T0189815

Topics on Continua

Sergio Macías

Topics on Continua

Second Edition

 Springer

Sergio Macías
Instituto de Matemáticas
Universidad Nacional Autónoma de México
Ciudad de México, México

The year of the former edition (2005) as well as the original title "Topics in Continua" and the original Publisher "Chapman & Hall/CRC".

ISBN 978-3-030-08127-0   ISBN 978-3-319-90902-8   (eBook)
https://doi.org/10.1007/978-3-319-90902-8

Mathematics Subject Classification (2010): 54B15, 54B20, 54C05, 54E35, 54E40, 54F15, 58E40

This Springer imprint is published by the registered company Springer International Publishing AG part of Springer Nature.
The registered company address is: Gewerbestrasse 11, 6330 Cham, Switzerland

To Elsa

León Felipe escribió un
tributo, no al héroe de la
historia, sino a su fiel
caballo Rocinante,
quien lo llevó en su lomo
por las tierras de España.

El héroe es, por supuesto,
Don Quijote de la Mancha:

"El Caballero de la Triste Figura"

¡Yo quería ese nombre!
pero me lo ganaron,
llegué a este mundo casi
trescientos cincuenta
años tarde. . .

Ya sólo me queda ser:

"El Caballero de la Triste Locura. . ."

S. M.

Preface to the Second Edition

After 12 years of the publication of *Topics on Continua* many things have happened. As it is well known, it is impossible to include everything. This Second Edition contains two new chapters which appear for the first time in a book, namely: *n-fold Hyperspace Suspensions* and *Induced Maps on n-fold Hyperspaces*. We include recent developments.

The first two chapters have very few modifications. In the first one, we prepare the way to prove the monotone-light factorization theorem, which appears later in chapter eight. We also add the notions of freely decomposable continuum and more concepts of aposyndesis. We include the notions of arc-smoothness of continua and arcwise decomposable continua too. For the second chapter we have not included much because of the two books on inverse limits and generalized inverse limits that appeared in 2012, namely: the book by Professors W. T. Ingram and William S. Mahavier *Inverse Limits: From Chaos to Continua*, Developments in Mathematics, Vol. 25, Springer, 2012 and the book by Professor W. T. Ingram *An Introduction to Inverse Limits with Set-valued Functions*, Springer Briefs in Mathematics, 2012. If the reader is interested in such topics, please refer to the mentioned books. We add the notions of confluent and weakly confluent maps to show that the bonding maps of an inverse limit are confluent if and only if the projection maps are confluent and the fact that each surjective map onto a chainable continuum is weakly confluent. By using inverse limits, it can be shown that the Cantor set is a topological group.

Chapter 3 has four new sections, namely: *Idempotency of $\mathcal{T}$, Three Decomposition Theorems, Examples*, and *$\mathcal{T}$-closed sets*. Throughout the chapter, we present characterizations of locally connected continua using the distinct forms of aposyndesis added in the first chapter. A sufficient condition for the idempotency on closed sets is given we also present an example showing that the condition is not necessary. We present a study of the relation between arc-smoothness and strict point $\mathcal{T}$-asymmetry. In particular, we show that Question 9.2.9 has a negative answer. We include results about the idempotency of $\mathcal{T}$ on products, cones, and suspensions. In particular, we prove that the first part of Question 9.2.3 has always a negative answer. We present three decomposition theorems using $\mathcal{T}$. In the strongest of the theorems, we obtain a continuous decomposition of the continuum with a locally connected quotient space and many of the elements of the decomposition are indecomposable continua. We present several classes of continua for which $\mathcal{T}$ is continuous and we study the family of $\mathcal{T}$-closed sets.

Chapter 4 remains essentially the same; we add three more consequences of the Property of Effros. The same happens with Chap. 5 where we include a few characterizations of the continuity of the set function $\mathcal{T}$ for homogeneous continua.

Chapter 6 has two new sections, namely: *Z-sets* and *Strong Size Maps*. Throughout the chapter, we include several bounds for the dimension of the n-fold hyperspace of certain classes of continua. We show that the n-fold hyperspaces are zero-dimensional aposyndetic. We give the correct statement and proof of Theorem 6.5.14. We give basic properties of Z-sets and sufficient conditions in order to show that the n-fold symmetric product of a continuum is a Z-set of the n-hyperspace of such continuum. We add several results that indicate when the n-fold symmetric product is a strong deformation retract of the m-fold hyperspace or of the hyperspace of closed sets. Also, we include properties of the continuum and the n-fold symmetric product when this is a retract of the m-fold hyperspace. We add a characterization of the graphs for which their n-fold hyperspace is a Cantor manifold. We also characterize the class of continua for which its n-fold hyperspace is a k-cell. We include results about suspensions and products related to the ones already given for cones. We end the chapter with a study of strong size maps, which are a nice generalization of Whitney maps to n-fold hyperspaces.

Chapter 7 is new. It is about hyperspace suspensions. We present most of what is known about n-fold hyperspace suspensions. We prove several properties of these spaces. We give sufficient conditions in order to obtain that n-fold hyperspace suspensions are contractible. We show that they are zero-dimensional aposyndetic. We study these hyperspaces when the continuum is locally connected. In particular, we give a sufficient condition to obtain that the n-fold hyperspace suspension of a locally connected continuum is the Hilbert cube. We characterize indecomposable continua by showing that their n-fold hyperspace suspensions are arcwise disconnected by removing two points. We present a description of the arc components of arcwise disconnected n-fold hyperspace suspensions when those two points are removed. We study properties of the n-fold hyperspace suspensions when they are homeomorphic to cones, suspensions, or products of continua. We present several results about the fixed point property of these hyperspaces. We study absolute n-fold hyperspace suspensions. We end this chapter by proving that hereditarily indecomposable continua have unique n-fold hyperspace suspensions.

Chapter 8 is also new. It is about induced maps between n-fold hyperspaces; these include hyperspace suspensions. We start with the definition of all the classes of maps that we study. Then we continue with general properties about the induced maps and present results about homeomorphisms, atomic maps, ε-maps, refinable maps, and almost monotone maps. We continue with results about confluent, monotone, open, light, and freely decomposable maps.

Chapter 9 (former Chap. 7), the last chapter, which is about questions, has two new sections (one for each of the new chapters), with questions on n-fold hyperspace suspensions and induced maps between n-fold hyperspaces.

I thank Javier Camargo for letting me include part of his dissertation in the second edition of the book.

I thank Ms. Elsa Arroyo for preparing all the pictures of the second edition of the book.

I thank the people at Springer, especially Professor Dr. Jan Holland, Ms. Anne Comment, Mr. Tilton Edward Stanley, Ms. Uma Periasamy and Ms. Kathleen Moriarty, for all their help.

Ciudad de México, México   Sergio Macías

Preface to the First Edition

My aim is to present four of my favorite topics in continuum theory: inverse limits, Professor Jones's set function $\mathcal{T}$, homogeneous continua, and n-fold hyperspaces.

Most topics treated in this book are not covered in Professor Sam B. Nadler Jr.'s book: *Continuum Theory: An Introduction*, Monographs and Textbooks in Pure and Applied Math., Vol. 158, Marcel Dekker, New York, Basel, Hong Kong, 1992.

The reader is assumed to have taken a one-year course on general topology.

The book has seven chapters. In Chap. 1, we include the basic background to be used in the rest of the book. The experienced readers may prefer to skip this chapter and jump right to the study of their favorite subject. This can be done without any problem. The topics of Chap. 1 are essentially independent of one another and can be read at any time.

Chapter 2 is for the most part about inverse limits of continua. We present the basic results on inverse limits. Some theorems are stated without proof in Professor W. Tom Ingram's book: *Inverse Limits*, Aportaciones Matemáticas, Textos # 18, Sociedad Matemática Mexicana, 2000. We show that the operation of taking inverse limits commutes with the operations of taking finite products, cones, and hyperspaces. We also include some applications of inverse limits.

In Chap. 3 we discuss Professor F. Burton Jones's set function $\mathcal{T}$. After giving the basic properties of this function, we present properties of continua in terms of $\mathcal{T}$, such as connectedness im kleinen, local connectedness, and semi-local connectedness. We also study continua for which the set function $\mathcal{T}$ is continuous. In the last section we present some applications of $\mathcal{T}$.

In Chap. 4 we start our study of homogeneous continua. We present a topological proof of a Theorem of Professor E. G. Effros given by F. D. Ancel. We include a brief introduction to topological groups and group actions.

Chapter 5 contains our main study of homogeneous continua. We present two Decomposition Theorems of such continua, whose proofs are applications of Professor Jones's set function $\mathcal{T}$ and Professor Effros's Theorem. These theorems have narrowed the study of homogeneous continua in such a way that they may hopefully be eventually classified. We also give examples of nontrivial homogeneous continua and their covering spaces.

In Chap. 6 we present most of what is known about n-fold hyperspaces. This chapter is slightly different from the other chapters because the proofs of many of the theorems are based on results in the literature that we do not prove; however, we give references to the appropriate places where proofs can be found. This chapter is a complement of the two existing books—Sam B. Nadler, Jr., *Hyperspaces of Sets: A Text with Research Questions*, Monographs and Textbooks in Pure and Applied Math., Vol. 49, Marcel Dekker, New York, Basel, 1978[1] and Alejandro Illanes and Sam B. Nadler, Jr., *Hyperspaces: Fundamentals and Recent Advances*, Monographs and Textbooks in Pure and Applied Math., Vol. 216, Marcel Dekker, New York, Basel, 1999, in which a thorough study of hyperspaces is done.

In Chap. 6, we also prove general properties of n-fold hyperspaces. In particular, we show that n-fold hyperspaces are unicoherent and finitely aposyndetic. We study the arcwise accessibility of points of the n-fold symmetric products from their complement in n-fold hyperspaces. We give a treatment of the points that arcwise disconnect n-fold hyperspaces of indecomposable continua. Then we study continua for which the operation of taking n-fold hyperspaces is continuous ($\mathcal{C}_n^*$-smoothness). We also investigate continua for which there exist retractions between their various hyperspaces. Next, we present some results about the n-fold hyperspaces of graphs. We end Chap. 6 by studying the relation between n-fold hyperspaces and cones over continua.

We end the book with a chapter (Chap. 7) containing open questions on each of the subjects presented in the book.

We include figures to illustrate definitions and aspects of proofs.

The book originates from two sources—class notes I took from the course on continuum theory given by Professor James T. Rogers, Jr. at Tulane University in the Fall Semester of 1988 and the one-year courses on continuum theory I have taught in the graduate program of mathematics at the Facultad de Ciencias of the Universidad Nacional Autónoma de México, since the spring of 1993. I thank all the students who have taken such courses.

I thank María Antonieta Molina and Juan Carlos Macías for letting me include part of their thesis in the book. Ms. Molina's thesis was based on two talks on the set function $\mathcal{T}$ given by Professor David P. Bellamy in the *IV Research Workshop on Topology*, celebrated in Oaxaca City, Oaxaca, México, November 14 through 16, 1996.

I thank Professors Sam B. Nadler, Jr. and James T. Rogers, Jr. for reading parts of the manuscript and making valuable suggestions. I also thank Ms. Gabriela Sanginés and Mr. Leonardo Espinosa for answering my questions about LaTeX, while I was typing this book.

I thank Professor Charles Hagopian and Marvi Hagopian for letting me use their living room to work on the book during my visit to California State University, Sacramento.

---

[1] This book has been reprinted in: Aportaciones Matemáticas de la Sociedad Matemática Mexicana, Serie Textos # 33, 2006.

I thank the Instituto de Matemáticas of the Universidad Nacional Autónoma de México and the Mathematics Department of West Virginia University, for the use of resources during the preparation of the book.

Finally, I thank the people at Marcel Dekker, Inc., especially Ms. Maria Allegra and Mr. Kevin Sequeira, who were always patient and helpful.

Ciudad de México, México   Sergio Macías

Contents

Chapter 1
Preliminaries

We gather some of the results of topology of metric spaces which will be useful for the rest of the book. We assume the reader is familiar with the notion of metric space and its elementary properties. We present the proofs of most of the results; we give an appropriate reference otherwise.

The topics reviewed in this chapter are: product topology, continuous decompositions, homotopy, fundamental group, geometric complexes, polyhedra, complete metric spaces, compacta, continua and hyperspaces.

1.1 Product Topology

The symbols $\mathbb{N}$, $\mathbb{Z}$, $\mathbb{Q}$, $\mathbb{R}$ and $\mathbb{C}$ denote the positive integers, integers, rational numbers, real numbers and complex numbers, respectively. The material of this section is taken from [10, 13, 17, 18, 25, 28].

The word *map* means a continuous function. A *compactum* is a compact metric space.

1.1.1 Definition Given a sequence, $\{X_n\}_{n=1}^{\infty}$, of nonempty sets, we define its *Cartesian product*, denoted by $\prod_{n=1}^{\infty} X_n$, as the set:

$$\prod_{n=1}^{\infty} X_n = \{(x_n)_{n=1}^{\infty} \mid x_n \in X_n \text{ for each } n \in \mathbb{N}\}.$$

For each $m \in \mathbb{N}$, there exists a function

$$\pi_m \colon \prod_{n=1}^{\infty} X_n \twoheadrightarrow X_m$$

defined by $\pi_m((x_n)_{n=1}^{\infty}) = x_m$. This function π_m is called the *mth-projection map*.

© Springer International Publishing AG, part of Springer Nature 2018
S. Macías, *Topics on Continua*, https://doi.org/10.1007/978-3-319-90902-8_1

1.1.2 Remark Given a metric space (X, d'), there exists a metric, d, which generates the same topology as d', with the property that $d(x, x') \leq 1$ for each pair of points x and x' of X. This metric d is called *bounded metric*. An example of such metric is given by $d(x, x') = \min\{1, d'(x, x')\}$.

1.1.3 Notation Given a metric space (X, d) and a subset A of X, $Cl_X(A)$, $Int_X(A)$ and $Bd_X(A)$ denote the closure, interior and boundary of A, respectively. We omit the subindex if there is no confusion. If ε is a positive real number, then the symbol $\mathcal{V}_\varepsilon^d(A)$ denotes the *open ball of radius ε about A*. If $A = \{x\}$, for some $x \in X$, we write $\mathcal{V}_\varepsilon^d(x)$ instead of $\mathcal{V}_\varepsilon^d(\{x\})$.

1.1.4 Definition If $\{(X_n, d_n)\}_{n=1}^\infty$ is a sequence of metric spaces, with bounded metrics, we define a metric ρ, for its Cartesian product as follows:

$$\rho((x_n)_{n=1}^\infty, (x'_n)_{n=1}^\infty) = \sum_{n=1}^\infty \frac{1}{2^n} d_n(x_n, x'_n).$$

1.1.5 Remark Since the metrics, d_n, in Definition 1.1.4 are bounded, ρ is well defined.

1.1.6 Lemma *If $\{(X_n, d_n)\}_{n=1}^\infty$ is a sequence of metric spaces, with bounded metrics, then ρ (Definition 1.1.4) is a metric and for each $m \in \mathbb{N}$, π_m is a continuous function.*

Proof The proof of the fact that ρ is, in fact, a metric is left to the reader.

Let $m \in \mathbb{N}$ be given. We show that π_m is continuous. Let $\varepsilon > 0$ and let $\delta = \frac{1}{2^m}\varepsilon$. If $(x_n)_{n=1}^\infty$ and $(x'_n)_{n=1}^\infty$ are two points of $\prod_{n=1}^\infty X_n$ such that $\rho((x_n)_{n=1}^\infty, (x'_n)_{n=1}^\infty) < \delta$, then, since $\frac{1}{2^m} d_m(x_m, x'_m) \leq \sum_{n=1}^\infty \frac{1}{2^n} d_n(x_n, x'_n)$, we have that $\frac{1}{2^m} d_m(x_m, x'_m) < \delta$. Hence,

$$d_m(x_m, x'_m) < 2^m \delta = \varepsilon.$$

Therefore, π_m is continuous.

Q.E.D.

1.1.7 Lemma *If $\{(X_n, d_n)\}_{n=1}^\infty$ is a sequence of metric spaces, with bounded metrics, then given $\varepsilon > 0$ and a point $(x_n)_{n=1}^\infty \in \prod_{n=1}^\infty X_n$, there exist $N \in \mathbb{N}$ and N positive real numbers, $\varepsilon_1, \dots, \varepsilon_N$, such that $\bigcap_{j=1}^N \pi_j^{-1}(\mathcal{V}_{\varepsilon_j}^{d_j}(x_j)) \subset \mathcal{V}_\varepsilon^\rho((x_n)_{n=1}^\infty)$.*

Proof Let $N \in \mathbb{N}$ be such that $\sum_{n=N+1}^\infty \frac{1}{2^n} < \frac{\varepsilon}{2}$. For each $j \in \{1, \dots, N\}$, let $\varepsilon_j = \frac{\varepsilon}{2^N}$. We assert that $\bigcap_{j=1}^N \pi_j^{-1}\left(\mathcal{V}_{\varepsilon_j}^{d_j}(x_j)\right) \subset \mathcal{V}_\varepsilon^\rho((x_n)_{n=1}^\infty)$. To see this, let $(y_n)_{n=1}^\infty \in \bigcap_{j=1}^N \pi_j^{-1}\left(\mathcal{V}_{\varepsilon_j}^{d_j}(x_j)\right)$. We want to see that $\rho((x_n)_{n=1}^\infty, (y_n)_{n=1}^\infty) < \varepsilon$.

Note that

$$\rho((x_n)_{n=1}^{\infty}, (y_n)_{n=1}^{\infty}) = \sum_{n=1}^{\infty} \frac{1}{2^n} d_n(x_n, y_n) =$$

$$\sum_{n=1}^{N} \frac{1}{2^n} d_n(x_n, y_n) + \sum_{n=N+1}^{\infty} \frac{1}{2^n} d_n(x_n, y_n) <$$

$$\sum_{n=1}^{N} \frac{1}{2^n} \frac{1}{2^N} \varepsilon + \frac{1}{2} \varepsilon = \left(1 - \frac{1}{2^N}\right) \frac{1}{2^N} \varepsilon + \frac{1}{2} \varepsilon \le \frac{1}{2} \varepsilon + \frac{1}{2} \varepsilon = \varepsilon.$$

Q.E.D.

1.1.8 Lemma *If $\{(X_n, d_n)\}_{n=1}^{\infty}$ is a sequence of metric spaces, with bounded metrics, then given a finite number of positive real numbers $\varepsilon_1, \ldots, \varepsilon_k$ and a point $(x_n)_{n=1}^{\infty} \in \prod_{n=1}^{\infty} X_n$, there exists $\varepsilon > 0$ such that $V_{\varepsilon}^{\rho}((x_n)_{n=1}^{\infty}) \subset \bigcap_{j=1}^{k} \pi_j^{-1}(V_{\varepsilon_j}^{d_j}(x_j))$.*

Proof Let $(x_n)_{n=1}^{\infty} \in \prod_{n=1}^{\infty} X_n$, and let $U = \bigcap_{j=1}^{k} \pi_j^{-1}\left(V_{\varepsilon_j}^{d_j}(x_j)\right)$. Take

$$\varepsilon = \min\left\{\frac{1}{2}\varepsilon_1, \ldots, \frac{1}{2^k}\varepsilon_k\right\}.$$

We show $V_{\varepsilon}^{\rho}((x_n)_{n=1}^{\infty}) \subset U$. Let $(y_n)_{n=1}^{\infty} \in V_{\varepsilon}^{\rho}((x_n)_{n=1}^{\infty})$. Then

$$\rho((x_n)_{n=1}^{\infty}, (y_n)_{n=1}^{\infty}) < \varepsilon, \text{ i.e., } \sum_{n=1}^{\infty} \frac{1}{2^n} d_n(x_n, y_n) < \varepsilon.$$

Hence, $\frac{1}{2^j} d_j(x_j, y_j) < \varepsilon \le \frac{1}{2^j}\varepsilon_j$ for each $j \in \{1, \ldots, k\}$. Thus, if $j \in \{1, \ldots, k\}$, then $d_j(x_j, y_j) < \varepsilon_j$. Therefore, $V_{\varepsilon}^{\rho}((x_n)_{n=1}^{\infty}) \subset U$.

Q.E.D.

1.1.9 Theorem *Let Z be a metric space. If $\{(X_n, d_n)\}_{n=1}^{\infty}$ is a sequence of metric spaces, then a function $f: Z \to \prod_{n=1}^{\infty} X_n$ is continuous if and only if $\pi_n \circ f$ is continuous for each $n \in \mathbb{N}$.*

Proof Clearly, if f is continuous, then $\pi_n \circ f$ is continuous for each $n \in \mathbb{N}$.

Suppose $\pi_n \circ f$ is continuous for each $n \in \mathbb{N}$. Let $\bigcap_{j=1}^{k} \pi_j^{-1}(U_j)$ be a basic open subset of $\prod_{n=1}^{\infty} X_n$. Since

$$f^{-1}\left(\bigcap_{j=1}^{k} \pi_j^{-1}(U_j)\right) = \bigcap_{j=1}^{k} f^{-1}(\pi_j^{-1}(U_j))$$

$$= \bigcap_{j=1}^{k} (\pi_j \circ f)^{-1}(U_j),$$

we have that $f^{-1}\left(\bigcap_{j=1}^{k} \pi_j^{-1}(U_j)\right)$ is open in Z. Hence, f is continuous.

Q.E.D.

1.1.10 Theorem *Let $\{X_n\}_{n=1}^{\infty}$ and $\{Y_n\}_{n=1}^{\infty}$ be two countable collections of metric spaces. Suppose that for each $n \in \mathbb{N}$, there exists a map $f_n: X_n \to Y_n$. Then the function $\prod_{n=1}^{\infty} f_n: \prod_{n=1}^{\infty} X_n \to \prod_{n=1}^{\infty} Y_n$ given by $\prod_{n=1}^{\infty} f_n((x_n)_{n=1}^{\infty}) = (f_n(x_n))_{n=1}^{\infty}$ is continuous.*

Proof For each $m \in \mathbb{N}$, let $\pi_m: \prod_{n=1}^{\infty} X_n \twoheadrightarrow X_m$ and $\pi_m': \prod_{n=1}^{\infty} Y_n \twoheadrightarrow Y_m$ be the projection maps.

Let $(x_n)_{n=1}^{\infty}$ be a point of $\prod_{n=1}^{\infty} X_n$, and let $m \in \mathbb{N}$. Then $\pi_m' \circ \prod_{n=1}^{\infty} f_n((x_n)_{n=1}^{\infty}) = \pi_m'\left((f_n(x_n))_{n=1}^{\infty}\right) = f_m(x_m) = f_m \circ \pi_m((x_n)_{n=1}^{\infty})$. Hence, by Theorem 1.1.9, $\prod_{n=1}^{\infty} f_n$ is continuous.

Q.E.D.

The following result is a particular case of *Tychonoff's Theorem*, which says that the Cartesian product of any family of compact topological spaces is compact. The proof of this theorem uses the *Axiom of Choice*. However, the case we show only uses the fact that compactness and sequential compactness are equivalent in metric spaces [17, Remark 3, p. 3].

1.1.11 Theorem *If $\{(X_n, d_n)\}_{n=1}^{\infty}$ is a sequence of compacta, then $\prod_{n=1}^{\infty} X_n$ is compact.*

Proof By Lemma 1.1.6, $\prod_{n=1}^{\infty} X_n$ is a metric space. We show that any sequence of points of $\prod_{n=1}^{\infty} X_n$ has a convergent subsequence.

Let $\{p^k\}_{k=1}^{\infty}$ be a sequence of points of $\prod_{n=1}^{\infty} X_n$, where $p^k = (p_n^k)_{n=1}^{\infty}$ for each $k \in \mathbb{N}$ (in this way, if we keep n fixed, $\{p_n^k\}_{k=1}^{\infty}$ is a sequence of points of X_n). Since (X_1, d_1) is sequentially compact, $\{p_1^k\}_{k=1}^{\infty}$ has a convergent subsequence $\{p_1^{k_j}\}_{j=1}^{\infty}$ converging to a point q_1 of X_1. Let us note that, implicitly, we have defined a subsequence $\{p^{k_j}\}_{j=1}^{\infty}$ of $\{p^k\}_{k=1}^{\infty}$.

Now, suppose, inductively, that for some $m \in \mathbb{N}$, we have defined a subsequence $\{p^{k_i}\}_{i=1}^{\infty}$ of $\{p^k\}_{k=1}^{\infty}$ such that $\{p_m^{k_i}\}_{i=1}^{\infty}$ converges to a point q_m of X_m. Since (X_{m+1}, d_{m+1}) is sequentially compact, $\{p_{m+1}^{k_i}\}_{i=1}^{\infty}$ has a convergent subsequence

$\{p_{m+1}^{k_{i_j}}\}_{j=1}^{\infty}$ such that it converges to a point q_{m+1} of X_{m+1}. Hence, by the Induction Principle, we have defined a sequence of subsequences of $\{p^k\}_{k=1}^{\infty}$ in such a way that each subsequence is a subsequence of the preceding one. Now, let $\Sigma = \{p^1, p^{k_2}, p^{k_{j_3}}, p^{k_{j_{i_4}}}, \ldots\}$. Clearly, Σ is a subsequence of $\{p^k\}_{k=1}^{\infty}$ which converges to the point $(q_n)_{n=1}^{\infty}$. Therefore, $\prod_{n=1}^{\infty} X_n$ is compact.

Q.E.D.

1.1.12 Definition Let $\mathcal{Q} = \prod_{n=1}^{\infty}[0, 1]_n$, where $[0, 1]_n = [0, 1]$, for each $n \in \mathbb{N}$. Then $\mathcal{Q}$ is called the *Hilbert cube*.

1.1.13 Theorem *The Hilbert cube is a connected compactum.*

Proof By Lemma 1.1.6, $\mathcal{Q}$ is a metric space. By Theorem 1.1.11, $\mathcal{Q}$ is compact. By [17, Theorem 11, p. 137], $\mathcal{Q}$ is connected.

Q.E.D.

1.1.14 Definition Let $f: X \to Y$ be a map between metric spaces. We say that f is an *embedding* if f is a homeomorphism onto $f(X)$.

1.1.15 Definition A map $f: X \to Y$ between metric spaces is said to be *closed* provided that for each closed subset K of X, $f(K)$ is closed in Y.

The next theorem says that there exists a "copy" of every compactum inside the Hilbert cube.

1.1.16 Theorem *If X is a compactum, then X can be embedded in the Hilbert cube $\mathcal{Q}$.*

Proof Let d be the metric of X. Without loss of generality, we assume that $\text{diam}(X) \leq 1$. Since X is a compactum, it contains a countable dense subset, $\{x_n\}_{n=1}^{\infty}$. Let $h: X \to \mathcal{Q}$ be given by $h(x) = (d(x, x_n))_{n=1}^{\infty}$. By Theorem 1.1.9, h is continuous. Clearly, h is one-to-one. Since X is compact and $\mathcal{Q}$ is metric, h is a closed map. Therefore, h is an embedding.

Q.E.D.

1.2 Continuous Decompositions

We present a method to construct "new" spaces from "old" ones by "shrinking" certain subsets to points. The preparation of this section is based on [10, 13, 18, 27].

1.2.1 Definition A *decomposition* of a set X is a collection of nonempty, pairwise disjoint sets whose union is X. The decomposition is said to be *closed* if each of its element is a closed subset of X.

1.2.2 Definition Let $\mathcal{G}$ be a decomposition of a metric space X. We define $X/\mathcal{G}$ as the set whose elements are the elements of the decomposition $\mathcal{G}$. $X/\mathcal{G}$ is called the *quotient space.* The function $q \colon X \twoheadrightarrow X/\mathcal{G}$, which sends each point x of X to the unique element G of $\mathcal{G}$ such that $x \in G$, is called the *quotient map.*

1.2.3 Remark Given a decomposition of a metric space X, note that $q(x) = q(y)$ if and only if x and y belong to the same element of $\mathcal{G}$. We give a topology to $X/\mathcal{G}$ in such a way that the function q is continuous and it is the biggest with this property.

1.2.4 Definition Let X be a metric space, let $\mathcal{G}$ be a decomposition of X and let $q \colon X \twoheadrightarrow X/\mathcal{G}$ be the quotient map. Then the topology

$$\mathcal{U} = \{U \subset X/\mathcal{G} \mid q^{-1}(U) \text{ is open in } X\}$$

is called the *quotient topology for $X/\mathcal{G}$.*

1.2.5 Remark Let $\mathcal{G}$ be a decomposition of a metric space X, and let $q \colon X \twoheadrightarrow X/\mathcal{G}$ be the quotient map. Then a subset U of $X/\mathcal{G}$ is open (closed, respectively) if and only if $q^{-1}(U)$ is an open (closed, respectively) subset of X.

1.2.6 Definition Let $f \colon X \twoheadrightarrow Y$ be a surjective map between metric spaces. Since f is a function, $\mathcal{G}_f = \{f^{-1}(y) \mid y \in Y\}$ is a decomposition of X. The function $\varphi_f \colon X/\mathcal{G}_f \to Y$ given by $\varphi_f(q(x)) = f(x)$ is of special interest. Note that φ_f is well defined; in fact, it is a bijection and the following diagram:

$$\begin{array}{ccc} X & \xrightarrow{\ f\ } & Y \\ & {}_{q}\searrow \quad \nearrow_{\varphi_f} & \\ & X/\mathcal{G}_f & \end{array}$$

is commutative.

The next lemma is a special case of the *Transgression Lemma* [27, 3.22].

1.2.7 Lemma *Let $f \colon X \twoheadrightarrow Y$ be a surjective map between metric spaces. If $X/\mathcal{G}_f$ has the quotient topology, then the function φ_f is continuous.*

Proof If U is an open subset of Y, then $\varphi_f^{-1}(U) = qf^{-1}(U)$. Since $q^{-1}\varphi_f^{-1}(U) = q^{-1}qf^{-1}(U) = f^{-1}(U)$ and $f^{-1}(U)$ is an open subset of X, we have, by the definition of quotient topology, that $\varphi_f^{-1}(U)$ is an open subset of $X/\mathcal{G}_f$. Therefore, φ_f is continuous.

Q.E.D.

1.2.8 Example Let $X = [0, 2\pi)$ and let $f \colon X \twoheadrightarrow \mathcal{S}^1$, where $\mathcal{S}^1$ is the unit circle, be given by $f(t) = \exp(t) = e^{it}$. Then f is a continuous bijection. Since $\mathcal{G}_f$ is, "essentially," X, it follows that $X/\mathcal{G}_f$ is homeomorphic to X. On the other hand, X is not homeomorphic to $\mathcal{S}^1$, since X is not compact and $\mathcal{S}^1$ is. Therefore, φ_f is not a homeomorphism.

1.2.9 Definition A map $f \colon X \to Y$ between metric spaces is said to be *open* provided that for each open subset K of X, $f(K)$ is open in Y.

The following theorem gives sufficient conditions to ensure that φ_f is a homeomorphism:

1.2.10 Theorem *Let $f: X \twoheadrightarrow Y$ be a surjective map between metric spaces. If f is open or closed, then $\varphi_f: X/\mathcal{G}_f \twoheadrightarrow Y$ is a homeomorphism.*

Proof Suppose f is an open map. Since φ is a bijective map, it is enough to show that φ_f is open. Let A be an open subset of $X/\mathcal{G}_f$. Since $\varphi_f(A) = fq^{-1}(A)$, $\varphi_f(A)$ is an open subset of Y. Therefore, φ_f is an open map.

The proof of the case when f is closed is similar.

Q.E.D.

Decompositions are also used to construct the cone and suspension over a given space.

1.2.11 Definition Let X be a metric space and let $\mathcal{G} = \{\{(x, t)\} \mid x \in X$ and $t \in [0, 1)\} \cup \{(X \times \{1\})\}$. Then $\mathcal{G}$ is a decomposition of $X \times [0, 1]$. The *cone over X*, denoted by $K(X)$, is the quotient space $(X \times [0, 1])/\mathcal{G}$. The element $\{X \times \{1\}\}$ of $(X \times [0, 1])/\mathcal{G}$ is called the *vertex* of the cone and it is denoted by v_X.

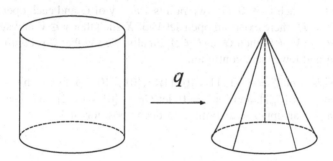

A proof of the following proposition may be found in [10, 5.2, p. 127].

1.2.12 Proposition *Let $f: X \rightarrow Y$ be a map between metric spaces. Then f induces a map $K(f): K(X) \rightarrow K(Y)$ by*

$$K(f)(\omega) = \begin{cases} v_Y, & \text{if } \omega = v_X \in K(X); \\ (f(x), t), & \text{if } \omega = (x, t) \in K(X) \setminus \{v_X\}. \end{cases}$$

1.2.13 Definition Let X be a metric space and let $\mathcal{G} = \{\{(x, t)\} \mid x \in X$ and $t \in (0, 1)\} \cup \{(X \times \{0\}), (X \times \{1\})\}$. Then $\mathcal{G}$ is a decomposition of $X \times [0, 1]$. The *suspension over X*, denoted by $\Sigma(X)$, is the quotient space $(X \times [0, 1])/\mathcal{G}$. The elements $\{X \times \{0\}\}$ and $\{X \times \{1\}\}$ of $(X \times [0, 1])/\mathcal{G}$ are called the *vertexes* of the suspension and are denoted by v^- and v^+, respectively.

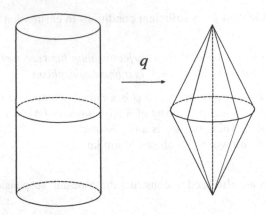

1.2.14 Definition Let X be a metric space and let $\mathcal{G}$ be a decomposition of X. We say that $\mathcal{G}$ is *upper semicontinuous* if for each $G \in \mathcal{G}$ and each open subset U of X such that $G \subset U$, there exists an open subset V of X such that $G \subset V$ and such that if $G' \in \mathcal{G}$ and $G' \cap V \neq \emptyset$, then $G' \subset U$. We say that $\mathcal{G}$ is *lower semicontinuous* provided that for each $G \in \mathcal{G}$ any two points x and y of G and each open set U of X such that $x \in U$, there exists an open set V of X such that $y \in V$ and such that if $G' \in \mathcal{G}$ and $G' \cap V \neq \emptyset$, then $G \cap U \neq \emptyset$. Finally, we say that $\mathcal{G}$ is *continuous* if $\mathcal{G}$ is both upper and lower semicontinuous.

1.2.15 Example Let $X = ([-1, 1] \times [0, 1]) \cup (\{0\} \times [0, 2])$. For each $t \in [-1, 1] \setminus \{0\}$, let $G_t = \{t\} \times [0, 1]$, and for $t = 0$, let $G_0 = \{0\} \times [0, 2]$. Let $\mathcal{G} = \{G_t \mid t \in [0, 1]\}$. Then $\mathcal{G}$ is an upper semicontinuous decomposition of X.

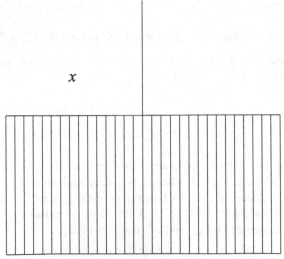

Upper semicontinuous decomposition

1.2.16 Example Let

$$X = ([0, 1] \times [0, 1)) \cup \left(\{1\} \times \left[0, \frac{1}{3} \right] \cup \{1\} \times \left[\frac{2}{3}, 1 \right] \right).$$

For each $t \in [0, 1)$, let $G_t = \{t\} \times [0, 1]$, let $G_1 = \{1\} \times \left[0, \frac{1}{3} \right]$ and let $G'_1 = \{1\} \times \left[\frac{2}{3}, 1 \right]$. Let $\mathcal{G} = \{G_t \mid t \in [0, 1]\} \cup \{G'_1\}$. Then $\mathcal{G}$ is a lower semicontinuous decomposition of X.

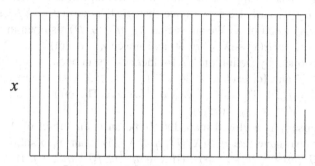

Lower semicontinuous decomposition

The following theorem gives us a way to obtain upper semicontinuous decompositions of compacta.

1.2.17 Theorem *Let $f : X \twoheadrightarrow Y$ be a surjective map between compacta. If $\mathcal{G}_f = \{f^{-1}(y) \mid y \in Y\}$, then $\mathcal{G}_f$ is an upper semicontinuous decomposition of X.*

Proof Let U be an open subset of X such that $f^{-1}(y) \subset U$. Note that $X \setminus U$ is a closed subset; hence, compact, of X. Then $f(X \setminus U)$ is a compact subset; hence, closed, of Y such that $y \notin f(X \setminus U)$. Thus, $Y \setminus f(X \setminus U)$ is an open subset of Y containing y.

If $V = f^{-1}(Y \setminus f(X \setminus U))$, then V is an open subset of X such that $f^{-1}(y) \subset V \subset U$. Since $V = \bigcup \{f^{-1}(y) \mid y \in Y \setminus f(X \setminus U)\}$, clearly V satisfies the required property of the definition of upper semicontinuous decomposition.

Q.E.D.

1.2.18 Remark Let us note that Theorem 1.2.17 is not true without the compactness of X. Let X be the Euclidean plane $\mathbb{R}^2$ and let $\pi : \mathbb{R}^2 \twoheadrightarrow \mathbb{R}$ be given by $\pi((x, y)) = x$. Then $\mathcal{G}_\pi$ is a decomposition of X which is not upper semicontinuous. To see this, let $U = \left\{ (x, y) \in X \mid x \neq 0 \text{ and } y < \frac{1}{x} \right\} \cup \{0\} \times \mathbb{R}$. Then U is an open set of X such that $\pi^{-1}(0) \subset U$, whose boundary is asymptotic to $\pi^{-1}(0)$. Hence, for each $t \in \mathbb{R} \setminus \{0\}$, $\pi^{-1}(t) \cap (X \setminus U) \neq \emptyset$.

The next theorem gives three other ways to think about upper semicontinuous decompositions.

1.2.19 Theorem *If X is a metric space and $\mathcal{G}$ is a decomposition of X, then the following conditions are equivalent:*

(a) *$\mathcal{G}$ is an upper semicontinuous decomposition;*
(b) *the quotient map $q: X \twoheadrightarrow X/\mathcal{G}$ is closed;*
(c) *if U is an open subset of X, then $W_U = \bigcup\{G \in \mathcal{G} \mid G \subset U\}$ is an open subset of X;*
(d) *if D is a closed subset of X, then $K_D = \bigcup\{G \in \mathcal{G} \mid G \cap D \neq \emptyset\}$ is a closed subset of X.*

Proof Suppose $\mathcal{G}$ is an upper semicontinuous decomposition. Let D be a closed subset of X. By Remark 1.2.5, we have that $q(D)$ is closed in $X/\mathcal{G}$ if and only if $q^{-1}(q(D))$ is closed in X. We show that $X \setminus q^{-1}(q(D))$ is open in X. Let $x \in X \setminus q^{-1}(q(D))$. Then $q(x) \in X/\mathcal{G} \setminus q(D)$ and, hence, $q^{-1}(q(x)) \subset X \setminus D$. Therefore, since $X \setminus D$ is open, by Definition 1.2.14, there exists an open set V of X such that $q^{-1}(q(x)) \subset V$ and for each $y \in V$, $q^{-1}(q(y)) \subset X \setminus D$. Clearly, $x \in V$ and $q(V) \subset X/\mathcal{G} \setminus q(D)$. Thus, $V \subset X \setminus q^{-1}(q(D))$. Therefore, $X \setminus q^{-1}(q(D))$ is open, since $x \in V \subset X \setminus q^{-1}(q(D))$.

Now, suppose q is a closed map. Let U be an open subset of X. Since q is a closed map, we have that $q^{-1}(X/\mathcal{G} \setminus q(X \setminus U))$ is an open subset of X such that $q^{-1}(X/\mathcal{G} \setminus q(X \setminus U)) = W_U$. (If $x \in q^{-1}(X/\mathcal{G} \setminus q(X \setminus U))$, then $q(x) \in X/\mathcal{G} \setminus q(X \setminus U)$. Hence, $q^{-1}(q(x)) \subset X \setminus q^{-1}(q(X \setminus U)) \subset X \setminus (X \setminus U)) = U$. Thus, $x \in W_U$. The other inclusion is obvious.)

Next, suppose W_U is open for each open subset U of X. Let D be a closed subset of X. Then $X \setminus D$ is open in X. Hence, $W_{X\setminus D}$ is open in X. Since, clearly, $K_D = X \setminus W_{X\setminus D}$, we have that K_D is closed.

Finally, suppose K_D is closed for each closed subset D of X. To see $\mathcal{G}$ is upper semicontinuous, let $G \in \mathcal{G}$ and let U be an open subset of X such that $G \subset U$. Note that $X \setminus U$ is a closed subset of X. Hence, $K_{X\setminus U}$ is a closed subset of X. Let $V = X \setminus K_{X\setminus U}$. Then V is open, $G \subset V \subset U$ and if $G' \in \mathcal{G}$ and $G' \cap V \neq \emptyset$, then $G' \subset V$. Therefore, $\mathcal{G}$ is upper semicontinuous.

                                                                                 Q.E.D.

1.2.20 Corollary *Let X be a metric space. If $\mathcal{G}$ is an upper semicontinuous decomposition of X, then the elements of $\mathcal{G}$ are closed.*

Proof Let $G \in \mathcal{G}$. Take $x \in G$ and let $q: X \twoheadrightarrow X/\mathcal{G}$ be the quotient map. Since X is a metric space, $\{x\}$ is closed in X. By Theorem 1.2.19, $q(\{x\})$ is closed in $X/\mathcal{G}$. Since q is continuous and $q^{-1}(q(\{x\})) = G$, G is a closed subset of X.

                                                                                 Q.E.D.

1.2.21 Theorem *If X is a compactum and $\mathcal{G}$ is an upper semicontinuous decomposition of X, then $X/\mathcal{G}$ has a countable basis.*

Proof Let $q: X \twoheadrightarrow X/\mathcal{G}$ be the quotient map. Since X is a compactum, it has a countable basis $\mathcal{U}$. Let

$$\mathcal{B} = \left\{ \bigcup_{j=1}^{n} U_j \;\middle|\; U_1, \ldots, U_n \in \mathcal{U} \text{ and } n \in \mathbb{N} \right\}.$$

Note that $\mathcal{B}$ is a countable family of open subsets of X.

Let

$$\boldsymbol{\mathcal{B}} = \{ X/\mathcal{G} \setminus q(X \setminus U) \mid U \in \mathcal{B} \}.$$

We see that $\boldsymbol{\mathcal{B}}$ is a countable basis for $X/\mathcal{G}$. Clearly, $\boldsymbol{\mathcal{B}}$ is a countable family of open subsets of $X/\mathcal{G}$. Let U be an open subset of $X/\mathcal{G}$ and let $x \in U$. Then $q^{-1}(U)$ is an open subset of X and $q^{-1}(x) \subset q^{-1}(U)$. Since $q^{-1}(x)$ is compact, there exist $U_1, \ldots, U_k \in \mathcal{U}$ such that $q^{-1}(x) \subset \bigcup_{j=1}^{k} U_j \subset q^{-1}(U)$. Let $U = \bigcup_{j=1}^{k} U_j$. Then $U \in \mathcal{B}$. Hence, $X/\mathcal{G} \setminus q(X \setminus U) \in \boldsymbol{\mathcal{B}}$. Also, $x \in X/\mathcal{G} \setminus q(X \setminus U) \subset U$. Therefore, $\boldsymbol{\mathcal{B}}$ is a countable basis for $X/\mathcal{G}$.

Q.E.D.

1.2.22 Corollary *If X is a compactum and $\mathcal{G}$ is an upper semicontinuous decomposition of X, then $X/\mathcal{G}$ is metrizable.*

Proof By Theorem 1.2.21, we have that $X/\mathcal{G}$ has a countable basis. By [16, Theorem 1, p. 241], it suffices to show that $X/\mathcal{G}$ is a Hausdorff space. Let x and y be two distinct points of $X/\mathcal{G}$. Then $q^{-1}(x)$ and $q^{-1}(y)$ are two disjoint closed subsets of X. Since X is a metric space, there exist two disjoint open subsets, U_1 and U_2, of X such that $q^{-1}(x) \subset U_1$ and $q^{-1}(y) \subset U_2$. Note that, by Theorem 1.2.19 (c), W_{U_1} and W_{U_2} are open subsets of X such that $q^{-1}(x) \subset W_{U_1} \subset U_1, q^{-1}(y) \subset W_{U_2} \subset U_2$, and $q(W_{U_1})$ and $q(W_{U_2})$ are open subsets of $X/\mathcal{G}$. Since $U_1 \cap U_2 = \emptyset$, $q(W_{U_1}) \cap q(W_{U_2}) = \emptyset$. Therefore, $X/\mathcal{G}$ is a Hausdorff space.

Q.E.D.

The next theorem gives a characterization of lower semicontinuous decompositions.

1.2.23 Theorem *Let X be a metric space and let $\mathcal{G}$ be a decomposition of X. Then $\mathcal{G}$ is lower semicontinuous if and only if the quotient map $q: X \twoheadrightarrow X/\mathcal{G}$ is open.*

Proof Suppose $\mathcal{G}$ is lower semicontinuous. Let U be an open subset of X. We show $q(U)$ is an open subset of $X/\mathcal{G}$. To this end, by Remark 1.2.5, we only need to prove that $q^{-1}(q(U))$ is an open subset of X.

Let $y \in q^{-1}(q(U))$. Then $q(y) \in q(U)$, and there exists a point x in U such that $q(x) = q(y)$. Since $\mathcal{G}$ is a lower semicontinuous decomposition, there exists an open subset V of X containing y such that if $G \in \mathcal{G}$ and $G \cap V \neq \emptyset$, then $G \cap U \neq \emptyset$. Hence, $V \subset q^{-1}(q(U))$. Therefore, q is open.

Now, suppose q is open. Let $G \in \mathcal{G}$. Take $x, y \in G$ and let U be an open subset of X such that $x \in U$. Since q is open, $V = q^{-1}(q(U))$ is an open subset of X such that $G \subset V$. In particular, $y \in V$. Let $G' \in \mathcal{G}$ be such that $G' \cap V \neq \emptyset$. Then $G' \subset V$. Thus, $q(G') \in q(U)$. Hence, there exists a point $u \in U$ such that $q(u) = q(G')$. Since $q^{-1}(q(G')) = G'$, $u \in G'$. Thus, $G' \cap U \neq \emptyset$. Therefore, $\mathcal{G}$ is lower semicontinuous.

Q.E.D.

The following corollary is a consequence of Theorems 1.2.19 and 1.2.23:

1.2.24 Corollary *Let X be a metric space and let $\mathcal{G}$ be a decomposition of X. Then $\mathcal{G}$ is continuous if and only if the quotient map is both open and closed.*

The following theorem gives us a necessary and sufficient condition on a map $f \colon X \twoheadrightarrow Y$ between compacta, to have that $\mathcal{G}_f = \{f^{-1}(y) \mid y \in Y\}$ is a continuous decomposition.

1.2.25 Theorem *Let X and Y be compacta and let $f \colon X \twoheadrightarrow Y$ be a surjective map. Then $\mathcal{G}_f = \{f^{-1}(y) \mid y \in Y\}$ is continuous if and only if f is open.*

Proof If $\mathcal{G}_f$ is a continuous decomposition of X, by Theorem 1.2.23, the quotient map $q \colon X \twoheadrightarrow X/\mathcal{G}_f$ is open. By Theorem 1.2.10, $\varphi_f \colon X/\mathcal{G}_f \twoheadrightarrow Y$ is a homeomorphism. Hence, $f = \varphi_f \circ q$ is an open map.

Now, suppose f is open. By Theorem 1.2.17, $\mathcal{G}_f$ is upper semicontinuous. Since $q = \varphi_f^{-1} \circ f$ and f is open, q is open. By Theorem 1.2.23, $\mathcal{G}_f$ is a lower semicontinuous decomposition. Therefore, $\mathcal{G}_f$ is continuous.

Q.E.D.

In the following definition a notion of convergence of sets is introduced.

1.2.26 Definition Let $\{X_n\}_{n=1}^{\infty}$ be a sequence of subsets of the metric space X. Then:

(1) the *limit inferior* of the sequence $\{X_n\}_{n=1}^{\infty}$ is defined as follows:

$$\liminf X_n = \{x \in X \mid \text{for each open subset } U \text{ of } X \text{ such that}$$

$$x \in U, \ U \cap X_n \neq \emptyset \text{ for each } n \in \mathbb{N}, \text{ save, possibly, finitely many}\}.$$

(2) the *limit superior* of the sequence $\{X_n\}_{n=1}^{\infty}$ is defined as follows:

$$\limsup X_n = \{x \in X \mid \text{for each open subset } U \text{ of } X \text{ such that}$$

$$x \in U, \ U \cap X_n \neq \emptyset \text{ for infinitely many indices } n \in \mathbb{N}\}.$$

Clearly, $\liminf X_n \subset \limsup X_n$. If $\liminf X_n = \limsup X_n = L$, then we say that the sequence $\{X_n\}_{n=1}^{\infty}$ is a *convergent sequence* with limit $L = \lim_{n \to \infty} X_n$.

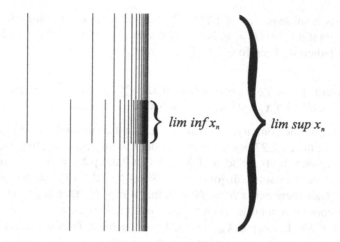

$$lim\ inf\ x_n \qquad lim\ sup\ x_n$$

1.2.27 Lemma *Let $\{X_n\}_{n=1}^{\infty}$ be a sequence of subsets of the metric space X. Then $\lim\inf X_n$ and $\lim\sup X_n$ are both closed subsets of X.*

Proof Let $x \in Cl(\lim\inf X_n)$. Let U be an open subset of X such that $x \in U$. Since $x \in Cl(\lim\inf X_n) \cap U$, we have that $\lim\inf X_n \cap U \neq \emptyset$. Hence, $U \cap X_n \neq \emptyset$ for each $n \in \mathbb{N}$, save, possibly, finitely many. Therefore, $x \in \lim\inf X_n$. The proof for $\lim\sup$ is similar.

Q.E.D.

The next theorem tells us that separable metric spaces behave like sequentially compact spaces using the notion of convergence just introduced.

1.2.28 Theorem *Each sequence $\{X_n\}_{n=1}^{\infty}$ of closed subsets of a separable metric space X has a convergent subsequence.*

Proof Let $\{U_m\}_{m=1}^{\infty}$ be a countable basis for X. Let $\{X_n^1\}_{n=1}^{\infty} = \{X_n\}_{n=1}^{\infty}$. Suppose, inductively, that we have defined the sequence $\{X_n^m\}_{n=1}^{\infty}$. We define the sequence $\{X_n^{m+1}\}_{n=1}^{\infty}$ as follows:

(1) If $\{X_n^m\}_{n=1}^{\infty}$ has a subsequence $\{X_{n_k}^m\}_{k=1}^{\infty}$ such that $\lim\sup X_{n_k}^m \cap U_m = \emptyset$, then let $\{X_n^{m+1}\}_{n=1}^{\infty}$ be such subsequence of $\{X_n^m\}_{n=1}^{\infty}$.

(2) If for each subsequence $\{X_{n_k}^m\}_{k=1}^{\infty}$ of $\{X_n^m\}_{n=1}^{\infty}$, we have that $\lim\sup X_{n_k}^m \cap U_m \neq \emptyset$, we define $\{X_n^{m+1}\}_{n=1}^{\infty}$ as $\{X_n^m\}_{n=1}^{\infty}$.

Since we have the subsequences $\{X_n^m\}_{n=1}^{\infty}$, let us consider the "diagonal subsequence" $\{X_m^m\}_{m=1}^{\infty}$. By construction, $\{X_m^m\}_{m=1}^{\infty}$ is a subsequence of $\{X_n\}_{n=1}^{\infty}$. We see that $\{X_m^m\}_{m=1}^{\infty}$ converges.

Let us assume that $\{X_m^m\}_{m=1}^{\infty}$ does not converge. Hence, there exists $p \in \lim\sup X_m^m \setminus \lim\inf X_m^m$. Let U_k be a basic open set such that $p \in U_k$ and $U_k \cap X_{m_\ell}^{m} = \emptyset$ for some subsequence $\{X_{m_\ell}^m\}_{\ell=1}^{\infty}$ of $\{X_m^m\}_{m=1}^{\infty}$ ($\lim\inf X_n$ is a closed subset of X by Lemma 1.2.27). Clearly, $\{X_{m_\ell}^{m_\ell}\}_{\ell=k}^{\infty}$ is a subsequence of $\{X_n^k\}_{n=1}^{\infty}$. Thus, $\{X_n^k\}_{n=1}^{\infty}$ satisfies condition (1), with k in place of m. Hence, $\lim\sup X_n^{k+1} \cap U_k = \emptyset$. Since

$\{X_m^m\}_{m=k+1}^{\infty}$ is a subsequence of $\{X_n^{k+1}\}_{n=1}^{\infty}$ and $\limsup X_m^m \subset \limsup X_n^{k+1}$, it follows that $\limsup X_m^m \cap U_k = \emptyset$. Now, recall that $p \in \limsup X_m^m \cap U_k$. Thus, we obtain a contradiction. Therefore, $\{X_m^m\}_{m=1}^{\infty}$ converges.

Q.E.D.

1.2.29 Theorem *Let X be a compactum. If $\{X_n\}_{n=1}^{\infty}$ is a sequence of connected subsets of X and $\liminf X_n \neq \emptyset$, then $\limsup X_n$ is connected.*

Proof Suppose, to the contrary, that $\limsup X_n$ is not connected. Since $\limsup X_n$ is closed, by Lemma 1.2.27, we assume, without loss of generality, that there exist two disjoint closed subsets A and B of X such that $\limsup X_n = A \cup B$. Since X is a metric space, there exist two disjoint open subsets U and V of X such that $A \subset U$ and $B \subset V$. Then there exists $N' \in \mathbb{N}$ such that if $n \geq N'$, then $X_n \subset U \cup V$. To show this, suppose it is not true. Then for each $n \in \mathbb{N}$, there exists $m_n > n$ such that $X_{m_n} \setminus (U \cup V) \neq \emptyset$. Let $x_{m_n} \in X_{m_n} \setminus (U \cup V)$ for each $n \in \mathbb{N}$. Since X is compact, without loss of generality, we assume that the sequence $\{x_{m_n}\}_{n=1}^{\infty}$ converges to a point x of X. Note that $x \in X \setminus (U \cup V)$ and, by construction, $x \in \limsup X_n$, a contradiction. Therefore, there exists $N' \in \mathbb{N}$ such that if $n \geq N'$, then $X_n \subset U \cup V$.

Since $\liminf X_n \neq \emptyset$ and $\liminf X_n \subset \limsup X_n$, we assume, without loss of generality, that $\liminf X_n \cap U \neq \emptyset$. Then there exists $N'' \in \mathbb{N}$ such that if $n \geq N''$, $U \cap X_n \neq \emptyset$. Let $N = \max\{N', N''\}$. Hence, if $n \geq N$, then $X_n \subset U \cup V$ and $U \cap X_n \neq \emptyset$. Since X_n is connected for every $n \in \mathbb{N}$, $X_n \cap V = \emptyset$ for each $n \geq N$, a contradiction. Therefore, $\limsup X_n$ is connected.

Q.E.D.

The following theorem gives us a characterization of an upper semicontinuous decomposition of a compactum in terms of limits inferior and superior.

1.2.30 Theorem *Let X be a compactum, with metric d. Then a decomposition $\mathcal{G}$ of X is upper semicontinuous if and only if $\mathcal{G}$ is a closed decomposition and for each sequence $\{X_n\}_{n=1}^{\infty}$ of elements of $\mathcal{G}$ and each element Y of $\mathcal{G}$ such that $\liminf X_n \cap Y \neq \emptyset$, then $\limsup X_n \subset Y$.*

Proof Suppose $\mathcal{G}$ is an upper semicontinuous decomposition of X. By Corollary 1.2.20, $\mathcal{G}$ is a closed decomposition. Let $\{X_n\}_{n=1}^{\infty}$ be a sequence of elements of $\mathcal{G}$ and let Y be an element of $\mathcal{G}$ such that $\liminf X_n \cap Y \neq \emptyset$.

Suppose there exists $p \in \limsup X_n \setminus Y$. Since Y is closed and p is not an element of Y, there exists an open set W of X such that $p \in W$ and $Cl(W) \cap Y = \emptyset$. Let $U = X \setminus Cl(W)$. Since $\mathcal{G}$ is upper semicontinuous, there exists an open set V of X such that $Y \subset V$ and if $G \in \mathcal{G}$ such that $G \cap V \neq \emptyset$, $G \subset U$.

Let $q \in \liminf X_n \cap Y$. Then $q \in \liminf X_n \cap V$. Hence, $V \cap X_n \neq \emptyset$ for each $n \in \mathbb{N}$, save, possibly, finitely many. Thus, $W \cap X_n = \emptyset$ for each $n \in \mathbb{N}$, save, possibly, finitely many. This contradicts the fact that $p \in W \cap \limsup X_n$. Therefore, $\limsup X_n \subset Y$.

Now, suppose $\mathcal{G}$ is a closed decomposition and let Y be an element of $\mathcal{G}$. Suppose that if $\{X_n\}_{n=1}^{\infty}$ is a sequence of elements of $\mathcal{G}$ such that $\liminf X_n \cap Y \neq \emptyset$, then $\limsup X_n \subset Y$.

To see $\mathcal{G}$ is upper semicontinuous, let U be an open subset of X such that $Y \subset U$. For each $n \in \mathbb{N}$, let $V_n = \mathcal{V}_{\frac{1}{n}}^d(Y)$. Suppose that for each $n \in \mathbb{N}$, there exists an element X_n of $\mathcal{G}$ such that $X_n \cap V_n \neq \emptyset$ and $X_n \not\subset U$. For each $n \in \mathbb{N}$, let $p_n \in X_n \cap V_n$. Since X is compact, $\{p_n\}_{n=1}^\infty$ has a convergent subsequence $\{p_{n_k}\}_{k=1}^\infty$. Let p be the point of convergence of $\{p_{n_k}\}_{k=1}^\infty$. Note that $p \in \liminf X_{n_k} \cap Y$. Hence, $\limsup X_{n_k} \subset Y$.

For each $k \in \mathbb{N}$, let $q_k \in X_{n_k} \setminus U$. Since X is compact, the sequence $\{q_{n_k}\}_{k=1}^\infty$ has a convergent subsequence $\{q_{n_{k_\ell}}\}_{\ell=1}^\infty$. Let q be the point of convergence of $\{q_{n_{k_\ell}}\}_{\ell=1}^\infty$. Note that $q \notin Y$ and $q \in \limsup X_{n_{k_\ell}} \subset \limsup X_{n_k}$, a contradiction. Therefore, $\mathcal{G}$ is upper semicontinuous.

Q.E.D.

As an application of Theorem 1.2.30, we have the following result which says that the components of the elements of an upper semicontinuous decomposition form another upper semicontinuous decomposition.

1.2.31 Theorem *Let X be a compactum and let $\mathcal{G}$ be an upper semicontinuous decomposition of X. If $\mathcal{D} = \{D \mid D$ is a component of G, for some $G \in \mathcal{G}\}$, then $\mathcal{D}$ is an upper semicontinuous decomposition of X.*

Proof Clearly, the elements of $\mathcal{D}$ are closed in X. Let $\{D_n\}_{n=1}^\infty$ be a sequence of elements of $\mathcal{D}$ and let $D \in \mathcal{D}$ be such that $\liminf D_n \cap D \neq \emptyset$. Let $G \in \mathcal{G}$ be such that $D \subset G$ and for each $n \in \mathbb{N}$, let $G_n \in \mathcal{G}$ be such that $D_n \subset G_n$. Since $\liminf D_n \subset \liminf G_n$ and $\liminf D_n \cap D \neq \emptyset$, $\liminf G_n \cap G \neq \emptyset$. Hence, by the upper semicontinuity of $\mathcal{G}$, $\limsup G_n \subset G$ (Theorem 1.2.30). Thus, since $\limsup D_n \subset \limsup G_n$, $\limsup D_n \subset G$.

By hypothesis, each D_n is connected. Then, by Theorem 1.2.29, $\limsup D_n$ is connected. Since D is a component of G, $\limsup D_n \subset G$ and $\limsup D_n \cap D \neq \emptyset$, $\limsup D_n \subset D$. Therefore, by Theorem 1.2.30, $\mathcal{D}$ is an upper semicontinuous decomposition.

Q.E.D.

The following theorem gives us a characterization of a continuous decomposition of a compactum in terms of limits inferior and superior.

1.2.32 Theorem *Let X be a compactum, with metric d. Then a decomposition $\mathcal{G}$ of X is continuous if and only if $\mathcal{G}$ is a closed decomposition and for each sequence $\{X_n\}_{n=1}^\infty$ of elements of $\mathcal{G}$ and each element Y of $\mathcal{G}$ such that $\liminf X_n \cap Y \neq \emptyset$, then $\limsup X_n = Y$.*

Proof Suppose $\mathcal{G}$ is a continuous decomposition of X. By Corollary 1.2.20, $\mathcal{G}$ is a closed decomposition. Let $\{X_n\}_{n=1}^\infty$ be a sequence of elements of $\mathcal{G}$ and let Y be an element of $\mathcal{G}$ such that $\liminf X_n \cap Y \neq \emptyset$. By Theorem 1.2.30, $\limsup X_n \subset Y$. Suppose there exists $p \in Y \setminus \limsup X_n$. Let U be an open subset of X such that $p \in U$ and $U \cap \limsup X_n = \emptyset$ (by Lemma 1.2.27, $\limsup X_n$ is closed). Let $q \in \limsup X_n \subset Y$. Since $\mathcal{G}$ is a lower semicontinuous decomposition, there exists an open set V of X such that $q \in V$ and if $G \in \mathcal{G}$ and $G \cap V \neq \emptyset$, then $G \cap U \neq \emptyset$.

Since $q \in \limsup X_n$ and V is an open set containing q, $V \cap X_n \neq \emptyset$ for infinitely many indices $n \in \mathbb{N}$. Hence, $U \cap X_n \neq \emptyset$ for infinitely many indices $n \in \mathbb{N}$. Then $U \cap \limsup X_n \neq \emptyset$, a contradiction. Therefore, $\limsup X_n = Y$.

Now, suppose that $\mathcal{G}$ is a closed decomposition of X such that for each sequence $\{X_n\}_{n=1}^{\infty}$ of elements of $\mathcal{G}$ and each element Y of $\mathcal{G}$ such that $\liminf X_n \cap Y \neq \emptyset$, then $\limsup X_n = Y$. By Theorem 1.2.30, $\mathcal{G}$ is upper semicontinuous.

Let Y be an element of $\mathcal{G}$. Let p and q be two points of Y, and let U be an open subset of X such that $p \in U$. For each $n \in \mathbb{N}$, let $V_n = \mathcal{V}_{\frac{1}{n}}^d(q)$. Suppose that for each $n \in \mathbb{N}$, there exists an element X_n of $\mathcal{G}$ such that $X_n \cap V_n \neq \emptyset$ and $X_n \cap U = \emptyset$. Hence, $\limsup X_n \cap U = \emptyset$. For each $n \in \mathbb{N}$, let $q_n \in X_n \cap V_n$. Clearly, $\{q_n\}_{n=1}^{\infty}$ converges to q. Thus, $q \in \liminf X_n \cap Y$. By hypothesis, $Y = \limsup X_n$. Hence, $Y \cap U \neq \emptyset$, a contradiction. Therefore, $\mathcal{G}$ is a continuous decomposition.

Q.E.D.

1.3   Homotopy and Fundamental Group

We introduce the fundamental group of a metric space. We show that the fundamental group of the unit circle $\mathcal{S}^1$ is isomorphic to the group of integers $\mathbb{Z}$. This section is based on the following [1, 10, 13, 19, 29].

We assume that the reader is familiar with the elementary concepts of group theory. The reader may find more than enough information about groups in [29].

1.3.1 Definition   Given a metric space X and two points x and y of X, a *path* joining x and y is a map $\alpha \colon [0, 1] \to X$ such that $\alpha(0) = x$ and $\alpha(1) = y$. In this case, x and y are the *end points* of α. When α is one-to-one, α is called an *arc*. Some times, we identify a path or an arc α with its image $\alpha([0, 1])$.

1.3.2 Definition   Let X and Y be metric spaces. Let $g, f \colon X \to Y$ be two maps. We say that *f is homotopic to g* provided that there exists a map $G \colon X \times [0, 1] \to Y$ such that $G((x, 0)) = f(x)$ and $G((x, 1)) = g(x)$ for each $x \in X$. The map G is called a *homotopy between f and g*. If for each $t \in [0, 1]$, $G(\cdot, t) \colon X \twoheadrightarrow Y$ is a homeomorphism, G is called an *isotopy between f and g*.

1.3.3 Example   Let $g, f \colon X \to Y$ be two constant maps between metric spaces, say $g(x) = q$ and $f(x) = p$ for each $x \in X$. Then f and g are homotopic if and only if p and q both belong to the same path component of Y. To show this, suppose first that p and q belong to the same path component of Y. Then there exists a map $\alpha \colon [0, 1] \to Y$ such that $\alpha(0) = p$ and $\alpha(1) = q$. Hence, $G \colon X \times [0, 1] \to Y$ given by $G((x, t)) = \alpha(t)$ is a homotopy between f and g. Now, suppose f and g are homotopic. Then there exists a map $G' \colon X \times [0, 1] \to Y$ such that $G'((x, 0)) = f(x) = p$ and $G'((x, 1)) = g(x) = q$ for each $x \in X$. Let $x_0 \in X$. Then the map $\beta \colon [0, 1] \to Y$ given by $\beta(t) = G'((x_0, t))$ is a path such that $\beta(0) = p$ and $\beta(1) = q$.

1.3.4 Example Let X be a metric space. If $g, f : X \to \mathbb{R}^n$ are two maps, for some $n \in \mathbb{N}$. Then f and g are homotopic. To see this, let $G : X \times [0, 1] \to \mathbb{R}^n$ be given by $G((x, t)) = (1 - t) f(x) + t g(x)$. Then G is a homotopy between f and g.

The next theorem is known as *Borsuk's Homotopy Extension Theorem*. A proof of this result may be found in [13, Theorem 4–4].

1.3.5 Theorem *Let A be a closed subset of a separable metric space X, and let $g', f' : A \to S^n$ be homotopic maps of A into the n-sphere S^n. If there exists a map $f : X \to S^n$ such that $f|_A = f'$ (i.e., f extends f'), then there also exists a map $g : X \to S^n$ such that $g|_A = g'$, and f and g may be chosen to be homotopic too.*

1.3.6 Theorem *Let Y be a closed subset of a compactum X. If $f : X \to S^1$ is a map such that $f|_Y : Y \to S^1$ is homotopic to a constant map, then there exists an open set U of X such that $Y \subset U$ and $f|_U : U \to S^1$ is homotopic to a constant map.*

Proof Let $f : X \to S^1$ be a map such that $f|_Y : Y \to S^1$ is homotopic to a constant map $g' : Y \to S^1$ given by $g'(y) = b$ for every $y \in Y$.

Since f is a map which extends $f|_Y$ to X, by Theorem 1.3.5, there exists a map $g : X \to S^1$ such that $g|_Y = g'$ and f and g are homotopic.

Let V be a proper open subset of S^1 such that $b \in V$. By continuity of g, there exists an open set U of X such that $Y \subset U$ and $g(U) \subset V$. Since V is an arc, it is easy to see that $g|_U$ is homotopic to a constant map. Since f and g are homotopic, $f|_U$ is also homotopic to a constant map.

Q.E.D.

1.3.7 Theorem *Let X be a compactum. If $f : X \twoheadrightarrow S^1$ is a map not homotopic to a constant map, then there exists a closed connected subset C of X such that $f|_C : C \twoheadrightarrow S^1$ is not homotopic to a constant map and for each proper closed subset C' of C, $f|_{C'} : C \to S^1$ is homotopic to a constant map.*

Proof Let $\mathcal{C}$ be the collection of all closed subsets, K, of X such that $f|_K : K \twoheadrightarrow S^1$ is not homotopic to a constant map. Since $X \in \mathcal{C}, \mathcal{C} \neq \emptyset$. Define the following partial order on $\mathcal{C}$. If $K_1, K_2 \in \mathcal{C}$, then $K_1 > K_2$ if $K_1 \subset K_2$. Let $\mathcal{K}$ be a (set theoretic) chain of elements of $\mathcal{C}$. We show that $\mathcal{K}$ has an upper bound. Let $Y = \bigcap_{K \in \mathcal{K}} K$. We assert that $Y \in \mathcal{C}$. If Y does not belong to $\mathcal{C}$, then $f|_Y : Y \to S^1$ is homotopic to a constant map. Hence, by Theorem 1.3.6 there exists an open subset U of X such that $f|_U : U \to S^1$ is homotopic to a constant map. Note that $\{X \setminus K \mid K \in \mathcal{K}\}$ is an open cover of $X \setminus U$. Then there exist $K_1, \ldots, K_m \in \mathcal{K}$ such that $X \setminus U \subset \bigcup_{j=1}^m X \setminus K_j$. Hence, $K = \bigcap_{j=1}^m K_j \subset U$. Since $\mathcal{K}$ is a chain, $K \in \mathcal{K}$. Since $K \subset U$, $f|_K : K \to S^1$ is homotopic to a constant map, a contradiction to the fact that $K \in \mathcal{C}$. Therefore, $Y \in \mathcal{C}$ and Y is an upper bound for $\mathcal{K}$. Thus, by Kuratowski–Zorn Lemma, there exists an element $C \in \mathcal{C}$ such that $f|_C : C \twoheadrightarrow S^1$ is not homotopic to a constant map and for each proper closed subset C' of C, $f|_{C'} : C \to S^1$ is homotopic to a constant map. It remains to show that C is connected.

Suppose C is not connected. Thus, there exist two disjoint closed subsets C_1 and C_2 of X such that $C = C_1 \cup C_2$. Since C is a maximal element of $\mathcal{C}$, $f|_{C_j} \colon C_j \to S^1$ is homotopic to a constant map $g_j \colon C_j \to S^1$ by a homotopy H_j such that $H_j((c, 0)) = (f|_{C_j})(c)$ and $H_j((c, 1)) = g_j(c)$ for each $c \in C_j$ and $j \in \{1, 2\}$. Since S^1 is arcwise connected, we assume, without loss of generality, that g_1 and g_2 have the same image $\{b\} \subset S^1$. Let $g \colon C \to S^1$ be given by $g(c) = b$ for each $c \in C$. Let $H \colon C \times [0, 1] \to S^1$ be given by $H((c, t)) = H_j((c, t))$ if $c \in C_j$, $j \in \{1, 2\}$. Then H is a homotopy such that $H((c, 0)) = (f|_{C_j})(c) = (f|_C)(c)$ and $H((c, 1)) = g_j(c) = b$ for every $c \in C$. Thus, $f|_C$ is homotopic to a constant map, a contradiction. Therefore, C is connected.

Q.E.D.

1.3.8 Theorem *Let X and Y be metric spaces. The relation of homotopy is an equivalence relation in the set of maps between X and Y. The equivalence classes of this equivalence relation are called homotopy classes.*

Proof Let $f \colon X \to Y$ be a map. Then $G \colon X \times [0, 1] \to Y$, given by $G((x, t)) = f(x)$, is a homotopy between f and f. Hence, the relation is reflexive.

Now, let $g, f \colon X \to Y$ be two maps and suppose f is homotopic to g. Then there exists a homotopy $H \colon X \times [0, 1] \to Y$ such that $H((x, 0)) = f(x)$ and $H((x, 1)) = g(x)$ for each $x \in X$. Hence, the map $K \colon X \times [0, 1] \to Y$ given by $K((x, t)) = H((x, 1 - t))$ is a homotopy between g and f, since $K((x, 0)) = g(x)$ and $K((x, 1)) = f(x)$ for each $x \in X$. Thus, the relation is symmetric.

Finally, let $h, g, f \colon X \to Y$ be three maps and suppose that f is homotopic to g and g is homotopic to h. Then there exist two homotopies $J, L \colon X \times [0, 1] \to Y$ such that $J((x, 0)) = f(x)$, $J((x, 1)) = g(x)$, $L((x, 0)) = g(x)$ and $L((x, 1)) = h(x)$ for each $x \in X$. Thus, the map $R \colon X \times [0, 1] \to Y$ given by

$$
R((x, t)) = \begin{cases} J((x, 2t)), & \text{if } t \in \left[0, \dfrac{1}{2}\right]; \\[2mm] L((x, 2t - 1)), & \text{if } t \in \left[\dfrac{1}{2}, 1\right], \end{cases}
$$

is a homotopy between f and h, since for each $x \in X$, $R((x, 0)) = f(x)$ and $R((x, 1)) = h(x)$. Hence, the relation is transitive.

Q.E.D.

1.3.9 Notation If $f \colon X \to Y$ is a map, then the homotopy class to which f belongs is denoted by $[f]$.

1.3.10 Definition A metric space X is said to be *contractible* provided that the identity map, 1_X, of X is homotopic to a constant map g. We say that X is *locally contractible at p* if for each neighborhood U of p in X, there exist a neighborhood V of p in X and a homotopy $G \colon V \times [0, 1] \to U$ such that $G((x, 0)) = x$ and $G((x, 1)) = x_0$ for each $x \in V$ and some $x_0 \in U$. The metric space X is *locally contractible* if it is locally contractible at each of its points.

The following two theorems present some consequences of the contractibility of a space.

1.3.11 Theorem *If X is a contractible metric space, then X is path connected.*

Proof Since X is contractible, there exists a map $G: X \times [0, 1] \to X$ such that $G((x, 0)) = x$ and $G((x, 1)) = p$ for each $x \in X$ and some point p of X. Let x and y be two points of X. Then the map $\alpha: [0, 1] \to X$ given by

$$
\alpha(t) = \begin{cases} G((x, 2t)), & \text{if } t \in \left[0, \dfrac{1}{2}\right]; \\ G((y, 2 - 2t)), & \text{if } t \in \left[\dfrac{1}{2}, 1\right], \end{cases}
$$

is a path such that $\alpha(0) = x$ and $\alpha(1) = y$. Therefore, X is path connected.

Q.E.D.

1.3.12 Theorem *Let X and Y be metric spaces, where Y is arcwise connected. If either X or Y is contractible and if $g, f: X \to Y$ are two maps, then f and g are homotopic.*

Proof Suppose Y is contractible. Then there exists a map $G: Y \times [0, 1] \twoheadrightarrow Y$ such that $G((y, 0)) = y$ and $G((y, 1)) = y_0$ for each $y \in Y$ and some point y_0 of Y. Then the map $K: X \times [0, 1] \to Y$ given by

$$
K((x, t)) = \begin{cases} G((f(x), 2t)), & \text{if } t \in \left[0, \dfrac{1}{2}\right]; \\ G((g(x), 2 - 2t)), & \text{if } t \in \left[\dfrac{1}{2}, 1\right], \end{cases}
$$

is a homotopy between f and g.

Now, assume X is contractible. Then there exists a map $H: X \times [0, 1] \twoheadrightarrow X$ such that $H((x, 0)) = x$ and $H((x, 1)) = x_0$ for each $x \in X$ and some point $x_0 \in X$. Let $\alpha: [0, 1] \to Y$ be a path such that $\alpha(0) = f(x_0)$ and $\alpha(1) = g(x_0)$. Then the map $L: X \times [0, 1]) \to Y$ given by

$$
L((x, t)) = \begin{cases} f \circ H((x, 3t)), & \text{if } t \in \left[0, \dfrac{1}{3}\right]; \\ \alpha(3t - 1), & \text{if } t \in \left[\dfrac{1}{3}, \dfrac{2}{3}\right]; \\ g \circ H((x, 3t - 2)), & \text{if } t \in \left[\dfrac{2}{3}, 1\right], \end{cases}
$$

is a homotopy between f and g.

Q.E.D.

Now, we consider a particular case of homotopy; namely, we study the homotopies between paths.

1.3.13 Definition Let X be a metric space. We say that two paths $\beta, \alpha\colon [0, 1] \to X$ are *homotopic relative to* $\{0, 1\}$ provided that there exists a homotopy $G\colon [0, 1] \times [0, 1] \to X$ such that $G((s, 0)) = \alpha(s)$, $G((s, 1)) = \beta(s)$, $G((0, t)) = \alpha(0) = \beta(0)$ and $G((1, t)) = \alpha(1) = \beta(1)$ for each $s, t \in [0, 1]$.

In order to define the fundamental group we need the following definitions:

1.3.14 Definition Let X be a metric space and let x_0 be a point in X. The pair (X, x_0) is called *pointed space*.

1.3.15 Definition Let X be a metric space. We say that a path $\alpha\colon [0, 1] \to X$ is *closed* provided that $\alpha(0) = \alpha(1)$.

1.3.16 Notation If (X, x_0) is a pointed space and $\alpha\colon [0, 1] \to X$ is a closed path, we assume that $\alpha(0) = \alpha(1) = x_0$. The point x_0 is called the *base of the closed path*.

1.3.17 Definition Let X be a metric space. Given two closed paths $\beta, \alpha\colon [0, 1] \to X$ such that $\alpha(0) = \beta(0)$, we define their *product*, denoted by $\alpha * \beta$, as the closed path given by:

$$(\alpha * \beta)(t) = \begin{cases} \alpha(2t), & \text{if } t \in \left[0, \dfrac{1}{2}\right]; \\ \beta(2t - 1), & \text{if } t \in \left[\dfrac{1}{2}, 1\right]. \end{cases}$$

1.3.18 Definition Let X be a metric space. Given a closed path $\alpha\colon [0, 1] \to X$, we define its *inverse*, denoted by α^{-1}, as the closed path $\alpha^{-1}\colon [0, 1] \to X$ given by $\alpha^{-1}(t) = \alpha(1 - t)$.

1.3.19 Theorem *Let X be a metric space. If $\beta', \beta, \alpha', \alpha\colon [0, 1] \to X$ are closed paths such that $\alpha(0) = \alpha'(0) = \beta(0) = \beta'(0)$ and such that α is homotopic to α' relative to $\{0, 1\}$ and β is homotopic to β' relative to $\{0, 1\}$, then $\alpha * \beta$ is homotopic to $\alpha' * \beta'$ relative to $\{0, 1\}$ and α^{-1} is homotopic to $(\alpha')^{-1}$ relative to $\{0, 1\}$.*

Proof Since α is homotopic to α' relative to $\{0, 1\}$ and β is homotopic to β' relative to $\{0, 1\}$, there exist two homotopies $K, G\colon [0, 1] \times [0, 1] \to X$ such that $G((s, 0)) = \alpha(s)$, $G((s, 1)) = \alpha'(s)$, $G((0, t)) = G((1, t)) = \alpha(0) = \alpha'(0)$, $K((s, 0)) = \beta(s)$, $K((s, 1)) = \beta'(s)$, and $K((0, t)) = K((1, t)) = \beta(0) = \beta'(0)$. Let $L\colon [0, 1] \times [0, 1] \to X$ be given by

$$L((s, t)) = \begin{cases} G((2s, t)), & \text{if } s \in \left[0, \dfrac{1}{2}\right]; \\ K((2s - 1, t)), & \text{if } s \in \left[\dfrac{1}{2}, 1\right]. \end{cases}$$

Then L is the required homotopy between $\alpha * \beta$ and $\alpha' * \beta'$, relative to $\{0, 1\}$.

Next, let $R: [0, 1] \times [0, 1] \to X$ be given by $R((s, t)) = G((1 - s, t))$. Then R is the required homotopy between α^{-1} and $(\alpha')^{-1}$, relative to $\{0, 1\}$.

$$\textbf{Q.E.D.}$$

We are ready to define the fundamental group of a metric space.

1.3.20 Definition Let (X, x_0) be a pointed space. The *fundamental group of* (X, x_0), denoted by $\pi_1(X, x_0)$, is the family of all homotopy classes of closed paths whose base is x_0. The group operation is given by $[\alpha] * [\beta] = [\alpha * \beta]$.

1.3.21 Remark If (X, x_0) is a pointed space, then, by Theorem 1.3.19, the operation defined on $\pi_1(X, x_0)$ is well defined. Clearly, the identity element of $\pi_1(X, x_0)$ is the homotopy class of the constant path "x_0."

1.3.22 Notation Let (X, x_0) and (Y, y_0) be two pointed spaces. By a *map,* $f: (X, x_0) \to (Y, y_0)$, *between the pointed spaces* (X, x_0) *and* (Y, y_0), we mean a map $f: X \to Y$ such that $f(x_0) = y_0$.

1.3.23 Lemma *Let* $f: (X, x_0) \to (Y, y_0)$ *be a map between pointed spaces. If* α *and* β *are two closed paths whose base is* x_0 *which are homotopic relative to* $\{0, 1\}$, *then* $f \circ \alpha$ *and* $f \circ \beta$ *are two closed paths whose base is* y_0 *which are homotopic relative to* $\{0, 1\}$.

Proof Clearly, $f \circ \alpha$ and $f \circ \beta$ are two closed paths whose base is y_0.

Since α and β are homotopic relative to $\{0, 1\}$, there exists a homotopy $G: [0, 1] \times [0, 1] \to X$ such that $G((s, 0)) = \alpha(s)$, $G((s, 1)) = \beta(s)$ and $G((0, t)) = G((1, t)) = x_0$ for each $s, t \in [0, 1]$. Then the map $K: [0, 1] \times [0, 1] \to Y$ given by $K((s, t)) = f(G((s, t)))$ is a homotopy between $f \circ \alpha$ and $f \circ \beta$ relative to $\{0, 1\}$.

$$\textbf{Q.E.D.}$$

1.3.24 Definition If $f: (X, x_0) \to (Y, y_0)$ is a map between pointed spaces, then f *induces a homomorphism* $\pi_1(f): \pi_1(X, x_0) \to \pi_1(Y, y_0)$ given by $\pi_1(f)([\alpha]) = [f \circ \alpha]$.

1.3.25 Remark Let $f: (X, x_0) \to (Y, y_0)$ be a map between pointed spaces. If α and β are two closed paths whose base is x_0, then, clearly, $f \circ (\alpha * \beta) = (f \circ \alpha) * (f \circ \beta)$. Hence, the induced map defined in Definition 1.3.24 is a well defined group homomorphism.

The next lemma says that the induced map of a composition is the composition of the induced maps.

1.3.26 Lemma *If* $f: (X, x_0) \to (Y, y_0)$ *and* $g: (Y, y_0) \to (Z, z_0)$ *are maps between pointed spaces, then* $\pi_1(g \circ f) = \pi_1(g) \circ \pi_1(f)$.

Proof Let $[\alpha] \in \pi_1(X, x_0)$. Then $\pi_1(g \circ f)([\alpha]) = [(g \circ f) \circ \alpha] = [g \circ (f \circ \alpha)] = \pi_1(g)([f \circ \alpha]) = \pi_1(g)(\pi_1(f)([\alpha])) = (\pi_1(g) \circ \pi_1(f))([\alpha])$.

$$\textbf{Q.E.D.}$$

In order to show that the fundamental group of the unit circle $\mathcal{S}^1$ is isomorphic to $\mathbb{Z}$, we associate to each closed path α in $\mathcal{S}^1$ a number $\eta(\alpha)$, which is called the *degree of* α, in such a way that two closed paths are homotopic if and only if they have the same degree. We use the exponential map $\exp\colon \mathbb{R} \twoheadrightarrow \mathcal{S}^1$ given by $\exp(t) = e^{it}$.

1.3.27 Remark Recall that given $z \in \mathcal{S}^1$, $\exp^{-1}(z) = \{t + 2\pi n \mid n \in \mathbb{Z}\}$, where t is any real number such that $\exp(t) = z$.

1.3.28 Notation Let A be a nonempty subset of $\mathbb{R}$ and let $t \in \mathbb{R}$. Then $A + t = \{a + t \mid a \in A\}$.

1.3.29 Lemma *The exponential map is open.*

Proof Let U be an open subset of $\mathbb{R}$, and let $F = \mathcal{S}^1 \setminus \exp(U)$. We show that F is closed in $\mathcal{S}^1$.

Note that $\exp^{-1}(\exp(U)) = \bigcup \{U + 2\pi n \mid n \in \mathbb{Z}\}$, which is an open subset of $\mathbb{R}$. Hence, its complement, $\exp^{-1}(F)$, is closed in $\mathbb{R}$. Since for each $t \in \exp^{-1}(F)$, there exists $t' \in [0, 2\pi]$ such that $\exp(t') = \exp(t)$, $F = \exp(\exp^{-1}(F)) = \exp(\exp^{-1}(F) \cap [0, 2\pi])$. Since $\exp^{-1}(F) \cap [0, 2\pi]$ is compact, F is compact. Thus, F is closed in $\mathcal{S}^1$.

Q.E.D.

1.3.30 Corollary *If $t \in \mathbb{R}$, then the restriction,*

$$\exp|_{(t,t+2\pi)}\colon (t, t + 2\pi) \twoheadrightarrow \mathcal{S}^1 \setminus \{\exp(t)\},$$

of the exponential map to $(t, t + 2\pi)$ is a homeomorphism onto $\mathcal{S}^1 \setminus \{\exp(t)\}$.

1.3.31 Theorem *Let $\alpha\colon [0, 1] \to \mathcal{S}^1$ be a closed path whose base is $(1, 0)$. Then there exists a unique map $\alpha^\star\colon [0, 1] \to \mathbb{R}$ such that $\alpha^\star(0) = 0$ and $\alpha(t) = \exp(\alpha^\star(t))$. The map $\alpha^\star$ is called the lifting of α beginning at 0.*

Proof First, suppose that $\alpha([0, 1]) \neq \mathcal{S}^1$. Let A be the component of $\exp^{-1}(\alpha([0, 1]))$ containing 0. Then $\exp|_A\colon A \twoheadrightarrow \alpha([0, 1])$ is a homeomorphism (Corollary 1.3.30). Hence, the map $\alpha^\star\colon [0, 1] \to \mathbb{R}$ given by $\alpha^\star(t) = (\exp|_A)^{-1}(\alpha(t))$ is the required map.

Next, suppose $\alpha([0, 1]) = \mathcal{S}^1$. Let $t_0 = 0 < t_1 < \cdots < t_{n-1} < t_n = 1$ be a subdivision of $[0, 1]$ such that $\alpha([t_{j-1}, t_j]) \neq \mathcal{S}^1$ for each $j \in \{1, \ldots, n\}$.

Let A_0 be the component of $\exp^{-1}(\alpha([0, t_1]))$ containing 0. Then, as before, $\exp|_{A_0}\colon A_0 \twoheadrightarrow \alpha([0, t_1])$ is a homeomorphism. Let $\alpha_0^*\colon [0, t_1] \to \mathbb{R}$ be given by $\alpha_0^*(t) = (\exp|_{A_0})^{-1}(\alpha(t))$. Let A_1 be the component of $\exp^{-1}(\alpha([t_1, t_2]))$ containing $\alpha_0^*(t_1)$, and let $\alpha_1^*\colon [t_1, t_2] \to \mathbb{R}$ be given by $\alpha_1^*(t) = (\exp|_{A_1})^{-1}(\alpha(t))$. Repeating this process, for each $j \in \{0, \ldots, n-1\}$, we define maps $\alpha_j^*\colon [t_j, t_{j+1}] \to \mathbb{R}$ given by $\alpha_j^*(t) = (\exp|_{A_j})^{-1}(\alpha(t))$.

Let $\alpha^\star\colon [0, 1] \to \mathbb{R}$ be given by $\alpha^\star(t) = \alpha_j^*(t)$ if $t \in [t_j, t_{j+1}]$. Then $\alpha^\star(0) = 0$ and $\alpha(t) = \exp(\alpha^\star(t))$.

To see α^* is unique, suppose $\beta \colon [0, 1] \to \mathbb{R}$ is another map such that $\beta(0) = 0$ and $\alpha(t) = \exp(\beta(t))$. Hence, $\exp(\alpha^*(t)) = \exp(\beta(t))$ for each $t \in [0, 1]$. Consider the map $\gamma \colon [0, 1] \to \mathbb{R}$ given by $\gamma(t) = \frac{\alpha^*(t) - \beta(t)}{2\pi}$. Then $\gamma([0, 1]) \subset \mathbb{Z}$. Since $\alpha^*(0) = \beta(0)$, $\gamma([0, 1]) = \{0\}$. Therefore, $\alpha^*(t) = \beta(t)$ for each $t \in [0, 1]$.

$$\text{Q.E.D.}$$

1.3.32 Remark If in Theorem 1.3.31 we do not require that the map α^* satisfies that $\alpha^*(0) = 0$, we may have many liftings of the map α. However, any other lifting α' of α satisfies that $\alpha'(t) = \alpha^*(t) + 2\pi k$ for some $k \in \mathbb{Z}$ and each $t \in [0, 1]$.

1.3.33 Definition Let $\alpha \colon [0, 1] \to S^1$ be a closed path, and let α' be a lifting of α. Then

$$\eta(\alpha) = \frac{\alpha'(1) - \alpha'(0)}{2\pi}$$

is an integer, and it is called the *degree of α*.

1.3.34 Remark Observe that for each closed path α, the definition of $\eta(\alpha)$ does not depend on the lifting of α. Since for any two liftings α' and α'' of α, we have that $\alpha'(1) - \alpha'(0) = \alpha''(1) - \alpha''(0)$ by Remark 1.3.32. Intuitively, the degree of a closed path "counts" the number of times that the closed path wraps $[0, 1]$ around S^1.

1.3.35 Theorem *Let $\beta, \alpha \colon [0, 1] \to S^1$ be two closed paths whose base is $(1, 0)$. Then:*

(1) $\eta(\alpha * \beta) = \eta(\alpha) + \eta(\beta)$;
(2) If α and β are homotopic relative to $\{0, 1\}$ if and only if $\eta(\alpha) = \eta(\beta)$;
(3) Given $k \in \mathbb{Z}$, there exists a closed path γ, whose base is $(1, 0)$, such that $\eta(\gamma) = k$.

Proof Let $\alpha^* \colon [0, 1] \to \mathbb{R}$ be a lifting of α such that $\alpha^*(0) = 0$ (Theorem 1.3.31), and let $\beta^* \colon [0, 1] \to \mathbb{R}$ be a lifting of β such that $\beta^*(0) = \alpha^*(1)$. Define $\alpha^* * \beta^* \colon [0, 1] \to \mathbb{R}$ by

$$(\alpha^* * \beta^*)(s) = \begin{cases} \alpha^*(2s), & \text{if } s \in \left[0, \dfrac{1}{2}\right]; \\ \beta^*(2s - 1), & \text{if } s \in \left[\dfrac{1}{2}, 1\right]. \end{cases}$$

Then it is easy to see that $\alpha^* * \beta^*$ is a lifting of $\alpha * \beta$. Since $2\pi \eta(\alpha * \beta) = (\alpha^* * \beta^*)(1) - (\alpha^* * \beta^*)(0) = \beta^*(1) - \alpha^*(0) = (\beta^*(1) - \beta^*(0)) + (\alpha^*(1) - \alpha^*(0)) = 2\pi(\eta(\beta) + \eta(\alpha))$, we have that $\eta(\alpha * \beta) = \eta(\alpha) + \eta(\beta)$.

Suppose $\eta(\alpha) = \eta(\beta)$. Let α' and β' be liftings of α and β, respectively, such that $\alpha'(0) = \beta'(0) = 0$. Since $\eta(\alpha) = \eta(\beta)$, $\alpha'(1) - \alpha'(0) = \beta'(1) - \beta'(0)$. In particular, $\alpha'(1) = \beta'(1)$. By Example 1.3.4, the map $G \colon [0, 1] \times [0, 1] \to \mathbb{R}$ given by $G((s, t)) = (1 - t)\alpha'(s) + t\beta'(s)$ is a homotopy between α' and β'. Note that for each $t \in [0, 1]$, $G((1, t)) - G((0, t)) = (1 - t)\left[\alpha'(1) - \alpha'(0)\right] + t\left[\beta'(1) - \beta'(0)\right] =$

$(1 - t)2\pi\eta(\alpha) + t2\pi\eta(\beta) = 2\pi\eta(\alpha)$. Hence, the map $K = \exp \circ G$ is a homotopy between α and β such that $K((0, t)) = K((1, t)) = (1, 0)$. Thus, K is a homotopy between α and β relative to $\{0, 1\}$.

Now, assume α and β are homotopic relative to $\{0, 1\}$. First, suppose that $\|\alpha(s) - \beta(s)\| < 2$ for each $s \in [0, 1]$, i.e., $\alpha(s)$ and $\beta(s)$ are never antipodal points. Let α'' and β'' be liftings of α and β, respectively, such that $\alpha''(0) = \beta''(0) = 0$. Since $\|\alpha(s) - \beta(s)\| < 2$, $|\alpha''(s) - \beta''(s)| < \pi$ for each $s \in [0, 1]$. Hence, $2\pi|\eta(\alpha) - \eta(\beta)| = |\alpha''(1) - \alpha''(0) - \beta''(1) + \beta''(0)| \le |\alpha''(1) - \beta''(1)| + |\alpha''(0) - \beta''(0)| < \pi + \pi = 2\pi$. Thus, $|\eta(\alpha) - \eta(\beta)| = 0$, i.e., $\eta(\alpha) = \eta(\beta)$. Next, assume there exists an $s \in [0, 1]$ such that $\|\alpha(s) - \beta(s)\| = 2$. Let $H: [0, 1] \times [0, 1] \to S^1$ be a homotopy between α and β relative to $\{0, 1\}$. Since H is uniformly continuous, there exists $\delta > 0$ such that if $|t - t'| < \delta$, then $\|H((s, t)) - H((s, t'))\| < 2$ for each $s \in [0, 1]$. Let $t_0 = 0 < t_1 < \cdots < t_{k-1} < t_k = 1$ be a subdivision of $[0, 1]$ such that $t_j - t_{j-1} < \delta$ for each $j \in \{1, \dots, k\}$. Let $\alpha_j: [0, 1] \to S^1$ be given by $\alpha_j(s) = H((s, t_j))$ for each $j \in \{0, \dots, k\}$. Note that $\alpha_0 = \alpha$ and $\alpha_k = \beta$. By construction, $\|\alpha_{j-1}(s) - \alpha_j(s)\| < 2$ for each $j \in \{1, \dots, k\}$. Then, applying the above argument, we obtain that $\eta(\alpha) = \eta(\alpha_0) = \eta(\alpha_1) = \cdots = \eta(\alpha_k) = \eta(\beta)$.

Finally, let $k \in \mathbb{Z}$. Define $\gamma: [0, 1] \to S^1$ by $\gamma(s) = \exp(2\pi k s)$. Then γ is a closed path whose base is $(1, 0)$. Note that the map $\gamma^\star: [0, 1] \to \mathbb{R}$ given by $\gamma^\star(s) = 2\pi k s$ is a lifting of γ such that $\gamma^\star(0) = 0$. Hence, $\eta(\gamma) = \frac{\gamma^\star(1) - \gamma^\star(0)}{2\pi} = k$.

Q.E.D.

Now, we are ready to show that the fundamental group of the unit circle S^1 is isomorphic to $\mathbb{Z}$.

1.3.36 Theorem *The fundamental group of the unit circle S^1 is isomorphic to $\mathbb{Z}$.*

Proof Let $\Sigma: \pi_1(S^1) \twoheadrightarrow \mathbb{Z}$ be given by $\Sigma([\alpha]) = \eta(\alpha)$. Note that, by Theorem 1.3.35 (2), Σ is well defined and by (3) of the same Theorem, Σ is, indeed, a surjection. By Theorem 1.3.35 (1), Σ is a homomorphism. Finally, by Theorem 1.3.35 (2), Σ is one-to-one. Therefore, Σ is an isomorphism.

Q.E.D.

Next, we define the degree of a map between simple closed curves.

1.3.37 Definition Let $f: S^1 \to S^1$ be a map. The *degree of f*, denoted by $\deg(f)$, is defined as follows. Consider the induced map $\pi_1(f): \pi_1(S^1) \to \pi_1(S^1)$. Then $\deg(f) = (\Sigma \circ \pi_1(f))(\Sigma^{-1}(1))$, where Σ is defined in Theorem 1.3.36.

1.3.38 Remark Let $f: S^1 \to S^1$ be a map. If $\deg(f) = 0$, then f is homotopic to a constant map [10, Theorem 7.4, p. 352].

1.3.39 Lemma *If $g, f: S^1 \to S^1$ are two maps, then $\deg(g \circ f) = \deg(g) \cdot \deg(f)$.*

Proof Recall that, by Lemma 1.3.26, $\pi_1(g \circ f) = \pi_1(g) \circ \pi_1(f)$.

By definition,

$$\deg(g \circ f) = [\Sigma \circ \pi_1(g \circ f)] (\Sigma^{-1}(1)) =$$

$$[\Sigma \circ (\pi_1(g) \circ \pi_1(f))] (\Sigma^{-1}(1)) = [\Sigma \circ \pi_1(g)] \left(\pi_1(f)(\Sigma^{-1}(1))\right) =$$

$$[\Sigma \circ \pi_1(g)] \left(\pi_1(f)(\Sigma^{-1}(1)) * \Sigma^{-1}(1)\right) =$$

$$\left[\Sigma \circ \pi_1(f)(\Sigma^{-1}(1))\right] \cdot \left[\Sigma \circ \pi_1(g)(\Sigma^{-1}(1))\right] = \deg(f) \cdot \deg(g).$$

Q.E.D.

We end this section with the following Theorem.

1.3.40 Theorem *A map* $f : S^1 \to S^1$ *is homotopic to a constant map if and only if there exists a map* $\tilde{f} : S^1 \to \mathbb{R}$ *such that* $f = \exp \circ \tilde{f}$.

Proof If $\tilde{f} : S^1 \to \mathbb{R}$ exists such that $f = \exp \circ \tilde{f}$, then $\tilde{f}$ is homotopic to a constant map since $\mathbb{R}$ is contractible (Theorem 1.3.12). Hence, $f = \exp \circ \tilde{f}$ is also homotopic to a constant map.

Now, suppose f is homotopic to a constant map. Without loss of generality, we assume that $f((1, 0)) = (1, 0)$.

Since f is homotopic to a constant map, $\deg(f) = 0$. Let $g : [0, 1] \twoheadrightarrow S^1$ be given by $g(t) = \exp(2\pi t)$. Since f is homotopic to a constant map, $f \circ g$ is homotopic to a constant map. By Theorem 1.3.35 (2), $\eta(f \circ g) = 0$. Let $\xi : [0, 1] \to \mathbb{R}$ be a map such that $\exp \circ \xi = f \circ g$; i.e., ξ is a lifting of $f \circ g$ (Theorem 1.3.31). Since $\eta(f \circ g) = 0$, $\xi(0) = \xi(1)$. Hence, the function $\tilde{f} : S^1 \to \mathbb{R}$ given by $\tilde{f}(z) = \xi(g^{-1}(z))$ is well defined, continuous, and satisfies that $(\exp \circ \tilde{f})(z) = (\exp \circ \xi)(g^{-1}(z)) = (f \circ g)(g^{-1}(z)) = f(z)$ for each $z \in S^1$.

Q.E.D.

1.4 Geometric Complexes and Polyhedra

This is a very small section. We present the definitions of polyhedra and the nerve of a finite collection of sets, which are used in Chap. 2. We use [10, 13, 18, 24] to produce this section.

1.4.1 Definition We say that $r + 1$ points, $\{x_0, \ldots, x_r\}$, of $\mathbb{R}^n$ are *affinely independent* provided that the set $\{x_1 - x_0, \ldots, x_r - x_0\}$ is linearly independent.

The following theorem gives us an alternative way, in terms of linear algebra, to see affinely independent subsets of $\mathbb{R}^n$.

1.4.2 Theorem *The set of points $\{x_0, \ldots, x_r\}$ of $\mathbb{R}^n$ is affinely independent if and only if each time*

$$\sum_{j=0}^{r} \xi_j x_j = 0 \text{ and } \sum_{j=0}^{r} \xi_j = 0,$$

where $\xi_j \in \mathbb{R}$, we have that $\xi_j = 0$ for each $j \in \{0, \ldots, r\}$.

Proof Suppose $\{x_0, \ldots, x_r\}$ is affinely independent. Let $\xi_0, \ldots, \xi_r$ be real numbers such that $\sum_{j=0}^{r} \xi_j x_j = 0$ and $\sum_{j=0}^{r} \xi_j = 0$.

Since $\sum_{j=0}^{r} \xi_j = 0$, $\xi_0 = -\sum_{j=1}^{r} \xi_j$. Then:

$$0 = \sum_{j=0}^{r} \xi_j x_j = \xi_0 x_0 + \sum_{j=1}^{r} \xi_j x_j =$$

$$\left(-\sum_{j=1}^{r} \xi_j\right) x_0 + \sum_{j=1}^{r} \xi_j x_j = \sum_{j=1}^{r} \xi_j (x_j - x_0).$$

Since $\{x_1 - x_0, \ldots, x_r - x_0\}$ is linearly independent, $\xi_1 = \cdots = \xi_r = 0$. Since $\xi_0 = -\sum_{j=1}^{r} \xi_j$, $\xi_0 = 0$.

Next, suppose $\{x_0, \ldots, x_r\}$ satisfies the hypothesis of the theorem. Let $\xi_1, \ldots, \xi_r$ be real numbers such that $\sum_{j=1}^{r} \xi_j (x_j - x_0) = 0$. Note that $\sum_{j=1}^{r} \xi_j (x_j - x_0) = \sum_{j=1}^{r} \xi_j x_j - \left(\sum_{j=1}^{r} \xi_j\right) x_0$. Let $\xi_0 = -\left(\sum_{j=1}^{r} \xi_j\right)$. Then $\sum_{j=1}^{r} \xi_j x_j - \left(\sum_{j=1}^{r} \xi_j\right) x_0 = \sum_{j=0}^{r} \xi_j x_j$ and $\sum_{j=0}^{r} \xi_j = 0$. Then, by hypothesis, $\xi_0 = \cdots = \xi_r = 0$. In particular, $\xi_1 = \cdots = \xi_r = 0$. Thus, $\{x_1 - x_0, \ldots, x_r - x_0\}$ is linearly independent. Therefore, $\{x_0, \ldots, x_r\}$ is affinely independent.

Q.E.D.

1.4.3 Definition Let $\{x_0, \ldots, x_r\}$ be an affinely independent subset of $\mathbb{R}^n$. We define the *(geometric) r-simplex generated by* $\{x_0, \ldots, x_r\}$, denoted by $[x_0, \ldots, x_r]$, as the following subset of $\mathbb{R}^n$:

$$[x_0, \ldots, x_r] = \left\{ \sum_{j=0}^{r} \xi_j x_j \mid \text{for each } j \in \{0, \ldots, r\}, \right.$$

$$\left. \xi_j \in [0, 1] \text{ and } \sum_{j=0}^{r} \xi_j = 1 \right\}.$$

The set $\{x_0, \ldots, x_r\}$ is called the *set of vertexes* of the r-simplex. The point $x = \sum_{j=0}^{r} \frac{1}{r+1} x_j$ is called the *barycenter* of the r-simplex. Each s-simplex generated by $s + 1$ points taken from $\{x_0, \ldots, x_r\}$ is called an *s-face* of $[x_0, \ldots, x_r]$.

1.4.4 Remark It is easy to see that a 0-simplex is a point, a 1-simplex is a line segment, a 2-simplex is a triangle and a 3-simplex is a tetrahedron.

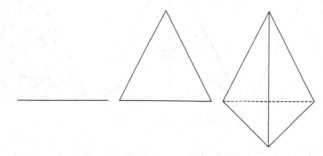

Simplexes of dimension one, two and three

1.4.5 Definition A *(geometric) complex*, $\mathcal{K}$, is a finite collection of simplexes in $\mathbb{R}^n$ such that:

(1) each face of a simplex in $\mathcal{K}$ also belongs to $\mathcal{K}$ and
(2) the intersection of any two simplexes is either empty or it is a face of both simplexes.

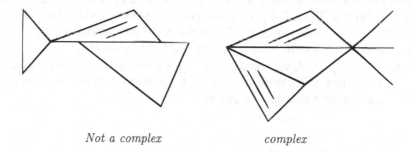

Not a complex   *complex*

1.4.6 Definition A *polyhedron* in $\mathbb{R}^n$ is the union of the simplexes of a geometric complex.

1.4.7 Definition The *barycentric subdivision of an r-simplex* $\mathcal{R}$ is a (geometric) complex $\mathcal{K}$ obtained as follows:

(a) If $r = 0$, then $\mathcal{K}$ is just $\mathcal{R}$.
(b) Suppose $r > 0$. If $\mathcal{S}_0, \ldots, \mathcal{S}_r$ are the $(r-1)$-faces of $\mathcal{R}$ and if x_b is the barycenter of $\mathcal{R}$, then $\mathcal{K}$ consists of all r-simplexes generated by x_b and the vertexes of all $(r-1)$-simplexes of the barycentric subdivision of $\mathcal{S}_j$ for each $j \in \{0, \ldots, r\}$. The *barycentric subdivision of a complex* $\mathcal{K}_0$ is a (geometric) complex $\mathcal{K}_1$ consisting of all the simplexes obtained by the barycentric subdivision of each simplex of $\mathcal{K}_0$.

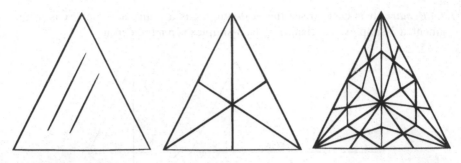

Barycentric subdivisions

1.4.8 Remark It is well known that a complex $\mathcal{K}_1$ obtained from a complex $\mathcal{K}_0$ by barycentric subdivision is, in fact, a complex [24, 15.2]. Some times the barycentric subdivision is performed several times; i.e., one can apply the barycentric subdivision to $\mathcal{K}_1$ to obtain a complex $\mathcal{K}_2$, and then apply the barycentric subdivision to $\mathcal{K}_2$ to obtain a complex $\mathcal{K}_3$, etc. It is also possible to show that the diameters of the simplexes of the complex $\mathcal{K}_m$, obtained after m barycentric subdivisions, tend to zero as m tends to infinity [24, 15.4].

1.4.9 Definition Let X be a compactum and let $\mathcal{U} = \{U_1, \ldots, U_n\}$ be a finite collection of subsets of X. Then the *nerve* of $\mathcal{U}$, denoted by $\mathcal{N}(\mathcal{U})$, is the complex defined as follows. To each $U_j \in \mathcal{U}$ associate the point e_j of $\mathbb{R}^n$, where $e_j = (0, \ldots, 0, 1, 0 \ldots, 0)$ (the number "1" appears in the jth coordinate). Hence, the vertexes of the complex are $\{e_1, \ldots, e_n\}$. For each subfamily $\{U_{j_1}, \ldots, U_{j_k}\}$ of $\mathcal{U}$ such that $\bigcap_{\ell=1}^{k} U_{j_\ell} \neq \emptyset$, we consider the simplex whose vertexes are the points $\{e_{j_1}, \ldots, e_{j_k}\}$. Then $\mathcal{N}(\mathcal{U})$ consists of all such possible simplexes. We denote by $\mathcal{N}^\star(\mathcal{U})$ the polyhedron associated to $\mathcal{N}(\mathcal{U})$.

1.5   Complete Metric Spaces

We present results about complete metric spaces based on [2, 10, 13, 16].

1.5.1 Definition Let X be a metric space, with metric d. A sequence, $\{x_n\}_{n=1}^{\infty}$, of elements of X is said to be a *Cauchy sequence* provided that for each $\varepsilon > 0$, there exists $N \in \mathbb{N}$ such that if $n, m \geq N$, then $d(x_n, x_m) < \varepsilon$.

1.5.2 Definition A metric space X, with metric d, is said to be *complete* provided that every Cauchy sequence of elements of X converges to a point of X. We say that X is *topologically complete* if there exists an equivalent metric d_1 for X such that (X, d_1) is a complete space.

The following results give some basic properties of complete metric spaces.

1.5.3 Lemma *Let X be a complete metric space. If F is a closed subset of X, then F is complete.*

Proof Let $\{x_n\}_{n=1}^{\infty}$ be a Cauchy sequence of elements of F. Since X is complete, there exists a point $x \in X$ such that $\{x_n\}_{n=1}^{\infty}$ converges to x. Since F is a closed subset of X, $x \in F$. Therefore, F is complete.

Q.E.D.

1.5.4 Proposition *Let X be a metric space. If Y is a subset of X and Y is complete, then Y is closed in X.*

Proof Let $x \in Cl(Y)$. Then there exists a sequence $\{y_n\}_{n=1}^{\infty}$ of elements of Y converging to x. Since a convergent sequence is a Cauchy sequence, and Y is complete, we have that $x \in Y$. Therefore, Y is closed in X.

Q.E.D.

1.5.5 Proposition *Let X be a complete metric space, with metric d. If $\{F_n\}_{n=1}^{\infty}$ is a sequence of closed subsets of X such that $F_{n+1} \subset F_n$ and $\lim_{n\to\infty} diam(F_n) = 0$, then there exists a point $x \in X$ such that $\bigcap_{n=1}^{\infty} F_n = \{x\}$.*

Proof For each $n \in \mathbb{N}$, let $x_n \in F_n$. Then $\{x_n\}_{n=1}^{\infty}$ is a sequence of points of X such that for each $N \in \mathbb{N}$, if $n, m \geq N$, then $x_n, x_m \in F_N$. Thus, since $\lim_{n\to\infty} diam(F_n) = 0$, $\{x_n\}_{n=1}^{\infty}$ is a Cauchy sequence. Hence, there exists $x \in X$ such that $\lim_{n\to\infty} x_n = x$. Note that, since for every $N \in \mathbb{N}$, $\{x_n\}_{n=N}^{\infty} \subset F_N$, $x \in \bigcap_{n=1}^{\infty} F_n$. Now, suppose there exists $y \in \bigcap_{n=1}^{\infty} F_n \setminus \{x\}$. Then for each $n \in \mathbb{N}$, $0 < d(x, y) \leq diam\left(\bigcap_{n=1}^{\infty} F_n\right) \leq diam(F_n)$, a contradiction. Therefore, $\bigcap_{n=1}^{\infty} F_n = \{x\}$.

Q.E.D.

1.5.6 Definition A subset of a metric space X is said to be a G_δ set provided that it is a countable intersection of open sets.

1.5.7 Lemma *Let X be a metric space, with metric d. If F is a closed subset of X, then F is a G_δ subset of X.*

Proof It follows from the fact that $F = \bigcap_{n=1}^{\infty} V_{\frac{1}{n}}^d(F)$.

Q.E.D.

The following theorem is due to Mazurkiewicz, a proof of which may be found in [10, Theorem 8.3, p. 308]:

1.5.8 Theorem *Let Y be a complete metric space. Then a nonempty subset A of Y is topologically complete if and only if A is a G_δ subset of Y.*

1.5.9 Lemma *If X is a complete metric space, with metric d, and $\{D_n\}_{n=1}^{\infty}$ is sequence of dense open subsets of X, then $\bigcap_{n=1}^{\infty} D_n$ is a dense subset of X.*

Proof Let U be an open subset of X. We show that $U \cap \left(\bigcap_{n=1}^{\infty} D_n\right) \neq \emptyset$.

Since D_1 is a dense subset of X, $D_1 \cap U \neq \emptyset$. In fact, $D_1 \cap U$ is an open subset of X. Hence, there exist $x_1 \in D_1 \cap U$ and $\varepsilon_1 > 0$ such that $\varepsilon_1 < 1$ and $Cl(V_{\varepsilon_1}^d(x_1)) \subset D_1 \cap U$. Since D_2 is a dense open subset of X and $V_{\varepsilon_1}^d(x_1)$ is open, $D_2 \cap V_{\varepsilon_1}^d(x_1) \neq \emptyset$. So, there exist $x_2 \in D_2 \cap V_{\varepsilon_1}^d(x_1)$ and $\varepsilon_2 > 0$ such that $\varepsilon_2 < \frac{1}{2}$ and $Cl(V_{\varepsilon_2}^d(x_2)) \subset D_2 \cap V_{\varepsilon_1}^d(x_1)$. If we continue with this process, we construct a decreasing family of closed subset of X, whose diameters tend to zero. Hence, by Proposition 1.5.5, there exists a point $x \in X$ such that $\{x\} = \bigcap_{n=1}^{\infty} Cl(V_{\varepsilon_n}^d(x_n)) \subset (\bigcap_{n=1}^{\infty} D_n) \cap U$.

Since $\bigcap_{n=1}^{\infty} Cl(V_{\varepsilon_n}^d(x_n)) \subset (\bigcap_{n=1}^{\infty} D_n) \cap U$, we obtain that $U \cap (\bigcap_{n=1}^{\infty} D_n) \neq \emptyset$.

Q.E.D.

1.5.10 Definition Let X be a metric space. A subset A of X is said to be *nowhere dense* provided that $Int(Cl(A)) = \emptyset$.

1.5.11 Definition Let X be a metric space. A subset A of X is said to be of the *first category* if it is the countable union of nowhere dense subsets of X. A subset of X that is not of the first category is said to be of the *second category*.

The following result is known as the *Baire Category Theorem*:

1.5.12 Theorem *If X is a complete metric space, then X is of the second category.*

Proof Suppose X is of the first category. Then there exists a sequence $\{A_n\}_{n=1}^{\infty}$ of nowhere dense subsets of X such that $X = \bigcup_{n=1}^{\infty} A_n$. Since, for each $n \in \mathbb{N}$, A_n is nowhere dense, each set $X \setminus Cl(A_n)$ is an open dense subset of X. By Lemma 1.5.9, $\bigcap_{n=1}^{\infty} (X \setminus Cl(A_n)) \neq \emptyset$.

Since $X = \bigcup_{n=1}^{\infty} A_n$, we have that

$$\bigcap_{n=1}^{\infty} (X \setminus Cl(A_n)) = X \setminus \bigcup_{n=1}^{\infty} Cl(A_n) \subset X \setminus \bigcup_{n=1}^{\infty} A_n = X \setminus X = \emptyset,$$

a contradiction. Therefore, X is of the second category.

Q.E.D.

The following theorem is due to Hausdorff, a proof of which may be found in [2, Appendix]:

1.5.13 Theorem *Let X and Y be metric spaces. If X is complete and $f \colon X \twoheadrightarrow Y$ is a surjective open map, then Y is topologically complete.*

1.6 Compacta

We give basic properties of compacta. We construct the Cantor set and present some of its properties. The material of this section comes from [16, 18, 27, 30].

A proof of the following theorem may be found in [18, Corolário 2, p. 212].

1.6.1 Theorem *If X is a compactum, then X is complete.*

1.6.2 Lemma *Let X be a compactum, and let A be a closed subset of X with a finite number of components. If $x \in Int(A)$ and C is the component of A such that $x \in C$, then $x \in Int(C)$.*

Proof Let $C, C_1, \ldots, C_n$ be the components of A. Since $x \in Int(A)$, there exists an open subset U of X such that $x \in U \subset A$. Since A is closed in X, each of $C, C_1, \ldots, C_n$ is closed in X. Let $V = U \cap \left(X \setminus \bigcup_{j=1}^{n} C_j \right)$. Then V is an open subset of X such that $x \in V \subset A$ and $V \cap \left(\bigcup_{j=1}^{n} C_j \right) = \emptyset$. Thus, $V \subset C$. Therefore, $x \in Int(C)$.

Q.E.D.

1.6.3 Definition Let X be a metric space. A subset Y of X is said to be *perfect* if Y is closed and every point of Y is a limit point of Y.

1.6.4 Example We construct the Cantor set and prove some of its properties. Let $C_0 = [0, 1]$. Remove $\left(\frac{1}{3}, \frac{2}{3} \right)$, and let $C_1 = \left[0, \frac{1}{3} \right] \cup \left[\frac{2}{3}, 1 \right]$. Remove the middle thirds of these intervals, and let $C_2 = \left[0, \frac{1}{9} \right] \cup \left[\frac{2}{9}, \frac{1}{3} \right] \cup \left[\frac{2}{3}, \frac{7}{9} \right] \cup \left[\frac{8}{9}, 1 \right]$. Continuing in this way, we obtain a sequence, $\{C_n\}_{n=0}^{\infty}$, of compact sets such that for each $n \in \mathbb{N}$, $C_{n+1} \subset C_n$ and C_n is the union of 2^n intervals, $I_{n,0}, \ldots, I_{n,2^n-1}$, of length 3^{-n}. The set

$$C = \bigcap_{n=0}^{\infty} C_n$$

is called the *Cantor set*. C is clearly compact and nonempty.

No interval of the form

$$(*) \qquad \left(\frac{3k+1}{3^m}, \frac{3k+2}{3^m} \right),$$

where $k, m \in \mathbb{N}$, has a point in common with C. Since every interval (x, y) contains an interval of the form $(*)$, if $3^{-m} < \frac{y-x}{6}$, C does not contain a nondegenerate interval.

We show that C is perfect. Let $x \in C$, and let A be any interval containing x. Let I_n be the interval in C_n that contains x. Let n large enough, so that $I_n \subset A$. Let x_n be an end point of I_n such that $x_n \neq x$. It follows from the construction that $x_n \in C$. Hence, x is a limit point of C. Therefore, C is perfect.

It follows, from the construction and the fact that C does not contain any nondegenerate interval, that C is totally disconnected. In Chap. 2, we present a characterization of the Cantor set as the only totally disconnected and perfect compactum.

1.6.5 Definition Let X be a metric space, and let $\mathcal{U}$ be an open cover of X. We say that a number $\lambda > 0$ is a *Lebesgue number for the open cover* $\mathcal{U}$ provided that if A is a nonempty subset of X such that $\operatorname{diam}(A) < \lambda$, then there exists $U \in \mathcal{U}$ such that $A \subset U$.

1.6.6 Theorem *If X is a compactum, with metric d, and $\mathcal{U}$ is an open cover of X, then there exists a Lebesgue number for $\mathcal{U}$.*

Proof Suppose that no such number exists. Then for each $n \in \mathbb{N}$, there exists a nonempty subset A_n of X such that $\operatorname{diam}(A_n) < \frac{1}{n}$ and A_n is not contained in any element of $\mathcal{U}$. Let $x_n \in A_n$. Since X is compact, without loss of generality, we assume that the sequence $\{x_n\}_{n=1}^{\infty}$ converges to a point $x \in X$.

Since $\mathcal{U}$ covers X, there exists $U \in \mathcal{U}$ such that $x \in U$. Hence, there exists $\varepsilon > 0$ such that $\mathcal{V}_\varepsilon^d(x) \subset U$ (U is open in X). Let $n \in \mathbb{N}$ be such that $\frac{1}{n} < \frac{\varepsilon}{2}$ and $d(x_n, x) < \frac{\varepsilon}{2}$. Then for each $y \in A_n$,

$$ d(y, x) \le d(y, x_n) + d(x_n, x) < \frac{1}{n} + \frac{\varepsilon}{2} < \varepsilon. $$

Hence, $A_n \subset \mathcal{V}_\varepsilon^d(x) \subset U$, a contradiction. Therefore, there exists a Lebesgue number for $\mathcal{U}$.

Q.E.D.

The following lemma is very useful in continuum theory.

1.6.7 Lemma *Let Z be a compactum and let $\{X_n\}_{n=1}^{\infty}$ be a sequence of closed subsets of Z such that $X_{n+1} \subset X_n$ for each $n \in \mathbb{N}$. If U is an open subset of Z such that $\bigcap_{n=1}^{\infty} X_n \subset U$, then there exists $N \in \mathbb{N}$ such that $X_n \subset U$ for each $n \ge N$.*

Proof Since U is an open subset of Z, $Z \setminus U$ is closed in Z. Note that, $Z \setminus U \subset Z \setminus \bigcap_{n=1}^{\infty} X_n = \bigcup_{n=1}^{\infty}(Z \setminus X_n)$. Hence, $\{Z \setminus X_n\}_{n=1}^{\infty}$ is an open cover of $Z \setminus U$. Since $Z \setminus U$ is compact, there exist $n_1, \ldots, n_k \in \mathbb{N}$ such that $Z \setminus U \subset \bigcup_{j=1}^{k}(Z \setminus X_{n_j})$. Thus, $\bigcap_{j=1}^{k} X_{n_j} \subset U$. Let $N = \max\{n_1, \ldots, n_k\}$. Then $X_N = \bigcap_{j=1}^{k} X_{n_j}$, and $X_N \subset U$. Therefore, $X_n \subset U$ for each $n \ge N$.

Q.E.D.

The following result is known as *The Cut Wire Theorem*; it is very useful in continuum theory. A proof of this result may be found in [27, 5.2].

1.6.8 Theorem *Let X be a compactum and let A and B be closed subsets of X. If no connected subset of X intersects both A and B, then there exist two disjoint closed subsets X_1 and X_2 of X such that $A \subset X_1$, $B \subset X_2$ and $X = X_1 \cup X_2$.*

1.7 Continua

We define the type of spaces we are more interested in, namely, continua, and give some of its main properties. We use [3, 7, 9, 11, 12, 14, 27], to prepare this section.

1.7.1 Definition A *continuum* is a connected compactum. A *subcontinuum* is a continuum contained in some metric space.

The following theorem provides a method to construct continua.

1.7.2 Theorem *Let Z be a compactum and let $\{X_n\}_{n=1}^{\infty}$ be a sequence of subcontinua of Z such that $X_{n+1} \subset X_n$ for each $n \in \mathbb{N}$. If $X = \bigcap_{n=1}^{\infty} X_n$, then X is a subcontinuum of Z.*

Proof Clearly, X is a closed; hence, compact, subset of Z. Suppose X is not connected. Then there exist two disjoint closed subsets A and B of Z such that $X = A \cup B$. Since Z is a metric space, there exist two disjoint open subsets U and V of Z such that $A \subset U$ and $B \subset V$. Hence, $X \subset U \cup V$. By Lemma 1.6.7, there exists $N \in \mathbb{N}$, such that $X_N \subset U \cup V$. Since X_N is connected, either $X_N \subset U$ or $X_N \subset V$. Assume, without loss of generality, that $X_N \subset U$. Since $X \subset X_N \subset U$ and $X = A \cup B$, we have that $B \subset U$, a contradiction. Therefore, X is connected. Hence, X is a subcontinuum of Z.

Q.E.D.

1.7.3 Theorem *If X is a continuum and $\mathcal{G}$ is an upper semicontinuous decomposition of X, then $X/\mathcal{G}$ is a continuum.*

Proof By Corollary 1.2.22, $X/\mathcal{G}$ is a metric space. Since continuous images of compact and connected spaces are compact and connected, $X/\mathcal{G}$ is a continuum.

Q.E.D.

The following definition is due to Davis and Doyle and they used it in their study of invertible continua [9]:

1.7.4 Definition Let X be a continuum, and let $x \in X$. We say that X *is almost connected im kleinen at x* provided that for each open subset U of X containing x, there exists a subcontinuum W of X such that $Int(W) \neq \emptyset$ and $W \subset U$. We say that X is *almost connected im kleinen* if it is connected im kleinen at each of its points.

1.7.5 Example Let X be the cone over the closure of the harmonic sequence $\{0\} \cup \left\{\frac{1}{n}\right\}_{n=1}^{\infty}$. Then X is called *harmonic fan*. It is easy to see that X is almost connected im kleinen at each of its points.

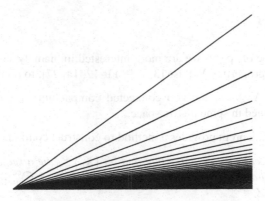

Almost connected im kleinen

1.7.6 Definition Let X be a continuum, and let $x \in X$. We say that X *is connected im kleinen at* x if for each closed subset F of X such that $F \subset X \setminus \{x\}$, there exists a subcontinuum W of X such that $x \in Int(W) \subset W \subset X \setminus F$.

1.7.7 Example Let X be a sequence of harmonic fans converging to a point p; see picture below. Then X is connected im kleinen at p but X is not locally connected at that point.

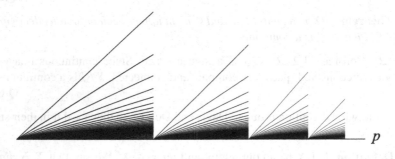

Connected im kleinen at p

1.7.8 Remark Note that we do not define connectedness im kleinen globally because, by Theorem 1.7.12, a continuum is connected im kleinen globally if and only if it is locally connected (Definition 1.7.10).

The way we defined connectedness im kleinen in Definition 1.7.6 is not the usual one. This definition provides the author the answer to the fact that aposyndesis (Definition 1.7.15) is a generalization of connectedness im kleinen. The following theorem presents the usual definition of this concept:

1.7.9 Theorem *If X is a continuum and $x \in X$, then the following are equivalent:*

(a) *X is connected im kleinen at x.*
(b) *For each open subset U of X such that $x \in U$, there exists an open subset V of X such that $x \in V \subset U$ and with the property that for each $y \in V$, there exists a connected subset C_y of X such that $\{x, y\} \subset C_y \subset U$.*
(c) *For each open subset U of X such that $x \in U$, there exists a subcontinuum W of X such that $x \in Int(W) \subset W \subset U$.*

Proof Suppose X is connected im kleinen at x. We show (b). Let U be an open subset of X such that $x \in U$. Then $X \setminus U$ is a closed subset of X not containing x. By hypothesis, there exists a subcontinuum W of X such that $x \in Int(W) \subset W \subset X \setminus (X \setminus U) = U$. Thus, $V = Int(W)$ satisfies the required properties.

Next, suppose (b). We show (c). Let U be an open subset of X such that $x \in U$. Let U' be an open subset of X such that $x \in U' \subset Cl(U') \subset U$. By hypothesis, there exists an open subset V of X such that $x \in V \subset U'$ with the property that for each $y \in V$, there exists a connected subset C_y of X such that $\{x, y\} \subset C_y \subset U'$. Let $W = Cl\left(\bigcup_{y \in V} C_y\right)$. Then W is a subcontinuum of X such that $x \in V \subset W \subset Cl(U') \subset U$.

Finally, suppose (c). We show X is connected im kleinen at x. Let F be a closed subset of X such that $F \subset X \setminus \{x\}$. Then $X \setminus F$ is an open subset of X containing x. By hypothesis, there exists a subcontinuum W of X such that $x \in Int(W) \subset W \subset X \setminus F$. Therefore, X is connected im kleinen at x.

Q.E.D.

1.7.10 Definition Let X be a continuum, and let $x \in X$. We say that *X is locally connected at x* provided that for each open subset U of X such that $x \in U$, there exists a connected open subset V of X such that $x \in V \subset U$. We say *X is locally connected* if it is locally connected at each of its points.

1.7.11 Lemma *A continuum X is locally connected if and only if the components of the open subsets of X are open.*

Proof Suppose X is locally connected. Let U be an open subset of X, and let C be a component of U. For each point $x \in C$, there exists an open connected set V_x such that $x \in V_x \subset U$. Then $C \cup V_x$ is a connected subset of U. Hence, $V_x \subset C$. Therefore, each point of C is an interior point of C. Thus, C is open.

Next, suppose the components of open sets are open. Let $x \in X$, and let U be an open subset of X such that $x \in U$. By hypothesis, the component, C, of U containing x is open. Then C is an open connected set such that $x \in C \subset U$. Therefore, X is locally connected.

Q.E.D.

1.7.12 Theorem *A continuum X is connected im kleinen at each of its points if and only if X is locally connected.*

Proof Clearly, if X is locally connected, then X is connected im kleinen at each of its points.

Suppose X is connected im kleinen at each of its points. Let U be an open subset of X, and let C be a component of U. If $x \in C$, then there exists a subcontinuum W of X such that $x \in Int(W) \subset W \subset U$, by Theorem 1.7.9. Since W is a connected subset of U and $W \cap C \neq \emptyset$, $W \subset C$. Then x is an interior point of C. Thus, C is open. Therefore, by Lemma 1.7.11, X is locally connected.

$$\textbf{Q.E.D.}$$

The notions of semi-aposyndesis and aposyndesis resemble those of T_0 and T_1 topological spaces, respectively.

1.7.13 Definition Let X be a continuum, and let $p, q \in X$. We say that X *is semi-aposyndetic at p and q* provided that there exists a subcontinuum W of X such that $\{p, q\} \cap Int(W) \neq \emptyset$ and $\{p, q\} \setminus W \neq \emptyset$. X is *semi-aposyndetic* if it is semi-aposyndetic at each pair of its points.

1.7.14 Example Let X be the harmonic fan (Example 1.7.5). Then X is semi-aposyndetic at each pair of its points.

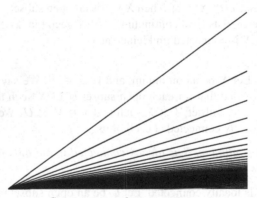

Semi–aposyndetic continuum

1.7.15 Definition Let X be a continuum, and let $p, q \in X$. We say that X *is aposyndetic at p with respect to q* provided that there exists a subcontinuum W of X such that $p \in Int(W) \subset W \subset X \setminus \{q\}$. Now, X is *aposyndetic at p* if X is aposyndetic at p with respect to each point of $X \setminus \{p\}$. We say that X *is aposyndetic* provided that X is aposyndetic at each of its points.

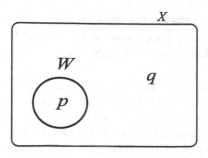

Aposyndetic at p with respect to q

1.7.16 Definition Let X be a continuum, and let $p \in X$. We say that X is *semi-locally connected at p* provided that for each open subset U of X such that $p \in U$, there exists an open subset V of X such that $p \in V \subset U$ and $X \setminus V$ has a finite number of components. We say X *is semi-locally connected* if X is semi-locally connected at each of its points.

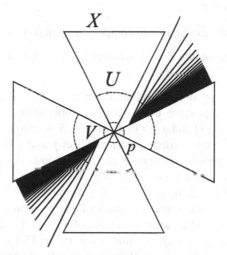

Semi–locally connected continuum

Even though aposyndesis and semi-local connectedness seem to be different concepts, it turns out that globally they are equivalent:

1.7.17 Theorem *A continuum X is aposyndetic if and only if it is semi-locally connected.*

Proof Suppose X is aposyndetic. Let $x \in X$. We show that X is semi-locally connected at x. Let U be an open subset of X such that $x \in U$. Since X is aposyndetic, for each $y \in X \setminus U$, there exists a subcontinuum W_y of X such that $y \in Int(W_y) \subset W_y \subset X \setminus \{x\}$. Since $X \setminus U$ is compact, there exists

$y_1, \ldots, y_m \in X \setminus U$ such that $X \setminus U \subset \bigcup_{j=1}^{m} Int(W_{y_j})$. Let $V = X \setminus \left(\bigcup_{j=1}^{m} W_{y_j} \right)$.
Then V is an open subset of X such that $x \in V \subset U$ and $X \setminus V$ has a finite number
of components. Therefore, X is semi-locally connected, since x is an arbitrary point
of X.

Now, suppose X is semi-locally connected. Let $x, y \in X$. We show that X is
aposyndetic at x with respect to y. Let U be an open subset of X such that $y \in U \subset$
$X \setminus \{x\}$. Let U' be an open subset of X such that $y \in U' \subset Cl(U') \subset U$. Since X is
semi-locally connected, there exists an open subset V of X such that $y \in V \subset U'$
and $X \setminus V$ has a finite number of components. By Lemma 1.6.2, x is contained in
the interior of the component of $X \setminus V$ containing x. Hence, X is aposyndetic at x
with respect to y.

Therefore, X is aposyndetic.

Q.E.D.

Another concept related to aposyndesis is free decomposability.

1.7.18 Definition A continuum X is *freely decomposable* if for each pair of distinct
points p and q of X, there exist two subcontinua P and Q of X such that $X = P \cup Q$,
$p \in P \setminus Q$ and $q \in Q \setminus P$.

The next theorem shows that free decomposability and aposyndesis coincide.

1.7.19 Theorem *A continuum X is aposyndetic if and only if X is freely decomposable.*

Proof Suppose X is freely decomposable. Let p and q be two distinct points of X.
Since X is freely decomposable, there exist two subcontinua P and Q of X such
that $X = P \cup Q$, $p \in P \setminus Q$ and $q \in Q \setminus P$. Hence, X is an aposyndetic continuum.

Now, assume X is an aposyndetic continuum. Let p and q be two distinct points
of X. Since X is aposyndetic, there exist two subcontinua A and B such that $p \in$
$Int(A) \subset A \subset X \setminus \{q\}$ and $q \in Int(B) \subset B \subset X \setminus \{p\}$. Let U and V open subsets
of X such that $U \subset A$, $V \subset B$, $p \in U \subset X \setminus B$ and $q \in V \subset X \setminus A$. Let A_p be
the component of $X \setminus V$ that contains p and let B_q be the component of $X \setminus U$ that
contains q. Note that A_p and B_q are subcontinua of X, $p \in A_p \setminus B_q$ and $q \in B_q \setminus A_p$.
We show that $X = A_p \cup B_q$. Suppose this is not true and let $L = X \setminus (A_p \cup B_q)$.
Since $p \in U \subset A \subset X \setminus V$ and $q \in V \subset B \subset X \setminus U$, we have that $A \subset A_p$ and
$B \subset B_q$. Hence, $U \cup V \subset A_p \cup B_q$. Thus, $L \subset X \setminus (U \cup V) = (X \setminus U) \cap (X \setminus V)$.
This implies that $Cl(L) \subset X \setminus V$. Note that $A_p \cup Cl(L)$ is a closed subset of $X \setminus V$
that contains A_p properly. Since A_p is a component of $X \setminus V$, $A_p \cup Cl(L)$ is not
connected. Hence, there exist two disjoint closed subsets R_1 and S_1 of X such that
$A_p \cup Cl(L) = R_1 \cup S_1$. Without loss of generality, we assume that $A_p \subset R_1$. Thus,
$S_1 \subset Cl(L)$. Since $Cl(L) \subset X \setminus U$, we have that $B_q \cup S_1 \subset X \setminus U$. Observe that
$B_q \cup S_1$ is a closed subset of $X \setminus U$ containing B_q properly. Since B_q is a component
of $X \setminus U$, $B_q \cup S_1$ is not connected. Thus, there exist two disjoint closed subsets R_2
and S_2 of X such that $B_q \cup S_1 = R_2 \cup S_2$. Without loss of generality, we assume
that $B_q \subset R_2$. Hence, $S_2 \subset S_1$. Since $R_1 \cap S_1 = \emptyset$ and $R_1 \cap S_2 = \emptyset$, we obtain that

$(R_1 \cup R_2) \cap S_2 = \emptyset$. Since $X = A_p \cup B_q \cup L$, we have that $X = A_p \cup B_q \cup Cl(L)$. Also, since $A_p \cup Cl(L) = R_1 \cup S_1$, $X = R_1 \cup S_1 \cup B_q$. Moreover, since $S_1 \cup B_q = R_2 \cup S_2$, we obtain that $X = (R_1 \cup R_2) \cup S_2$. A contradiction to the connectedness of X. Therefore, $X = A_p \cup B_q$ and X is freely decomposable.

Q.E.D.

The notion of aposyndesis may be extended as follows:

1.7.20 Definition A continuum X is *aposyndetic at p with respect to the subset K of X* if there exists a subcontinuum W of X such that $p \in Int(W) \subset W \subset X \setminus K$.

1.7.21 Definition A continuum X is *continuum aposyndetic at p* if X is aposyndetic at p with respect to each subcontinuum of X not containing p. X is *continuum aposyndetic* provided that X is continuum aposyndetic at each of its points.

1.7.22 Definition A continuum X is *freely decomposable with respect to points and continua* if for each subcontinuum C of X and each point $a \in X \setminus C$, there exist two subcontinua A and B of X such that $X = A \cup B$, $a \in A \setminus B$ and $C \subset B \setminus A$.

1.7.23 Lemma *Let X be a continuum, and let A be a subcontinuum of X such that $X \setminus A$ is not connected. If U and V are nonempty disjoint open subsets of X such that $X \setminus A = U \cup V$, then $A \cup U$ and $A \cup V$ are subcontinua of X.*

Proof Since $X \setminus (A \cup U) = V$, $A \cup U$ is closed, hence, compact. Similarly $A \cup V$ is compact.

We show $A \cup U$ is connected. To see this, suppose $A \cup U$ is not connected. Then there exist two nonempty disjoint closed subsets K and L of X such that $A \cup U = K \cup L$. Since A is connected, without loss of generality, we assume that $A \subset K$. Note that, in this case, $L \subset U$. Hence, $L \cap Cl(V) = \emptyset$. Thus, $X = L \cup (K \cup Cl(V))$, a contradiction, since L and $K \cup Cl(V)$ are disjoint closed subsets of X. Therefore, $A \cup U$ is connected. Similarly, $A \cup V$ is connected.

Q.E.D.

1.7.24 Definition A continuum X is *decomposable* provided that it can be written as the union of two of its proper subcontinua. We say X is *indecomposable* if it is not decomposable. We say X is *hereditarily decomposable (indecomposable)* if each nondegenerate subcontinuum of X is decomposable (indecomposable, respectively).

1.7.25 Lemma *A continuum X is decomposable if and only if X contains a proper subcontinuum with nonempty interior.*

Proof Suppose X is a decomposable continuum. Then there exist two proper subcontinua, A and B, of X such that $X = A \cup B$. Note that $X \setminus B$ is an open set contained in A. Therefore, $Int(A) \neq \emptyset$.

Now, assume A is a proper subcontinuum of X with nonempty interior. If $X \setminus A$ is connected, then $Cl(X \setminus A)$ is a proper subcontinuum of X and $X = A \cup Cl(X \setminus A)$. Hence, X is decomposable.

Suppose that $X \setminus A$ is not connected. Then there exist two nonempty disjoint open subsets U and V of X such that $X \setminus A = U \cup V$. By Lemma 1.7.23, $A \cup U$

and $A \cup V$ are subcontinua of X, and $X = (A \cup U) \cup (A \cup V)$. Therefore, X is decomposable.

Q.E.D.

1.7.26 Corollary *A continuum X is indecomposable if and only if each proper subcontinuum of X has empty interior.*

The following result is known as *The Boundary Bumping Theorem*. It has many applications in continuum theory and in hyperspaces. A proof of this theorem may be found in [27, 5.4].

1.7.27 Theorem *Let X be a continuum and let U be a nonempty, proper, open subset of X. If K is a component of $Cl(U)$, then $K \cap Bd(U) \neq \emptyset$.*

The following corollary is also very useful; a proof of it may be found in [27, 5.5].

1.7.28 Corollary *Let X be a nondegenerate continuum. If A is a proper subcontinuum of X and U is an open subset of X such that $A \subset U$, then there exists a subcontinuum B of X such that $A \subsetneq B \subset U$.*

1.7.29 Lemma *If X is a continuum such that each of its proper subcontinuum is indecomposable, then X is indecomposable. Hence, X is hereditarily indecomposable.*

Proof Suppose X is decomposable. Then there exist two proper subcontinua A and B of X such that $X = A \cup B$. Note that $X \setminus B$ is an open subset of X contained in A. On the other hand, there exists a proper subcontinuum, H, of X containing A (Corollary 1.7.28). Hence, H is an indecomposable continuum containing a proper subcontinuum with nonempty interior, which is impossible (Corollary 1.7.26). Therefore, X is indecomposable.

Q.E.D.

1.7.30 Definition A continuum X *is irreducible between two of its points* if no proper subcontinuum of X contains both points. A continuum is *irreducible* if it is irreducible between two of its points.

The following results present some of the properties of irreducible continua.

1.7.31 Theorem *Let X be an irreducible continuum between a and b. If C is a subcontinuum of X such that $X \setminus C$ is not connected, then $X \setminus C$ is the union of two open and connected sets, one containing a and the other containing b. Moreover, if $a \in C$, then $X \setminus C$ is connected.*

Proof Suppose $X \setminus C$ is not connected. Then there exist two nonempty disjoint open subsets U and V of X such that $X \setminus C = U \cup V$. By Lemma 1.7.23, $A = C \cup U$ and $B = C \cup V$ are subcontinua of X such that $X = A \cup B$, $A \cap B = C$, $A \neq X$ and $B \neq X$.

Since X is irreducible between a and b, $\{a, b\} \cap C = \emptyset$. If $\{a, b\} \cap C \neq \emptyset$, then either A or B is a proper subcontinuum of X containing $\{a, b\}$, a contradiction. Therefore, $\{a, b\} \cap C = \emptyset$. We assume that $a \in U$ and $b \in V$.

Since A and B are proper subcontinua of X, neither A nor B may contain $\{a, b\}$.

Since A is a proper subcontinuum of X and $a \in A$, we assert that $V = X \setminus A$ is connected. To see this, suppose $X \setminus A$ is not connected. Then there exist two nonempty disjoint open subsets K and L such that $X \setminus A = K \cup L$. Since $b \in X \setminus A$, we may assume that $b \in K$. Then, by Lemma 1.7.23, $A \cup K$ is subcontinuum of X which is proper and satisfies that $\{a, b\} \subset A \cup K$, a contradiction. Therefore, $V = X \setminus A$ is connected. Similarly, $U = X \setminus B$ is connected.

A similar argument shows that if $a \in C$, then $X \setminus C$ is connected.

Q.E.D.

1.7.32 Lemma *Let X be an irreducible decomposable continuum. If Y and Z are two disjoint subcontinua of X, then $X \setminus (Y \cup Z)$ has at most three components.*

Proof Since X is irreducible, by Theorem 1.7.31, $X \setminus Y$ has at most two components, U and V, such that they are connected open subsets of X, $Z \subset U$ and V may be empty.

Similarly, we assume that $X \setminus Z = H \cup K$, where H and K are connected open subsets of X such that $Y \subset H$ and K may be empty. Let $R = (U \setminus Z) \cap (H \setminus Y)$. Note that R is an open subset of X, $Cl(R) \cap Y \neq \emptyset$ and $Cl(R) \cap Z \neq \emptyset$. We assert that R is connected. To show this, assume R is not connected. Let C be a component of R. By Theorem 1.7.27, $Cl(C) \cap (Y \cup Z) \neq \emptyset$. If $Cl(C) \cap Y \neq \emptyset$ and $Cl(C) \cap Z \neq \emptyset$, then $V \cup Y \cup Cl(C) \cup Z \cup K$ is a subcontinuum of X containing the points of irreducibility of X. Thus, $X = V \cup Y \cup Cl(C) \cup Z \cup K$. Since $R \cap (V \cup Y \cup Z \cup K) = \emptyset$, it follows that $R \subset Cl(C)$. Hence, R is connected ($C \subset R \subset Cl(C)$), a contradiction. Therefore, either $Cl(C) \cap Y = \emptyset$ or $Cl(C) \cap Z = \emptyset$. Let

$$\mathcal{A} = \{Cl(C) \mid C \text{ is a component of } R \text{ and } Cl(C) \cap Y \neq \emptyset\}$$

and

$$\mathcal{B} = \{Cl(C) \mid C \text{ is a component of } R \text{ and } Cl(C) \cap Z \neq \emptyset\}.$$

We claim that $\mathcal{A}$ and $\mathcal{B}$ are both nonempty. Suppose, to the contrary, that $\mathcal{B}$ is empty. Note that $V \cup Y \cup (\bigcup \mathcal{A})$ is a connected set and that $\bigcup \mathcal{A}$ is not connected. Then there exist two separated subsets J and L of $V \cup Y \cup (\bigcup \mathcal{A})$ such that $\bigcup \mathcal{A} = J \cup L$. Note that $V \cup Y \cup J$ and $V \cup Y \cup L$ are connected sets (Lemma 1.7.23). Hence, either $(V \cup Y \cup Cl(J)) \cap Z \neq \emptyset$ or $(V \cup Y \cup Cl(L)) \cap Z \neq \emptyset$. Suppose $(V \cup Y \cup Cl(J)) \cap Z \neq \emptyset$. Then $(V \cup Y \cup Cl(J)) \cup Z \cup K$ is a proper subcontinuum of X containing the points of irreducibility of X, a contradiction. Therefore, $\mathcal{B} \neq \emptyset$. Similarly, $\mathcal{A} \neq \emptyset$.

Since X is a continuum, $Cl(\bigcup \mathcal{A}) \cap Cl(\bigcup \mathcal{B}) \neq \emptyset$. Let $x \in Cl(\bigcup \mathcal{A}) \cap Cl(\bigcup \mathcal{B})$. Since $x \in Cl(\bigcup \mathcal{A})$, there exists a sequence $\{a_n\}_{n=1}^{\infty}$ of elements of $\bigcup \mathcal{A}$ converging to x. For each $n \in \mathbb{N}$, let $Cl(C_n) \in \mathcal{A}$ such that $a_n \in Cl(C_n)$.

By Theorem 1.8.5, we assume that the sequence $\{Cl(C_n)\}_{n=1}^{\infty}$ of subcontinua of X converges (in the Hausdorff metric) to a subcontinuum T of X. Note that $T \cap Y \neq \emptyset$ and $x \in T$. Similarly, there exists a subcontinuum T' of X such that $T' \cap Z \neq \emptyset$ and $x \in T'$. Consequently, $V \cup Y \cup T \cup T' \cup Z \cup K$ is a proper subcontinuum of X containing its points of irreducibility, a contradiction. Therefore, R is connected.

Now, observe that

$$X \setminus [(V \cup Y) \cup (Z \cup K)] = (X \setminus V) \cap (X \setminus Y) \cap (X \setminus Z) \cap (X \setminus K)$$

$$= (U \cap Y) \cap (X \setminus Y) \cap (X \setminus Z) \cap (H \cap Z)$$

$$= U \cap H = U \cap (X \setminus Y) \cap (X \setminus Z) \cap H$$

$$= U \cap (X \setminus Z) \cap (X \setminus Y) \cap H$$

$$= (U \setminus Z) \cap (H \setminus Y) = R.$$

Since $V \cap (Y \cup Z \cup K) = \emptyset$ and $K \cap (Z \cup Y \cup V) = \emptyset$, $X \setminus (Y \cup Z) = V \cup R \cup K$.

Q.E.D.

1.7.33 Definition A continuum X *is weakly irreducible* provided that the complement of each finite union of subcontinua of X has a finite number of components.

1.7.34 Theorem *If X is an irreducible continuum, then X is weakly irreducible.*

Proof Let $Z_1, \ldots, Z_n$ be a finite family of subcontinua of X such that $Z_j \cap Z_k = \emptyset$ if $j \neq k$. For each $j \in \{1, \ldots, n\}$, by Theorem 1.7.31, we assume that $X \setminus Z_j = U_j \cup V_j$, where U_j and V_j are open connected subsets of X. Without loss of generality, we suppose that $\bigcup_{j=2}^{n} Z_j \subset V_1$ and $\bigcup_{j=1}^{n-1} Z_j \subset U_n$. Hence, U_1 and V_n may be empty. We assume also that, for each $j \in \{2, \ldots, n-1\}$, $\bigcup_{k=1}^{j-1} Z_k \subset U_j$ and $\bigcup_{k=j+1}^{n} Z_k \subset V_j$. For each $j \in \{1, \ldots, n-1\}$, let $R_j = (V_j \setminus Z_{j+1}) \cap (U_{j+1} \setminus Z_j)$. By the proof of Lemma 1.7.32, R_j is a connected open subset of X. Note that $X \setminus \left(\bigcup_{j=1}^{n} Z_j \right) = U_1 \cup \left(\bigcup_{j=1}^{n-1} R_j \right) \cup V_n$. Thus, $X \setminus \left(\bigcup_{j=1}^{n} Z_j \right)$ has, at most, $n+1$ components. Therefore, X is weakly irreducible.

Q.E.D.

1.7.35 Definition A continuum X *is unicoherent* provided that for each pair A and B of subcontinua of X such that $X = A \cup B$, $A \cap B$ is connected. We say that X *is hereditarily unicoherent* if each subcontinuum of X is unicoherent.

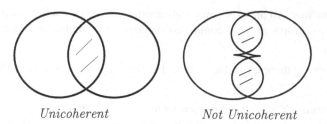

Unicoherent   *Not Unicoherent*

1.7.36 Lemma *If X is a subcontinuum of* $[0, 1]^n$ *and U is an open subset of* $[0, 1]^n$ *such that* $X \subset U$, *then there exists a polyhedron P such that* $X \subset Int(P) \subset P \subset U$.

Sketch of Proof Since $[0, 1]^n$ is a polyhedron, we can perform barycentric subdivisions to obtain complexes $\mathcal{K}_1, \mathcal{K}_2, \ldots$. For each $m \in \mathbb{N}$, let $\mathcal{K}_m^\star$ be the subcomplex of $\mathcal{K}_m$ consisting of all the simplexes of $\mathcal{K}_m$ which intersect X (we include all the faces of such simplexes as part of $\mathcal{K}_m^\star$). Let P_m be the polyhedron determined by $\mathcal{K}_m^\star$. Note that P_m is a continuum. (P_m is connected since it is the union of connected sets intersecting X and X itself is connected. Since P_m is a finite union of compact sets, P_m is compact.) Note that, by construction, $Int(P_{m+1}) \subset P_m$. Since the limit of the diameters of the simplexes of the complex $\mathcal{K}_m^\star$ tend to zero as m tends to infinity, $\bigcap_{m=1}^{\infty} Int(P_m) = \bigcap_{m=1}^{\infty} P_m = X$. By Lemma 1.6.7, there exists $m \in \mathbb{N}$ such that $P_m \subset U$.

Q.E.D.

Now, we present the definition of arc-smooth continuum. This notion is used to study continua which are strictly point $\mathcal{T}$-asymmetric (Definition 3.1.92).

1.7.37 Definition An arcwise connected continuum X is *arc-smooth at* $p \in X$ if there exists a map $\alpha \colon X \to \mathcal{C}(X)$ (Definition 1.8.1) such that $\alpha(p) = \{p\}$, $\alpha(x)$ is an arc in X joining p and x and if $y \in \alpha(x)$, then $\alpha(y) \subset \alpha(x)$. The continuum X is *arc-smooth* if there exists a point at which it is arc-smooth.

1.7.38 Theorem *If X is an arc-smooth continuum, then for each closed subset H of X, the set* $\bigcup_{x \in H} \alpha(x)$ *is a subcontinuum of X.*

Proof Suppose X is arc-smooth at p and let $M = \bigcup_{x \in H} \alpha(x)$. Note that for each $x \in H$, $p \in \alpha(x)$. Hence, M is a connected subset of X. Let $z \in Cl(M)$. Then there exists a sequence $\{z_n\}_{n=1}^{\infty}$ of points of M converging to z. Thus, for each $n \in \mathbb{N}$, there exists $x_n \in H$ such that $z_n \in \alpha(x_n)$. Since H is closed in X, without loss of generality, we assume that the sequence $\{x_n\}_{n=1}^{\infty}$ converges to a point $x \in H$. Since X is arc-smooth, α is continuous. In consequence, the sequence of arcs $\{\alpha(x_n)\}_{n=1}^{\infty}$ converges to the arc $\alpha(x)$. Note that, by the properties of α, the sequence of arcs $\{\alpha(z_n)\}_{n=1}^{\infty}$ converges to the arc $\alpha(z)$ and $z \in \alpha(x)$. Hence, $z \in M$ and M is closed. Therefore, M is a subcontinuum of X.

Q.E.D.

1.7.39 Definition An arcwise connected continuum X is *arcwise decomposable* if there exist two proper arcwise connected subcontinua A and B of X such that $X = A \cup B$.

The following theorem is due to David P. Bellamy and it is part of [3, Example II].

1.7.40 Theorem *If X is a continuum which is the continuous image of the cone over the Cantor set, then X is arcwise decomposable.*

Proof Let $\mathfrak{F}_{\mathcal{C}}$ be the cone over the Cantor set, and let $f : \mathfrak{F}_{\mathcal{C}} \twoheadrightarrow X$ be a surjective map. By the Kuratowski–Zorn lemma, there exists a subcontinuum K of $\mathfrak{F}_{\mathcal{C}}$ such that $f(K) = X$ and $f(L) \neq X$ for any proper subcontinuum L of K. Let v be the vertex of $\mathfrak{F}_{\mathcal{C}}$. If $K \setminus \{v\}$ is not connected, then K is the union of two arcwise connected proper subcontinua. If $K \setminus \{v\}$ is connected, then K is an arc. In either case, there exist two arcwise connected proper subcontinua K_1 and K_2 of K such that $K = K_1 \cup K_2$. Hence, by the minimality of K, $f(K_1)$ and $f(K_2)$ are arcwise connected proper subcontinua of X such that $X = f(K_1) \cup f(K_2)$. Therefore, X is arcwise decomposable.

Q.E.D.

1.8   Hyperspaces

We give the definition of the main hyperspaces associated to a continuum. We present some of their elementary properties. For this section, we use [4–6, 8, 15, 20–23, 26].

1.8.1 Definition Given a compactum X, we define its *hyperspaces* as the following sets:

$$2^X = \{A \subset X \mid A \text{ is closed and nonempty}\},$$

and for each $n \in \mathbb{N}$

$$C_n(X) = \{A \in 2^X \mid A \text{ has at most } n \text{ components}\},$$

$$\mathcal{F}_n(X) = \{A \in 2^X \mid A \text{ has at most } n \text{ points}\}.$$

$\mathcal{F}_n(X)$ is called *n-fold symmetric product* and $C_n(X)$ is called *n-fold hyperspace*.

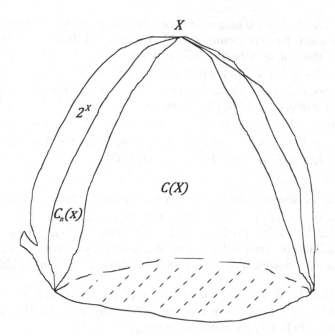

1.8.2 Remark Given a continuum X, we agree that $\mathcal{C}(X) = \mathcal{C}_1(X)$. Let us observe that for each $n \in \mathbb{N}$,

$$\mathcal{F}_n(X) \subset \mathcal{C}_n(X),$$

$$\mathcal{C}_n(X) \subset \mathcal{C}_{n+1}(X),$$

and that

$$\mathcal{F}_n(X) \subset \mathcal{F}_{n+1}(X).$$

The n-fold symmetric products were defined by Borsuk and Ulam [4]. These hyperspaces have been studied by many people. Some recent results and references about n-fold symmetric products may be found in [5, 20, 21] and [6]. It is not clear who defined or where the n-fold hyperspaces were defined. A study of the n-fold hyperspaces is presented in Chap. 6.

1.8.3 Theorem *Let X be a continuum with metric d. Then the map $\mathcal{H}: 2^X \times 2^X \to [0, \infty)$ given by:*

$$\mathcal{H}(A, B) = \inf\{\varepsilon > 0 \mid A \subset \mathcal{V}_\varepsilon^d(B) \ \text{and} \ B \subset \mathcal{V}_\varepsilon^d(A)\}$$

is a metric for 2^X. It is called the Hausdorff metric.

Proof We show the triangle inequality. The other two properties are obvious.

Let A, B and C be three elements of 2^X. We show that $\mathcal{H}(A, C) \leq \mathcal{H}(A, B) + \mathcal{H}(B, C)$. To this end, let α be a positive real number. Let $\beta_A = \mathcal{H}(A, B) + \frac{\alpha}{2}$ and $\beta_B = \mathcal{H}(B, C) + \frac{\alpha}{2}$. Let us observe that: $A \subset \mathcal{V}_{\beta_A}^d(B)$ and $B \subset \mathcal{V}_{\beta_B}^d(C)$. Then, given $a \in A$, there exists $b \in B$ such that $d(a, b) < \mathcal{H}(A, B) + \frac{\alpha}{2}$. For this b, there exists $c \in C$ such that $d(b, c) < \mathcal{H}(B, C) + \frac{\alpha}{2}$. Hence, $d(a, c) < \mathcal{H}(A, B) + \mathcal{H}(B, C) + \alpha$. Therefore, if $\beta = \mathcal{H}(A, B) + \mathcal{H}(B, C) + \alpha$, then $A \subset \mathcal{V}_\beta^d(C)$. A similar argument shows that $C \subset \mathcal{V}_\beta^d(A)$. Since α is an arbitrary positive number, it follows, from the definition of $\mathcal{H}$, that $\mathcal{H}(A, C) \leq \mathcal{H}(A, B) + \mathcal{H}(B, C)$.

$$\textbf{Q.E.D.}$$

Let X be a continuum. Note that given a sequence, $\{Y_n\}_{n=1}^\infty$, of elements of 2^X, we have two types of convergence of this sequence, namely, the one given in Definition 1.2.26 (the limit belongs to 2^X by Lemma 1.2.27), and the other one given by the Hausdorff metric. It is known that both types of convergence coincide, i.e., we have the following theorem, a proof of which may be found in [26, (0.7)]:

1.8.4 Theorem *Let X be a continuum, and let $\{Y_n\}_{n=1}^\infty$ be a sequence of elements of 2^X. Then $\{Y_n\}_{n=1}^\infty$ converges to Y in the sense of Definition 1.2.26 if and only if $\{Y_n\}_{n=1}^\infty$ converges to Y in the Hausdorff metric.*

1.8.5 Theorem *If X is a compactum, then 2^X and $\mathcal{C}(X)$ are compact.*

Proof First, we show that 2^X is compact. By Theorem 1.8.4, it is enough to prove that every sequence of elements in 2^X has a convergent subsequence in the sense of Definition 1.2.26. This is already done in Theorem 1.2.28. We just need to mention that the limit of the subsequence constructed in Theorem 1.2.28 is closed by Lemma 1.2.27, and, since X is compact, such limit is nonempty.

To see that $\mathcal{C}(X)$ is compact, it suffices to show that $\mathcal{C}(X)$ is closed in 2^X. This follows from Theorem 1.2.29.

$$\textbf{Q.E.D.}$$

Regarding n-fold symmetric products, we have the following results:

1.8.6 Lemma *Let X be a continuum with metric d, and let $n \in \mathbb{N}$. If D_n denotes the metric on X^n given by*

$$D_n((x_1, \ldots, x_n), (x_1', \ldots, x_n')) = \max\{d(x_1, x_1'), \ldots, d(x_n, x_n')\},$$

then the function $f_n \colon X^n \to \mathcal{F}_n(X)$ given by

$$f_n((x_1, \ldots, x_n)) = \{x_1, \ldots, x_n\}$$

is surjective and satisfies the following inequality:

$$\mathcal{H}(f_n((x_1, \ldots, x_n)), f_n((x_1', \ldots, x_n'))) \leq$$

$$D_n((x_1, \ldots, x_n), (x_1', \ldots, x_n')),$$

for each $(x_1, \ldots, x_n)$ and $(x_1', \ldots, x_n')$ in X^n.

Proof Clearly the map f_n is surjective. Let $(x_1, \ldots, x_n)$ and $(x_1', \ldots, x_n')$ be two points of X^n. Assume that

$$D_n((x_1, \ldots, x_n), (x_1', \ldots, x_n')) = r$$

and let $\varepsilon > 0$ be given. Hence, $D_n((x_1, \ldots, x_n), (x_1', \ldots, x_n')) < r + \varepsilon$. This implies that for each $j \in \{1, \ldots, n\}$, $d(x_j, x_j') < r + \varepsilon$. Thus, for each $x_j \in f_n((x_1, \ldots, x_n))$, we have that $x_j' \in f_n((x_1', \ldots, x_n'))$ and $d(x_j, x_j') < r + \varepsilon$. This shows that

$$f_n((x_1, \ldots, x_n)) \subset \mathcal{V}_{r+\varepsilon}^d(f_n((x_1', \ldots, x_n'))).$$

Similarly, $f_n((x_1', \ldots, x_n')) \subset \mathcal{V}_{r+\varepsilon}^d(f_n((x_1, \ldots, x_n)))$. Thus,

$$\mathcal{H}(f_n((x_1, \ldots, x_n)), f_n((x_1', \ldots, x_n'))) \leq r + \varepsilon.$$

Since the ε is arbitrary, we obtain that

$$\mathcal{H}(f_n((x_1, \ldots, x_n)), f_n((x_1', \ldots, x_n'))) \leq r.$$

Q.E.D.

As an immediate consequence of lemma 1.8.6, we have the following Corollary:

1.8.7 Corollary *Let X be a continuum and let $n \in \mathbb{N}$. Then the function $f_n \colon X^n \twoheadrightarrow \mathcal{F}_n(X)$ given by:*

$$f_n((x_1, \ldots, x_n)) = \{x_1, \ldots, x_n\}$$

is continuous.

1.8.8 Corollary *If X is a continuum, then $\mathcal{F}_n(X)$ is a continuum for each $n \in \mathbb{N}$.*

1.8.9 Corollary *If X is a continuum, then 2^X is a continuum.*

Proof By Corollary 1.8.8, $\mathcal{F}_n(X)$ is a continuum for each $n \in \mathbb{N}$. Then $\mathcal{F}(X) = \bigcup_{n=1}^{\infty} \mathcal{F}_n(X)$ is a connected subset of 2^X. We show that $\mathcal{F}(X)$ is dense in 2^X. To this end, let ε be a positive real number. Since X is compact, there exist $x_1, \ldots, x_m \in X$ such that $X \subset \bigcup_{n=1}^{m} \mathcal{V}_{\varepsilon}^d(x_n)$. Then $\{x_1, \ldots, x_m\} \in \mathcal{F}(X)$, $X \subset \mathcal{V}_{\varepsilon}^d(\{x_1, \ldots, x_m\})$

and, clearly, $\{x_1, \ldots, x_m\} \subset \mathcal{V}_\varepsilon^d(X)$. Hence, $\mathcal{H}(\{x_1, \ldots, x_m\}, X) < \varepsilon$. Therefore, $\mathcal{F}(X)$ is dense in 2^X. Thus, 2^X is connected. By Theorem 1.8.5, 2^X is a compactum. Therefore, 2^X is a continuum.

Q.E.D.

The following theorem (see [26, (1.13)]) is a better result than Corollary 1.8.9:

1.8.10 Theorem *If X is a continuum, then 2^X and $\mathcal{C}(X)$ are both arcwise connected continua.*

Recall that given a compactum X, we defined 2^X as the family of all nonempty closed subsets of X with the Hausdorff metric $\mathcal{H}$. By Theorem 1.8.5, 2^X is a compactum. Hence, we define 2^{2^X} as the family of all nonempty closed subsets of 2^X with the Hausdorff metric $\mathcal{H}^2$. The following lemma, a proof of which may be found in [26, (1.48)], gives a nice map from 2^{2^X} onto 2^X:

1.8.11 Lemma *Let X be a compactum. Let $\sigma: 2^{2^X} \to 2^X$ be given by $\sigma(\mathcal{A}) = \bigcup\{A \mid A \in \mathcal{A}\}$. Then σ is well defined and satisfies the following inequality:*

$$\mathcal{H}(\sigma(\mathcal{A}_1), \sigma(\mathcal{A}_2)) \leq \mathcal{H}^2(\mathcal{A}_1, \mathcal{A}_2)$$

for each $\mathcal{A}_1$ and $\mathcal{A}_2$ in 2^{2^X}. In particular, σ is continuous.

1.8.12 Corollary *Let X be a continuum. Then $\mathcal{C}_n(X)$ is an arcwise connected continuum for each $n \in \mathbb{N}$.*

Proof Let $n \in \mathbb{N}$. Since $\mathcal{C}(X)$ is an arcwise connected continuum (Theorem 1.8.10), $\mathcal{F}_n(\mathcal{C}(X))$ is a continuum (Corollary 1.8.8). It is easy to see that, in fact, $\mathcal{F}_n(\mathcal{C}(X))$ is arcwise connected. Let σ be the union map. Then $\sigma\,(\mathcal{F}_n(\mathcal{C}(X))) = \mathcal{C}_n(X)$. Hence, $\mathcal{C}_n(X)$ is an arcwise connected continuum.

Q.E.D.

1.8.13 Notation Let X be a compactum. Given a finite collection of nonempty subsets of X, $U_1, U_2, \ldots, U_m$, we define
$$\langle U_1, \ldots, U_m \rangle =$$

$$\left\{ A \in 2^X \;\middle|\; A \subset \bigcup_{k=1}^{m} U_k \text{ and } A \cap U_k \neq \emptyset \text{ for each } k \in \{1, \ldots, m\} \right\}.$$

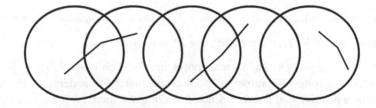

A proof of the following theorem may be found in [26, (0.11)]:

1.8.14 Theorem *Let X be a compactum. If*

$$\mathcal{B} = \{\langle U_1, \ldots, U_n \rangle \mid U_1, \ldots, U_n \text{ are open subsets of } X \text{ and } n \in \mathbb{N}\},$$

then $\mathcal{B}$ is a basis for a topology of 2^X.

1.8.15 Definition The topology for 2^X given by Theorem 1.8.14 is called the *Vietoris topology*.

It is known that the topology induced by the Hausdorff metric and the Vietoris topology coincide (see [26, (0.13)]):

1.8.16 Theorem *Let X be a compactum. Then the topology induced by the Hausdorff metric and the Vietoris topology for 2^X are the same.*

1.8.17 Definition Let X be a compactum. A *Whitney map* is a continuous function $\mu: 2^X \twoheadrightarrow [0, 1]$ such that:

(1) $\mu(X) = 1$,
(2) $\mu(\{x\}) = 0$ for every $x \in X$ and
(3) $\mu(A) < \mu(B)$ if $A, B \in 2^X$ and $A \subsetneq B$.

The fibres, $\mu^{-1}(t)$, of a Whitney map are called *Whitney levels*.

1.8.18 Remark Whitney maps exist; constructions of them may be found in [26] and [15].

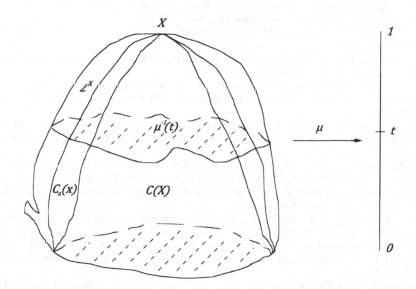

1.8.19 Definition Let X be a compactum, and let $A, B \in 2^X$. An *order arc from A to B* is a one-to-one map $\alpha: [0, 1] \to 2^X$ such that $\alpha(0) = A$, $\alpha(1) = B$ and for each $s, t \in [0, 1]$ such that $s < t$, $\alpha(s) \subsetneq \alpha(t)$.

The following theorem tells us when an order arc exists. A proof of it may be found in [26, (1.8)]:

1.8.20 Theorem *Let X be a compactum, and let A, $B \in 2^X$. Then there exists an order arc from A to B if and only if $A \subset B$ and each component of B intersects A.*

References

1. M. Aguilar, S. Gitler and C. Prieto, *Algebraic Topology From a Homotopical Viewpoint*, Universitext, Springer-Verlag, New York, Inc., 2002.
2. F. D. Ancel, An Alternative Proof and Applications of a Theorem of E. G. Effros, Michigan J. 34 (1987), 39–55.
3. D. P. Bellamy, The cone over the Cantor set-continuous maps from both directions, in: *Proceedings of the Topology Conference*, J. W. Rogers, Jr., ed., Emory University, Atlanta GA, (1970), 8–25.
4. K. Borsuk and S. Ulam, On Symmetric Products of Topological Spaces, Bull. Amer. Math. Soc., 37 (1931), 875–882.
5. E. Castañeda, A Unicoherent Continuum for Which its Second Symmetric Product is not Unicoherent, Topology Proceedings, 23 (1998), 61–67.
6. E. Castañeda, *Productos Simétricos*, Tesis Doctoral, Facultad de Ciencias, U. N. A. M., 2003. (Spanish)
7. J. J. Charatonik, On fans, Dissertationes Math. (Rozprawy Mat.), 54 (1967), 1–37.
8. J. J. Charatonik, A. Illanes and S. Macías, Induced Mappings on the Hyperspaces $C_n(X)$ of a Continuum X, Houston J. Math., 28 (2002), 781–805.
9. H. S. Davis, P. H. Doyle, Invertible Continua, Portugal. Math., 26 (1967), 487–491.
10. J. Dugundji, *Topology*, Allyn and Bacon, Inc., Boston, 1966.
11. J. B. Fugate, G. R. Gordh, Jr. and L. Lum, Arc-smooth Continua, Trans. Amer. Math. Soc., 265 (1981), 545–561.
12. G. R. Gordh, Jr. and C. B. Hughes, On Freely Decomposable Mappings of Continua, Glasnik Math., 14 (34) (1979), 137–146.
13. J. G. Hocking and G. S. Young, *Topology*, Dover Publications, Inc., New York, 1988.
14. F. B. Jones, Aposyndetic Continua and Certain Boundary Problems, Amer. J. Math., 53 (1941), 545–553.
15. A. Illanes and S. B. Nadler, Jr., *Hyperspaces: Fundamentals and Recent Advances*, Monographs and Textbooks in Pure and Applied Math., Vol. 216, Marcel Dekker, New York, Basel, 1999.
16. K. Kuratowski, *Topology*, Vol. I, Academic Press, New York, N. Y., 1966.
17. K. Kuratowski, *Topology*, Vol. II, Academic Press, New York, N. Y., 1968.
18. E. L. Lima, *Espaços Métricos*, terceira edição, Instituto de Matemática Pura e Aplicada, CNPq, (Projeto Euclides), 1977. (Portuguese)
19. E. L. Lima, *Grupo Fundamental e Espaços de Recobrimento*, Instituto de Matemática Pura e Aplicada, CNPq, (Projeto Euclides), 1993. (Portuguese)
20. S. Macías, On Symmetric Products of Continua, Topology Appl., 92 (1999), 173–182.
21. S. Macías, Aposyndetic Properties of Symmetric Products of Continua, Topology Proc., 22 (1997), 281–296.
22. S. Macías, On the Hyperspaces $C_n(X)$ of a Continuum X, Topology Appl., 109 (2001), 237–256.
23. S. Macías, On the Hyperspaces $C_n(X)$ of a Continuum X, II, Topology Proc., 25 (2000), 255–276.

24. J. Munkres, *Elements of Algebraic Topology*, Addison-Wesley Publishing Company, Inc., Redwood City, California, 1984.
25. J. Munkres, *Topology*, second edition, Prentice Hall, Upper Saddle River, NJ, 2000.
26. S. B. Nadler, Jr., *Hyperspaces of Sets,* Monographs and Textbooks in Pure and Applied Math., Vol. 49, Marcel Dekker, New York, Basel, 1978. Reprinted in: Aportaciones Matemáticas de la Sociedad Matemática Mexicana, Serie Textos # 33, 2006.
27. S. B. Nadler, Jr., *Continuum Theory: An Introduction,* Monographs and Textbooks in Pure and Applied Math., Vol. 158, Marcel Dekker, New York, Basel, Hong Kong, 1992.
28. S. B. Nadler, Jr., *Continuum Theory*, class notes, Fall Semester of 1999, Math. 481, West Virginia University.
29. J. J. Rotman, *An Introduction to the Theory of Groups*, fourth edition, Graduate Texts in Mathematics, Vol. 148, Springer-Verlag, New York, Inc., 1995.
30. W. Rudin, *Principles of Mathematical Analysis*, third edition, International Series in Pure and Applied Mathematics, McGraw-Hill, Inc., New York, 1976.

Chapter 2
Inverse Limits and Related Topics

We present basic results about inverse limits and related topics. An excellent treatment of inverse limits, distinct from the one given here, was written by W. Tom Ingram [13]. There are two excellent recent books on inverse limits one written by W. Tom Ingram and William S. Mahavier [15] and the other one written by W. Tom Ingram [14].

First, we present some basic results of inverse limits. Then a construction and a characterization of the Cantor set are given using inverse limits. With this technique, it is shown that a compactum is a continuous image of the Cantor set. Next, we show that inverse limits commute with the operation of taking finite products, cones and hyperspaces. We give some properties of chainable continua. We study circularly chainable and $\mathcal{P}$-like continua. We end the chapter presenting several properties of universal maps and AH-essential maps.

2.1 Inverse Limits

We introduce inverse limits of metric spaces. We prove basic properties of inverse limits of compacta and continua and we follow [1, 3–5, 7, 9, 17, 21–23, 25].

2.1.1 Definition Let $\{X_n\}_{n=1}^{\infty}$ be a countable collection of metric spaces. For each $n \in \mathbb{N}$, let $f_n^{n+1} \colon X_{n+1} \to X_n$ be a map. Then the sequence $\{X_n, f_n^{n+1}\}$ of metric spaces and maps is called an *inverse sequence*. The maps f_n^{n+1} are called *bonding maps*.

$$X_1 \xleftarrow{f_1^2} X_2 \longleftarrow \cdots \longleftarrow X_{n-1} \xleftarrow{f_{n-1}^n} X_n \xleftarrow{f_n^{n+1}} X_{n+1} \longleftarrow \cdots$$

2.1.2 Notation If $m, n \in \mathbb{N}$ and $n > m$, then $f_m^n = f_m^{m+1} \circ \cdots \circ f_{n-1}^n$ and $f_n^n = 1_{X_n}$, where 1_{X_n} denotes the identity map on X_n.

© Springer International Publishing AG, part of Springer Nature 2018
S. Macías, *Topics on Continua*, https://doi.org/10.1007/978-3-319-90902-8_2

2.1.3 Definition If $\{X_n, f_n^{n+1}\}$ is an inverse sequence of metric spaces, then the *inverse limit of* $\{X_n, f_n^{n+1}\}$, denoted by $\varprojlim\{X_n, f_n^{n+1}\}$ or X_∞, is the subspace of the topological product $\prod_{n=1}^{\infty} X_n$ given by

$$\varprojlim\{X_n, f_n^{n+1}\} = \left\{ (x_n)_{n=1}^{\infty} \in \prod_{n=1}^{\infty} X_n \; \middle| \; f_n^{n+1}(x_{n+1}) = x_n \right.$$

$$\text{for each } n \in \mathbb{N} \Big\}.$$

2.1.4 Remark If $\{X_n, f_n^{n+1}\}$ is an inverse sequence of metric spaces, then, by Lemma 1.1.6, its inverse limit, X_∞, is a metric space.

2.1.5 Definition Let $\{X_n, f_n^{n+1}\}$ be an inverse sequence of metric spaces with inverse limit X_∞. Recall that there exist maps $\pi_m \colon \prod_{n=1}^{\infty} X_n \twoheadrightarrow X_m$ called projection maps. For each $m \in \mathbb{N}$, let $f_m = \pi_m|_{X_\infty}$. Then f_m is a continuous function, since it is a restriction of a map (Lemma 1.1.6), and it is called a *projection map*.

2.1.6 Remark Let $\{X_n, f_n^{n+1}\}$ be an inverse sequence of metric spaces with inverse limit X_∞. Even though the maps π_m are surjective, the projection maps f_m are not, in general, surjective. However, if all the bonding maps f_n^{n+1} are surjective, then the projection maps are surjective, and vice versa. Note that, by definition, for each $n \in \mathbb{N}$, $f_n^{n+1} \circ f_{n+1} = f_n$; i.e., the following diagram

$$\begin{array}{ccc} & X_\infty & \\ f_{n+1} \swarrow & & \searrow f_n \\ X_{n+1} & \xrightarrow{\; f_n^{n+1} \;} & X_n \end{array}$$

is commutative.

2.1.7 Definition Let $\{X_n, f_n^{n+1}\}$ be an inverse sequence of metric spaces. For each $m \in \mathbb{N}$, let

$$S_m = \left\{ (x_n)_{n=1}^{\infty} \in \prod_{n=1}^{\infty} X_n \; \middle| \; f_k^{k+1}(x_{k+1}) = x_k, \; 1 \le k < m \right\}.$$

2.1.8 Proposition *Let* $\{X_n, f_n^{n+1}\}$ *be an inverse sequence of compacta whose inverse limit is* X_∞. *Then:*

(1) For each $m \in \mathbb{N}$, S_m *is homeomorphic to* $\prod_{n=m}^{\infty} X_n$.
(2) $\{S_m\}_{m=1}^{\infty}$ *is a decreasing sequence of compacta and* $X_\infty = \bigcap_{m=1}^{\infty} S_m$. *In particular,* $X_\infty \ne \emptyset$.
(3) If for each $n \in \mathbb{N}$, X_n *is a continuum, then* X_∞ *is a continuum.*

Proof We show (1). To this end, let $h \colon S_m \to \prod_{n=m}^{\infty} X_n$ be given by $h((x_n)_{n=1}^{\infty}) = (x_n)_{n=m}^{\infty}$. Since $\pi_n \circ h$ is continuous for each $n \geq m$, by Theorem 1.1.9, h is continuous.

Now, let $g \colon \prod_{n=m}^{\infty} X_n \to S_m$ be given by $g\left((x_n)_{n=m}^{\infty}\right) = (y_n)_{n=1}^{\infty}$, where $y_n = x_n$ if $n \geq m$ and $y_n = f_n^m(x_m)$ if $n < m$. Since $\pi_n \circ g$ is continuous for each $n \in \mathbb{N}$, by Theorem 1.1.9, g is continuous.

Note that $g \circ h = 1_{S_m}$ and $h \circ g = 1_{\prod_{n=m}^{\infty} X_n}$. Therefore, h is a homeomorphism.

We see (2). By definition, $S_{m+1} \subset S_m$ for each $m \in \mathbb{N}$. By (1), each S_m is a compactum (Theorem 1.1.11). Clearly, $X_\infty = \bigcap_{m=1}^{\infty} S_m$.

We prove (3). Note that (3) follows from (2) and Theorem 1.7.2.

Q.E.D.

2.1.9 Proposition *Let $\{X_n, f_n^{n+1}\}$ be an inverse sequence of compacta with inverse limit X_∞. For each $n \in \mathbb{N}$, let*

$$\mathcal{B}_n = \left\{ f_n^{-1}(U_n) \mid U_n \text{ is an open subset of } X_n \right\}.$$

If $\mathcal{B} = \bigcup_{n=1}^{\infty} \mathcal{B}_n$, then $\mathcal{B}$ is a basis for the topology of X_∞.

Proof Since X_∞ is a subspace of a topological product, a basic open subset of X_∞ is of the form

$$\bigcap_{j=1}^{k} f_{n_j}^{-1}(U_{n_j}),$$

where U_{n_j} is an open subset of X_{n_j}. Without loss of generality, we assume that $n_k = \max\{n_1, \ldots, n_k\}$. Let $U = \bigcap_{j=1}^{k} \left(f_{n_j}^{n_k}\right)^{-1}(U_{n_j})$. Then U is an open subset of X_{n_k} and

$$f_{n_k}^{-1}(U) = f_{n_k}^{-1}\left(\bigcap_{j=1}^{k} (f_{n_j}^{n_k})^{-1}(U_{n_j})\right) =$$

$$\bigcap_{j=1}^{k} \left(f_{n_j}^{n_k} \circ f_{n_k}\right)^{-1}(U_{n_j}) =$$

$$\bigcap_{j=1}^{k} f_{n_j}^{-1}(U_{n_j}).$$

Q.E.D.

The next corollary says that projection maps and bonding maps share the property of being open.

2.1.10 Corollary *Let $\{X_n, f_n^{n+1}\}$ be an inverse sequence of compacta with surjective bonding maps, whose inverse limit is X_∞. Then the projection maps are open if and only if all the bonding maps are open.*

Proof Suppose all the bonding maps are open. Let $m \in \mathbb{N}$. Let $\mathcal{U}$ be an open subset of X_∞, and let $(x_n)_{n=1}^\infty \in \mathcal{U}$. By Proposition 2.1.9, there exist $N \in \mathbb{N}$ and an open subset U_N of X_N such that

$$(x_n)_{n=1}^\infty \in f_N^{-1}(U_N) \subset \mathcal{U}.$$

To see that f_m is open, we consider two cases.

Suppose first that $N \geq m$. Then, by Remark 2.1.6,

$$x_m = f_m((x_n)_{n=1}^\infty) \in f_m\left(f_N^{-1}(U_N)\right) =$$

$$f_m^N f_N(f_N^{-1}(U_N)) = f_m^N(U_N) \subset f_m(\mathcal{U}).$$

Hence, since the bonding maps are open, $x_m = f_m((x_n)_{n=1}^\infty)$ is an interior point of $f_m(\mathcal{U})$. Therefore, $f_m(\mathcal{U})$ is open since $(x_n)_{n=1}^\infty$ is an arbitrary point of $\mathcal{U}$.

Now suppose that $m > N$. Then, by Remark 2.1.6,

$$x_m = f_m((x_n)_{n=1}^\infty) \in f_m\left(f_N^{-1}(U_N)\right) =$$

$$f_m f_m^{-1}(f_N^m)^{-1}(U_N) = (f_N^m)^{-1}(U_N) \subset f_m(\mathcal{U}).$$

Hence, since the bonding maps are continuous, $x_m = f_m((x_n)_{n=1}^\infty)$ is an interior point of $f_m(\mathcal{U})$. Therefore, $f_m(\mathcal{U})$ is open since $(x_n)_{n=1}^\infty$ is an arbitrary point of $\mathcal{U}$.

Next, suppose all the projection maps are open. Let $n \in \mathbb{N}$, and let U_{n+1} be an open subset of X_{n+1}. Since $f_n^{n+1}(U_{n+1}) = f_n f_{n+1}^{-1}(U_{n+1})$ (Remark 2.1.6) and the fact that the projection maps are continuous and open, $f_n^{n+1}(U_{n+1})$ is an open subset of X_n. Therefore, f_n^{n+1} is open.

Q.E.D.

2.1.11 Definition Let X and Y be continua. A surjective map $f : X \twoheadrightarrow Y$ is said to be *monotone* if $f^{-1}(y)$ is connected for each $y \in Y$.

The following lemma gives a very useful characterization of monotone maps.

2.1.12 Lemma *Let X and Y be continua. If $f : X \twoheadrightarrow Y$ is a surjective map, then f is monotone if and only if $f^{-1}(C)$ is connected for each connected subset C of Y.*

Proof Suppose f is monotone and let C be a subset of Y such that $f^{-1}(C)$ is not connected. Then there exist two nonempty subsets A and B of X such that $f^{-1}(C) = A \cup B$, $Cl_X(A) \cap B = \emptyset$ and $A \cap Cl_X(B) = \emptyset$. Observe that if $y \in C$ and $f^{-1}(y) \cap A \neq \emptyset$, then $f^{-1}(y) \subset A$ ($f^{-1}(y)$ is connected). Let $M = \{y \in C \mid f^{-1}(y) \subset A\}$. Hence, $A = f^{-1}(M)$. Similarly, if $N = \{y \in C \mid f^{-1}(y) \subset B\}$, then $B = f^{-1}(N)$. Clearly, $C = M \cup N$. Suppose there exists $y \in Cl_Y(M) \cap N$. Since $y \in N$, $f^{-1}(y) \subset B$. Since $y \in Cl_Y(M)$, there exists a sequence $\{y_n\}_{n=1}^\infty$ of points of M converging to y. This implies that $f^{-1}(y_n) \subset A$ for every $n \in \mathbb{N}$. Let $x_n \in f^{-1}(y_n)$. Since X is a compactum, without loss of generality, we assume that $\{x_n\}_{n=1}^\infty$ converges to a point x. Note that $x \in Cl_X(A)$ and, by the continuity of f, $f(x) = y$. Thus, $x \in Cl_X(A) \cap f^{-1}(y) \subset Cl_X(A) \cap B$, a contradiction. Hence, $Cl_Y(M) \cap N = \emptyset$. Similarly, $M \cap Cl_Y(N) = \emptyset$. Therefore, C is not connected.

The other implication is obvious.

Q.E.D.

The next result says that projection maps and bonding maps share the property of being monotone.

2.1.13 Proposition *Let $\{X_n, f_n^{n+1}\}$ be an inverse sequence of continua with surjective bonding maps and whose inverse limit is X_∞. Then the bonding maps are monotone if and only if the projection maps are monotone.*

Proof Suppose the bonding maps are monotone. Let $m \in \mathbb{N}$. We show f_m is monotone. To this end, let $x_m \in X_m$. For each $n \in \mathbb{N}$, let

$$W_n = \begin{cases} (f_m^n)^{-1}(x_m), & \text{if } n \geq m; \\ \{f_n^m(x_m)\}, & \text{if } n < m. \end{cases}$$

Then $\{W_n, f_n^{n+1}|_{W_{n+1}}\}$ is an inverse sequence of continua. Hence, by Proposition 2.1.8 (3), $W = \varprojlim\{W_n, f_n^{n+1}|_{W_{n+1}}\}$ is a subcontinuum of X_∞. We assert that $W = f_m^{-1}(x_m)$. To see this, let $(y_n)_{n=1}^\infty \in f_m^{-1}(x_m)$. Then $y_m = x_m$, $y_n = f_n^m(x_m)$ for each $n < m$, and $y_n \in (f_m^n)^{-1}(x_m) = W_n$ for each $n > m$. Hence, $(y_n)_{n=1}^\infty \in W$. Therefore, $f_m^{-1}(x_m) \subset W$.

Since $f_m(W) = W_m = \{x_m\}$, $W \subset f_m^{-1}(x_m)$. Therefore, $W = f_m^{-1}(x_m)$ and f_m is monotone.

Now, suppose all the projection maps are monotone. Let $m \in \mathbb{N}$. We prove that f_m^{m+1} is monotone. To see this, let $x_m \in X_m$. By Remark 2.1.6, $(f_m^{m+1})^{-1}(x_m) = f_{m+1}f_m^{-1}(x_m)$. Hence, $(f_m^{m+1})^{-1}(x_m)$ is connected since the projection maps are monotone and continuous. Therefore, f_m^{m+1} is monotone.

Q.E.D.

2.1.14 Corollary *Let $\{X_n, f_n^{n+1}\}$ be an inverse sequence of locally connected continua with surjective bonding maps and whose inverse limit is X_∞. If all the bonding maps are monotone, then X_∞ is locally connected.*

Proof Let $(x_n)_{n=1}^\infty \in X_\infty$, and let $\mathcal{U}$ be an open subset of X_∞ such that $(x_n)_{n=1}^\infty \in \mathcal{U}$. By Proposition 2.1.9, there exist $N \in \mathbb{N}$ and an open subset U_N of X_N such that $(x_n)_{n=1}^\infty \in f_N^{-1}(U_N) \subset \mathcal{U}$. Since X_N is locally connected, there exists a connected open subset V_N of X_N such that $x_N \in V_N \subset U_N$. Hence, $(x_n)_{n=1}^\infty \in f_N^{-1}(V_N) \subset f_N^{-1}(U_N) \subset \mathcal{U}$. Since the bonding maps are monotone, by Proposition 2.1.13, f_N

is monotone. Hence, $f_N^{-1}(V_N)$ is a connected (Lemma 2.1.12) open subset of X_∞. Therefore, X_∞ is locally connected.

Q.E.D.

The next proposition says that when the projection maps are surjective, this property is no longer true for the restriction of such maps to proper closed subsets.

2.1.15 Proposition *Let $\{X_n, f_n^{n+1}\}$ be an inverse sequence with inverse limit X_∞. If Y is a proper closed subset of X_∞, then there exists $N \in \mathbb{N}$, such that $f_m(Y) \neq X_m$ for each $m \geq N$.*

Proof Let $(x_n)_{n=1}^\infty \in X_\infty \setminus Y$. By Proposition 2.1.9, there exist $N \in \mathbb{N}$ and an open subset U_N of X_N such that

$$(x_n)_{n=1}^\infty \in f_N^{-1}(U_N) \subset X_\infty \setminus Y.$$

Hence, $f_N(Y) \subset X_N \setminus U_N$. Also, if $m > N$, then $f_m(Y) \subset X_m \setminus \left(f_N^m\right)^{-1}(U_N)$. To see this, let $z_m \in f_m(Y)$. Then there exists $(y_n)_{n=1}^\infty \in Y$ such that $f_m((y_n)_{n=1}^\infty) = z_m$. Since $Y \subset X_\infty$, $f_N^m(z_m) = f_N((y_n)_{n=1}^\infty)$. Thus, $z_m \in (f_N^m)^{-1}\left(f_N((y_n)_{n=1}^\infty)\right) \subset (f_N^m)^{-1}(f_N(Y))$. Since $f_N(Y) \subset X_N \setminus U_N$,

$$f_m(Y) \subset (f_N^m)^{-1}(f_N(Y)) \subset (f_N^m)^{-1}(X_N \setminus U_N) \subset X_m \setminus (f_N^m)^{-1}(U_N).$$

Therefore, $f_m(Y) \neq X_m$.

Q.E.D.

The following proposition gives us the image of the inverse limit under a projection map in terms of the factor spaces and the bonding maps.

2.1.16 Proposition *Let $\{X_n, f_n^{n+1}\}$ be an inverse sequence of compacta whose inverse limit is X_∞. Then, for each $n \in \mathbb{N}$, $f_n(X_\infty) = \bigcap_{m=n+1}^\infty f_n^m(X_m)$.*

Proof Let $n \in \mathbb{N}$. Since, for each $m > n$, $f_n = f_n^m \circ f_m$, $f_n(X_\infty) = f_n^m \circ f_m(X_\infty) \subset f_n^m(X_m)$. Therefore, $f_n(X_\infty) \subset \bigcap_{m=n+1}^\infty f_n^m(X_m)$.

Now, let $x_n \in \bigcap_{m=n+1}^\infty f_n^m(X_m)$. We show there exists an element of X_∞ whose nth coordinate is x_n. To this end, let $K = \prod_{k=1}^{n-1} X_k \times \{x_n\} \times \prod_{k=n+1}^\infty X_k$. Recall that, by Proposition 2.1.8 (2), $X_\infty = \bigcap_{m=1}^\infty S_m$ and $\{S_m\}_{m=1}^\infty$ is a decreasing sequence. Let $m_0 \in \mathbb{N}$. Then $K \cap S_{m_0} \neq \emptyset$. To see this, let $m_1 > \max\{n, m_0\}$. Since $x_n \in \bigcap_{m=n+1}^\infty f_n^m(X_m)$, $(f_n^{m_1})^{-1}(x_n) \neq \emptyset$. Let $x_{m_1} \in (f_n^{m_1})^{-1}(x_n)$, and let $x_{m_0} = f_{m_0}^{m_1}(x_{m_1})$. Now, let $(y_k)_{k=1}^\infty \in \prod_{k=1}^\infty X_k$ be such that $y_{m_1} = x_{m_1}$ and $y_k = f_k^{m_1}(x_{m_1})$ for each $k \in \{1, \ldots, m_1 - 1\}$. Note that $(y_k)_{k=1}^\infty \in K \cap S_{m_0}$. Thus, the family $\{K\} \cup \{S_m\}_{m=1}^\infty$ has the finite intersection property. Therefore, $K \cap X_\infty \neq \emptyset$. Hence, there exists an element of X_∞ whose nth coordinate is x_n.

Q.E.D.

2.1.17 Proposition *Let $\{X_n, f_n^{n+1}\}$ be an inverse sequence of compacta, whose inverse limit is X_∞. For each $n \in \mathbb{N}$, let A_n be a closed subset of X_n, and* *suppose that* $f_n^{n+1}(A_{n+1}) \subset A_n$. *Then* $\{A_n, f_n^{n+1}|_{A_{n+1}}\}$ *is an inverse sequence and* $\varprojlim\{A_n, f_n^{n+1}|_{A_{n+1}}\} = \bigcap_{n=1}^{\infty} f_n^{-1}(A_n)$.

Proof Clearly, $\{A_n, f_n^{n+1}|_{A_{n+1}}\}$ is an inverse sequence of compacta.

Observe that, since $f_n^{n+1}(A_{n+1}) \subset A_n$, $f_{n+1}^{-1}(A_{n+1}) \subset f_n^{-1}(A_n)$ for every $n \in \mathbb{N}$. Hence, $\{f_n^{-1}(A_n)\}_{n=1}^{\infty}$ is a decreasing sequence of compacta.

Let $a = (a_n)_{n=1}^{\infty} \in \varprojlim\{A_n, f_n^{n+1}|_{A_{n+1}}\}$. Since $f_n(a) = a_n \in A_n$, $a \in f_n^{-1}(A_n)$ for every $n \in \mathbb{N}$. Thus, $a \in \bigcap_{n=1}^{\infty} f_n^{-1}(A_n)$. Therefore, $\varprojlim\{A_n, f_n^{n+1}|_{A_{n+1}}\} \subset \bigcap_{n=1}^{\infty} f_n^{-1}(A_n)$.

Next, let $b = (b_n)_{n=1}^{\infty} \in \bigcap_{n=1}^{\infty} f_n^{-1}(A_n)$. Then $f_n(b) = b_n \in A_n$ for each $n \in \mathbb{N}$. Hence, $b \in \prod_{n=1}^{\infty} A_n$. Since $b \in X_{\infty}$, $f_n^{n+1}(b_{n+1}) = b_n$ for every $n \in \mathbb{N}$. This implies that $b \in \varprojlim\{A_n, f_n^{n+1}|_{A_{n+1}}\}$. Thus, $\bigcap_{n=1}^{\infty} f_n^{-1}(A_n) \subset \varprojlim\{A_n, f_n^{n+1}|_{A_{n+1}}\}$.

Therefore, $\bigcap_{n=1}^{\infty} f_n^{-1}(A_n) = \varprojlim\{A_n, f_n^{n+1}|_{A_{n+1}}\}$.

Q.E.D.

2.1.18 Definition Let $\{X_n, f_n^{n+1}\}$ be an inverse sequence of continua. Then $\{X_n, f_n^{n+1}\}$ is called an *indecomposable inverse sequence* provided that, for each $n \in \mathbb{N}$, whenever A_{n+1} and B_{n+1} are subcontinua of X_{n+1} such that $X_{n+1} = A_{n+1} \cup B_{n+1}$, we have that $f_n^{n+1}(A_{n+1}) = X_n$ or $f_n^{n+1}(B_{n+1}) = X_n$.

The motivation of the name "indecomposable inverse sequence" is given in the following result:

2.1.19 Theorem *Let* $\{X_n, f_n^{n+1}\}$ *be an indecomposable inverse sequence whose inverse limit is* X_{∞}. *Then* X_{∞} *is an indecomposable continuum.*

Proof By Proposition 2.1.8 (3), X_{∞} is a continuum. Now, suppose X_{∞} is decomposable. Then there exist two proper subcontinua A and B of X_{∞} such that $X_{\infty} = A \cup B$. By Proposition 2.1.15, there exists $n \in \mathbb{N}$ such that if $m \geq n$, then $f_m(A) \neq X_m$ and $f_m(B) \neq X_m$. Since, by definition, the bonding maps are surjective, the projection maps are surjective too (Remark 2.1.6). Hence,

$$X_{n+2} = f_{n+2}(X_{\infty}) = f_{n+2}(A) \cup f_{n+2}(B).$$

This implies that

$$X_{n+1} = f_{n+1}^{n+2}(X_{n+2}) = f_{n+1}^{n+2} \circ f_{n+2}(A) \cup f_{n+1}^{n+2} \circ f_{n+2}(B).$$

By hypothesis $f_{n+1}^{n+2} \circ f_{n+2}(A) = X_{n+1}$ or $f_{n+1}^{n+2} \circ f_{n+2}(B) = X_{n+1}$. Thus, $f_{n+1}(A) = X_{n+1}$ or $f_{n+1}(B) = X_{n+1}$, contradicting the election of n. Therefore, X_{∞} is indecomposable.

Q.E.D.

2.1.20 Proposition *Let* $\{X_n, f_n^{n+1}\}$ *be an inverse sequence of metric spaces whose inverse limit is* X_{∞}. *If* A *is a closed subset of* X_{∞}, *then the double sequence*

$\{f_n(A), f_n^{n+1}|_{f_{n+1}(A)}\}$ *is an inverse sequence with surjective bonding maps and*

$$(*) \quad \varprojlim \left\{ f_n(A), f_n^{n+1}|_{f_{n+1}(A)} \right\} = A = \left[\prod_{n=1}^{\infty} f_n(A) \right] \cap X_\infty.$$

Proof By Remark 2.1.6, for each $n \in \mathbb{N}$, $f_n^{n+1} \circ f_{n+1} = f_n$. It follows that $\{f_n(A), f_n^{n+1}|_{f_{n+1}(A)}\}$ is an inverse sequence with surjective bonding maps.

Now we show $(*)$. First observe that

$$\varprojlim \left\{ f_n(A), f_n^{n+1}|_{f_{n+1}(A)} \right\} = \left[\prod_{n=1}^{\infty} f_n(A) \right] \cap X_\infty.$$

Also observe that

$$A \subset \left[\prod_{n=1}^{\infty} f_n(A) \right] \cap X_\infty.$$

Thus, we need to show

$$\left[\prod_{n=1}^{\infty} f_n(A) \right] \cap X_\infty \subset A.$$

Let $\varepsilon > 0$, and let $y = (y_n)_{n=1}^{\infty} \in \left[\prod_{n=1}^{\infty} f_n(A) \right] \cap X_\infty$. Now, let $N \in \mathbb{N}$ be such that $\sum_{n=N+1}^{\infty} \frac{1}{2^n} < \varepsilon$.

Since for each $n \in \mathbb{N}$, $y_n \in f_n(A)$, there exists $a^{(n)} = \left(a_m^{(n)} \right)_{m=1}^{\infty} \in A$ such that $f_n\left(a^{(n)} \right) = y_n$. Now observe that

$$\rho(a^{(N)}, y) = \sum_{n=1}^{\infty} \frac{1}{2^n} d_n \left(a_n^{(N)}, y_n \right) =$$

$$\sum_{n=N+1}^{\infty} \frac{1}{2^n} d_n \left(a_n^{(N)}, y_n \right) < \varepsilon.$$

Since ε is arbitrary, $y \in Cl(A) = A$.

$$\text{Q.E.D.}$$

As a consequence of Propositions 2.1.17 and 2.1.20, we have the following corollary:

2.1.21 Corollary *Let $\{X_n, f_n^{n+1}\}$ be an inverse sequence of compacta, whose inverse limit is X_∞. If A is a closed subset of X_∞, then*

$$A = \bigcap_{n=1}^{\infty} f_n^{-1}(f_n(A)).$$

2.1.22 Definition Let X and Y be continua. A surjective map $f: X \twoheadrightarrow Y$ is *confluent* provided that for each subcontinuum Q of Y and each component K of $f^{-1}(Q)$, we have that $f(K) = Q$.

2.1.23 Theorem *Let $\{X_n, f_n^{n+1}\}$ be an inverse sequence of continua with surjective bonding maps and whose inverse limit is X_∞. Then the bonding maps are confluent if and only if the projection maps are confluent.*

Proof Suppose the projection maps are confluent. Then, since for each $n \in \mathbb{N}$, $f_n^{n+1} \circ f_{n+1} = f_n$, by Theorem 8.1.13, f_n^{n+1} is confluent.

Suppose the bonding maps are confluent. Let $m \in \mathbb{N}$, let Q_m be a subcontinuum of X_m and let C_m be a component of $f_m^{-1}(Q_m)$. Let K_{m+1} be the component of $(f_m^{m+1})^{-1}(Q_m)$ containing $f_{m+1}(C_m)$. Let K_{m+2} be the component of $(f_{m+1}^{m+2})^{-1}(K_{m+1})$ which contains $f_{m+2}(C_m)$. In general, for $n > m$, let K_n be the component of $(f_{n-1}^n)^{-1}(K_{n-1})$ containing $f_n(C_m)$. For $n \le m$, let $K_n = f_n^m(Q_m)$. Then $\{K_n, f_n^{n+1}|_{K_{n+1}}\}$ is an inverse sequence of continua with surjective bonding maps (each f_n^{n+1} is confluent). Let $K = \varprojlim\{K_n, f_n^{n+1}|_{K_{n+1}}\}$. Hence, K is a subcontinuum of X_∞ (Proposition 2.1.8), $f_m(K) = Q_m$, and $K \cap C_m \neq \emptyset$. Since C_m is a component of $f_m^{-1}(Q_m)$, $K \subset C_m$. Thus, $f_m(C_m) = Q_m$. Therefore, f_m is confluent.

Q.E.D.

2.1.24 Proposition *Let $\{X_n, f_n^{n+1}\}$ be an inverse sequence of metric spaces whose inverse limit is X_∞. If A and B are two closed subsets of X_∞, $C = A \cap B$ and $C_n = f_n(A) \cap f_n(B)$ for each $n \in \mathbb{N}$, then $C = \varprojlim\{C_n, f_n^{n+1}|_{C_{n+1}}\}$.*

Proof Let $x = (x_n)_{n=1}^{\infty} \in \varprojlim\{C_n, f_n^{n+1}|_{C_{n+1}}\}$. Then, by definition, $x_n \in C_n = f_n(A) \cap f_n(B)$ for each $n \in \mathbb{N}$. Hence, $x \in \varprojlim\{f_n(A), f_n^{n+1}|_{f_{n+1}(A)}\} = A$ and $x \in \varprojlim\{f_n(B), f_n^{n+1}|_{f_{n+1}(B)}\} = B$ (by $(*)$ of Proposition 2.1.20). Thus, $x \in C$.

Now, let $y = (y_n)_{n=1}^{\infty} \in C = A \cap B$. Then for each $n \in \mathbb{N}$, $y_n \in f_n(A)$ and $y_n \in f_n(B)$. Thus, $y_n \in C_n$ for each $n \in \mathbb{N}$. Therefore, $y \in \varprojlim\{C_n, f_n^{n+1}|_{C_{n+1}}\}$.

Q.E.D.

2.1.25 Definition Let $\{X_n, f_n^{n+1}\}$ be an inverse sequence of arcs (i.e., for each $n \in \mathbb{N}$, X_n is an arc) with surjective bonding maps. Then the inverse limit, X_∞, of $\{X_n, f_n^{n+1}\}$ is called an *arc-like continuum*.

Propositions 2.1.20 and 2.1.24 have the following corollaries:

2.1.26 Corollary *If X is an arc-like continuum, then each nondegenerate subcontinuum of X is arc-like.*

Proof Let X be an arc-like continuum, and let A be a nondegenerate subcontinuum of X. Since X is an arc-like continuum, there exists an inverse sequence $\{X_n, f_n^{n+1}\}$ of arcs such that $X = \varprojlim\{X_n, f_n^{n+1}\}$. By $(*)$ of Propositions 2.1.20, we have that $A = \varprojlim\{f_n(A), f_n^{n+1}|_{f_{n+1}(A)}\}$. Therefore, A is an arc-like continuum.

$$\textbf{Q.E.D.}$$

2.1.27 Corollary *If $\{X_n, f_n^{n+1}\}$ is an inverse sequence of unicoherent continua, with surjective bonding maps, whose inverse limit is X_∞, then X_∞ is a unicoherent continuum.*

Proof Let A and B be two subcontinua of X_∞ such that $X_\infty = A \cup B$. Since the bonding maps are surjective, by Remark 2.1.6, the projection maps are surjective too. Hence, for each $n \in \mathbb{N}$, $X_n = f_n(A) \cup f_n(B)$. By the unicoherence of the spaces, $f_n(A) \cap f_n(B)$ is connected for every $n \in \mathbb{N}$. Thus, by Propositions 2.1.20 and 2.1.24, $\{f_n(A) \cap f_n(B), f_n^{n+1}|_{f_{n+1}(A) \cap f_{n+1}(B)}\}$ is an inverse sequence of continua whose inverse limit is a continuum (Proposition 2.1.8 (3)). Since $A \cap B = \varprojlim\{f_n(A) \cap f_n(B), f_n^{n+1}|_{f_{n+1}(A) \cap f_{n+1}(B)}\}$ (Proposition 2.1.24) $A \cap B$ is connected. Therefore, X_∞ is unicoherent.

$$\textbf{Q.E.D.}$$

2.1.28 Corollary *Let $\{X_n, f_n^{n+1}\}$ be an inverse sequence of hereditarily unicoherent continua. If X_∞ is the inverse limit of $\{X_n, f_n^{n+1}\}$, then X_∞ is a hereditarily unicoherent continuum.*

Proof Let A and B be two subcontinua of X_∞. If $A \cap B = \emptyset$, then $A \cap B$ is connected. Thus, suppose that $A \cap B \neq \emptyset$. Then for each $n \in \mathbb{N}$, $f_n(A)$ and $f_n(B)$ are two subcontinua of X_n such that $f_n(A) \cap f_n(B) \neq \emptyset$. Since each X_n is hereditarily unicoherent, $f_n(A) \cap f_n(B)$ is connected. Hence, by Propositions 2.1.20 and 2.1.24, $\{f_n(A) \cap f_n(B), f_n^{n+1}|_{f_{n+1}(A) \cap f_{n+1}(B)}\}$ is an inverse sequence of continua whose inverse limit is a continuum (Proposition 2.1.8 (3)). Since $A \cap B = \varprojlim\{f_n(A) \cap f_n(B), f_n^{n+1}|_{f_{n+1}(A) \cap f_{n+1}(B)}\}$ (Proposition 2.1.24) $A \cap B$ is connected. Therefore, X_∞ is hereditarily unicoherent.

$$\textbf{Q.E.D.}$$

2.1.29 Remark Note that in Corollary 2.1.28 we do not require the bonding maps to be surjective.

2.1.30 Corollary *Each arc-like continuum is hereditarily unicoherent.*

Proof The corollary follows from the easy fact that an arc is hereditarily unicoherent and from Corollary 2.1.28.

$$\textbf{Q.E.D.}$$

2.1.31 Corollary *If X is an arc-like continuum, then X does not contain a simple closed curve.*

2.1.32 Remark Let $\{X_n, f_n^{n+1}\}$ be an inverse sequence of arcs, with surjective bonding maps, whose inverse limit is X_∞. Note that, in this case, S_m is homeomorphic to the Hilbert cube Q (being a countable product of arcs). Since $X_\infty = \bigcap_{m=1}^\infty S_m$ (Proposition 2.1.8 (2)), X_∞ may be written as a countable intersection of Hilbert cubes.

The next theorem tells us a way to define a map from a metric space into an inverse limit of compacta.

2.1.33 Theorem *Let Y be a metric space. Let $\{X_n, f_n^{n+1}\}$ be a sequence of compacta. If for each $n \in \mathbb{N}$, there exists a map $h_n: Y \to X_n$ such that $f_n^{n+1} \circ h_{n+1} = h_n$, then there exists a map $h_\infty: Y \to X_\infty$ such that $f_n \circ h_\infty = h_n$ for each $n \in \mathbb{N}$. The map h_∞ is called induced map, and it is denoted, also, by $\varprojlim\{h_n\}$.*

Proof Let $h_\infty: Y \to X_\infty$ be given by $h_\infty(y) = (h_n(y))_{n=1}^\infty$. Since $f_n^{n+1} \circ h_{n+1} = h_n$ for each $n \in \mathbb{N}$, h_∞ is well defined. Clearly, $f_n \circ h_\infty = h_n$. Hence, by Theorem 1.1.9, h_∞ is continuous.

Q.E.D.

2.1.34 Theorem *Let Y be a metric space. Let $\{X_n, f_n^{n+1}\}$ be a sequence of compacta. Suppose that for each $n \in \mathbb{N}$, there exists a map $h_n: Y \to X_n$ such that $f_n^{n+1} \circ h_{n+1} = h_n$. If h_m is one-to-one for some $m \in \mathbb{N}$, then the induced map h_∞ is one-to-one.*

Proof Suppose h_m is one-to-one for some $m \in \mathbb{N}$. Let $y, y' \in Y$ be such that $y \neq y'$. Since h_m is one-to-one, $h_m(y) \neq h_m(y')$. Hence, $h_\infty(y) \neq h_\infty(y')$. Therefore, h_∞ is one-to-one.

Q.E.D.

2.1.35 Theorem *Let Y be a metric space. Let $\{X_n, f_n^{n+1}\}$ be a sequence of compacta. Suppose that for each $n \in \mathbb{N}$, there exists a map $h_n: Y \to X_n$ such that $f_n^{n+1} \circ h_{n+1} = h_n$. If each h_n is surjective, then $h_\infty(Y)$ is dense in X_∞. In particular, if Y is compact, then h_∞ is surjective.*

Proof Let $f_n^{-1}(U_n)$ be a basic open set of X_∞, where U_n is an open subset of X_n (Proposition 2.1.9). Since h_n is surjective, there exists $y \in Y$ such that $h_n(y) \in U_n$. Hence, $h_\infty(y) \in f_n^{-1}(U_n)$. Therefore, $h_\infty(Y)$ is dense in X_∞.

Q.E.D.

2.1.36 Definition For each $n \in \mathbb{N}$, let $X_n = \mathcal{S}^1$, and let $f_n^{n+1}: X_{n+1} \twoheadrightarrow X_n$ be given by $f_n^{n+1}(z) = z^2$ (complex number multiplication). Let $\Sigma_2 = \varprojlim\{X_n, f_n^{n+1}\}$. Then Σ_2 is called the *dyadic solenoid*. Note that, by Theorem 2.1.19, Σ_2 is an indecomposable continuum.

The following example shows that an induced map may not be surjective.

2.1.37 Example Let Σ_2 be the dyadic solenoid. For each $n \in \mathbb{N}$, let $h_n \colon \mathbb{R} \to S^1$ be given by $h_n(t) = \exp\left(\frac{2\pi t}{2^{n-1}}\right)$. Let $n \in \mathbb{N}$. Then $f_n^{n+1} \circ h_{n+1}(t) = f_n^{n+1}\left(\exp\left(\frac{2\pi t}{2^n}\right)\right) = \left(\exp\left(\frac{2\pi t}{2^n}\right)\right)^2 = \exp\left(\frac{2\pi t}{2^{n-1}}\right) = h_n(t)$. Hence, $f_n^{n+1} \circ h_{n+1} = h_n$ for each $n \in \mathbb{N}$. By Theorem 2.1.33, there exists the induced map $h_\infty \colon \mathbb{R} \to \Sigma_2$. Observe that h_∞ is not surjective since $h_\infty(\mathbb{R})$ is an arcwise connected subset of Σ_2. But, since Σ_2 is indecomposable, it is not arcwise connected (This follows easily from [23, 11.15]).

The next two propositions present some expected results.

2.1.38 Proposition *Let $\{X_n\}_{n=1}^\infty$ be a sequence of compacta such that $X_{n+1} \subset X_n$ for each $n \in \mathbb{N}$. If $f_n^{n+1} \colon X_{n+1} \to X_n$ is the inclusion map, then $\varprojlim\{X_n, f_n^{n+1}\}$ is homeomorphic to $\bigcap_{n=1}^\infty X_n$.*

Proof Note that if $(x_n)_{n=1}^\infty \in \varprojlim\{X_n, f_n^{n+1}\}$, then $x_n = x_1$ for every $n \in \mathbb{N}$ and $x_1 \in \bigcap_{n=1}^\infty X_n$.

For each $n \in \mathbb{N}$, let $h_n \colon \bigcap_{m=1}^\infty X_m \to X_n$ be given by $h_n(x) = x$; i.e., h_n is the inclusion map. Clearly $f_n^{n+1} \circ h_{n+1} = h_n$ for every $n \in \mathbb{N}$. By Theorem 2.1.33, there exists the induced map $h_\infty \colon \bigcap_{n=1}^\infty X_n \to \varprojlim\{X_n, f_n^{n+1}\}$. Since h_1 is one-to-one, h_∞ is one-to-one (Theorem 2.1.34). Now, if $(x_n)_{n=1}^\infty \in \varprojlim\{X_n, f_n^{n+1}\}$, then $h_\infty(x_1) = (x_n)_{n=1}^\infty$. Hence, h_∞ is surjective. Since $\bigcap_{n=1}^\infty X_n$ is compact, h_∞ is a homeomorphism.

Q.E.D.

2.1.39 Proposition *Let $\{X_n, f_n^{n+1}\}$ be an inverse sequence of compacta whose inverse limit is X_∞. If for each $n \in \mathbb{N}$, X_n is homeomorphic to a compactum Y and all the bonding maps are homeomorphisms, then X_∞ is homeomorphic to Y.*

Proof Let $h_1 \colon Y \to X_1$ be any homeomorphism. For $n \geq 2$, let $h_{n+1} = (f_n^{n+1})^{-1} \circ h_n$. Observe that, by definition, $f_n^{n+1} \circ h_{n+1} = h_n$ for each $n \in \mathbb{N}$. Then, by Theorem 2.1.33, there exists the induced map $h_\infty \colon Y \to X_\infty$. Since each h_n is a homeomorphism and Y is compact, by Theorems 2.1.34 and 2.1.35, h_∞ is a homeomorphism.

Q.E.D.

The next theorem says that certain type of subsequences of an inverse sequence converge to the same limit.

2.1.40 Theorem *Let $\{X_n, f_n^{n+1}\}$ be an inverse sequence of compacta, with surjective bonding maps, whose inverse limit is X_∞, and let $\{m(n)\}_{n=1}^\infty$ be an increasing subsequence of $\mathbb{N}$. If $Y_\infty = \varprojlim\{Y_n, g_n^{n+1}\}$, where $Y_n = X_{m(n)}$ and $g_n^{n+1} = f_{m(n)}^{m(n+1)}$ for each $n \in \mathbb{N}$, then Y_∞ is homeomorphic to X_∞.*

Proof For each $k \in \mathbb{N}$, let $h_k \colon X_\infty \to Y_k$ be given by $h_k((x_n)_{n=1}^\infty) = x_{m(k)}$; i.e., $h_k = f_{m(k)}$. Hence, h_k is continuous and $g_k^{k+1} \circ h_{k+1} = h_k$. Then, by

Theorem 2.1.33, there exists the induced map $h_\infty \colon X_\infty \to Y_\infty$. Since the bonding maps are surjective, by Remark 2.1.6, the projections are surjective. In particular, each h_k is surjective. Thus, h_∞ is surjective (Theorem 2.1.35). To see h_∞ is one-to-one, it is enough to observe that if $(x_n)_{n=1}^\infty$ and $(x_n')_{n=1}^\infty$ are two distinct points of X_∞, then there exists $N \in \mathbb{N}$ such that $x_n \neq x_n'$ for each $n \geq N$.

$$\textbf{Q.E.D.}$$

2.1.41 Notation We use the notation $\{X_n, f_n^{n+1}\}_{n=N}^\infty$ to denote the inverse subsequence obtained from $\{X_n, f_n^{n+1}\}$ by removing the first $N - 1$ factor spaces.

Next, we present some consequences of Theorem 2.1.40. First, we need the following definition:

2.1.42 Definition A continuum X is said to be a *triod*, provided that there exists a subcontinuum M of X such that $X \setminus M = K_1 \cup K_2 \cup K_3$, where each $K_j \neq \emptyset$, $j \in \{1, 2, 3\}$, and they are mutually separated; i.e., $Cl_X(K_j) \cap K_\ell = \emptyset$, $j, \ell \in \{1, 2, 3\}$ and $j \neq \ell$.

2.1.43 Corollary *If X is an arc-like continuum, then X does not contain a triod.*

Proof Since X is an arc-like continuum, there exists an inverse sequence $\{X_n, f_n^{n+1}\}$ of arcs whose inverse limit is X.

Suppose Y is a triod contained in X. Then there exists a subcontinuum M of Y such that $Y \setminus M = K_1 \cup K_2 \cup K_3$, where each $K_j \neq \emptyset$, $j \in \{1, 2, 3\}$, and they are mutually separated. Note that $M \cup K_j$ is a continuum (Lemma 1.7.23).

Let $n \in \mathbb{N}$. Then $f_n(Y) = f_n(M \cup K_1 \cup K_2 \cup K_3) = f_n(M) \cup f_n(K_1) \cup f_n(K_2) \cup f_n(K_3)$. Since X_n is an arc, there exists $j_n \in \{1, 2, 3\}$ such that $f_n(M \cup K_{j_n}) \subset f_n(M \cup K_{\ell_1}) \cup f_n(M \cup K_{\ell_2})$, $\ell_1, \ell_2 \in \{1, 2, 3\} \setminus \{j_n\}$ and $\ell_1 \neq \ell_2$. Since the set of positive integers is infinite and we only have three choices, by Theorem 2.1.40, we assume, without loss of generality, that $f_n(M \cup K_3) \subset f_n(M \cup K_1) \cup f_n(M \cup K_2)$ for each $n \in \mathbb{N}$.

Note that, by $(*)$ of Proposition 2.1.20,

$$M \cup K_j = \varprojlim\{f_n(M \cup K_j), f_n^{n+1}|_{f_{n+1}(M \cup K_j)}\}.$$

Since $f_n(M \cup K_3) \subset f_n(M \cup K_1) \cup f_n(M \cup K_2)$ for each $n \in \mathbb{N}$, it follows that $M \cup K_3 \subset (M \cup K_1) \cup (M \cup K_2) = M \cup K_1 \cup K_2$, a contradiction. Therefore, X does not contain a triod.

$$\textbf{Q.E.D.}$$

2.1.44 Definition Let $\{X_n, f_n^{n+1}\}$ be an inverse sequence of simple closed curves with surjective bonding maps. Then the inverse limit, X_∞, of such an inverse sequence is called a *circle-like continuum*.

2.1.45 Corollary *If X is a circle-like continuum, then each nondegenerate proper subcontinuum of X is arc-like.*

Proof Since X is circle-like, there exists an inverse sequence $\{X_n, f_n^{n+1}\}$ of simple closed curves whose inverse limit is X.

Let Z be a nondegenerate proper subcontinuum of X. Note that by $(*)$ of Proposition 2.1.20, $Z = \varprojlim\{f_n(Z), f_n^{n+1}|_{f_{n+1}(Z)}\}$. By Proposition 2.1.15, there exists $N \in \mathbb{N}$ such that $f_n(Z) \neq X_n$ for each $n \geq N$. Hence, $f_n(Z)$ is an arc for every $n \geq N$.

Applying Theorem 2.1.40, we have that Z is homeomorphic to $\varprojlim\{f_n(Z),$ $f_n^{n+1}|_{f_{n+1}(Z)}\}_{n=N}^{\infty}$. Therefore, Z is an arc-like continuum.

Q.E.D.

2.1.46 Corollary *If X is a circle-like continuum, then X does not contain a triod.*

2.1.47 Theorem *Let $\{X_n, f_n^{n+1}\}$ be an inverse sequence of compacta whose inverse limit is X_∞. If for each $n \in \mathbb{N}$, X_n has at most k components, for some $k \in \mathbb{N}$, then X_∞ has at most k components.*

Proof Suppose X_∞ has more than k components. Let $\mathcal{C}_1, \ldots, \mathcal{C}_{k+1}$ be $k+1$ distinct components of X_∞. For each $n \in \mathbb{N}$, let us consider the set $\{f_n(\mathcal{C}_1), \ldots, f_n(\mathcal{C}_{k+1})\}$. Since X_n has at most k components, there exist $i_n, j_n \in \{1, \ldots, k+1\}$ such that $f_n(\mathcal{C}_{i_n})$ and $f_n(\mathcal{C}_{j_n})$ are contained in the same component of X_n. Since $\{1, \ldots, k + 1\}$ is a finite set, there exist $i_0, j_0 \in \{1, \ldots, k + 1\}$ and a subsequence $\{n_\ell\}_{\ell=1}^{\infty}$ of $\mathbb{N}$ such that for each $\ell \in \mathbb{N}$, $f_{n_\ell}(\mathcal{C}_{i_0})$ and $f_{n_\ell}(\mathcal{C}_{j_0})$ are contained in the same component Y_{n_ℓ} of X_{n_ℓ}. Note that, by Theorem 2.1.40 and Proposition 2.1.20, $\mathcal{C}_{i_0}$ is homeomorphic to $\mathcal{C}_{i_0}' = \varprojlim\{f_{n_\ell}(\mathcal{C}_{i_0}), f_{n_\ell}^{n_{\ell+1}}|_{f_{n_{\ell+1}}(\mathcal{C}_{i_0})}\}$ and $\mathcal{C}_{j_0}$ is homeomorphic to $\mathcal{C}_{j_0}' = \varprojlim\{f_{n_\ell}(\mathcal{C}_{j_0}), f_{n_\ell}^{n_{\ell+1}}|_{f_{n_{\ell+1}}(\mathcal{C}_{j_0})}\}$.

Let $Y_\infty = \varprojlim\{Y_{n_\ell}, f_{n_\ell}^{n_{\ell+1}}|_{Y_{n_{\ell+1}}}\}$. Then, by Proposition 2.1.8 and Theorem 2.1.40, Y_∞ is a subcontinuum of $X_\infty' = \varprojlim\{X_{n_\ell}, f_{n_\ell}^{n_{\ell+1}}\}$, which is homeomorphic to X_∞ (Theorem 2.1.40). Note that $\mathcal{C}_{i_0}' \cup \mathcal{C}_{j_0}' \subset Y_\infty$, a contradiction. Therefore, X_∞ has at most k components.

Q.E.D.

The next theorem gives us a way to define a map between inverse limits.

2.1.48 Theorem *Let $\{X_n, f_n^{n+1}\}$ and $\{Y_n, g_n^{n+1}\}$ be inverse sequences of compacta, whose inverse limits are X_∞ and Y_∞, respectively. If for each $n \in \mathbb{N}$, there exists a map $k_n \colon X_n \to Y_n$ such that $k_n \circ f_n^{n+1} = g_n^{n+1} \circ k_{n+1}$, then there exists a map $k_\infty \colon X_\infty \to Y_\infty$ such that $g_n \circ k_\infty = k_n \circ f_n$. The map k_∞ is called induced map, and it is denoted, also, by $\varprojlim\{k_n\}$.*

Proof For each $n \in \mathbb{N}$, let $h_n \colon X_\infty \to Y_n$ be given by $h_n = k_n \circ f_n$. Note that $g_n^{n+1} \circ h_{n+1} = h_n$. Hence, by Theorem 2.1.33, there exists the induced map $h_\infty \colon X_\infty \to Y_\infty$. Let $k_\infty = h_\infty$. Then $g_n \circ k_\infty = g_n \circ h_\infty = h_n = k_n \circ f_n$.

Q.E.D.

2.1.49 Theorem *Let* $\{X_n, f_n^{n+1}\}$ *and* $\{Y_n, g_n^{n+1}\}$ *be inverse sequences of compacta, whose inverse limits are* X_∞ *and* Y_∞, *respectively. Suppose that for each* $n \in \mathbb{N}$, *there exists a map* $k_n : X_n \to Y_n$ *such that* $k_n \circ f_n^{n+1} = g_n^{n+1} \circ k_{n+1}$. *If all the maps* k_n *are one-to-one, then the induced map* k_∞ *is one-to-one.*

Proof Let $(x_\ell)_{\ell=1}^\infty$ and $(x_\ell')_{\ell=1}^\infty$ be two distinct points of X_∞. Then there exists $m \in \mathbb{N}$ such that $x_m \neq x_m'$. Since k_m is one-to-one, $k_m(x_m) \neq k_m(x_m')$. Hence, $k_\infty((x_\ell)_{\ell=1}^\infty) = (k_n \circ f_n((x_\ell)_{\ell=1}^\infty))_{n=1}^\infty = (k_n(x_n))_{n=1}^\infty \neq (k_n(x_n'))_{n=1}^\infty = k_\infty((x_\ell')_{\ell=1}^\infty)$. Therefore, k_∞ is one-to-one.

Q.E.D.

2.1.50 Theorem *Let* $\{X_n, f_n^{n+1}\}$ *and* $\{Y_n, g_n^{n+1}\}$ *be inverse sequences of compacta, with surjective bonding maps, whose inverse limits are* X_∞ *and* Y_∞, *respectively. Suppose that for each* $n \in \mathbb{N}$, *there exists a map* $k_n : X_n \to Y_n$ *such that* $k_n \circ f_n^{n+1} = g_n^{n+1} \circ k_{n+1}$. *If all the maps* k_n *are surjective, then the induced map* k_∞ *is surjective.*

Proof Let $g_m^{-1}(V_m)$ be a basic open set in Y_∞, where V_m is an open subset of Y_m (Proposition 2.1.9). Since k_m is surjective, there exists $z_m \in X_m$ such that $k_m(z_m) \in V_m$. Since for each $n \in \mathbb{N}$, f_n^{n+1} is surjective, the projection maps, f_n, are surjective (Remark 2.1.6). Hence, there exists $(x_n)_{n=1}^\infty \in X_\infty$ such that $f_m((x_n)_{n=1}^\infty) = z_m$. Then $g_m \circ k_\infty((x_n)_{n=1}^\infty) = k_m \circ f_m((x_n)_{n=1}^\infty) = k_m(z_m) \in V_m$. Thus, $k_\infty((x_n)_{n=1}^\infty) \in g_m^{-1}(V_m)$. Therefore, $k_\infty(X_\infty)$ is dense in Y_∞. Since X_∞ is compact, $k_\infty(X_\infty) = Y_\infty$.

Q.E.D.

2.1.51 Theorem *Let* $\{X_n, f_n^{n+1}\}$ *and* $\{Y_n, g_n^{n+1}\}$ *be inverse sequences of compacta, whose inverse limits are* X_∞ *and* Y_∞, *respectively. Suppose that for each* $n \in \mathbb{N}$, $k_n : Y_n \twoheadrightarrow X_n$ *and* $h_{n+1} : X_{n+1} \twoheadrightarrow Y_n$ *are surjective maps such that* $k_n \circ h_{n+1} = f_n^{n+1}$ *and* $h_{n+1} \circ k_{n+1} = g_n^{n+1}$. *Then there exists a map* $h_\infty : X_\infty \to Y_\infty$ *such that* $h_\infty = k_\infty^{-1}$, *where* $k_\infty = \varprojlim\{k_n\}$. *In particular,* k_∞ *is a homeomorphism. Also,* X_∞ *and* Y_∞ *are homeomorphic.*

Proof Let $h_\infty : X_\infty \to Y_\infty$ be given by $h_\infty((x_n)_{n=1}^\infty) = (h_n(x_n))_{n=2}^\infty$. By Theorem 1.1.9, h_∞ is continuous.

Now, let $(y_n)_{n=1}^\infty \in Y_\infty$. Then

$$h_\infty \circ k_\infty((y_n)_{n=1}^\infty) = h_\infty((k_n(y_n))_{n=1}^\infty) = (h_n k_n(y_n))_{n=2}^\infty$$
$$= (g_{n-1}^n(y_n))_{n=2}^\infty = (y_n)_{n-1}^\infty;$$

i.e., $h_\infty \circ k_\infty = 1_{Y_\infty}$.

Next, let $(x_n)_{n=1}^\infty \in X_\infty$. Then

$$k_\infty \circ h_\infty((x_n)_{n=1}^\infty) = k_\infty((h_n(x_n))_{n=2}^\infty) = (k_{n-1} h_n(x_n))_{n=2}^\infty$$
$$= (f_{n-1}^n(x_n))_{n=2}^\infty = (x_n)_{n=1}^\infty;$$

i.e., $k_\infty \circ h_\infty = 1_{X_\infty}$.

Therefore, $h_\infty = k_\infty^{-1}$.

Q.E.D.

2.1.52 Theorem *Let $\{X_n, f_n^{n+1}\}$ and $\{Y_n, g_n^{n+1}\}$ be inverse sequences of continua, whose inverse limits are X_∞ and Y_∞, respectively. Suppose that for each $n \in \mathbb{N}$, $k_n \colon X_n \twoheadrightarrow Y_n$ is a confluent map such that $k_{n+1} \circ f_n^{n+1} = g_n^{n+1} \circ k_n$. Then the induced map k_∞ is confluent.*

Proof Let Q be a subcontinuum of Y_∞ and let K be a component of $k_\infty^{-1}(Q)$. Note that $\{g_n(Q), g_n^{n+1}|_{g_{n+1}(Q)}\}$ and $\{f_n(K), f_n^{n+1}|_{f_{n+1}(K)}\}$ are inverse sequences of continua, $Q = \varprojlim\{g_n(Q), g_n^{n+1}|_{g_{n+1}(Q)}\}$ and $K = \varprojlim\{f_n(K), f_n^{n+1}|_{f_{n+1}(K)}\}$ (Proposition 2.1.20). Observe that $k_n(f_n(K)) \subset g_n(Q)$. Let L_n be the component of $k_n^{-1}(g_n(Q))$ which contains $f_n(K)$. Then $f_n^{n+1}(L_{n+1}) \subset L_n$ for each $n \in \mathbb{N}$. Hence, $\{L_n, f_n^{n+1}|_{L_{n+1}}\}$ is an inverse sequence of continua. Let $L = \varprojlim\{L_n, f_n^{n+1}|_{L_{n+1}}\}$. Then, by Proposition 2.1.8, L is a subcontinuum of X_∞ such that $K \subset L$ and $k_\infty(L) = Q$. Since K is a component of $k_\infty^{-1}(Q)$, $K = L$ and $k_\infty(K) = Q$. Therefore, k_∞ is confluent.

Q.E.D.

2.1.53 Theorem *Let $\{X_n, f_n^{n+1}\}$ and $\{Y_n, g_n^{n+1}\}$ be inverse sequences of continua, whose inverse limits are X_∞ and Y_∞, respectively. Suppose that for each $n \in \mathbb{N}$, $k_n \colon X_n \twoheadrightarrow Y_n$ is a monotone map such that $k_{n+1} \circ f_n^{n+1} = g_n^{n+1} \circ k_n$. Then the induced map k_∞ is monotone.*

Proof Let $(y_n)_{n=1}^\infty$ be a point of Y_∞. Since each k_n is monotone, $L_n = k_n^{-1}(y_n)$ is a subcontinuum of X_n. Then $f_n^{n+1}(L_{n+1}) \subset L_n$ for each $n \in \mathbb{N}$. Hence, $\{L_n, f_n^{n+1}|_{L_{n+1}}\}$ is an inverse sequence of continua. Let $L = \varprojlim\{L_n, f_n^{n+1}|_{L_{n+1}}\}$. Then, by Proposition 2.1.8, L is a subcontinuum of X_∞, and $L \subset k_\infty^{-1}((y_n)_{n=1}^\infty)$. Let $(x_n)_{n=1}^\infty \in k_\infty^{-1}((y_n)_{n=1}^\infty)$. Then $k_n(x_n) = y_n$; and hence, $x_n \in L_n$ for every $n \in \mathbb{N}$. Thus, $(x_n)_{n=1}^\infty \in L$. Therefore, $k_\infty^{-1}((y_n)_{n=1}^\infty) = L$, and k_∞ is monotone.

Q.E.D.

Next, we prove a theorem due to Anderson and Choquet, originally proved in [1], which tells us when we can embed an inverse limit in a compactum.

2.1.54 Theorem *Let X be a compactum, with metric d. Let $\{X_n, f_n^{n+1}\}$ be an inverse sequence of closed subsets of X with surjective bonding maps. Assume:*

(1) For each $\varepsilon > 0$, there exists $k \in \mathbb{N}$ such that for all $x \in X_k$,

$$diam \left[\bigcup_{j \geq k} \left(f_k^j \right)^{-1} (x) \right] < \varepsilon;$$

(2) For each $n \in \mathbb{N}$ and each $\delta > 0$, there exists $\delta' > 0$ such that whenever $j > n$ and $x, y \in X_j$ such that $d(f_n^j(x), f_n^j(y)) > \delta$, then $d(x, y) > \delta'$.

Then $\varprojlim\{X_n, f_n^{n+1}\}$ is homeomorphic to $\bigcap_{n=1}^{\infty} \left[Cl \left(\bigcup_{m \geq n} X_m \right) \right]$. In particular, if $X_n \subset X_{n+1}$ for each $n \in \mathbb{N}$, then $\varprojlim\{X_n, f_n^{n+1}\}$ is homeomorphic to $Cl \left(\bigcup_{n=1}^{\infty} X_n \right)$.

Proof Let $X_\infty = \varprojlim\{X_n, f_n^{n+1}\}$. If $x = (x_n)_{n=1}^{\infty} \in X_\infty$, then, by (1), we have that $(x_n)_{n=1}^{\infty}$ is a Cauchy sequence in X. Thus, $(x_n)_{n=1}^{\infty}$ converges to a point in X, which we call $h(x)$. Hence, we have defined a function $h: X_\infty \to X$. We see that h is an embedding.

First, we show h is continuous. Let $x = (x_n)_{n=1}^{\infty} \in X_\infty$, and let $\varepsilon > 0$. Now, choose $k \in \mathbb{N}$ as guaranteed by (1). Let $\ell = k + 1$, and let

$$V = f_\ell^{-1}(V_\varepsilon^d(h(x)) \cap X_\ell).$$

Since $x \in X_\infty$ and k satisfies (1), we have from the definition of h that $d(x_\ell, h(x)) < \varepsilon$. Hence, $x_\ell \in V_\varepsilon^d(h(x)) \cap X_\ell$. Thus, $x \in V$. Next, observe that, since $V_\varepsilon^d(h(x)) \cap X_\ell$ is open in X_ℓ, V is open in X_∞. Now, let $y = (y_n)_{n=1}^{\infty} \in V$. For the same reasons as for x, $d(y_\ell, h(y)) < \varepsilon$. Since $y \in V$, $y_\ell \in V_\varepsilon^d(h(x))$. Hence,

$$d(h(y), h(x)) \leq d(h(y), y_\ell) + d(y_\ell, h(x)) < 2\varepsilon.$$

Therefore, h is continuous.

To prove h is one-to-one, let $x = (x_n)_{n=1}^{\infty}$ and $y = (y_n)_{n=1}^{\infty}$ be two distinct points of X_∞. Then there exists $n \in \mathbb{N}$, such that $x_n \neq y_n$. Let $\delta = \frac{d(x_n, y_n)}{2}$. Then, by (2), there exists $\delta' > 0$ such that whenever $j > n$, then, since

$$d(f_n^j(x_j), f_n^j(y_j)) = d(x_n, y_n) = 2\delta > \delta$$

we have that $d(x_j, y_j) > \delta'$. Therefore, $h(x) \neq h(y)$.

It remains to see that $h(X_\infty) = \bigcap_{n=1}^{\infty} \left[Cl \left(\bigcup_{m \geq n} X_m \right) \right]$. For convenience, let

$$Z = \bigcap_{n=1}^{\infty} \left[Cl \left(\bigcup_{m \geq n} X_m \right) \right].$$

It is clear, from the definition of h, that $h(X_\infty) \subset Z$. To show $h(X_\infty) = Z$, we prove that $h(X_\infty)$ is dense in Z.

For this purpose, let $z \in Z$ and let $\varepsilon > 0$. Let $k \in \mathbb{N}$ be as guaranteed by (1). Since $z \in Z$, there exists $m \geq k$ such that $d(z, p) < \varepsilon$ for some $p \in X_m$. Since the bonding maps are surjective, by Remark 2.1.6, there exists $x = (x_n)_{n=1}^{\infty} \in X_\infty$ such that $f_m(x) = f_m((x_n)_{n=1}^{\infty}) = x_m = p$.

Note that, since $m \geq k$, condition (1) holds for m in place of k. Hence, it follows that $d(z, h(x)) < 2\varepsilon$. Thus, $h(X_\infty)$ is dense in Z. Therefore, since X_∞ is compact, $h(X_\infty) = Z$.

Q.E.D.

We end this section proving that any continuum is homeomorphic to an inverse limit of polyhedra.

2.1.55 Theorem *Each continuum is homeomorphic to an inverse limit of polyhedra.*

Proof Let X be a continuum. By Theorem 1.1.16, we assume that X is contained in the Hilbert cube Q.

For each $n \in \mathbb{N}$, let d_n denote the Euclidean metric on $[0, 1]^n$, and let $\pi_n \colon Q \to [0, 1]^n$ and $\pi_n^{n+1} \colon [0, 1]^{n+1} \to [0, 1]^n$ be the projection maps. By Lemma 1.7.36, there exists a polyhedron P_1 in $[0, 1]$ such that $\pi_1(X) \subset Int(P_1) \subset P_1 \subset \mathcal{V}_1^{d_1}(\pi_1(X))$. Note that $\pi_2(X) \subset (\pi_1^2)^{-1}(Int(P_1))$. Hence, by Lemma 1.7.36, there exists a polyhedron P_2 in $[0, 1]^2$ such that $\pi_2(X) \subset Int(P_2) \subset P_2 \subset \mathcal{V}_{\frac{1}{2}}^{d_2}(\pi_2(X))$ and $\pi_1^2(P_2) \subset Int(P_1)$. Continue in this way to define a sequence, $\{P_n\}_{n=1}^{\infty}$, of polyhedra such that $\pi_n(X) \subset Int(P_n) \subset P_n \subset \mathcal{V}_{\frac{1}{n}}^{d_n}(\pi_n(X))$ and $\pi_n^{n+1}(P_{n+1}) \subset Int(P_n)$. Hence, $\{P_n, \pi_n^{n+1}|_{P_{n+1}}\}$ is an inverse sequence. Let $P_{\infty} = \varprojlim\{P_n, \pi_n^{n+1}|_{P_{n+1}}\}$. We show that X is homeomorphic to P_{∞}.

For each $n \in \mathbb{N}$, let $h_n \colon X \to P_n$ be given by $h_n = \pi_n|_X$. Observe that, by construction, if $x \in X$, then $\pi_n^{n+1}(\pi_{n+1}(x)) = \pi_n(x)$ for each $n \in \mathbb{N}$. Thus, $\pi_n^{n+1}(h_{n+1}(x)) = h_n(x)$ for every $n \in \mathbb{N}$. Hence, the sequence $\{h_n\}_{n=1}^{\infty}$ induces a map $h_{\infty} \colon X \to P_{\infty}$ (Theorem 2.1.33). To see h_{∞} is one-to-one, let x and x' be two distinct points of X. Then there exists $n \in \mathbb{N}$, such that $\pi_n(x) \neq \pi_n(x')$; i.e., $h_n(x) \neq h_n(x')$. Thus, $h_{\infty}(x) \neq h_{\infty}(x')$. Therefore, h_{∞} is one-to-one.

Clearly, if z_1 and z_2 are two points of $[0, 1]^{k+1}$, then

$$d_k(\pi_k^{k+1}(z_1), \pi_k^{k+1}(z_2)) \leq d_{k+1}(z_1, z_2).$$

Now, let $\overline{p} = (\overline{p}_k)_{k=1}^{\infty} \in P_{\infty}$, and let $n \in \mathbb{N}$. Then, by Proposition 2.1.16, $\overline{p}_n \in \bigcap_{m=n+1}^{\infty} \pi_n^m(P_m)$. Since $h_n(X) = \pi_n^{n+1} \circ h_{n+1}(X)$, $d_n(\overline{p}_n, h_n(X)) \leq d_{n+1}(\overline{p}_{n+1}, h_{n+1}(X))$. Hence,

$$d_n(\overline{p}_n, h_n(X)) \leq d_m(\overline{p}_m, h_m(X)) \leq \frac{1}{m}$$

for each $m > n$. Therefore, $\overline{p}_n \in \pi_n(X) = h_n(X)$.

Finally, suppose h_{∞} is not surjective. Then there exists a point $\overline{p} = (\overline{p}_k)_{k=1}^{\infty} \in P_{\infty} \setminus h_{\infty}(X)$. By Proposition 2.1.9, there exist $N \in \mathbb{N}$ and an open subset U_N of P_N such that $(\overline{p}_k)_{k=1}^{\infty} \in (\pi_N|_{P_{\infty}})^{-1}(U_N) \subset P_{\infty} \setminus h_{\infty}(X)$. Hence, $\overline{p}_N \in U_N \setminus h_N(X)$, a contradiction with the preceding paragraph. Therefore, h_{∞} is surjective.

Q.E.D.

Note that in Theorem 2.1.55 the bonding maps are not surjective. The following result tells us that each continuum can be written as an inverse limit of polyhedra with surjective bonding maps; a proof of this may be found in [17, Theorem 2].

2.1.56 Theorem *Each continuum is homeomorphic to an inverse limit of polyhedra with surjective bonding maps.*

2.2 Inverse Limits and the Cantor Set

The purpose of this section is to characterize the Cantor set C (Example 1.6.4) and to show that every compactum is a continuous image of C. The material for this section comes from [5, 6, 9, 12].

We begin by showing that C is an inverse limit of finite discrete spaces. First, we need the following definition:

2.2.1 Definition If $x \in \mathbb{R}$, then $[x] = \max\{n \in \mathbb{Z} \mid n \leq x\}$ is called the *greatest integer function*.

2.2.2 Theorem *For each $n \in \mathbb{N}$, let $X_n = \{0, 1, \ldots, 2^n - 1\}$ with the discrete topology, and define $f_n^{n+1} \colon X_{n+1} \to X_n$ by $f_n^{n+1}(x) = \left[\frac{x}{2}\right]$. If $X_\infty = \varprojlim\{X_n, f_n^{n+1}\}$, then X_∞ is homeomorphic to C.*

Proof For each $n \in \mathbb{N}$, let $h_n \colon C \to X_n$ be given by $h_n(x) = k$ if $x \in I_{n,k}$ (Example 1.6.4). Clearly, h_n is surjective for each $n \in \mathbb{N}$. To see h_n is continuous, it is enough to observe that for each $k \in X_n$, $h_n^{-1}(k) = I_{n,k} \cap C$, which is an open subset of C.

Let $n \in \mathbb{N}$. Observe that, by the construction of C, $I_{n+1,2\ell} \cup I_{n+1,2\ell+1} \subset I_{n,\ell}$. Hence, if $x \in C$ and $h_{n+1}(x) = k$, then $h_n(x) = \left[\frac{k}{2}\right]$. Consequently, if $x \in C$, then $f_n^{n+1} \circ h_{n+1}(x) = \left[\frac{h_{n+1}(x)}{2}\right] = h_n(x)$. Therefore, $f_n^{n+1} \circ h_{n+1} = h_n$. Thus, by Theorem 2.1.33, there exists the induced map $h_\infty \colon C \to X_\infty$. By Theorem 2.1.35, h_∞ is surjective.

To see h_∞ is one-to-one, let x and x' be two distinct points of C. Then there exist $n \in \mathbb{N}$ and $j, k \in \{0, \ldots, 2^n - 1\}$ such that $j \neq k$, $x \in I_{n,j}$ and $x' \in I_{n,k}$. Hence, $h_n(x) \neq h_n(x')$. Thus, h_∞ is one-to-one.

Therefore, h_∞ is a homeomorphism.

Q.E.D.

Now, we present some results about totally disconnected compacta.

2.2.3 Proposition *If X is a totally disconnected compactum, then the family of all open and closed subsets of X forms a basis for the topology of X.*

Proof Let U be an open subset of X, and let $x \in U$. Since X is totally disconnected, no connected subset of X intersects both $\{x\}$ and $X \setminus U$. By Theorem 1.6.8, there exist two disjoint open and closed subsets X_1 and X_2 of X such that $X = X_1 \cup X_2$, $\{x\} \subset X_1$ and $X \setminus U \subset X_2$. Hence, $x \in X_1 \subset U$.

Q.E.D.

2.2.4 Definition Let X be a metric space. If $\mathcal{U}$ is a family of subsets of X, then the *mesh of* $\mathcal{U}$, denoted by mesh($\mathcal{U}$), is defined by:

$$\text{mesh}(\mathcal{U}) = \sup\{\text{diam}(U) \mid U \in \mathcal{U}\}.$$

2.2.5 Definition Let X be a metric space. If $\mathcal{U}$ and $\mathcal{V}$ are two coverings of X, then $\mathcal{V}$ is a *refinement of* $\mathcal{U}$ provided that for each $V \in \mathcal{V}$, there exists $U \in \mathcal{U}$ such that $V \subset U$. If $\mathcal{V}$ refines $\mathcal{U}$, we write $\mathcal{V} \prec \mathcal{U}$.

2.2.6 Theorem *If X is a totally disconnected compactum, then there exists a sequence $\{\mathcal{U}_n\}_{n=1}^{\infty}$ of finite coverings of X such that for each $n \in \mathbb{N}$:*

(i) $\mathcal{U}_{n+1} \prec \mathcal{U}_n$;
(ii) mesh($\mathcal{U}_n$) $< \frac{1}{n}$;
(iii) if $U \in \mathcal{U}_n$, then U is open and closed in X; and
(iv) if $U, V \in \mathcal{U}_n$ and $U \neq V$, then $U \cap V = \emptyset$.

Proof We construct $\mathcal{U}_1$ first. By Proposition 2.2.3, for each $x \in X$, there exists an open and closed subset U_x of X such that $x \in U_x$ and diam(U_x) < 1. Hence, $\{U_x \mid x \in X\}$ is an open cover of X. Since X is compact, there exist $x_{11}, \ldots, x_{1k_1} \in X$ such that $X = \bigcup_{j=1}^{k_1} U_{x_{1j}}$. The sets $U_{x_{11}}, \ldots, U_{x_{1k_1}}$ are not necessarily pairwise disjoint. But this can be fixed. Let $U_{11} = U_{x_{11}}$ and for $j \in \{2, \ldots, k_1\}$, let $U_{1j} = U_{x_{1j}} \setminus \bigcup_{\ell=1}^{j-1} U_{1\ell}$. Then $\mathcal{U}_1 = \{U_{11}, \ldots, U_{1k_1}\}$ is a finite family of pairwise disjoint open and closed subsets of X covering X.

Let λ_1 be a Lebesgue number of the open covering $\mathcal{U}_1$ (Theorem 1.6.6). Without loss of generality, we assume that $\lambda_1 < \frac{1}{2}$. By Proposition 2.2.3, for each $x \in X$, there exists an open and closed subset V_x of X such that $x \in V_x$ and diam(V_x) $< \lambda_1$. Hence, the family $\{V_x \mid x \in X\}$ forms an open cover of X. Since X is compact, there exist $x_{21}, \ldots, x_{2k_2} \in X$ such that $X = \bigcup_{j=1}^{k_2} V_{x_{2j}}$. Again, the elements of $\{V_{x_{21}}, \ldots, V_{x_{2k_2}}\}$ may not be pairwise disjoint. Hence, let $U_{21} = V_{x_{21}}$ and for each $j \in \{2, \ldots, k_2\}$, let $U_{2j} = V_{x_{2j}} \setminus \bigcup_{\ell=1}^{j-1} V_{2\ell}$. Then $\mathcal{U}_2 = \{U_{21}, \ldots, U_{2k_2}\}$ is a finite family of pairwise disjoint open and closed subsets of X covering X. Note that, by construction, $\mathcal{U}_2 \prec \mathcal{U}_1$.

Proceeding by induction, we obtain the result.

Q.E.D.

2.2.7 Theorem *If X is a totally disconnected compactum, then there exists an inverse sequence $\{X_n, f_n^{n+1}\}$ of finite discrete spaces such that $\varprojlim\{X_n, f_n^{n+1}\}$ is homeomorphic to X.*

Proof Let $\{\mathcal{U}_n\}_{n=1}^{\infty}$ be a sequence of finite coverings given by Theorem 2.2.6. For each $n \in \mathbb{N}$, let $X_n = \mathcal{U}_n$. Give X_n the discrete topology. We define the bonding maps f_n^{n+1} as follows: If $U_{n+1} \in X_{n+1}$, then $f_n^{n+1}(U_{n+1}) = U_n$, where U_n is the unique element of $\mathcal{U}_n = X_n$ such that $U_{n+1} \subset U_n$. Hence, $\{X_n, f_n^{n+1}\}$ is an inverse sequence. Let $X_\infty = \varprojlim\{X_n, f_n^{n+1}\}$.

For each $n \in \mathbb{N}$, let $h_n \colon X \to X_n$ be given by $h_n(x) = U$, where U is the unique element of X_n such that $x \in U$. Let $n \in \mathbb{N}$. Note that, by construction, h_n is surjective and $f_n^{n+1} \circ h_{n+1} = h_n$. Also note that if $U \in X_n$, then $h_n^{-1}(U) = U$. Then h_n is continuous.

By Theorem 2.1.33, there exists the induced map $h_\infty \colon X \to X_\infty$. Since X is compact and each h_n is surjective, by Theorem 2.1.35, h_∞ is surjective.

To see h_∞ is one-to-one, let x and y be two distinct points of X. Since $\lim\limits_{n \to \infty} \mathrm{mesh}(\mathcal{U}_n) = 0$, there exist $n \in \mathbb{N}$ and $U, V \in \mathcal{U}_n$ such that $x \in U$ and $y \in V$. Hence, $h_n(x) \neq h_n(y)$, since $U \cap V = \emptyset$. Therefore, h_∞ is one-to-one.

Q.E.D.

2.2.8 Lemma *If X is a totally disconnected and perfect compactum, then for each $n \in \mathbb{N}$, there exist n nonempty pairwise disjoint open and closed subsets of X whose union is X.*

Proof The proof is done by induction on n. For $n = 1$, X itself is open and closed. Let $n \geq 2$ and assume that there exist n nonempty pairwise disjoint open and closed subsets, $U_1, \ldots, U_n$, of X such that $X = \bigcup_{j=1}^{n} U_j$. Let $x \in U_n$. Since X is perfect, $U_n \neq \{x\}$. By Proposition 2.2.3, there exists an open and closed subset of X such that $x \in V \subset U_n$. Hence, $X = U_1 \cup \cdots \cup U_{n-1} \cup V \cup (U_n \setminus V)$.

Q.E.D.

We are ready to prove the characterization of the Cantor set, $\mathcal{C}$, mentioned above.

2.2.9 Theorem *If X and Y are two totally disconnected and perfect compacta, then X and Y are homeomorphic. In particular, X and Y are homeomorphic to $\mathcal{C}$ (Example 1.6.4).*

Proof By Proposition 2.2.3 and Lemma 2.2.8, there exist two open covers $\mathcal{U}_1$ and $\mathcal{U}_1'$ of X and Y, respectively, such that the following four conditions hold:

(i)   The elements of each cover are open and closed.
(ii) The elements of each cover are pairwise disjoint.
(iii) The mesh of each cover is less than one.
(iv) Both covers have the same cardinality.

Let $h_1 \colon \mathcal{U}_1 \to \mathcal{U}_1'$ be any bijection. Apply Proposition 2.2.3 to obtain refinements $\mathcal{U}_2$ and $\mathcal{U}_2'$ of $\mathcal{U}_1$ and $\mathcal{U}_1'$, respectively, of mesh less than $\frac{1}{2}$ and satisfying conditions (i) and (ii) above.

By Lemma 2.2.8, we assume that if $U_1 \in \mathcal{U}_1$, then U_1 and $h_1(U_1)$ contain the same number of elements of $\mathcal{U}_2$ and $\mathcal{U}_2'$, respectively. It follows that there exists a bijection $h_2 \colon \mathcal{U}_2 \to \mathcal{U}_2'$ such that if $U_2 \in \mathcal{U}_2$, $U_1 \in \mathcal{U}_1$ and $U_2 \subset U_1$, then $h_2(U_2) \subset h_1(U_1)$. This process can be repeated inductively to obtain two inverse sequences, $\{\mathcal{U}_n, f_n^{n+1}\}$ and $\{\mathcal{U}_n', (f')_n^{n+1}\}$, and a sequence $\{h_n\}_{n=1}^{\infty}$ of bijections such that we have the following infinite ladder:

$$\cdots \longleftarrow \mathcal{U}_{n-1} \xleftarrow{f^n_{n-1}} \mathcal{U}_n \xleftarrow{f^{n+1}_n} \mathcal{U}_{n+1} \longleftarrow \cdots : \mathcal{U}_\infty$$

$$\downarrow h_{n-1} \qquad \downarrow h_n \qquad \downarrow h_{n+1} \qquad \downarrow h_\infty$$

$$\cdots \longleftarrow \mathcal{U}'_{n-1} \xleftarrow[(f')^n_{n-1}]{} \mathcal{U}'_n \xleftarrow[(f')^{n+1}_n]{} \mathcal{U}'_{n+1} \longleftarrow \cdots : \mathcal{U}'_\infty$$

where each $\mathcal{U}_n$ and each $\mathcal{U}'_n$ have the discrete topology, and the functions f^{n+1}_n and $(f')^{n+1}_n$ are defined as in the proof of Theorem 2.2.7. Since each h_n is a homeomorphism, by Theorems 2.1.49 and 2.1.50, h_∞ is a homeomorphism. It follows from the proof of Theorem 2.2.7 that $\mathcal{U}_\infty$ and $\mathcal{U}'_\infty$ are homeomorphic to X and Y, respectively. Therefore, X and Y are homeomorphic.

Q.E.D.

As an application of Theorem 2.2.9, we have the following:

2.2.10 Theorem *For each $n \in \mathbb{N}$, let*

$$L_n = \left\{ \cos\left(\frac{2\pi k}{2^n}\right) + i \sin\left(\frac{2\pi k}{2^n}\right) \ \Big| \ k \in \{0, \ldots, 2^n - 1\} \right\};$$

i.e., L_n is the collection of 2^nth roots of unity, with the discrete topology. Let $f^{n+1}_n \colon L_{n+1} \to L_n$ be given by $f^{n+1}_n(z) = z^2$. Then $\varprojlim\{L_n, f^{n+1}_n\}$ is homeomorphic to $\mathcal{C}$.

Proof Let $L_\infty = \varprojlim\{L_n, f^{n+1}_n\}$. Note that for each $z_n \in L_n$, $\left(f^{n+1}_n\right)^{-1}(z_n)$ has exactly two points. Also observe that L_∞ is nonempty and compact, by Proposition 2.1.8. Since each f^{n+1}_n is surjective, each f_n is surjective also, by Remark 2.1.6. Let $(z_n)^\infty_{n=1}$ and $(w_n)^\infty_{n=1}$ be two distinct points of L_∞. Then there exists a positive integer N such that $z_N \neq w_N$. Hence, since f_N is surjective and L_N is discrete, $f^{-1}_N(\{z_N\})$ and $f^{-1}_N(\{w_N\})$ are two disjoint open and closed subsets of L_∞ containing $(z_n)^\infty_{n=1}$ and $(w_n)^\infty_{n=1}$, respectively. Thus, L_∞ is totally disconnected.

Let $\mathcal{U}$ be an open subset of L_∞ and let $(z_n)^\infty_{n=1} \in \mathcal{U}$. Hence, there exists $m \in \mathbb{N}$ such that the basic open set $f^{-1}_m(\{z_m\})$ is contained in $\mathcal{U}$, by Proposition 2.1.9. Now, let $w_{m+1} \in \left(f^{m+1}_m\right)^{-1}(z_m) \setminus \{z_{m+1}\}$. Next, let $w_{m+2} \in \left(f^{m+1}_m\right)^{-1}(w_{m+1})$. In general, if $n > m$, let $w_n \in \left(f^n_{n-1}\right)^{-1}(w_{n-1})$; and if $n < m$, then let $w_n = f^m_n(z_m)$. Note that $(w_n)^\infty_{n=1} \in L_\infty$. In fact, $(w_n)^\infty_{n=1} \in f^{-1}_m(\{z_m\}) \subset \mathcal{U} \setminus \{(z_n)^\infty_{n=1}\}$. Therefore, L_∞ is perfect. Since L_∞ is compact, totally disconnected and perfect, by Theorem 2.2.9, L_∞ is homeomorphic to $\mathcal{C}$.

Q.E.D.

2.2.11 Remark Note that the space L_n, of 2^nth roots of unity, is a group [12, pp. 294–295]. Since in Theorem 2.2.10, we are considering L_n with the discrete topology, L_n is a topological group (see Definition 4.1.1). Observe that each of the maps f^{n+1}_n is a group homomorphism. Hence, it is not difficult to prove that

$L_\infty = \varprojlim\{L_n, f_n^{n+1}\}$ is a topological group, where the operations are defined coordinatewise. Therefore, the Cantor set $\mathcal{C}$ is a topological group.

2.2.12 Lemma *If X is a totally disconnected compactum, then $X \times \mathcal{C}$ is homeomorphic to $\mathcal{C}$.*

Proof Let X be a totally disconnected compactum. By Theorem 2.2.9, we only need to show that $X \times \mathcal{C}$ is perfect. Let $(x, t) \in X \times \mathcal{C}$, and let $U \times V$ be a basic open set of $X \times \mathcal{C}$ such that $(x, t) \in U \times V$. Since $\mathcal{C}$ is perfect (Example 1.6.4), there exists $t' \in V \setminus \{t\}$. Hence, $(x, t') \in U \times V$. Therefore, $X \times \mathcal{C}$ is perfect.

Q.E.D.

The characterization of $\mathcal{C}$ in Theorem 2.2.9 allows us to show that any compactum is a continuous image of $\mathcal{C}$.

2.2.13 Theorem *If X is a compactum, with metric d, then there exists a surjective map $f : \mathcal{C} \twoheadrightarrow X$ of the Cantor set onto X.*

Proof Since X is compact, using Theorem 1.6.6 (Lebesgue numbers), there exists a sequence $\{\mathcal{U}_n\}_{n=1}^\infty$ of finite covers of X such that, for each $n \in \mathbb{N}$, the following three conditions are satisfied:

(i) each $U \in \mathcal{U}_n$ is the closure of an open subset of X;
(ii) $\mathrm{mesh}(\mathcal{U}_n) < \frac{1}{2^n}$; and
(iii) $\mathcal{U}_{n+1} \prec \mathcal{U}_n$.

Let $\mathcal{U}_1 = \{U_{11}, \ldots, U_{1k_1}\}$. The elements of $\mathcal{U}_1$ may not be pairwise disjoint. We use the following trick. For each $i \in \{1, \ldots, k_1\}$, let $V_{1i} = U_{1i} \times \{i\}$. Then we say that a subset $W \times \{i\}$ is open in V_{1i} if W is an open subset of U_{1i}. Let $\mathcal{V}_1 = \bigcup_{i=1}^{k_1} V_{1i}$. We define a metric d_1 on $\mathcal{V}_1$ as follows:

$$d_1((x, i), (y, j)) = \begin{cases} d(x, y), & \text{if } i - j; \\ 1, & \text{if } i \neq j. \end{cases}$$

Now, let $\mathcal{U}_2 = \{U_{21}, \ldots, U_{2k_2}\}$. For each $U_{2j} \in \mathcal{U}_2$, let $i \in \{1, \ldots, k_1\}$ be such that $U_{2j} \subset U_{1i}$. Then let $V_{2ij} = U_{2j} \times \{i\} \times \{j\}$ (whenever $U_{2j} \subset U_{1i}$). Again, we say that a subset $W \times \{i\} \times \{j\}$ is open in V_{2ij} if W is open in U_{2j}. Let $\mathcal{V}_2 = \bigcup_{i,j} V_{2ij}$. We define a metric d_2 in a similar way as we defined d_1 on $\mathcal{V}_1$. Finally, let $f_1^2 : \mathcal{V}_2 \to \mathcal{V}_1$ be given by $f_1^2((u, i, j)) = (u, i)$. Clearly, f_1^2 is continuous.

Now, let $\mathcal{U}_3 = \{U_{31}, \ldots, U_{3k_3}\}$. For each $U_{3k} \in \mathcal{U}_3$, let $j \in \{1, \ldots, k_2\}$ and $i \in \{1, \ldots, k_1\}$ be such that $U_{3k} \subset U_{2j} \subset U_{1i}$. Then define $V_{3ijk} = U_{3k} \times \{i\} \times \{j\} \times \{k\}$. As before, a subset $W \times \{i\} \times \{j\} \times \{k\}$ is open in V_{3ijk} if W is open in U_{3k}. Let $\mathcal{V}_3 = \bigcup_{i,j,k} V_{3ijk}$. We define a metric on $\mathcal{V}_3$ in the same way we defined d_1 on $\mathcal{V}_1$. Finally, let $f_2^3 : \mathcal{V}_3 \to \mathcal{V}_2$ be given by $f_2^3((u, i, j, k)) = (u, i, j)$. Clearly, f_2^3 is continuous. Although it is notationally complicated, the general inductive step is clear now. Hence, we obtain an inverse sequence $\{\mathcal{V}_n, f_n^{n+1}\}$ of compacta. Let $\mathcal{V}_\infty = \varprojlim\{\mathcal{V}_n, f_n^{n+1}\}$. Then $\mathcal{V}_\infty$ is a compactum.

A second inverse sequence, $\{X_n, g_n^{n+1}\}$, may be constructed letting for each $n \in \mathbb{N}$, $X_n = X$ and $g_n^{n+1} = 1_X$. It follows, from Proposition 2.1.39, that X is homeomorphic to $\varprojlim\{X_n, g_n^{n+1}\}$.

For each $n \in \mathbb{N}$, let $h_n \colon \mathcal{V}_n \to X_n$ be given by $h_n(u, i, j, \ldots, p) = u$. Thus, the following diagram

$$\cdots \longleftarrow \mathcal{V}_{n-1} \xleftarrow{f_{n-1}^n} \mathcal{V}_n \xleftarrow{f_n^{n+1}} \mathcal{V}_{n+1} \longleftarrow \cdots : \mathcal{V}_\infty$$
$$\downarrow h_{n-1} \quad\quad \downarrow h_n \quad\quad \downarrow h_{n+1} \quad\quad \downarrow h_\infty$$
$$\cdots \longleftarrow X_{n-1} \xleftarrow{g_{n-1}^n} X_n \xleftarrow{g_n^{n+1}} X_{n+1} \longleftarrow \cdots : X_\infty$$

is commutative, where $X_\infty = \varprojlim\{X_n, g_n^{n+1}\}$ and $h_\infty = \varprojlim\{h_n\}$. Since each h_n is surjective and each $\mathcal{V}_n$ is a compactum, by Theorem 2.1.50, h_∞ is surjective.

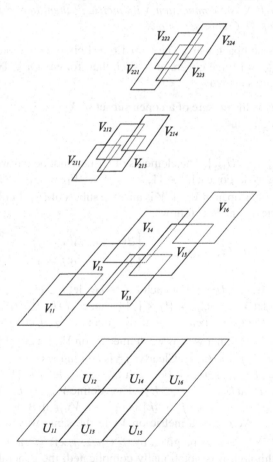

Now, we show that $\mathcal{V}_\infty$ is totally disconnected. To this end, we prove that for any two distinct points of $\mathcal{V}_\infty$, there exists an open and closed subset of $\mathcal{V}_\infty$ containing one of them and not containing the other.

Let $x = (x_n)_{n=1}^{\infty}$ and $y = (y_n)_{n=1}^{\infty}$ be two distinct points of $\mathcal{V}_{\infty}$. Then there exists $n_0 \in \mathbb{N}$ such that $x_{n_0} \neq y_{n_0}$ and n_0 is the smallest with this property.

First, suppose that $n_0 = 1$; i.e., $x_1 \neq y_1$. Suppose, also, that $x_1 = (u, i)$ and $y_1 = (u', i')$. If $u \neq u'$, then there exist $m \in \mathbb{N}$ and $U_{m\ell}, U_{m\ell'} \in \mathcal{U}_m$ such that $u \in U_{m\ell}$, $u' \in U_{m\ell'}$ and $U_{m\ell} \cap U_{m\ell'} = \emptyset$. Then the corresponding sets $V_{mij...\ell}$ and $V_{mij...\ell'}$ are disjoint and both are open and closed subsets of V_m. Hence, $f_m^{-1}(V_{mij...\ell})$ is an open and closed subset of $\mathcal{V}_{\infty}$ containing x and not containing y.

If $u = u'$, then $i \neq i'$. Hence, $x_1 \in V_{1i}$ and $y_1 \in V_{1i'}$. Note that V_{1i} and $V_{1i'}$ are disjoint open and closed subsets of $\mathcal{V}_1$. Then $f_1^{-1}(V_{1i})$ is an open and closed subset of $\mathcal{V}_{\infty}$ containing x and not containing y.

Now, suppose $n_0 \geq 2$. For simplicity, we assume that $n_0 = 3$. Then $x_3 \neq y_3$, $x_3 = (u, i, j, k)$ and $y = (u, i, j, k')$, with $k \neq k'$. Hence, $x_3 \in V_{3ijk}$ and $y_3 \in V_{3ijk'}$. Note that V_{3ijk} and $V_{3ijk'}$ are disjoint open and closed subsets of V_3. Then $f_3^{-1}(V_{ijk})$ is an open and closed subset of $\mathcal{V}_{\infty}$ containing x and not containing y.

Therefore, $\mathcal{V}_{\infty}$ is totally disconnected.

Even though $\mathcal{V}_{\infty}$ may not be perfect, by Lemma 2.2.12, $\mathcal{V}_{\infty} \times \mathcal{C}$ is perfect. Hence, by Theorem 2.2.9, $\mathcal{V}_{\infty} \times \mathcal{C}$ is homeomorphic to $\mathcal{C}$. Thus, we have the following maps:

(a) A homeomorphism $\xi : \mathcal{C} \to \mathcal{V}_{\infty} \times \mathcal{C}$;
(b) A surjective map $\zeta : \mathcal{V}_{\infty} \times \mathcal{C} \to \mathcal{V}_{\infty}$ given by $\zeta((v, c)) = v$ (the projection map);
(c) A surjective map $h_{\infty} : \mathcal{V}_{\infty} \to X_{\infty}$; and
(d) A homeomorphism $\varphi : X_{\infty} \to X$.

Therefore, $f = \varphi \circ h_{\infty} \circ \zeta \circ \xi$ is a surjective map from $\mathcal{C}$ onto X.

$$\text{Q.E.D.}$$

2.3 Inverse Limits and Other Operations

We show that taking inverse limits commute with products, cones and hyperspaces. To this end, we use [5, 9, 23].

2.3.1 Theorem *Let $\{X_n, f_n^{n+1}\}$ and $\{Y_n, g_n^{n+1}\}$ be inverse sequences of compacta, with surjective bonding maps, whose inverse limits are X_{∞} and Y_{∞}, respectively. Then $\{X_n \times Y_n, f_n^{n+1} \times g_n^{n+1}\}$ is an inverse sequence and $\varprojlim\{X_n \times Y_n, f_n^{n+1} \times g_n^{n+1}\}$ is homeomorphic to $X_{\infty} \times Y_{\infty}$.*

Proof Let $n \in \mathbb{N}$. By Theorem 1.1.10, the function $f_n \times g_n$ is continuous.

For each $n \in \mathbb{N}$, let $h_n : X_{\infty} \times Y_{\infty} \to X_n \times Y_n$ be given by $h_n = f_n \times g_n$. Then $(f_n^{n+1} \times g_n^{n+1}) \circ h_{n+1} = h_n$. Hence, by Theorem 2.1.33, there exists the induced map

$$h_{\infty} : X_{\infty} \times Y_{\infty} \to \varprojlim\{X_n \times Y_n, f_n^{n+1} \times g_n^{n+1}\}.$$

Since all the bonding maps are surjective, h_n is surjective for each $n \in \mathbb{N}$. Then h_∞ is surjective by Theorem 2.1.35. Clearly, h_∞ is one-to-one.

Q.E.D.

The next theorem shows that taking inverse limits and taking cones commute.

2.3.2 Theorem *Let $\{X_n, f_n^{n+1}\}$ be an inverse sequence of compacta whose inverse limit is X_∞. Then $\{K(X_n), K(f_n^{n+1})\}$ is an inverse sequence of cones with induced bonding maps and $K(X_\infty)$ is homeomorphic to $\varprojlim \{K(X_n), K(f_n^{n+1})\}$.*

Proof By Proposition 1.2.12, $K(f_n^{n+1})$ is a map from $K(X_{n+1})$ into $K(X_n)$ for each $n \in \mathbb{N}$. Hence, $\{K(X_n), K(f_n^{n+1})\}$ is an inverse sequence of continua. Let $Y_\infty = \varprojlim\{K(X_n), K(f_n^{n+1})\}$.

Note that for each $n \in \mathbb{N}$, by Proposition 1.2.12, $K(f_n)\colon K(X_\infty) \to K(X_n)$ is a map. Clearly, $K(f_n^{n+1}) \circ K(f_{n+1}) = K(f_n)$ for every $n \in \mathbb{N}$. Hence, by Theorem 2.1.33, there exists an induced map $K(f)_\infty\colon K(X_\infty) \to Y_\infty$. To see $K(f)_\infty$ is a homeomorphism, we show it is a bijection.

Note that $K(f)_\infty(v_{X_\infty}) = (v_{X_n})_{n=1}^\infty$. If $((x_n, t))_{n=1}^\infty \in Y_\infty \setminus \{(v_{X_n})_{n=1}^\infty\}$, then $((x_n)_{n=1}^\infty, t)$ is a point of $K(X_\infty)$ satisfying that $K(f)_\infty(((x_n)_{n=1}^\infty, t)) = ((x_n, t))_{n=1}^\infty$. Hence, $K(f)_\infty$ is surjective.

To see $K(f)_\infty$ is one-to-one, note that $(K(f)_\infty)^{-1}((v_{X_n})_{n=1}^\infty) = \{v_{X_\infty}\}$. Now, if $((x_n)_{n=1}^\infty, t)$ and $((x_n')_{n=1}^\infty, t')$ are two points of $K(X_\infty)$ such that

$$K(f)_\infty(((x_n)_{n=1}^\infty, t)) = K(f)_\infty(((x_n')_{n=1}^\infty, t')),$$

then $((x_n, t))_{n=1}^\infty = ((x_n', t'))_{n=1}^\infty$. Hence, $x_n = x_n'$ for each $n \in \mathbb{N}$, and $t = t'$. Thus, $((x_n)_{n=1}^\infty, t) = ((x_n')_{n=1}^\infty, t')$. Consequently, $K(f)_\infty$ is one-to-one.

Therefore, $K(f)_\infty$ is a homeomorphism.

Q.E.D.

The following lemma gives a base for the hyperspace of subcompacta of an inverse limit in terms of the open subsets of the factor spaces and the projection maps.

2.3.3 Lemma *Let $\{X_n, f_n^{n+1}\}$ be an inverse sequence of compacta whose inverse limit is X_∞. For each $j \in \mathbb{N}$, let*

$$\mathcal{B}_j = \{\langle f_j^{-1}(U_1), \ldots, f_j^{-1}(U_k)\rangle \mid U_1, \ldots, U_k \text{ are open subsets of } X_j\}.$$

If $\mathcal{B} = \bigcup_{j=1}^\infty \mathcal{B}_j$, then $\mathcal{B}$ is a base for the Vietoris topology for 2^{X_∞}.

Proof By Proposition 2.1.9 and Theorem 1.8.14,

$$\mathcal{B}^* = \{\langle f_{n(1)}^{-1}(U_{n(1)}), \ldots, f_{n(m)}^{-1}(U_{n(m)})\rangle \mid U_{n(\ell)} \text{ is an open}$$

$$\text{subset of } X_{n(\ell)} \text{ for each } \ell \in \{1, \ldots, m\}, m \in \mathbb{N}\}$$

is a base for the Vietoris topology for 2^{X_∞}. We show that $\mathcal{B} = \mathcal{B}^*$. Clearly, $\mathcal{B} \subset \mathcal{B}^*$. Let $\langle f_{n(1)}^{-1}(U_{n(1)}), \ldots, f_{n(m)}^{-1}(U_{n(m)}) \rangle \in \mathcal{B}^*$, and let

$$k = \max\{n(1), \ldots, n(m)\}.$$

Note that, since $f_{n(\ell)}^k \circ f_k = f_{n(\ell)}$, we have that for every $\ell \in \{1, \ldots, m\}$, $f_{n(\ell)}^{-1}(U_{n(\ell)}) = f_k^{-1}(f_{n(\ell)}^k)^{-1}(U_{n(\ell)})$.

For each $\ell \in \{1, \ldots, m\}$, let $V_{n(\ell)} = (f_{n(\ell)}^k)^{-1}(U_{n(\ell)})$. Then, since the bonding maps are continuous, each $V_{n(\ell)}$ is an open subset of X_k. Hence,

$$\langle f_k^{-1}(V_{n(1)}), \ldots, f_k^{-1}(V_{n(m)}) \rangle \in \mathcal{B}.$$

Since $f_k^{-1}(V_{n(\ell)}) = f_{n(\ell)}^{-1}(U_{n(\ell)})$ for each $\ell \in \{1, \ldots, m\}$,

$$\langle f_k^{-1}(V_{n(1)}), \ldots, f_k^{-1}(V_{n(m)}) \rangle = \langle f_{n(1)}^{-1}(U_{n(1)}), \ldots, f_{n(m)}^{-1}(U_{n(m)}) \rangle.$$

Hence, $\langle f_{n(1)}^{-1}(U_{n(1)}), \ldots, f_{n(m)}^{-1}(U_{n(m)}) \rangle \in \mathcal{B}$.
Therefore, $\mathcal{B} = \mathcal{B}^*$.

Q.E.D.

2.3.4 Theorem *Let $\{X_n, f_n^{n+1}\}$ be an inverse sequence of continua whose inverse limit is X_∞. Then the following hold.*

(1) 2^{X_∞} is homeomorphic to $\varprojlim \left\{ 2^{X_n}, 2^{f_n^{n+1}} \right\}$;

(2) $\mathcal{C}_m(X_\infty)$ is homeomorphic to $\varprojlim \{\mathcal{C}_m(X_n), \mathcal{C}_m(f_n^{n+1})\}$ for each $m \in \mathbb{N}$; and

(3) $\mathcal{F}_m(X_\infty)$ is homeomorphic to $\varprojlim \{\mathcal{F}_m(X_n), \mathcal{F}_m(f_n^{n+1})\}$ for each $m \in \mathbb{N}$.

Furthermore, there exists a homeomorphism

$$h \colon \varprojlim \left\{ 2^{X_n}, 2^{f_n^{n+1}} \right\} \twoheadrightarrow 2^{X_\infty}$$

such that for each $m \in \mathbb{N}$,

$$h \left(\varprojlim \left\{ \mathcal{C}_m(X_n), \mathcal{C}_m(f_n^{n+1}) \right\} \right) = \mathcal{C}_m(X_\infty)$$

and

$$h \left(\varprojlim \left\{ \mathcal{F}_m(X_n), \mathcal{F}_m(f_n^{n+1}) \right\} \right) = \mathcal{F}_m(X_\infty).$$

Proof We define the homeomorphism h after making some observations about the points of $\varprojlim \left\{ 2^{X_n}, 2^{f_n^{n+1}} \right\}$.

Let $(A_n)_{n=1}^\infty \in \varprojlim \left\{ 2^{X_n}, 2^{f_n^{n+1}} \right\}$. Then, by definition, $2^{f_n^{n+1}}(A_{n+1}) = A_n$. Thus, $\{A_n, f_n^{n+1}|_{A_{n+1}}\}$ is an inverse sequence with surjective bonding maps.

Since, for each $n \in \mathbb{N}$, $A_n \subset X_n$, we have that $\varprojlim\{A_n, f_n^{n+1}|_{A_{n+1}}\} \subset \varprojlim\{X_n, f_n^{n+1}\} = X_\infty$. Hence, $\varprojlim\{A_n, f_n^{n+1}|_{A_{n+1}}\} \in 2^{X_\infty}$.

Now, let $h \colon \varprojlim \left\{ 2^{X_n}, 2^{f_n^{n+1}} \right\} \twoheadrightarrow 2^{X_\infty}$ be given by

$$h((A_n)_{n=1}^\infty) = \varprojlim\{A_n, f_n^{n+1}|_{A_{n+1}}\}.$$

By the above considerations, h is well defined.

Let $K \in 2^{X_\infty}$. Then K is a closed subset of X_∞. Hence, by $(*)$ of Proposition 2.1.20, $K = \varprojlim\{f_n(K), f_n^{n+1}|_{f_{n+1}(K)}\}$. Note that $f_n(K) \in 2^{X_n}$ and $2^{f_n^{n+1}}(f_{n+1}(K)) = f_n(K)$ for each $n \in \mathbb{N}$. Thus, $(f_n(K))_{n=1}^\infty \in \varprojlim \left\{ 2^{X_n}, 2^{f_n^{n+1}} \right\}$, and $h((f_n(K))_{n=1}^\infty) = K$. Therefore, h is surjective.

Next, let $(A_n)_{n=1}^\infty$ and $(B_n)_{n=1}^\infty$ be two elements of $\varprojlim \left\{ 2^{X_n}, 2^{f_n^{n+1}} \right\}$ such that $h((A_n)_{n=1}^\infty) = h((B_n)_{n=1}^\infty)$. We prove that $A_k = B_k$ for every $k \in \mathbb{N}$.

Let $k \in \mathbb{N}$, and let $p \in A_k$. Since $\{A_n, f_n^{n+1}|_{A_{n+1}}\}$ is an inverse sequence with surjective bonding maps, there exists a point $(x_n)_{n=1}^\infty \in h((A_n)_{n=1}^\infty)$ such that $x_k = p$ (Remark 2.1.6). Since $h((A_n)_{n=1}^\infty) = h((B_n)_{n=1}^\infty)$, $(x_n)_{n=1}^\infty \in h((B_n)_{n=1}^\infty)$. Hence, $x_k \in B_k$; i.e., $p \in B_k$. Therefore, $A_k \subset B_k$. A similar argument shows that $B_k \subset A_k$. Thus, $A_k = B_k$. Therefore, h is one-to-one.

Now, we see that h is continuous. Let $\pi_j \colon \varprojlim \left\{ 2^{X_n}, 2^{f_n^{n+1}} \right\} \to 2^{X_j}$ be the projection map for every $j \in \mathbb{N}$.

Let $\langle f_j^{-1}(U_1), \ldots, f_j^{-1}(U_k) \rangle \in \mathcal{B}$ (Lemma 2.3.3). We show that

$$h^{-1}(\langle f_j^{-1}(U_1), \ldots, f_j^{-1}(U_k) \rangle)$$

is an open subset of $\varprojlim \left\{ 2^{X_n}, 2^{f_n^{n+1}} \right\}$. To this end, it suffices to prove that

$$h^{-1}(\langle f_j^{-1}(U_1), \ldots, f_j^{-1}(U_k) \rangle) = \pi_j^{-1}(\langle U_1, \ldots, U_k \rangle).$$

First observe that if $(A_n)_{n=1}^\infty \in \varprojlim \left\{ 2^{X_n}, 2^{f_n^{n+1}} \right\}$, then $f_j(h((A_n)_{n=1}^\infty)) = A_j$ for each $j \in \mathbb{N}$. Hence,

$$h^{-1}(\langle f_j^{-1}(U_1), \ldots, f_j^{-1}(U_k) \rangle) =$$

$$\left\{ (A_n)_{n=1}^\infty \in \varprojlim \left\{ 2^{X_n}, 2^{f_n^{n+1}} \right\} \; \middle| \; h((A_n)_{n=1}^\infty) \subset \bigcup_{\ell=1}^{k} f_j^{-1}(U_\ell) \text{ and} \right.$$

$$\left. h((A_n)_{n=1}^{\infty}) \cap f_j^{-1}(U_\ell) \neq \emptyset, \text{ for each } \ell \in \{1, \ldots, k\} \right\} =$$

$$\left\{ (A_n)_{n=1}^{\infty} \in \varprojlim \left\{ 2^{X_n}, 2^{f_n^{n+1}} \right\} \;\middle|\; h((A_n)_{n=1}^{\infty}) \subset f_j^{-1}\left(\bigcup_{\ell=1}^{k} U_\ell \right) \text{ and} \right.$$

$$\left. h((A_n)_{n=1}^{\infty}) \cap f_j^{-1}(U_\ell) \neq \emptyset, \text{ for each } \ell \in \{1, \ldots, k\} \right\} =$$

$$\left\{ (A_n)_{n=1}^{\infty} \in \varprojlim \left\{ 2^{X_n}, 2^{f_n^{n+1}} \right\} \;\middle|\; f_j(h((A_n)_{n=1}^{\infty})) \subset \bigcup_{\ell=1}^{k} U_\ell \text{ and} \right.$$

$$\left. f_j(h((A_n)_{n=1}^{\infty})) \cap U_\ell \neq \emptyset, \text{ for each } \ell \in \{1, \ldots, k\} \right\} =$$

$$\left\{ (A_n)_{n=1}^{\infty} \in \varprojlim \left\{ 2^{X_n}, 2^{f_n^{n+1}} \right\} \;\middle|\; A_j \subset \bigcup_{\ell=1}^{k} U_\ell \text{ and} \right.$$

$$\left. A_j \cap U_\ell \neq \emptyset, \text{ for each } \ell \in \{1, \ldots, k\} \right\} =$$

$$\pi_j^{-1}(\langle U_1, \ldots, U_k \rangle).$$

Therefore, h is continuous. Hence, h is a homeomorphism.
Observe that by the definition of h and by Theorem 2.1.47, it follows that

$$h\left(\varprojlim \left\{ \mathcal{C}_m(X_n), \mathcal{C}_m(f_n^{n+1}) \right\} \right) = \mathcal{C}_m(X_\infty)$$

and that

$$h\left(\varprojlim \left\{ \mathcal{F}_m(X_n), \mathcal{F}_m(f_n^{n+1}) \right\} \right) = \mathcal{F}_m(X_\infty)$$

for every $m \in \mathbb{N}$.

Q.E.D.

2.4   Chainable Continua

We discuss an important class of continua, namely, chainable continua. Besides presenting some of its properties, we prove that it coincides with the class of arc-like continua. The material of this section is based on [4, 7, 16, 25].

2.4.1 Definition Let X be a compactum. A *chain* $\mathcal{U}$ *in* X is a finite sequence, $U_1, \ldots, U_n$, of subsets of X such that $U_i \cap U_j \neq \emptyset$ if and only if $|i - j| \leq 1$ for each $i, j \in \{1, \ldots, n\}$. Each U_j is called a *link* of $\mathcal{U}$. If each link of $\mathcal{U}$ is open, then $\mathcal{U}$ is called an *open chain*. If $\varepsilon > 0$, $\mathcal{U}$ is an open chain and the mesh($\mathcal{U}$) $< \varepsilon$, then $\mathcal{U}$ is called an ε-*chain*.

2.4.2 Remark Let us observe that we do not require the links of a chain to be connected. Hence, a chain may look like in the following picture:

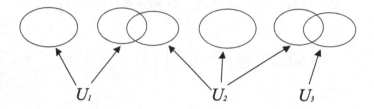

2.4.3 Definition A continuum X is said to be *chainable* provided that for each $\varepsilon > 0$, there exists an ε-chain covering X. If $x_1, x_2 \in X$, then X *is chainable from* x_1 *to* x_2 if for each $\varepsilon > 0$, there exists an ε-chain $\mathcal{U} = \{U_1, \ldots, U_n\}$ covering X such that $x_1 \in U_1$ and $x_2 \in U_n$.

2.4.4 Example The unit interval $[0, 1]$ is chainable from 0 to 1.

0   1

2.4.5 Example Let

$$X = \{0\} \times [-1, 1] \cup \left\{ \left(x, \sin\left(\frac{1}{x}\right) \right) \in \mathbb{R}^2 \ \middle|\ x \in \left(0, \frac{2}{\pi}\right] \right\}.$$

Then X is called the *topologist sine curve*. Note that X is a chainable continuum from $(0, -1)$ to $\left(\frac{2}{\pi}, 1\right)$.

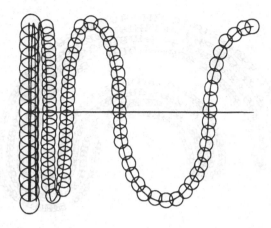

2.4.6 Example Let X be the topologist sine curve and let X' be the reflection of X in $\mathbb{R}^2$ with respect to the line $x = \frac{2}{\pi}$. Let $Z = X \cup X'$. Then Z is a chainable continuum from $(0, -1)$ to $\left(\frac{4}{\pi}, -1\right)$.

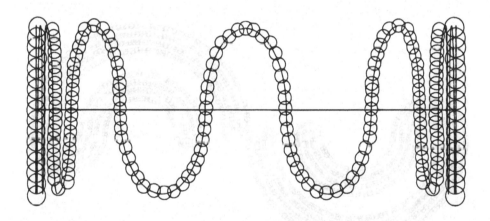

2.4.7 Example The *Knaster continuum*, $\mathcal{K}$, is defined in the following way. The continuum consists of

(a) all semi-circles in $\mathbb{R}^2$ with nonnegative ordinates, with center at the point $\left(\frac{1}{2}, 0\right)$ and passing through every point of the Cantor set $\mathcal{C}$.

(b) all semi-circles in $\mathbb{R}^2$ with nonpositive ordinates, which have for each $n \in \mathbb{N}$, the center at the point $\left(\frac{5}{2 \cdot 3^n}, 0\right)$ and passing through each point of the Cantor set $\mathcal{C}$ lying in the interval $\left[\frac{2}{3^n}, \frac{1}{3^{n-1}}\right]$.

Then $\mathcal{K}$ is a chainable continuum.

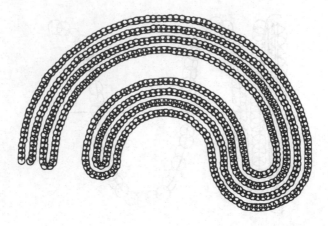

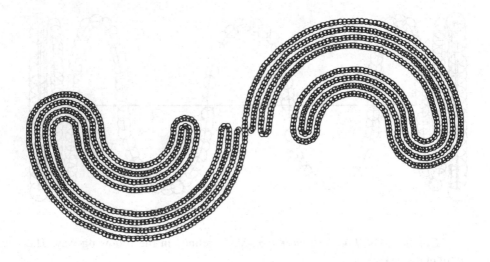

2.4.8 Remark It is well known that $\mathcal{K}$ is an indecomposable continuum [16, Remark to Theorem 8, p. 213].

2.4.9 Example Let $\mathcal{K}'$ be the reflection of $\mathcal{K}$ in $\mathbb{R}^2$ with respect to the origin $(0, 0)$. Let $\mathcal{M} = \mathcal{K} \cup \mathcal{K}'$. Then $\mathcal{M}$ is a chainable continuum.

The next theorem shows that the property of being chainable is hereditary; i.e., each nondegenerate subcontinuum has the property.

2.4.10 Theorem *If X is a chainable continuum and K is a subcontinuum of X, then K is chainable.*

Proof Let $\varepsilon > 0$. Since X is chainable, there exists an ε-chain $\mathcal{U} = \{U_1, \ldots, U_n\}$ covering X. Let $i = \min\{k \in \{1, \ldots, n\} \mid K \cap U_k \neq \emptyset\}$, and let $j = \max\{k \in$

$\{1, \ldots, n\} \mid K \cap U_k \neq \emptyset\}$. We show that $\mathcal{U}' = \{U_i \cap K, \ldots, U_j \cap K\}$ is an ε-chain covering K. Clearly, $\text{mesh}(\mathcal{U}') < \varepsilon$.

Suppose $\mathcal{U}'$ is not a chain. Then there exists $\ell \in \{i, \ldots, j-1\}$ such that $(U_\ell \cap K) \cap (U_{\ell+1} \cap K) = \emptyset$. Hence,

$$K \subset \left(\bigcup_{m=i}^{\ell} (U_m \cap K) \right) \bigcup \left(\bigcup_{m=\ell+1}^{j} (U_m \cap K) \right)$$

and

$$\left(\bigcup_{m=i}^{\ell} (U_m \cap K) \right) \bigcap \left(\bigcup_{m=\ell+1}^{j} (U_m \cap K) \right) = \emptyset.$$

This contradicts the fact that K is connected since $K \cap U_i \neq \emptyset$ and $K \cap U_j \neq \emptyset$. Thus, $\mathcal{U}'$ is a chain.

Therefore, K is chainable.

Q.E.D.

The following lemma is known as the *Shrinking Lemma*.

2.4.11 Lemma *Let X be a compactum. If $\mathcal{V} = \{V_1, \ldots, V_m\}$ is a finite open cover of X, then there exists an open cover $\mathcal{U} = \{U_1, \ldots, U_m\}$ such that $Cl(U_j) \subset V_j$ for each $j \in \{1, \ldots, m\}$.*

Proof The construction of $\mathcal{U}$ is done inductively.

Let $F_1 = X \setminus \left(\bigcup_{j=2}^{m} V_j \right)$. Then $F_1 \subset V_1$, and F_1 is a closed subset of X. Since X is a metric space, there exists an open subset U_1 of X such that $F_1 \subset U_1 \subset Cl(U_1) \subset V_1$.

Suppose U_{k-1} has been defined for each $k < m$. Let

$$F_k = X \setminus \left(\bigcup_{j=1}^{k-1} U_j \cup \bigcup_{j=k+1}^{m} V_j \right).$$

Then $F_k \subset V_k$, and F_k is a closed subset of X. Since X is a metric space, there exists an open subset U_k of X such that $F_k \subset U_k \subset Cl(U_k) \subset V_k$. Thus, we finish the inductive step.

Now, let $\mathcal{U} = \{U_1, \ldots, U_m\}$. Then $\mathcal{U}$ is a family of m open subsets of X. We show $\mathcal{U}$ covers X. To this end, let $x \in X$. Then x belongs to finitely many elements of $\mathcal{V}$, say $V_{k_1}, \ldots, V_{k_n}$. Let $k = \max\{k_1, \ldots, k_n\}$. Now, $x \in X \setminus V_\ell$ for every $\ell > k$, and hence, if $x \in X \setminus U_j$ for any $j < k$, then $x \in F_k \subset U_k$. Thus, in any case, $x \in U_j$ for some $j \in \{1, \ldots, m\}$. Therefore, $\mathcal{U}$ covers X.

Q.E.D.

2.4.12 Definition Let X be a compactum. A cover $\mathcal{U}$ of X is said to be *essential* provided that no proper subfamily of $\mathcal{U}$ covers X.

2.4.13 Lemma *If X is a chainable continuum, then there exists a sequence $\{\mathcal{U}_n\}_{n=1}^{\infty}$ of essential covers of X such that for each $n \in \mathbb{N}$, the following conditions hold:*

(a) *$\mathcal{U}_n$ is a $\frac{1}{2^n}$-chain with the property that disjoint links have disjoint closures; and*
(b) *The closure of the union of any three consecutive links of $\mathcal{U}_{n+1}$ is contained in one link of $\mathcal{U}_n$.*

Proof Note that if $\mathcal{U} = \{U_1, \ldots, U_n\}$ is a chain covering X is not essential, by the definition of chain, then $U_1 \subset U_2$ and/or $U_n \subset U_{n-1}$. Since one can always modify such a chain $\mathcal{U}$ so that its end links contain a point not in any other link (by simply removing U_1 and/or U_n if necessary), we may assume that the chain is essential.

The construction of the coverings is done inductively.

Since X is chainable, there exists a $\frac{1}{2}$-chain $\mathcal{V}_1 = \{V_{11}, \ldots, V_{1,m_1}\}$ covering X. By Lemma 2.4.11, there exists an open cover $\mathcal{U}_1 = \{U_{11}, \ldots, U_{1m_1}\}$ such that $Cl(U_{1j}) \subset V_{1j}$ for each $j \in \{1, \ldots, m_1\}$. Then $\mathcal{U}_1$ is a $\frac{1}{2}$-chain covering X with the property that disjoint links have disjoint closures.

Suppose that, for some $n \in \mathbb{N}$, we have constructed open covers $\mathcal{U}_1, \ldots, \mathcal{U}_n$ of X satisfying conditions (a) and (b) of the statement above. We construct $\mathcal{U}_{n+1}$ as follows.

Let λ_{n+1} be a Lebesgue number for the cover $\mathcal{U}_n$ (Theorem 1.6.6). Let $\alpha < \min\left\{\frac{1}{3}\lambda_{n+1}, \frac{1}{2^{n+1}}\right\}$. Since X is chainable, there exists an α-chain $\mathcal{V}_{n+1} = \{V_{n+11}, \ldots, V_{n+1m_{n+1}}\}$ covering X. Now, by Lemma 2.4.11, there exists an open cover $\mathcal{U}_{n+1} = \{U_{n+11}, \ldots, U_{n+1m_{n+1}}\}$ such that $Cl(U_{n+1j}) \subset V_{n+1j}$. Then $\mathcal{U}_{n+1}$ is a $\frac{1}{2^{n+1}}$-chain covering X with the property that disjoint links have disjoint closures.

We need to see that the closure of the union of three consecutive links of $\mathcal{U}_{n+1}$ is contained in a link of $\mathcal{U}_n$. Let U_{n+1j}, U_{n+1j+1} and U_{n+1j+2} be three consecutive links of $\mathcal{U}_{n+1}$. Since $\text{diam}(Cl(U_{n+1j}) \cup Cl(U_{n+1j+1}) \cup Cl(U_{n+1j+2})) = \text{diam}(U_{n+1j} \cup U_{n+1j+1} \cup U_{n+1j+2}) \leq 3\alpha < \lambda_{n+1}$, there exists a link U_{nk} of $\mathcal{U}_n$ such that $Cl(U_{n+1j}) \cup Cl(U_{n+1j+1}) \cup Cl(U_{n+1j+2}) \subset U_{nk}$ (since λ_{n+1} is a Lebesgue number for $\mathcal{U}_n$).

In this way, we finish the inductive step. Therefore, the existence of the sequence of open coverings is proven.

Q.E.D.

2.4.14 Definition Let X be a chainable continuum. A sequence $\{\mathcal{U}_n\}_{n=1}^{\infty}$ of essential covers satisfying the conditions of Lemma 2.4.13 is called a *defining sequence of chains for X*.

2.4.15 Definition Let X be a metric space, and let $f \colon X \to X$ be a map. We say that *f has a fixed point* if there exists $x \in X$ such that $f(x) = x$.

2.4.16 Definition Let X be a metric space. We say that *X has the fixed point property* provided that for each map $f \colon X \to X$, f has a fixed point.

It is a well known fact from calculus that the unit interval $[0, 1]$ has the fixed point property. Hamilton [8] has shown that this property is shared by all chainable continua. To prove this result, we need the following lemma:

2.4.17 Lemma *Let X be a compactum, with metric d, and let $f: X \to X$ be a map. If for each $\varepsilon > 0$, there exists a point x_ε in X such that $d(f(x_\varepsilon), x_\varepsilon) < \varepsilon$, then f has a fixed point.*

Proof By hypothesis, for each $n \in \mathbb{N}$, there exists $x_n \in X$ such that $d(x_n, f(x_n)) < \frac{1}{n}$. Since X is compact, without loss of generality, we assume that the sequence $\{x_n\}_{n=1}^\infty$ converges to a point $x \in X$. Since f is continuous, the sequence $\{f(x_n)\}_{n=1}^\infty$ converges to $f(x)$.

Let $\varepsilon > 0$. Then there exists $N \in \mathbb{N}$ such that

$$d(x_N, x) < \frac{\varepsilon}{3}, \ d(f(x_N), f(x)) < \frac{\varepsilon}{3} \text{ and } \frac{1}{N} < \frac{\varepsilon}{3}.$$

Hence,

$$d(x, f(x)) \leq d(x, x_N) + d(x_N, f(x_N)) + d(f(x_N), f(x)) <$$

$$\frac{\varepsilon}{3} + \frac{1}{N} + \frac{\varepsilon}{3} < \frac{2}{3}\varepsilon + \frac{\varepsilon}{3} = \varepsilon.$$

Since ε is arbitrary, $d(x, f(x)) = 0$. Thus, $f(x) = x$. Therefore, f has a fixed point
                                                                                    Q.E.D.

2.4.18 Theorem *If X is a chainable continuum, with metric d, then X has the fixed point property.*

Proof Let $f: X \to X$ be a map. Let $\{\mathcal{U}_n\}_{n=1}^\infty$ be a defining sequence of chains for X. For each $k \in \mathbb{N}$, we assume that $\mathcal{U}_k = \{U_{k1}, \ldots, U_{kn_k}\}$.

Let $\varepsilon > 0$. By Lemma 2.4.17, we need to find a point $x_\varepsilon \in X$ such that $d(x_\varepsilon, f(x_\varepsilon)) < \varepsilon$.

Let $k \in \mathbb{N}$ be such that $\frac{1}{2^k} < \varepsilon$. Consider the chain $\mathcal{U}_k$ and define the following subsets of X:

$$A = \{x \in X \mid \text{if } x \in Cl(U_{kj}) \text{ and } f(x) \in Cl(U_{ki}), \text{ then } j < i\};$$

$$B = \{x \in X \mid x, f(x) \in Cl(U_{ki}) \text{ for some } i \in \{1, \ldots, n_k\}\};$$

$$C = \{x \in X \mid \text{if } x \in Cl(U_{kj}) \text{ and } f(x) \in Cl(U_{ki}), \text{ then } j > i\}.$$

We show that $B \neq \emptyset$. To this end, suppose $B = \emptyset$. Let $x \in X \setminus A$, and let λ be a Lebesgue number of $\mathcal{U}_k$ (Theorem 1.6.6). Since X is compact, f is uniformly continuous. Hence, there exists $\delta > 0$ such that $\delta < \lambda$ and such that if $y, z \in X$ and $d(y, z) < \delta$, then $d(f(y), f(z)) < \lambda$. Let $y \in X$ be such that $d(y, x) < \delta$.

Since $\delta < \lambda$, there exists $U_{km} \in \mathcal{U}_k$ such that $x, y \in U_{km} \subset Cl(U_{km})$. Since $x \in Cl(U_{km}) \cap (X \setminus A)$, then $f(x) \in Cl(U_{kj})$, where $j < m$. Now, since $d(f(y), f(x)) < \lambda$, $f(y) \in Cl(U_{kn})$, where $n \leq m$. Since $B = \emptyset$, $n < m$. Hence, $y \in X \setminus A$. Consequently, $X \setminus A$ is an open subset of X. Therefore, A is closed. Similarly, C is a closed subset of X. Since $X = A \cup C$ and $A \cap C = \emptyset$, we obtain a contradiction. Therefore, $B \neq \emptyset$.

Q.E.D.

The next concept is used to define classes of continua.

2.4.19 Definition Let $f : X \twoheadrightarrow Y$ be a surjective map between metric spaces, and let $\varepsilon > 0$. We say that f is an ε-*map* provided that for each $y \in Y$, $\text{diam}(f^{-1}(y)) < \varepsilon$.

The following lemma says that ε-maps between compacta behave in the same way with sets of positive small diameter.

2.4.20 Lemma *Let X and Y be compacta, with metrics d and d', respectively. Let $\varepsilon > 0$. If $f : X \twoheadrightarrow Y$ is an ε-map, then there exists $\delta > 0$ such that $\text{diam}(f^{-1}(U)) < \varepsilon$ for each subset U of Y with $\text{diam}(U) < \delta$.*

Proof First note that since $\text{diam}(U) = \text{diam}(Cl(U))$ for any set U, it suffices to prove the lemma for closed sets.

Suppose the lemma is not true. Then for each $n \in \mathbb{N}$, there exists a closed subset U_n of Y such that $\text{diam}(U_n) < \frac{1}{n}$ and $\text{diam}(f^{-1}(U_n)) \geq \varepsilon$. Let $x_n, x'_n \in f^{-1}(U_n)$ be such that $d(x_n, x'_n) = \text{diam}(f^{-1}(U_n))$. Since X is compact, without loss of generality, we assume that there exist two points $x, x' \in X$ be such that $\lim_{n \to \infty} x_n = x$ and $\lim_{n \to \infty} x'_n = x'$. Note that $d(x, x') \geq \varepsilon$.

Now, by continuity,

$$d'(f(x), f(x')) = \lim_{n \to \infty} d'(f(x_n), f(x'_n)) \leq$$

$$\lim_{n \to \infty} \text{diam}(U_n) \leq \lim_{n \to \infty} \frac{1}{n} = 0,$$

a contradiction to the fact that f is an ε-map. Therefore, the lemma is true.

Q.E.D.

2.4.21 Definition A continuum X is said to be *like an arc* provided that for each $\varepsilon > 0$, there exists an ε-map $f : X \twoheadrightarrow [0, 1]$.

The next theorem gives us the equality of the class of chainable continua and the class of arc-like continua. The proof we present is due to James T. Rogers, Jr.

2.4.22 Theorem *If X is a continuum with metric d, then the following are equivalent:*

(1) X is chainable.
(2) X is arc-like.
(3) X is like an arc.

Proof Suppose X is a chainable continuum. We show X is arc-like.

Let $\{\mathcal{U}_n\}_{n=1}^{\infty}$ be a defining sequence of chains for X, Lemma 2.4.13. For every $n \in \mathbb{N}$, we use the notation $\mathcal{U}_n = \{U_{n,0}, \ldots, U_{n,k(n)}\}$.

For each $n \in \mathbb{N}$, let $X_n = [0, 1]$. Divide X_n into $k(n)$ equal subintervals with vertexes $v_{n,0} = 0, \ldots, v_{n,k(n)} = 1$. Note that there exists a one-to-one correspondence between the vertexes of the subintervals of X_n and the links of the chain $\mathcal{U}_n$.

We define the functions $f_n^{n+1} \colon X_{n+1} \twoheadrightarrow X_n$ as follows:

$$f_n^{n+1}(v_{n+1,m}) = \begin{cases} v_{n,j}, & \text{if } U_{n,j} \text{ is the only link of } \mathcal{U}_n \\ & \text{containing } U_{n+1,m}; \\ \dfrac{v_{n,j} + v_{n,j+1}}{2}, & \text{if } U_{n+1,m} \subset U_{n,j} \cap U_{n,j+1}, \end{cases}$$

and extend f_n^{n+1} linearly over X_{n+1}. Since the chains are essential, the function f_n^{n+1} is well defined for every $n \in \mathbb{N}$. Also, all these functions are continuous and surjective.

Let $X_\infty = \varprojlim\{X_n, f_n^{n+1}\}$. Then X_∞ is an arc-like continuum. We show X_∞ is homeomorphic to X. To this end, let $h_n \colon X \to \mathcal{C}(X_n)$ be given by:

$$h_n(x) = \begin{cases} \{v_{n,j}\}, & \text{if } U_{n,j} \text{ is the only link of } \mathcal{U}_n \text{ containing } x; \\ [v_{n,j}, v_{n,j+1}], & \text{if } x \in U_{n,j} \cap U_{n,j+1}. \end{cases}$$

Note that for each $n \in \mathbb{N}$, $f_n^{n+1}(h_{n+1}(x)) \subset h_n(x)$. To see this, we consider six cases.

If $x \in U_{n+1,m} \setminus (U_{n+1,m-1} \cup U_{n+1,m+1})$ and $U_{n+1,m} \subset U_{n,j} \setminus (U_{n,j-1} \cup U_{n,j+1})$, then $f_n^{n+1}(h_{n+1}(x)) = f_n^{n+1}(\{v_{n+1,m}\}) = \{v_{n,j}\} = h_n(x)$.

If $x \in U_{n+1,m} \setminus (U_{n+1,m-1} \cup U_{n+1,m+1})$ and $U_{n+1,m} \subset U_{n,j} \cap U_{n,j+1}$, then $f_n^{n+1}(h_{n+1}(x)) = f_n^{n+1}(\{v_{n+1,m}\}) = \left\{\frac{v_{n,j}+v_{n,j+1}}{2}\right\} \subset [v_{n,j}, v_{n,j+1}] = h_n(x)$.

If $x \in U_{n+1,m} \setminus (U_{n+1,m-1} \cup U_{n+1,m+1})$, $U_{n+1,m} \subset U_{n,j}$ and $U_{n+1,m} \cap U_{n,j+1} \neq \emptyset$, then $f_n^{n+1}(h_{n+1}(x)) = f_n^{n+1}(\{v_{n+1,m}\}) = \{v_{n,j}\} \subset [v_{n,j}, v_{n,j+1}] = h_n(x)$.

If $x \in U_{n+1,m} \cap U_{n+1,m+1}$ and $U_{n+1,m} \cap U_{n+1,m+1} \subset U_{n,j} \setminus (U_{n,j-1} \cup U_{n,j+1})$, then $f_n^{n+1}(h_{n+1}(x)) = f_n^{n+1}([v_{n+1,m}, v_{n+1,m+1}]) = \{v_{n,j}\} = h_n(x)$.

If $x \in U_{n+1,m} \cap U_{n+1,m+1}$ and $U_{n+1,m} \cup U_{n+1,m+1} \subset U_{n,j} \cap U_{n,j+1}$, then $f_n^{n+1}(h_{n+1}(x)) = f_n^{n+1}([v_{n+1,m}, v_{n+1,m+1}]) = \left\{\frac{v_{n,j}+v_{n,j+1}}{2}\right\} \subset [v_{n,j}, v_{n,j+1}] = h_n(x)$.

If $x \in U_{n+1,m} \cap U_{n+1,m+1}, U_{n+1,m} \cup U_{n+1,m+1} \subset U_{n,j}$ and $U_{n+1,m+1} \cap U_{n,j+1} \neq \emptyset$, then $f_n^{n+1}(h_{n+1}(x)) = f_n^{n+1}([v_{n+1,m}, v_{n+1,m+1}]) = \left[v_{n,j}, \frac{v_{n,j}+v_{n,j+1}}{2}\right] \subset [v_{n,j}, v_{n,j+1}] = h_n(x)$.

Therefore, $f_n^{n+1}(h_{n+1}(x)) \subset h_n(x)$. Note that this implies that $\{f_n^{-1}(h_n(x))\}_{n=1}^{\infty}$ is a decreasing sequence of closed subsets of X_∞. Thus, $\bigcap_{n=1}^{\infty} f_n^{-1}(h_n(x)) \neq \emptyset$.

Now, we are ready to define the homeomorphism $g \colon X \twoheadrightarrow X_\infty$ as follows: $g(x)$ is the unique point in $\bigcap_{n=1}^{\infty} f_n^{-1}(h_n(x))$.

Since $\lim_{n\to\infty} k(n) = \infty$ and the projection maps are $\frac{1}{2^n}$-maps, we have that $\lim_{n\to\infty} \operatorname{diam}(f_n^{-1}(h_n(x))) = 0$. Hence, g is well defined.

Now, we show that g is one-to-one. Let $x, x' \in X$ be such that $x \neq x'$. Let $\mathcal{U}_n$ be an element of the defining sequence of chains for X with the property that x and x' belong to different links of $\mathcal{U}_n$. In this case, $h_n(x) \cap h_n(x') = \emptyset$. Hence, $g(x) \neq g(x')$.

To see g is surjective, let $y = (y_n)_{n=1}^{\infty} \in X_\infty$. Let $\mathcal{M}_n \colon X_\infty \to \mathcal{C}(X_n)$ be given by

$$\mathcal{M}_n(y) = \begin{cases} \{v_{n,j}\}, & \text{if } y_n = v_{n,m}; \\ [v_{n,j}, v_{n,j+1}], & \text{if } y_n \in (v_{n,m}, v_{n,m+1}). \end{cases}$$

Let R_n be the union of the links of $\mathcal{U}_n$ whose vertexes are in $\mathcal{M}_n(y)$. Then R_n consists of one link or it is the union of two consecutive links of $\mathcal{U}_n$. Let $\{x\} = \bigcap_{n=1}^{\infty} R_n = \bigcap_{n=1}^{\infty} Cl(R_n)$ (recall that $Cl(R_{n+1}) \subset R_n$ for every $n \in \mathbb{N}$). We prove that $g(x) = y$. Let $n \in \mathbb{N}$. Then either $R_n = U_{n,j}$ or $R_n = U_{n,j} \cup U_{n,j+1}$, and $x \in R_n$. Hence, either $h_n(x) = \{v_{n,j}\}$ or $h_n(x) = [v_{n,j}, v_{n,j+1}]$. In either case, $h_n(x) = \mathcal{M}_n(y)$. Thus, $y \in f_n^{-1}(\mathcal{M}_n(y)) = f_n^{-1}(h_n(x))$. Hence, $y \in \bigcap_{n=1}^{\infty} f_n^{-1}(h_n(x))$. Therefore, $g(x) = y$.

Now, we see that g is continuous. To this end, by Theorem 1.1.9, it is enough to prove that for each $n \in \mathbb{N}$, $f_n \circ g$ is continuous. Let $n \in \mathbb{N}$ and let $\varepsilon > 0$. Let $\ell > n$ be such that for each $m \in \{1, \ldots, k(\ell) - 2\}$, $\operatorname{diam}(f_n^{\ell}([v_{\ell,m}, v_{\ell,m+2}])) < \varepsilon$. Let $\lambda > 0$ be a Lebesgue number for the cover $\mathcal{U}_\ell$ (Theorem 1.6.6). Hence, if $x, x' \in X$ and $d(x, x') < \lambda$, then there exists $j \in \{1, \ldots, k(\ell)\}$ such that $x, x' \in U_{\ell,j}$.

Note that $f_\ell \circ g(U_{\ell,j}) \subset [v_{\ell,m}, v_{\ell,m+2}]$ for some $m \in \{1, \ldots, k(\ell) - 2\}$. Hence, $\operatorname{diam}(f_n \circ g(U_{\ell,j})) = \operatorname{diam}(f_n^{\ell} \circ f_\ell \circ g(U_{\ell,j})) < \varepsilon$. Thus, $d_n(f_n \circ g(x), f_n \circ g(x')) < \varepsilon$, where d_n is the metric on X_n. Therefore, $f_n \circ g$ is continuous.

Next, suppose X is arc-like. We show that X is like an arc. Since X is arc-like, there exists an inverse sequence $\{X_n, f_n^{n+1}\}$, where $X_n = [0, 1]$ and f_n^{n+1} is surjective for each $n \in \mathbb{N}$, and whose inverse limit is X. Let $\varepsilon > 0$, and let $N \in \mathbb{N}$ be such that $\sum_{n=N+1}^{\infty} \frac{1}{2^n} < \varepsilon$. To see f_N is an ε-map, let $z \in X_N$. Let $x = (x_n)_{n=1}^{\infty}$

and $y = (y_n)_{n=1}^{\infty}$ be two points of $f_N^{-1}(z)$. Then

$$\rho(x, y) = \sum_{n=1}^{\infty} \frac{1}{2^n} d_n(x_n, y_n) = \sum_{n=N+1}^{\infty} \frac{1}{2^n} d_n(x_n, y_n) \leq \sum_{n=N+1}^{\infty} \frac{1}{2^n} < \varepsilon.$$

Therefore, $\operatorname{diam}(f_N^{-1}(z)) < \varepsilon$.

Finally, suppose X is like an arc. To see X is chainable, let $f : X \twoheadrightarrow [0, 1]$ be an ε-map. By Lemma 2.4.20, there exists $\delta > 0$ such that if U is a subset of $[0, 1]$ and $\operatorname{diam}(U) < \delta$, then $\operatorname{diam}(f^{-1}(U)) < \varepsilon$.

Let $n \in \mathbb{N}$ be such that $\frac{1}{n} < \frac{\delta}{2}$. Divide $[0, 1]$ into n equal parts and let

$$\mathcal{U} = \left\{ f^{-1}\left(\left[0, \frac{1}{n}\right)\right), f^{-1}\left(\left(0, \frac{2}{n}\right)\right), f^{-1}\left(\left(\frac{1}{n}, \frac{3}{n}\right)\right), \ldots, \right.$$

$$\left. f^{-1}\left(\left(\frac{n-2}{n}, 1\right)\right), f^{-1}\left(\left(\frac{n-1}{n}, 1\right]\right) \right\}.$$

Since $\left\{\left[0, \frac{1}{n}\right), \left(0, \frac{2}{n}\right), \left(\frac{1}{n}, \frac{3}{n}\right), \ldots, \left(\frac{n-2}{n}, 1\right), \left(\frac{n-1}{n}, 1\right]\right\}$ is a δ-chain of $[0, 1]$, $\mathcal{U}$ is an ε-chain covering X. Since ε is arbitrary, therefore, X is chainable.

Q.E.D.

2.4.23 Definition A map $f : X \rightarrow Y$ between continua is said to be *weakly confluent* provided that for every subcontinuum Z of Y, there exists a subcontinuum W of X such that $f(W) = Z$.

As an application of Theorem 2.4.22, we show that every map onto a chainable continuum is weakly confluent. First, we prove the special case of the unit interval.

2.4.24 Lemma *If Z is a continuum and $f : Z \twoheadrightarrow [0, 1]$ is a surjective map, then f is weakly confluent.*

Proof Let $[a, b]$ be a subcontinuum of $[0, 1]$. Let $K = \{(z, f(z)) \mid z \in Z\}$. Note that K is a subcontinuum of $Z \times [0, 1]$. Suppose that $K \cap (Z \times [a, b])$ does not contain an irreducible continuum between $K \cap (Z \times \{a\})$ and $K \cap (Z \times \{b\})$. Then, by Theorem 1.6.8, there exist two nonempty closed subsets P and R of $K \cap (Z \times [a, b])$ such that $K \cap (Z \times [a, b]) = P \cup R$, $K \cap (Z \times \{a\}) \subset P$ and $K \cap (Z \times \{b\}) \subset R$. Let

$$P' = P \cup \left[\pi_{[0,1]}^{-1}([0, a]) \cap K\right] \text{ and } R' = R \cup \left[\pi_{[0,1]}^{-1}([b, 1]) \cap K\right],$$

where $\pi_{[0,1]} : Z \times [0, 1] \twoheadrightarrow [0, 1]$ is the projection map. Observe that this implies that $K = P' \cup R'$, which is a contradiction, because P' and R' are nonempty, closed and disjoint subsets of K, and K is connected. Thus, there exists an irreducible subcontinuum L of $K \cap (Z \times [a, b])$ between $K \cap (Z \times \{a\})$ and $K \cap (Z \times \{b\})$. Let

$\pi_Z \colon Z \times [0, 1] \twoheadrightarrow Z$ be the projection map. Note that $f(\pi_Z(L))$ is a subcontinuum of $[a, b]$ such that $\{a, b\} \subset f(\pi_Z(L))$. Hence, $f(\pi_Z(L)) = [a, b]$. Therefore, f is weakly confluent.

Q.E.D.

2.4.25 Theorem *Let X be a chainable continuum and let Z be a continuum. If $f \colon Z \twoheadrightarrow X$ is a surjective map, then f is weakly confluent.*

Proof Since X is chainable, by Theorem 2.4.22, X is homeomorphic to $\varprojlim\{[0, 1], g_n^{n+1}\}$, where each g_n^{n+1} is a surjective map. We assume that $X = \varprojlim\{[0, 1], g_n^{n+1}\}$.

Let Q be a subcontinuum of X. Observe that, by Proposition 2.1.20, $Q = \varprojlim\{g_n(Q), g_n^{n+1}|_{g_{n+1}(Q)}\}$, where the bonding maps are surjective. Then, by Lemma 2.4.24, each of the maps $g_n \circ f$ is weakly confluent. Hence, since $g_n(Q)$ is a subcontinuum of $[0, 1]$, for each $n \in \mathbb{N}$, there exists a subcontinuum K_n of Z such that $g_n \circ f(K_n) = g_n(Q)$. Since the hyperspace of subcontinua, $\mathcal{C}(Z)$, of Z is compact (Theorem 1.8.5), there exists a subsequence $\{K_{n_\ell}\}_{\ell=1}^\infty$ of $\{K_n\}_{n=1}^\infty$ converging to a subcontinuum K of Z. Since, by Theorem 2.1.40, X is homeomorphic to $\varprojlim\{[0, 1], g_{n_\ell}^{n_\ell+1}\}$, without loss of generality, we assume that K is the limit of $\{K_n\}_{n=1}^\infty$.

Let $q \in Q$. For each $n \in \mathbb{N}$, there exists $z_n \in K_n$ such that $g_n \circ f(z_n) = g_n(q)$. Let z be a limit point of the sequence $\{z_n\}_{n=1}^\infty$. We prove that $f(z) = q$. Suppose that $f(z) \neq q$. Then there exists $N \in \mathbb{N}$ such that $g_N \circ f(z) \neq g_N(q)$. Thus, there exists an open subset U of $[0, 1]$ such that $g_N \circ f(z) \in U$ and $g_N(q) \in [0, 1] \setminus U$. Hence, $f^{-1}(g_N^{-1}(U))$ is an open subset of Z containing z. Since z is a limit point of $\{z_n\}_{n=1}^\infty$, there exists $m \in \mathbb{N}$ such that $m > N$ and $z_m \in f^{-1}(g_N^{-1}(U))$. As a consequence of this, we have

$$g_N(q) = g_N^m \circ g_m(q) = g_N^m \circ g_m \circ f(z_m) = g_N \circ f(z_m) \in U,$$

a contradiction. Therefore, $f(z) = q$ and $Q \subset f(K)$.

Next, let $z \in K$. Since $\{K_n\}_{n=1}^\infty$ converges to K, for each $n \in \mathbb{N}$, there exists $z_n \in K_n$ such that the sequence $\{z_n\}_{n=1}^\infty$ converges to z. Let $n \in \mathbb{N}$. Then $g_n \circ f(z_n) \in g_n(Q)$. For every $n \in \mathbb{N}$, let $x_n \in g_n^{-1}(g_n \circ f(z_n)) \cap Q$. Let x be a limit point of the sequence $\{x_n\}_{n=1}^\infty$. We show that $f(z) = x$. Suppose that $f(z) \neq x$. Then there exists $N \in \mathbb{N}$ such that $g_N \circ f(z) \neq g_N(x)$. Hence, there exist two disjoint open subsets V and W of $[0, 1]$ such that $g_N \circ f(z) \in V$ and $g_N(x) \in W$. Note that $f^{-1}(g_N^{-1}(V))$ is an open subset of Z containing z and $g_N^{-1}(W)$ is an open subset of X containing x. Since $\{z_n\}_{n=1}^\infty$ converges to z, there exists $N_1 \in \mathbb{N}$ such that $N_1 \geq N$ and if $n \geq N_1$, then $z_n \in f^{-1}(g_N^{-1}(V))$. Since x is a limit point of $\{x_n\}_{n=1}^\infty$, there exists $m \geq N_1$ such that $x_m \in g_N^{-1}(W)$; i.e., $g_N(x_m) \in W$. Thus, we have that

$$g_N(x_m) = g_N^m \circ g_m(x_m) = g_N^m \circ g_m \circ f(z_m) = g_N \circ f(z_m) \in V,$$

a contradiction, since $V \cap W = \emptyset$. Hence, $f(z) = x$ and $f(K) \subset Q$. Therefore, $f(K) = Q$ and f is weakly confluent.

Q.E.D.

2.5 Circularly Chainable and $\mathcal{P}$-Like Continua

We present basic facts about circularly chainable and $\mathcal{P}$-like continua. For this, we follow [2, 8, 17, 23, 26].

2.5.1 Definition Let X be a compactum. A *circular chain* $\mathcal{U}$ in X is a finite sequence, $U_0, \ldots, U_n$, of subsets of X such that $U_i \cap U_j \neq \emptyset$ if and only if $|i - j| \leq 1$ or $i, j \in \{0, n\}$. Each U_j is called a *link* of $\mathcal{U}$. If each link of $\mathcal{U}$ is open, then $\mathcal{U}$ is called an *open circular chain*. If $\varepsilon > 0$, $\mathcal{U}$ is an open circular chain and the mesh($\mathcal{U}$) $< \varepsilon$, then $\mathcal{U}$ is called a *circular ε-chain*.

2.5.2 Definition A continuum X is *circularly chainable* provided that for each $\varepsilon > 0$, there exists a circular ε-chain covering X.

2.5.3 Example The unit circle S^1 is a circularly chainable continuum.

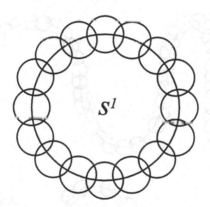

2.5.4 Example Let X be the topologist sine curve (see Example 2.4.5). Join the points $(0, -1)$ and $\left(\frac{2}{\pi}, 1\right)$ with an arc, Y, in $\mathbb{R}^2$. Then $W = X \cup Y$ is called the *Warsaw circle*, and it is a circularly chainable continuum.

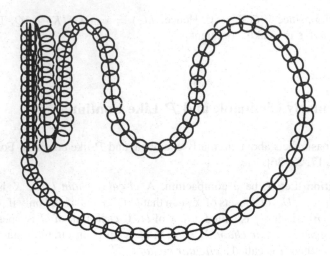

2.5.5 Example The *double Warsaw circle* consists of two copies of the topologist sine curve joined by two arcs (see picture below). It is also a circularly chainable continuum.

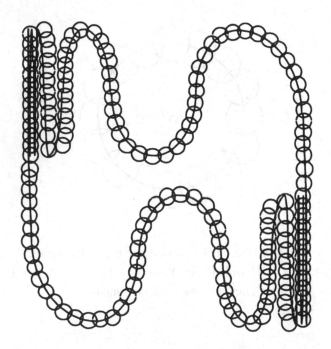

2.5.6 Example Some chainable continua are circularly chainable. For instance, let X be either $\mathcal{K}$, the Knaster continuum (Example 2.4.7), or $\mathcal{M}$ (Example 2.4.9).

Then X is a circularly chainable continuum. If $\varepsilon > 0$ and $\mathcal{U} = \{U_0, \ldots, U_n\}$ is an $\frac{\varepsilon}{2}$-chain, then $\mathcal{C} = \{U_0, \ldots, U_n, U_0 \cup U_n\}$ is a circular ε-chain.

2.5.7 Definition A continuum X is said to be *like a circle* if for each $\varepsilon > 0$, there exists an ε-map from X onto $\mathcal{S}^1$.

The following theorem is analogous to Theorem 2.4.22, for circularly chainable continua, and it is a special case of Theorem 2.5.13:

2.5.8 Theorem *Let X be a continuum. Then the following are equivalent:*

(1) X is a circularly chainable continuum.
(2) X is a circle-like continuum.
(3) X is a like a circle continuum.

The next theorem says that each circle-like continuum may be written as an inverse limit of circles where the bonding maps have nonnegative degree.

2.5.9 Theorem *If X is a circle-like continuum, then there exists an inverse sequence $\{X_n, f_n^{n+1}\}$ such that $\varprojlim \{X_n, f_n^{n+1}\}$ is homeomorphic to X, where each $X_n = \mathcal{S}^1$ and such that the bonding maps satisfy one of the following three conditions:*

(a) for each $n \in \mathbb{N}$, $\deg(f_n^{n+1}) = 0$;
(b) for each $n \in \mathbb{N}$, $\deg(f_n^{n+1}) = 1$;
(c) for each $n \in \mathbb{N}$, $\deg(f_n^{n+1}) > 1$.

Proof Since X is circle-like, there exists an inverse sequence $\{Y_n, r_n^{n+1}\}$ of circles with surjective bonding maps such that $\varprojlim \{Y_n, r_n^{n+1}\}$ is homeomorphic to X.

First, we show that X is homeomorphic to an inverse limit of circles where all the bonding maps have nonnegative degrees. To this end, we construct a new inverse sequence $\{Y_n, g_n^{n+1}\}$ and maps $h_n : Y_n \to Y_n$ such that $h_n \circ r_n^{n+1} = g_n^{n+1} \circ h_{n+1}$ and $\deg(g_n^{n+1}) \geq 0$ for every $n \in \mathbb{N}$.

Let $h_1 = 1_{Y_1}$. If $\deg(r_1^2) \geq 0$, then $g_1^2 = r_1^2$, and $h_2 = 1_{Y_2}$. If $\deg(r_1^2) < 0$, then $h_2 = \aleph$, where $\aleph$ is the antipodal map, and $g_1^2 = r_1^2 \circ \aleph$. Hence, $h_1 \circ r_1^2 = g_1^2 \circ h_2$, and $\deg(g_1^2) \geq 0$.

Next, suppose inductively that we have constructed $h_1, \ldots, h_k$, where either $h_j = 1_{Y_j}$ or $h_j = \aleph$, $j \in \{1, \ldots, k\}$, and $g_1^2, \ldots, g_{k-1}^k$, where $\deg(g_j^{j+1}) \geq 0$, $j \in \{1, \ldots, k-1\}$, such that $h_{j-1} \circ r_{j-1}^j = g_{j-1}^j \circ h_j$ for every $j \in \{2, \ldots, k\}$. To define h_{k+1} and g_k^{k+1} we consider four cases.

If $h_k = 1_{Y_k}$ and $\deg(r_k^{k+1}) \geq 0$, then let $h_{k+1} = 1_{Y_{k+1}}$ and let $g_k^{k+1} = r_k^{k+1}$.

If $h_k = 1_{Y_k}$ and $\deg(r_k^{k+1}) < 0$, then let $h_{k+1} = \aleph$ and let $g_k^{k+1} = r_k^{k+1} \circ \aleph$.

If $h_k = \aleph$ and $\deg(r_k^{k+1}) \geq 0$, then let $h_{k+1} = \aleph$ and let $g_k^{k+1} = \aleph \circ r_k^{k+1} \circ \aleph$.

If $h_k = \aleph$ and $\deg(r_k^{k+1}) < 0$, then let $h_{k+1} = 1_{Y_{k+1}}$ and let $g_k^{k+1} = \aleph \circ r_k^{k+1}$.

In each case, $h_k \circ r_k^{k+1} = g_k^{k+1} \circ h_{k+1}$ and $\deg(g_k^{k+1}) \geq 0$. In this way we finish the inductive step.

So, we have constructed an inverse sequence $\{Y_n, g_n^{n+1}\}$, where $\deg(g_n^{n+1}) \geq 0$ for all $n \in \mathbb{N}$. By Theorems 2.1.48, 2.1.49 and 2.1.50, X is homeomorphic to $\varprojlim\{Y_n, g_n^{n+1}\}$.

Now, suppose there exists a subsequence $\{m(n)\}_{n=1}^{\infty}$ of $\mathbb{N}$ such that each map $g_{m(n)}^{m(n)+1}$ has degree zero. Then, by Lemma 1.3.39, $\deg\left(g_{m(n)}^{m(n+1)}\right) = 0$ for each $n \in \mathbb{N}$. Thus, each bonding map of the inverse sequence $\left\{Y_{m(n)}, g_{m(n)}^{m(n+1)}\right\}$ has degree zero and, by Theorem 2.1.40, $\varprojlim\left\{Y_{m(n)}, g_{m(n)}^{m(n+1)}\right\}$ is homeomorphic to X. Therefore, taking $X_n = Y_{m(n)}$ and $f_n^{n+1} = g_{m(n)}^{m(n+1)}$, $\{X_n, f_n^{n+1}\}$ is an inverse sequence of circles with surjective bonding maps with degree zero, whose inverse limit is homeomorphic to X.

Next, suppose none of the bonding maps g_n^{n+1} has degree zero. Hence, we assume that for each $n \in \mathbb{N}$, $\deg(g_n^{n+1}) > 0$.

If only finitely many bonding maps have degree greater than one, then taking a subsequence, by Theorem 2.1.40, we assume that all the bonding maps have degree one. Thus, X is homeomorphic to the inverse limit of such subsequence.

Finally, suppose there exists a subsequence $\{m(n)\}_{n=1}^{\infty}$ of $\mathbb{N}$ such that each map $g_{m(n)}^{m(n)+1}$ has degree greater than one. Then, by Lemma 1.3.39, $\deg\left(g_{m(n)}^{m(n+1)}\right) > 1$ for each $n \in \mathbb{N}$. Thus, each bonding map of the inverse sequence $\left\{Y_{m(n)}, g_{m(n)}^{m(n+1)}\right\}$ has degree greater than one and, by Theorem 2.1.40, $\varprojlim\left\{Y_{m(n)}, g_{m(n)}^{m(n+1)}\right\}$ is homeomorphic to X. Therefore, taking $X_n = Y_{m(n)}$ and $f_n^{n+1} = g_{m(n)}^{m(n+1)}$, $\{X_n, f_n^{n+1}\}$ is an inverse sequence of circles with surjective bonding maps with degree greater than one, whose inverse limit is homeomorphic to X.

Q.E.D.

The following theorem tells us that each circle-like continuum, expressed as an inverse sequence of circles with bonding maps having degree zero, is arc-like too.

2.5.10 Theorem *Let $\{X_n, f_n^{n+1}\}$ be an inverse sequence of circles with surjective bonding maps having degree zero. If $X_\infty = \varprojlim\{X_n, f_n^{n+1}\}$, then X_∞ is an arc-like continuum.*

Proof For each $n \in \mathbb{N}$, let $p_n \colon \mathbb{R} \to \mathcal{S}^1$ be given by $p_n(t) = \exp(2\pi t)$. Since $\deg(f_n^{n+1}) = 0$ for every $n \in \mathbb{N}$, each map f_n^{n+1} is homotopic to a constant map (Remark 1.3.38). Hence, by Theorem 1.3.40, there exists a map $h_{n+1} \colon X_{n+1} \to \mathbb{R}$ such that $h_{n+1}((1, 0)) = 0$ and $f_n^{n+1} = p_n \circ h_{n+1}$ for every $n \in \mathbb{N}$.

For each $n \in \mathbb{N}$, let $Y_n = h_{n+1}(X_{n+1})$. Note that each Y_n is an interval of length at least one (the bonding maps f_n^{n+1} are surjective). For each $n \in \mathbb{N}$, let $g_n^{n+1} \colon Y_{n+1} \twoheadrightarrow Y_n$ be given by $g_n^{n+1} = h_{n+1} \circ (p_n|_{Y_{n+1}})$. Note that g_n^{n+1} is surjective since it is the composition of surjective maps. Hence, $\{Y_n, g_n^{n+1}\}$ is an inverse sequence of arcs with surjective bonding maps whose inverse limit, Y_∞, is homeomorphic to X_∞ (Theorem 2.1.51). Therefore, X_∞ is an arc-like continuum.

Q.E.D.

2.5.11 Remark It is known that circle-like continua expressed as an inverse limit of circles with bonding maps having degree one are planar continua which separate the plane. If the degree of the bonding maps are greater than one, then such continua are not planar [2].

Next, we present a generalization of arc-like and circle-like continua.

2.5.12 Definition Let $\mathcal{P}$ be a collection of continua. If the continuum X is homeomorphic to an inverse limit of elements of $\mathcal{P}$ with surjective bonding maps, then X is said to be $\mathcal{P}$-like. If $\mathcal{P}$ consists of just one element P, then X is said to be P-like.

A proof of the following theorem may be found in [17]. Note that it generalizes Theorems 2.4.22 and 2.5.8:

2.5.13 Theorem *Let $\mathcal{P}$ be a collection of polyhedra, and let X be a continuum. Then the following are equivalent:*

(a) *For each $\varepsilon > 0$, there exists a finite open cover $\mathcal{U}$ of X such that* mesh$(\mathcal{U}) < \varepsilon$ *and the polyhedron, $\mathcal{N}^*(\mathcal{U})$, associated with the nerve of $\mathcal{U}$, is homeomorphic to an element of $\mathcal{P}$;*
(b) *X is a $\mathcal{P}$-like continuum; and*
(c) *For each $\varepsilon > 0$, there exists an ε-map from X onto an element of $\mathcal{P}$.*

2.5.14 Corollary *If X is a continuum, then there exists a countable family $\mathcal{P}$ of polyhedra such that X is a $\mathcal{P}$-like continuum.*

Proof For each $n \in \mathbb{N}$, let $\mathcal{U}_n$ be a finite open cover of X such that mesh$(\mathcal{U}_n) < \frac{1}{n}$. Let $\mathcal{P} = \{\mathcal{N}^*(\mathcal{U}_n)\}_{n=1}^{\infty}$. Then, by Theorem 2.5.13, X is a $\mathcal{P}$-like continuum.

Q.E.D.

2.5.15 Lemma *Let $\mathcal{P}$ and $\mathcal{R}$ be two families of continua. If X is a $\mathcal{P}$-like continuum and each element of $\mathcal{P}$ is an $\mathcal{R}$-like continuum, then X is an $\mathcal{R}$-like continuum.*

Proof Let $\varepsilon > 0$. By Theorem 2.5.13, there exist an element $P \in \mathcal{P}$ and an ε-map $f \colon X \twoheadrightarrow P$. By Lemma 2.4.20, there exists $\delta > 0$ such that if U is a subset of P and $\operatorname{diam}(U) < \delta$, then $\operatorname{diam}(f^{-1}(U)) < \varepsilon$. Since P is an $\mathcal{R}$-like continuum, there exist an element $R \in \mathcal{R}$ and a δ-map $g \colon P \twoheadrightarrow R$. Hence, $g \circ f \colon X \twoheadrightarrow R$ is an ε-map from X onto R. Since the ε is arbitrary, by Theorem 2.5.13, X is $\mathcal{R}$-like.

Q.E.D.

2.5.16 Theorem *Let $T = [-1, 1] \times \{0\} \cup \{0\} \times [-1, 0] \subset \mathbb{R}^2$. Then $[0, 1]$ is T-like.*

Proof By Theorem 2.5.13, it suffices to show that for each $\varepsilon > 0$, there exists an ε-map from $[0, 1]$ onto T.

Let $\varepsilon > 0$. Let $f : [0, 1] \twoheadrightarrow T$ be given by

$$f(t) = \begin{cases} \left(0, \dfrac{t + \varepsilon - 1}{1 - \varepsilon}\right), & \text{if } t \in [0, 1 - \varepsilon]; \\[3mm] \left(-\dfrac{2}{\varepsilon}(t + \varepsilon - 1), 0\right), & \text{if } t \in \left[1 - \varepsilon, 1 - \dfrac{\varepsilon}{2}\right]; \\[3mm] \left(\dfrac{1}{\varepsilon}(\varepsilon + 4(t - 1)), 0\right), & \text{if } t \in \left[1 - \dfrac{\varepsilon}{2}, 1\right]. \end{cases}$$

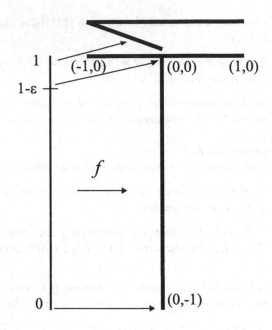

Then f is an ε-map.

Q.E.D.

2.5.17 Corollary *Let* $T = [-1, 1] \times \{0\} \cup \{0\} \times [-1, 0] \subset \mathbb{R}^2$. *If* X *is an arc-like continuum, then* X *is* T-*like.*

Proof It follows easily from Theorem 2.5.16 and Lemma 2.5.15.

Q.E.D.

The following theorem is a slight modification of the Hahn–Mazurkiewicz Theorem:

2.5.18 Theorem *Let* X *be a locally connected continuum. If* x *and* y *are two points of* X, *then there exists a surjective map* $f : [0, 1] \twoheadrightarrow X$ *such that* $f(0) = x$ *and* $f(y) = 1$.

Proof Since X is locally connected, by Hahn–Mazurkiewicz Theorem [23, 8.14], there exists a surjective map $g\colon [0, 1] \twoheadrightarrow X$. Let s_x and s_y be points of $[0, 1]$ such that $g(s_x) = x$ and $g(s_y) = y$. Let $h\colon [0, 1] \to [0, 1]$ be given by

$$
h(t) = \begin{cases}
s_x - 3s_x t, & \text{if } t \in \left[0, \dfrac{1}{3}\right]; \\[2mm]
3t - 1, & \text{if } t \in \left[\dfrac{1}{3}, \dfrac{2}{3}\right]; \\[2mm]
3(s_y - 1)t - 2s_y + 3, & \text{if } t \in \left[\dfrac{2}{3}, 1\right].
\end{cases}
$$

Then h is a surjective map such that $h(0) = s_x$ and $h(1) = s_y$. Hence, $f = g \circ h$ is a surjective map from $[0, 1]$ onto X such that $f(0) = x$ and $f(1) = y$.

Q.E.D.

The proof of the following theorem is due to Ray L. Russo.

2.5.19 Theorem *Let $n \in \mathbb{N}$. If $\mathcal{B} = \{x \in \mathbb{R}^n \mid ||x|| \leq 1\}$, then each arc-like continuum is $\mathcal{B}$-like.*

Proof By Lemma 2.5.15, it suffices to show that $[0, 1]$ is $\mathcal{B}$-like. Let $\varepsilon > 0$. Let $m \in \mathbb{N}$ be such that $\frac{1}{m} < \frac{\varepsilon}{2}$. For each $k \in \{1, \ldots, m\}$, let

$$
E_k = \left\{x \in \mathbb{R}^n \mid 1 - \frac{k}{m} \leq ||x|| \leq 1 - \frac{k-1}{m}\right\}.
$$

Note that E_m is the closed ball of radius $\frac{1}{m}$ and E_k is an annular region, $k \in \{1, \ldots, m - 1\}$. Note, also, that $\mathcal{B} = \bigcup_{k=1}^{m} E_k$.

Divide $[0, 1]$ into m equal subintervals of length $\frac{1}{m}$. By Theorem 2.5.18, there exists a surjective map $g_1\colon \left[0, \frac{1}{m}\right] \twoheadrightarrow E_1$ such that $g_1\left(\frac{1}{m}\right) \in E_1 \cap E_2$. By the same theorem, there exists a surjective map $g_2\colon \left[\frac{1}{m}, \frac{2}{m}\right] \twoheadrightarrow E_2$ such that $g_2\left(\frac{1}{m}\right) = g_1\left(\frac{1}{m}\right)$ and $g_2\left(\frac{2}{m}\right) \in E_2 \cap E_3$. In general, we have a surjective map $g_k\colon \left[\frac{k-1}{m}, \frac{k}{m}\right] \twoheadrightarrow E_k$ such that $g_{k-1}\left(\frac{k-1}{m}\right) = g_k\left(\frac{k-1}{m}\right)$ for each $k \in \{2, \ldots, m\}$.

Let $f\colon [0, 1] \twoheadrightarrow \mathcal{B}$ be given by $f(t) = g_k(t)$ if $t \in \left[\frac{k-1}{m}, \frac{k}{m}\right]$. Then f is an ε-map. Hence, by Theorem 2.5.13, $[0, 1]$ is $\mathcal{B}$-like.

Q.E.D.

2.6   Universal and AH-Essential Maps

We study universal maps and AH-essential maps. We present some of their properties. In particular, we show that these maps are the same, when the range space is an n-cell. To this end, we follow [10, 11, 18–20, 24, 27–30].

We begin this section with the following Theorem:

2.6.1 Theorem *Let* $\{X_n, f_n^{n+1}\}$ *be an inverse sequence of continua with the fixed point property, and surjective bonding maps. For each* $n \in \mathbb{N}$, *let* $h_n \colon X_n \to X_n$ *be a map such that* $f_n^{n+1} \circ h_{n+1} = h_n \circ f_n^{n+1}$. *If* $X_\infty = \varprojlim\{X_n, f_n^{n+1}\}$, *then* $h_\infty = \varprojlim\{h_n\}$ *has a fixed point.*

Proof Let $\varepsilon > 0$. Let $m \in \mathbb{N}$ be such that $\frac{1}{2^m} < \varepsilon$.

Since each X_m has the fixed point property, there exists $x_m \in X_m$ such that $h_m(x_m) = x_m$. Since all the bonding maps are surjective, by Remark 2.1.6, the projection maps are surjective too. Hence, there exists a point $z = (z_n)_{n=1}^\infty \in X_\infty$ such that $f_m(z) = z_m = x_m$.

Since $f_n^{n+1} \circ h_{n+1} = h_n \circ f_n^{n+1}$ for each $n \in \mathbb{N}$,

$$h_\infty(z) = (z_1, z_2, \ldots, z_{m-1}, x_m, h_{m+1}(z_{m+1}), \ldots).$$

Hence,

$$\rho\left(h_\infty(z), z\right) = \sum_{n=m+1}^{\infty} \frac{1}{2^n} d(h_n(z_n), z_n) \leq \sum_{n=m+1}^{\infty} \frac{1}{2^n} = \frac{1}{2^m} < \varepsilon.$$

Therefore, by Lemma 2.4.17, h_∞ has a fixed point.

Q.E.D.

2.6.2 Remark Let us observe that Theorem 2.6.1 does not imply that the inverse limit of continua with the fixed point property has the fixed point property; compare with Theorem 2.6.14.

2.6.3 Definition A map $f \colon X \to Y$ between metric spaces is called *universal* provided that for any other map $g \colon X \to Y$, there exists a point $x \in X$ such that $f(x) = g(x)$.

As an easy consequence of the definition of a universal map we have the following two propositions:

2.6.4 Proposition *Let* $f \colon X \to Y$ *be a universal map between metric spaces. Then the following hold:*

(1) f is surjective.
(2) Y has the fixed point property.
(3) If X_0 is a subspace of X and $f|_{X_0} \colon X_0 \to Y$ is universal, then f is universal.

(4) If $g: Y \to Z$ is a map between metric spaces and $g \circ f$ is universal, then g is universal.

2.6.5 Proposition A metric space X has the fixed point property if and only if the identity map, 1_X, is universal.

2.6.6 Lemma Let X and Y be metric spaces, and let $f: X \twoheadrightarrow Y$ be a universal map. If $g: W \twoheadrightarrow X$ and $h: Y \twoheadrightarrow Z$ are homeomorphisms, then $f \circ g$ and $h \circ f$ are universal.

Proof We show $f \circ g$ is universal. The proof for $h \circ f$ is similar.

Let $\ell: W \to Y$ be a map. Note that $\ell \circ g^{-1}: X \to Y$. Since f is universal, there exists $x \in X$ such that $(\ell \circ g^{-1})(x) = f(x)$. Hence, $\ell(g^{-1}(x)) = (f \circ g)(g^{-1}(x))$. Therefore, $f \circ g$ is universal.

Q.E.D.

2.6.7 Proposition For each $n \in \mathbb{N}$, let $f_n: X \to Y$ be a map between compacta. Suppose the sequence $\{f_n\}_{n=1}^{\infty}$ converges uniformly to a map $f: X \to Y$. If each f_n is universal, then f is universal.

Proof Let $g: X \to Y$ be a map. Since for each $n \in \mathbb{N}$, f_n is universal, there exists $x_n \in X$ such that $f_n(x_n) = g(x_n)$ for every $n \in \mathbb{N}$. Since X is a compactum, the sequence $\{x_n\}_{n=1}^{\infty}$ has a convergent subsequence $\{x_{n_k}\}_{k=1}^{\infty}$. Let x be the limit point of $\{x_{n_k}\}_{k=1}^{\infty}$. Then $f(x) = g(x)$. Therefore, f is universal.

Q.E.D.

2.6.8 Theorem Let X be a continuum. If $f: X \twoheadrightarrow [0, 1]$ is a surjective map, then f is universal.

Proof Let $g: X \to [0, 1]$ be a map. Note that the sets

$$\{x \in X \mid f(x) \le g(x)\} \text{ and } \{x \in X \mid f(x) \ge g(x)\}$$

are nonempty closed subsets of X whose union is X. Hence, these sets have a point x in common. Thus, $f(x) = g(x)$. Therefore, f is universal.

Q.E.D.

2.6.9 Theorem Let $\{Y_n, g_n^{n+1}\}$ be an inverse sequence of compacta whose inverse limit is Y_∞, and let X be a compactum. For each $n \in \mathbb{N}$, let $h_n: X \to Y_n$ be a map such that $g_n^{n+1} \circ h_{n+1} = h_n$. If each h_n is universal, then the induced map $h_\infty = \varprojlim\{h_n\}$ is universal.

Proof Let $k\colon X \to Y_\infty$ be a map. Since each h_n is universal, for each $n \in \mathbb{N}$, there exists a point $x_n \in X$ such that $(g_n \circ k)(x_n) = h_n(x_n)$. Since $h_n(x_n) = (g_n \circ h_\infty)(x_n)$,

$$\rho(h_\infty(x_n), k(x_n)) = \sum_{\ell=n+1}^{\infty} \frac{1}{2^\ell} d_\ell((g_\ell \circ h_\infty)(x_n), (g_\ell \circ k)(x_n))$$

$$\leq \sum_{\ell=n+1}^{\infty} \frac{1}{2^\ell} = \frac{1}{2^n}$$

for every $n \in \mathbb{N}$.

Since X is a compactum, without loss of generality, we assume that the sequence $\{x_n\}_{n=1}^{\infty}$ converges to a point $x \in X$. Hence, $\rho(h_\infty(x), k(x)) = \lim_{n\to\infty} \rho(h_\infty(x_n), k(x_n)) \leq \lim_{n\to\infty} \frac{1}{2^n} = 0$. Thus, $h_\infty(x) = k(x)$. Therefore, h_∞ is universal.

Q.E.D.

2.6.10 Corollary *Let $\{Y_n, g_n^{n+1}\}$ be an inverse sequence of arcs, with surjective bonding maps, whose inverse limit is Y_∞. If X is a continuum and $f\colon X \twoheadrightarrow Y_\infty$ is a surjective map, then f is universal.*

Proof Note that $f = \varprojlim\{g_n \circ f\}$. Since each Y_n is an arc, by Theorem 2.6.8, each $g_n \circ f$ is universal. Hence, by Theorem 2.6.9, f is universal.

Q.E.D.

2.6.11 Corollary *If X is an arc-like continuum, then X has the fixed point property.*

2.6.12 Definition Let Y be a metric space. We say that Y is an *absolute retract* (*absolute neighborhood retract*), provided that for any metric space X, and for any closed subset A of X, every map $f\colon A \to Y$ can be extended over Y (over a neighborhood (depending on f) of A in X).

2.6.13 Remark The concepts defined in Definition 2.6.12 are *absolute extensor* and *absolute neighborhood extensor*. It is known that these concepts are equivalent to the original definition of an absolute retract and an absolute neighborhood retract [20, 1.5.2].

The following theorem gives us a sufficient condition to obtain an inverse limit with the fixed point property.

2.6.14 Theorem *Let $\{X_n, f_n^{n+1}\}$ be an inverse sequence of compact absolute neighborhood retracts, where all the bonding maps $f_n^m\colon X_m \twoheadrightarrow X_n$ are universal. Then $X_\infty = \varprojlim\{X_n, f_n^{n+1}\}$ has the fixed point property.*

Proof First, we prove each projection map $f_n \colon X_\infty \twoheadrightarrow X_n$ is universal. To this end, let $n \in \mathbb{N}$. Let $(a_\ell)_{\ell=1}^\infty \in \prod_{\ell=1}^\infty X_\ell$. For each $m \in \mathbb{N}$, let $i_m \colon X_m \to \prod_{\ell=1}^\infty X_\ell$ be given by

$$i_m(x) = (f_1^m(x), f_2^m(x), \ldots, f_{m-1}^m(x), x, a_{m+1}, a_{m+2}, \ldots).$$

Note that $i_m \colon X_m \twoheadrightarrow i_m(X_m) \subset S_m$ (Definition 2.1.7) is a homeomorphism. Also note that given $z \in X_m$, $i_m(z)$ is not necessarily an element of X_∞.

Let $k \colon X_\infty \to X_n$ be a map. We show there exists $x = (x_\ell)_{\ell=1}^\infty \in X_\infty$ such that $k(x) = f_n(x)$. Since X_n is an absolute neighborhood retract, there exist an open set G of $\prod_{\ell=1}^\infty X_\ell$ containing X_∞ and a map $\hat{k} \colon G \to X_n$ extending k.

Since $X_\infty = \bigcap_{m=1}^\infty S_m$ (Proposition 2.1.8), by Lemma 1.6.7, there exists $M \in \mathbb{N}$ such that $i_m(X_m) \subset S_m \subset G$ for each $m \geq M$. Without loss of generality, we assume that $M \geq n$.

Let $m \geq M$. Then, by Lemma 2.6.6, $f_n^m \circ i_m^{-1}$ is universal. Hence, since $i_m(X_m) \subset G$, there exists $x_m \in X_m$ such that $f_n^m \circ i_m^{-1}(i_m(x_m)) = f_n^m(x_m) = \hat{k} \circ i_m(x_m)$. Note that this is true for each $m \geq M$. Since $\prod_{\ell=1}^\infty X_\ell$ is compact, without loss of generality, we assume that the sequence $\{i_m(x_m)\}_{m=M}^\infty$ converges to a point $x' = (x_\ell')_{\ell=1}^\infty \in X_\infty$.

Note that the bonding maps are surjective, since they are universal (Proposition 2.6.4 (1)). Hence, the projection maps are surjective (Remark 2.1.6). Thus, for each $m \geq M$, there exists $z^m = (z_\ell^m)_{\ell=1}^\infty \in X_\infty$ such that $f_m(z^m) = x_m$. Note that $\lim_{m \to \infty} z^m = x'$. In consequence:

$$k(x') = \hat{k}(x') = \hat{k}(\lim_{m \to \infty} i_m(x_m)) = \lim_{m \to \infty} \hat{k} \circ i_m(x_m) = \lim_{m \to \infty} f_n^m(x_m) =$$

$$\lim_{m \to \infty} f_n^m f_m(z^m) = \lim_{m \to \infty} f_n(z^m) = f_n\left(\lim_{m \to \infty} z^m\right) = f_n(x').$$

Therefore, f_n is universal.

Now, observe that $\lim_{\leftarrow}\{f_n\} = 1_{X_\infty}$. Hence, by Theorem 2.6.9, 1_{X_∞} is universal. Therefore, by Proposition 2.6.5, X_∞ has the fixed point property.

Q.E.D.

A proof of the following theorem may be found in [20, 3.6.11].

2.6.15 Theorem *Each polyhedron is an absolute neighborhood retract.*

2.6.16 Corollary *Let $\{X_n, f_n^{n+1}\}$ be an inverse sequence of polyhedra, where all the bonding maps $f_n^m \colon X_m \twoheadrightarrow X_n$ are universal. Then $X_\infty = \lim_{\leftarrow}\{X_n, f_n^{n+1}\}$ has the fixed point property.*

2.6.17 Definition Let X be a metric space and let A be a subset of X. We say that A is a *retract* of X if there exists a map $r \colon X \twoheadrightarrow A$ such that $r(a) = a$ for each $a \in A$. The map r is called a *retraction*.

2.6.18 Corollary *Let $\{X_n, f_n^{n+1}\}$ be an inverse sequence of absolute neighborhood retracts with the fixed point property. If each bonding map is a retraction, then $\varprojlim\{X_n, f_n^{n+1}\}$ has the fixed point property.*

Proof By Theorem 2.6.14, it suffices to show that each bonding map $f_n^m : X_m \to X_n$ is universal.

Since each bonding map f_n^m is a retraction, $f_n^m|_{X_n} : X_n \to X_n$ is the identity map 1_{X_n}. Since every X_n has the fixed point property, by Proposition 2.6.5, 1_{X_n} is universal. Hence, $f_n^m|_{X_n} = 1_{X_n}$ is universal. Therefore, by (3) of Proposition 2.6.4, f_n^m is universal.

Q.E.D.

From now on, we identify the manifold boundary, $\partial([0,1]^n)$, of the n-cell $[0,1]^n$ with the unit $(n-1)$-dimensional sphere S^{n-1}.

2.6.19 Theorem *Let X be a separable metric space such that every map $g : X \to S^{n-1}$ is homotopic to a constant map. Then a map $f : X \twoheadrightarrow [0,1]^n$ is universal if and only if the restriction $f|_{f^{-1}(S^{n-1})} : f^{-1}(S^{n-1}) \twoheadrightarrow S^{n-1}$ is not homotopic to a constant map.*

Proof Let $f : X \twoheadrightarrow [0,1]^n$ be a map and suppose that $f|_{f^{-1}(S^{n-1})}$ is homotopic to a constant map. Hence, by Theorem 1.3.5, there exists a map $f' : X \twoheadrightarrow S^{n-1}$ such that $f'|_{f^{-1}(S^{n-1})} = f|_{f^{-1}(S^{n-1})}$. Let $g : X \to S^{n-1} \subset [0,1]^n$ be given by $g(x) = -f'(x)$. Then $f(x) \neq g(x)$ for any $x \in X$. Therefore, f is not universal.

Next, suppose $f : X \twoheadrightarrow [0,1]^n$ is not universal. Then there exists a map $g : X \to [0,1]^n$ such that $f(x) \neq g(x)$ for any $x \in X$. Let $f' : X \twoheadrightarrow S^{n-1}$ be the map satisfying the following equation:

$$\|f'(x) - f(x)\| + \|f(x) - g(x)\| = \|f'(x) - g(x)\|.$$

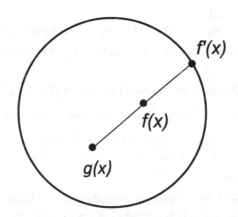

Then f' is map such that $f'|_{f^{-1}(S^{n-1})} = f|_{f^{-1}(S^{n-1})}$. Note that, by hypothesis, f' is homotopic to a constant map.

Q.E.D.

2.6.20 Definition Let $n = \{n_k\}_{k=1}^{\infty}$ be a sequence of positive integers. For each $k \in \mathbb{N}$, let $f_k^{k+1} : S^1 \twoheadrightarrow S^1$ be given by $f_k^{k+1}(z) = z^{n_k}$. The *n-solenoid*, denoted by Σ_n, is $\varprojlim\{X_k, f_k^{k+1}\}$, where $X_k = S^1$ for every $k \in \mathbb{N}$. Note that, by Theorem 2.1.19, Σ_n is an indecomposable continuum.

2.6.21 Theorem *Let* $n = \{n_k\}_{k=1}^{\infty}$ *be a sequence of positive integers. Then the cone over the n-solenoid,* Σ_n, *has the fixed point property.*

Proof Let $\Sigma_n = \varprojlim\{X_k, f_k^{k+1}\}$, where every $X_k = S^1$. By Theorem 2.3.2, $K(\Sigma_n)$ is homeomorphic to $\varprojlim\{K(X_k), K(f_k^{k+1})\}$. It is well known that $K(S^1)$ is homeomorphic to $\{x \in \mathbb{R}^2 \mid ||x|| \leq 1\}$ (which is, in turn, homeomorphic to $[0, 1]^2$). Since $K(\Sigma_n)$ is contractible, then every map from $K(\Sigma_n)$ into S^1 is homotopic to a constant map (Theorem 1.3.12).

Observe that for each $k \in \mathbb{N}$, $K(f_k)^{-1}(S^1) = \Sigma_n \times \{0\}$. Hence, each $K(f_k)|_{K(f_k)^{-1}(S^1)} : K(f_k)^{-1}(S^1) \to S^1$ is not homotopic to a constant map. Thus, by Theorem 2.6.19, every $K(f_k)$ is universal. Since the induced map $\varprojlim\{K(f_k)\} = 1_{K(\Sigma_n)}$, by Theorem 2.6.9, $1_{K(\Sigma_n)}$ is universal. Therefore, by Proposition 2.6.5, $K(\Sigma_n)$ has the fixed point property.

Q.E.D.

2.6.22 Remark Let Σ_n be a solenoid. Since James T. Rogers, Jr. showed that the hyperspace of subcontinua, $C(\Sigma_n)$, of Σ_n is homeomorphic to $K(\Sigma_n)$ [27, Theorem 2], we have, by Theorem 2.6.21, that $C(\Sigma_n)$ has the fixed point property. Hence, Σ_n is an example of a continuum without the fixed point property (it is not difficult to show this) such that its cone and its hyperspace of subcontinua have the fixed point property.

2.6.23 Definition Let X be a metric space. A map $f : X \twoheadrightarrow [0, 1]^n$ is said to be *AH-inessential* provided that there exists a map $g : X \to S^{n-1}$ such that $f|_{f^{-1}(S^{n-1})} = g|_{f^{-1}(S^{n-1})}$. The map f is *AH-essential* if it is not *AH*-inessential.

2.6.24 Lemma *Let* X *be a separable metric space. If* $f : X \twoheadrightarrow [0, 1]^n$ *is AH-essential, then* $f|_{f^{-1}(S^{n-1})} : f^{-1}(S^{n-1}) \twoheadrightarrow S^{n-1}$ *is not homotopic to a constant map.*

Proof Suppose $f|_{f^{-1}(S^{n-1})}$ is homotopic to a constant map g'. Then, clearly, g' can be extended to a constant map g defined on X. Hence, by Theorem 1.3.5, there exists a map $\hat{f} : X \to S^{n-1}$ such that $\hat{f}|_{f^{-1}(S^{n-1})} = f|_{f^{-1}(S^{n-1})}$ and $\hat{f}$ is homotopic to g. Hence, f is *AH*-inessential.

Q.E.D.

2.6.25 Corollary *Let X be a compactum. If $f: X \twoheadrightarrow [0, 1]^2$ is AH-essential, then there exists a component K of $f^{-1}(S^1)$ such that $f|_K: K \to S^1$ is not homotopic to a constant map.*

Proof By Lemma 2.6.24, $f|_{f^{-1}(S^1)}$ is not homotopic to a constant map. Hence, by Theorem 1.3.7, there exists a subcontinuum C of $f^{-1}(S^1)$ such that $f|_C$ is not homotopic to a constant map. Let K be the component of $f^{-1}(S^1)$ such that $C \subset K$. Then $f|_K: K \twoheadrightarrow S^1$ is not homotopic to a constant map.

Q.E.D.

2.6.26 Theorem *Let X be a metric space. Then the map $f: X \twoheadrightarrow [0, 1]^n$ is AH-essential if and only if f is universal.*

Proof Suppose f is not universal. Then there exists a map $g: X \to [0, 1]^n$ such that $f(x) \neq g(x)$ for any $x \in X$. Let $f': X \to S^{n-1}$ be the map satisfying the equation

$$\|f'(x) - f(x)\| + \|f(x) - g(x)\| = \|f'(x) - g(x)\|.$$

Then f' is an extension of $f|_{f^{-1}(S^{n-1})}$ to X. Therefore, f is AH-inessential.

Next, suppose f is AH-inessential. Then there exists a map $g: X \to S^{n-1}$ such that $g|_{f^{-1}(S^{n-1})} = f|_{f^{-1}(S^{n-1})}$. Let $h: X \to S^{n-1}$ be given by $h(x) = -g(x)$. Hence, $f(x) \neq h(x)$ for any $x \in X$. Therefore, f is not universal.

Q.E.D.

Now, our goal is to prove Mazurkiewicz's theorem about the existence of indecomposable subcontinua in continua of dimension at least two.

A proof of the following theorem may be found in [24, 18.6].

2.6.27 Theorem *Let X be a separable metric space and let $n \in \mathbb{N}$. Then $dim(X) \geq n$ if and only if there exists an AH-essential map of X to $[0, 1]^n$.*

2.6.28 Lemma *Let X be a continuum, let $f: X \twoheadrightarrow [0, 1]^2$ be an AH-essential map, and let J be a simple closed curve contained in $[0, 1]^2$. If H is the bounded complementary domain (in $\mathbb{R}^2$) of J, then $f|_{f^{-1}(H \cup J)}: f^{-1}(H \cup J) \twoheadrightarrow H \cup J$ is an AH-essential map.*

Proof Since $H \cup J$ is homeomorphic to $[0, 1]^2$, by Theorem 2.6.26, it suffices to show that $f|_{f^{-1}(H \cup J)}: f^{-1}(H \cup J) \to H \cup J$ is universal.

Let $g: f^{-1}(H \cup J) \to H \cup J$ be a map. Since $H \cup J$ is an absolute retract [20, 1.5.5], there exists a map $\hat{g}: X \to H \cup J$ such that $\hat{g}|_{f^{-1}(H \cup J)} = g$. Since f is a universal map (Theorem 2.6.26), there exists $x \in X$ such that $f(x) = \hat{g}(x)$. Note that $f(x) \in H \cup J$. Hence, $x \in f^{-1}(H \cup J)$. Thus, $\hat{g}(x) = g(x)$. Therefore, $f|_{f^{-1}(H \cup J)}$ is a universal map.

Q.E.D.

2.6.29 Corollary *Let* X *be a continuum, and let* $f: X \twoheadrightarrow [0, 1]^2$ *be an AH-essential map. If* J *is a simple closed curve contained in* $[0, 1]^2$, *then there exists a subcontinuum* K *of* X *such that* $f(K) = J$.

Proof Note that, by Lemmas 2.6.28 and 2.6.24, $f|_{f^{-1}(J)}: f^{-1}(J) \twoheadrightarrow J$ is not homotopic to a constant map. Hence, by Corollary 2.6.25, there exists a component K of $f^{-1}(J)$ such that $f|_K: K \twoheadrightarrow J$ is not homotopic to a constant map. In particular, $f(K) = J$.

Q.E.D.

A proof of the following theorem may be found in [18, p. 157].

2.6.30 Theorem *If* C *is a subcontinuum of* $[0, 1]^2$, *then there exists a sequence* $\{J_n\}_{n=1}^{\infty}$ *of simple closed curves such that* $\lim_{n \to \infty} J_n = C$.

2.6.31 Theorem *Let* X *be a continuum. If* $f: X \twoheadrightarrow [0, 1]^2$ *is an AH-essential map, then* f *is weakly confluent.*

Proof Let C be a subcontinuum of $[0, 1]^2$. Then, by Theorem 2.6.30, there exists a sequence $\{J_n\}_{n=1}^{\infty}$ of simple closed curves converging to C. By Corollary 2.6.29, for each $n \in \mathbb{N}$, there exists a subcontinuum K_n of X such that $f(K_n) = J_n$. Since $\mathcal{C}(X)$ is compact (Theorem 1.8.5), without loss of generality, we assume that the sequence $\{K_n\}_{n=1}^{\infty}$ converges to a subcontinuum K of X. Therefore, by Corollary 8.2.3, $f(K) = C$.

Q.E.D.

As a consequence of Theorems 2.6.27 and 2.6.31, we have the following:

2.6.32 Corollary *If* X *is a continuum of dimension greater than one, then there exists a weakly confluent map* $f: X \twoheadrightarrow [0, 1]^2$.

Now, we are ready to show Mazurkiewicz's Theorem.

2.6.33 Theorem *If* X *is a continuum of dimension greater than one, then* X *contains an indecomposable continuum.*

Proof Since the dimension of X is greater than one, by Corollary 2.6.32, there exists a weakly confluent map $f: X \twoheadrightarrow [0, 1]^2$. Let K be an indecomposable subcontinuum (like Knaster's continuum, Example 2.4.7) of $[0, 1]^2$. Since f is weakly confluent, there exists a subcontinuum Y' of X such that $f(Y') = K$. Using Kuratowski–Zorn Lemma, there exists a subcontinuum Y of Y' such that $f(Y) = K$ and for each proper subcontinuum C of Y, $f(C) \neq K$.

We assert that Y is indecomposable. Suppose this is not true. Then there exist two proper subcontinua A and B of Y such that $Y = A \cup B$. Since A and B are proper subcontinua of Y, $f(A) \neq K$ and $f(B) \neq K$. Note that $K = f(Y) = f(A \cup B) = f(A) \cup f(B)$. Hence, K is decomposable, a contradiction. Therefore, Y is an indecomposable subcontinuum of X.

Q.E.D.

2.6.34 Corollary *Each hereditarily decomposable continuum is of dimension one.*

To finish this section, we describe a family of plane continua constructed by Z. Waraszkiewicz, which are known as the Waraszkiewicz spirals [30] and present some applications.

Let $\mathcal{M}$ be the following subset of $\mathbb{R}^2$, defined in polar coordinates (e denotes the real exponential map):

$$\mathcal{M} = \{(1, \theta) \mid \theta \geq 0\} \cup \{(1 + e^{-\theta}, \theta) \mid \theta \geq 0\} \cup \{(1 + e^{-\theta}, -\theta) \mid \theta \geq 0\}.$$

Hence, $\mathcal{M}$ is the union of the unit circle $\mathcal{S}^1$, a ray R^+ spiraling in a counterclockwise direction onto $\mathcal{S}^1$, and a ray R^- spiraling in a clockwise direction onto $\mathcal{S}^1$.

2.6.35 Definition The *Waraszkiewicz spirals* are subcontinua of $\mathcal{M}$. Each spiral is homeomorphic to a compactification of $[0, 1)$ with remainder $\mathcal{S}^1$. To determine a Waraszkiewicz spiral, begin at the point with Cartesian coordinates $(2, 0)$ and follow R^+ or R^-. At any time that the positive x-axis is crossed, feel free to change rays. This procedure determines the uncountable collection of continua called the Waraszkiewicz spirals.

Z. Waraszkiewicz gave a proof of the following result in [30]. Ray L. Russo has improved the proof [29]:

2.6.36 Theorem *No continuum can be mapped onto all Waraszkiewicz spirals.*

2.6.37 Corollary *Given a countable collection of continua, there exists a Waraszkiewicz spiral W such that no member of this collection can be mapped onto W.*

Proof Suppose the result is not true. Then there exists a countable family $\{X_n\}_{n=1}^{\infty}$ of continua such that for each Waraszkiewicz spiral W, there exists $n \in \mathbb{N}$ such that X_n can be mapped onto W.

Without loss of generality, we assume that the members of the family $\{X_n\}_{n=1}^{\infty}$ are pairwise disjoint and $\lim_{n \to \infty} \operatorname{diam}(X_n) = 0$. For each $n \in \mathbb{N}$, let $x_n \in X_n$. Identify all the points of the sequence $\{x_n\}_{n=1}^{\infty}$ to a point ω. In this way we obtain an infinite wedge. We call this quotient space X. Note that, by construction, X is a continuum.

Observe that, for each $n \in \mathbb{N}$, there exists a retraction $r_n \colon X \twoheadrightarrow X_n$ given by

$$r_n(x) = \begin{cases} \omega, & \text{if } x \in X_m \text{ and } n \neq m; \\ x, & \text{if } x \in X_n. \end{cases}$$

Hence, X can be mapped onto all the Waraszkiewicz spirals, a contradiction to Theorem 2.6.36.

$$\text{Q.E.D.}$$

2.6.38 Definition A continuum X is *hereditarily equivalent* provided that X is homeomorphic to each of its nondegenerate subcontinua.

2.6.39 Corollary *Each hereditarily equivalent continuum has dimension one.*

Proof Let X be a continuum of dimension greater than one. By Corollary 2.6.32, there exists a weakly confluent map $f \colon X \twoheadrightarrow [0, 1]^2$. By Theorem 2.6.36, there exists a Waraszkiewicz spiral W such that X cannot be mapped onto W. We assume that $W \subset [0, 1]^2$. Since f is weakly confluent, there exists a subcontinuum Y of X such that $f(Y) = W$. Hence, Y and X are not homeomorphic. Therefore, X is not hereditarily equivalent.

$$\text{Q.E.D.}$$

2.6.40 Corollary *If X is a continuum of dimension greater than one, then X contains uncountably many nonhomeomorphic subcontinua.*

Proof Suppose X has countably many nonhomeomorphic subcontinua $\{X_n\}_{n=1}^{\infty}$. By Corollary 2.6.37, there exists a Waraszkiewicz spiral W such that X_n cannot be mapped onto W for any $n \in \mathbb{N}$. We assume that $W \subset [0, 1]^2$.

Since X is a continuum of dimension greater than one, by Corollary 2.6.32, there exists a weakly confluent map $f \colon X \twoheadrightarrow [0, 1]^2$. Hence, there exists a subcontinuum Y of X such that $f(Y) = W$. This implies that Y is not homeomorphic to X_n for any $n \in \mathbb{N}$, a contradiction. Therefore, X contains uncountably many nonhomeomorphic subcontinua.

Q.E.D.

An argument for the following result may be found in [28, p. 484].

2.6.41 Theorem *For each Waraszkiewicz spiral W, there exists a plane continuum $\widehat{W}$ such that each nondegenerate subcontinuum of $\widehat{W}$ can be mapped onto W.*

2.6.42 Definition A metric space X is *homogeneous* provided that for each pair of points $x, y \in X$, there exists a homeomorphism $h \colon X \twoheadrightarrow X$ such that $h(x) = y$.

We end this chapter showing that homogeneous hereditarily indecomposable continua are one-dimensional.

2.6.43 Theorem *If X is a hereditarily indecomposable continuum of dimension greater than one, then X is not homogeneous.*

Proof Let x be a point of X, and let $[\![x, X]\!]$ be the family of all subcontinua of X containing x, with the topology generated by the Hausdorff metric (Theorem 1.8.3). Note that $[\![x, X]\!]$ is the image of an order arc. Hence, $[\![x, X]\!]$ is homeomorphic to $[0, 1]$. Let $\{X_n\}_{n=1}^{\infty}$ be a countable dense subset of $[\![x, X]\!]$.

By Corollary 2.6.37, there exists a Waraszkiewicz spiral W such that X_n cannot be mapped onto W for any $n \in \mathbb{N}$. By Theorem 2.6.41, there exists a continuum $\widehat{W}$ in $[0, 1]^2$ such that each map from X_n into $\widehat{W}$ is constant for each $n \in \mathbb{N}$.

Suppose X is homogeneous. Since X is of dimension greater than one, by Corollary 2.6.32, there exists a weakly confluent map $f \colon X \twoheadrightarrow [0, 1]^2$. Hence, there exists a subcontinuum Z of X such that $f(Z) = \widehat{W}$. Since X is homogeneous, we assume that $x \in Z$; i.e., $Z \in [\![x, X]\!]$. Since $\{X_n\}_{n=1}^{\infty}$ is dense in $[\![x, X]\!]$, there exists a subsequence $\{X_{n_k}\}_{k=1}^{\infty}$ of $\{X_n\}_{n=1}^{\infty}$ such that $\lim_{k \to \infty} X_{n_k} = Z$ and $X_{n_k} \subset Z$ for every $k \in \mathbb{N}$. Since for each $k \in \mathbb{N}$, $x \in X_{n_k}$, $f(X_{n_k}) = \{f(x)\}$. This implies the contradiction that $f(Z) = \{f(x)\}$. Therefore, X is not homogeneous.

Q.E.D.

References

1. R. D. Anderson and G. Choquet, A Plane Continuum no Two of Whose Nondegenerate Subcontinua are Homeomorphic: An Application of Inverse Limits, Proc. Amer. Math. Soc., 10 (1959), 347–353.
2. R H Bing, Embedding Circle-like Continua in the Plane, Canad. J. Math., 14 (1962), 113–128.
3. C. E. Capel, Inverse Limit Spaces, Duke Math. J. 21 (1954), 233–245.

4. J. J. Charatonik and W. J. Charatonik, On Projection and Limit Mappings of Inverse Systems of Compact Spaces, Topology Appl., 16 (1983), 1–9.
5. C. O. Christenson and W. L. Voxman, *Aspects of Topology*, Monographs and Textbooks in Pure and Applied Math., Vol. 39, Marcel Dekker, New York, Basel, 1977.
6. J. Dugundji, *Topology*, Allyn and Bacon, Inc., Boston, 1966.
7. J. Grispolakis and E. D. Tymchatyn, On Confluent Mappings and Essential Mappings—A Survey, Rocky Mountain J. Math., 11 (1981), 131–153.
8. O. H. Hamilton, A Fixed Point Theorem for Pseudo-arcs and Certain Other Metric Continua, Proc. Amer. Math. Soc., 2 (1951), 173–174.
9. J. G. Hocking and G. S. Young, *Topology*, Dover Publications, Inc., New York, 1988.
10. W. Holsztyński, Universal Mappings and Fixed Point Theorems, Bull. Acad. Polon. Sci. Sér. Sci. Math. Astronom. Phys., 15 (1967), 433–438.
11. W. Holsztyński, A Remark on the Universal Mappings of 1-dimensional Continua, Bull. Acad. Polon. Sci. Sér. Sci. Math. Astronom. Phys., 15 (1967), 547–549.
12. T. W. Hugerford, *Algebra*, Graduate Texts in Mathematics, Vol. 73, Springer-Verlag, New York, Inc., 1987.
13. W. T. Ingram, *Inverse Limits*, Aportaciones Matemáticas, Textos # 15, Sociedad Matemática Mexicana, 2000.
14. W. T. Ingram *An Introduction to Inverse Limits with Set-valued Functions*, Springer Briefs in Mathematics, 2012.
15. W. T. Ingram and William S. Mahavier *Inverse Limits: From Chaos to Continua*, Developments in Mathematics, Vol. 25, Springer, 2012.
16. K. Kuratowski, *Topology*, Vol. II, Academic Press, New York, 1968.
17. S. Mardešić and J. Segal, ε-mappings onto Polyhedra, Trans. Amer. Math. Soc., 109 (1963), 146–164.
18. S. Mazurkiewicz, Sur les Continus Absolutment Indécomposables, Fund. Math., 16 (1930), 151–159.
19. S. Mazurkiewicz, Sur L'existence des Continus Indécomposables, Fund. Math., 25 (1935), 327–328.
20. J. van Mill, *Infinite-Dimensional Topology*, North Holland, Amsterdam, 1989.
21. S. B. Nadler, Jr., Multicoherence Techniques Applied to Inverse Limits, Trans. Amer. Math. Soc., 157 (1971), 227–234.
22. S. B. Nadler, Jr., *Hyperspaces of Sets,* Monographs and Textbooks in Pure and Applied Math., Vol. 49, Marcel Dekker, New York, Basel, 1978. Reprinted in: Aportaciones Matemáticas de la Sociedad Matemática Mexicana, Serie Textos # 33, 2006.
23. S. B. Nadler, Jr., *Continuum Theory: An Introduction,* Monographs and Textbooks in Pure and Applied Math., Vol. 158, Marcel Dekker, New York, Basel, Hong Kong, 1992.
24. S. B. Nadler, Jr., *Dimension Theory: An Introduction with Exercises*, Aportaciones Matemáticas, Textos # 18, Sociedad Matemática Mexicana, 2002.
25. D. R. Reed, Confluent and Related Mappings, Colloq. Math., 29 (1974), 233–239.
26. J. T. Rogers, Jr., The Pseudo-circle is not Homogeneous, Trans. Amer. Math. Soc. 148 (1970), 417–428.
27. J. T. Rogers, Jr., Embedding the Hyperspaces of Circle-like Plane Continua, Proc. Amer. Math. Soc., 29 (1971), 165–168.
28. J. T. Rogers, Jr., Orbits of Higher-Dimensional Hereditarily Indecomposable Continua, Proc. Amer. Math. Soc., 95 (1985), 483–486.
29. R. L. Russo, Universal Continua, Fund. Math., 105 (1979), 41–60.
30. Z. Waraszkiewicz, Sur un Probléme de M. H. Hahn, Fund. Math., 18 (1932), 118–137.

Chapter 3
Jones's Set Function $\mathcal{T}$

We prove basic results about the set function $\mathcal{T}$ defined by F. Burton Jones [26]. We define this function on compacta and then we concentrate on continua. In particular, we present some of the well known properties (such as connectedness im kleinen, local connectedness, semi-local connectedness, etc.) using the set function $\mathcal{T}$. The notion of aposyndesis is the main motivation of Jones to define this function. We study the idempotency of $\mathcal{T}$ on products, cones and suspensions. We present some properties of a continuum assuming the continuity of the set function $\mathcal{T}$ and examples of classes of continua for which $\mathcal{T}$ is continuous. We give three decomposition theorems using $\mathcal{T}$. We also present some applications.

3.1 The Set Function $\mathcal{T}$

The material of this section is based on [2, 3, 5, 12, 14–17, 20, 22, 25, 36, 39, 41, 46, 48, 50].

3.1.1 Definition Given a compactum X, the *power set of X*, denoted by $\mathcal{P}(X)$, is:

$$\mathcal{P}(X) = \{A \mid A \subset X\}.$$

3.1.2 Remark Let X be a metric space. Let us note that if $A \in \mathcal{P}(X)$, then its closure satisfies:

$$Cl(A) = \{x \in X \mid \text{for each open subset } U \text{ of } X \text{ such that}$$

$$x \in U, \text{ we have that } U \cap A \neq \emptyset\}.$$

As we see below (Definition 3.1.3), this property is similar to the definition of the function $\mathcal{T}$.

© Springer International Publishing AG, part of Springer Nature 2018
S. Macías, *Topics on Continua*, https://doi.org/10.1007/978-3-319-90902-8_3

3.1.3 Definition Let X be a compactum. Define

$$\mathcal{T}\colon \mathcal{P}(X) \to \mathcal{P}(X)$$

by

$$\mathcal{T}(A) = \{x \in X \mid \text{for each subcontinuum } W \text{ of } X \text{ such that}$$

$$x \in Int(W), \text{ we have that } W \cap A \neq \emptyset\},$$

for every $A \in \mathcal{P}(X)$. The function $\mathcal{T}$ is called *Jones's set function $\mathcal{T}$*.

3.1.4 Remark In general, when working with the set function $\mathcal{T}$, we usually work with complements. Hence, for any compactum X and any $A \in \mathcal{P}(X)$, we have that:

$$\mathcal{T}(A) = X \setminus \{x \in X \mid \text{there exists a subcontinuum } W \text{ of } X$$

$$\text{such that } x \in Int(W) \subset W \subset X \setminus A\}.$$

3.1.5 Remark Let X be a compactum. If $A \in \mathcal{P}(X)$, then $A \subset \mathcal{T}(A)$, and $\mathcal{T}(A)$ is closed in X. Hence, the range of $\mathcal{T}$ is $2^X \cup \{\emptyset\}$ (see Definition 1.8.1); i.e.:

$$\mathcal{T}\colon \mathcal{P}(X) \to 2^X \cup \{\emptyset\}.$$

The following proposition gives a relation between aposyndesis and $\mathcal{T}$.

3.1.6 Proposition *Let X be a compactum, and let $A \in \mathcal{P}(X)$. If $x \in X \setminus \mathcal{T}(A)$, then X is aposyndetic at x with respect to each point of A.*

Proof Let $x \in X \setminus \mathcal{T}(A)$, and let $a \in A$. Since $x \in X \setminus \mathcal{T}(A)$, by Definition 3.1.3, there exists a subcontinuum W of X such that $x \in Int(W) \subset W \subset X \setminus A$. In particular, $W \subset X \setminus \{a\}$. Therefore, X is aposyndetic at x with respect to a.
                                                                              Q.E.D.

3.1.7 Proposition *Let X be a compactum. If $A, B \in \mathcal{P}(X)$ and $A \subset B$, then $\mathcal{T}(A) \subset \mathcal{T}(B)$.*

Proof Let $x \in X \setminus \mathcal{T}(B)$. Then there exists a subcontinuum W of X such that $x \in Int(W) \subset W \subset X \setminus B$. Since $X \setminus B \subset X \setminus A$, $x \in Int(W) \subset W \subset X \setminus A$. Therefore, $x \in X \setminus \mathcal{T}(A)$.
                                                                              Q.E.D.

3.1.8 Corollary *Let X be a compactum. If $A, B \in \mathcal{P}(X)$, then $\mathcal{T}(A) \cup \mathcal{T}(B) \subset \mathcal{T}(A \cup B)$.*

3.1.9 Remark Let us note that the reverse inclusion of Corollary 3.1.8 is, in general, not true (see Example 3.1.18). It is an open question to characterize continua X such that $\mathcal{T}(A \cup B) = \mathcal{T}(A) \cup \mathcal{T}(B)$ for each $A, B \in \mathcal{P}(X)$.

Let us see a couple of examples.

3.1.10 Example Let X be the Cantor set. Then $\mathcal{T}(\emptyset) = X$. To see this, suppose there exists a point $x \in X \setminus \mathcal{T}(\emptyset)$. Then there exists a subcontinuum W of X such that $x \in Int(W) \subset W \subset X \setminus \emptyset = X$. Since X is the Cantor set, X is totally disconnected and perfect. Hence, no subcontinuum of X has interior. Therefore, W cannot exist. Hence, $\mathcal{T}(\emptyset) = X$. Note that, by Proposition 3.1.7, $\mathcal{T}(A) = X$ for each $A \in \mathcal{P}(X)$.

3.1.11 Example Let $X = \{0\} \cup \left\{ \frac{1}{n} \right\}_{n=1}^{\infty}$. Then $\mathcal{T}(\emptyset) = \{0\}$. This follows from the fact that the only subcontinuum of X with empty interior is $\{0\}$. Hence, by Remark 3.1.5 and Proposition 3.1.7, $\mathcal{T}(A) = \{0\} \cup A$ for each $A \in \mathcal{P}(X)$.

3.1.12 Remark Note that if X is as in Example 3.1.10 or 3.1.11, then $\mathcal{T}|_{2^X} : 2^X \to 2^X$ is continuous, where 2^X has the topology given by the Hausdorff metric. In Sect. 3.3, we discuss the continuity of $\mathcal{T}$ on continua.

The following theorem gives us a characterization of compacta X for which $\mathcal{T}(\emptyset) = \emptyset$.

3.1.13 Theorem *Let X be a compactum. Then $\mathcal{T}(\emptyset) = \emptyset$ if and only if X has only finitely many components.*

Proof Suppose X has only finitely many components. Let $x \in X$, and let C be the component of X such that $x \in C$. Hence, C is a subcontinuum of X and, by Lemma 1.6.2, $x \in Int(C)$. Thus, each point of X is contained in the interior of a proper subcontinuum of X. Therefore, $\mathcal{T}(\emptyset) = \emptyset$.

Now, suppose $\mathcal{T}(\emptyset) = \emptyset$. Then for each point $x \in X$, there exists a subcontinuum W_x of X such that $x \in Int(W_x) \subset W_x \subset X \setminus \emptyset$. Hence, $\{Int(W_x) \mid x \in X\}$ is an open cover of X. Since X is compact, there exist $x_1, \ldots, x_n \in X$ such that $X = \bigcup_{j=1}^{n} Int(W_{x_j}) \subset \bigcup_{j=1}^{n} W_{x_j} \subset X$. Thus, X is the union of finitely many continua. Therefore, X has only finitely many components.

Q.E.D.

3.1.14 Corollary *If X is a continuum, then $\mathcal{T}(\emptyset) = \emptyset$.*

Let us see more examples.

3.1.15 Example Let X be the cone over the Cantor set, with vertex v_X and base B. Since the Cantor set is totally disconnected and perfect, it follows that the subcontinua of X having nonempty interior must contain v_X. Hence, $\mathcal{T}(\{v_X\}) = X$. If $r \in X \setminus \{v_X\}$, then $\mathcal{T}(\{r\})$ is the line segment from r to the base B. Also, $\mathcal{T}(B) = B$.

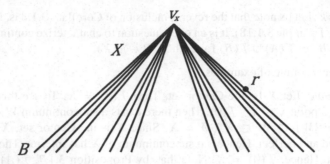

3.1.16 *Example* Let $Y = X \cup X'$, where X is the cone over the Cantor set, with vertex v_X and base B, and X' is another copy of the cone over the Cantor set, with vertex $v_{X'} \in B$. In this case, $\mathcal{T}(\{v_X\}) = X$, $\mathcal{T}(\{v_{X'}\}) = X'$ and $\mathcal{T}(\mathcal{T}(\{v_X\})) = \mathcal{T}^2(\{v_X\}) = Y$. Note that this implies, in general, that the function $\mathcal{T}$ is not idempotent (see Definition 3.1.57).

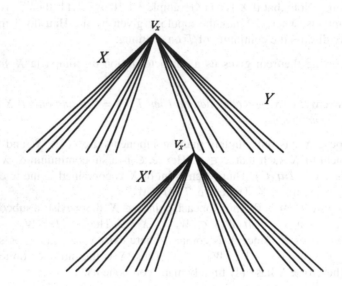

3.1.17 *Example* Let X be the topologist sine curve, see Example 2.4.5. Let $J = X \setminus \left\{ \left(x, \sin\left(\frac{1}{x}\right) \right) \,\middle|\, 0 < x \leq \frac{2}{\pi} \right\}$. Since X is not locally connected at any point of J, if $p \in J$, then $\mathcal{T}(\{p\}) = J$. Also, $\mathcal{T}(\{q\}) = \{q\}$ for each $q \in X \setminus J$.

The following example shows that, in general, $\mathcal{T}(A \cup B) = \mathcal{T}(A) \cup \mathcal{T}(B)$ does not hold.

3.1.18 Example Let X be the suspension over the harmonic sequence $\{0\} \cup \left\{\frac{1}{n}\right\}_{n=1}^{\infty}$, with vertices a and b. Note that for each $x \in X$, $\mathcal{T}(\{x\}) = \{x\}$. In particular, $\mathcal{T}(\{a\}) = \{a\}$ and $\mathcal{T}(\{b\}) = \{b\}$; however $\mathcal{T}(\{a, b\})$ consists of the limit segment from a to b.

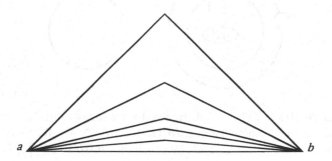

According to E. L. VandenBoss [46, p. 18], the following result is due to H. S. Davis.

3.1.19 Theorem *Let X be a compactum, and let A and B be nonempty closed subsets of X. Then the following are equivalent:*

(1) $\mathcal{T}(A) \cap B = \emptyset$.
(2) There exist two closed subsets M and N of X such that $A \subset Int(M)$, $B \subset Int(N)$ and $\mathcal{T}(M) \cap N = \emptyset$.

Proof Assume (1), we show (2). For each $b \in B$, there exists a subcontinuum W_b of X such that $b \in Int(W_b) \subset W_b \subset X \setminus A$. Since B is compact, there exist $b_1, \ldots, b_n \in B$ such that $B \subset \bigcup_{j=1}^{n} Int(W_{b_j}) \subset \bigcup_{j=1}^{n} W_{b_j} \subset X \setminus A$. Since X is a metric space, there exist two open sets U and V of X such that $A \subset U \subset Cl(U) \subset \left(X \setminus \bigcup_{j=1}^{n} W_{b_j}\right)$ and $B \subset V \subset Cl(V) \subset \bigcup_{j=1}^{n} Int(W_{b_j})$.

Let $M = Cl(U)$ and let $N = Cl(V)$. Then M and N are closed subsets of X such that $A \subset Int(M)$, $B \subset Int(N)$ and $\mathcal{T}(M) \cap N = \emptyset$.

The other implication follows easily from Proposition 3.1.7.

Q.E.D.

The following corollary is an easy consequence of proof of Theorem 3.1.19.

3.1.20 Corollary *Let X be a continuum, and let A be a closed subset of X. If $x \in X \setminus \mathcal{T}(A)$, then there exists an open subset U of X such that $A \subset U$ and $x \in X \setminus \mathcal{T}(Cl(U))$.*

3.1.21 Theorem *Let X be a continuum. If W is a subcontinuum of X, then $\mathcal{T}(W)$ is also a subcontinuum of X.*

Proof By Remark 3.1.5, $\mathcal{T}(W)$ is a closed subset of X.

Suppose $\mathcal{T}(W)$ is not connected. Then there exist two disjoint closed subsets A and B of X such that $\mathcal{T}(W) = A \cup B$. Since W is connected, without loss of generality, we assume that $W \subset A$. Since X is a metric space, there exists an open subset U of X such that $A \subset U$ and $Cl(U) \cap B = \emptyset$.

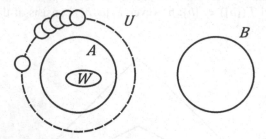

Note that $Bd(U) \cap \mathcal{T}(W) = \emptyset$. Hence, for each $z \in Bd(U)$, there exists a subcontinuum K_z of X such that $z \in Int(K_z) \subset K_z \subset X \setminus W$. Since $Bd(U)$ is compact, there exist $z_1, \ldots, z_n \in Bd(U)$ such that $Bd(U) \subset \bigcup_{j=1}^n Int(K_{z_j}) \subset \bigcup_{j=1}^n K_{z_j}$. Let $V = U \setminus \left(\bigcup_{j=1}^n K_{z_j} \right)$. Let $Y = X \setminus V = (X \setminus U) \cup \left(\bigcup_{j=1}^n K_{z_j} \right)$. By Theorem 1.7.27, Y has only a finite number of components. Observe that $B \subset X \setminus Cl(U) \subset X \setminus U \subset Y$. Hence, $B \subset Int(Y)$. Let $b \in B$, and let C be the component of Y such that $b \in C$. By Lemma 1.6.2, $b \in Int(C)$. By construction, $C \cap W = \emptyset$. Thus, $b \in X \setminus \mathcal{T}(W)$, a contradiction. Therefore, $\mathcal{T}(W)$ is connected.
Q.E.D.

The set function $\mathcal{T}$ may be used to define some of the properties of continua like connectedness im kleinen or local connectedness. In the following theorems we show the equivalence between both types of definitions.

The following theorem gives us a local definition of almost connectedness im kleinen using $\mathcal{T}$.

3.1.22 Theorem *Let X be a continuum, with metric d. If $p \in X$, then X is almost connected im kleinen at p if and only if for each $A \in \mathcal{P}(X)$ such that $p \in Int(\mathcal{T}(A))$, $p \in Cl(A)$.*

Proof Suppose X is almost connected im kleinen at p. Let $A \in \mathcal{P}(X)$ be such that $p \in Int(\mathcal{T}(A))$. Then there exists $N \in \mathbb{N}$ such that $\mathcal{V}_{\frac{1}{n}}^d(p) \subset Int(\mathcal{T}(A))$ for each $n \geq N$. Since X is almost connected im kleinen at p, for each $n \geq N$, there exists a subcontinuum W_n of X such that $Int(W_n) \neq \emptyset$ and $W_n \subset \mathcal{V}_{\frac{1}{n}}^d(p) \subset Cl(\mathcal{V}_{\frac{1}{n}}^d(p)) \subset \mathcal{T}(A)$. Hence, $W_n \cap A \neq \emptyset$ for each $n \geq N$. Let $x_n \in W_n \cap A$ for every $n \geq N$. Note that, by construction, the sequence $\{x_n\}_{n=N}^\infty$ converges to p and $\{x_n\}_{n=N}^\infty \subset A$. Therefore, $p \in Cl(A)$.

Now suppose that $p \in X$ satisfies that for each $A \in \mathcal{P}(X)$ such that $p \in Int(\mathcal{T}(A))$, $p \in Cl(A)$. Let U be an open subset of X such that $p \in U$. Let V an open subset of X such that $p \in V \subset Cl(V) \subset U$. If some component of $Cl(V)$ has nonempty interior, then X is almost connected im kleinen at p. Suppose, then, that all the components of $Cl(V)$ have empty interior. Let $A = Bd(V)$. Then A is a closed subset of X and $p \in X \setminus A$. We show that $V \subset \mathcal{T}(A)$. To this end, suppose there exists $x \in V \setminus \mathcal{T}(A)$. Thus, there exists a subcontinuum W of X such that $x \in Int(W) \subset W \subset X \setminus A$. Since all the components of $Cl(V)$ have empty interior, and W is a subcontinuum with nonempty interior, we have that $W \cap (X \setminus V) \neq \emptyset$ and, of course, $W \cap V \neq \emptyset$. Since W is connected, $W \cap Bd(V) = W \cap A \neq \emptyset$, a contradiction. Hence, $V \subset \mathcal{T}(A)$. Since $p \in V \subset \mathcal{T}(A)$, by hypothesis, $p \in Cl(A) = A$, a contradiction. Therefore, $Cl(V)$ has a component with nonempty interior, and X is almost connected im kleinen at p.

Q.E.D.

The next theorem gives us a global definition of almost connectedness im kleinen using $\mathcal{T}$.

3.1.23 Theorem *A continuum X is almost connected im kleinen if and only if for each closed subset, F, of X, $Int(F) = Int(\mathcal{T}(F))$.*

Proof Suppose X is almost connected im kleinen. Let F be a closed subset of X. Since $F \subset \mathcal{T}(F)$ (Remark 3.1.5), $Int(F) \subset Int(\mathcal{T}(F))$. Let $x \in Int(\mathcal{T}(F))$. Then, by Theorem 3.1.22, $x \in Cl(F) = F$. Hence, $Int(\mathcal{T}(F)) \subset F$. Therefore, $Int(\mathcal{T}(F)) \subset Int(F)$ and $Int(\mathcal{T}(F)) = Int(F)$.

Now, suppose that $Int(F) = Int(\mathcal{T}(F))$ for each closed subset F of X. Let $x \in X$, and let U be an open subset of X such that $x \in U$. Since $Int(X \setminus U) \cap U = \emptyset$, by hypothesis, $Int(\mathcal{T}(X \setminus U)) \cap U = \emptyset$. Hence, there exists $y \in U$ such that $y \in X \setminus \mathcal{T}(X \setminus U)$. Thus, there exists a subcontinuum K of X such that $y \in Int(K) \subset K \subset X \setminus (X \setminus U) = U$. Therefore, X is almost connected im kleinen at x.

Q.E.D.

3.1.24 Theorem *Let X be continuum. If $p \in X$, then X is connected im kleinen at p if and only if for each $A \in \mathcal{P}(X)$ such that $p \in \mathcal{T}(A)$, $p \in Cl(A)$.*

Proof Let $p \in X$. Assume X is connected im kleinen at p. Let $A \in \mathcal{P}(X)$ and suppose $p \in X \setminus Cl(A)$. Hence, $Cl(A)$ is a closed subset of $X \setminus \{p\}$. Since X is connected im kleinen at p, there exists a subcontinuum W of X such that $p \in Int(W) \subset W \subset X \setminus A$. Therefore, $p \in X \setminus \mathcal{T}(A)$.

Now, suppose that $p \in X$ satisfies that for each $A \in \mathcal{P}(X)$ such that $p \in \mathcal{T}(A)$, $p \in Cl(A)$. Let D be a closed subset of X such that $D \subset X \setminus \{p\}$. Since $p \in X \setminus D$, by hypothesis, $p \in X \setminus \mathcal{T}(D)$. Then there exists a subcontinuum K of X such that $p \in Int(K) \subset K \subset X \setminus D$. Therefore, X is connected im kleinen at p.

Q.E.D.

3.1.25 Corollary *Let X be continuum. If $p \in X$, then X is connected im kleinen at p if and only if for each closed subset A of X such that $p \in \mathcal{T}(A)$, $p \in A$.*

The following theorem follows from Definition 1.7.13:

3.1.26 Theorem *Let X be a continuum, and let $p, q \in X$. Then X is semi-aposyndetic at p and q if and only if either $p \in X \setminus \mathcal{T}(\{q\})$ or $q \in X \setminus \mathcal{T}(\{p\})$.*

The next theorem follows from the definition of aposyndesis:

3.1.27 Theorem *Let X be a continuum, and let $p, q \in X$. Then X is aposyndetic at p with respect to q if and only if $p \in X \setminus \mathcal{T}(\{q\})$.*

The following theorem is a global version of Theorem 3.1.27.

3.1.28 Theorem *A continuum X is aposyndetic if and only if $\mathcal{T}(\{p\}) = \{p\}$ for each $p \in X$.*

Proof Suppose X is aposyndetic. Let $p \in X$. Then for each $q \in X \setminus \{p\}$, X is aposyndetic at q with respect to p. Hence, by Remark 3.1.5 and Theorem 3.1.27, $q \in X \setminus \mathcal{T}(\{p\})$. Therefore, $\mathcal{T}(\{p\}) = \{p\}$.

Now, suppose $\mathcal{T}(\{x\}) = \{x\}$ for each $x \in X$. Let $p, q \in X$ such that $p \neq q$. Since $\mathcal{T}(\{q\}) = \{q\}$, $p \in X \setminus \mathcal{T}(\{q\})$. Thus, by Theorem 3.1.27, X is aposyndetic at p with respect to q. Since p and q are arbitrary points of X, X is aposyndetic.

Q.E.D.

3.1.29 Theorem *Let X be a continuum. If $p \in X$, then X is semi-locally connected at p if and only if $\mathcal{T}(\{p\}) = \{p\}$.*

Proof Suppose X is semi-locally connected at p. By Remark 3.1.5, $\{p\} \subset \mathcal{T}(\{p\})$. Let $q \in X \setminus \{p\}$, and let U be an open subset of X such that $p \in U$ and $q \in X \setminus Cl(U)$. Since X is semi-locally connected at p, there exists an open subset V of X such that $p \in V \subset U$ and $X \setminus V$ has only finitely many components. By Lemma 1.6.2, q belongs to the interior of the component of $X \setminus V$ containing q. Hence, $q \in X \setminus \mathcal{T}(\{p\})$. Therefore, $\mathcal{T}(\{p\}) = \{p\}$.

Now, suppose $\mathcal{T}(\{p\}) = \{p\}$. Let U be an open subset of X such that $p \in U$. Since $\mathcal{T}(\{p\}) = \{p\}$, for each $q \in X \setminus U$, there exists a subcontinuum W_q of X such that $q \in Int(W_q) \subset W_q \subset X \setminus \{p\}$. Note that $\{Int(W_q) \mid q \in X \setminus U\}$ is an open cover of $X \setminus U$. Since this set is compact, there exist $q_1, \ldots, q_n \in X \setminus U$ such that $X \setminus U \subset \bigcup_{j=1}^{n} Int(W_{q_j})$. Let $V = X \setminus \left(\bigcup_{j=1}^{n} W_{q_j} \right)$. Then V is an open subset of X such that $p \in V \subset U$ and $X \setminus V$ has only finitely many components. Therefore, X is semi-locally connected at p.

Q.E.D.

As a consequence of Theorems 3.1.28 and 3.1.29, we have:

3.1.30 Corollary *A continuum X is aposyndetic if and only if it is semi-locally connected.*

The next theorem gives us a characterization of local connectedness in terms of the set function $\mathcal{T}$.

3.1.31 Theorem *A continuum X is locally connected if and only if for each closed subset A of X, $\mathcal{T}(A) = A$.*

Proof Suppose X is locally connected. Let A be a closed subset of X. Then $X \setminus A$ is an open subset of X. Let $p \in X \setminus A$. Then there exists an open subset U of X such that $p \in U \subset Cl(U) \subset X \setminus A$. Since X is locally connected, there exists an open connected subset V of X such that $p \in V \subset U$. Hence, $Cl(V)$ is a subcontinuum of X such that $p \in V \subset Cl(V) \subset Cl(U) \subset X \setminus A$. Consequently, $p \in X \setminus \mathcal{T}(A)$. Thus, $\mathcal{T}(A) \subset A$. By Remark 3.1.5, $A \subset \mathcal{T}(A)$. Therefore, $\mathcal{T}(A) = A$.

Now, suppose $\mathcal{T}(A) = A$ for each closed subset of X. Let $p \in X$, and let U be an open subset of X such that $p \in U$. Then $X \setminus U$ is a closed subset of X such that $p \notin X \setminus U$. Since $X \setminus U = \mathcal{T}(X \setminus U)$, there exists a subcontinuum W of X such that $p \in Int(W) \subset W \subset X \setminus (X \setminus U) = U$. Thus, by Theorem 1.7.9, X is connected im kleinen at p. Since p is an arbitrary point of X, by Theorem 1.7.12, X is locally connected.

Q.E.D.

Theorem 3.1.31 may be strengthened as follows:

3.1.32 Theorem *Let X be a continuum. Then X is locally connected if and only if $\mathcal{T}(Y) = Y$ for each subcontinuum Y of X.*

Proof If X is locally connected, by Theorem 3.1.31, $\mathcal{T}(Y) = Y$ for each subcontinuum Y of X.

Now suppose that for every subcontinuum Y of X, $\mathcal{T}(Y) = Y$. We show that X is locally connected. To this end, by Lemma 1.7.11, it is enough to see the components of open subsets of X are open. Let V be an open subset of X. Let $p \in V$. Since $\mathcal{T}(\{p\}) = \{p\}$ and $X \setminus V$ is compact, there exist finitely many subcontinua contained in $X \setminus \{p\}$ whose interiors cover $X \setminus V$. Let $\mathcal{W} = \{W_1, \ldots, W_m\}$ be such a collection of smallest possible cardinality. Let $U = X \setminus \left(\bigcup_{j=1}^{m} W_j \right)$, and let P be the component of $Cl(U)$ such that $p \in P$. We show that $p \in Int(P) \subset P \subset V$. By Theorem 1.7.27, each component of $Cl(U)$ intersects $\bigcup_{j=1}^{m} W_j$. By the minimality of m, we assert that no component of $Cl(U)$ except P can intersect more than one of the W_j's. To see this, let P' be a component of $Cl(U)$, different from P, and suppose that P' intersects both W_j and W_k, where $j, k \in \{1, \ldots, m\}$ and $j \neq k$. Note that, in this case, $P' \cup W_j \cup W_j$ is a continuum. Let $\mathcal{W}' = \{W_1, \ldots, W_m, P' \cup W_j \cup W_k\} \setminus \{W_j, W_k\}$. Then $\mathcal{W}'$ has $m - 1$ elements and $X \setminus V \subset \left(\bigcup_{\substack{l \neq j \\ l \neq k}} W_l \right) \cup (P' \cup W_j \cup W_k)$, a contradiction to the minimality of m.

For each $j \in \{1, \ldots, m\}$, let A_j be the union of all the components of $Cl(U)$ which intersect W_j. Then $A_j \cap A_k \subset P$, for each $j, k \in \{1, \ldots, m\}$ such that $j \neq k$.

Now, since $p \notin \mathcal{T}(W_j) = W_j$, for each $j \in \{1, \ldots, m\}$, there exists a subcontinuum K_j of X such that $p \in Int(K_j) \subset K_j \subset X \setminus W_j$. Then $K_j \cap (A_j \setminus P) = \emptyset$. To see this, suppose there exists $x \in K_j \cap (A_j \setminus P)$. Let L be the component of $K_j \cap Cl(U)$ such that $x \in L$. By Theorem 1.7.27, L intersects the boundary of $K_j \cap Cl(U)$ in K_j. This boundary is contained in $\bigcup_{\ell=1}^{m} W_\ell$. (If $y \in Bd_{K_j}(K_j \cap Cl(U))$, then $y \in Cl(U)$. If $y \in U$, then $y \in K_j \cap U \subset K_j \cap Cl(U)$ and $(K_j \cap U) \cap (K_j \setminus (K_j \cap Cl(U))) = \emptyset$, a contradiction. Therefore, $y \in Cl(U) \setminus U \subset \bigcup_{\ell=1}^{m} W_\ell$.) Hence, there exists $k \in \{1, \ldots, m\} \setminus \{j\}$

such that $L \cap W_k \neq \emptyset$. But $L \subset M$, for some component M of $Cl(U)$, $M \neq P$ and $M \subset A_j$, so that $M \cap W_k = \emptyset$, a contradiction.

Thus, $\left(\bigcap_{j=1}^{m} K_j \right) \cap \left(\bigcup_{j=1}^{m} (A_j \setminus P) \right) = \emptyset$, and it follows that $\bigcap_{j=1}^{m} K_j \subset P$. Since $p \in Int \left(\bigcap_{j=1}^{m} K_j \right)$, $p \in Int(P)$. Therefore, P is a connected neighborhood of p in X contained in V. Since p is an arbitrary point of V, every component of V is open.

Q.E.D.

3.1.33 Remark Note that Theorem 3.1.32 gives a partial answer to Question 9.2.10.

Our next three results are consequence of Theorem 3.1.32.

3.1.34 Theorem *Let X be an aposyndetic continuum. If for each subcontinuum W of X, $X \setminus W$ has only finitely many components, then X is locally connected.*

Proof To show that X is locally connected, by Theorem 3.1.32, it suffices to prove that $\mathcal{T}(W) = W$ for every subcontinuum W of X. Let W be a subcontinuum of X, we know that $W \subset \mathcal{T}(W)$ (Remark 3.1.5). Let $x \in X \setminus W$. Since X is aposyndetic, for each $w \in W$, there exists a subcontinuum K_w of X such that $w \in Int(K_w) \subset K_w \subset X \setminus \{x\}$. Since W is compact, there exist $w_1, \dots, w_n \in W$ such that $W \subset \bigcup_{j=1}^{n} Int(K_{w_j}) \subset \bigcup_{j=1}^{n} K_{w_j} \subset X \setminus \{x\}$. Let $K = \bigcup_{j=1}^{n} K_{w_j}$. Then K is a subcontinuum of X such that $W \subset Int(K) \subset K \subset X \setminus \{x\}$. By hypothesis, $X \setminus K$ has only finitely many components. Let C be the component of $X \setminus K$ containing x. Then, since $X \setminus K$ is open and has only finitely many components, C is open in X. Thus, $Cl(C)$ is a subcontinuum of X such that $x \in Int(Cl(C)) \subset X \setminus Int(K) \subset X \setminus W$. Hence, $x \in X \setminus \mathcal{T}(W)$. Therefore, $\mathcal{T}(W) = W$, and X is locally connected.

Q.E.D.

3.1.35 Corollary *Let X be an aposyndetic continuum such that $\dim(\mathcal{C}(X)) < \infty$. Then X is locally connected.*

Proof Since $\dim(\mathcal{C}(X)) < \infty$, it follows that there exists $n \in \mathbb{N}$ such that $\mathcal{C}(X)$ does not contain an n-cell. Hence, using a similar proof to the one given for Theorem 6.1.11, we have that X does not contain n-ods. Note that this implies that the complement of each subcontinuum of X has at most $n - 1$ components. The corollary now follows from Theorem 3.1.34.

Q.E.D.

3.1.36 Theorem *Let X be an aposyndetic continuum. If $\mathcal{T}(W) = W$ for each subcontinuum W of X with nonempty interior, then X is locally connected.*

Proof Let A be a subcontinuum of X and let $x \in X \setminus A$. Since X is aposyndetic, for each $a \in A$, there exists a subcontinuum W_a of X such that $a \in Int(W_a) \subset W_a \subset X \setminus \{x\}$. Since A is compact, there exist $a_1, \dots, a_n \in A$ such that $A \subset \bigcup_{j=1}^{n} Int(W_{a_j})$. Let $W = \bigcup_{j=1}^{n} W_{a_j}$. Note that $A \subset W$, $Int(W) \neq \emptyset$ and $x \in X \setminus W$. Hence, by hypothesis, $\mathcal{T}(W) = W$. Since $A \subset W$, by Proposition 3.1.7,

$\mathcal{T}(A) \subset \mathcal{T}(W) = W$. Thus, $x \in X \setminus \mathcal{T}(A)$. Therefore, $\mathcal{T}(A) = A$. Now the theorem follows from Theorem 3.1.32.

Q.E.D.

3.1.37 Theorem *A continuum X is continuum aposyndetic if and only if X is locally connected.*

Proof Suppose X is continuum aposyndetic. It follows from the definition that if W is a subcontinuum of X, then $\mathcal{T}(W) = W$. Hence, by Theorem 3.1.32, X is locally connected.

The reverse implication is clear.

Q.E.D.

As a consequence of Theorem 3.1.37, we obtain:

3.1.38 Theorem *A continuum X is freely decomposable with respect to points and continua if and only if X is locally connected.*

Proof Suppose X is freely decomposable with respect to points and continua. Then X is continuum aposyndetic. Thus, by Theorem 3.1.37, X is locally connected.

Now, assume X is locally connected. Let C be a subcontinuum of X and let $a \in X \setminus C$. Consider the quotient space X/C. Then X/C is a locally connected continuum. Since locally connected continua are aposyndetic, by Theorem 1.7.19, X/C is freely decomposable. Let $q \colon X \to X/C$ be the quotient map and let χ be the point of X/C corresponding to $q(C)$. Note that q is a monotone map. Since X/C is freely decomposable, there exist two subcontinua Γ_1 and Γ_2 of X/C such that $X/C = \Gamma_1 \cup \Gamma_2$, $q(a) \in \Gamma_1 \setminus \Gamma_2$ and $\chi \in \Gamma_2 \setminus \Gamma_1$. Hence, $X = q^{-1}(\Gamma_1) \cup q^{-1}(\Gamma_2)$, $a \in q^{-1}(\Gamma_1) \setminus q^{-1}(\Gamma_2)$ and $C = q^{-1}(\chi) \subset q^{-1}(\Gamma_2) \setminus q^{-1}(\Gamma_1)$. Therefore, X is freely decomposable with respect to points and continua.

Q.E.D.

3.1.39 Theorem *Let X be a continuum. Then X is indecomposable if and only if $\mathcal{T}(\{p\}) = X$ for each $p \in X$. Hence, $\mathcal{T}(A) = X$ for each nonempty closed subset A of X.*

Proof Suppose X is indecomposable. By Corollary 1.7.26, each proper subcontinuum of X has empty interior. Therefore, $\mathcal{T}(\{p\}) = X$ for each $p \in X$.

Now, suppose X is decomposable. Then there exist two proper subcontinua A and B of X such that $X = A \cup B$. Let $a \in A \setminus B$, and let $b \in B \setminus A$. Then $b \in Int(B) \subset B \subset X \setminus \{a\}$. Hence, $b \in X \setminus \mathcal{T}(\{a\})$. Therefore, $\mathcal{T}(\{a\}) \neq X$.

Q.E.D.

3.1.40 Definition A continuum X *is $\mathcal{T}$-symmetric* if for each pair, A and B, of closed subsets of X, $A \cap \mathcal{T}(B) = \emptyset$ if and only if $B \cap \mathcal{T}(A) = \emptyset$. We say that X *is point $\mathcal{T}$-symmetric* if for each pair, p and q, of points of X, $p \notin \mathcal{T}(\{q\})$ if and only if $q \notin \mathcal{T}(\{p\})$.

3.1.41 Theorem *If X is a weakly irreducible continuum, then X is $\mathcal{T}$-symmetric.*

Proof Let A and B be two closed subsets of X. Suppose $A \cap \mathcal{T}(B) = \emptyset$. Hence, for each $a \in A$, there exists a subcontinuum W_a of X such that $a \in Int(W_a) \subset W_a \subset X \setminus B$. Since A is compact, there exist $a_1, \ldots, a_n \in A$ such that $A \subset \bigcup_{j=1}^{n} Int(W_{a_j}) \subset \bigcup_{j=1}^{n} W_{a_j} \subset X \setminus B$. This implies that $B \subset X \setminus \left(\bigcup_{j=1}^{n} W_{a_j} \right) \subset X \setminus \left(\bigcup_{j=1}^{n} Int(W_{a_j}) \right) \subset X \setminus A$.

Let $b \in B$. Since X is weakly irreducible, $X \setminus \left(\bigcup_{j=1}^{n} W_{a_j} \right)$ has only a finite number of components which are open subsets of X. Let C be the component of $X \setminus \left(\bigcup_{j=1}^{n} W_{a_j} \right)$ such that $b \in C$. Then $b \in C \subset Cl(C) \subset X \setminus \left(\bigcup_{j=1}^{n} Int(W_{a_j}) \right) \subset X \setminus A$. Thus, $b \in X \setminus \mathcal{T}(A)$. Therefore, $B \cap \mathcal{T}(A) = \emptyset$.

Similarly, if $B \cap \mathcal{T}(A) = \emptyset$, then $A \cap \mathcal{T}(B) = \emptyset$. Therefore, X is $\mathcal{T}$-symmetric.

Q.E.D.

3.1.42 Corollary *If X is an irreducible continuum, then X is $\mathcal{T}$-symmetric.*

Proof Let X be an irreducible continuum. By Theorem 1.7.34, X is weakly irreducible. Hence, by Theorem 3.1.41, X is $\mathcal{T}$-symmetric.

Q.E.D.

3.1.43 Theorem *Let X be a $\mathcal{T}$-symmetric continuum, and let p be a point of X. Then X is connected im kleinen at p if and only if X is semi-locally connected at p.*

Proof Suppose X is connected im kleinen at p. Let $q \in \mathcal{T}(\{p\})$. Since X is $\mathcal{T}$-symmetric, $p \in \mathcal{T}(\{q\})$. Thus, $p \in \{q\}$ (Corollary 3.1.25). Hence, $p = q$ and $\mathcal{T}(\{p\}) = \{p\}$. Therefore, by Theorem 3.1.29, X is semi-locally connected at p.

Now, suppose X is semi-locally connected at p. Then $\mathcal{T}(\{p\}) = \{p\}$ (Theorem 3.1.29). Let A be a closed subset of X such that $p \in \mathcal{T}(A)$. Hence, since X is $\mathcal{T}$-symmetric, $A \cap \mathcal{T}(\{p\}) \neq \emptyset$. Thus, $p \in A$. Therefore, by Corollary 3.1.25, X is connected im kleinen at p.

Q.E.D.

3.1.44 Definition A continuum X *is $\mathcal{T}$-additive* if for each pair, A and B, of closed subsets of X, $\mathcal{T}(A \cup B) = \mathcal{T}(A) \cup \mathcal{T}(B)$.

The following theorem gives us a sufficient condition for a continuum X to be $\mathcal{T}$-additive.

3.1.45 Theorem *Let X be a continuum. If for each point $x \in X$ and any two subcontinua W_1 and W_2 of X such that $x \in Int(W_j)$, $j \in \{1, 2\}$, there exists a subcontinuum W_3 of X such that $x \in Int(W_3)$ and $W_3 \subset W_1 \cap W_2$, then X is $\mathcal{T}$-additive.*

Proof Let A_1 and A_2 be closed subsets of X, and let $x \in X \setminus \mathcal{T}(A_1) \cup \mathcal{T}(A_2)$. Hence, there exist two subcontinua W_1 and W_2 of X such that $x \in Int(W_j) \subset W_j \subset X \setminus A_j$, $j \in \{1, 2\}$. By hypothesis, there exists a subcontinuum W_3 of X such that $x \in Int(W_3)$ and $W_3 \subset W_1 \cap W_2$. Thus, $x \in Int(W_3) \subset W_3 \subset X \setminus (A_1 \cup A_2)$. Hence, $x \in X \setminus \mathcal{T}(A_1 \cup A_2)$. Therefore, by Corollary 3.1.8, $\mathcal{T}(A_1 \cup A_2) = \mathcal{T}(A_1) \cup \mathcal{T}(A_2)$.

Q.E.D.

We need the following definition to give a characterization of $\mathcal{T}$-additive continua.

3.1.46 Definition Let X be a continuum. A *filterbase* $\aleph$ *in* X is a family $\aleph = \{A_\omega\}_{\omega \in \Omega}$ of subsets of X having two properties:

(a) For each $\omega \in \Omega$, $A_\omega \neq \emptyset$, and
(b) For each $\omega_1, \omega_2 \in \Omega$, there exists $\omega_3 \in \Omega$ such that $A_{\omega_3} \subset A_{\omega_1} \cap A_{\omega_2}$.

3.1.47 Lemma *Let X be a compactum. If Γ is a filterbase of closed subsets of X, then $\mathcal{T}\left(\bigcap\{G \mid G \in \Gamma\}\right) = \bigcap\{\mathcal{T}(G) \mid G \in \Gamma\}$.*

Proof By Proposition 3.1.7, $\mathcal{T}\left(\bigcap\{G \mid G \in \Gamma\}\right) \subset \bigcap\{\mathcal{T}(G) \mid G \in \Gamma\}$.

Let $p \in X \setminus \mathcal{T}\left(\bigcap\{G \mid G \in \Gamma\}\right)$. Then there exists a subcontinuum W of X such that $p \in Int(W) \subset W \subset X \setminus \left(\bigcap\{G \mid G \in \Gamma\}\right)$. Since W is compact, there exist $G_1, \ldots, G_n \in \Gamma$ such that $W \cap \left(\bigcap_{j=1}^{n} G_j\right) = \emptyset$. Since Γ is a filterbase, there exists $G \in \Gamma$ such that $G \subset \bigcap_{j=1}^{n} G_j$. Note that $W \cap G = \emptyset$. Thus, $p \in X \setminus \mathcal{T}(G)$. Hence, $p \in X \setminus \bigcap\{\mathcal{T}(G) \mid G \in \Gamma\}$. Therefore, $\mathcal{T}\left(\bigcap\{G \mid G \in \Gamma\}\right) = \bigcap\{\mathcal{T}(G) \mid G \in \Gamma\}$.
                                                                         Q.E.D.

3.1.48 Theorem *Let X be a continuum. Then X is $\mathcal{T}$-additive if and only if for each family Λ of closed subsets of X whose union is closed, $\mathcal{T}\left(\bigcup\{L \mid L \in \Lambda\}\right) = \bigcup\{\mathcal{T}(L) \mid L \in \Lambda\}$.*

Proof Suppose X is $\mathcal{T}$-additive. By Proposition 3.1.7,

$$\bigcup\{\mathcal{T}(L) \mid L \in \Lambda\} \subset \mathcal{T}\left(\bigcup\{L \mid L \in \Lambda\}\right).$$

Now, suppose $x \in X \setminus \bigcup\{\mathcal{T}(L) \mid L \in \Lambda\}$. Then for each $L \in \Lambda$, let $F(L) = \{A \subset X \mid A$ is closed and $L \subset Int(A)\}$. If $L = \emptyset$, then $\mathcal{T}(L) = \bigcap\{\mathcal{T}(A) \mid A \in F(L)\}$. (Clearly $\mathcal{T}(L) \subset \bigcap\{\mathcal{T}(A) \mid A \in F(L)\}$. Let $z \in X \setminus \mathcal{T}(L)$. Then there exists a subcontinuum W of X such that $z \in Int(W)$. Let A be any proper closed subset of $X \setminus W$. Then $A \in F(L)$ and $z \in X \setminus \mathcal{T}(A)$. Hence, $z \in X \setminus \bigcap\{\mathcal{T}(A) \mid A \in F(L)\}$.) If $L \neq \emptyset$, then $F(L)$ is a filterbase of closed subsets of X. Since $\bigcap\{A \mid A \in F(L)\} = L$, $\mathcal{T}(L) = \bigcap\{\mathcal{T}(A) \mid A \in F(L)\}$, by Lemma 3.1.47.

Hence, for each $L \in \Lambda$, $x \in X \setminus \bigcap\{\mathcal{T}(A) \mid A \in F(L)\}$; and thus, there exists, for each $L \in \Lambda$, $A_L \in F(L)$ such that $x \in X \setminus \mathcal{T}(A_L)$. Note that $\{Int(A_L) \mid L \in \Lambda\}$ is an open cover of $\bigcup\{L \mid L \in \Lambda\}$. Since this set is compact, there exist $L_1, \ldots, L_m \in \Lambda$ such that

$$\bigcup\{L \mid L \in \Lambda\} \subset \bigcup_{j=1}^{m} Int(A_{L_j}).$$

Since, by hypothesis and mathematical induction, $\mathcal{T}\left(\bigcup_{j=1}^{m} A_{L_j}\right) = \bigcup_{j=1}^{m} \mathcal{T}(A_{L_j})$, $\mathcal{T}\left(\bigcup\{L \mid L \in \Lambda\}\right) \subset \bigcup_{j=1}^{m} \mathcal{T}(A_{L_j})$. Now, since for every $j \in \{1, \ldots, m\}$,

$x \in X \setminus \mathcal{T}(A_{L_j})$, it follows that $x \in X \setminus \mathcal{T}\left(\bigcup\{L \mid L \in \Lambda\}\right)$. Thus, we have that $\mathcal{T}\left(\bigcup\{L \mid L \in \Lambda\}\right) \subset \bigcup\{\mathcal{T}(L) \mid L \in \Lambda\}$.

The other implication is obvious.

Q.E.D.

3.1.49 Theorem *Each $\mathcal{T}$-symmetric continuum is $\mathcal{T}$-additive.*

Proof Let X be a $\mathcal{T}$-symmetric continuum, and let A and B be two closed subsets of X. By Corollary 3.1.8, $\mathcal{T}(A) \cup \mathcal{T}(B) \subset \mathcal{T}(A \cup B)$.

Let $x \in \mathcal{T}(A \cup B)$. Then $\{x\} \cap \mathcal{T}(A \cup B) \neq \emptyset$. Since X is $\mathcal{T}$-symmetric, $\mathcal{T}(\{x\}) \cap (A \cup B) \neq \emptyset$. Hence, either $\mathcal{T}(\{x\}) \cap A \neq \emptyset$ or $\mathcal{T}(\{x\}) \cap B \neq \emptyset$. Thus, since X is $\mathcal{T}$-symmetric, $\{x\} \cap \mathcal{T}(A) \neq \emptyset$ or $\{x\} \cap \mathcal{T}(B) \neq \emptyset$; i.e., $x \in \mathcal{T}(A)$ or $x \in \mathcal{T}(B)$. Then $x \in \mathcal{T}(A) \cup \mathcal{T}(B)$. Therefore, X is $\mathcal{T}$-additive.

Q.E.D.

3.1.50 Theorem *If X is a hereditarily unicoherent continuum, then X is $\mathcal{T}$-additive.*

Proof Let X be a hereditarily unicoherent continuum, and let A and B be two closed subsets of X. By Corollary 3.1.8, $\mathcal{T}(A) \cup \mathcal{T}(B) \subset \mathcal{T}(A \cup B)$.

Let $x \in X \setminus (\mathcal{T}(A) \cup \mathcal{T}(B))$. Then $x \notin \mathcal{T}(A)$ and $x \notin \mathcal{T}(B)$. Hence, there exist subcontinua W_A and W_B of X such that $x \in Int(W_A) \cap Int(W_B) \subset W_A \cap W_B \subset X \setminus (A \cup B)$. Since X is hereditarily unicoherent, $W_A \cap W_B$ is a subcontinuum of X. Therefore, $x \in X \setminus \mathcal{T}(A \cup B)$.

Q.E.D.

The following corollary is a consequence of Theorem 3.1.48; however, we present a different proof based on Corollary 3.1.20.

3.1.51 Corollary *If X is a $\mathcal{T}$-additive continuum and if A is a closed subset of X, then $\mathcal{T}(A) = \bigcup_{a \in A} \mathcal{T}(\{a\})$.*

Proof Let A be a closed subset of the $\mathcal{T}$-additive continuum X. If $a \in A$, then $\{a\} \subset A$ and $\mathcal{T}(\{a\}) \subset \mathcal{T}(A)$, by Remark 3.1.5. Hence, $\bigcup_{a \in A} \mathcal{T}(\{a\}) \subseteq \mathcal{T}(A)$.

Now, let $x \in X$ be such that $x \in X \setminus \mathcal{T}(\{a\})$ for each $a \in A$. Then, by Corollary 3.1.20, for each $a \in A$, there exists an open subset U_a of X such that $a \in U_a$ and $x \in X \setminus \mathcal{T}(Cl(U_a))$. Since A is compact and $\{U_a\}_{a \in A}$ is an open cover of A, there exist $a_1, \ldots, a_n \in A$ such that $A \subset \bigcup_{j=1}^{n} U_{a_j}$. Since for each $j \in \{1, \ldots, n\}$, $x \in X \setminus \mathcal{T}(Cl(U_{a_j}))$, $x \in X \setminus \bigcup_{j=1}^{n} \mathcal{T}(Cl(U_{a_j}))$. Thus, $x \in X \setminus \mathcal{T}\left(\bigcup_{j=1}^{n} Cl(U_{a_j})\right)$ (since X is $\mathcal{T}$-additive). Hence, $x \in X \setminus \mathcal{T}(A)$.

Q.E.D.

3.1.52 Theorem *A continuum X is locally connected if and only if X is aposyndetic and $\mathcal{T}$-additive.*

Proof Suppose X is a locally connected continuum. Then, by Theorem 3.1.31, $\mathcal{T}(A) = A$ for each closed subset A of X. Hence, $\mathcal{T}(\{p\}) = \{p\}$ for each point

$p \in X$ and $\mathcal{T}(A \cup B) = A \cup B = \mathcal{T}(A) \cup \mathcal{T}(B)$ for each pair of closed subsets A and B of X. Therefore, X is aposyndetic (Theorem 3.1.28) and $\mathcal{T}$-additive.

Now, suppose X is aposyndetic and $\mathcal{T}$-additive. Let A be a closed subset of X. Since X is $\mathcal{T}$-additive, by Corollary 3.1.51, $\mathcal{T}(A) = \bigcup_{a \in A} \mathcal{T}(\{a\})$. Since X is aposyndetic, by Theorem 3.1.28, $\mathcal{T}(\{a\}) = \{a\}$ for each $a \in A$. Thus, $\mathcal{T}(A) = \bigcup_{a \in A} \mathcal{T}(\{a\}) = \bigcup_{a \in A} \{a\} = A$. Therefore, by Theorem 3.1.31, X is locally connected.

Q.E.D.

As a consequence of Theorems 3.1.41, 3.1.49 and 3.1.52, we have the following:

3.1.53 Corollary *If X is an aposyndetic weakly irreducible continuum, then X is locally connected.*

3.1.54 Corollary *If X is an aposyndetic hereditarily unicoherent continuum, then X is locally connected.*

Proof Let X be an aposyndetic hereditarily unicoherent continuum. Since X is hereditarily unicoherent, by Theorem 3.1.50, X is $\mathcal{T}$-additive. Then, since X is an aposyndetic $\mathcal{T}$-additive continuum, by Theorem 3.1.52, X is locally connected.

Q.E.D.

3.1.55 Theorem *A continuum X is $\mathcal{T}$-symmetric if and only if X is point $\mathcal{T}$-symmetric and $\mathcal{T}$-additive.*

Proof Suppose X is $\mathcal{T}$-symmetric. By Theorem 3.1.49, X is $\mathcal{T}$-additive. Clearly, X is point $\mathcal{T}$-symmetric.

Now, suppose X is point $\mathcal{T}$-symmetric and $\mathcal{T}$-additive. Let A and B be two closed subsets of X. Suppose that $A \cap \mathcal{T}(B) = \emptyset$ and $B \cap \mathcal{T}(A) \neq \emptyset$. Let $x \in B \cap \mathcal{T}(A)$. Then $x \in B$ and $x \in \mathcal{T}(A)$. Note that $x \in X \setminus A$ (if $x \in A$, then $x \in A \cap B \subset A \cap \mathcal{T}(B)$, a contradiction). Since X is $\mathcal{T}$-additive, $\mathcal{T}(A) = \bigcup_{a \in A} \mathcal{T}(\{a\})$. Thus, there exists $a \in A$ such that $x \in \mathcal{T}(\{a\})$. This implies that $a \in \mathcal{T}(\{x\})$ (since X is point $\mathcal{T}$-symmetric). Consequently, since $x \in B$, $a \in \mathcal{T}(\{x\}) \subset \mathcal{T}(B)$ (Proposition 3.1.7). Hence, $a \in A \cap \mathcal{T}(B)$, a contradiction to our assumption. Therefore, $B \cap \mathcal{T}(A) = \emptyset$, and X is $\mathcal{T}$-symmetric.

Q.E.D.

3.1.56 Theorem *Let X be a semi-aposyndetic irreducible continuum. Then X is homeomorphic to $[0, 1]$.*

Proof Since X is an irreducible continuum, by Corollary 3.1.42, X is $\mathcal{T}$-symmetric. Let $x, y \in X$. Since X is semi-aposyndetic, by Theorem 3.1.26, either $x \in X \setminus \mathcal{T}(\{y\})$ or $y \in X \setminus \mathcal{T}(\{x\})$. This implies that $x \in X \setminus \mathcal{T}(\{y\})$ if and only if $y \in X \setminus \mathcal{T}(\{x\})$, by the $\mathcal{T}$-symmetry of X. Hence, $\mathcal{T}(\{x\}) = \{x\}$ for each $x \in X$. Therefore, X is aposyndetic (Theorem 3.1.28).

Since X is $\mathcal{T}$-symmetric, X is $\mathcal{T}$-additive (Theorem 3.1.49). Since X is an aposyndetic $\mathcal{T}$-additive continuum, X is locally connected (Theorem 3.1.52). Thus, X is an irreducible locally connected continuum. Consequently, X is an irreducible arcwise connected continuum [23, Theorem 3–15]. Therefore, X is homeomorphic to $[0, 1]$.

Q.E.D.

3.1.57 Definition Let X be a continuum. We say that $\mathcal{T}$ *is idempotent on* X provided that $\mathcal{T}^2(A) = \mathcal{T}(A)$ for each subset A of X. We say that $\mathcal{T}$ *is idempotent on closed sets* if $\mathcal{T}^2(A) = \mathcal{T}(A)$ for each closed subset A of X.

3.1.58 Proposition *Let X be a continuum for which $\mathcal{T}$ is idempotent on closed sets, and let Z be a nonempty closed subset of X. If $\mathcal{T}(Z) = A \cup B$, where A and B are nonempty closed subsets of X, then $\mathcal{T}(A \cup B) = \mathcal{T}(A) \cup \mathcal{T}(B) = A \cup B$.*

Proof By the idempotency of $\mathcal{T}$ on closed sets, Remark 3.1.5 and Proposition 3.1.7, we have

$$\mathcal{T}(Z) = A \cup B \subset \mathcal{T}(A) \cup \mathcal{T}(B) \subset \mathcal{T}(A \cup B) = \mathcal{T}^2(Z) = \mathcal{T}(Z).$$

Q.E.D.

3.1.59 Proposition *Let X be a continuum for which $\mathcal{T}$ is idempotent on closed sets. If Z is a nonempty closed subset of X, then $\mathcal{T}(Z) = \bigcup\{\mathcal{T}(\{w\}) \mid w \in \mathcal{T}(Z)\}$.*

Proof Let $x \in \bigcup\{\mathcal{T}(\{w\}) \mid w \in \mathcal{T}(Z)\}$. Then there exists $w \in \mathcal{T}(Z)$ such that $x \in \mathcal{T}(\{w\})$. Since $\mathcal{T}$ is idempotent on closed sets, $\mathcal{T}(\{w\}) \subset \mathcal{T}^2(Z) = \mathcal{T}(Z)$. Hence, $x \in \mathcal{T}(Z)$, and $\bigcup\{\mathcal{T}(\{w\}) \mid w \in \mathcal{T}(Z)\} \subset \mathcal{T}(Z)$.

The other inclusion is clear.

Q.E.D.

3.1.60 Theorem *Let X be a continuum. Then $\mathcal{T}$ is idempotent on X if and only if for each subcontinuum W of X and each point $x \in Int(W)$, there exists a subcontinuum M of X such that $x \in Int(M) \subset M \subset Int(W)$.*

Proof Suppose $\mathcal{T}$ is idempotent on X. Let W be a subcontinuum of X, and let $x \in Int(W)$. Hence, $x \in X \backslash \mathcal{T}(X \backslash W)$. Since $\mathcal{T}$ is idempotent, $x \in X \backslash \mathcal{T}^2(X \backslash W)$. Thus, there exists a subcontinuum M of X such that $x \in Int(M) \subset M \subset X \backslash \mathcal{T}(X \backslash W) \subset X \backslash (X \backslash W) = W$. Therefore, $x \in Int(M) \subset M \subset Int(W)$.

Now, suppose the condition stated holds. We show $\mathcal{T}$ is idempotent. By Remark 3.1.5, for each $A \in \mathcal{P}(X)$, $\mathcal{T}(A) \subset \mathcal{T}^2(A)$. Let $B \in \mathcal{P}(X)$, and let $x \in X \backslash \mathcal{T}(B)$. Then there exists a subcontinuum W of X such that $x \in Int(W) \subset W \subset X \backslash B$. By hypothesis, there exists a subcontinuum M such that $x \in Int(M) \subset M \subset Int(W)$. Since $Int(W) \subset X \backslash \mathcal{T}(B)$, $x \in Int(M) \subset M \subset X \backslash \mathcal{T}(B)$. Hence, $x \in X \backslash \mathcal{T}^2(B)$. Therefore, $\mathcal{T}$ is idempotent.

Q.E.D.

3.1.61 Corollary *Let X be a continuum. If $\mathcal{T}$ is idempotent on X, W is a subcontinuum of X and K is a component of $Int(W)$, then K is open.*

Proof Let W be a subcontinuum of X, and let K be a component of $Int(W)$. If $x \in K$, then, by Theorem 3.1.60, there exists a subcontinuum M of X such that

$x \in Int(M) \subset Int(W)$. Hence, $M \subset K$ and x is an interior point of K. Therefore, K is open.

Q.E.D.

3.1.62 Definition Let X be a continuum. A subcontinuum M of X is a *continuum domain* if $M = Cl(Int(M))$. A continuum domain M is a *strong continuum domain* provided that $Int(M)$ is connected.

3.1.63 Corollary *Let X be a continuum. If $\mathcal{T}$ is idempotent on X, $x \in X$ and W is a subcontinuum of X such that $x \in Int(W)$, then there exists a strong continuum domain M of X such that $x \in Int(M) \subset M \subset W$.*

Proof Let W be a subcontinuum of X, and let $x \in Int(W)$. By Corollary 3.1.61, the component, K, of W containing x is open. Let $M = Cl(K)$. Then M is a strong continuum domain and $x \in Int(M) \subset M \subset W$.

Q.E.D.

3.1.64 Definition A metric space Z is *continuumwise connected* provided that for each pair of points z_1 and z_2 of Z, there exists a subcontinuum W of Z such that $\{z_1, z_2\} \subset W$.

3.1.65 Theorem *Let X be a continuum and let $A \in \mathcal{P}(X)$. If $\mathcal{T}^2(A) = \mathcal{T}(A)$, then the components of $X \setminus \mathcal{T}(A)$ are open and continuumwise connected.*

Proof If $A = \emptyset$ or $\mathcal{T}(A) = X$, then the result is clear. Let $A \in \mathcal{P}(X) \setminus \{\emptyset\}$ be such that $\mathcal{T}(A) \neq X$. Let L be a component of $X \setminus \mathcal{T}(A)$ and let $x \in L$. Since $\mathcal{T}^2(A) = \mathcal{T}(A)$, $x \in X \setminus \mathcal{T}^2(A)$. Hence, there exists a subcontinuum W of X such that $x \in Int(W) \subset W \subset X \setminus \mathcal{T}(A)$. Since L is a component of $X \setminus \mathcal{T}(A)$ and $L \cap W \neq \emptyset$, $W \subset L$. Thus, x is an interior point of L. Therefore, L is open.

We show that L is continuumwise connected. Let ℓ' and ℓ'' be two points of L. For each $\ell \in L$, there exists a subcontinuum W_ℓ of X such that $\ell \in Int(W_\ell) \subset W_\ell \subset X \setminus \mathcal{T}(A)$. Observe that $W_\ell \subset L$ for all $\ell \in L$. Also note that $\{Int(W_\ell) \mid \ell \in L\}$ forms an open cover of L, since L is connected, there exist $\ell_1, \ldots, \ell_n \in L$ such that $\{Int(W_{\ell_1}), \ldots, Int(W_{\ell_n})\}$ forms a chain such that $\ell' \in Int(W_{\ell_1})$ and $\ell'' \in Int(W_{\ell_n})$ [12, (2.F.2)]. Hence, $W = \bigcup_{j=1}^{n} W_{\ell_j}$ is a subcontinuum of X such that $\{\ell', \ell''\} \subset W \subset L$. Therefore, L is continuumwise connected.

Q.E.D.

A similar result to Corollary 3.1.63 is true when $\mathcal{T}$ is idempotent on closed sets:

3.1.66 Theorem *Let X be a continuum such that $\mathcal{T}$ is idempotent on closed sets. If $A \in 2^X$ and $x \in X \setminus \mathcal{T}(A)$, then there exists a strong continuum domain W of X such that $x \in Int(W) \subset W \subset X \setminus \mathcal{T}(A) \subset X \setminus A$.*

Proof Let $A \in 2^X$ and let $x \in X \setminus \mathcal{T}(A)$. Since $\mathcal{T}$ is idempotent on closed sets, we have that $x \in X \setminus \mathcal{T}^2(A)$. By Corollary 3.1.20, there exists an open subset U of X such that $\mathcal{T}(A) \subset U$ and $x \in X \setminus \mathcal{T}(Cl(U))$. Let L be the component of $X \setminus \mathcal{T}(Cl(U))$ containing x. Since $\mathcal{T}$ is idempotent on closed sets, by Theorem 3.1.65,

L is open. Let $W = Cl(L)$. Then W is a strong continuum domain of X such that $x \in Int(W) \subset W \subset X \setminus U \subset X \setminus \mathcal{T}(A)$.

Q.E.D.

3.1.67 Theorem *Let X be an aposyndetic continuum such that $\mathcal{T}$ is idempotent on closed sets. If $\mathcal{T}(W) = W$ for each strong continuum domain W of X, then X is locally connected.*

Proof Note that, by Theorem 3.1.32, it is enough to show that $\mathcal{T}(A) = A$ for each subcontinuum A of X. Clearly $\mathcal{T}(X) = X$. Let A be a proper subcontinuum of X. By Remark 3.1.5, $A \subset \mathcal{T}(A)$. Let $x \in X \setminus A$. Since X is aposyndetic, by Theorem 3.1.66, for each $a \in A$, there exists a strong continuum domain W_a of X such that $a \in Int(W_a) \subset W_a \subset X \setminus \{x\}$. Since A is compact, there exist $a_1, \ldots, a_n \in A$ such that $A \subset \bigcup_{j=1}^{n} Int(W_{a_j}) \subset \bigcup_{j=1}^{n} W_{a_j} \subset X \setminus \{x\}$. Observe that $\bigcup_{j=1}^{n} Int(W_{a_j})$ is a connected open set. Let $W = Cl\left(\bigcup_{j=1}^{n} Int(W_{a_j})\right)$. Then W is a strong continuum domain such that $A \subset W \subset X \setminus \{x\}$. By hypothesis, $\mathcal{T}(W) = W$. Hence, by Proposition 3.1.7, $\mathcal{T}(A) \subset \mathcal{T}(W) = W \subset X \setminus \{x\}$. Thus, $x \in X \setminus \mathcal{T}(A)$. Therefore, $\mathcal{T}(A) = A$, and X is locally connected.

Q.E.D.

The next theorem gives a sufficient condition to have $\mathcal{T}$ idempotent on closed sets.

3.1.68 Theorem *Let X be a continuum. If for each $A \in 2^X$ and each subcontinuum K of X such that $Int(K) \neq \emptyset$ and $A \cap K = \emptyset$, there exists a subcontinuum W of X such that $K \subset Int(W) \subset W \subset X \setminus A$, then $\mathcal{T}$ is idempotent on closed sets.*

Proof Suppose that for each $A \in 2^X$ and each subcontinuum K of X such that $Int(K) \neq \emptyset$ and $K \cap A = \emptyset$, there exists a subcontinuum W of X such that $K \subset Int(W) \subset W \subset X \setminus A$. Let $A \in 2^X$ and let $x \in X \setminus \mathcal{T}(A)$. Then there exists a subcontinuum K of X such that $x \in Int(K) \subset K \subset X \setminus A$. By hypothesis, there exists a subcontinuum W of X such that $K \subset Int(W) \subset W \subset X \setminus A$. This implies that $x \in X \setminus \mathcal{T}^2(A)$. Hence, $\mathcal{T}^2(A) \subset \mathcal{T}(A)$. By Remark 3.1.5 and Proposition 3.1.7, $\mathcal{T}(A) \subset \mathcal{T}^2(A)$. Therefore, $\mathcal{T}$ is idempotent on closed sets.

Q.E.D.

The next example shows that the converse of Theorem 3.1.68 is not true.

3.1.69 Example Let

$$X = (\{0\} \times [-1, 2]) \cup \left\{ \left(x, \sin\left(\frac{1}{x}\right) \right) \mid x \in \left(0, \frac{\pi}{2}\right] \right\},$$

let $K = \{0\} \times \left[0, \frac{3}{2}\right]$ and let $A = \{0\} \times \left[-\frac{1}{3}, -\frac{1}{2}\right]$. Then $\mathcal{T}$ is idempotent on closed sets, also, $Int(K) = \{0\} \times \left(1, \frac{3}{2}\right)$, $K \cap A = \emptyset$ and if W is a subcontinuum of X such that $K \subset Int(W)$, then $A \subset W$.

The next theorem is an application of Theorem 3.1.68.

3.1.70 Theorem *If X is a continuum with the property of Kelley (Definition 6.1.18), then $\mathcal{T}$ is idempotent on closed sets.*

Proof Let $A \in 2^X$. Let K be a subcontinuum of X such that $Int(K) \neq \emptyset$ and $K \cap A = \emptyset$.

Let $x \in Int(K)$ and let $\varepsilon > 0$ be such that $\varepsilon < d(K, A)$ and $\mathcal{V}_\varepsilon^d(x) \subset Int(K)$. Let $\delta > 0$ be a Kelley number for this ε. Since K is compact, there exist $k_1, \ldots, k_n \in K$ such that $K \subset \bigcup_{j=1}^n \mathcal{V}_\delta^d(k_j)$.

Fix $j \in \{1, \ldots, n\}$. Then for each $k \in \mathcal{V}_\delta^d(k_j)$, there exists a subcontinuum W_k of X such that $k \in W_k$ and $\mathcal{H}(K, W_k) < \varepsilon$. This implies that $W_k \cap \mathcal{V}_\varepsilon^d(x) \neq \emptyset$. Hence, $K \cup W_k$ is connected. Let

$$W_j = Cl\left(\bigcup_{k \in \mathcal{V}_\delta^d(k_j)} W_k\right).$$

Then W_j is a subcontinuum of X such that $\mathcal{V}_\delta^d(k_j) \subset W_j$. Let $W = \bigcup_{j=1}^n W_j$. Thus, W is a subcontinuum of X and $K \subset Int(W) \subset W \subset X \setminus A$. Therefore, by Theorem 3.1.68, $\mathcal{T}$ is idempotent on closed sets.

$$\textbf{Q.E.D.}$$

3.1.71 Theorem *Let X be a continuum. If $\mathcal{T}$ is idempotent on X and $\mathcal{T}(\{p, q\})$ is a continuum for all $p, q \in X$, then X is indecomposable.*

Proof Suppose X is decomposable. Then, by Corollary 3.1.63, there exists a strong continuum domain W of X. Let p_0 and q_0 be any two points in $X \setminus Int(W)$. Note that, by Theorem 3.1.60, $\mathcal{T}(\{p_0, q_0\}) \cap Int(W) = \emptyset$.

Since $\mathcal{T}(\{p_0, q_0\})$ is connected and $\mathcal{T}(\{p_0, q_0\}) \cap Int(W) = \emptyset$, p_0 and q_0 belong to the same component of $X \setminus Int(W)$. Thus, $X \setminus Int(W)$ is a continuum. By Theorem 3.1.60, there exists a subcontinuum M, with nonempty interior, such that $M \subset Int(X \setminus Int(W)) \subset X \setminus W$. Then let $K = X \setminus (Int(W) \cup Int(M))$, and let p_1 and q_1 be any two points of K. By Theorem 3.1.60, $\mathcal{T}(\{p_1, q_1\}) \subset K$. Hence, K is a continuum also. Now, let $p \in Int(M)$ and let $q \in Int(W)$. Observe that, by Theorem 3.1.60, $\mathcal{T}(\{p, q\}) \cap Int(K) = \emptyset$. Then $\mathcal{T}(\{p, q\}) \subset X \setminus Int(K) \subset W \cup M$. Hence, $\mathcal{T}(\{p, q\})$ is not a continuum, a contradiction.

$$\textbf{Q.E.D.}$$

3.1.72 Lemma *Let X be a compactum. If A is a closed subset of X, then $p \in X \setminus \mathcal{T}(A)$ if and only if there exist a subcontinuum W and an open subset Q of X such that $p \in Int(W) \cap Q$, $Bd(Q) \cap \mathcal{T}(A) = \emptyset$ and $W \cap A \cap Q = \emptyset$.*

Proof Let $p \in X \setminus \mathcal{T}(A)$. Then there exists a subcontinuum W of X such that $p \in Int(W) \subset W \subset X \setminus A$. Since X is a metric space, there exists an open subset Q of X such that $p \in Q \subset Cl(Q) \subset Int(W)$. Hence, $Bd(Q) \cap \mathcal{T}(A) = \emptyset$ and $W \cap A \cap Q = \emptyset$.

Now, suppose there exist a continuum W and an open subset Q of X such that $p \in Int(W) \cap Q$, $Bd(Q) \cap \mathcal{T}(A) = \emptyset$ and $W \cap A \cap Q = \emptyset$. Since $Bd(Q)$ is a compact set and $Bd(Q) \cap \mathcal{T}(A) = \emptyset$, there exist a finite collection, $W_1, \ldots, W_n$, of subcontinua of X such that $Bd(Q) \subset \bigcup_{j=1}^{n} Int(W_j) \subset \bigcup_{j=1}^{m} W_j \subset X \setminus A$. If $W \subset Q$, then, clearly, $p \in X \setminus \mathcal{T}(A)$.

Assume $W \setminus Q \neq \emptyset$. By Theorem 1.7.27, the closure of each component of $W \cap Q$ must intersect at least one of the W_j's, since $Bd(Q) \subset \bigcup_{j=1}^{m} W_j$. Hence, $H = (W \cap Q) \cup \left(\bigcup_{j=1}^{m} W_j \right)$ has only a finite number of components. Let K be the component of H such that $p \in K$. Since $p \in Int(W) \cap Q$, by Lemma 1.6.2, $p \in Int(K)$, and, of course, $K \cap A \subset H \cap A = \emptyset$. Thus, $p \in X \setminus \mathcal{T}(A)$.

Q.E.D.

3.1.73 Lemma *Let X be a compactum, and let A be a subset of X. If $\mathcal{T}(A) = M \cup N$, where M and N are disjoint closed sets, then $\mathcal{T}(A \cap M) = M \cup \mathcal{T}(\emptyset)$.*

Proof Suppose $p \in \mathcal{T}(A \cap M) \setminus (M \cup \mathcal{T}(\emptyset))$. Since $p \in X \setminus \mathcal{T}(\emptyset)$, there exists a subcontinuum W of X such that $p \in Int(W)$. Since X is a metric space, there exists an open subset Q of X such that $N \subset Q$ and $Cl(Q) \cap M = \emptyset$. Note that $p \in Int(W) \cap Q$ and $Bd(Q) \cap \mathcal{T}(A \cap M) \subset Bd(Q) \cap \mathcal{T}(A) = \emptyset$, and $W \cap (A \cap M) \cap Q \subset Q \cap M = \emptyset$. Then, by Lemma 3.1.72, $p \in X \setminus \mathcal{T}(A \cap M)$; thus, contradicting the assumption.

Now, suppose $p \in (M \cup \mathcal{T}(\emptyset)) \setminus \mathcal{T}(A \cap M)$. Since $p \in X \setminus \mathcal{T}(A \cap M)$ and $\emptyset \subset A \cap M$, $p \in X \setminus \mathcal{T}(\emptyset)$. Hence, $p \in M$. Since X is a metric space, there exists an open subset Q of X such that $M \subset Q$ and $Cl(Q) \cap N = \emptyset$. Since $p \in X \setminus \mathcal{T}(A \cap M)$, there exists a subcontinuum W of X such that $p \in Int(W) \subset W \subset X \setminus (A \cap M)$. Observe that $p \in Int(W) \cap Q$ and $Bd(Q) \cap \mathcal{T}(A) = \emptyset$. Since $Q \cap N = \emptyset$, $W \cap A \cap Q = W \cap (A \cap M) \cap Q = \emptyset$. Hence, by Lemma 3.1.72, $p \in X \setminus \mathcal{T}(A) \subset X \setminus M$; thus, contradicting our assumption.

Q.E.D.

3.1.74 Theorem *Let X be a compactum, and let A be a closed subset of X. If K is a component of $\mathcal{T}(A)$, then $\mathcal{T}(A \cap K) = K \cup \mathcal{T}(\emptyset)$.*

Proof Let K be a component of $\mathcal{T}(A)$. Let $\Gamma(K) = \{L \subset \mathcal{T}(A) \mid K \subset L$ and L is open and closed in $\mathcal{T}(A)\}$. Note that the collection of $\{A \cap L \mid L \in \Gamma(K)\}$ fails to be a filterbase if for some $L \in \Gamma(K)$, $A \cap L = \emptyset$. In this case the conclusion of Lemma 3.1.47 holds (an argument similar to the one given in the proof of Theorem 3.1.48 shows this). Note that Lemma 3.1.73 remains true even if $A \cap M = \emptyset$. Hence, by Lemma 3.1.73, for each $L \in \Gamma(K)$, $\mathcal{T}(A \cap L) = L \cup \mathcal{T}(\emptyset)$. Therefore, by Lemma 3.1.47, the following sequence of equalities establishes the theorem:

$\mathcal{T}(A \cap K) = \mathcal{T}\left(\bigcap \{A \cap L \mid L \in \Gamma(K)\}\right) = \bigcap \{\mathcal{T}(A \cap L) \mid L \in \Gamma(K)\} = \bigcap \{L \cup \mathcal{T}(\emptyset) \mid L \in \Gamma(K)\} = \bigcap \{L \mid L \in \Gamma(K)\} \cup \mathcal{T}(\emptyset) = K \cup \mathcal{T}(\emptyset)$. (The fact that $\bigcap \{L \mid L \in \Gamma(K)\} = K$ follows from [23, Theorem 2–14].)

Q.E.D.

The following corollary says that for each nonempty closed subset A of a continuum X, the components of $\mathcal{T}(A)$ are also in the image of $\mathcal{T}$.

3.1.75 Corollary *Let X be a continuum, and let A be a closed subset of X. If K is a component of $\mathcal{T}(A)$, then $K = \mathcal{T}(A \cap K)$.*

Proof Let A be a closed subset of X and let K be a component of $\mathcal{T}(A)$. By Theorem 3.1.74, $\mathcal{T}(A \cap K) = K \cup \mathcal{T}(\emptyset)$. Since $\mathcal{T}(\emptyset) = \emptyset$, by Corollary 3.1.14, $K = \mathcal{T}(A \cap K)$.

Q.E.D.

3.1.76 Corollary *If X is a continuum and A is a nonempty closed subset of X, then $\mathcal{T}(A)$ has at most as many components as A has.*

3.1.77 Corollary *Let X be a continuum and let $n \in \mathbb{N}$. If $A \in \mathcal{C}_n(X)$, then $\mathcal{T}(A) \in \mathcal{C}_n(X)$.*

3.1.78 Corollary *Let X be a continuum, and let W_1 and W_2 be subcontinua of X. If $\mathcal{T}(W_1 \cup W_2) \neq \mathcal{T}(W_1) \cup \mathcal{T}(W_2)$, then $\mathcal{T}(W_1 \cup W_2)$ is a continuum.*

Proof Suppose $\mathcal{T}(W_1 \cup W_2)$ is not connected. Then there exist two disjoint closed subsets A and B of X such that $\mathcal{T}(W_1 \cup W_2) = A \cup B$. By Lemma 3.1.73 and Corollary 3.1.14, $\mathcal{T}((W_1 \cup W_2) \cap A) = A$ and $\mathcal{T}((W_1 \cup W_2) \cap B) = B$. Suppose $W_1 \subset A$. If $W_2 \subset A$, then $A = \mathcal{T}((W_1 \cup W_2) \cap A) = \mathcal{T}(W_1 \cup W_2)$, a contradiction. Thus, $W_2 \subset B$. Hence, $\mathcal{T}(W_1) = A$ and $\mathcal{T}(W_2) = B$, which implies that $\mathcal{T}(W_1 \cup W_2) = \mathcal{T}(W_1) \cup \mathcal{T}(W_2)$, a contradiction. Therefore, $\mathcal{T}(W_1 \cup W_2)$ is connected.

Q.E.D.

3.1.79 Corollary *Let X be a continuum. If $p, q \in X$ are such that $\mathcal{T}(\{p, q\}) \neq \mathcal{T}(\{p\}) \cup \mathcal{T}(\{q\})$, then $\mathcal{T}(\{p, q\})$ is a continuum.*

The following theorem gives relationships between the images of maps and the images of $\mathcal{T}$. We subscript $\mathcal{T}$ to differentiate the continua on which $\mathcal{T}$ is defined.

3.1.80 Theorem *Let X and Y be continua, and let $f : X \twoheadrightarrow Y$ be a surjective map. If $A \in \mathcal{P}(X)$ and $B \in \mathcal{P}(Y)$, then the following hold:*

(a) $\mathcal{T}_Y(B) \subset f\mathcal{T}_X f^{-1}(B)$.
(b) If f is monotone, then $f\mathcal{T}_X(A) \subset \mathcal{T}_Y f(A)$ and $\mathcal{T}_X f^{-1}(B) \subset f^{-1}\mathcal{T}_Y(B)$.
(c) If f is monotone, then $\mathcal{T}_Y(B) = f\mathcal{T}_X f^{-1}(B)$.
(d) If f is open, then $f^{-1}\mathcal{T}_Y(B) \subset \mathcal{T}_X f^{-1}(B)$.
(e) If f is monotone and open, then $f^{-1}\mathcal{T}_Y(B) = \mathcal{T}_X f^{-1}(B)$.

Proof We show (a). Let $y \in Y \setminus f\mathcal{T}_X f^{-1}(B)$. Then $f^{-1}(y) \cap \mathcal{T}_X f^{-1}(B) = \emptyset$. Thus, for each $x \in f^{-1}(y)$, there exists a subcontinuum W_x of X such that $x \in Int(W_x) \subset W_x \subset X \setminus f^{-1}(B)$. Since $f^{-1}(y)$ is compact, there exist $x_1, \ldots, x_n \in f^{-1}(y)$ such that $f^{-1}(y) \subset \bigcup_{j=1}^{n} Int(W_{x_j})$. Note that $\left(\bigcup_{j=1}^{n} W_{x_j}\right) \cap f^{-1}(B) = \emptyset$, and for each $j \in \{1, \ldots, n\}$, $f^{-1}(y) \cap W_{x_j} \neq \emptyset$. Hence, $f\left(\bigcup_{j=1}^{n} W_{x_j}\right) \cap B = $

$\emptyset$ and $f\left(\bigcup_{j=1}^{n} W_{x_j}\right)$ is a continuum ($y \in f(W_{x_j}) \cap f(W_{x_k})$ for every $j, k \in \{1, 2, \ldots, n\}$).

Observe that $Y \setminus f\left(X \setminus \bigcup_{j=1}^{n} Int(W_{x_j})\right)$ is an open set of Y and it is contained in $f\left(\bigcup_{j=1}^{n} W_{x_j}\right)$. To see this, note that, since $\bigcup_{j=1}^{n} Int(W_{x_j}) \subset \bigcup_{j=1}^{n} W_{x_j}$, $X \setminus \bigcup_{j=1}^{n} W_{x_j} \subset X \setminus \bigcup_{j=1}^{n} Int(W_{x_j})$. Then $f\left(X \setminus \bigcup_{j=1}^{n} W_{x_j}\right) \subset f\left(X \setminus \bigcup_{j=1}^{n} Int(W_{x_j})\right)$ and, consequently, $Y \setminus f\left(X \setminus \bigcup_{j=1}^{n} Int(W_{x_j})\right) \subset Y \setminus f\left(X \setminus \bigcup_{j=1}^{n} W_{x_j}\right)$. Since f is a surjection, $Y \setminus f\left(X \setminus \bigcup_{j=1}^{n} Int(W_{x_j})\right) \subset f\left(X \setminus \left(X \setminus \bigcup_{j=1}^{n} W_{x_j}\right)\right) = f\left(\bigcup_{j=1}^{n} W_{x_j}\right)$. Therefore, $Y \setminus f\left(X \setminus \bigcup_{j=1}^{n} Int(W_{x_j})\right) \subset f\left(\bigcup_{j=1}^{n} W_{x_j}\right)$.

Also, since $f^{-1}(y) \subset \bigcup_{j=1}^{n} Int(W_{x_j})$, we have that $\{y\} \cap f\left(X \setminus \bigcup_{j=1}^{n} Int(W_{x_j})\right) = \emptyset$. Hence, $y \in Y \setminus f\left(X \setminus \bigcup_{j=1}^{n} Int(W_{x_j})\right)$. Therefore, $y \in Y \setminus \mathcal{T}_Y(B)$.

We prove (b). First, we see $f\mathcal{T}_X(A) \subset \mathcal{T}_Y f(A)$. Let $y \in Y \setminus \mathcal{T}_Y f(A)$. Then there exists a subcontinuum W of Y such that $y \in Int(W) \subset W \subset Y \setminus f(A)$. It follows that $f^{-1}(y) \subset f^{-1}(Int(W)) \subset f^{-1}(W) \subset f^{-1}(Y) \setminus f^{-1}f(A) \subset X \setminus A$. Hence, $f^{-1}(y) \subset X \setminus A$. Since f is monotone, $f^{-1}(W)$ is a subcontinuum of X (Lemma 2.1.12). Thus, since $f^{-1}(y) \subset f^{-1}(Int(W)) \subset f^{-1}(W) \subset X \setminus A$, $f^{-1}(y) \cap \mathcal{T}_X(A) = \emptyset$. Therefore, $y \in Y \setminus f\mathcal{T}_X(A)$.

Now, we show $\mathcal{T}_X f^{-1}(B) \subset f^{-1}\mathcal{T}_Y(B)$. Let $x \in X \setminus f^{-1}\mathcal{T}_Y(B)$. Then $f(x) \in Y \setminus \mathcal{T}_Y(B)$. Hence, there exists a subcontinuum W of Y such that $f(x) \in Int(W) \subset W \subset Y \setminus B$. This implies that $x \in f^{-1}(f(x)) \subset f^{-1}(Int(W)) \subset f^{-1}(W) \subset X \setminus f^{-1}(B)$. Since f is monotone, $f^{-1}(W)$ is a subcontinuum of X. Therefore, $x \in X \setminus \mathcal{T}_X f^{-1}(B)$.

We prove (c). By (a), $\mathcal{T}_Y(B) \subset f\mathcal{T}_X f^{-1}(B)$. Since $\mathcal{T}_X f^{-1}(B) \subset f^{-1}\mathcal{T}_Y(B)$, by (b), $f\mathcal{T}_X f^{-1}(B) \subset ff^{-1}\mathcal{T}_Y(B) = \mathcal{T}_Y(B)$. Therefore, $f\mathcal{T}_X f^{-1}(B) = \mathcal{T}_Y(B)$.

We show (d). Let $x \in X \setminus \mathcal{T}_X f^{-1}(B)$. Then there exists a subcontinuum W of X such that $x \in Int(W) \subset W \subset X \setminus f^{-1}(B)$. Hence, $f(x) \in f(Int(W)) \subset f(W) \subset Y \setminus B$. Since f is open, $f(Int(W))$ is an open subset of Y. Therefore, $f(x) \in Y \setminus \mathcal{T}_Y(B)$.

Note that (e) follows directly from (b) and (d).

Q.E.D.

The following four theorems are applications of Theorem 3.1.80.

3.1.81 Theorem *Let X and Y be continua, and let $f : X \twoheadrightarrow Y$ be a surjective map. If X is locally connected, then Y is locally connected.*

Proof By Theorem 3.1.31, it suffices to show that $\mathcal{T}_Y(B) = B$ for each closed subset B of Y. Let B be a closed subset of Y. By Remark 3.1.5, $B \subset \mathcal{T}_Y(B)$. By Theorem 3.1.80 (a), $\mathcal{T}_Y(B) \subseteq f\mathcal{T}_X f^{-1}(B)$. Since X is locally connected,

$\mathcal{T}_X f^{-1}(B) = f^{-1}(B)$. Hence, $\mathcal{T}_Y(B) \subseteq ff^{-1}(B) = B$. Therefore, $\mathcal{T}_Y(B) = B$, and Y is locally connected.

Q.E.D.

3.1.82 Theorem *The monotone image of an indecomposable continuum is an indecomposable continuum.*

Proof Let $f: X \twoheadrightarrow Y$ be a monotone map, where X is an indecomposable continuum. Let $y \in Y$. Then, by Theorem 3.1.80 (b), $\mathcal{T}_X f^{-1}(\{y\}) \subset f^{-1}\mathcal{T}_Y(\{y\})$. Since X is indecomposable, by Theorem 3.1.39, $\mathcal{T}_X f^{-1}(\{y\}) = X$. Hence, $Y = f(X) \subset ff^{-1}\mathcal{T}_Y(\{y\}) \subset Y$. Thus, $\mathcal{T}_Y(\{y\}) = Y$. Therefore, by Theorem 3.1.39, Y is indecomposable.

Q.E.D.

3.1.83 Theorem *The monotone image of a $\mathcal{T}$-symmetric continuum is $\mathcal{T}$-symmetric.*

Proof Let X be a $\mathcal{T}$-symmetric continuum, and let $f: X \twoheadrightarrow Y$ be a monotone map. Let A and B be two closed subsets of Y such that $A \cap \mathcal{T}_Y(B) = \emptyset$. Then, by Theorem 3.1.80 (c), $A \cap f\mathcal{T}_X f^{-1}(B) = \emptyset$. Thus, $f^{-1}(A) \cap \mathcal{T}_X f^{-1}(B) = \emptyset$. Since X is $\mathcal{T}$-symmetric, $\mathcal{T}_X f^{-1}(A) \cap f^{-1}(B) = \emptyset$. Hence, $f(\mathcal{T}_X f^{-1}(A) \cap f^{-1}(B)) = f\mathcal{T}_X f^{-1}(A) \cap B = \mathcal{T}_Y(A) \cap B = \emptyset$. Therefore, Y is $\mathcal{T}$-symmetric.

Q.E.D.

3.1.84 Theorem *The monotone image of a $\mathcal{T}$-additive continuum is $\mathcal{T}$-additive.*

Proof Let X be a $\mathcal{T}$-additive continuum, and let $f: X \twoheadrightarrow Y$ be a monotone map. Let A and B be two closed subsets of Y. Then, by Theorem 3.1.80 (c), $\mathcal{T}_Y(A \cup B) = f\mathcal{T}_X f^{-1}(A \cup B) = f\mathcal{T}_X(f^{-1}(A) \cup f^{-1}(B)) = f(\mathcal{T}_X f^{-1}(A) \cup \mathcal{T}_X f^{-1}(B)) = f\mathcal{T}_X f^{-1}(A) \cup f\mathcal{T}_X f^{-1}(B) = \mathcal{T}_Y(A) \cup \mathcal{T}_Y(B)$.

Q.E.D.

Now, we present some relations between $\mathcal{T}$ and inverse limits due to H. S. Davis [15].

3.1.85 Lemma *Let $\{X_n, f_n^{n+1}\}$ be an inverse sequence of continua whose inverse limit is X. Let $p = (p_n)_{n=1}^{\infty} \in X$, and let A be a closed subset of X. If there exist $N \in \mathbb{N}$, two open subsets U_N and V_N of X_N and an inverse sequence $\{W_n, f_n^{n+1}|_{W_{n+1}}\}$ of subcontinua such that $p_N \in U_N$, $f_N(A) \subset V_N$ and for each $n \geq N$,*

$$(f_N^n)^{-1}(U_N) \subset W_n \subset X_n \setminus (f_N^n)^{-1}(V_N),$$

then $p \in X \setminus \mathcal{T}(A)$.

Proof Let $W = \varprojlim\{W_n, f_n^{n+1}|_{W_{n+1}}\}$. Then W is a subcontinuum of X, by Proposition 2.1.8. Note that $f_N^{-1}(U_N) \subset W$ since, for each $n \geq N$, $(f_N^n)^{-1}(U_N) \subset Cl((f_N^n)^{-1}(U_N)) \subset Cl(f_n^{-1}(f_N^n)^{-1}(U_N)) \subset W_n$ and $W = \bigcap_{n=N}^{\infty} f_n^{-1}(W_n)$ (Corollary 2.1.21). Since $p \in f_N^{-1}(U_N)$, $p \in Int(W)$.

Since, for every $n \geq N$, $f_n^{-1}(W_n) \subset X \setminus f_n^{-1}(f_N^n)^{-1}(V_N) = X \setminus f_N^{-1}(V_N) = f_N^{-1}(X_N \setminus V_N) \subset X \setminus A$, $\bigcap_{n=N}^{\infty} f_n^{-1}(W_n) \subset X \setminus A$. Thus, $W \cap A = \emptyset$. Therefore, $p \in X \setminus \mathcal{T}(A)$.

$$\text{Q.E.D.}$$

3.1.86 Lemma *Let $\{X_n, f_n^{n+1}\}$ be an inverse sequence of continua whose inverse limit is X. Let $p = (p_n)_{n=1}^{\infty} \in X$, and let A be a closed subset of X. If $p \in X \setminus \mathcal{T}(A)$, then there exist $N \in \mathbb{N}$, two open subsets U_N and V_N of X_N and an inverse sequence $\{W_n, f_n^{n+1}|_{W_{n+1}}\}$ of subcontinua such that $p_N \in U_N$, $f_N(A) \subset V_N$ and for every $n \geq N$,*

$$(f_N^n)^{-1}(U_N) \cap f_n(X) \subset W_n \subset f_n(X) \setminus (f_N^n)^{-1}(V_N).$$

Proof Since $p \in X \setminus \mathcal{T}(A)$, there exists a subcontinuum W of X such that $p \in Int(W) \subset W \subset X \setminus A$. By Proposition 2.1.9, there exist $N(p) \in \mathbb{N}$ and an open subset $U_{N(p)}$ of $X_{N(p)}$ such that $p \in f_{N(p)}^{-1}(U_{N(p)}) \subset Int(W)$. Similarly, for each $a = (a_n)_{n=1}^{\infty} \in A$, there exist $N(a) \in \mathbb{N}$ and an open subset $U_{N(a)}$ of $X_{N(a)}$ such that $a \in f_{N(a)}^{-1}(U_{N(a)}) \subset X \setminus W$. Since A is compact, there exist $a^1, \ldots, a^m \in A$ such that $A \subset \bigcup_{j=1}^{m} f_{N(a^j)}^{-1}(U_{N(a^j)})$. Let $N = \max\{N(p), N(a^1), \ldots, N(a^m)\}$. Define $U_N = (f_{N(p)}^N)^{-1}(U_{N(p)})$ and $V_N = \bigcup_{j=1}^{m}(f_{N(a^j)}^N)^{-1}(U_{N(a^j)})$.

For each $n \in \mathbb{N}$, let $W_n = f_n(W)$. Let $n \geq N$, and let $z_n \in (f_N^n)^{-1}(U_N) \cap f_n(X)$. Since $z_n \in f_n(X)$, there exists $z \in X$ such that $f_n(z) = z_n$. Since $f_{N(p)}^n(z_n) \in U_{N(p)}$, $f_{N(p)}(z) \in U_{N(p)}$. Therefore,

$$z \in f_{N(p)}^{-1}(U_{N(p)}) \subset Int(W) \subset W.$$

Thus, $z_n = f_n(z) \in f_n(W) = W_n$. Hence, $(f_N^n)^{-1}(U_N) \cap f_n(X) \subset W_n$.

Now, let $w_n \in W_n = f_n(W)$. Then there exists $w \in W$ such that $f_n(w) = w_n$. Clearly, $w_n \in f_n(X)$. Suppose $w_n \in (f_N^n)^{-1}(V_N)$. Then there exists $j \in \{1, \ldots, m\}$ such that $f_{N(a^j)}^n(w_n) \in U_{N(a^j)}$. Hence, $f_{N(a^j)}(w) \in U_{N(a^j)}$. Therefore, $w \in X \setminus W$. This contradiction establishes that $W_n \subset f_n(X) \setminus (f_N^n)^{-1}(V_N)$.

$$\text{Q.E.D.}$$

3.1.87 Theorem *Let $\{X_n, f_n^{n+1}\}$ be an inverse sequence of continua whose inverse limit is X. Let $p = (p_n)_{n=1}^{\infty} \in X$, and let A be a closed subset of X. Then the following are equivalent:*

(a) $p \in X \setminus \mathcal{T}(A)$,
(b) there exist $N \in \mathbb{N}$, two open subsets U_N and V_N of X_N and an inverse sequence $\{W_n, f_n^{n+1}|_{W_{n+1}}\}$ of subcontinua such that $p_N \in U_N$, $f_N(A) \subset V_N$ and for every $n \geq N$,

$$(f_N^n)^{-1}(U_N) \cap f_n(X) \subset W_n \subset f_n(X) \setminus (f_N^n)^{-1}(V_N).$$

Proof Suppose (a). Then Lemma 3.1.86 shows (b).

Now, suppose (b). Note that if $W = \underleftarrow{\lim}\{W_n, f_n^{n+1}|_{W_{n+1}}\}$, then for each $n \in \mathbb{N}$, $f_n(W) = f_n(X) \cap W_n$, $f_n(W)$ is a continuum and $W = \underleftarrow{\lim}\{f_n(W), f_n^{n+1}|_{f_{n+1}(W)}\}$ (Proposition 2.1.20).

The relations in Lemma 3.1.85 imply that $(f_N^n)^{-1}(U_N) \cap f_n(X) \subset W_n \subset f_n(X) \setminus (f_N^n)^{-1}(V_N)$.

Q.E.D.

3.1.88 Theorem *Let $\{X_n, f_n^{n+1}\}$ be an inverse sequence of continua, with surjective bonding maps, whose inverse limit is X. Let $p = (p_n)_{n=1}^{\infty} \in X$, and let A be a closed subset of X. Then the following are equivalent:*

(a) $p \in X \setminus \mathcal{T}(A)$,
(b) there exist $N \in \mathbb{N}$, two open subsets U_N and V_N of X_N and an inverse sequence $\{W_n, f_n^{n+1}|_{W_{n+1}}\}$ of subcontinua such that $p_N \in U_N$, $f_N(A) \subset V_N$ and for every $n \geq N$,

$$(f_N^n)^{-1}(U_N) \subset W_n \subset X_n \setminus (f_N^n)^{-1}(V_N).$$

Proof Note that, since the bonding maps are surjective, the projection maps are surjective also (Remark 2.1.6). Thus, the theorem now follows directly from Theorem 3.1.87

Q.E.D.

3.1.89 Corollary *Let $\{X_n, f_n^{n+1}\}$ be an inverse sequence of continua, with surjective bonding maps, whose inverse limit is X. Let A be a closed subset of X, and let $N_0 \in \mathbb{N}$. Then*

$$\bigcap_{n=N_0}^{\infty} f_n^{-1}(\mathcal{T}_{X_n}(f_n(A))) \subset \mathcal{T}_X(A).$$

Proof Let $p = (p_n)_{n=1}^{\infty} \in X \setminus \mathcal{T}_X(A)$. By Theorem 3.1.88, there exist $N \in \mathbb{N}$, two open subsets U_N and V_N of X_N and an inverse sequence $\{W_n, f_n^{n+1}|_{W_{n+1}}\}$ of subcontinua such that $p_N \in U_N$, $f_N(A) \subset V_N$ and for every $n \geq N$,

$$(f_N^n)^{-1}(U_N) \subset W_n \subset X_n \setminus (f_N^n)^{-1}(V_N).$$

Let $m \geq \max\{N, N_0\}$. Then $p_m \in (f_N^m)^{-1}(U_N) \subset W_m \subset X_m \setminus (f_N^m)^{-1}(V_N) \subset X_m \setminus f_m(A)$. Thus, $p_m \in X_m \setminus \mathcal{T}_{X_m}(f_m(A))$. Hence, $p \in X \setminus f_m^{-1}(\mathcal{T}_{X_m}(f_m(A)))$. Therefore, $p \in X \setminus \bigcap_{n=N_0}^{\infty} f_n^{-1}(\mathcal{T}_{X_n}(f_n(A)))$.

Q.E.D.

3.1.90 Corollary *Let $\{X_n, f_n^{n+1}\}$ be an inverse sequence of continua, with surjective bonding maps, whose inverse limit is X. Let A be a closed subset of X, and let $N_0 \in \mathbb{N}$. If the bonding maps are monotone, then*

$$\bigcap_{n=N_0}^{\infty} f_n^{-1}(\mathcal{T}_{X_n}(f_n(A))) = \mathcal{T}_X(A).$$

Proof Let $p = (p_n)_{n=1}^{\infty} \in X \setminus \bigcap_{n=N_0}^{\infty} f_n^{-1}(\mathcal{T}_{X_n}(f_n(A)))$. Then there exist $N \geq N_0$ such that $p_N \in X_N \setminus \mathcal{T}_{X_N}(f_N(A))$. Thus, there exist a subcontinuum W_N of X_N and two open subsets U_N and V_N of X_N such that $p_N \in U_N \subset W_N \subset X_N \setminus V_N \subset X_N \setminus f_N(A)$.

Now, for each $n \in \mathbb{N}$, let

$$W_n = \begin{cases} (f_N^n)^{-1}(W_N), & \text{if } n \geq N; \\ f_n^N(W_N), & \text{if } n \leq N. \end{cases}$$

Since the bonding maps are monotone, W_n is a continuum for each $n \in \mathbb{N}$ (Lemma 2.1.12). By construction, $\{W_n, f_n^{n+1}|_{W_{n+1}}\}$ is an inverse sequence of subcontinua and $(f_N^n)^{-1}(U_N) \subset W_n \subset X_n \setminus (f_N^n)^{-1}(V_N)$. Hence, by Theorem 3.1.88, $p \in X \setminus \mathcal{T}_X(A)$. The other inclusion is given by Corollary 3.1.89.

Q.E.D.

The next corollary is a consequence of Davis's work:

3.1.91 Corollary *Let $\{X_n, f_n^{n+1}\}$ be an inverse sequence of aposyndetic continua, with surjective bonding maps, whose inverse limit is X. If the bonding maps are monotone, then X is aposyndetic.*

Proof Let $p = (p_n)_{n=1}^{\infty} \in X$. By Corollary 3.1.90, we have that $\mathcal{T}_X(\{p\}) = \bigcap_{n=1}^{\infty} f_n^{-1}(\mathcal{T}_{X_n}(\{p_n\}))$. Since each factor space is aposyndetic, by Theorem 3.1.28, $\bigcap_{n=1}^{\infty} f_n^{-1}(\mathcal{T}_{X_n}(\{p_n\})) = \bigcap_{n=1}^{\infty} f_n^{-1}(\{p_n\}) = \{p\}$ (Corollary 2.1.21). Therefore, $\mathcal{T}_X(\{p\}) = \{p\}$. Since p is arbitrary, by Theorem 3.1.28, X is aposyndetic.

Q.E.D.

We turn our attention to strictly point $\mathcal{T}$-asymmetric continua.

3.1.92 Definition A continuum X is *strictly point $\mathcal{T}$-asymmetric* if for any two distinct points p and q of X with $p \in \mathcal{T}(\{q\})$, we have that $q \in X \setminus \mathcal{T}(\{p\})$.

3.1.93 Theorem *If X is an arc-smooth continuum, then X is strictly point $\mathcal{T}$-asymmetric.*

Proof Let X be an arc-smooth continuum at the point p. Let $\alpha \colon X \to \mathcal{C}(X)$ be the map given by the definition of arc-smoothness (Definition 1.7.37). Let x and y be two distinct points of X such that $x \in \mathcal{T}(\{y\})$. Then $y \in W$ for each subcontinuum W of X such that $x \in Int(W)$. We show that $y \in \alpha(x)$. To this end, we prove first that either $\alpha(x) \subset \alpha(y)$ or $\alpha(y) \subset \alpha(x)$. Suppose that there exists a point

$z \in \alpha(x) \cap \alpha(y) \setminus \{x, y\}$ such that $\alpha(z) = \alpha(x) \cap \alpha(y)$. Let $r > 0$ be such that $Cl(V_r^d(x)) \cap \{z, y\} = \emptyset$ and $Cl(V_r^d(y)) \cap \{z, x\} = \emptyset$. By Theorem 1.7.38, $W = \bigcup_{w \in Cl(V_r^d(x))} \alpha(w)$ is a subcontinuum of X which contains x in its interior. Since $x \in \mathcal{T}(\{y\})$, $y \in W$. Let $N \in \mathbb{N}$ be such that $\frac{1}{n} < r$ for each $n \geq N$. Then, for every $n \geq N$, there exists $x_n \in V_{\frac{1}{n}}^d(x)$ such that there exists $y_n \in V_{\frac{1}{n}}^d(y)$ such that $y_n \in \alpha(x_n)$ because if for every $w \in V_{\frac{1}{n}}^d(x)$, there does not exist such $w' \in V_{\frac{1}{n}}^d(y)$, then, by Theorem 1.7.38, $W_n = \bigcup_{w \in Cl(V_{\frac{1}{n}}^d(x))} \alpha(w)$ would be a subcontinuum of X which contains x in its interior and such that $W_n \cap V_{\frac{1}{n}}^d(y) = \emptyset$, a contradiction to the fact that $x \in \mathcal{T}(\{y\})$. Hence, there exist a sequence $\{x_n\}_{n=N}^{\infty}$ which converges to x and a sequence $\{y_n\}_{n=N}^{\infty}$ converging to y such that $y \in \alpha(x_n)$ for all $n \geq N$. Since $y \in \lim_{n \to \infty} \alpha(x_n)$, we have that $\lim_{n \to \infty} \alpha(x_n) \neq \alpha(x)$, a contradiction to the continuity of α. Thus, either $\alpha(x) \subset \alpha(y)$ or $\alpha(y) \subset \alpha(x)$.

Now, assume that $x \in \alpha(y)$. Let $\ell > 0$ be such that $y \in X \setminus V_\ell^d(x)$. By Theorem 1.7.38, $W' = \bigcup_{w \in Cl(V_\ell^d(x))} \alpha(w)$ is a subcontinuum of X such that $x \in Int(W')$. A similar argument to the one given in the previous paragraph shows that there exist a sequence $\{x_n\}_{n=1}^{\infty}$ which converges to x and a sequence $\{y_n\}_{n=1}^{\infty}$ converging to y such that $y_n \in \alpha(x_n)$ for every $n \in \mathbb{N}$. Since $y \in \lim_{n \to \infty} \alpha(x_n)$, we obtain that $\lim_{n \to \infty} \alpha(x_n) \neq \alpha(x)$, a contradiction to the continuity of α. Thus, $y \in \alpha(x)$.

Finally, let $s > 0$ be such that $Cl(V_s^d(x)) \cap Cl(V_s^d(y)) = \emptyset$. Let

$$W_y = \bigcup_{w \in Cl(V_s^d(y))} \alpha(w).$$

Then, by Theorem 1.7.38, we have that W_y is a subcontinuum of X which contains y in its interior. Since $y \in \alpha(x)$, we obtain that $x \in X \setminus W_y$. Thus, $y \in X \setminus \mathcal{T}(\{x\})$. Therefore, X is strictly point $\mathcal{T}$-asymmetric.

Q.E.D.

Note that by [20, Theorem II-4-B], a one-dimensional arc-smooth continuum is a smooth dendroid (Definition 6.7.10). Hence, as a consequence of this and Theorem 3.1.93, we have:

3.1.94 Corollary *If X is a smooth dendroid, then X is strictly point $\mathcal{T}$-asymmetric.*

The following example shows that the converse of Theorem 3.1.93 is not true even for dendroids. This gives a negative answer to Question 9.2.9. We need the following notation:

3.1.95 Notation For each pair of points x and y of $\mathbb{R}^n$, let $[\![x, y]\!] = \{(1 - t)x + ty \mid t \in [0, 1]\}$.

3.1.96 Example Let X be the following subcontinuum of $\mathbb{R}^2$:

$$X = \bigcup_{n=1}^{\infty}\left[\!\!\left[(-1,0),\left(0,\frac{1}{n}\right)\right]\!\!\right] \cup [(-1,0),(1,0)] \cup \bigcup_{n=1}^{\infty}\left[\!\!\left[(1,0),\left(0,-\frac{1}{n}\right)\right]\!\!\right].$$

Then X is a strictly $\mathcal{T}$-asymmetric dendroid which is not smooth.

The converse of Theorem 3.1.93 is true for the class of fans (Definition 6.7.9). In order to prove this, we need the following lemma.

3.1.97 Lemma *Let X be a fan with top τ and let p be a point of $X \setminus \{\tau\}$. If $\{p_n\}_{n=1}^{\infty}$ is a sequence of points of X converging to p, then the arc τp is contained in $\liminf \tau p_n$.*

Proof Suppose that there exists a point $t \in \tau p \setminus \liminf \tau p_n$. Then there exist an open subset U of X such that $t \in U$ and a sequence $\{n_k\}_{k=1}^{\infty}$ of positive integers such that $\tau p_{n_k} \cap U = \emptyset$ for each $k \in \mathbb{N}$. Note that by Lemma 1.2.27 and Theorem 1.2.29, $\limsup \tau p_{n_k}$ is a subcontinuum of X. Hence, since for every $k \in \mathbb{N}$, $\tau \in \tau p_{n_k}$ and the sequence $\{p_{n_k}\}_{k=1}^{\infty}$ converges to p, we obtain that $\tau p \subset \limsup \tau p_{n_k}$. Thus, $t \in \limsup \tau p_{n_k}$. Then there exists a subsequence $\{n_{k_\ell}\}_{\ell=1}^{\infty}$ of $\{n_k\}_{k=1}^{\infty}$ such that $U \cap \tau p_{n_{k_\ell}} \neq \emptyset$ for all $\ell \in \mathbb{N}$, a contradiction to our assumption. Therefore, $\tau p \subset \limsup \tau p_n$.

Q.E.D.

3.1.98 Theorem *A fan X is strictly point $\mathcal{T}$-asymmetric if and only if it is smooth.*

Proof Let X be a fan with top τ. If X is smooth, then by [20, Theorem II-4-B] and Theorem 3.1.93, X is strictly point $\mathcal{T}$-asymmetric.

Suppose X is not smooth. Then there exist a point $p \in X$ and a sequence $\{p_n\}_{n=1}^{\infty}$ of points of X such that $\{p_n\}_{n=1}^{\infty}$ converges to p but $\limsup \tau p_n \neq \liminf \tau p_n$. Hence, by Lemma 3.1.97, $\tau p \subset \liminf \tau p_n$. Thus, $\limsup \tau p_n \neq \tau p$. Let $q \in \limsup \tau p_n \setminus \tau p$ and let e_p be the end point of the leg of X which contains p. Then there exists a subsequence $\{p_{n_k}\}_{k=1}^{\infty}$ of $\{p_n\}_{n=1}^{\infty}$ such that for every $k \in \mathbb{N}$, there exists $q_{n_k} \in \tau p_{n_k}$ in such a way that the sequence $\{q_{n_k}\}_{k=1}^{\infty}$ converges to q, and $p_{n_k} \in X \setminus \tau e_p$ for all $k \in \mathbb{N}$. We consider two cases.

Case (1) $p \in \tau q$.

Let W be a subcontinuum of X containing p in its interior. Since it is easy to see that subcontinua of dendroids are dendroids, W is arcwise connected. Since $p \in Int(W)$, there exists $K \in \mathbb{N}$ such that $p_{n_k} \in Int(W)$ for each $k \geq K$. Since W is arcwise connected and $q_{n_k} \in \tau p_{n_k}$ for each $k \in \mathbb{N}$, we have that $q_{n_k} \in W$ for every $k \geq K$. Thus, $q \in W$. This implies that $p \in \mathcal{T}(\{q\})$. Now if W' is a subcontinuum of X containing q in its interior, there exists $K' \in \mathbb{N}$ such that $q_{n_k} \in Int(W')$ for each $k \geq K'$. Since $q_{n_k} \in \tau p_{n_k}$ and W' is arcwise connected, the arcs τq and τq_{n_k} are contained in W' for every $k \geq K'$. Since $p \in \tau q$, $p \in W'$. Thus, $q \in \mathcal{T}(\{p\})$. Therefore, X is not strictly point $\mathcal{T}$-asymmetric.

Case (2) $p \in X \setminus \tau q$.

Let $r \in \tau q \setminus \{\tau, q\}$. By Lemma 1.2.27 and Theorem 1.2.29, $\limsup q_{n_k} p_{n_k}$ is a subcontinuum of X which contains q and p. Thus, $qp \subset \limsup q_{n_k} p_{n_k}$ ($\limsup q_{n_k} p_{n_k}$ is arcwise connected). Hence, there exists a sequence $\{r_{n_k}\}_{k=1}^{\infty}$ converging to r such that $r_{n_k} \in q_{n_k} p_{n_k}$ for every $k \in \mathbb{N}$. Thus, r is in the same situation as p in Case (1); i.e., $q_{n_k} \in \tau r_{n_k}$ and the sequences $\{q_{n_k}\}_{k=1}^{\infty}$ and $\{r_{n_k}\}_{k=1}^{\infty}$ converge to q and r, respectively. Hence, $q \in \mathcal{T}(\{r\})$ and $r \in \mathcal{T}(\{q\})$. Therefore, X is not strictly point $\mathcal{T}$ asymmetric.

Q.E.D.

3.2 Idempotency of $\mathcal{T}$

In 1980, David Bellamy asked: *If X and Y are indecomposable continua, is $\mathcal{T}$ idempotent on $X \times Y$? Even for only the closed sets of $X \times Y$?* (see Question 9.2.3). We prove that the set function $\mathcal{T}$ is not idempotent on the product of two indecomposable continua. This gives a negative answer to the first question, the second one remains open. We also prove that $\mathcal{T}$ is never idempotent on the cone and suspension over an indecomposable continuum. We give an example of a decomposable continuum Z such that $\mathcal{T}$ is not idempotent on the family of closed sets of either $Z \times [0, 1]$ or $K(Z)$ or $\Sigma(Z)$. To this end, we use [13, 35, 42].

We begin with a characterization of locally connected continua.

3.2.1 Theorem *Let X be a continuum such that $\mathcal{T}$ is idempotent on closed sets. Then X is locally connected if and only if $\mathcal{T}|_{\mathcal{C}(X)} \colon \mathcal{C}(X) \twoheadrightarrow \mathcal{C}(X)$ is surjective.*

Proof First note that, by Remark 3.1.5, $A \subset \mathcal{T}(A)$ for all $A \in \mathcal{P}(X)$. Thus, $\mathcal{T}(\{x\}) = \{x\}$ for all $x \in X$. Hence, X is aposyndetic, by Theorem 3.1.28.

Let $A \in \mathcal{C}(X)$. Since $\mathcal{T}|_{\mathcal{C}(X)}$ is surjective, there exists $A' \in \mathcal{C}(X)$ such that $\mathcal{T}(A') = A$. Thus, since $\mathcal{T}$ is idempotent on closed sets, $\mathcal{T}(A) = \mathcal{T}^2(A') = \mathcal{T}(A') = A$. Therefore, by Theorem 3.1.32, X is locally connected.

The converse implication is clear since for locally connected continua X, $\mathcal{T}$ is the identity map on 2^X, by Theorem 3.1.31. In particular, $\mathcal{T}|_{\mathcal{C}(X)}$ is the identity map on $\mathcal{C}(X)$.

Q.E.D.

3.2.2 Corollary *Let X be a continuum such that $\mathcal{T}$ is idempotent on closed sets. Then X is locally connected if and only if $\mathcal{T} \colon 2^X \twoheadrightarrow 2^X$ is surjective.*

3.2.3 Theorem *Let X and Y be continua such that for each subcontinuum W of X, $\mathcal{T}_X^2(W) = \mathcal{T}_X(W)$, and let $f \colon X \twoheadrightarrow Y$ be a monotone map. If for each subcontinuum Z of Y, there exists a subcontinuum W of X such that $\mathcal{T}_X(W) = f^{-1}(Z)$, then Y is locally connected.*

Proof Let Z be a subcontinuum of Y. Since f is monotone, by Theorem 3.1.80 (c), we have that $\mathcal{T}_Y(Z) = f\mathcal{T}_X f^{-1}(Z)$. By hypothesis, there exists a subcontinuum W of X such that $\mathcal{T}_X(W) = f^{-1}(Z)$. Hence, since $\mathcal{T}_X^2(W) = \mathcal{T}_X(W)$, we have that $f\mathcal{T}_X f^{-1}(Z) = f\mathcal{T}_X\mathcal{T}_X(W) = f\mathcal{T}_X(W) = ff^{-1}(Z) = Z$. Thus, $\mathcal{T}_Y(Z) = Z$. Therefore, by Theorem 3.1.32, Y is locally connected.

$$\textbf{Q.E.D.}$$

Next, we prove that the set function $\mathcal{T}$ is not idempotent on the product of two continua one of which is indecomposable. In particular, $\mathcal{T}$ is not idempotent on the product of two indecomposable continua. We need the following:

3.2.4 Lemma *Let X be an indecomposable continuum, let $x_0 \in X$ and let Y be a continuum. If W is a subcontinuum of $X \times Y$ such that $Int_{X \times Y}(W) \neq \emptyset$, then $W \cap (\{x_0\} \times Y) \neq \emptyset$.*

Proof Let $\pi_X \colon X \times Y \twoheadrightarrow X$ be the projection map. Note that $\pi_X(W)$ is a subcontinuum of X such that $Int_X(\pi_X(W)) \neq \emptyset$. Thus, since X is indecomposable, $\pi_X(W) = X$, by Corollary 1.7.26. Let $w \in W$ be such that $\pi_X(w) = x_0$. Then $w \in W \cap (\{x_0\} \times Y)$. Therefore, $W \cap (\{x_0\} \times Y) \neq \emptyset$.

$$\textbf{Q.E.D.}$$

3.2.5 Corollary *If X is an indecomposable continuum, $x_0 \in X$ and Y is a continuum, then $\mathcal{T}_{X \times Y}(\{x_0\} \times Y) = X \times Y$.*

3.2.6 Theorem *If X is an indecomposable continuum and Y is a continuum, then $\mathcal{T}_{X \times Y}$ is not idempotent on $X \times Y$.*

Proof Let $(x_0, y_0) \in X \times Y$. Let $A = (\{x_0\} \times Y) \setminus \{(x_0, y_0)\}$, and let $(x, y) \in (X \times Y) \setminus A$, where $y \neq y_0$. Let U be an open subset of X such that $x \in U \subset Cl_X(U) \subset X \setminus \{x_0\}$. Let

$$W = (Cl_X(U) \times Y) \cup (X \times \{y_0\}).$$

Then W is a subcontinuum of $X \times Y$ such that $(x, y) \in Int_{X \times Y}(W) \subset W \subset (X \times Y) \setminus A$. Hence, $\mathcal{T}_{X \times Y}(A) \neq X \times Y$. Since $\mathcal{T}_{X \times Y}(A)$ is closed, $Cl_{X \times Y}(A) \subset \mathcal{T}_{X \times Y}(A)$. Since A is dense in $\{x_0\} \times Y$, $Cl_{X \times Y}(A) = \{x_0\} \times Y$. Hence, by Corollary 3.2.5, $X \times Y \subset \mathcal{T}_{X \times Y}(\{x_0\} \times Y) \subset \mathcal{T}_{X \times Y}^2(A)$. Therefore, $\mathcal{T}_{X \times Y}$ is not idempotent on $X \times Y$.

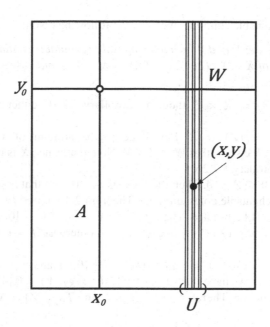

Q.E.D.

3.2.7 Corollary *If X and Y are indecomposable continua, then $\mathcal{T}_{X \times Y}$ is not idempotent on $X \times Y$.*

3.2.8 Remark Observe that Corollary 3.2.7 gives a negative answer to the first question of Question 9.2.3. The second question remains open.

3.2.9 Theorem *Let X and Y be continua. If Z is a closed subset of $X \times Y$ such that $\pi_X(Z) \neq X$ and $\pi_Y(Z) \neq Y$, where π_X and π_Y are the projection maps, then $\mathcal{T}_{X \times Y}(Z) \subset \pi_X(Z) \times \pi_Y(Z)$.*

Proof Let $(x, y) \in (X \times Y) \setminus (\pi_X(Z) \times \pi_Y(Z))$. Without loss of generality, we assume that $x \in X \setminus \pi_X(Z)$. Since $\pi_X(Z)$ is closed in X, there exists an open set U of X such that $x \in U \subset Cl_X(U) \subset X \setminus \pi_X(Z)$. Let $y' \in Y \setminus \pi_Y(Z)$. Let $W = (Cl_X(U) \times Y) \cup (X \times \{y'\})$. Then W is a subcontinuum of $X \times Y$ such that $(x, y) \in Int_{X \times Y}(W) \subset W \subset (X \times Y) \setminus \pi_X(Z) \times \pi_Y(Z)$. Therefore, $\mathcal{T}_{X \times Y}(Z) \subset \pi_X(Z) \times \pi_Y(Z)$.

Q.E.D.

3.2.10 Corollary *Let X and Y be continua. If Z is a closed subset of $X \times Y$ such that $\mathcal{T}_{X \times Y}(Z) = X \times Y$, then either $\pi_X(Z) = X$ or $\pi_Y(Z) = Y$.*

3.2.11 Corollary *Let X and Y be continua. If A and B are nonempty proper closed subsets of X and Y, respectively, then $\mathcal{T}_{X \times Y}(A \times B) = A \times B$.*

Regarding chainable continua, we have the following result:

3.2.12 Theorem *Let X and Y be indecomposable chainable continua, and let Z be a subcontinuum of $X \times Y$. Then $\mathcal{T}_{X \times Y}(Z) = X \times Y$ if and only if $\pi_X(Z) = X$ or $\pi_Y(Z) = Y$.*

Proof If $\mathcal{T}_{X \times Y}(Z) = X \times Y$, then, by Corollary 3.2.10, either $\pi_X(Z) = X$ or $\pi_Y(Z) = Y$.

Now, suppose $\pi_X(Z) = X$. Let W be a subcontinuum of $X \times Y$ such that $Int_{X \times Y}(W) \neq \emptyset$. We show that $W \cap Z \neq \emptyset$. Note that, since X is indecomposable, $\pi_X(W) = X$, Corollary 1.7.26.

Suppose that $W \cap Z = \emptyset$. Then there exists $\varepsilon > 0$ such that $\varepsilon < d(W, Z)$. Note that $\pi_Y(Z)$ is a chainable continuum, by Theorem 2.4.10. Let $f \colon X \twoheadrightarrow [0, 1]$ and $g \colon \pi_Y(Z) \twoheadrightarrow [0, 1]$ be ε-maps. Then $f \times g \colon X \times \pi_Y(Z) \twoheadrightarrow [0, 1] \times [0, 1]$ is an ε-map. Hence, $(f \times g)(Z) \cap (f \times g)(W) = \emptyset$, otherwise $f \times g$ would not be an ε-map.

Since $\pi_X((f \times g)(W)) = f(\pi_X(W)) = [0, 1]$ and $\pi_Y((f \times g)(Z)) = g(\pi_Y((Z)) = [0, 1]$, we have that $(f \times g)(Z) \cap (f \times g)(W) \neq \emptyset$ [42, Theorem 130, p. 158], a contradiction. Therefore, $W \cap Z \neq \emptyset$, and $\mathcal{T}_{X \times Y}(Z) = X \times Y$.

Q.E.D.

As a consequence of Theorems 3.2.9 and 3.2.12, we have:

3.2.13 Corollary *Let X and Y be indecomposable chainable continua. If Z is a subcontinuum of $X \times Y$, then either $\mathcal{T}_{X \times Y}(Z) = X \times Y$ or $\mathcal{T}_{X \times Y}(Z) \subset \pi_X(Z) \times \pi_Y(Z)$.*

Now, we turn our attention to cones, $q \colon X \times [0, 1] \twoheadrightarrow K(X)$ denotes the quotient map.

3.2.14 Lemma *Let X be an indecomposable continuum and let $x_0 \in X$. If W is a subcontinuum of $K(X)$ such that $Int_{K(X)}(W) \neq \emptyset$, then $W \cap q(\{x_0\} \times [0, 1]) \neq \emptyset$.*

Proof If $v_X \in W$, then it is clear that $W \cap q(\{x_0\} \times Y) \neq \emptyset$. Suppose that $v_X \notin W$. Then, since q is monotone, $q^{-1}(W)$ is a subcontinuum of $X \times [0, 1]$ such that $Int_{X \times [0,1]}(q^{-1}(W)) \neq \emptyset$. By Lemma 3.2.4, $q^{-1}(W) \cap (\{x_0\} \times [0, 1]) \neq \emptyset$. Therefore, $W \cap q(\{x_0\} \times [0, 1]) \neq \emptyset$.

Q.E.D.

3.2.15 Corollary *If X is an indecomposable continuum and $x_0 \in X$, then $\mathcal{T}_{K(X)}q(\{x_0\} \times [0, 1]) = K(X)$.*

3.2.16 Theorem *If X is an indecomposable, then $\mathcal{T}_{K(X)}$ is not idempotent on $K(X)$.*

Proof Let $(x_0, t_0) \in X \times (0, 1)$. Let $A = (X \times [0, 1]) \setminus \{(x_0, t_0)\}$, and let $(x, t) \in (X \times (0, 1)) \setminus A$, where $t \neq t_0$. Construct W as in the proof of Theorem 3.2.6 in such a way that $W \cap (X \times \{1\}) = \emptyset$. Then $q(x, t) \in Int_{K(X)}(q(W)) \subset q(W) \subset K(X) \setminus q(A)$. Thus, $\mathcal{T}_{K(X)}q(A) \neq K(X)$. Since $q(A)$ is dense in $q(\{x_0\} \times Y)$,

$Cl_{K(X)}(q(A)) = q(\{x_0\} \times Y)$. Hence, by Corollary 3.2.15, $K(X) \subset \mathcal{T}_{K(X)}q(\{x_0\} \times Y) \subset \mathcal{T}^2_{K(X)}q(A)$. Therefore, $\mathcal{T}_{K(X)}$ is not idempotent on $K(X)$.

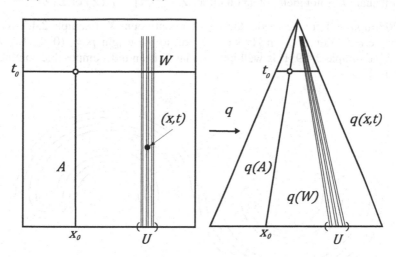

Q.E.D.

Next, we consider suspensions. The results about suspensions are similar to the ones about cones, we state them and present the necessary changes. Here, $q : X \times [0, 1] \twoheadrightarrow \Sigma(X)$ denotes the quotient map.

3.2.17 Lemma *Let X be an indecomposable continuum and let $x_0 \in X$. If W is a subcontinuum of $\Sigma(X)$ such that $Int_{\Sigma(X)}(W) \neq \emptyset$, then $W \cap q(\{x_0\} \times [0, 1]) \neq \emptyset$.*

Proof If either $v^+ \in W$ or $v^- \in W$, then it is clear that $W \cap q(\{x_0\} \times [0, 1]) \neq \emptyset$. Suppose that $\{v^+, v^-\} \cap W = \emptyset$. Then, since q is monotone, $q^{-1}(W)$ is a subcontinuum of $X \times [0, 1]$ such that $Int_{X \times [0,1]}(q^{-1}(W)) \neq \emptyset$. By Lemma 3.2.4, $q^{-1}(W) \cap (\{x_0\} \times [0, 1]) \neq \emptyset$. Therefore, $W \cap q(\{x_0\} \times [0, 1]) \neq \emptyset$.

Q.E.D.

3.2.18 Corollary *If X is an indecomposable continuum and $x_0 \in X$, then $\mathcal{T}_{\Sigma(X)}q(\{x_0\} \times [0, 1]) = \Sigma(X)$.*

3.2.19 Theorem *If X is an indecomposable, then $\mathcal{T}_{\Sigma(X)}$ is not idempotent on $\Sigma(X)$.*

Proof Let $(x_0, t_0) \in X \times (0, 1)$. Let $A = (X \times [0, 1]) \setminus \{(x_0, t_0)\}$, and let $(x, t) \in (X \times (0, 1)) \setminus A$, where $t \neq t_0$. Construct W as in the proof of Theorem 3.2.6 in such a way that $W \cap (X \times \{1\} \cup X \times \{0\}) = \emptyset$. Then $q(x, t) \in Int_{\Sigma(X)}(q(W)) \subset q(W) \subset \Sigma(X) \setminus q(A)$. Thus, $\mathcal{T}_{\Sigma(X)}q(A) \neq \Sigma(X)$. Since $q(A)$ is dense in $q(\{x_0\} \times [0, 1])$, $Cl_{\Sigma(X)}(q(A)) = q(\{x_0\} \times [0, 1])$. Hence, by Corollary 3.2.18, $\Sigma(X) \subset \mathcal{T}_{\Sigma(X)}q(\{x_0\} \times Y) \subset \mathcal{T}^2_{\Sigma(X)}q(A)$. Therefore, $\mathcal{T}_{\Sigma(X)}$ is not idempotent on $\Sigma(X)$.

Q.E.D.

To end this section, we give a decomposable continuum Z such that $\mathcal{T}$ is idempotent on the family of closed sets of neither $Z \times [0, 1]$ nor $K(Z)$ nor $\Sigma(Z)$. In particular, $\mathcal{T}$ is not idempotent on either $Z \times [0, 1]$ or $K(Z)$ or $\Sigma(Z)$.

3.2.20 Example Let us consider the Knaster continuum X, Example 2.4.7, and let X' be the reflection of X in $\mathbb{R}^2$ with respect to the origin $p = (0, 0)$. Let $Z = X \cup X'$, Example 2.4.9. It is well known that X is an indecomposable continuum, Remark 2.4.8.

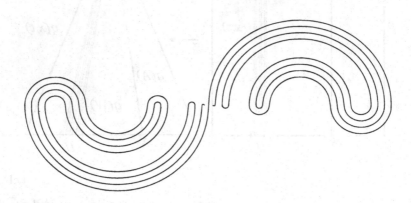

To see that $\mathcal{T}_{Z \times [0,1]}$ is not idempotent on the family of closed subsets of $Z \times [0, 1]$, let us note that, using Corollary 3.2.5, we have that $\mathcal{T}_{Z \times [0,1]}(\{p\} \times [0, 1]) = Z \times [0, 1]$. Let $z \in X \setminus \{p\}$. Then, using again Corollary 3.2.5, we obtain that $\mathcal{T}_{Z \times [0,1]}(\{z\} \times [0, 1]) = X \times [0, 1]$. In particular, $\{p\} \times [0, 1] \subset \mathcal{T}_{Z \times [0,1]}(\{z\} \times [0, 1])$. Hence,

$$\mathcal{T}_{Z \times [0,1]}\left(\mathcal{T}_{Z \times [0,1]}(\{z\} \times [0, 1])\right) = \mathcal{T}_{Z \times [0,1]}(X \times [0, 1]) = Z \times [0, 1].$$

Therefore, $\mathcal{T}_{Z \times [0,1]}$ is not idempotent on the family of closed subsets of $Z \times [0, 1]$.

To show that $\mathcal{T}_{K(Z)}$ is not idempotent on the family of closed subsets of $K(Z)$, let us observe that, using Corollary 3.2.15, we obtain that $\mathcal{T}_{K(Z)}(q(\{p\} \times [0, 1])) = K(Z)$. Let $z \in X \setminus \{p\}$. Then, using Corollary 3.2.15, we have that $\mathcal{T}_{K(Z)}(q(\{z\} \times [0, 1])) = K(X)$. Observe that $q(\{p\} \times [0, 1]) \subset \mathcal{T}_{K(Z)}(q(\{z\} \times [0, 1])$. Thus,

$$\mathcal{T}_{K(Z)}\left(\mathcal{T}_{K(Z)}(q(\{z\} \times [0, 1]))\right) = \mathcal{T}_{K(Z)}(K(X)) = K(Z).$$

Therefore, $\mathcal{T}_{K(Z)}$ is not idempotent on the family of closed subsets of $K(Z)$.

To prove that $\mathcal{T}_{\Sigma(Z)}$ is not idempotent on the family of closed subsets of $\Sigma(Z)$, let us note that, using Corollary 3.2.18, we obtain that $\mathcal{T}_{\Sigma(Z)}(q(\{p\} \times [0, 1])) = \Sigma(Z)$. Let $z \in X \setminus \{p\}$. Then, by Corollary 3.2.18, we have that $\mathcal{T}_{\Sigma(Z)}(q(\{z\} \times [0, 1])) = \Sigma(X)$. Hence, $q(\{p\} \times [0, 1]) \subset \mathcal{T}_{\Sigma(Z)}(q(\{z\} \times [0, 1]))$. Thus,

$$\mathcal{T}_{\Sigma(Z)}\left(\mathcal{T}_{\Sigma(Z)}(q(\{z\} \times [0, 1]))\right) = \mathcal{T}_{\Sigma(Z)}(\Sigma(X)) = \Sigma(Z).$$

Therefore, $\mathcal{T}_{\Sigma(Z)}$ is not idempotent on the family of closed subsets of $\Sigma(Z)$.

3.3 Continuity of $\mathcal{T}$

Let X be a continuum. By Remark 3.1.5, the image of any subset of X under $\mathcal{T}$ is a closed subset of X. Then we may restrict the domain of $\mathcal{T}$ to the hyperspace, 2^X, of nonempty closed subsets of X. Since 2^X has a topology, we may ask if $\mathcal{T}: 2^X \to 2^X$ is continuous. The answer to this question is negative, as can be easily seen from Example 3.1.15. On the other hand, by Theorems 3.1.31 and 3.1.39, $\mathcal{T}$ is continuous for locally connected continua and for indecomposable continua, respectively. In this section we present some results related to the continuity of $\mathcal{T}$. In particular, we show that if a continuum X is almost connected im kleinen at each of its points and $\mathcal{T}$ is continuous, then X is locally connected. The results of this section are taken from [1, 2, 32] and [50].

We begin with the following theorem which says that $\mathcal{T}$ is always upper semicontinuous.

3.3.1 Theorem *Let X be a continuum. If U is an open subset of X, then $\mathcal{U} = \{A \in 2^X \mid \mathcal{T}(A) \subset U\}$ is open in 2^X.*

Proof Let U be an open subset of X and let $\mathcal{U} = \{A \in 2^X \mid \mathcal{T}(A) \subset U\}$. We show $\mathcal{U}$ is open in 2^X. Let $B \in Cl_{2^X}(2^X \setminus \mathcal{U})$. Then there exists a sequence $\{B_n\}_{n=1}^{\infty}$ of elements of $2^X \setminus \mathcal{U}$ converging to B. Note that for each $n \in \mathbb{N}$, $\mathcal{T}(B_n) \cap (X \setminus U) \neq \emptyset$. Let $x_n \in \mathcal{T}(B_n) \cap (X \setminus U)$. Without loss of generality, we assume that $\{x_n\}_{n=1}^{\infty}$ converges to a point $x \in X$. Note that $x \in X \setminus U$. We assert that $x \in \mathcal{T}(B)$. Suppose $x \in X \setminus \mathcal{T}(B)$. Then there exists a subcontinuum W of X such that $x \in Int(W) \subset W \subset X \setminus B$. Since $\{B_n\}_{n=1}^{\infty}$ converges to B and $\{x_n\}_{n=1}^{\infty}$ converges to x, there exists $N \in \mathbb{N}$ such that for each $n \geq N$, $B_n \subset X \setminus W$ and $x_n \in Int(W)$. Let $n \geq N$. Then $x \in Int(W) \subset W \subset X \setminus B_n$. This implies that $x_n \in X \setminus \mathcal{T}(B_n)$, a contradiction. Thus, $x \in \mathcal{T}(B) \cap (X \setminus U)$. Hence, $B \in 2^X \setminus \mathcal{U}$. Therefore, $2^X \setminus \mathcal{U}$ is closed in 2^X.

$\qquad$ **Q.E.D.**

The next two theorems give sufficient conditions for $\mathcal{T}$ to be continuous for a continuum.

3.3.2 Theorem *Let X and Z be continua, where Z is locally connected. If $f: X \twoheadrightarrow$ Z is a surjective, monotone and open map such that for each proper subcontinuum W of X, $f(W) \neq Z$, then $\mathcal{T}_X$ is continuous for X.*

Proof Let A be a closed subset of X. Since Z is locally connected, $\mathcal{T}_Z(f(A)) = f(A)$ (Theorem 3.1.31). Hence, $f\mathcal{T}_X f^{-1}f(A) = f(A)$, by Theorem 3.1.80 (c). Thus,

$$f^{-1}f\mathcal{T}_X f^{-1}f(A) = f^{-1}f(A)$$

and

$$\mathcal{T}_X f^{-1}f(A) \subset f^{-1}f(A).$$

Then $\mathcal{T}_X f^{-1}f(A) = f^{-1}f(A)$, by Remark 3.1.5. Since $A \subset f^{-1}f(A)$, $\mathcal{T}_X(A) \subset \mathcal{T}_X f^{-1}f(A) = f^{-1}f(A)$ (the inclusion holds by Proposition 3.1.7).

Suppose there exists $x \in f^{-1}f(A) \setminus \mathcal{T}_X(A)$. Then there exists a subcontinuum W of X such that $x \in Int(W) \subset W \subset X \setminus A$. Since f is open, $f(x) \in Int(f(W))$. For each $z \in Z \setminus f(Int(W))$, there exists a subcontinuum M_z of Z such that $z \in Int(M_z) \subset M_z \subset Z \setminus \{f(x)\}$ ($\mathcal{T}_Z(\{z\}) = \{z\}$). Since $Z \setminus f(Int(W))$ is compact, there exist $z_1, \ldots, z_n \in Z \setminus f(Int(W))$ such that $Z \setminus f(Int(W)) \subset \bigcup_{j=1}^{n} Int(M_{z_j}) \subset \bigcup_{j=1}^{n} M_{z_j}$. By choosing n as small as possible, we may assume that $M_{z_j} \cap f(W) \neq \emptyset$ for each $j \in \{1, \ldots, n\}$. Then $Z = f(W) \cup \left(\bigcup_{j=1}^{n} M_{z_j} \right)$, and $f^{-1}f(W) \cup \left(\bigcup_{j=1}^{n} f^{-1}(M_{z_j}) \right) = X$.

For each $j \in \{1, \ldots, n\}$, let $q_j \in f(W) \cap M_{z_j}$. Then there exists $p_j \in W$ such that $f(p_j) = q_j$. Since $p_j \in f^{-1}(M_{z_j})$, $W \cap f^{-1}(M_{z_j}) \neq \emptyset$, $j \in \{1, \ldots, n\}$. Let $Y = W \cup \left(\bigcup_{j=1}^{n} f^{-1}(M_{z_j}) \right)$. Then Y is a subcontinuum of X. We assert that $Y \neq X$. To see this, recall that $x \in f^{-1}f(A)$. Thus, there exists $y \in A$ such that $f(y) = f(x)$. Then $y \in X \setminus W$ and $f(y) \in Z \setminus \left(\bigcup_{j=1}^{n} M_{z_j} \right)$ (recall the construction of the M_{z_j}'s). Hence, $y \in X \setminus \left(\bigcup_{j=1}^{n} f^{-1}(M_{z_j}) \right)$. However, $f(Y) = Z$, contrary to our hypothesis. Therefore, $\mathcal{T}_X(A) = f^{-1}f(A)$; i.e., $\mathcal{T}_X = \Im(f) \circ 2^f$.

Since f is continuous and open, $\Im(f)$ and 2^f are continuous (Theorems 8.5.5 and 8.2.2, respectively). Therefore, $\mathcal{T}_X$ is continuous.

Q.E.D.

3.3.3 Definition Let $f: X \to Z$ be a map. We say that f is $\mathcal{T}_{XZ}$-*continuous* provided that always $f\mathcal{T}_X(A) \subset \mathcal{T}_Z f(A)$ for every $A \in \mathcal{P}(X)$, or equivalently, $f^{-1}\mathcal{T}_Z(B) \supset \mathcal{T}_X f^{-1}(B)$ for every $B \in \mathcal{P}(Z)$.

3.3.4 Theorem *Let X be a continuum for which $\mathcal{T}_X$ is continuous, and let Z be a continuum. If $f: X \twoheadrightarrow Z$ is a $\mathcal{T}_{XZ}$-continuous surjective open map, then $\mathcal{T}_Z$ is continuous for Z.*

Proof Since f is $\mathcal{T}_{XZ}$-continuous, by Theorem 3.1.80 (a),

$$f\mathcal{T}_X f^{-1}(B) = \mathcal{T}_Z(B)$$

for each $B \in \mathcal{P}(Z)$. Thus, $\mathcal{T}_Z = 2^f \circ \mathcal{T}_X \circ \Im(f)$. Since f is continuous and open, $\Im(f)$ and 2^f are continuous (Theorems 8.5.5 and 8.2.2, respectively). Hence, $\mathcal{T}_Z$ is a composition of three maps. Therefore, $\mathcal{T}_Z$ is continuous.

Q.E.D.

The next lemma says that given a continuum X for which $\mathcal{T}|_{\mathcal{F}_1(X)}$ is continuous, the family $\{\mathcal{T}(\{x\}) \mid x \in X\}$ resembles an upper semicontinuous decomposition.

3.3.5 Lemma *Let X be a continuum, with metric d, for which $\mathcal{T}|_{\mathcal{F}_1(X)}$ is continuous. If U is an open subset of X and K is a closed subset of X, then:*

(1) $A = \{x \in X \mid \mathcal{T}(\{x\}) \subset U\}$ is open in X.
(2) $B = \{x \in X \mid \mathcal{T}(\{x\}) \cap U \neq \emptyset\}$ is open in X.
(3) $C = \{x \in X \mid \mathcal{T}(\{x\}) \cap K \neq \emptyset\}$ is closed in X.
(4) $D = \{x \in X \mid \mathcal{T}(\{x\}) \subset K\}$ is closed in X.

Proof Let $x \in A$. Then $\mathcal{T}(\{x\}) \subset U$, and there exists $\varepsilon > 0$ such that $\mathcal{V}_\varepsilon^d(\mathcal{T}(\{x\})) \subset U$. Let $\delta > 0$ be given by the uniform continuity of $\mathcal{T}|_{\mathcal{F}_1(X)}$ for ε. Let $w \in \mathcal{V}_\delta^d(x)$. Then $\{w\} \in \mathcal{V}_\delta^{\mathcal{H}}(\{x\})$. Thus, $\mathcal{H}(\mathcal{T}(\{w\}), \mathcal{T}(\{x\})) < \varepsilon$. Hence, $\mathcal{T}(\{w\}) \subset \mathcal{V}_\varepsilon^d(\mathcal{T}(\{x\})) \subset U$. This implies that $w \in A$. Therefore, A is open.

Let $x \in B$. Then $\mathcal{T}(\{x\}) \cap U \neq \emptyset$. Let $y \in \mathcal{T}(\{x\}) \cap U$ and let $\varepsilon > 0$ such that $\mathcal{V}_\varepsilon^d(y) \subset U$. Let $\delta > 0$ be given by the uniform continuity of $\mathcal{T}|_{\mathcal{F}_1(X)}$ for ε. Let $z \in \mathcal{V}_\delta^d(x)$. Then $\{z\} \in \mathcal{V}_\delta^{\mathcal{H}}(\{x\})$. Hence, $\mathcal{H}(\mathcal{T}(\{z\}), \mathcal{T}(\{x\})) < \varepsilon$. This implies that there exists $z' \in \mathcal{T}(\{z\})$ such that $d(z', y) < \varepsilon$. Thus, $\mathcal{T}(\{z\}) \cap U \neq \emptyset$. Hence, $z \in B$. Therefore, B is open.

Now, let Z be either C or D. Let $x \in Cl(Z)$. Then there exists a sequence $\{x_n\}_{n=1}^\infty$ of elements of Z converging to x. Since $\mathcal{T}|_{\mathcal{F}_1(X)}$ is continuous, $\lim_{n \to \infty} \mathcal{T}(\{x_n\}) = \mathcal{T}(\{x\})$.

If Z is C, then for each $n \in \mathbb{N}$, there exists $k_n \in \mathcal{T}(\{x_n\}) \cap K$. Hence, without loss of generality, we assume that $\{k_n\}_{n=1}^\infty$ converges to a point k. Since $\lim_{n \to \infty} \mathcal{T}(\{x_n\}) = \mathcal{T}(\{x\})$ and K is closed, $k \in \mathcal{T}(\{x\}) \cap K$. Thus, $x \in C$. Therefore, C is closed.

If Z is D, then for each $n \in \mathbb{N}$, $\mathcal{T}(\{x_n\}) \subset K$. Since $\lim_{n \to \infty} \mathcal{T}(\{x_n\}) = \mathcal{T}(\{x\})$ and K is closed, $\mathcal{T}(\{x\}) \subset K$. Hence, $x \in D$. Therefore, D is closed.

Q.E.D.

3.3.6 Notation Let X be a compactum. If A is any subset of X, then

$$CL(A) = \{B \in 2^X \mid B \subset A\}.$$

The following theorem tells us that the idempotency of $\mathcal{T}$ is related to its continuity.

3.3.7 Theorem *If X is a continuum for which $\mathcal{T}$ is continuous, then $\mathcal{T}$ is idempotent on X.*

Proof Let W be a subcontinuum of X, and let $x \in Int(W)$. Since $X \setminus Int(W)$ is a closed subset of X, it is easy to see that $CL(X \setminus Int(W))$ is closed in 2^X. Then $\mathcal{T}^{-1}(CL(X \setminus Int(W)))$ is a closed subset of 2^X by the continuity of $\mathcal{T}$. Note that

$$CL(X \setminus W) \subset \mathcal{T}^{-1}(CL(X \setminus Int(W)))$$

because $X \setminus W \subset X \setminus Int(W)$, and Remark 3.1.5.

Since, for $W \neq X$, $X \setminus Int(W)$ is a limit point of $CL(X \setminus W)$ (recall that $Cl(X \setminus W) = X \setminus Int(W)$), it follows that $\mathcal{T}(X \setminus Int(W)) \subset X \setminus Int(W)$. Then there exists a subcontinuum M of X such that $x \in Int(M) \subset M \subset Int(W)$. If $W = X$, then it suffices to choose $M = X$. Thus, in either case, the theorem follows from Theorem 3.1.60.

Q.E.D.

As a consequence of Theorems 3.2.2 and 3.3.7, we have the following:

3.3.8 Corollary *If X is a continuum for which $\mathcal{T}$ is continuous and $\mathcal{T} \colon 2^X \twoheadrightarrow 2^X$ is surjective, then X is locally connected.*

3.3.9 Theorem *If X is a point $\mathcal{T}$-symmetric continuum for which $\mathcal{T}$ is continuous, then $\mathcal{T}(\{p, q\}) = \mathcal{T}(\{p\}) \cup \mathcal{T}(\{q\})$ for every $p, q \in X$.*

Proof Let $p \in X$. For each $x \in X$, define

$$A(x) = \{y \in X \mid \mathcal{T}(\{x, y\}) \in \mathcal{C}(X)\}$$

and

$$B(x) = \{y \in X \mid \mathcal{T}(\{x, y\}) = \mathcal{T}(\{x\}) \cup \mathcal{T}(\{y\})\}.$$

We assert that $A(x)$ and $B(x)$ are both closed for every $x \in X$. Let $x \in X$, and let $Z(x) \in \{A(x), B(x)\}$. If $y \in Cl(Z(x))$, then there exists a sequence $\{y_n\}_{n=1}^{\infty}$ of elements of $Z(x)$ converging to y. Hence, the sequence $\{\{x, y_n\}\}_{n=1}^{\infty}$ converges to $\{x, y\}$ (note that for each $z, w \in X$, $\{z, w\} = \sigma(\{\{z\}, \{w\}\})$). Since σ is continuous (Lemma 1.8.11),

$$\lim_{n \to \infty} \{x, y_n\} = \lim_{n \to \infty} \sigma(\{\{x\}, \{y_n\}\}) = \sigma(\{\{x\}, \{y\}\}) = \{x, y\}.$$

Hence, by the continuity of $\mathcal{T}$, the sequence $\{\mathcal{T}(\{x, y_n\})\}_{n=1}^{\infty}$ converges to $\mathcal{T}(\{x, y\})$.

If $Z(x)$ is $A(x)$, then, since $\mathcal{C}(X)$ is compact (Theorem 1.8.5), $\mathcal{T}(\{x, y\}) \in \mathcal{C}(X)$. If $Z(x)$ is $B(x)$, then $\mathcal{T}(\{x, y\}) = \lim_{n \to \infty} \mathcal{T}(\{x, y_n\}) = \lim_{n \to \infty} [\mathcal{T}(\{x\}) \cup \mathcal{T}(\{y_n\})] = \mathcal{T}(\{x\}) \cup \lim_{n \to \infty} \mathcal{T}(\{y_n\}) = \mathcal{T}(\{x\}) \cup \mathcal{T}(\{y\})$ (the third equality follows from the continuity of the union map, σ). Hence, $A(x)$ and $B(x)$ are both closed subsets of X. Note that, by Corollary 3.1.79, $X = A(p) \cup B(p)$.

Let $x \in \mathcal{T}(\{p\})$. Then $\mathcal{T}(\{x\}) \subset \mathcal{T}^2(\{p\}) = \mathcal{T}(\{p\})$ (by Proposition 3.1.7 and the fact that $\mathcal{T}$ is idempotent, by Theorem 3.3.7). Since X is point $\mathcal{T}$-symmetric and $x \in \mathcal{T}(\{p\})$, $p \in \mathcal{T}(\{x\})$. Hence, $\mathcal{T}(\{p\}) \subset \mathcal{T}(\{x\})$. Therefore, $\mathcal{T}(\{p\}) = \mathcal{T}(\{x\})$. Also, since $\{p, x\} \subset \mathcal{T}(\{p\})$, $\mathcal{T}(\{p, x\}) \subset \mathcal{T}(\{p\})$. Thus, $\mathcal{T}(\{p, x\}) = \mathcal{T}(\{p\})$. Since $\mathcal{T}(\{p\})$ is a continuum (Theorem 3.1.21), $x \in A(p) \cap B(p)$. Now, let $x \in A(p) \cap B(p)$. Then $\mathcal{T}(\{p, x\}) = \mathcal{T}(\{p\}) \cup \mathcal{T}(\{x\})$, and this set is a continuum. Hence, $\mathcal{T}(\{p\}) \cap \mathcal{T}(\{x\}) \neq \emptyset$. Let $y \in \mathcal{T}(\{p\}) \cap \mathcal{T}(\{x\})$. Then $x \in \mathcal{T}(\{y\}) \subset \mathcal{T}^2(\{p\}) = \mathcal{T}(\{p\})$. Therefore, $A(p) \cap B(p) = \mathcal{T}(\{p\})$.

Now, suppose $B(p) \neq \mathcal{T}(\{p\})$. Let $y \in A(p)$, and let $x \in B(p) \setminus \mathcal{T}(\{p\})$ be arbitrary points. Then $\mathcal{T}(\{p, x\}) = \mathcal{T}(\{p\}) \cup \mathcal{T}(\{x\})$ and $\mathcal{T}(\{p\}) \cap \mathcal{T}(\{x\}) = \emptyset$. Hence, $\mathcal{T}(\{x\}) \cap A(p) = \emptyset$. (Suppose there exists $z \in \mathcal{T}(\{x\}) \cap A(p)$, then, as before, $\mathcal{T}(\{z\}) = \mathcal{T}(\{x\})$. Since $\{p, z\} \subset \mathcal{T}(\{p\}) \cup \mathcal{T}(\{z\}) = \mathcal{T}(\{p\}) \cup \mathcal{T}(\{x\}) \subset \mathcal{T}(\{p, x\})$, $\mathcal{T}(\{p, z\}) \subset \mathcal{T}^2(\{p, x\}) = \mathcal{T}(\{p, x\})$. Similarly, we have that $\mathcal{T}(\{p, x\}) \subset \mathcal{T}(\{p, z\})$. Hence, $\mathcal{T}(\{p, z\}) = \mathcal{T}(\{p, x\})$. Since $z \in A(p)$, $\mathcal{T}(\{p, z\})$ is connected. Thus, $\mathcal{T}(\{p, x\})$ is connected, a contradiction.) Let U be an open set such that $\mathcal{T}(\{x\}) \subset U$ and $Cl(U) \cap A(p) = \emptyset$. Now, let $q \in Bd(U)$. Then $q \in (X \setminus A(p)) \cap (X \setminus \mathcal{T}(\{x\}))$. Hence, $q \in X \setminus \mathcal{T}(\{p, x\})$. Since $\mathcal{T}$ is idempotent, $q \in X \setminus \mathcal{T}^2(\{p, x\})$. Thus, there exists a subcontinuum W of X such that $q \in Int(W) \subset W \subset X \setminus \mathcal{T}(\{p, x\})$. Then $W \subset B(p)$, since otherwise $(W \cap A(p)) \cup (W \cap B(p))$ would be a separation of W (recall that $A(p) \cap B(p) = \mathcal{T}(\{p\})$ and $W \cap \mathcal{T}(\{p\}) = \emptyset$). Therefore, $y \in X \setminus W$, and $q \in X \setminus \mathcal{T}(\{x, y\})$ (if $q \in \mathcal{T}(\{x, y\})$, then $\{x, y\} \cap W \neq \emptyset$, a contradiction). Since q is an arbitrary point of $Bd(U)$,

$$\mathcal{T}(\{x, y\}) = (\mathcal{T}(\{x, y\}) \cap U) \cup (\mathcal{T}(\{x, y\}) \cap (X \setminus Cl(U))),$$

where $\mathcal{T}(\{x, y\}) \cap U$ and $\mathcal{T}(\{x, y\}) \cap (X \setminus Cl(U))$ are separated. Thus, $\mathcal{T}(\{x, y\})$ is not connected. Hence, $\mathcal{T}(\{x, y\}) = \mathcal{T}(\{x\}) \cup \mathcal{T}(\{y\})$, by Corollary 3.1.79. So, $x \in B(y)$. Thus, $B(p) \setminus \mathcal{T}(\{p\}) \subset B(y)$.

Note that $p \in Cl(B(p) \setminus \mathcal{T}(\{p\}))$. If not, $p \in Int(A(p))$, and since $A(p)$ is a continuum and $\mathcal{T}$ is idempotent (Theorem 3.3.7), there exists a continuum M such that $p \in Int(M) \subset M \subset Int(A(p))$ (Theorem 3.1.60). Hence, M misses some point $w \in \mathcal{T}(\{p\})$, and $p \in X \setminus \mathcal{T}(\{w\})$, contradicting the point $\mathcal{T}$-symmetry of X. Thus, $p \in Cl(B(p) \setminus \mathcal{T}(\{p\}))$.

Since $B(y)$ is closed and $p \in Cl(B(p) \setminus \mathcal{T}(\{p\}))$, it follows that $p \in B(y)$, or that $y \in B(p)$. But $y \in A(p)$, so that $y \in \mathcal{T}(\{p\})$, and $A = \mathcal{T}(\{p\})$. By contrapositive, if $A(p) \neq \mathcal{T}(\{p\})$, then $B(p) = \mathcal{T}(\{p\})$. So, for each $p \in X$, either $A(p) = X$ or $B(p) = X$. Suppose there exists $p \in X$ such that $A(p) = X$. Let $q \in X$ be arbitrary. If $q \in \mathcal{T}(\{p\})$, then $A(q) = A(p) = X$. (As we did before, if $z \in A(p) \cup A(q)$, then, since $\mathcal{T}(\{p\}) = \mathcal{T}(\{q\})$, $\mathcal{T}(\{p, z\}) = \mathcal{T}(\{q, z\})$.) Suppose $q \in X \setminus \mathcal{T}(\{p\})$. Then $q \in A(p)$, so $p \in A(q)$. Since $p \in X \setminus \mathcal{T}(\{q\})$, $A(q) \neq \mathcal{T}(\{q\})$. Hence, $A(q) = X$. Thus, either $A(p) = X$ for every $p \in X$ or $B(p) = X$ for every $p \in X$. If $B(p) = X$ for every $p \in X$, the theorem is proved. So, suppose $A(p) = X$

for all $p \in X$. Then for each $p, q \in X$, $\mathcal{T}(\{p, q\})$ is a continuum. Hence, by Theorem 3.1.71, X is indecomposable. Thus, by Theorem 3.1.39, $B(p) = X$ for all $p \in X$ in this case also.

Q.E.D.

3.3.10 Theorem *If X is a point $\mathcal{T}$-symmetric continuum for which $\mathcal{T}$ is continuous, then X is $\mathcal{T}$-additive.*

Proof Let $\mathcal{F}(X) = \bigcup_{n=1}^{\infty} \mathcal{F}_n(X)$. Then $\mathcal{F}(X)$ is the family of all finite subsets of X.

We show first that for each $A \in \mathcal{F}(X)$, $\mathcal{T}(A) = \bigcup_{a \in A} \mathcal{T}(\{a\})$. Suppose this is not true. Then there exists an $M \in \mathcal{F}(X)$ such that $\mathcal{T}(M) \neq \bigcup_{p \in M} \mathcal{T}(\{p\})$. Take M with the smallest cardinality. By Theorem 3.3.9, M has at least three elements. We assert that $\mathcal{T}(M)$ is a continuum. If $\mathcal{T}(M)$ is not connected, then there exist two disjoint closed subsets of A and B of X such that $\mathcal{T}(M) = A \cup B$. Then, by Lemma 3.1.73, Corollary 3.1.14 and the minimality of M:

$$\mathcal{T}(M) = A \cup B = \mathcal{T}(M \cap A) \cup \mathcal{T}(M \cap B)$$

$$= \bigcup_{p \in M \cap A} \mathcal{T}(\{p\}) \cup \bigcup_{p \in M \cap B} \mathcal{T}(\{p\}) = \bigcup_{p \in M} \mathcal{T}(\{p\}),$$

contrary to the choice of M. Moreover, if $p, q \in M$ are distinct points, $\mathcal{T}(\{p\}) \cap \mathcal{T}(\{q\}) = \emptyset$; otherwise, by point $\mathcal{T}$-symmetry and idempotency, $\mathcal{T}(\{p\}) = \mathcal{T}(\{q\})$ and then, since $M \subset \mathcal{T}(M \setminus \{p\})$:

$$\mathcal{T}(M) \subset \mathcal{T}^2(M \setminus \{p\}) = \mathcal{T}(M \setminus \{p\}) = \bigcup_{r \in M \setminus \{p\}} \mathcal{T}(\{r\}) \subset \mathcal{T}(M).$$

Thus, $\mathcal{T}(M) = \bigcup_{p \in M} \mathcal{T}(\{p\})$. This also contradicts the choice of M.

Now, let $p \in M$ be arbitrary, and let $N = M \setminus \{p\}$. Then N has at least two elements. Since N has cardinality smaller than the cardinality of M, $\mathcal{T}(N) = \bigcup_{r \in N} \mathcal{T}(\{r\})$; i.e., $\mathcal{T}(N)$ is a finite union of pairwise disjoint subcontinua. Hence, $\mathcal{T}(N)$ is not a continuum. Now, define

$$L = \left\{ x \in X \mid \mathcal{T}(N \cup \{x\}) = \bigcup_{r \in N} \mathcal{T}(\{r\}) \cup \mathcal{T}(\{x\}) \right\}$$

and

$$K = \{x \in X \mid \mathcal{T}(N \cup \{x\}) \in \mathcal{C}(X)\}.$$

Note that $N \subset L$ and $p \in K$. Thus, $L \neq \emptyset$ and $K \neq \emptyset$. We claim that L and K are closed subsets of X. To see this, let Z be either L or K. Let $x \in Cl(Z)$. Then there exists a sequence $\{x_n\}_{n=1}^{\infty}$ of elements of Z converging to x.

If Z is L, since $\mathcal{T}$ and the union map σ (Lemma 1.8.11) are continuous,

$$\mathcal{T}(N \cup \{x\}) = \lim_{n \to \infty} \mathcal{T}(N \cup \{x_n\}) = \lim_{n \to \infty} \left(\bigcup_{r \in N} \mathcal{T}(\{r\}) \cup \mathcal{T}(\{x_n\}) \right) =$$

$$\bigcup_{r \in N} \mathcal{T}(\{r\}) \cup \lim_{n \to \infty} \mathcal{T}(\{x_n\}) = \bigcup_{r \in N} \mathcal{T}(\{r\}) \cup \mathcal{T}(\{x\}).$$

Hence, $x \in L$. Therefore, L is closed.

If Z is K, since $\mathcal{T}$ and the union map σ are continuous, the sequence $\{\mathcal{T}(N \cup \{x_n\})\}_{n=1}^{\infty}$ is a sequence of continua converging to $\mathcal{T}(N \cup \{x\})$ and $\mathcal{C}(X)$ is closed in 2^X (Theorem 1.8.5), then $\mathcal{T}(N \cup \{x\}) \in \mathcal{C}(X)$. Hence, $x \in K$. Therefore, K is closed.

If $y \in K \cap L$, then $\mathcal{T}(N \cup \{y\}) = \bigcup_{r \in N} \mathcal{T}(\{r\}) \cup \mathcal{T}(\{y\})$ and this set is connected. Thus, $\mathcal{T}(\{y\}) \cap \mathcal{T}(\{r\}) \neq \emptyset$ for each $r \in N$. Then, by point $\mathcal{T}$-symmetry and idempotency, $\mathcal{T}(\{y\}) = \mathcal{T}(\{r\})$ for every $r \in N$, a contradiction to the fact that for $r_1, r_2 \in N$, if $r_1 \neq r_2$, then $\mathcal{T}(\{r_1\}) \neq \mathcal{T}(\{r_2\})$). Thus, $K \cap L = \emptyset$.

Let $x \in X \setminus L$. Then $\mathcal{T}(N \cup \{x\})$ is a continuum; otherwise, there exist two disjoint closed subsets A and B of X such that $\mathcal{T}(N \cup \{x\}) = A \cup B$. Hence, by Lemma 3.1.73 and Corollary 3.1.14:

$$\mathcal{T}(N \cup \{x\}) = A \cup B = \mathcal{T}((N \cup \{x\}) \cap A) \cup \mathcal{T}((N \cup \{x\}) \cap B)$$

$$= \bigcup_{r \in (N \cup \{x\}) \cap A} \mathcal{T}(\{r\}) \cup \bigcup_{r \in (N \cup \{x\}) \cap B} \mathcal{T}(\{r\}) = \bigcup_{r \in N} \mathcal{T}(\{r\}) \cup \mathcal{T}(\{x\}),$$

a contradiction to the fact that $x \in X \setminus L$. Therefore, $\mathcal{T}(N \cup \{x\})$ is a continuum. Thus, $x \in K$. Hence, $X = K \cup L$, and K and L are disjoint closed sets, a contradiction to the fact that X is connected.

Therefore, $\mathcal{T}(A) = \bigcup_{a \in A} \mathcal{T}(\{a\})$ for each $A \in \mathcal{F}(X)$.

Now, let $B \in 2^X$, and let $\varepsilon > 0$. Let $\delta > 0$ be given by the uniform continuity of $\mathcal{T}$ for $\frac{\varepsilon}{2}$. Since $\mathcal{F}(X)$ is dense in 2^X (proof of Corollary 1.8.9), there exists $A \in \mathcal{F}(X)$ such that $\mathcal{H}(B, A) < \delta$.

Since $\mathcal{H}(B, A) < \delta$, $\mathcal{H}^2(\mathcal{F}_1(B), \mathcal{F}_1(A)) < \delta$. Hence, by Theorem 8.2.2, $\mathcal{H}^2(2^{\mathcal{T}}(\mathcal{F}_1(B)), 2^{\mathcal{T}}(\mathcal{F}_1(A))) < \frac{\varepsilon}{2}$. This implies, by Lemma 1.8.11, that

$$\mathcal{H}\left(\bigcup_{b \in B} \mathcal{T}(\{b\}), \bigcup_{a \in A} \mathcal{T}(\{a\}) \right) < \frac{\varepsilon}{2}.$$

Therefore:

$$\mathcal{H}\left(\mathcal{T}(B), \bigcup_{b \in B} \mathcal{T}(\{b\})\right) \leq$$

$$\mathcal{H}(\mathcal{T}(B), \mathcal{T}(A)) + \mathcal{H}\left(\bigcup_{b \in B} \mathcal{T}(\{b\}), \bigcup_{a \in A} \mathcal{T}(\{a\})\right) < \frac{\varepsilon}{2} + \frac{\varepsilon}{2} = \varepsilon.$$

Since ε is arbitrary, $\mathcal{T}(B) = \bigcup_{b \in B} \mathcal{T}(\{b\})$. Now, the fact that X is $\mathcal{T}$-additive follows easily.

Q.E.D.

As a consequence of Theorems 3.1.55 and 3.3.10, we have the following:

3.3.11 Corollary *If X is a continuum for which $\mathcal{T}$ is continuous, then X is point $\mathcal{T}$-symmetric if and only if X is $\mathcal{T}$-symmetric.*

Proof Clearly, if X is $\mathcal{T}$-symmetric, then X is point $\mathcal{T}$-symmetric.

Suppose $\mathcal{T}$ is continuous for X and X is point $\mathcal{T}$-symmetric. Hence, X is $\mathcal{T}$-additive (Theorem 3.3.10). Thus, since X is point $\mathcal{T}$-symmetric and $\mathcal{T}$-additive, X is $\mathcal{T}$-symmetric (Theorem 3.1.55).

Q.E.D.

3.3.12 Theorem *If X is a continuum for which $\mathcal{T}$ is continuous, W is a subcontinuum of X with nonempty interior, and U is an open subset of X such that $W \subset U$, then there exists a point $p \in X$ such that $\mathcal{T}(\{p\}) \subset U$.*

Proof If $X \setminus W$ is connected, let $p \in Int(W)$. Then $Cl(X \setminus W)$ is a continuum with every point outside W in its interior such that $Cl(X \setminus W) \subset X \setminus \{p\}$. Hence, $\mathcal{T}(\{p\}) \subset W \subset U$.

Suppose $X \setminus W$ is not connected. Thus, there exist two nonempty disjoint open sets M and N of X such that $X \setminus W = M \cup N$. Then if $x \in M$, $\mathcal{T}(\{x\}) \subset Cl(M)$, since $W \cup N$ is a continuum (Lemma 1.7.23) having every point outside $Cl(M)$ in its interior and such that $W \cup N \subset X \setminus \{x\}$. Similarly, if $x \in N$, $\mathcal{T}(\{x\}) \subset Cl(N)$. Now, let

$$A = \{x \in X \mid \mathcal{T}(\{x\}) \cap M \cap (X \setminus U) \neq \emptyset\}$$

and

$$B = \{x \in X \mid \mathcal{T}(\{x\}) \cap M \neq \emptyset\}.$$

Observe that $A \subset B$, $B \neq \emptyset$ (since $M \subset B$), B is open (Lemma 3.3.5 (2)) while A is closed. (Note that $X \setminus A = \{x \in X \mid \mathcal{T}(\{x\}) \subset (X \setminus M) \cup U\}$. Since $Cl(M) \setminus M \subset U$, $(X \setminus M) \cup U$ is open. Thus, $X \setminus A$ is open, by Lemma 3.3.5 (1)). Since $N \cap B = \emptyset$, $B \neq X$. Thus, $A \neq B$ since X is connected. Let $x \in B \setminus A$. Then $\mathcal{T}(\{x\}) \cap M \neq \emptyset$

and $\mathcal{T}(\{x\}) \cap M \cap (X \backslash U) = \emptyset$. Let $p \in \mathcal{T}(\{x\}) \cap M$. Then $\mathcal{T}(\{p\}) \subset Cl(M) \cap \mathcal{T}(\{x\})$ ($\mathcal{T}$ is idempotent by Theorem 3.3.7); in particular:

$$\mathcal{T}(\{p\}) \cap (X \backslash U) \subset Cl(M) \cap \mathcal{T}(\{x\}) \cap (X \backslash U) = \emptyset,$$

since $Cl(M) \backslash M \subset U$. Thus, $\mathcal{T}(\{p\}) \subset U$.

Q.E.D.

3.3.13 Theorem *If X is a $\mathcal{T}$-additive continuum for which $\mathcal{T}$ is continuous and W is a continuum domain of X, then $\mathcal{T}(W) = W$.*

Proof Let $L = \{p \in X \mid \mathcal{T}(\{p\}) \subset W\}$. Let $x \in W$, and let M be a subcontinuum of X such that $x \in Int(M)$. Then $Int(M) \cap Int(W) \neq \emptyset$. Let $y \in Int(M) \cap Int(W)$. Then, since $\mathcal{T}$ is idempotent (Theorem 3.3.7),

$$y \in X \backslash (\mathcal{T}(X \backslash Int(M)) \cup \mathcal{T}(X \backslash Int(W)))$$

by Theorem 3.1.60. By additivity:

$$y \in X \backslash \mathcal{T}((X \backslash Int(M)) \cup (X \backslash Int(W)))$$

and

$$y \in X \backslash \mathcal{T}(X \backslash (Int(M) \cap Int(W))).$$

Hence, there exists a subcontinuum N of X such that $y \in Int(N)$ and $N \subset Int(M) \cap Int(W)$. Then, by Theorems 3.3.12 and 3.1.60, there exists $p \in N$ such that $p \in Int(N)$ and $\mathcal{T}(\{p\}) \subset N$. Thus, $\mathcal{T}(\{p\}) \subset W$ so that $p \in L$. Hence, $M \cap L \neq \emptyset$ and $x \in \mathcal{T}(L)$. Since x is an arbitrary point of W, $W \subset \mathcal{T}(L)$. By definition of L and additivity, $\mathcal{T}(L) \subset W$. Thus, $\mathcal{T}(W) = \mathcal{T}^2(L) = \mathcal{T}(L) = W$.

Q.E.D.

The following theorem says that for continua for which $\mathcal{T}$ is continuous $\mathcal{T}$-additivity is equivalent to $\mathcal{T}$-symmetry.

3.3.14 Theorem *If X is a continuum for which $\mathcal{T}$ is continuous, then X is $\mathcal{T}$-additive if and only if X is $\mathcal{T}$-symmetric.*

Proof If X is $\mathcal{T}$-symmetric, by Theorem 3.1.49, X is $\mathcal{T}$-additive.

Suppose X is $\mathcal{T}$-additive. Let A and B be two closed subsets of X such that $A \cap \mathcal{T}(B) = \emptyset$. Then, by the definition of $\mathcal{T}$, compactness and Corollary 3.1.63, there exists a finite collection $\{W_j\}_{j=1}^n$ such that W_j is a continuum domain, for each $j \in \{1, \ldots, n\}$, $A \subset \bigcup_{j=1}^n Int(W_j)$ and $B \cap \left(\bigcup_{j=1}^n W_j\right) = \emptyset$. Then, by additivity and Theorem 3.3.13, $\mathcal{T}\left(\bigcup_{j=1}^n W_j\right) = \bigcup_{j=1}^n W_j$. Hence, $\mathcal{T}(A) \subset \bigcup_{j=1}^n W_j$. Therefore, $\mathcal{T}(A) \cap B = \emptyset$.

Q.E.D.

As a consequence of Corollary 3.3.11 and Theorem 3.3.14, we have the following:

3.3.15 Corollary *If X is a continuum for which $\mathcal{T}$ is continuous, then the following are equivalent:*

(1) X is point $\mathcal{T}$-symmetric;
(2) X is $\mathcal{T}$-symmetric;
(3) X is $\mathcal{T}$-additive.

3.3.16 Corollary *If X is an aposyndetic continuum for which $\mathcal{T}$ is continuous, then X is locally connected.*

Proof Since X is aposyndetic, by Theorem 3.1.28, X is point $\mathcal{T}$-symmetric. Hence, by Corollary 3.3.15, X is $\mathcal{T}$-additive. Thus, by Theorem 3.1.52, X is locally connected.

$$\textbf{Q.E.D.}$$

3.3.17 Theorem *If X is a continuum for which $\mathcal{T}$ is continuous and X is almost connected im kleinen at $p \in X$, then X is semi-locally connected at p.*

Proof Let

$$\mathcal{L} = \{A \in 2^X \mid p \in Int(A)\}.$$

Note that $X \in \mathcal{L}$. Hence, $\mathcal{L} \neq \emptyset$. By the almost connectedness im kleinen of X and Theorem 3.3.12, the set $B(A) = \{x \in X \mid \mathcal{T}(\{x\}) \subset A\}$ is nonempty for each $A \in \mathcal{L}$. By continuity of $\mathcal{T}$, $B(A)$ is closed (Lemma 3.3.5 (4)) for each $A \in \mathcal{L}$. Hence, $\{B(A) \mid A \in \mathcal{L}\}$ is a filterbase of closed subsets, and $\bigcap_{A \in \mathcal{L}} B(A) \neq \emptyset$. But,

$$\bigcap_{A \in \mathcal{L}} B(A) \subset \bigcap_{A \in \mathcal{L}} A = \{p\}.$$

Thus, $\mathcal{T}(\{p\}) \subset \bigcap_{A \in \mathcal{L}} A = \{p\}$. Therefore, by Theorem 3.1.29, X is semi-locally connected at p.

$$\textbf{Q.E.D.}$$

3.3.18 Theorem *Let X be a continuum. If $\mathcal{T}$ is both additive and continuous for X and $p \in X$, then the following are equivalent:*

(1) X is connected im kleinen at p.
(2) X is almost connected im kleinen at p.
(3) X is semi-locally connected at p.

Proof Note that, by Corollary 3.3.15, X is $\mathcal{T}$-symmetric. Thus, by Theorem 3.1.43, X is connected im kleinen at p if and only if X is semi-locally connected at p.

Clearly, if X is connected im kleinen at p, then X is almost connected im kleinen at p.

By Theorem 3.3.17, if X is almost connected im kleinen at p, then X is semi-locally connected at p.

Q.E.D.

3.3.19 Theorem *Let X be a continuum for which $\mathcal{T}$ is continuous. Then X is semi-locally connected if and only if X is locally connected.*

Proof First note that if X is either semi-locally connected or locally connected, then $\mathcal{T}(\{p\}) = \{p\}$ for each $p \in X$ (Theorems 3.1.29 and 3.1.31, respectively). Hence, X is point $\mathcal{T}$-symmetric. Thus, by Corollary 3.3.15, X is $\mathcal{T}$-symmetric. Therefore, by Theorems 3.1.43 and 1.7.12, X is semi-locally connected if and only if X is locally connected.

Q.E.D.

As a consequence of Theorems 3.3.17 and 3.3.19, we have the following:

3.3.20 Corollary *Let X be a continuum for which $\mathcal{T}$ is continuous. If X is almost connected im kleinen at each of its points, then X is locally connected.*

3.4 Three Decomposition Theorems

We present three Decomposition Theorems, one for the class of point $\mathcal{T}$-symmetric continua X for which $\mathcal{T}$ is idempotent on singletons; the second one is for the class of point $\mathcal{T}$-symmetric continua X for which $\mathcal{T}|_{\mathcal{F}_1(X)}$ is continuous and for which $\mathcal{T}$ is idempotent on singletons ; and the third for the class of point $\mathcal{T}$-symmetric continua for which the set function $\mathcal{T}$ is continuous. We restrict ourselves to decomposable nonlocally connected continua; for it is well known that $\mathcal{T}$ is a constant map on indecomposable continua, Theorem 3.1.39, and $\mathcal{T}$ is the identity map on locally connected continua, Theorem 3.1.31. For this section, we follow [7, 27, 33, 36, 43, 50].

The proof of the following theorem is similar to the one given for Theorem 3.3.1.

3.4.1 Theorem *Let X be a continuum. If*

$$\mathcal{G} = \{\mathcal{T}(\{x\}) \mid x \in X\}$$

is a decomposition of X, then $\mathcal{G}$ is upper semicontinuous. Hence, $X/\mathcal{G}$ is a continuum.

Proof Let $x_0 \in X$ and let U be an open subset of X such that $\mathcal{T}(\{x_0\}) \subset U$. Let $K = \{x \in X \mid \mathcal{T}(\{x\}) \cap (X \setminus U) \neq \emptyset\}$. We show that K is closed in X. Let $x \in Cl(K)$. Then there exists a sequence $\{x_n\}_{n=1}^{\infty}$ of points of K converging to x. Since each $x_n \in K$, there exists $z_n \in \mathcal{T}(\{x_n\}) \setminus U$ for every positive integer n. Without loss of generality, we assume that the sequence $\{z_n\}_{n=1}^{\infty}$ converges to a point z. Note that $z \in X \setminus U$. We prove that $z \in \mathcal{T}(\{x\})$. To this end, suppose that $z \in X \setminus \mathcal{T}(\{x\})$. Then there exists a subcontinuum H of X such that $z \in Int(H) \subset H \subset X \setminus \{x\}$. Since $\{z_n\}_{n=1}^{\infty}$ converges to z and $\{x_n\}_{n=1}^{\infty}$ converges to x, there exists a positive integer N such that $z_n \in Int(H)$ and $x_n \in X \setminus H$ for all $n \geq N$. This implies that $z_n \in X \setminus \mathcal{T}(\{x_n\})$ for each $n \geq N$. A contradiction. Thus, $z \in \mathcal{T}(\{x\}) \setminus U$, and K is closed. Let $V = X \setminus K$. Then $V = \{x \in X \mid \mathcal{T}(\{x\}) \subset U\}$, $\mathcal{T}(\{x_0\}) \subset V$ and V is open in X. Therefore, $\mathcal{G}$ is upper semicontinuous. By Theorem 1.7.3, $X/\mathcal{G}$ is a continuum.

Q.E.D.

3.4.2 Definition Let X be a continuum. We say that $\mathcal{T}$ is *idempotent on singletons* if $\mathcal{T}(\mathcal{T}(\{x\})) = \mathcal{T}(\{x\})$ for all $x \in X$.

3.4.3 Corollary *Let X be a continuum such that $\mathcal{T}_X$ is idempotent on singletons. If*

$$\mathcal{G} = \{\mathcal{T}_X(\{x\}) \mid x \in X\}$$

is a decomposition of X, then $X/\mathcal{G}$ is an aposyndetic continuum.

Proof By Theorem 3.4.1, $X/\mathcal{G}$ is a continuum. Let $q \colon X \twoheadrightarrow X/\mathcal{G}$ be the quotient map. By Theorem 3.1.21, q is a monotone map. Let $\chi \in X/\mathcal{G}$ and let $x \in X$ be such that $\mathcal{T}_X(\{x\}) = q^{-1}(\chi)$. Since q is monotone, by Theorem 3.1.80 (c), $\mathcal{T}_{X/\mathcal{G}}(\{\chi\}) = q\mathcal{T}_X q^{-1}(\chi) = q\mathcal{T}_X(\mathcal{T}_X(\{x\})) = q\mathcal{T}_X(\{x\}) = qq^{-1}(\chi) = \{\chi\}$, the third equality is true because $\mathcal{T}_X$ is idempotent on singletons. Therefore, $X/\mathcal{G}$ is an aposyndetic continuum, by Theorem 3.1.28.

Q.E.D.

3.4.4 Lemma *Let X be a continuum and let $z \in X$. If*

$$\mathcal{G} = \{\mathcal{T}(\{x\}) \mid x \in X\}$$

is a decomposition and W is a subcontinuum of X such that $\mathcal{T}(\{z\}) \cap Int(W) \neq \emptyset$, then $\mathcal{T}(\{z\}) \subset W$.

Proof Let W be a subcontinuum of X and let z be a point of X such that $\mathcal{T}(\{z\}) \cap Int(W) \neq \emptyset$. Let $x \in \mathcal{T}(\{z\}) \cap Int(W)$ and suppose there exists $y \in \mathcal{T}(\{z\}) \setminus W$. Thus, $x \in Int(W) \subset W \subset X \setminus \{y\}$; i.e., $x \notin \mathcal{T}(\{y\})$. Since $\mathcal{G}$ is a decomposition, $\mathcal{T}(\{x\}) = \mathcal{T}(\{z\}) = \mathcal{T}(\{y\})$, a contradiction. Therefore, $\mathcal{T}(\{z\}) \subset W$.

Q.E.D.

The following definition is based on a property obtained by Bellamy and Lum in [7, Lemma 5].

3.4.5 Definition Let X be a continuum and let $z \in X$. We say that $\mathcal{T}(\{z\})$ *has property BL* provided that $\mathcal{T}(\{z\}) \subset \mathcal{T}(\{x\})$ for each $x \in \mathcal{T}(\{z\})$.

3.4.6 Lemma *Let X be a decomposable continuum for which $\mathcal{T}$ is idempotent on singletons, and let $z \in X$. If $\mathcal{T}(\{z\})$ has property BL, then $\mathcal{T}(\{w\}) = \mathcal{T}(\{z\})$ for every $w \in \mathcal{T}(\{z\})$.*

Proof Let $z \in X$ be such that $\mathcal{T}(\{z\})$ has property BL, and let $w \in \mathcal{T}(\{z\})$. Since $w \in \mathcal{T}(\{z\})$ and $\mathcal{T}$ is idempotent on singletons, $\mathcal{T}(\{w\}) \subset \mathcal{T}(\mathcal{T}(\{z\})) = \mathcal{T}(\{z\})$.

Hence, $\mathcal{T}(\{w\}) \subset \mathcal{T}(\{z\})$. Since $\mathcal{T}(\{z\})$ has property BL, $\mathcal{T}(\{z\}) \subset \mathcal{T}(\{w\})$. Therefore, $\mathcal{T}(\{w\}) = \mathcal{T}(\{z\})$.

$$Q.E.D.$$

3.4.7 Corollary *Let X be a decomposable continuum for which $\mathcal{T}$ is idempotent on singletons. If z_1 and z_2 are two points of X such that both $\mathcal{T}(\{z_1\})$ and $\mathcal{T}(\{z_2\})$ have property BL, then either $\mathcal{T}(\{z_1\}) = \mathcal{T}(\{z_2\})$ or $\mathcal{T}(\{z_1\}) \cap \mathcal{T}(\{z_2\}) = \emptyset$.*

Proof Let z_1 and z_2 be points of X such that both $\mathcal{T}(\{z_1\})$ and $\mathcal{T}(\{z_2\})$ have property BL. Suppose $\mathcal{T}(\{z_1\}) \cap \mathcal{T}(\{z_2\}) \neq \emptyset$. Let $z_3 \in \mathcal{T}(\{z_1\}) \cap \mathcal{T}(\{z_2\})$. Then, by Lemma 3.4.6, $\mathcal{T}(\{z_1\}) = \mathcal{T}(\{z_3\}) = \mathcal{T}(\{z_2\})$.

$$Q.E.D.$$

The proof following lemma is very similar to the proof of Proposition 3.1.59.

3.4.8 Lemma *Let X be a decomposable continuum for which $\mathcal{T}$ is idempotent on singletons. If $z \in X$, then*

$$\mathcal{T}(\{z\}) = \bigcup \{\mathcal{T}(\{w\}) \mid w \in \mathcal{T}(\{z\})\}.$$

Proof Let $z \in X$ and let $w_0 \in \mathcal{T}(\{z\})$. Since $\mathcal{T}$ is idempotent on singletons $\mathcal{T}(\{w_0\}) \subset \mathcal{T}(\mathcal{T}(\{z\})) = \mathcal{T}(\{z\})$. Hence, $\bigcup \{\mathcal{T}(\{w\}) \mid w \in \mathcal{T}(\{z\})\} \subset \mathcal{T}(\{z\})$. The other inclusion is clear.

$$Q.E.D.$$

The proof of the following theorem is based on a technique of Bellamy and Lum [7, Lemma 5].

3.4.9 Theorem *If X is a decomposable continuum for which $\mathcal{T}$ is idempotent on singletons, then for each $x \in X$, there exists $z \in \mathcal{T}(\{x\})$ such that $\mathcal{T}(\{z\})$ has property BL.*

Proof Let $x \in X$. By Lemma 3.4.8, we have that

$$\mathcal{T}(\{x\}) = \bigcup \{\mathcal{T}(\{w\}) \mid w \in \mathcal{T}(\{z\})\}.$$

Let $\mathcal{G}_x = \{\mathcal{T}(\{w\}) \mid w \in \mathcal{T}(\{x\})\}$. Partially order $\mathcal{G}_x$ by inclusion. Let $\mathcal{H} = \{\mathcal{T}(\{w_\lambda\})\}_{\lambda \in \Lambda}$ be a (set theoretic) chain of elements of $\mathcal{G}_x$. We show that $\mathcal{H}$ has a lower bound in $\mathcal{G}_x$.

Since $\mathcal{H}$ is a chain of continua (Theorem 3.1.21), $\bigcap_{\lambda \in \Lambda} \mathcal{T}(\{w_\lambda\}) \neq \emptyset$. Let $w_0 \in \bigcap_{\lambda \in \Lambda} \mathcal{T}(\{w_\lambda\})$. Then

$$\mathcal{T}(\{w_0\}) \subset \mathcal{T}\left(\bigcap_{\lambda \in \Lambda} \mathcal{T}(\{w_\lambda\})\right) \subset \bigcap_{\lambda \in \Lambda} \mathcal{T}(\mathcal{T}(\{w_\lambda\})) =$$

$$\bigcap_{\lambda \in \Lambda} \mathcal{T}(\{w_\lambda\}) \subset \mathcal{T}(\{x\}).$$

Hence, by Kuratowski–Zorn Lemma, there exists $z \in \mathcal{T}(\{x\})$ such that $\mathcal{T}(\{z\})$ is a minimal element; i.e., each $w \in \mathcal{T}(\{z\})$ satisfies that $\mathcal{T}(\{z\}) \subset \mathcal{T}(\{w\})$. Therefore, $\mathcal{T}(\{z\})$ has property BL.

Q.E.D.

Now we are ready to prove our first decomposition theorem.

3.4.10 Theorem *Let X be a decomposable point $\mathcal{T}_X$-symmetric continuum for which $\mathcal{T}_X$ is idempotent on singletons. Then*

$$\mathcal{G} = \{\mathcal{T}_X(\{x\}) \mid x \in X\}$$

is an upper semicontinuous decomposition of X such that the quotient space $X/\mathcal{G}$ is an aposyndetic continuum.

Proof Let $x \in X$. By Theorem 3.4.9, there exists $z \in \mathcal{T}_X(\{x\})$ such that $\mathcal{T}_X(\{z\})$ has property BL. Since X is point $\mathcal{T}_X$-symmetric and $z \in \mathcal{T}_X(\{x\})$, we have that $x \in \mathcal{T}_X(\{z\})$. Thus, since $\mathcal{T}_X(\{z\})$ has property BL, $\mathcal{T}_X(\{x\}) = \mathcal{T}_X(\{z\})$, by Lemma 3.4.6. Hence, $\mathcal{T}_X(\{x\})$ has property BL. Therefore, by Corollary 3.4.7,

$$\mathcal{G} = \{\mathcal{T}_X(\{x\}) \mid x \in X\}$$

is a decomposition of X. Hence, by Theorem 3.4.1, $\mathcal{G}$ is upper semicontinuous and, by Corollary 3.4.3, $X/\mathcal{G}$ is an aposyndetic continuum.

Q.E.D.

A proof of the following result may be found in [27, 2.1]. The theorem is originally proved by E. Dyer.

3.4.11 Theorem *Let X and Y be nondegenerate continua. If $f : X \twoheadrightarrow Y$ is a surjective, monotone and open map, then there exists a dense G_δ subset W of Y having the following property: for each $y \in W$, for each subcontinuum B of $f^{-1}(y)$, for each $x \in Int_{f^{-1}(y)}(B)$ and for each neighborhood U of B in X, there exist a proper subcontinuum Z of X containing B and a neighborhood V of y in Y such that $x \in Int_X(Z)$, $(f|_Z)^{-1}(V) \subset U$ and $f|_Z : Z \twoheadrightarrow Y$ is a monotone surjective map.*

Now we are ready to prove our second decomposition theorem.

3.4.12 Theorem *Let X be a decomposable point $\mathcal{T}_X$-symmetric continuum for which $\mathcal{T}_X|_{\mathcal{F}_1(X)}$ is continuous and for which $\mathcal{T}_X$ is idempotent on singletons. Then*

$$\mathcal{G} = \{\mathcal{T}_X(\{x\}) \mid x \in X\}$$

is a continuous decomposition of X such that the quotient space $X/\mathcal{G}$ is an aposyndetic continuum. Moreover, all the elements of $\mathcal{G}$ are nowhere dense in X; also there exists a dense G_δ subset $\mathcal{W}$ of $X/\mathcal{G}$ such that if $q(z) \in \mathcal{W}$, then $\mathcal{T}_X(\{z\})$ is an indecomposable continuum, where $q : X \twoheadrightarrow X/\mathcal{G}$ is the quotient map.

Proof By Theorem 3.4.10, $\mathcal{G}$ is an upper semicontinuous decomposition of X such that $X/\mathcal{G}$ is an aposyndetic continuum. Since $\mathcal{T}_X|_{\mathcal{F}_1(X)}$ is continuous, in fact, $\mathcal{G}$ is a continuous decomposition of X. Note that, by Theorem 3.1.21, q is a monotone map. Also, since $\mathcal{G}$ is a continuous decomposition, by Corollary 1.2.24, q is an open map. Since q is an open map, all the elements of $\mathcal{G}$ are clearly nowhere dense in X.

Since the quotient map q is surjective, monotone and open, let $\mathcal{W}$ be the dense G_δ subset of $X/\mathcal{G}$ given by Theorem 3.4.11. Let $\omega \in \mathcal{W}$ and let $z \in X$ be such that $q(z) = \omega$. Suppose $\mathcal{T}_X(\{z\})$ is decomposable. Then there exists a subcontinuum K of $\mathcal{T}_X(\{z\})$ such that $Int_{\mathcal{T}_X(\{z\})}(K) \neq \emptyset$. Let $x \in Int_{\mathcal{T}_X(\{z\})}(K)$ and let U be an open subset of X such that $\mathcal{T}_X(\{z\}) \subset U$. By Theorem 3.4.11, there exist a proper subcontinuum Z of X containing $\mathcal{T}_X(\{z\})$ and a neighborhood $\mathcal{V}$ of ω in $X/\mathcal{G}$ such that $x \in Int_X(Z)$, $(q|_Z)^{-1}(\mathcal{V}) \subset U$ and $q|Z: Z \twoheadrightarrow X/\mathcal{G}$ is a monotone surjection. Let $x' \in X \setminus Z$. On one hand, by Lemma 3.4.4, $\mathcal{T}_X(\{x'\}) \cap Z = \emptyset$. On the other hand, since $q|_Z$ is surjective, we have that $\mathcal{T}_X(\{x'\}) \cap Z \neq \emptyset$, a contradiction. Therefore, $\mathcal{T}_X(\{z\})$ is indecomposable.

Q.E.D.

Before stating our third decomposition theorem, let us observe the following:

3.4.13 Theorem *Let X be a continuum for which $\mathcal{T}_X$ is continuous. If*

$$\mathcal{G} = \{\mathcal{T}_X(\{x\}) \mid x \in X\}$$

is a decomposition of X, then $\mathcal{G}$ is continuous, $X/\mathcal{G}$ is a locally connected continuum and $\mathcal{T}_X\left(2^X\right)$ is homeomorphic to $2^{X/\mathcal{G}}$. (In particular, $\mathcal{T}_X\left(2^X\right)$ is homeomorphic to the Hilbert cube.) Moreover, all the elements of $\mathcal{G}$ are nowhere dense in X; also there exists a dense G_δ subset $\mathcal{W}$ of $X/\mathcal{G}$ such that if $q(z) \in \mathcal{W}$, then $\mathcal{T}_X(\{z\})$ is an indecomposable continuum, where $q: X \twoheadrightarrow X/\mathcal{G}$ is the quotient map.

Proof Observe that, since $\mathcal{G}$ is a decomposition, X is point $\mathcal{T}_X$-symmetric. Also, since $\mathcal{T}_X$ is continuous, $\mathcal{G}$ is a continuous decomposition; and, by Theorem 3.3.7, $\mathcal{T}_X$ is idempotent. Hence, by Theorem 3.4.10, $X/\mathcal{G}$ is an aposyndetic continuum.

Let $q: X \twoheadrightarrow X/\mathcal{G}$ be the quotient map. As in Theorem 3.4.12, q is a monotone, open and surjective map. Note that $q^{-1}\mathcal{T}_{X/\mathcal{G}}(\Gamma) = \mathcal{T}_X q^{-1}(\Gamma)$ for each subset Γ of $X/\mathcal{G}$ (Theorem 3.1.80 (e)). Hence, q is a $\mathcal{T}_X X/\mathcal{G}$-continuous surjective open map. Since $\mathcal{T}_X$ is continuous, $\mathcal{T}_{X/\mathcal{G}}$ is continuous too (Theorem 3.3.4). Since aposyndetic continua for which $\mathcal{T}$ is continuous are locally connected (Corollary 3.3.16), we have that $X/\mathcal{G}$ is locally connected.

To see that $\mathcal{T}_X\left(2^X\right)$ is homeomorphic to $2^{X/\mathcal{G}}$, let $\Im(q): 2^{X/\mathcal{G}} \to 2^X$ be given by $\Im(q)(\Gamma) = q^{-1}(\Gamma)$. By Theorem 8.5.5, $\Im(q)$ is continuous. Note that $2^q \circ \Im(q) = 1_{2^{X/\mathcal{G}}}$. In particular, $\Im(q): 2^{X/\mathcal{G}} \to \Im(q)\left(2^{X/\mathcal{G}}\right)$ is a homeomorphism. Thus, it is enough to show that $\mathcal{T}_X\left(2^X\right) = \Im(q)\left(2^{X/\mathcal{G}}\right)$.

Let $\Gamma \in 2^{X/\mathcal{G}}$. Then $\mathcal{T}_X\left(\Im(q)(\Gamma)\right) = \mathcal{T}_X q^{-1}(\Gamma) = q^{-1}\mathcal{T}_{X/\mathcal{G}}(\Gamma) = q^{-1}(\Gamma) = \Im(q)(\Gamma)$; the second equality is true by Theorem 3.1.80 (e), and the second the last equality is valid by Theorem 3.1.31. Thus, $\Im(q)(\Gamma) \in \mathcal{T}_X\left(2^X\right)$ and $\Im(q)\left(2^{X/\mathcal{G}}\right) \subset \mathcal{T}_X\left(2^X\right)$.

Now, let $K \in \mathcal{T}_X\left(2^X\right)$. Then there exists $A \in 2^X$ such that $\mathcal{T}_X(A) = K$. We prove that $K = \Im(q)(q(A))$. Note that $q^{-1}(q(A)) = \bigcup\{q^{-1}(q(a)) \mid a \in A\} = \bigcup\{\mathcal{T}_X(\{a\}) \mid a \in A\}$. Since $\mathcal{G}$ is a decomposition, X is point $\mathcal{T}_X$-symmetric. Hence, X is $\mathcal{T}_X$-additive (Theorem 3.3.10). Since X is $\mathcal{T}_X$-additive, $\bigcup\{\mathcal{T}_X(\{a\}) \mid a \in A\} = \mathcal{T}_X(A)$ (Corollary 3.1.51). Thus, $\Im(q)(q(A)) = \mathcal{T}_X(A) = K$, $K \in \Im(q)\left(2^{X/\mathcal{G}}\right)$ and $\mathcal{T}_X\left(2^X\right) \subset \Im(q)\left(2^{X/\mathcal{G}}\right)$.

Therefore, $\mathcal{T}_X\left(2^X\right) = \Im(q)\left(2^{X/\mathcal{G}}\right)$. Since $\Im(q)\left(2^{X/\mathcal{G}}\right)$ is homeomorphic to $2^{X/\mathcal{G}}$, we see that $\mathcal{T}_X\left(2^X\right)$ is homeomorphic to $2^{X/\mathcal{G}}$. Since $2^{X/\mathcal{G}}$ is homeomorphic to the Hilbert cube [43, (1.97)], $\mathcal{T}_X\left(2^X\right)$ is homeomorphic to the Hilbert cube.

Now the theorem follows from Theorem 3.4.12.

Q.E.D.

As a consequence of Theorems 3.4.12 and 3.4.13, we have our third decomposition theorem:

3.4.14 Theorem *Let X be a point $\mathcal{T}_X$-symmetric continuum for which $\mathcal{T}_X$ is continuous. Then*

$$\mathcal{G} = \{\mathcal{T}_X(\{x\}) \mid x \in X\}$$

is a continuous decomposition of X such that $X/\mathcal{G}$ is a locally connected continuum and $\mathcal{T}_X\left(2^X\right)$ is homeomorphic to $2^{X/\mathcal{G}}$. (In particular, $\mathcal{T}_X\left(2^X\right)$ is homeomorphic to the Hilbert cube.) Moreover, all the elements of $\mathcal{G}$ are nowhere dense in X; also there exists a dense G_δ subset $\mathcal{W}$ of $X/\mathcal{G}$ such that if $q(z) \in \mathcal{W}$, then $\mathcal{T}_X(\{z\})$ is an indecomposable continuum, where $q \colon X \twoheadrightarrow X/\mathcal{G}$ is the quotient map.

3.4.15 Corollary *Let X be a hereditarily decomposable $\mathcal{T}$-additive continuum. Then $\mathcal{T}$ is continuous if and only if X is locally connected.*

Proof If X is locally connected, then $\mathcal{T}$ is the identity map (Theorem 3.1.31). Hence, $\mathcal{T}$ is continuous.

Suppose X is a hereditarily decomposable $\mathcal{T}$-additive continuum. Note that if X is also aposyndetic, then, by Theorem 3.1.52, X is locally connected.

Assume X is not aposyndetic. Since X is $\mathcal{T}$-additive and $\mathcal{T}$ is continuous, by Corollary 3.3.15, we obtain that X is point $\mathcal{T}$-symmetric. Thus, by Theorem 3.4.14, $\mathcal{G} = \{\mathcal{T}_X(\{x\}) \mid x \in X\}$ is a continuous decomposition of X. Since X is not aposyndetic, by Theorem 3.1.28, there exists $x_0 \in X$ such that $\mathcal{T}(\{x_0\})$ is nondegenerate. Then, by the continuity of $\mathcal{T}$, there exists an open subset U of X such that $x_0 \in U$ and for each $x \in U$, $\mathcal{T}(\{x\})$ is nondegenerate. Thus, by Theorem 3.4.14, there exists $x_1 \in U$ such that $\mathcal{T}(\{x_1\})$ is indecomposable, a contradiction. Hence, X is aposyndetic. Therefore, X is locally connected.

Q.E.D.

3.4.16 Corollary *Let X be a hereditarily decomposable and hereditarily unicoherent continuum. Then $\mathcal{T}$ is continuous if and only if X is locally connected.*

Proof Since X is hereditarily unicoherent, X is $\mathcal{T}$-additive, by Theorem 3.1.50. Now the corollary follows form Corollary 3.4.15.

Q.E.D.

3.4.17 Remark Note that Theorem 3.4.14 gives a partial answer to Question 9.2.2.

3.4.18 Remark Let us observe that, since for continua X for which $\mathcal{T}$ is continuous, being $\mathcal{T}$-additive is equivalent to being point $\mathcal{T}$-symmetric (Corollary 3.3.15), by Theorem 3.4.14, Question 9.2.1 is equivalent to Question 9.2.2.

3.5 Examples of Continua for Which $\mathcal{T}$ Is Continuous

In 1970, David P. Bellamy asked in [1] if any nonlocally connected continuum for which $\mathcal{T}$ is continuous had to be indecomposable. In 1983, Bellamy [3, Remark 2, p. 10] answered the question in the negative by showing that the circle of pseudo-arcs (Definition 3.5.5) is a decomposable nonlocally connected continuum for which $\mathcal{T}$ is continuous.

We give several families of decomposable nonlocally connected continua for which $\mathcal{T}$ is continuous. In order to do this, we use [1, 3, 9, 10, 18, 28, 29, 31, 34, 40, 44, 45, 47].

We start with the following definition.

3.5.1 Definition The *pseudo-arc* is the only hereditarily indecomposable chainable continuum.

3.5.2 Remark R H Bing proved that any two hereditarily indecomposable chainable continua are homeomorphic [9, Theorem 21].

A proof of the next theorem is due to Wayne Lewis and may be found in [29].

3.5.3 Theorem *If X is a one-dimensional continuum, then there exists a one-dimensional continuum $\widehat{X}$ such that $\widehat{X}$ has a terminal (Definition 5.1.9) continuous decomposition into pseudo-arcs such that the decomposition space is homeomorphic to X.*

3.5.4 Remark Observe that the continua constructed by Wayne Lewis do not contain arcs. Hence, they are not locally connected and, by construction, the continuum $\widehat{X}$ is decomposable if and only if X is decomposable.

3.5.5 Definition The *circle of pseudo-arcs* is the space $\widehat{X}$, obtained from Theorem 3.5.3 when X is the circle.

Now we are ready to present our first family of examples.

3.5.6 Theorem *If X is a locally connected one-dimensional continuum, then the set function $\mathcal{T}_{\widehat{X}}$ is continuous for the continuum $\widehat{X}$ given in Theorem 3.5.3.*

Proof Let X be a one-dimensional locally connected continuum, and let $\widehat{X}$ be the one-dimensional continuum given in Theorem 3.5.3. Since $\widehat{X}$ admits a continuous decomposition into pseudo-arcs, the quotient map $q: \widehat{X} \twoheadrightarrow X$ is monotone, open and surjective. Let W be a proper subcontinuum of $\widehat{X}$, let $\hat{x} \in \widehat{X} \setminus W$, and let $P_{\hat{x}}$ be the pseudo-arc of the decomposition that contains $\hat{x}$. Since $P_{\hat{x}}$ is a terminal subcontinuum of $\widehat{X}$, either $W \subset P_{\hat{x}}$ or $W \cap P_{\hat{x}} = \emptyset$. In any case, $q(W)$ is a proper subcontinuum of X. Therefore, by Theorem 3.3.2, $\mathcal{T}_{\widehat{X}}$ is continuous.

                                                                                              Q.E.D.

3.5.7 Remark What Bellamy did to show the continuity of $\mathcal{T}_X$ in Theorem 3.3.2 is to prove that $\mathcal{T}_X = \Im(f) \circ 2^f$; i.e., $\mathcal{T}_X(A) = f^{-1} \circ f(A)$ for each $A \in 2^X$. Thus, for the continuum $\widehat{X}$ of Theorem 3.5.6, $\mathcal{T}_{\widehat{X}}(\{\hat{x}\}) = P_{\hat{x}}$. This implies that $\widehat{X}$ is point $\mathcal{T}$-symmetric. Hence, $\widehat{X}$ is $\mathcal{T}$-additive by Theorem 3.3.10.

The following example shows that the requirement in Theorem 3.3.2, that the images of proper subcontinua of the domain are proper subcontinua of the range is essential. The existence of the continuum $\widetilde{Z}$ is due to Karen Villarreal [47]:

3.5.8 Example Let Z be the circle of pseudo-arcs. Then there exists a two-dimensional aposyndetic nonlocally connected homogeneous continuum $\widetilde{Z} \subset Z \times Z$ such that if π_1 is the projection map of $Z \times Z$ onto the first factor, then $\pi_1|_{\widetilde{Z}}: \widetilde{Z} \twoheadrightarrow Z$ is open, monotone and surjective. Let $\tilde{q}: Z \twoheadrightarrow S^1$ be the quotient map from Z onto the unit circle. Then $\tilde{q} \circ \pi_1|_{\widetilde{Z}}: \widetilde{Z} \twoheadrightarrow S^1$ is open, monotone and surjective. It is easy to see, from the construction of $\widetilde{Z}$, that $\Delta_Z = \{(z, z) \mid z \in Z\}$ is a proper subcontinuum of $\widetilde{Z}$ such that $(\tilde{q} \circ \pi_1|_{\widetilde{Z}})(\Delta_Z) = S^1$. Observe that, since $\widetilde{Z}$ is aposyndetic and nonlocally connected, $\mathcal{T}_{\widetilde{Z}}$ is not continuous, by Corollary 3.3.16.

Let us note the following question:

3.5.9 Question Let X be a nonlocally connected decomposable continuum for which the set function $\mathcal{T}$ is continuous. Does X admit a terminal continuous decomposition into pseudo-arcs such that the decomposition space is locally connected?

The answer to this question is negative as can be seen in the following:

3.5.10 Example Let Z be the circle of pseudo-arcs, and let $q: Z \twoheadrightarrow S^1$ be the quotient map. Let $A = \{(a, b) \in S^1 \mid a \geq 0 \text{ and } b \geq 0\}$. Let $\mathcal{L} = \{q^{-1}((a, b)) \mid (a, b) \in A\} \cup \{\{\hat{x}\} \mid \hat{x} \in Z \text{ and } q(\hat{x}) \in S^1 \setminus A\}$. Then $\mathcal{L}$ is an upper semicontinuous decomposition of Z such that $Z/\mathcal{L}$ is a continuum containing an arc. Let $q_{\mathcal{L}}: Z \twoheadrightarrow Z/\mathcal{L}$ be the quotient map. Note that for each $\chi \in Z/\mathcal{L}$, there exists $z \in S^1$ such that $q_{\mathcal{L}}^{-1}(\chi) \subset q^{-1}(z)$. Hence, by [44, 3.22], the function $f: Z/\mathcal{L} \twoheadrightarrow S^1$ given by $f = q \circ q_{\mathcal{L}}^{-1}$ is well defined and continuous. Since q is monotone and open, f is monotone and open. Let $\mathcal{W}$ be a proper subcontinuum of $Z/\mathcal{L}$. Then $q_{\mathcal{L}}^{-1}(\mathcal{W})$ is a proper subcontinuum of Z. Thus, $f(\mathcal{W}) = q\left(q_{\mathcal{L}}^{-1}(\mathcal{W})\right)$ is a proper subcontinuum of S^1. Therefore, by Theorem 3.3.2, $\mathcal{T}_{Z/\mathcal{L}}$ is continuous for

$Z/\mathcal{L}$. Since $Z/\mathcal{L}$ contains an arc, there does not exist a continuous decomposition of $Z/\mathcal{L}$ into pseudo-arcs.

3.5.11 Definition A continuum X is of *type* λ provided that X is irreducible and each indecomposable subcontinuum of X has empty interior.

3.5.12 Remark By [45, Theorem 10, p. 15], a continuum X is of type λ if and only if it admits a monotone upper semicontinuous decomposition $\mathcal{G}$ such that each element of $\mathcal{G}$ is nowhere dense and $X/\mathcal{G}$ is an arc. Each element of $\mathcal{G}$ is called a *layer* of X.

Following Mohler and Oversteegen [40], we give the following definition:

3.5.13 Definition A continuum X of type λ for which $\mathcal{G}$ (Remark 3.5.12) is continuous, is a *continuously irreducible continuum*.

The next theorem characterizes the class of irreducible continua for which $\mathcal{T}$ is continuous:

3.5.14 Theorem *Let X be an irreducible continuum. Then $\mathcal{T}$ is continuous for X if and only if X is continuously irreducible. Moreover, $\mathcal{G} = \{\mathcal{T}(\{x\}) \mid x \in X\}$ is the finest continuous monotone decomposition of X such that $X/\mathcal{G}$ is an arc.*

Proof Suppose X is an irreducible continuum for which $\mathcal{T}$ is continuous. Since X is irreducible, X is $\mathcal{T}$-symmetric, by Corollary 3.1.42. Hence, X is point $\mathcal{T}$-symmetric. Since $\mathcal{T}$ is also continuous, we have that

$$\mathcal{G} = \{\mathcal{T}(\{x\}) \mid x \in X\}$$

is a monotone continuous decomposition of X such that $X/\mathcal{G}$ is a locally connected continuum, and the elements of $\mathcal{G}$ are nowhere dense, by Theorem 3.4.14. Since X is irreducible and the quotient map $q\colon X \twoheadrightarrow X/\mathcal{G}$ is monotone, $X/\mathcal{G}$ is an irreducible continuum [28, Theorem 3, p. 192]. Hence, $X/\mathcal{G}$ is an arc by Theorem 3.1.56. Thus, X is a continuum of type λ, by Remark 3.5.12. We identify $X/\mathcal{G}$ with $[0, 1]$.

Let $\mathbb{G}$ be the finest upper semicontinuous decomposition of X each element of which is nowhere dense and $X/\mathbb{G}$ is an arc. Let $q'\colon X \twoheadrightarrow X/\mathbb{G} = [0, 1]$ be the quotient map. Let $x, z \in X$ be such that $q'(x) \neq q'(z)$. Then there exists a subinterval A of $[0, 1]$ such that $q'(z) \in Int_{[0,1]}(A)$ and $q'(x) \in [0, 1] \setminus A$. Hence, $z \in Int_X((q')^{-1}(A))$ and $x \in X \setminus (q')^{-1}(A)$. Thus, $z \in X \setminus \mathcal{T}_X(\{x\})$. Therefore, $\mathcal{T}_X(\{x\}) \subset (q')^{-1}(q'(x))$. Let $w \in (q')^{-1}(q'(x))$, and suppose W is a subcontinuum of X such that $w \in Int_X(W)$. Then $(q')^{-1}(q'(x)) \subset W$ [45, Theorem 5, p. 10]. Hence, $w \in \mathcal{T}_X(\{x\})$ and $\mathcal{T}_X(\{x\}) = (q')^{-1}(q'(x))$. Thus, $\mathbb{G} = \{\mathcal{T}_X(\{x\}) \mid x \in X\} = \mathcal{G}$ is the finest upper semicontinuous monotone decomposition of X such that $X/\mathbb{G}$ is an arc. Since $\mathcal{T}_X$ is continuous, $\mathcal{G}$ is continuous. Therefore, X is a continuously irreducible continuum.

Suppose X is a continuously irreducible continuum. Let $\mathbb{G}$ be the finest continuous monotone decomposition of X such that $X/\mathbb{G}$ is an arc, and let $q\colon X \twoheadrightarrow [0, 1]$ be the quotient map. Then q is monotone and open. Since X is continuously irreducible, if Z is a proper subcontinuum of X, then $q(Z)$ is a proper subcontinuum of $[0, 1]$. Hence, by Theorem 3.3.2, $\mathcal{T}_X$ is continuous. In fact, $\mathcal{T}_X(A) = q^{-1}(q(A))$ for every nonempty closed subset A of X. In particular, $\mathcal{T}_X(\{x\}) = q^{-1}(q(x))$ for each $x \in X$, and $\mathbb{G} = \{\mathcal{T}_X(\{x\}) \mid x \in X\}$.

Q.E.D.

Let us observe that, so far, all known examples of decomposable nonlocally connected continua X for which the set function $\mathcal{T}$ is continuous have the property that there exist many points $x \in X$ such that $\mathcal{T}(\{x\})$ is a pseudo-arc (Theorem 3.5.6, Example 3.5.10 and Theorem 3.5.14). In Theorem 3.5.17 we present a new family of one-dimensional nonlocally connected continua for which the set function $\mathcal{T}$ is continuous which do not contain pseudo-arcs.

The following theorem is due to Mohler and Oversteegen [40, Corollary 1.1 and Theorem 2.1].

3.5.15 Theorem *Let X be any continuously irreducible one-dimensional continuum. Then there exist a continuously irreducible one-dimensional continuum $\widehat{X}$ such that every nondegenerate subcontinuum of $\widehat{X}$ contains an arc, and an atomic map $g \colon \widehat{X} \twoheadrightarrow X$ (Definition 8.1.1).*

As a consequence of Theorem 3.5.15 and Theorem 3.5.14, we have:

3.5.16 Theorem *For each one-dimensional continuously irreducible continuum X, there exist a one-dimensional continuously irreducible continuum $\widehat{X}$ such that $\mathcal{T}_{\widehat{X}}(\{\hat{x}\})$ does not contain a pseudo-arc for any $\hat{x} \in \widehat{X}$, and an atomic map $g \colon \widehat{X} \twoheadrightarrow X$ (Definition 8.1.1).*

Let $\mathcal{Z}$ be the class of one-dimensional continuously irreducible continua and let $\widehat{\mathcal{Z}} = \{\widehat{X} \mid X \in \mathcal{Z}\}$, where $\widehat{X}$ is given in Theorem 3.5.16. Hence, we have the following:

3.5.17 Theorem *The class $\widehat{\mathcal{Z}}$, defined above, is a class of one-dimensional nonlocally connected continua $\widehat{X}$ for which the set function $\mathcal{T}_{\widehat{X}}$ is continuous such that $\mathcal{T}_{\widehat{X}}(\{\hat{x}\})$ does not contain a pseudo-arc for any $\hat{x} \in \widehat{X}$. In particular, no element of $\widehat{\mathcal{Z}}$ contains a pseudo-arc.*

3.5.18 Theorem *Let X be a continuum and let $\mathcal{G}$ be a monotone, terminal and continuous decomposition of X (Definition 5.1.9). Then there exists an embedding of $C(X/\mathcal{G})$ into $C(X)$.*

Proof Let $q \colon X \twoheadrightarrow X/\mathcal{G}$ be the quotient map. Note that q is open, by Corollary 1.2.24, and monotone. Let $\Im(q) \colon C(X/\mathcal{G}) \to C(X)$ be given by $\Im(q)(\Gamma) = q^{-1}(\Gamma)$. Since q is monotone, $\Im(q)$ is well defined. Since q is open, $\Im(q)$ is continuous, by Theorem 8.5.5. We show that $\Im(q)$ is one-to-one. To this end, let Γ_1 and Γ_2 be two distinct elements of $C(X/\mathcal{G})$. Without loss of generality, we assume that there exists $\chi \in \Gamma_1 \setminus \Gamma_2$. Then, since $\mathcal{G}$ is a monotone terminal decomposition, we have that $q^{-1}(\chi) \subset q^{-1}(\Gamma_1) \setminus q^{-1}(\Gamma_2)$. Thus, $q^{-1}(\Gamma_1) \neq q^{-1}(\Gamma_2)$. Hence, $\Im(q)(\Gamma_1) \neq \Im(q)(\Gamma_2)$. Therefore, $\Im(q)$ is an embedding.

Q.E.D.

As a consequence of Theorem 3.5.18, we have:

3.5.19 Corollary *Let X be a continuum and let $\mathcal{G}$ be a monotone, terminal and continuous decomposition of X. Then $\dim(\mathcal{C}(X/\mathcal{G})) \leq \dim(\mathcal{C}(X))$.*

3.5.20 Theorem *Let X be a continuum and let $\mathcal{G}$ be a monotone, terminal and continuous decomposition of X. If $X/\mathcal{G}$ is an aposyndetic continuum and $\dim(\mathcal{C}(X)) < \infty$, then $\mathcal{T}$ is continuous for X.*

Proof Since $\dim(\mathcal{C}(X)) < \infty$, by Corollary 3.5.19, $\dim(\mathcal{C}(X/\mathcal{G})) < \infty$. Hence, since $X/\mathcal{G}$ is aposyndetic, by Corollary 3.1.35, $X/\mathcal{G}$ is locally connected. Let $q \colon X \twoheadrightarrow X/\mathcal{G}$ be the quotient map. Since $\mathcal{G}$ is a monotone, terminal and continuous decomposition of X, q is a monotone open (Corollary 1.2.24) and for every proper subcontinuum W of X, $q(W) \neq X$. Therefore, by Theorem 3.3.2, $\mathcal{T}$ is continuous for X.

Q.E.D.

3.6 $\mathcal{T}$-Closed Sets

We introduce the family of $\mathcal{T}$-closed sets of a continuum X and present its main properties. We also present a characterization of $\mathcal{T}$-closed sets. The material of this section comes from [6, 18, 37, 38].

3.6.1 Definition Given a continuum X, a subset A of X is a $\mathcal{T}$-*closed set* provided that $\mathcal{T}(A) = A$. We denote the family of $\mathcal{T}$-closed sets of a continuum X by $\mathfrak{T}(X)$. Note that $X \in \mathfrak{T}(X)$.

3.6.2 Theorem *If X is a continuum, then $\mathfrak{T}(X)$ is a G_δ subset of 2^X.*

Proof Given $\varepsilon > 0$ define:

$$\Theta_\varepsilon = \{A \in 2^X \mid \mathcal{T}(A) \subset \mathcal{V}_\varepsilon(A)\}.$$

Note that if $A \in \Theta_\varepsilon$, then $\mathcal{H}(A, \mathcal{T}(A)) < \varepsilon$. If $A \in \Theta_\varepsilon$, then define:

$$\mathcal{M}_\varepsilon(A) = \bigcup_{0 < s < 1} \left(\mathcal{V}_{s\varepsilon}^{\mathcal{H}}(A) \cap \{B \in 2^X \mid \mathcal{T}(B) \subset \mathcal{V}_{(1-s)\varepsilon}(A)\} \right).$$

Note that $A \in \mathcal{M}_\varepsilon(A)$ and $\mathcal{M}_\varepsilon(A)$ is an open subset of 2^X (the second set is open in 2^X since $\mathcal{T}$ is upper semicontinuous by Theorem 3.3.1).

We show that $\mathcal{M}_\varepsilon(A) \subset \Theta_\varepsilon$. Let $B \in \mathcal{M}_\varepsilon(A)$. Then there exists $s \in (0, 1)$ such that $\mathcal{H}(B, A) < s\varepsilon$ and $\mathcal{T}(B) \subset \mathcal{V}_{(1-s)\varepsilon}(A)$. We need to prove that $\mathcal{T}(B) \subset \mathcal{V}_\varepsilon(B)$. Let $x \in \mathcal{T}(B)$. Then, since $\mathcal{T}(B) \subset \mathcal{V}_{(1-s)\varepsilon}(A)$, there exists $a \in A$ such that $d(x, a) < (1 - s)\varepsilon$. Hence, since $\mathcal{H}(B, A) < s\varepsilon$, there exists $b \in B$ such that $d(b, a) < s\varepsilon$. By the triangle inequality, we obtain that $d(x, b) \leq d(x, a) +$

$d(a, b) < (1 - s)\varepsilon + s\varepsilon = \varepsilon$. Thus, $x \in V_\varepsilon(B)$, and $\mathcal{T}(B) \subset V_\varepsilon(B)$. Hence, A is an interior point of Θ_ε. Therefore, Θ_ε is open.

Observe that

$$\mathfrak{T}(X) = \bigcap_{n=1}^{\infty} \Theta_{\frac{1}{n}}.$$

Therefore, $\mathfrak{T}(X)$ is a G_δ subset of 2^X.

Q.E.D.

The following example shows that there exist continua X such that $\mathfrak{T}(X)$ is not connected.

3.6.3 Example Let X be an irreducible continuum such that X is the union of two sequences of indecomposable continua, $\{Y_n\}_{n=1}^{\infty}$ and $\{Z_n\}_{n=1}^{\infty}$, both converging to the singleton $\{p\}$. Then $|\mathfrak{T}(X)| = 4$, namely:

$$\mathfrak{T}(X) = \left\{ X, \{p\}, \bigcup_{n=1}^{\infty} Y_n \cup \{p\}, \bigcup_{n=1}^{\infty} Z_n \cup \{p\} \right\}.$$

Therefore, $\mathfrak{T}(X)$ is not connected.

The following example shows that there exists a continuum X such that $\mathcal{T}$ is idempotent and $\mathfrak{T}(X)$ is not closed in 2^X.

3.6.4 Example Let X be the topologist sine curve (see Example 2.4.5). For each positive integer n, let

$$A_n = \left\{ \left(\frac{2}{\pi(2n + 1)}, \sin\left(\frac{\pi(2n + 1)}{2} \right) \right) \right\}$$

and let $A_0 = \{(0, 0)\}$. Then $\{A_n\}_{n=1}^{\infty}$ converges to A_0, $\mathcal{T}(A_n) = A_n$ for all n and $\mathcal{T}(A_0) = \{0\} \times [-1, 1]$. Therefore, $\mathfrak{T}(X)$ is not closed in 2^X. It is easy to see that $\mathcal{T}$ is idempotent on closed sets.

Note that the following theorem is a consequence of Corollary 3.1.75, we present another proof here.

3.6.5 Theorem *Let X be a continuum. If $A \in \mathfrak{T}(X)$ and K is a component of A, then $K \in \mathfrak{T}(X)$.*

Proof Let $A \in \mathfrak{T}(X)$ and let K be a component of A. Then $\mathcal{T}(K) \subset \mathcal{T}(A) = A$, and $\mathcal{T}(K)$ is connected by Theorem 3.1.21. Since K is a component of A, $\mathcal{T}(K) \subset K$. Therefore, $K \in \mathfrak{T}(X)$.

Q.E.D.

3.6.6 Theorem *Let X be a continuum. If $A \in \mathfrak{T}(X)$, then for each separation $P \cup Q$ of $X \setminus A$, we have that $A \cup P \in \mathfrak{T}(X)$.*

Proof Let $A \in \mathfrak{T}(X)$ and let $P \cup Q$ be a separation of $X \setminus A$. Let $x \in X \setminus (A \cup P)$. Then $x \in Q$. Since $x \in X \setminus A$ and $A \in \mathfrak{T}(X)$, there exists a subcontinuum W of X such that $x \in Int(W) \subset W \subset X \setminus A$. Since $W \subset P \cup Q$, $Q \cap W \neq \emptyset$ and P and Q are separated, we obtain that $W \subset Q$. Hence, $W \cap (A \cup P) = \emptyset$. Thus, $x \in X \setminus \mathcal{T}(A \cup P)$. Therefore, $A \cup P \in \mathfrak{T}(X)$.

Q.E.D.

A similar argument to the one given for Theorem 3.6.6 proves the following:

3.6.7 Theorem *Let X be a continuum. If $A \in \mathfrak{T}(X)$ and K is a component of $Cl(X \setminus A)$, then $A \cup K \in \mathfrak{T}(X)$.*

The following simple theorem is very important for the family of minimal $\mathcal{T}$-closed sets.

3.6.8 Theorem *Let X be a continuum. If $\{A_\lambda\}_{\lambda \in \Lambda}$ is a family of elements of $\mathfrak{T}(X)$, then $\bigcap_{\lambda \in \Lambda} A_\lambda \in \mathfrak{T}(X)$.*

Proof Since

$$\bigcap_{\lambda \in \Lambda} A_\lambda \subset \mathcal{T}\left(\bigcap_{\lambda \in \Lambda} A_\lambda\right) \subset \bigcap_{\lambda \in \Lambda} \mathcal{T}(A_\lambda) = \bigcap_{\lambda \in \Lambda} A_\lambda,$$

we have that $\bigcap_{\lambda \in \Lambda} A_\lambda \in \mathfrak{T}(X)$.

Q.E.D.

The following example shows that the union of two $\mathcal{T}$-closed sets is not necessarily a $\mathcal{T}$-closed set.

3.6.9 Example Let X be the suspension of the Cantor ternary set. Let v^+ and v^- be the vertexes of X. Then it is easy to see that $\mathcal{T}(\{v^+\}) = \{v^+\}$, $\mathcal{T}(\{v^-\}) = \{v^-\}$ and $\mathcal{T}(\{v^+, v^-\}) = X$.

Note that the characterization of aposyndetic and locally connected continua in terms of the set function $\mathcal{T}$ given in Theorems 3.1.28, 3.1.31, and 3.1.32, can be written in terms of the family of $\mathcal{T}$-closed sets as follows:

3.6.10 Theorem *A continuum X is aposyndetic if and only if $\mathcal{F}_1(X) \subset \mathfrak{T}(X)$.*

3.6.11 Theorem *A continuum X is locally connected if and only if $\mathfrak{T}(X) = 2^X$.*

3.6.12 Theorem *A continuum X is locally connected if and only if $\mathcal{C}(X) \subset \mathfrak{T}(X)$.*

From Theorem 3.1.39, we obtain:

3.6.13 Theorem *If X is an indecomposable, then $\mathfrak{T}(X) = \{X\}$.*

The next example shows that the converse of Theorem 3.6.13 is not true:

3.6.14 Example Let X be two Knaster continua glued by its end point (see Example 2.4.9). Then X is a decomposable continuum and $\mathfrak{T}(X) = \{X\}$.

3.6.15 Theorem *Let X and Y be continua and let $f: X \twoheadrightarrow Y$ be a monotone map. If $B \in \mathfrak{T}(Y)$, then $f^{-1}(B) \in \mathfrak{T}(X)$.*

Proof Let $x \in X \setminus f^{-1}(B)$. Then $f(x) \in Y \setminus B$. Since $B \in \mathfrak{T}(Y)$, there exists a subcontinuum W of Y such that $f(x) \in Int_Y(W) \subset W \subset Y \setminus B$. Hence, since f is monotone, $f^{-1}(W)$ is a subcontinuum of X and $x \in f^{-1}(x) \subset Int_X(f^{-1}(W)) \subset f^{-1}(W) \subset f^{-1}(Y \setminus B) \subset X \setminus f^{-1}(B)$. Thus, $x \in X \setminus \mathcal{T}f^{-1}(B)$. Therefore, $f^{-1}(B) \in \mathfrak{T}(X)$.

Q.E.D.

As a consequence of Theorem 3.6.15, we have:

3.6.16 Corollary *Let X and Y be continua and let $f: X \twoheadrightarrow Y$ be a monotone map. Then $|\mathfrak{T}(Y)| \le |\mathfrak{T}(X)|$.*

We present a result in which we have the equality of the cardinalities of $\mathcal{T}$-closed sets (Theorem 3.6.23). We need the following:

3.6.17 Lemma *Let X be a continuum. If W is a proper terminal subcontinuum (Definition 5.1.9) of X, then $Int_X(W) = \emptyset$.*

Proof Suppose W is a proper terminal subcontinuum of X and $Int_X(W) \ne \emptyset$. Note that $Bd_X(Int_X(W)) \subset W$. Let $x \in X \setminus W$ and let C be the component of $X \setminus Int_X(W)$ containing x. By Theorem 1.7.27, $C \cap Bd_X(Int_X(W)) \ne \emptyset$. Hence, $C \cap W \ne \emptyset$, and $C \setminus W \ne \emptyset$. Since W is a terminal subcontinuum of X, $W \subset C$, a contradiction to the fact that $C \cap Int_X(W) = \emptyset$. Therefore, $Int_X(W) = \emptyset$.

Q.E.D.

3.6.18 Lemma *Let X and Y be continua, where Y is aposyndetic. If $f: X \twoheadrightarrow Y$ is a monotone map, then $\mathcal{T}_X(\{x\}) \subset f^{-1}(f(x))$ for each $x \in X$.*

Proof Let $x \in X$ and let $x' \in X \setminus f^{-1}(f(x))$. This implies that $f(x) \ne f(x')$. Hence, since Y is aposyndetic, there exists a subcontinuum W of Y such that $f(x') \in Int_Y(W) \subset W \subset Y \setminus \{f(x)\}$. Thus, $x' \in f^{-1}(f(x')) \subset Int_X(f^{-1}(W)) \subset f^{-1}(W) \subset X \setminus f^{-1}(f(x)) \subset X \setminus \{x\}$. Since f is monotone, $f^{-1}(W)$ is a subcontinuum of X, Theorem 2.1.12. It follows that $x' \in X \setminus \mathcal{T}_X(\{x\})$. Therefore, $\mathcal{T}_X(\{x\}) \subset f^{-1}(f(x))$.

Q.E.D.

3.6.19 Lemma *Let X and Y be continua and let $f: X \twoheadrightarrow Y$ be an atomic map (Definition 8.1.1). Then for every $x \in X$, $f^{-1}(f(x)) \subset \mathcal{T}_X(\{x\})$.*

Proof Let $x \in X$ and let $z \in X \setminus \mathcal{T}_X(\{x\})$. Then there exists a subcontinuum W of X such that $x \in Int_X(W) \subset W \subset X \setminus \{x\}$. Since $f^{-1}(f(x))$ is a terminal subcontinuum of X, Theorem 8.1.25, by Lemma 3.6.17, we have that $f^{-1}(f(x)) \cap W = \emptyset$. In particular, $z \in X \setminus f^{-1}(f(x))$. Therefore, $f^{-1}(f(x)) \subset \mathcal{T}_X(\{x\})$.

Q.E.D.

Since atomic maps are monotone, Theorem 8.1.24, as a consequence of Lemmas 3.6.19 and 3.6.18, we have:

3.6.20 Corollary *Let X and Y be continua, where Y is aposyndetic, and let $f: X \twoheadrightarrow Y$ be an atomic map. Then for every $x \in X$, $\mathcal{T}_X(\{x\}) = f^{-1}(f(x))$.*

3.6.21 Lemma *Let X and Y be continua and let $f: X \twoheadrightarrow Y$ be an atomic map. If $A \in \mathfrak{T}(X)$, then $A = f^{-1}(f(A))$.*

Proof Let $A \in \mathfrak{T}(X)$. We know that $A \subset f^{-1}(f(A))$. Let $x \in f^{-1}(f(A))$. There exists $a \in A$ such that $f(a) = f(x)$. Thus, by Lemma 3.6.19, $x \in f^{-1}(f(a)) \subset \mathcal{T}_X(\{a\}) \subset \mathcal{T}(A) = A$. Therefore, $A = f^{-1}(f(A))$.

$$\text{Q.E.D.}$$

3.6.22 Theorem *Let X and Y be continua and let $f: X \twoheadrightarrow Y$ be an atomic map. If $A \in \mathfrak{T}(X)$, then $f(A) \in \mathfrak{T}(Y)$.*

Proof Let $A \in \mathfrak{T}(X)$. By Theorem 3.1.80 (a), we have that $\mathcal{T}_Y(f(A)) \subset f\mathcal{T}_X f^{-1}(f(A))$. By Lemma 3.6.21, $A = f^{-1}(f(A))$. Thus, we obtain that $\mathcal{T}_Y(f(A)) \subset f\mathcal{T}_X(A) = f(A)$. Hence, $\mathcal{T}_Y(f(A)) = f(A)$. Therefore, $f(A) \in \mathfrak{T}(Y)$.

$$\text{Q.E.D.}$$

3.6.23 Theorem *Let X and Y be continua and let $f: X \twoheadrightarrow Y$ be an atomic map. Then $|\mathfrak{T}(X)| = |\mathfrak{T}(Y)|$.*

Proof Note that atomic maps are monotone, Theorem 8.1.24. Also observe that by Theorem 3.6.22, we have that $|\mathfrak{T}(X)| \leq |\mathfrak{T}(Y)|$. By Corollary 3.6.16, we know that $|\mathfrak{T}(X)| \geq |\mathfrak{T}(Y)|$. Therefore, $|\mathfrak{T}(X)| = |\mathfrak{T}(Y)|$.

$$\text{Q.E.D.}$$

The following example shows that there exist monotone maps, that are not atomic, between decomposable continua with the same cardinality of $\mathcal{T}$-closed sets.

3.6.24 Example Let Z be the Knaster indecomposable continuum, Example 2.4.7, let X be the union of three copies of Z glued by their endpoints and let Y be two copies of Z also glued by their endpoints. Let v be the common point of the two copies of Z in Y. Let $f: X \twoheadrightarrow Y$ be the map that sends two of the copies of Z in X homeomorphically to the two copies of Z in Y and sends the third copy of Z in X to $\{v\}$. Then f is a monotone map, f is not an atomic map, and $|\mathfrak{T}(X)| = |\mathfrak{T}(Y)| = 1$.

3.6.25 Theorem *Let X be a type λ continuum and let $q: X \twoheadrightarrow [0, 1]$ be the quotient map of the finest monotone upper semicontinuous decomposition of X. Then $A \in \mathfrak{T}(X)$ if and only if there exists a nonempty closed subset B of $[0, 1]$ such that $A = q^{-1}(B)$.*

Proof Let B be a closed subset of $[0, 1]$ and let $x \in X \backslash q^{-1}(B)$. Then $q(x) \in [0, 1] \backslash B$. Thus, there exists an interval $[a, b]$ such that $q(x) \in (a, b)$ and $[a, b] \cap B = \emptyset$. This implies, since q is monotone, that $q^{-1}([a, b])$ is a subcontinuum of X such that $x \in Int(q^{-1}([a, b]))$ and $q^{-1}([a, b]) \cap q^{-1}(B) = \emptyset$. Therefore, $\mathcal{T}(q^{-1}(B)) = q^{-1}(B)$, and $q^{-1}(B) \in \mathfrak{T}(X)$.

Let $A \in \mathfrak{T}(X)$. Since X is a type λ continuum, by [18, Theorem 5.2], $\mathcal{G} = \{\mathcal{T}^2(\{x\}) \mid x \in X\}$ is the finest monotone upper semicontinuous decomposition of X. Thus, since $A \in \mathfrak{T}(X)$, $q^{-1}q(A) = \bigcup\{\mathcal{T}^2(\{a\}) \mid a \in A\} \subset \mathcal{T}^2(A) = A$. Therefore, $A = q^{-1}q(A)$.

$$\text{Q.E.D.}$$

3.6.26 Theorem *Let X and Y be continua, let $f \colon X \twoheadrightarrow Y$ be an atomic map, and let A be a nonempty closed subset of X. Then $A \in \mathfrak{T}(X)$ if and only if there exists a nonempty closed subset B of Y such that $B \in \mathfrak{T}(Y)$ and $A = f^{-1}(B)$.*

Proof Assume that $A \in \mathfrak{T}(X)$. By Lemma 3.6.21, $A = f^{-1}(f(A))$, and by Theorem 3.6.22, $f(A) \in \mathfrak{T}(Y)$. Hence, if $B = f(A)$, we are done.

Suppose that there exists $B \in \mathfrak{T}(Y)$ and $A = f^{-1}(B)$. Since atomic maps are monotone (Theorem 8.1.24) by Theorem 3.6.15, $A = f^{-1}(B) \in \mathfrak{T}(X)$.

$$\text{Q.E.D.}$$

The next example shows that the monotone image of a $\mathcal{T}$-closed set is not necessarily $\mathcal{T}$-closed.

3.6.27 Example Let

$$Z = (\{0\} \times [-1, 2]) \cup \left\{ \left(x, \sin\left(\frac{1}{x}\right) \right) \mid x \in \left(0, \frac{\pi}{2}\right] \right\}$$

and let X be as in Example 3.6.4. Note that $X \subset Z$. Let $f \colon Z \twoheadrightarrow X$ be given by

$$f((x, y)) = \begin{cases} (x, y), & \text{if } (x, y) \in X; \\ (0, 1), & \text{if } (x, y) \in Z \setminus X. \end{cases}$$

Then f is a monotone retraction. Note that $A = \{0\} \times \left[\frac{3}{2}, 2\right] \in \mathfrak{T}(Z)$, $f(A) = \{(0, 1)\}$ and $\mathcal{T}_X(\{(0, 1)\}) = \{0\} \times [-1, 1]$. Hence, $f(A) \notin \mathfrak{T}(X)$.

3.6.28 Theorem *Let X and Y be continua and let $f \colon X \twoheadrightarrow Y$ be a monotone map. If A is a subset of X, then $f \mathcal{T}_X^\alpha(A) \subset \mathcal{T}_Y^\alpha f(A)$ for all ordinals α.*

Proof The proof is done by transfinite induction. The case $\alpha = 1$ follows from part (b) of Theorem 3.1.80. Now, suppose that $f \mathcal{T}_X^\alpha(A) \subset \mathcal{T}_Y^\alpha f(A)$ for some ordinal α. Then $f \mathcal{T}_X^{\alpha+1}(A) \subset \mathcal{T}_Y^\alpha f \mathcal{T}_X(A) \subset \mathcal{T}_Y^{\alpha+1} f(A)$.

Let γ be a limit ordinal and suppose that for each $\alpha < \gamma$, $f\mathcal{T}_X^\alpha(A) \subset \mathcal{T}_Y^\alpha f(A)$. Then $\bigcup_{\alpha<\gamma} f\mathcal{T}_X^\alpha(A) \subset \bigcup_{\alpha<\gamma} \mathcal{T}_Y^\alpha f(A)$. Hence, since f is a closed map,

$$f\mathcal{T}^\gamma(A) = f\left(Cl\left(\bigcup_{\alpha<\gamma} \mathcal{T}_X^\alpha(A)\right)\right) = Cl\left(f\left(\bigcup_{\alpha<\gamma} \mathcal{T}_X^\alpha(A)\right)\right) =$$

$$Cl\left(\bigcup_{\alpha<\gamma} f\mathcal{T}_X^\alpha(A)\right) \subset Cl\left(\bigcup_{\alpha<\gamma} \mathcal{T}_Y^\alpha f(A)\right) = \mathcal{T}_Y^\gamma f(A).$$

Therefore, $f\mathcal{T}_X^\alpha(A) \subset \mathcal{T}_Y^\alpha f(A)$ for all ordinals α.

Q.E.D.

Now, we give a characterization of the elements of the family of $\mathcal{T}$-closed sets. We present two necessary conditions for a set to be $\mathcal{T}$-closed that, in general, are not sufficient, but we prove that these conditions are sufficient for the class of continua with the property of Kelley.

3.6.29 Theorem *Let X be a continuum and let A be a closed subset of X. Then $A \in \mathfrak{T}(X)$ if and only if for every open subset U of X containing A, there exists an open subset V of X such that $A \subset V \subset U$ and $X \setminus V$ has only finitely many components.*

Proof Suppose $A \in \mathfrak{T}(X)$ and let U be an open subset of X such that $A \subset U$. Since $A \in \mathfrak{T}(X)$, for each $x \in X \setminus U$, there exists a subcontinuum W_x of X such that $x \in Int(W_x) \subset W_x \subset X \setminus A$. Since $X \setminus U$ is compact, there exist $x_1, \dots, x_n \in X \setminus U$ such that $X \setminus U \subset \bigcup_{j=1}^n Int(W_{x_j})$. Let $V = \bigcup_{j=1}^n (X \setminus W_{x_j})$. Then $A \subset V \subset U$ and $X \setminus V = \bigcup_{j=1}^n W_{x_j}$. Hence, $X \setminus V$ has only finitely many components.

Now, assume the condition stated is satisfied. Let $x \in X \setminus A$. Since A is closed, there exist an open subsets U of X such that $A \subset U$ and $x \in X \setminus U$. By our assumption, there exists an open subset V of X such that $A \subset V \subset U$ and $X \setminus V$ has only finitely many components. Let L be the component of $X \setminus V$ containing x. By Lemma 1.6.2, $x \in Int(L)$. Thus, $x \in X \setminus \mathcal{T}(A)$. Therefore, $A \in \mathfrak{T}(X)$.

Q.E.D.

As a consequence of this, we obtain the following:

3.6.30 Corollary *Let X be a continuum and let $A \in \mathfrak{T}(X)$. If $Z = X \times \{0, 1\}/R$, where R is the equivalence relation that identifies $(a, 0)$ with $(a, 1)$ for all $a \in A$, then $q(A \times \{0, 1\}) \in \mathfrak{T}(Z)$, where $q: X \times \{0, 1\} \twoheadrightarrow Z$ is the quotient map.*

Proof Let U be an open subset of Z containing $q(A \times \{0, 1\})$. Then $q^{-1}(U)$ is an open subset of $X \times \{0, 1\}$ containing $A \times \{0, 1\}$. Since $A \in \mathfrak{T}(X)$, by Theorem 3.6.29, there exists an open subset V of X such that $A \times \{0, 1\} \subset V \times \{0, 1\} \subset q^{-1}(U)$ and $(X \times \{0, 1\}) \setminus (V \times \{0, 1\})$ has only finitely many components. Then $q(V \times \{0, 1\})$ is an open subset of Z, $q(A \times \{0, 1\}) \subset q(V \times \{0, 1\}) \subset U$ and $Z \setminus q(V \times \{0, 1\})$

has only finitely many components. Therefore, by Theorem 3.6.29, $q(A \times \{0, 1\}) \in \mathfrak{T}(Z)$.

Q.E.D.

Note the following property of $\mathcal{T}$-closed sets, which is a consequence of Theorem 3.1.65:

3.6.31 Theorem *Let X be a continuum. If $A \in \mathfrak{T}(X)$, then every component of $X \setminus A$ is open and continuumwise connected.*

The following example shows that the converse of Theorem 3.6.31 is not true. First, we introduce the following:

3.6.32 Notation If χ_1 and χ_2 are two elements of $\mathbb{R}^n$, $\overline{\chi_1 \chi_2}$ denotes the convex in $\mathbb{R}^n$ arc joining χ_1 and χ_2.

3.6.33 Example For each positive integer n, let $A_n = \overline{(0,0)(1, \frac{1}{n})}$, let $A_0 = \overline{(0,0)(1,0)}$ and let $Z = \bigcup_{n=0}^{\infty} A_n$ (Z is homeomorphic to the harmonic fan, Example 1.7.5). For each positive integer n, note that the point $(\frac{1}{2^n}, \frac{1}{2^n n}) \in A_n$. Given a positive integer n, let B_n a semi-circle in $\mathbb{R}^3$ joining the points $(\frac{1}{2^n}, \frac{1}{2^n n})$ and $(\frac{1}{2^n}, 0)$ and such that $B_n \cap Z = \{(\frac{1}{2^n}, \frac{1}{2^n n}), (\frac{1}{2^n}, 0)\}$. Let $X = Z \cup (\bigcup_{n=1}^{\infty} B_n)$. Note that $\{(0, 0)\}$ is a closed subset of X such that $X \setminus \{(0, 0)\}$ is open and arcwise connected, but $\mathcal{T}(\{(0, 0)\}) = A_0$. Hence, $\{(0, 0)\}$ is not a $\mathcal{T}$-closed set of X. Note that $\mathcal{T}$ is idempotent on closed sets for X, but $\mathcal{T}$ is not idempotent for X.

The converse of Theorem 3.6.31 is true for continua with the property of Kelley.

3.6.34 Theorem *Let X be a continuum with the property of Kelley. Then $A \in \mathfrak{T}(X)$ if and only if each component of $X \setminus A$ is open and continuumwise connected.*

Proof If $A \in \mathfrak{T}(X)$, then, by Theorem 3.6.31, each component of $X \setminus A$ is open and continuumwise connected.

Suppose A is a closed subset of X such that each component of $X \setminus A$ is open and continuumwise connected. Let $x_0 \in X \setminus A$ and let L be the component of $X \setminus A$ containing x_0. For each positive integer n, let W_n be the component of $X \setminus \mathcal{V}_{\frac{1}{n}}(A)$ containing x_0. Then $W_n \subset L$. Let $x_1 \in L$. Since L is continuumwise connected, there exists a subcontinuum K of X such that $\{x_0, x_1\} \subset K \subset L$. Hence, there exists a positive integer n such that $K \subset W_n$. Thus, $L = \bigcup_{n=1}^{\infty} W_n$. Note that $W_n \subset W_{n+1}$ for every n. Since L is open, by the Baire Category Theorem, there exists m such that $Int(W_m) \neq \emptyset$. Let $w \in Int(W_m)$. Let $\varepsilon > 0$ be such that $\mathcal{V}_{\varepsilon}(w) \subset W$ and $d(W_m, A) > \varepsilon$. By the proof of Theorem 3.1.70, there exists a subcontinuum K_0 of X such that $\mathcal{V}_{\delta}(x_0) \subset K_0$. By construction, we see that $K_0 \cap A = \emptyset$. Thus, $x_0 \in X \setminus \mathcal{T}(A)$. Therefore, $A \in \mathfrak{T}(X)$.

Q.E.D.

Since homogeneous continua have the property of Kelley (Theorem 4.2.36), from of Theorem 3.6.34, we have:

3.6.35 Corollary *Let X be a homogeneous continuum. Then $A \in \mathfrak{T}(X)$ if and only if each component of $X \setminus A$ is open and continuumwise connected.*

Next, we consider minimal $\mathcal{T}$-closed sets.

3.6.36 Definition Let X be a continuum and let A be a closed subset of X. Then A is a *minimal $\mathcal{T}$-closed set of X* provided that $A \in \mathfrak{T}(X)$ and $A' \notin \mathfrak{T}(X)$ for each proper closed subset A' of A. The family of minimal $\mathcal{T}$-closed sets of X is denoted by $\mathfrak{MT}(X)$.

As a consequence of Theorem 3.6.8, we have:

3.6.37 Theorem *Let X be a continuum. If $A \in \mathfrak{MT}(X)$, then $A \in \mathcal{C}(X)$.*

Let us note the following consequence of Theorem 3.6.10:

3.6.38 Theorem *A continuum X is aposyndetic if and only if $\mathcal{F}_1(X) = \mathfrak{MT}(X)$.*

In the next example we present a continuum X such that all the elements of $\mathfrak{MT}(X)$ are nondegenerate.

3.6.39 Example Let Z be the cone over the ternary Cantor set, with vertex v_Z. Remove each point of the base of Z and compactify each ray with an arc as a remainder, obtaining a space homeomorphic to the topologist $\sin\left(\frac{1}{x}\right)$-curve (Example 3.6.4) in such a way that we obtain a continuum X. Then $\mathfrak{MT}(X)$ consists of the family of limit bars of all the topologist $\sin\left(\frac{1}{x}\right)$-curves.

The next example shows that there exists a continuum X such that $\mathfrak{MT}(X)$ is not closed in $\mathcal{C}(X)$.

3.6.40 Example Let X be the continuum of Example 3.6.39. Shrink the limit bar of the topologist $\sin\left(\frac{1}{x}\right)$-curve of the leg corresponding to 0 and identify this point with the vertex v_Z, obtaining a continuum Y. Then $\{v_Z\} \in Cl_{\mathcal{C}(Y)}(\mathfrak{MT}(Y))$ but $\mathcal{T}_Y(\{v_Z\}) = Y$.

3.7   Applications

We present several applications of the set function $\mathcal{T}$; these are taken from [4, 8, 11, 19, 21, 30, 41, 43, 44, 49].

We begin with a couple of definitions.

3.7.1 Definition A continuum X *is m-aposyndetic* provided that for each subset K of X with m points, $\mathcal{T}(K) = K$. We say X *is countable closed aposyndetic* if for each countable closed subset K of X, $\mathcal{T}(K) = K$.

3.7.2 Definition Let X be a continuum, and let A and B be two nonempty disjoint subsets of X. A subset C of X is a *separator of A and B* (or *separates X between A*

and B) if $X \setminus C = U \cup V$, where U and V are separated (i.e., $Cl(U) \cap V = \emptyset$ and $U \cap Cl(V) = \emptyset$), $A \subset U$ and $B \subset V$.

3.7.3 Theorem *Let X be a continuum. If $A \in \mathcal{P}(X)$, then $\mathcal{T}(A)$ intersects each closed separator of A and some point of $\mathcal{T}(A)$.*

Proof Suppose the theorem is not true. Then there exist $p \in \mathcal{T}(A)$ and a closed subset B of X such that $X \setminus B = H \cup K$, $A \subset H$, $p \in K$, where H and K are disjoint open subsets of X and $B \cap \mathcal{T}(A) = \emptyset$. Hence, for each $b \in B$, there exists a subcontinuum W_b of X such that $b \in Int(W_b) \subset W_b \subset X \setminus A$. Since B is compact, there exist $b_1, \ldots, b_n \in B$ such that $B \subset \bigcup_{j=1}^{n} Int(W_{b_j})$.

Since $Cl(K) \setminus K \subset B$, by Theorem 1.7.27, $Y = K \cup \bigcup_{j=1}^{n} W_{b_j}$ only has a finite number of components. Note that Y is closed in X since it is the union of a finite number of continua and $Cl(K) \setminus K \subset Y$. Since $p \in K$ and $K \subset Y$, $p \in Y$. Let C be the component of Y such that $p \in C$. By Lemma 1.6.2, $p \in Int(C)$. This implies that $p \in X \setminus \mathcal{T}(A)$, a contradiction. Therefore, $B \cap \mathcal{T}(A) \neq \emptyset$.

Q.E.D.

Now, we present several consequences of Theorem 3.7.3.

3.7.4 Corollary *Let X be a continuum. If $A \in \mathcal{P}(X)$, then each component of $\mathcal{T}(A)$ intersects $Cl(A)$.*

Proof Suppose the corollary is not true. Then there exists a component C of $\mathcal{T}(A)$ such that $C \cap Cl(A) = \emptyset$. Hence, by Theorem 1.6.8, there exist two disjoint closed subsets H and K of $\mathcal{T}(A)$ such that $\mathcal{T}(A) = H \cup K$, $Cl(A) \subset H$ and $C \subset K$. Since X is a metric space, there exist two disjoint open subsets U and V of X such that $H \subset U$ and $K \subset V$. Let $B = X \setminus (U \cup V)$. Then B is a closed separator of A and C and $B \cap \mathcal{T}(A) = \emptyset$, a contradiction to Theorem 3.7.3.

Q.E.D.

3.7.5 Corollary *Let X be a continuum, and let A and B be two closed subsets of X. If K is a component of $\mathcal{T}(A \cup B)$ such that $K \setminus (\mathcal{T}(A) \cup \mathcal{T}(B)) \neq \emptyset$, then $K \cap A \neq \emptyset$ and $K \cap B \neq \emptyset$.*

Proof Since X is a continuum, by Corollary 3.1.14, $\mathcal{T}(\emptyset) = \emptyset$. By Corollary 3.7.4, $K \cap (A \cup B) \neq \emptyset$. Now, by Theorem 3.1.74, $\mathcal{T}((A \cup B) \cap K) = K$. Since $K \setminus (\mathcal{T}(A) \cup \mathcal{T}(B)) \neq \emptyset$, $(A \cup B) \cap K$ meets both A and B. Thus, $K \cap A \neq \emptyset$ and $K \cap B \neq \emptyset$.

Q.E.D.

3.7.6 Corollary *Let X be a continuum, and let A and B be closed subsets of X. If $\mathcal{T}(A \cup B) \neq \mathcal{T}(A) \cup \mathcal{T}(B)$, then there exists a subcontinuum K of X such that $K \subset \mathcal{T}(A \cup B)$, $K \cap A \neq \emptyset$ and $K \cap B \neq \emptyset$.*

Proof Let K be a component of $\mathcal{T}(A \cup B)$ such that $K \setminus (\mathcal{T}(A) \cup \mathcal{T}(B)) \neq \emptyset$ and apply Corollary 3.7.5.

Q.E.D.

3.7.7 Corollary *If X is a continuum and A is a closed subset of X such that $\mathcal{T}(A)$ is totally disconnected then $\mathcal{T}(A) = A$.*

Proof Let A be a closed subset of X such that $\mathcal{T}(A)$ is a totally disconnected subset of X. By Remark 3.1.5, $A \subset \mathcal{T}(A)$. By Corollary 3.7.4, each component of $\mathcal{T}(A)$ intersects $Cl(A) = A$. Since all the components of $\mathcal{T}(A)$ are singletons, $\mathcal{T}(A) \subset A$.

Q.E.D.

The following theorem tells us that $\mathcal{T}$ of a product of continua behaves like the identity map at the product of two proper closed subsets. This result is already proved in Corollary 3.2.11, but we include another proof here.

3.7.8 Theorem *Let X and Y be continua. If A and B are proper closed subsets of X and Y, respectively, then $\mathcal{T}(A \times B) = A \times B$.*

Proof Let A and B be two proper closed subsets of X and Y, respectively. By Remark 3.1.5, $A \times B \subset \mathcal{T}(A \times B)$.

Let $(x, y) \in (X \times Y) \setminus (A \times B)$; without loss of generality, we assume that $x \in X \setminus A$. Then there exists an open subset U of X such that $x \in U \subset Cl(U) \subset X \setminus A$. Hence, $(Cl(U) \times Y) \cap (A \times B) = \emptyset$.

Let $z \in Y \setminus B$. Then $(X \times \{z\}) \cap (A \times B) = \emptyset$. Thus, $(x, y) \in (Cl(U) \times Y) \cup (X \times \{z\}) \subset (X \times Y) \setminus (A \times B)$. Since $(Cl(U) \times Y) \cup (X \times \{z\})$ is a continuum containing (x, y) in its interior, $(x, y) \in (X \times Y) \setminus \mathcal{T}(A \times B)$. Therefore, $\mathcal{T}(A \times B) = A \times B$.

Q.E.D.

As a consequence of Theorem 3.1.28 and Theorem 3.7.8, we have the following corollary:

3.7.9 Corollary *If X and Y are continua, then $X \times Y$ is an aposyndetic continuum.*

3.7.10 Theorem *Let X and Y be continua. If A and B are two closed totally disconnected subsets of X and Y, respectively, then for each closed subset K of $A \times B$, $\mathcal{T}(K) = K$.*

Proof Let K be a closed subset of $A \times B$. By Theorem 3.7.8, $\mathcal{T}(A \times B) = A \times B$. Since K is a closed subset of $A \times B$, by Proposition 3.1.7, $\mathcal{T}(K) \subset \mathcal{T}(A \times B) = A \times B$. Hence, $\mathcal{T}(K)$ is totally disconnected. By Corollary 3.7.7, $\mathcal{T}(K) = K$.

Q.E.D.

Now we are ready to prove our application of $\mathcal{T}$ to Cartesian products.

3.7.11 Theorem *Let X and Y be continua. If K is a countable closed subset of $X \times Y$, then $\mathcal{T}(K) = K$. In particular, $X \times Y$ is m-aposyndetic for each $m \in \mathbb{N}$.*

Proof Let K be a countable closed subset of $X \times Y$. Let $\pi_X : X \times Y \twoheadrightarrow X$ and $\pi_Y : X \times Y \twoheadrightarrow Y$ be the projection maps. Since K is countable and closed, $\pi_X(K)$ and $\pi_Y(K)$ are countable closed subsets of X and Y, respectively. Then $K \subset \pi_X(K) \times \pi_Y(K)$. Thus, by Theorem 3.7.10, $\mathcal{T}(K) = K$.

Q.E.D.

Next, we present an application of $\mathcal{T}$ to cones and suspensions similar to the one to products.

3.7.12 Theorem *If X is a continuum and K is a countable closed subset of $K(X)$, then $\mathcal{T}_{K(X)}(K) = K$.*

Proof Let K be a countable closed subset of $K(X)$. We assume first that $v_X \notin K$. Then $q^{-1}(K)$ is a countable closed subset of $X \times [0, 1]$. Hence, by Theorem 3.7.11, $\mathcal{T}_{X \times [0,1]}q^{-1}(K) = q^{-1}(K)$. Since q is monotone, by Theorem 3.1.80 (c), we have that $\mathcal{T}_{K(X)}(K) = q\mathcal{T}_{X \times [0,1]}q^{-1}(K)$. Therefore, $\mathcal{T}_{K(X)}(K) = qq^{-1}(K) = K$.

If $v_X \in K$ and v_X is an isolated point of K, then $K \setminus \{v_X\}$ is a closed subset of $K(X)$. Then, by the previous paragraph, $\mathcal{T}_{K(X)}(K \setminus \{v_X\}) = K \setminus \{v_X\}$. Hence, it is easy to see that $\mathcal{T}_{K(X)}(K) = K$.

Next, we assume that $v_X \in K$. Then $q^{-1}(K) = (X \times \{1\}) \cup \{(x_n, t_n)\}_{n=1}^{\infty}$, also suppose that $Cl_{X \times [0,1]}(\{(x_n, t_n)\}_{n=1}^{\infty}) \setminus \{(x_n, t_n)\}_{n=1}^{\infty} \subset X \times \{1\}$. Let $(x, t) \in (X \times [0, 1]) \setminus q^{-1}(K)$. We assume that $t \in (0, 1)$, the case $t = 0$ is similar. Since $q^{-1}(K)$ is closed and $X \times [0, 1]$ is metric, there exist an open set U of X such that $x \in U$ and an $\varepsilon > 0$ such that $0 < t - \varepsilon < t + \varepsilon < 1$ and $Cl_X(U) \times [t - \varepsilon, t + \varepsilon] \subset (X \times [0, 1]) \setminus q^{-1}(K)$. Let $s \in [t - \varepsilon, t + \varepsilon] \setminus \{t_n\}_{n=1}^{\infty}$, and let $W = (X \times \{s\}) \cup (Cl_X(U) \times [t - \varepsilon, t + \varepsilon])$. Then W is a subcontinuum of $X \times [0, 1]$ such that $(x, t) \in Int_{X \times [0,1]}(W) \subset W \subset (X \times [0, 1]) \setminus q^{-1}(K)$. Hence, $(x, t) \in (X \times [0, 1]) \setminus \mathcal{T}_{X \times [0,1]}q^{-1}(K)$. Thus, $\mathcal{T}_{X \times [0,1]}q^{-1}(K) = q^{-1}(K)$. Since q is monotone, by Theorem 3.1.80 part (c), we have that $\mathcal{T}_{K(X)}(K) = q\mathcal{T}_{X \times [0,1]}q^{-1}(K)$. Therefore, $\mathcal{T}_{K(X)}(K) = qq^{-1}(K) = K$.

Q.E.D.

The proof of the following theorem is similar to the one given in Theorem 3.7.12, we present the details.

3.7.13 Theorem *If X is a continuum and K is a countable closed subset of $\Sigma(X)$, then $\mathcal{T}_{\Sigma(X)}(K) = K$.*

Proof Let K be a countable closed subset of $\Sigma(X)$. Let v^+ and v^- be the vertexes of $\Sigma(X)$. We assume first that $\{v^+, v^-\} \cap K = \emptyset$. Then $q^{-1}(K)$ is a countable closed subset of $X \times [0, 1]$. Hence, by Theorem 3.7.11, $\mathcal{T}_{X \times [0,1]}q^{-1}(K) = q^{-1}(K)$. Since q is monotone, by Theorem 3.1.80 (c), $\mathcal{T}_{\Sigma(X)}(K) = q\mathcal{T}_{X \times [0,1]}q^{-1}(K)$. Therefore, $\mathcal{T}_{\Sigma(X)}(K) = qq^{-1}(K) = K$.

Next, we assume that $v^+ \in K$ and $v^- \in \Sigma(X) \setminus K$. If v^+ is an isolated point of K, a similar argument to the one given in Theorem 3.7.12, shows that $\mathcal{T}_{\Sigma(X)}(K) = K$. Suppose that $q^{-1}(K) = (X \times \{1\}) \cup \{(x_n, t_n)\}_{n=1}^{\infty}$, also assume that $Cl_{X \times [0,1]}(\{(x_n, t_n)\}_{n=1}^{\infty}) \setminus \{(x_n, t_n)\}_{n=1}^{\infty} \subset X \times \{1\}$. Let $(x, t) \in (X \times [0, 1]) \setminus q^{-1}(K)$. Without loss of generality, we assume that $t \neq 0$. Since $q^{-1}(K)$ is closed and $X \times [0, 1]$ is metric, there exist an open set U of X such that $x \in U$ and an $\varepsilon > 0$ such that $0 < t - \varepsilon < t + \varepsilon < 1$ and $Cl_X(U) \times [t - \varepsilon, t + \varepsilon] \subset (X \times [0, 1]) \setminus q^{-1}(K)$. Let $s \in [t - \varepsilon, t + \varepsilon] \setminus \{t_n\}_{n=1}^{\infty}$, and let $W = (X \times \{s\}) \cup (Cl_X(U) \times [t - \varepsilon, t + \varepsilon])$.

Then W is a subcontinuum of $X \times [0, 1]$ such that $(x, t) \in Int_{X \times [0,1]}(W) \subset W \subset (X \times [0, 1]) \setminus q^{-1}(K)$. Hence, $(x, t) \in (X \times [0, 1]) \setminus \mathcal{T}_{X \times [0,1]} q^{-1}(K)$. Thus, $\mathcal{T}_{X \times [0,1]} q^{-1}(K) = q^{-1}(K)$. Since q is monotone, by Theorem 3.1.80 (c), we have that $\mathcal{T}_{\Sigma(X)}(K) = q \mathcal{T}_{X \times [0,1]} q^{-1}(K)$. Therefore, $\mathcal{T}_{\Sigma(X)}(K) = q q^{-1}(K) = K$.

Now, suppose that $\{v^+, v^-\} \subset K$. If either v^+ or v^- or both are isolated points of K, a similar argument to the one given in Theorem 3.7.12, shows that $\mathcal{T}_{\Sigma(X)}(K) = K$. Assume that $q^{-1}(K) = (X \times \{1\}) \cup (X \times \{0\}) \cup \{(x_n, t_n)\}_{n=1}^\infty \cup \{(x'_n, t'_n)\}_{n=1}^\infty$. Also suppose that $Cl_{X \times [0,1]}(\{(x_n, t_n)\}_{n=1}^\infty) \setminus \{(x_n, t_n)\}_{n=1}^\infty \subset X \times \{1\}$ and $Cl_{X \times [0,1]}(\{(x'_n, t'_n)\}_{n=1}^\infty) \setminus \{(x'_n, t'_n)\}_{n=1}^\infty \subset X \times \{0\}$. Let $(x, t) \in (X \times (0, 1)) \setminus q^{-1}(K)$. Since $X \times [0, 1]$ is metric, there exist an open set U of X and an $\varepsilon > 0$ such that $0 < t - \varepsilon < t + \varepsilon < 1$ and $Cl_X(U) \times [t - \varepsilon, t + \varepsilon] \subset (X \times [0, 1]) \setminus q^{-1}(K)$. Let $s \in [t - \varepsilon, t + \varepsilon] \setminus (\{t_n\}_{n=1}^\infty \cup \{t'_n\}_{n=1}^\infty)$ and let $W = (X \times \{s\}) \cup (Cl_X(U) \times [t - \varepsilon, t + \varepsilon])$. Then W is a subcontinuum of $X \times [0, 1]$ such that $(x, t) \in Int_{X \times [0,1]}(W) \subset W \subset (X \times [0, 1]) \setminus q^{-1}(K)$. Hence, $(x, t) \in (X \times [0, 1]) \setminus \mathcal{T}_{X \times [0,1]} q^{-1}(K)$. Thus, $\mathcal{T}_{X \times [0,1]} q^{-1}(K) = q^{-1}(K)$. Since q is monotone, by Theorem 3.1.80 (c), we have that $\mathcal{T}_{\Sigma(X)}(K) = q \mathcal{T}_{X \times [0,1]} q^{-1}(K)$. Therefore, $\mathcal{T}_{\Sigma(X)}(K) = q q^{-1}(K) = K$.

Q.E.D.

3.7.14 Corollary *Let Y and Z be continua. If $X = Y \times Z$ is a continuum for which $\mathcal{T}$ is continuous, then X is locally connected. In particular, Y and Z are locally connected.*

Proof Let A be a nonempty closed subset of X. By the proof of Corollary 1.8.9, there exists a sequence $\{K_m\}_{m=1}^\infty$ of finite subsets of X converging to A. Since $\mathcal{T}$ is continuous, $\{\mathcal{T}(K_m)\}_{m=1}^\infty$ converges to $\mathcal{T}(A)$. By Theorem 3.7.11, $\mathcal{T}(K_m) = K_m$ for each $m \in \mathbb{N}$. Hence, $\mathcal{T}(A) = A$. Since A is an arbitrary nonempty closed subset of X, by Theorem 3.1.31, X is locally connected. Note that Y and Z are locally connected by Theorem 3.1.81.

Q.E.D.

As a consequence of Corollary 3.7.14 and Theorem 3.1.81, we obtain:

3.7.15 Corollary *If X is a continuum such that $\mathcal{T}$ is continuous for $K(X)$, then $K(X)$ is locally connected. In particular, X is locally connected.*

3.7.16 Corollary *If X is a continuum such that $\mathcal{T}$ is continuous for $\Sigma(X)$, then $\Sigma(X)$ is locally connected. In particular, X is locally connected.*

Our application of $\mathcal{T}$ to symmetric products is the following:

3.7.17 Theorem *Let X be a continuum and let $n \geq 2$ be an integer. If Γ is a countable closed subset of $\mathcal{F}_n(X)$, then $\mathcal{T}_{\mathcal{F}_n(X)}(\Gamma) = \Gamma$. In particular, $\mathcal{F}_n(X)$ is m-aposyndetic for each $m \in \mathbb{N}$.*

Proof Let Γ be a countable closed subset of $\mathcal{F}_n(X)$. Then the function $f_n \colon X^n \to \mathcal{F}_n(X)$ is a continuous surjection (Lemma 1.8.6). Since for each $\chi \in \mathcal{F}_n(X)$, $f_n^{-1}(\chi)$ is finite and Γ is countable and closed, $f_n^{-1}(\Gamma) = \bigcup_{\gamma \in \Gamma} f_n^{-1}(\gamma)$

is a countable closed subset of X^n. Hence, by Theorem 3.7.11 and induction, $\mathcal{T}_{X^n} f_n^{-1}(\Gamma) = f_n^{-1}(\Gamma)$. This implies that $f_n \mathcal{T}_{X^n} f_n^{-1}(\Gamma) = f_n f_n^{-1}(\Gamma) = \Gamma$. Thus, by Theorem 3.1.80 (a), $\mathcal{T}_{\mathcal{F}_n(X)}(\Gamma) \subset f_n \mathcal{T}_{X^n} f_n^{-1}(\Gamma)$. Hence, $\mathcal{T}_{\mathcal{F}_n(X)}(\Gamma) \subset \Gamma$. Since $\Gamma \subset \mathcal{T}_{\mathcal{F}_n(X)}(K)$, by Remark 3.1.5, $\mathcal{T}_{\mathcal{F}_n(X)}(\Gamma) = \Gamma$.

Q.E.D.

3.7.18 Corollary *Let X be a continuum and let $n \geq 2$ be an integer. If $\mathcal{T}_{\mathcal{F}_n(X)}$ is continuous for $\mathcal{F}_n(X)$, then X is locally connected.*

Proof As in the proof of Corollary 3.7.14, $\mathcal{T}_{\mathcal{F}_n(X)}(\Gamma) = \Gamma$ for each closed subset Γ of $\mathcal{F}_n(X)$. Hence, $\mathcal{F}_n(X)$ is locally connected, by Theorem 3.1.31. Therefore, X is locally connected by [30, Lemma 2].

Q.E.D.

3.7.19 Remark Let us observe that Corollaries 3.7.14 and 3.7.18 follow from Theorems 3.7.11 and 3.7.17 and Corollary 3.3.16. We decided to present different proofs.

Regarding the continuity of $\mathcal{T}$ on the other hyperspaces, we have the following Theorem:

3.7.20 Theorem *Let X be a continuum. If*

$$\mathcal{L} \in \left\{ 2^X, \mathcal{C}_n(X) \ (n \in \mathbb{N}) \right\}$$

and $\mathcal{T}_{\mathcal{L}}$ is continuous for $\mathcal{L}$, then X is locally connected.

Proof Suppose $\mathcal{T}_{\mathcal{L}}$ is continuous for $\mathcal{L}$. Since $\mathcal{L}$ is aposyndetic ([21, Theorem 1] for 2^X, and Corollary 6.3.3 for $\mathcal{C}_n(X)$), by Corollary 3.3.16, $\mathcal{L}$ is locally connected. Hence, X is locally connected ([43, (1.92)] for 2^X, and Theorem 3.6.11 for $\mathcal{C}_n(X)$).

Q.E.D.

The set function $\mathcal{T}$ may be used to show that continua are not contractible. First, we show the following lemma:

3.7.21 Lemma *Suppose X is a continuum, A is a closed subset of X and $r \in \mathcal{T}_X(A)$. If $H : X \times [0, 1] \twoheadrightarrow X$ is a homotopy such that $H((x, 0)) = x$ for each $x \in X$ and such that $H((r, 1)) \in X \setminus \mathcal{T}_X(A)$, then $H((r, t)) \in A$ for some $t \in [0, 1]$.*

Proof Let $f : X \to X$ be given by $f(x) = H((x, 1))$. Suppose the result is not true. Then $H^{-1}(A) \cap (\{r\} \times [0, 1]) = \emptyset$. Thus, there exists an open subset U of X such that $r \in U$ and $H^{-1}(A) \cap (Cl(U) \times [0, 1]) = \emptyset$. Since $f(r) = H((r, 1)) \in X \setminus \mathcal{T}_X(A)$, there exists a subcontinuum W of X such that $f(r) \in Int(W) \subset W \subset X \setminus A$. Let $V = U \cap f^{-1}(Int(W))$. Then $r \in V$, $H^{-1}(A) \cap (Cl(V) \times [0, 1]) = \emptyset$ and $f(Cl(V)) \subset W$. Let $L = (X \times \{0\}) \cup (Cl(V) \times [0, 1])$. Then L is a subcontinuum of $X \times [0, 1]$. Consider the disjoint union $L \cup W$. Let K be the quotient space of $L \cup W$ obtained by identifying each $(v, 1) \in Cl(V) \times \{1\}$ with $f(v)$, and let $q : L \cup W \twoheadrightarrow K$ be the quotient map.

Define $G\colon K \twoheadrightarrow X$ by

$$G(q(\omega)) = \begin{cases} \omega, & \text{if } \omega \in W; \\ H(\omega), & \text{if } \omega \in L. \end{cases}$$

Note that G is, in fact, surjective since $G(q((x,0))) = x$ for each $x \in X$. By Theorem 3.1.80 (a), $\mathcal{T}_X(A) \subset G\mathcal{T}_K G^{-1}(A)$. In particular, $r \in G\mathcal{T}_K G^{-1}(A)$. Hence, $G^{-1}(r) \cap \mathcal{T}_K G^{-1}(A) \neq \emptyset$. However, since $A \cap W = \emptyset$ and $H^{-1}(A) \cap (Cl(V) \times [0,1]) = \emptyset$, it follows that $G^{-1}(A) = q(A \times \{0\})$. Thus, $G^{-1}(r) \cap \mathcal{T}_K(q(A \times \{0\})) \neq \emptyset$.

Let $J = q(W \cup (Cl(V) \times [0,1]))$. Then J is a subcontinuum of K, and

$$M = q(W \cup (Cl(V) \times (0,1]) \cup (V \times [0,1])) \subset Int_K(J).$$

Since $G^{-1}(r) \cap q(X \times \{0\}) = q(\{(r,0)\})$, $G^{-1}(r) \subset M \subset Int_K(J)$. Since $q(A \times \{0\}) \cap J = \emptyset$, we have that $G^{-1}(r) \cap \mathcal{T}_K(q(A \times \{0\})) = \emptyset$, a contradiction. Therefore, there exists $t \in [0,1]$ such that $H((r,t)) \in A$.

Q.E.D.

The next theorem gives a sufficient condition for a continuum not to be contractible.

3.7.22 Theorem *If X is a continuum and A and B are closed subsets of X such that $A \cap \mathcal{T}(B) = \emptyset$, $\mathcal{T}(A) \cap B = \emptyset$ and $\mathcal{T}(A) \cap \mathcal{T}(B) \neq \emptyset$, then X is not contractible.*

Proof Suppose X is contractible. Then there exists a homotopy $H\colon X \times [0,1] \twoheadrightarrow X$ such that $H((x,0)) = x$ and $H((x,1)) = p$ for each $x \in X$ and some $p \in X$. Without loss of generality, we assume that $p \in A$. Let $r \in \mathcal{T}(A) \cap \mathcal{T}(B)$. Since $H((r,1)) \in A$ and $H((r,1)) \in X \setminus \mathcal{T}(B)$, by Lemma 3.7.21, there exists $t \in [0,1]$ such that $H((r,t)) \in B$. Let t_A and t_B be the smallest elements of $[0,1]$ such that $H((r,t_A)) \in A$ and $H((r,t_B)) \in B$. Thus, either $t_A < t_B$ or $t_A > t_B$, since $t_A = t_B$ is impossible. If $t_A < t_B$, applying Lemma 3.7.21 to $H|_{X \times [0,t_A]}$, we obtain an element $t' < t_A$ such that $H((r,t')) \in B$, a contradiction. Similarly, if $t_A > t_B$, applying Lemma 3.7.21 to $H|_{X \times [0,t_B]}$, we obtain an element $t'' < t_B$ such that $H((r,t'')) \in A$, a contradiction.

Therefore, X is not contractible.

Q.E.D.

Now, we present some connectivity properties of $\mathcal{T}$.

3.7.23 Lemma *Let X be a continuum. If S is a nonempty totally disconnected closed subset of X such that there exists a point $p \in \mathcal{T}(S) \setminus S$ and such that for each proper subset S' of S, $p \in X \setminus \mathcal{T}(S')$, then $\mathcal{T}(S)$ is connected.*

Proof Let S be a totally disconnected closed subset of X satisfying the properties stated. Let S_0 be a nonempty subset of S which is both open and closed in S. Since $p \in X \setminus \mathcal{T}(S \setminus S_0)$, there exists a subcontinuum W of X such that $p \in Int(W) \subset$

$W \subset X \setminus (S \setminus S_0)$. Let $\{U_n\}_{n=1}^{\infty}$ and $\{V_n\}_{n=1}^{\infty}$ be decreasing sequences of open subsets of X such that for each $n \in \mathbb{N}$, $S \setminus S_0 \subset U_n$, $S_0 \subset V_n$, $U_1 \cap Cl(V_1) = U_1 \cap W = Cl(V_1) \cap \{p\} = \emptyset$, $S \setminus S_0 = \bigcap_{n=1}^{\infty} U_n$ and $S_0 = \bigcap_{n=1}^{\infty} V_n$.

For each $n \in \mathbb{N}$, let C_n be the component of $X \setminus U_n$ such that $W \subset C_n$. Since p does not belong to the interior of the component of $C_n \setminus V_n$ in which it lies, there exists a sequence $\{D_n^j\}_{j=1}^{\infty}$ of distinct components of $C_n \setminus V_n$ such that $p \in D_n = \lim_{j \to \infty} D_n^j$. Since, by Theorem 1.7.27, $D_n^j \cap Cl(V_n) \neq \emptyset$, for each $n \in \mathbb{N}$, D_n is a continuum, $D_n \subset C_n \setminus V_n$, $p \in D_n$ and $D_n \cap Cl(V_n) \neq \emptyset$. Let $D = \lim_{n \to \infty} D_n$. Then, since $\mathcal{C}(X)$ is compact (Theorem 1.8.5), D is a continuum containing p and $D \cap S_0 \neq \emptyset$.

We assert that $D \subset \mathcal{T}(S)$. Suppose this is not true. Then there exist $q \in D \setminus \mathcal{T}(S)$ and a subcontinuum W' of X such that $q \in Int(W') \subset W' \subset X \setminus S$. It follows that there exists $N_1 \in \mathbb{N}$ such that for $n > N_1$, $W' \subset X \setminus (U_n \cup V_n)$. Since $q \in Int(W') \cap D$, there exists $N_2 \in \mathbb{N}$ such that for $n > N_2$, $(Int(W')) \cap D_n \neq \emptyset$. Let $m > \max\{N_1, N_2\}$. Since $(Int(W')) \cap D_m \neq \emptyset$, there exists $N_3 \in \mathbb{N}$ such that for $j > N_3$, $(Int(W')) \cap D_m^j \neq \emptyset$. Let $j > \max\{N_1, N_2, N_3\}$. Then $D_m^j \subset C_m$ and $W' \cap C_m \neq \emptyset$. Thus, $W' \subset C_m \setminus V_m$, a contradiction, since $j > \max\{N_1, N_2, N_3\}$ and $(Int(W')) \cap D_m^j \neq \emptyset$. Therefore, $D \subset \mathcal{T}(S)$.

Now, let $s \in S$. Then there exists a sequence $\{S_n\}_{n=1}^{\infty}$ of subsets of S such that for each $n \in \mathbb{N}$, S_n is open and closed in S, and $\{s\} = \bigcap_{n=1}^{\infty} S_n$. By the previous construction, for each $n \in \mathbb{N}$, there exists a subcontinuum A_n of $\mathcal{T}(S)$ such that $p \in A_n$ and $A_n \cap S_n \neq \emptyset$. Since $\mathcal{C}(X)$ is compact (Theorem 1.8.5), without loss of generality, we assume that the sequence $\{A_n\}_{n=1}^{\infty}$ converges. Let $A_s = \lim_{n \to \infty} A_n$. Then A_s is a subcontinuum of $\mathcal{T}(S)$ and $\{p, s\} \subset A_s$.

Let $A = \bigcup_{s \in S} A_s$. Then A is connected (for each $s \in S$, $p \in A_s$) and $S \subset A \subset \mathcal{T}(S)$. For each $x \in \mathcal{T}(S)$, let C_x be the component of $\mathcal{T}(S)$ such that $x \in C_x$. By Corollary 3.7.4, $C_x \cap S \neq \emptyset$. Thus, $C_x \cap A \neq \emptyset$ for each $x \in \mathcal{T}(S)$, and it follows that

$$\mathcal{T}(S) = A \cup \left(\bigcup_{x \in \mathcal{T}(S)} C_x \right)$$

is connected.

Q.E.D.

The following theorem shows that we may remove the "totally disconnected" hypothesis in Lemma 3.7.23.

3.7.24 Theorem *Let X be a continuum. Suppose A is a nonempty closed subset of X, and there exists a point $p \in \mathcal{T}_X(A) \setminus A$ such that for each proper closed subset A' of A, $p \in X \setminus \mathcal{T}_X(A')$. Then $\mathcal{T}_X(A)$ is connected.*

Proof If $\mathcal{T}_X(A) = X$, there is nothing to do. Then suppose $\mathcal{T}_X(A)$ is a proper closed subset of X. Let $\{A_\lambda\}_{\lambda \in \Lambda}$ be the family of components of A. Then $\mathcal{G} = \{A_\lambda\}_{\lambda \in \Lambda} \cup$

$\{\{x\} \mid x \in X \setminus A\}$ is an upper semicontinuous decomposition of X and $X/\mathcal{G}$ is a continuum (Theorem 1.7.3). Let $q: X \twoheadrightarrow X/\mathcal{G}$ be the quotient map. Note that q is monotone.

Since A is a closed subset of X, $q(A) = q\left(\bigcup_{\lambda \in \Lambda} A_\lambda\right) = \bigcup_{\lambda \in \Lambda} q(A_\lambda)$ is a closed totally disconnected subset of $X/\mathcal{G}$. Thus, $\mathcal{T}_{X/\mathcal{G}}(q(A))$ is connected, by Lemma 3.7.23. Since q is monotone, by Theorem 3.1.80 (b), we have that $\mathcal{T}_X(A) = \mathcal{T}_X(q^{-1}q(A)) \subset q^{-1}\mathcal{T}_{X/\mathcal{G}}(q(A))$. By Theorem 3.1.80 (c), we have $q^{-1}\mathcal{T}_{X/\mathcal{G}}(q(A)) = q^{-1}q\mathcal{T}_X(q^{-1}(q(A))) = q^{-1}q\mathcal{T}_X(A)$. Since $A \subset \mathcal{T}(A)$, $q^{-1}q\mathcal{T}_X(A) = \mathcal{T}_X(A)$. Hence, $q^{-1}\mathcal{T}_{X/\mathcal{G}}(q(A)) = \mathcal{T}_X(A)$. Therefore, $\mathcal{T}_X(A)$ is connected.

Q.E.D.

3.7.25 Theorem *Let X be a continuum. Suppose A is a nonempty closed subset of X and there exists $x \in \mathcal{T}(A) \setminus A$. Then there exists a closed subset D of X such that:*

(a) $D \subset A$,
(b) $x \in \mathcal{T}(D)$,
(c) if E is a nonempty closed subset of X and $E \subsetneq D$, then $x \in X \setminus \mathcal{T}(E)$, and
(d) $\mathcal{T}(D)$ is a continuum.

Proof Let $\{B_n\}_{n=1}^{\infty}$ be a decreasing sequence of nonempty closed subsets of X such that for each $n \in \mathbb{N}$, $x \in \mathcal{T}(B_n)$. Let $B = \bigcap_{n=1}^{\infty} B_n$. Then B is a nonempty closed subset of X. If $x \in X \setminus \mathcal{T}(B)$, then there exists a subcontinuum W of X such that $x \in Int(W) \subset W \subset X \setminus B$. Since W is compact, there exists $N \in \mathbb{N}$ such that $W \subset \bigcup_{n=1}^{N}(X \setminus B_n)$, a contradiction to the fact that $x \in \mathcal{T}(B_N)$. Hence, $x \in \mathcal{T}(B)$. By the Brouwer Reduction Theorem (see [49, (11.1), p. 17]), there exists a minimal element D satisfying conditions (a), (b) and (c). Condition (d) follows from Theorem 3.7.24.

Q.E.D.

In [44, 5.12] it is shown that if X is a continuum and x is a point at which X is not connected im kleinen, then there exists a subcontinuum K of X such that $x \in K$ and X is not connected im kleinen at any of the points of K. In the following theorem, we show that something similar happens with the images of $\mathcal{T}$.

3.7.26 Theorem *Let X be a continuum. Suppose A is a nonempty closed subset of X such that there exists $x \in \mathcal{T}(A) \setminus A$. Then there exists a nondegenerate subcontinuum K of X such that $x \in K \subset \mathcal{T}(A)$.*

Proof Let A be a nonempty closed subset of X and let $x \in \mathcal{T}(A) \setminus A$. By Theorem 3.7.25, there exists a nonempty closed subset D of X satisfying conditions (a), (b), (c) and (d). Let V be an open subset of X such that $x \in V \subset Cl(V) \subset X \setminus A$. Let H be the component of $V \cap \mathcal{T}(D)$ containing x, and let $K = Cl(H)$. Since $V \cap \mathcal{T}(D)$ is open in $\mathcal{T}(D)$, by Theorem 1.7.27, K intersects $Bd(V)$. Thus, K is a nondegenerate subcontinuum of X contained in $\mathcal{T}(A)$.

Q.E.D.

The following theorem characterizes local connectedness in unicoherent continua.

3.7.27 Theorem *Let X be a unicoherent continuum. Then X is locally connected if and only if for each nonempty closed subset C of X which separates X between two points x and y, there exists a component E of C which separates X between x and y.*

Proof First, suppose that X is unicoherent and locally connected. Let C be a nonempty closed subset of X which separates X between x and y. Let A be the component of $X \setminus C$ containing x, and let B be the component of $X \setminus Cl(A)$ containing y. Since X is locally connected, by Lemma 1.7.11, A and B are open subsets of X. Also, $X \setminus (Cl(A) \cap Cl(B)) \subset (X \setminus Cl(B)) \cup B, x \in X \setminus Cl(B)$, and $y \in B$. Thus, $Cl(A) \cap Cl(B)$ is a closed subset of X which separates X between x and y.

Let $\{S_\lambda\}_{\lambda \in \Lambda}$ be the family of components of $X \setminus Cl(A)$ such that $S_\lambda \cap B = \emptyset$ for each $\lambda \in \Lambda$. Then, by Theorem 1.7.27, $Cl(A) \cup \left(\bigcup_{\lambda \in \Lambda} Cl(S_\lambda)\right)$ is connected. Note that $X = Cl(A) \cup Cl\left(\bigcup_{\lambda \in \Lambda} Cl(S_\lambda)\right) \cup Cl(B)$. Since X is unicoherent, $Cl(A) \cap Cl(B) = \left(Cl(A) \cup Cl\left(\bigcup_{\lambda \in \Lambda} Cl(S_\lambda)\right)\right) \cap Cl(B)$ is a continuum. Let E be the component of C that contains $Cl(A) \cap Cl(B)$. If x and y belong to the same component of $X \setminus E \subset X \setminus (Cl(A) \cap Cl(B))$, then x and y are in the same component of $X \setminus (Cl(A) \cap Cl(B))$. Since this is not true, E separates X between x and y.

Now, suppose X is a unicoherent continuum satisfying the conditions stated, and suppose X is not locally connected; hence, not connected im kleinen at some point $p \in X$. Then, by Theorem 3.1.24, there exists a nonempty closed subset A of X such that $p \in \mathcal{T}(A) \setminus A$. By Theorem 3.7.25, there exists a nonempty closed subset B of X satisfying conditions (a), (b), (c) and (d). Let $x \in B$ and let U_0 be an open subset of X such that $p \in U_0 \subset Cl(U_0) \subset X \setminus B$. Then $Bd(U_0)$ is a closed set which separates X between x and p. Hence, there exists a component N of $Bd(U_0)$ which separates X between x and p.

Let U and V be disjoint open sets such that $X \setminus N = U \cup V$, $p \in U$ and $x \in V$. Then, by Lemma 1.7.23, $H = N \cup U$ and $K = N \cup V$ are continua, and $X = H \cup K$. Since $x \in B \cap K$, then $B \cap K \neq \emptyset$. If $B \cap H = \emptyset$, then $p \in X \setminus \mathcal{T}(B)$, which is contrary to (b) of Theorem 3.7.25. Thus, $B \cap H \neq \emptyset$. Since $B \cap K$ and $B \cap H$ are two nonempty closed proper subsets of B, it follows from (c) of Theorem 3.7.25 that there exist two subcontinua L_1 and L_2 of X such that $p \in Int(L_1) \subset L_1 \subset X \setminus (B \cap H)$ and $p \in Int(L_2) \subset L_2 \subset X \setminus (B \cap K)$. Again, if $L_1 \cap (B \cap K) = \emptyset$ or $L_2 \cap (B \cap H) = \emptyset$, it follows that $p \in X \setminus \mathcal{T}(B)$. Thus, assume that $L_1 \cap (B \cap K) \neq \emptyset$ and $L_2 \cap (B \cap H) \neq \emptyset$. Let $X_1 = L_1 \cup K$ and $X_2 = L_2 \cup H$. Then X_1 and X_2 are continua and $X = X_1 \cup X_2$. Now, since X is unicoherent, $X_1 \cap X_2$ is a continuum. Note that $p \in Int(X_1 \cap X_2) \subset (X_1 \cap X_2) \subset X \setminus B$. Hence, $p \in X \setminus \mathcal{T}(B)$, which is contrary to condition (b) of Theorem 3.7.25. Therefore, X is locally connected.

$\hfill$ **Q.E.D.**

References

1. D. P. Bellamy, Continua for Which the Set Function $\mathcal{T}$ is Continuous, Trans. Amer. Math. Soc., 1511 (1970), 581–587.
2. D. P. Bellamy, Set Functions and Continuous Maps, in *General Topology and Modern Analysis*, (L. F. McAuley and M. M. Rao, eds.), Academic Press, (1981), 31–38.
3. D. P. Bellamy, Some Topics in Modern Continua Theory, in *Continua Decompositions Manifolds*, (R H Bing, W. T. Eaton and M. P. Starbird, eds.), University of Texas Press, (1983), 1–26.
4. D. P. Bellamy and J. J. Charatonik, The Set Function $\mathcal{T}$ and Contractibility of Continua, Bull. Acad. Polon. Sci. Sér. Sci. Math. Astronom. Phys., 25 (1977), 47–49.
5. D. P. Bellamy and H. S. Davis, Continuum Neighborhoods and Filterbases, Proc. Amer. Math. Soc., 27 (1971), 371–374.
6. D. P. Bellamy, L. Fernández and S. Macías, On $\mathcal{T}$-closed Sets, Topology Appl., 195 (2015), 209–225. (Special issue honoring the memory of Professor Mary Ellen Rudin.)
7. D. P. Bellamy and L. Lum, The Cyclic Connectivity of Homogeneous Arcwise Connected Continua, Trans. Amer. Math. Soc., 266 (1981), 389–396.
8. D. E. Bennet, A Characterization of Locally Connectedness By Means of the Set Function $\mathcal{T}$, Fund. Math. 86 (1974), 137–141.
9. R H Bing, Concerning Hereditarily Indecomposable Continua, Pacific J. Math., 1 (1951), 43–51.
10. R H Bing and F. B. Jones, Another Homogeneous Plane Continuum, Trans. Amer. Math. Soc., 90 (1959), 171–192.
11. K. Borsuk and S. Ulam, On Symmetric Products of Topological Spaces, Bull. Amer. Math. Soc., 37 (1931), 875–882.
12. C. O. Christenson and W. L. Voxman, *Aspects of Topology*, Monographs and Textbooks in Pure and Applied Math., Vol. 39, Marcel Dekker, New York, Basel, 1977.
13. H. Cook, W. T. Ingram, and A. Lelek, *A List of Problems Known as Houston Problem Book*, in *Continua with The Houston Problem Book*, H. Cook, W. T. Ingram, K. T. Kuperberg, A. Lelek, and P. Minc, editors. Lecture Notes in Pure and Applied Mathematics, Vol. 170, Marcel Dekker, New York, Basel, Hong Kong (1995), 365–398.
14. H. S. Davis, A Note on Connectedness im Kleinen, Proc. Amer. Math. Soc., 19 (1968), 1237–1241.
15. H. S. Davis, Relationships Between Continuum Neighborhoods in Inverse Limit Spaces and Separations in Inverse Limit Sequences, Proc. Amer. Math. Soc., 64 (1977), 149–153.
16. H. S. Davis and P. H. Doyle, Invertible Continua, Portugal. Math., 26 (1967), 487–491.
17. L. Fernández, On strictly point $\mathcal{T}$-asymmetric continua, Topology Proc., 35 (2010), 91–96.
18. L. Fernández and S. Macías, The Set Functions $\mathcal{T}$ and $\mathcal{K}$ and Irreducible Continua, Colloq. Math., 121 (2010), 79–91.
19. R. W. FitzGerald, The Cartesian Product of Non-degenerate Compact Continua is n-point Aposyndetic, Topology Conference (Arizona State Univ., Tempe, Ariz., 1967), Arizona State University, Tempe, Ariz., (1968), 324–326.
20. J. B. Fugate, G. R. Gordh, Jr. and L. Lum, Arc-smooth Continua, Trans. Amer. Math. Soc., 265 (1981), 545–561.
21. J. T. Goodykoontz, Jr., Aposyndetic Properties of Hyperspaces, Pacific J. Math., 47 (1973), 91–98.
22. G. R. Gordh, Jr. and C. B. Hughes, On Freely Decomposable Mappings of Continua, Glasnik Math., 14 (34) (1979), 137–146.
23. J. G. Hocking and G. S. Young, *Topology*, Dover Publications, Inc., New York, 1988.
24. F. B. Jones, Concerning the Boundary of a Complementary Domain of a Continuous Curve, Bull. Amer. Math. Soc., 45 (1939), 428–435.
25. F. B. Jones, Aposyndetic Continua and Certain Boundary Problems, Amer. J. Math., 53 (1941), 545–553.

26. F. B. Jones, Concerning Nonaposyndetic Continua, Amer. J. Math., 70 (1948), 403–413.
27. J. Krasinkiewicz, On Two Theorems of Dyer, Colloq. Math., 50 (1986), 201–208.
28. K. Kuratowski, *Topology*, Vol. II, Academic Press, New York, N. Y., 1968.
29. W. Lewis, Continuous Curves of Pseudo-arcs, Houston J. Math., 11 (1985), 91–99.
30. S. Macías, Aposyndetic Properties of Symmetric Products of Continua, Topology Proc., 22 (1997), 281–296.
31. S. Macías, A Class of One-dimensional Nonlocally Connected Continua for Which the Set Function $\mathcal{T}$ is Continuous, Houston J. Math., 32 (2006), 161–165.
32. S. Macías, Homogeneous Continua for Which the Set Function $\mathcal{T}$ is Continuous, Topology Appl., 153 (2006), 3397–3401.
33. S. Macías, A Decomposition Theorem for a Class of Continua for Which the Set Function $\mathcal{T}$ is Continuous, Colloq. Math., 109 (2007), 163–170.
34. S. Macías, On Continuously Irreducible Continua, Topology Appl., 156 (2009), 2357–2363.
35. S. Macías, On the Idempotency of the Set Function $\mathcal{T}$, Houston J. Math., 37 (2011), 1297–1305.
36. S. Macías, A Note on The Set Functions $\mathcal{T}$ and $\mathcal{K}$, Topology Proc., 46 (2015), 55–65.
37. S. Macías, On Continuously Type A' θ-continua, JP Journal Geometry and Topology, 18 (2015), 1–14.
38. S. Macías, Atomic Maps and $\mathcal{T}$-closed Sets, Topology Proc., 50 (2017), 97–100.
39. S. Macías and S. B. Nadler, Jr., On Hereditarily Decomposable Homogeneous Continua, Topology Proc., 34 (2009), 131–145.
40. L. Mohler and L. G. Oversteegen, On the Structure of Tranches in Continuously Irreducible Continua, Colloq. Math., 54 (1987), 23–28.
41. M. A. Molina, *Algunos Aspectos Sobre la Función $\mathcal{T}$ de Jones*, Tesis de Licenciatura, Facultad de Ciencias, U. N. A. M., 1998. (Spanish)
42. R. L. Moore, *Foundations of Point Set Theory*, Rev. Ed., Amer. Math. Soc. Colloq. Publ., Vol. 13, Amer. Math. Soc., Providence, Rhode Island, 1962.
43. S. B. Nadler, Jr., *Hyperspaces of Sets*, Monographs and Textbooks in Pure and Applied Math., Vol. 49, Marcel Dekker, New York, Basel, 1978. Reprinted in: Aportaciones Matemáticas de la Sociedad Matemática Mexicana, Serie Textos # 33, 2006.
44. S. B. Nadler, Jr., *Continuum Theory: An Introduction,* Monographs and Textbooks in Pure and Applied Math., Vol. 158, Marcel Dekker, New York, Basel, Hong Kong, 1992.
45. E. S. Thomas, Jr., Monotone Decompositions of Irreducible Continua, Dissertationes Math. (Rozprawy Mat.), 50 (1966), 1–74.
46. E. L. VandenBoss, *Set Functions and Local Connectivity*, Ph. D. Dissertation, Michigan State University (1970). University Microfilms # 71-11997.
47. K. Villarreal, Fibered Products of Homogeneous Continua, Trans. Amer. Math. Soc., 338 (1993), 933–939.
48. G. T. Whyburn, Semi-locally-connected Sets, Amer. J. Math., 61 (1939), 733–749.
49. G. T. Whyburn, *Analytic Topology,* Amer. Math. Soc. Colloq. Publ., Vol. 28, Amer. Math. Soc., Providence, R. I., 1942.
50. S. Willard, *General Topology*, Addison-Wesley Publishing Co., 1970.

Chapter 4
A Theorem of E. G. Effros

We present a topological proof of a Theorem by E. G. Effros [4] and a consequence of it, due to C. L. Hagopian [5], which has been very useful in the theory of homogeneous continua. We present the proof of Effros's result given by Fredric G. Ancel [1].

Before showing Effros's Theorem, we present some background on topological groups and actions of topological groups on metric spaces.

4.1 Topological Groups

We introduce topological groups and give some of its elementary properties. To this end, we use [3, 6, 11].

4.1.1 Definition We say that a group G, which has a topology τ, is a *topological group* provided that the group operations are continuous; i.e., the functions given by:

$$\pi: G \times G \to G \quad \pi((g_1, g_2)) = g_1 \cdot g_2$$

and

$$\xi: G \to G \quad \xi(g) = g^{-1}$$

are continuous.

4.1.2 Proposition *Let G be a group with a topology τ. Then G is a topological group if and only if the function $\omega: G \times G \to G$ given by $\omega((g_1, g_2)) = g_1 \cdot g_2^{-1}$ is continuous.*

© Springer International Publishing AG, part of Springer Nature 2018
S. Macías, *Topics on Continua*, https://doi.org/10.1007/978-3-319-90902-8_4

Proof Suppose G is a topological group. Observe that $\omega = \pi \circ (1_G \times \xi)$, where 1_G denotes the identity map of G. Hence, ω is continuous.

Now, suppose that ω is continuous. Note that $\xi(g) = \omega((e_G, g))$, where e_G is the identity element of G. Thus, ξ is continuous. Since $\pi = \omega \circ (1_G \times \xi)$, π is continuous.

Q.E.D.

4.1.3 Notation Let G be a group. If H and K are nonempty subsets of G, then define

$$H \cdot K = \{h \cdot k \mid h \in H \text{ and } k \in K\}$$

and

$$H^{-1} = \{h^{-1} \mid h \in H\}.$$

If $H = \{h\}$, then we write $h \cdot K$ instead of $\{h\} \cdot K$. Similarly, we write $H \cdot k$ instead of $H \cdot \{k\}$, when $K = \{k\}$. We define, inductively, $H^{n+1} = H \cdot H^n$ for each $n \in \mathbb{N}$.

4.1.4 Notation If G is a topological group and $g \in G$, then $\mathcal{N}(g)$ denotes the family of all neighborhoods of g in G. Also, e_G denotes the identity element of G.

4.1.5 Remark If G is a topological group, then we may describe the continuity of the group operations as follows: Let g_1 and g_2 be two elements of G. For the continuity of π: for each $U \in \mathcal{N}(g_1 \cdot g_2)$, there exist $V \in \mathcal{N}(g_1)$ and $W \in \mathcal{N}(g_2)$ such that $V \cdot W \subset U$. For the continuity of ξ: for each $U \in \mathcal{N}(g_1^{-1})$, there exists $V \in \mathcal{N}(g_1)$ such that $V^{-1} \subset U$.

4.1.6 Example It is easy to see that $(\mathbb{R}^n, +)$, $(\mathbb{C}, +)$, $(\mathbb{C} \setminus \{0\}, \cdot)$ and $(\mathcal{S}^1, \cdot)$ are topological groups with the usual operations and the usual topologies.

4.1.7 Proposition *Let G be a topological group. If g_0 is a fixed element of G, then the functions $\psi_{g_0}, \varphi_{g_0} : G \to G$ given by $\psi_{g_0}(g) = g_0 \cdot g$ and $\varphi_{g_0}(g) = g \cdot g_0$ are homeomorphisms.*

Proof Note that ψ_{g_0} and φ_{g_0} are restrictions of the operation π to $\{g_0\} \times G$ and $G \times \{g_0\}$, respectively. Hence, both functions are continuous. Also note that $\psi_{g_0} \circ \psi_{g_0^{-1}} = \psi_{g_0^{-1}} \circ \psi_{g_0} = 1_G$ and that $\varphi_{g_0} \circ \varphi_{g_0^{-1}} = \varphi_{g_0^{-1}} \circ \varphi_{g_0} = 1_G$. Therefore, ψ_{g_0} and φ_{g_0} are homeomorphisms, since, clearly, $\psi_{g_0^{-1}}$ and $\varphi_{g_0^{-1}}$ are continuous functions.

Q.E.D.

4.1.8 Definition Let G be a topological group. The maps ψ_{g_0} and φ_{g_0} of Proposition 4.1.7 are called *left translation by g_0* and *right translation by g_0*, respectively.

4.1.9 Corollary *If G is a topological group and ψ_{g_0} and φ_{g_0} are the left and right translations by g_0, respectively, then $\psi_{g_0}^{-1} = \psi_{g_0^{-1}}$ and $\varphi_{g_0}^{-1} = \varphi_{g_0^{-1}}$.*

4.1.10 Proposition *Let G be a topological group. Then the map $\xi: G \to G$ given by $\xi(g) = g^{-1}$ is a homeomorphism.*

Proof By Definition 4.1.1, ξ is continuous. Since $\xi \circ \xi = 1_G$, ξ is a homeomorphism.

Q.E.D.

4.1.11 Corollary *Each topological group is a homogeneous space.*

Proof Let g_1 and g_2 be two elements of G. Let $\zeta: G \to G$ be given by $\zeta(g) = g \cdot g_1^{-1} \cdot g_2$. Note that $\zeta(g_1) = g_2$ and that $\zeta = \varphi_{g_1^{-1} \cdot g_2}$. Therefore, ζ is a homeomorphism.

Q.E.D.

4.1.12 Proposition *Let G be a topological group, and let $\mathcal{N}^\circ(e_G)$ be a local basis of open sets of G containing e_G. Let $\mathcal{L}_g = \{g \cdot U \mid U \in \mathcal{N}^\circ(e_G)\}$ and let $\mathcal{R}_g = \{U \cdot g \mid U \in \mathcal{N}^\circ(e_G)\}$, for each $g \in G$. If $\mathcal{L} = \bigcup_{g \in G} \mathcal{L}_g$ and $\mathcal{R} = \bigcup_{g \in G} \mathcal{R}_g$, then $\mathcal{L}$ and $\mathcal{R}$ both form a basis for the topology of G.*

Proof Let W be an open subset of G and let g_0 be an element of W. Since $\psi_{g_0^{-1}}$ is a homeomorphism (Proposition 4.1.7), $\psi_{g_0^{-1}}(W) = g_0^{-1} \cdot W$ is an open subset of G such that $e_G \in g_0^{-1} \cdot W$. Since $\mathcal{N}^\circ(e_G)$ is a local basis of e_G, there exists $U \in \mathcal{N}^\circ(e_G)$ such that $U \subset g_0^{-1} \cdot W$. Hence, $\psi_{g_0}(e_G) \in \psi_{g_0}(U) \subset \psi_{g_0}(g_0^{-1} \cdot W)$; i.e., $g_0 \in g_0 \cdot U \subset (g_0 \cdot g_0^{-1}) \cdot W = W$. Therefore, $\mathcal{L}$ is a basis for the topology of G.

The proof of the fact that $\mathcal{R}$ forms a basis for the topology of G is similar.

Q.E.D.

4.1.13 Definition Let G be a topological group and let S be a nonempty subset of G. We say that S is *symmetric* if $S = S^{-1}$. The family of symmetric neighborhoods of e_G is denoted by $\mathcal{N}^*(e_G)$; i.e., $\mathcal{N}^*(e_G) = \{V \in \mathcal{N}(e_G) \mid V = V^{-1}\}$.

The next three propositions say that each of the families of symmetric neighborhoods, powers of neighborhoods and closure of neighborhoods of the identity element of a topological group forms a local basis.

4.1.14 Proposition *If G is a topological group and if $U \in \mathcal{N}(e_G)$, then there exists $V \in \mathcal{N}^*(e_G)$ such that $V \subset U$.*

Proof Let $U \in \mathcal{N}(e_G)$. Since ξ is a homeomorphism (Proposition 4.1.10), $\xi(U) \in \mathcal{N}(e_G)$. Note that $\xi(U) = U^{-1}$. Let $V = U \cap U^{-1}$. Then, clearly, $V \in \mathcal{N}^*(e_G)$ and $V \subset U$.

Q.E.D.

4.1.15 Proposition *If G is a topological group and $U \in \mathcal{N}(e_G)$, then for each $n \in \mathbb{N}$, there exists $V \in \mathcal{N}(e_G)$ such that $V^n \subset U$.*

Proof Let $U \in \mathcal{N}(e_G)$. We do the proof by induction over n. For $n = 1$, let $V = U$.

Let $n \geq 2$ and suppose there exists $W \in \mathcal{N}(e_G)$ such that $W^n \subset U$. We show there exists $V \in \mathcal{N}(e_G)$ such that $V^{n+1} \subset U$.

Since the operation π is continuous and $e_G \cdot e_G = e_G$, there exist $V_1, V_2 \in \mathcal{N}(e_G)$ such that $V_1 \cdot V_2 \subset W$. Let $V = V_1 \cap V_2$. Then $V \in \mathcal{N}(e_G)$ and $V^2 \subset W$. Thus, $V^{n+1} = V^2 \cdot V^{n-1} \subset W \cdot W^{n-1} = W^n \subset U$.

$$\textbf{Q.E.D.}$$

4.1.16 Proposition *If G is a topological group and $U \in \mathcal{N}(e_g)$, then there exists $V \in \mathcal{N}(e_G)$ such that $Cl(V) \subset U$.*

Proof Let $U \in \mathcal{N}(e_G)$. By Proposition 4.1.15, there exists $W \in \mathcal{N}(e_G)$ such that $W^2 \subset U$. By Proposition 4.1.14, there exists $V \in \mathcal{N}^*(e_G)$ such that $V \subset W$. Hence, $V^2 \subset U$. Let $g \in Cl(V)$. Then $g \cdot V \in \mathcal{N}(g)$ (Proposition 4.1.12). Thus, $g \cdot V \cap V \neq \emptyset$. This implies that there exist $g_1, g_2 \in V$ such that $g \cdot g_1 = g_2$. Hence, $g = g_2 \cdot g_1^{-1} \in V \cdot V^{-1} = V^2 \subset U$. Therefore, $Cl(V) \subset U$.

$$\textbf{Q.E.D.}$$

4.1.17 Proposition *Let G be a topological group. Let S, T, U and R be nonempty subsets of G, and let $g_0 \in G$. Then the following statements are true:*

(1) If U is an open subset of G, then the sets: $g_0 \cdot U$, $U \cdot g_0$, U^{-1}, $R \cdot U$ and $U \cdot R$ are open subsets of G.

(2) If S is a closed subset of G, then the sets: $g_0 \cdot S$, $S \cdot g_0$ and S^{-1} are closed subsets of G.

(3) If S and T are compact subsets of G, then the sets: $S \cdot T$ and S^{-1} are compact subsets of G.

(4) If S is a compact subset of G and T is a closed subset of G, then $S \cdot T$ and $T \cdot S$ are closed subsets of G.

(5) $Cl(S) = \bigcap_{W \in \mathcal{N}(e_G)} S \cdot W$ and $Cl(S) = \bigcap_{W \in \mathcal{N}(e_G)} W \cdot S$.

Proof We show (1) is true. Note that, since the maps ψ_{g_0}, φ_{g_0} (Proposition 4.1.7) and ξ (Proposition 4.1.10) are homeomorphisms, $g_0 \cdot U$, $U \cdot g_0$ and U^{-1} are open subsets of G.

Since $R \cdot U = \bigcup_{r \in R} r \cdot U$ and $U \cdot R = \bigcup_{r \in R} U \cdot r$, $R \cdot U$ and $U \cdot R$ are open subsets of G.

The proof of (2) is similar to the one given in (1).

Note that (3) follows from the continuity of the operations π and ξ and the fact that S and T are compact sets.

We see (4) is true. To this end, we show that $T \cdot S$ is closed. The proof of the fact that $S \cdot T$ is closed is similar.

To see $T \cdot S$ is closed, we show $G \setminus (T \cdot S)$ is an open subset of G. Let $g \in G \setminus (T \cdot S)$. Then for each $s \in S$, $T \cdot s$ is a closed subset of G (part (2)). Thus, there exists $U_s \in \mathcal{N}^*(e_G)$ such that $g \cdot U_s \cap T \cdot s = \emptyset$. By Proposition 4.1.15, there exists $W_s \in \mathcal{N}(e_G)$ such that $W_s^2 \subset U_s$. By Proposition 4.1.14, there exists $V_s \in \mathcal{N}^*(e_G)$ such that $V_s \subset W_s$. Hence, for each $s \in S$, $V_s^2 \subset U_s$. Note that $g \cdot V_s \cap (T \cdot s) \cdot V_s = \emptyset$ for every $s \in S$.

Now, observe that the family of neighborhoods $\{s \cdot V_s \mid s \in S\}$ covers S. Since S is compact, there exist $s_1, \ldots, s_n \in S$ such that $S \subset \bigcup_{j=1}^n s_j \cdot V_{s_j}$. Let $W' = \bigcap_{j=1}^n V_{s_j}$. Then $W' \in \mathcal{N}^*(e_G)$, $g \cdot W' \in \mathcal{N}(g)$ and $g \cdot W' \cap (T \cdot s_j) \cdot V_{s_j} = \emptyset$, for each $j \in \{1, \ldots, n\}$. Thus, $g \cdot W' \cap T \cdot S = \emptyset$. Therefore, $G \setminus (T \cdot S)$ is open.

We prove (5) is satisfied. Let $W \in \mathcal{N}(e_G)$. Then there exists $V \in \mathcal{N}^*(e_G)$ such that $V \subset W$. Note that $S \subset S \cdot V \subset S \cdot W$. Let $g \in Cl(S)$. Since $V \in \mathcal{N}(e_G)$, $g \cdot V \cap S \neq \emptyset$. Thus, there exist $v \in V$ and $s \in S$ such that $g \cdot v = s$. Hence, $g = s \cdot v^{-1} \in S \cdot V^{-1} = S \cdot V \subset S \cdot W$. Therefore, $Cl(S) \subset \bigcap_{W \in \mathcal{N}(e_G)} S \cdot W$.

Next, let $g \in \bigcap_{W \in \mathcal{N}(e_G)} S \cdot W$, and let $V \in \mathcal{N}(g)$. We show that $V \cap S \neq \emptyset$, which implies that $g \in Cl(S)$. To this end, note that $V^{-1} \cdot g \in \mathcal{N}(e_G)$. Then $g \in S \cdot (V^{-1} \cdot g)$. Thus, there exist $s \in S$ and $v \in V$ such that $g = s \cdot v^{-1} \cdot g$. Hence, $s = v$. Therefore, $S \cap V \neq \emptyset$.

The proof of the other equality is similar.

$$\text{Q.E.D.}$$

4.2 Group Actions and a Theorem of Effros

We present elementary properties of an action of a topological group on a topological space. We also give a topological proof of a Theorem of E. G. Effros (Theorem 4.2.25). The material for this section is taken from [1–5, 7–10].

4.2.1 Definition An *action of a topological group G on a metric space X* is a map $\theta \colon G \times X \to X$ such that:

(1) $\theta(g, \theta(g', x)) = \theta(g \cdot g', x)$ for each $g', g \in G$ and each $x \in X$; and
(2) $\theta(e_G, x) = x$ for each $x \subset X$.

4.2.2 Notation Instead of $\theta(g, x)$, we write $g \cdot x$. In this way, the properties (1) and (2) of Definition 4.2.1 become:

(1) $g \cdot (g' \cdot x) = (g \cdot g') \cdot x$; and
(2) $e_G \cdot x = x$.

4.2.3 Remark Let G be a topological group acting on a metric space X. Note that for each $g \in G$, the function $\eta_g \colon X \twoheadrightarrow X$, given by $\eta_g(x) = g \cdot x$, is a homeomorphism, whose inverse is the map $\eta_{g^{-1}}$.

4.2.4 Example

(1) If G is a topological group, then G acts on G^n, for each $n \in \mathbb{N}$.
(2) If G is $\mathbb{R} \setminus \{0\}$ (or G is $\mathbb{C} \setminus \{0\}$), then G acts on $\mathbb{R}^n$ (or on $\mathbb{C}^n$, respectively).
(3) If G is $\mathcal{S}^1$, then G acts on $\mathbb{C}$.

4.2.5 Notation Let G be a topological group acting on a metric space X. If K is a nonempty subset of G and $x \in X$, then $K \cdot x = \{g \cdot x \mid g \in K\}$.

4.2.6 Definition Let G be a topological group acting on a metric space X. We say that G *acts transitively on* X, if $G \cdot x = X$ for each $x \in X$. We say that G *acts micro-transitively on* X if for each $x \in X$ and each $U \in \mathcal{N}(e_G)$, $U \cdot x$ is a neighborhood of x in X.

4.2.7 Remark From now on, G denotes a topological group whose topology is given by a complete metric acting on a metric space X. We call such a group a *complete metric group*.

4.2.8 Definition Let G be a complete metric group acting on a metric space X. For each $x \in X$, define the map $\gamma_x \colon G \twoheadrightarrow X$ by $\gamma_x(g) = g \cdot x$ for every $g \in G$.

4.2.9 Lemma *If G is a complete metric group acting transitively on a metric space X, then the following are equivalent:*

(a) G acts micro-transitively on X;
(b) γ_x is an open map for each $x \in X$;
(c) γ_x is an open map for some $x \in X$.

Proof First assume (a). We show (b). Let x be a point of X and let U be an open subset of G. Take $g \in U$. Note that $g^{-1} \cdot U$ is an open subset of G such that $e_G \in g^{-1} \cdot U$ (Proposition 4.1.7). Since G acts micro-transitively on X, $(g^{-1} \cdot U) \cdot x$ is a neighborhood of x in X. Hence, $U \cdot x = \gamma_x(U)$ is a neighborhood of $g \cdot x = \gamma_x(g)$. Therefore, γ_x is an open map.

Obviously, if (b) is true, then (c) is true. Assume (c). We prove (b). Suppose γ_x is an open map for some $x \in X$. Let $z \in X$. We see that γ_z is also an open map. Since G acts transitively on X, there exists $g_0 \in G$ such that $z = g_0 \cdot x$. Since φ_{g_0} is a homeomorphism (Proposition 4.1.7) and $\gamma_z = \gamma_x \circ \varphi_{g_0}$, γ_z is an open map.

Finally, assume (b). We see (a). Let $x \in X$ and let $U \in \mathcal{N}(e_g)$. Since γ_x is an open map, $\gamma_x(Int_G(U)) = (Int_G(U)) \cdot x$ is an open subset of X such that $x \in (Int_G(U)) \cdot x$. Hence, $U \cdot x$ is a neighborhood of x in X. Therefore, G acts micro-transitively on X.

Q.E.D.

4.2.10 Definition Let G be a complete metric group acting on a metric space X. We say that X is *G-countably covered* if for each $x \in X$ and each $U \in \mathcal{N}(e_G)$, there exists a sequence $\{h_n\}_{n=1}^{\infty}$ of homeomorphisms of X such that the family $\{h_n(U \cdot x)\}_{n=1}^{\infty}$ covers X.

4.2.11 Definition Let G be a complete metric group acting on a metric space X. We say that G *acts weakly micro-transitively on* X if for each $x \in X$ and each $U \in \mathcal{N}(e_G)$, $Cl(U \cdot x)$ is a neighborhood of x in X.

The following lemma provides examples of metric spaces which are G-countably covered.

4.2.12 Lemma *Let G be a complete metric group acting transitively on a metric space X. If G is separable, then X is G-countably covered.*

Proof Suppose G is separable. Let $x \in X$ and let $U \in \mathcal{N}(e_G)$. Then the collection $\{g \cdot U \mid g \in G\}$ covers G. Since G is a separable metric space, there exists a sequence, $\{g_n\}_{n=1}^{\infty}$, of elements of G such that the sequence of open sets $\{g_n \cdot U\}_{n=1}^{\infty}$ (Proposition 4.1.7) covers G. Because G acts transitively on X, $\{(g_n \cdot U) \cdot x\}_{n=1}^{\infty}$ covers X. Note that for each $n \in \mathbb{N}$, $(g_n \cdot U) \cdot x = \eta_{g_n}(U \cdot x)$ and each η_{g_n} is a homeomorphism (Remark 4.2.3). Therefore, X is G-countably covered.

Q.E.D.

4.2.13 Lemma *Let G be a complete metric group acting on a metric space X. If X is of the second category and is G-countably covered, then G acts weakly microtransitively on X.*

Proof Assume X is of the second category and is G-countably covered. Let $x \in X$, and let $U \in \mathcal{N}(e_G)$. We show that $Cl(U \cdot x)$ is a neighborhood of x in X.

Since G is a topological group, there exists $V \in \mathcal{N}(e_G)$ such that $V^{-1} \cdot V \subset U$ (Proposition 4.1.2). By hypothesis, there exists a sequence, $\{h_n\}_{n=1}^{\infty}$, of homeomorphisms of X such that the family $\{h_n(V \cdot x)\}_{n=1}^{\infty}$ covers X. Since X is of the second category, there exists $n \in \mathbb{N}$ such that $Cl(h_n(V \cdot x))$ has nonempty interior. Hence, $Cl(V \cdot x)$ has nonempty interior. Since any open nonempty subset of $Cl(V \cdot x)$ must intersect $V \cdot x$, it follows that there exists $g \in V$ such that $g \cdot x \in Int(Cl(V \cdot x))$. Thus, $x \in g^{-1} \cdot (Int(Cl(V \cdot x))) = Int\left(Cl((g^{-1} \cdot V) \cdot x)\right) \subset Int\left(Cl((V^{-1} \cdot V) \cdot x)\right) \subset Int(Cl(U \cdot x))$. Therefore, $Cl(U \cdot x)$ is a neighborhood of x in X.

Q.E.D.

Let G be a complete metric group acting transitively on a metric space X. If G also acts micro-transitively on X, then G acts weakly micro-transitively on X. The next lemma shows that the converse is also true.

4.2.14 Lemma *Let G be a complete metric group acting transitively on a metric space X. If G acts weakly micro-transitively on X, then G acts micro-transitively on X.*

Proof Suppose G acts weakly micro-transitively on X. Let ρ_X be a metric on X and let ρ_G be a complete metric on G.

Let $x_0 \in X$ and let $U \in \mathcal{N}(e_G)$. We show that $U \cdot x_0$ is a neighborhood of x_0 in X. To this end, let $U_0 \in \mathcal{N}(e_G)$ be such that $(Cl(U_0))^{-1} \cdot (Cl(U_0)) \subset U$ (Propositions 4.1.2 and 4.1.16). Since G acts weakly micro-transitively on X, there exists an open subset M_0 of X such that $x_0 \in M_0 \subset Cl(U_0 \cdot x_0)$. We prove that $M_0 \subset U \cdot x_0$.

Let $y_0 \in M_0$; we must find $g \in U$ such that $g \cdot x_0 = y_0$. At this point we follow a technique introduced by T. Homma [7]. What we do is to move from x_0 to y_0 and from y_0 to x_0, planning ahead each movement so that the next movement is possible and is close to e_G.

First, let $V_0 = U_0$. Invoke the weak micro-transitivity action of G on X to obtain an open subset N_0 such that $y_0 \in N_0 \subset Cl(V_0 \cdot y_0)$.

We construct eight sequences, namely:

two sequences, $\{g_n\}_{n=1}^{\infty}$ and $\{h_n\}_{n=1}^{\infty}$, of elements of G,
two sequences, $\{x_n\}_{n=0}^{\infty}$ and $\{y_n\}_{n=0}^{\infty}$, of elements of X,
two sequences, $\{U_n\}_{n=0}^{\infty}$ and $\{V_n\}_{n=0}^{\infty}$, of neighborhoods of e_G in G,

and

two sequences, $\{M_n\}_{n=0}^{\infty}$ and $\{N_n\}_{n=0}^{\infty}$, of open subsets of X.

These eight sequences are constructed to satisfy the following thirteen properties:

(1_n) $g_n \in U_{n-1}$.   (2_n) $h_n \in V_{n-1}$.
(3_n) $x_n = g_n \cdot x_{n-1}$.   (4_n) $y_n = h_n \cdot y_{n-1}$.
(5_n) $x_n \in N_{n-1}$.   (6_n) $y_n \in M_n$.
(7_n) $U_n \cdot g_n \cdot \ldots \cdot g_1 \subset U_0$.   ($8_n$) $V_n \cdot h_n \cdot \ldots \cdot h_1 \subset V_0$.
(9_n) $\mathrm{diam}(U_n \cdot g_n \cdot \ldots \cdot g_1) < \frac{1}{2^n}$.   ($10_n$) $\mathrm{diam}(V_n \cdot h_n \cdot \ldots \cdot h_1) < \frac{1}{2^n}$.
(11_n) $x_n \in M_n \subset Cl(U_n \cdot x_n)$.   ($12_n$) $y_n \in N_n \subset Cl(V_n \cdot y_n)$.
(13_n) $\mathrm{diam}(M_n) < \frac{1}{n}$.

The construction of these eight sequences is done by induction. We already have chosen U_0, V_0, x_0, y_0, M_0 and N_0. Let $n \geq 1$ and inductively assume that for $k \in \{1, \ldots, n-1\}$, we have chosen $g_k, h_k, x_k, y_k, U_k, V_k, M_k$ and N_k in such a way that $(1_k), \ldots, (13_k)$ are satisfied.

Note that from (6_{n-1}), (12_{n-1}) and (11_{n-1}), we have that

$$y_{n-1} \in M_{n-1} \cap N_{n-1} \subset Cl(U_{n-1} \cdot x_{n-1}).$$

Hence, there exists $g_n \in U_{n-1}$ such that $g_n \cdot x_{n-1} \in N_{n-1}$; so (1_n) holds. Let $x_n = g_n \cdot x_{n-1}$. Then (3_n) and (5_n) also hold. From (7_{n-1}) and (1_n) follows that $g_n \cdot g_{n-1} \cdot \ldots \cdot g_1 \in U_0$. Thus, we may choose $U_n \in \mathcal{N}(e_G)$ in such a way that (7_n) and (9_n) are satisfied. Since G acts weakly micro-transitively on X, there exists an open subset M_n of X such that satisfies (11_n) and (13_n). So far, we have chosen g_n, x_n, U_n and M_n in such a way that (1_n), (3_n), (5_n), (7_n), (9_n), (11_n) and (13_n) hold.

From (5_n), (11_n) and (12_{n-1}), we conclude that

$$x_n \in N_{n-1} \cap M_n \subset Cl(V_{n-1} \cdot y_{n-1}).$$

Hence, there exists $h_n \in V_{n-1}$ such that $h_n \cdot y_{n-1} \in M_n$, so (2_n) holds. Let $y_n = h_n \cdot y_{n-1}$. Then (4_n) and (6_n) are also satisfied. From (8_{n-1}) and (2_n), we obtain that $h_n \cdot h_{n-1} \cdot \ldots \cdot h_1 \in V_0$. Thus, we may choose $V_n \in \mathcal{N}(e_G)$ such that (8_n) and (10_n) hold. Since G acts weakly micro-transitively on X, we may find an open subset N_n of X satisfying (12_n). So, we have completed the inductive construction of the eight sequences.

For each $n \geq 1$, let $\tilde{g}_n = g_n \cdot \ldots \cdot g_1$ and let $\tilde{h}_n = h_n \cdot \ldots \cdot h_1$. From (1_n), (2_n), (9_n) and (10_n), we conclude that $\{\tilde{g}_n\}_{n=1}^{\infty}$ and $\{\tilde{h}_n\}_{n=1}^{\infty}$ are Cauchy sequences with respect to the metric ρ_G of G. Since ρ_G is a complete metric, it follows that

$\{\tilde{g}_n\}_{n=1}^{\infty}$ and $\{\tilde{h}_n\}_{n=1}^{\infty}$ converge to elements g and h of G, respectively. It follows from (7_n) and (8_n) that $\{\tilde{g}_n\}_{n=1}^{\infty} \subset U_0$ and $\{\tilde{h}_n\}_{n=1}^{\infty} \subset V_0$. Therefore, $g \in Cl(U_0)$ and $h \in Cl(V_0) = Cl(U_0)$. By the election of U_0, we have that $h^{-1} \cdot g \in U$.

It follows from (3_n) and (4_n) that for each $n \geq 1$, $x_n = \tilde{g}_n \cdot x_0$ and $y_n = \tilde{h}_n \cdot y_0$. Hence, the sequence $\{x_n\}_{n=1}^{\infty}$ converges to $g \cdot x_0$, and the sequence $\{y_n\}_{n=1}^{\infty}$ converges to $h \cdot y_0$. Since for each $n \in \mathbb{N}$, it follows from (6_n) and (11_n) that both x_n and y_n belong to M_n, then, from (13_n), we have that $\rho_X(x_n, y_n) < \frac{1}{n}$. Thus, $g \cdot x_0 = h \cdot y_0$. Therefore, $h^{-1} \cdot g \in U$ and $(h^{-1} \cdot g) \cdot x_0 = y_0$.

Q.E.D.

4.2.15 Theorem *Let G be a complete metric group acting transitively on a metric space X. If G is separable, then the following are equivalent:*

(a) G acts micro-transitively on X.
(b) X is topologically complete.
(c) X is of the second category.

Proof First, suppose G acts micro-transitively on X. To see X has a complete metric, let $x \in X$ and consider the map $\gamma_x \colon G \twoheadrightarrow X$ (Definition 4.2.8). Since G acts transitively on X, γ_x is surjective. Since G acts micro-transitively on X, by Lemma 4.2.9, γ_x is an open map. Hence, by Theorem 1.5.13, X has a complete metric.

By the Baire Category Theorem (Theorem 1.5.12), if X has a complete metric, then X is of the second category.

Finally, suppose X is of the second category. Since G is separable, by Lemma 4.2.12, X is G-countably covered. Hence, since X is of the second category and is G-countably covered, by Lemma 4.2.13, G acts weakly micro-transitively on X. Therefore, by Lemma 4.2.14, G acts micro-transitively on X.

Q.E.D.

Now, we present some definitions needed for Effros's Theorem.

4.2.16 Definition Let G be a complete metric group acting on a metric space X. Then $G \cdot x$ is called the *orbit of x under the action of G on X*. Note that distinct orbits are disjoint. The set

$$\{G \cdot x \mid x \in X\}$$

of orbits is called the *orbit space determined by the action of G on X*, and it is denoted by X/G. We give X/G the quotient topology. The function $q \colon X \twoheadrightarrow X/G$ given by $q(x) = G \cdot x$ is the quotient map.

4.2.17 Lemma *If G is a complete metric group acting on a metric space X, then quotient map $q \colon X \twoheadrightarrow X/G$ is open.*

Proof It is enough to observe that if U is an open subset of X, then $q^{-1}(q(U)) = G \cdot U$. Since $G \cdot U = \bigcup\{g \cdot U \mid g \in G\} = \bigcup\{\eta_g(U) \mid g \in G\}$ and each η_g is a homeomorphism (Remark 4.2.3), $q^{-1}(q(U))$ is open in X.

Q.E.D.

4.2.18 Definition Let G be a complete metric group acting on a metric space X. If $x \in X$, then

$$G_x = \{g \in G \mid g \cdot x = x\}$$

is a subgroup of G called the *stabilizer subgroup of x*.

4.2.19 Remark Note that G_x acts on G by multiplication on the right. The orbit space of this action, G/G_x, is the set

$$G/G_x = \{g \cdot G_x \mid g \in G\}$$

and consists of all the left cosets of G_x. We give G/G_x the quotient topology. The function $q_x : G \rightarrow G/G_x$ given by $q_x(g) = g \cdot G_x$ is the quotient map.

4.2.20 Lemma *Let G be a complete metric group acting on a metric space X. If $x \in X$, then quotient map $q_x : G \twoheadrightarrow G/G_x$ is open.*

Proof It is enough to observe that if U is an open subset of G, then $q_x^{-1}(q_x(U)) = U \cdot G_x$. By Proposition 4.1.17 (1), $U \cdot G_x$ is open in G.

                                                                    Q.E.D.

4.2.21 Definition Let G be a complete metric group acting on a metric space X. If $x \in X$, define the function $\psi_x : G/G_x \rightarrow G \cdot x$ by $\psi_x(g \cdot G_x) = g \cdot x$.

In the following lemma we show that ψ_x is a well defined function which is a continuous bijection.

4.2.22 Lemma *Let G be a complete metric group acting on a metric space X. If $x \in X$, then the function $\psi_x : G/G_x \rightarrow G \cdot x$ of Definition 4.2.21 is a well defined continuous bijection. Moreover, the following diagram*

$$
\begin{array}{ccc}
 & G & \\
q_x \swarrow & & \searrow \gamma_x \\
G/G_x & \xrightarrow[\psi_x]{} & (G \cdot x)
\end{array}
$$

is commutative.

Proof Let $x \in X$. For $g', g \in G$, $g \cdot G_x = g' \cdot G_x$ if and only if $g \cdot x = g' \cdot x$. Hence, ψ_x is well defined and one-to-one. To see ψ_x is surjective, let $g \cdot x \in G \cdot x$. Then $\psi_x(g \cdot G_x) = g \cdot x$. Therefore, ψ_x is a bijection.

Next, let $g \in G$. Then $\psi_x \circ q_x(g) = \psi_x(q_x(g)) = \psi_x(g \cdot G_x) = g \cdot x = \gamma_x(g)$. Therefore, the diagram is commutative.

Finally, since $\psi_x \circ q_x = \gamma_x$, q_x is open (Lemma 4.2.20) and γ_x is continuous, we have that ψ_x is continuous.

                                                                    Q.E.D.

4.2.23 Lemma *Let G be a complete metric group acting on a metric space X. If $x \in X$, then G acts micro-transitively on the orbit $G \cdot x$ if and only if the map $\psi_x \colon G/G_x \to G \cdot x$ of Definition 4.2.21 is a homeomorphism.*

Proof By Lemma 4.2.22, ψ_x is continuous. Since $\psi_x \circ q_x = \gamma_x$ and q_x is continuous and open (Lemma 4.2.20), ψ_x is open if and only if γ_x is open. By Lemma 4.2.9, γ_x is open if and only if G acts micro-transitively on the orbit $G \cdot x$.

Q.E.D.

4.2.24 Lemma *If G is a complete metric group acting on a separable complete metric space X, then each orbit is a G_δ subset of X if and only if X/G is a T_0 space.*

Proof First, assume that each orbit is a G_δ subset of X. We show that X/G is a T_0 space. Let $x, y \in X$. Suppose that $q(x) \in Cl_{X/G}(\{q(y)\})$ and $q(y) \in Cl_{X/G}(\{q(x)\})$. We see that $q(x) = q(y)$.

We assert that $G \cdot x \subset Cl_X(G \cdot y)$. Indeed, suppose there exists $g \in G$ such that $g \cdot x \notin Cl_X(G \cdot y)$. Then there exists an open subset U of X such that $g \cdot x \in U$ and $U \cap G \cdot y = \emptyset$. Since the map $q \colon X \to X/G$ is open (Lemma 4.2.17), $q(U)$ is an open subset of X/G such that $q(g \cdot x) = q(x) \in q(U)$. Note that $q(G \cdot y) = \{q(y)\}$ and $q(y) \notin q(U)$. A contradiction to the fact that $q(x) \in Cl_{X/G}(\{q(y)\})$. Therefore, $G \cdot x \subset Cl_X(G \cdot y)$. Similarly, $G \cdot y \subset Cl_X(G \cdot x)$. We conclude that $G \cdot x \cup G \cdot y \subset Cl_X(G \cdot x) \cap Cl_X(G \cdot y)$.

Since $G \cdot x$ is dense in $Cl_X(G \cdot x)$ and $G \cdot y$ is dense in $Cl_X(G \cdot y)$, $G \cdot x$ and $G \cdot y$ both are dense in $Cl_X(G \cdot x) \cap Cl_X(G \cdot y)$. By hypothesis, $G \cdot x$ and $G \cdot y$ are both G_δ subsets of $Cl_X(G \cdot x) \cap Cl_X(G \cdot y)$. Since $Cl_X(G \cdot x) \cap Cl_X(G \cdot y)$ is a closed subset of X and X is a complete metric space, $Cl_X(G \cdot x) \cap Cl_X(G \cdot y)$ has a complete metric (Lemma 1.5.3). By Lemma 1.5.9, $G \cdot x \cap G \cdot y$ is dense in $Cl_X(G \cdot x) \cap Cl_X(G \cdot y)$. In particular, $G \cdot x \cap G \cdot y \neq \emptyset$. Hence, $G \cdot x = G \cdot y$ and $q(x) = q(y)$. Therefore, X/G is a T_0 space.

Next, suppose X/G is a T_0 space. Since X is a separable metric space, X has a countable basis $\{U_n\}_{n=1}^{\infty}$. We assert that $\{q(U_n)\}_{n=1}^{\infty}$ is a countable basis for X/G. Clearly, this family is countable. Since the map $q \colon X \to X/G$ is open (Lemma 4.2.17), $q(U_n)$ is an open subset of X/G for each $n \in \mathbb{N}$. Furthermore, if $x \in X$ and V is an open subset of X/G such that $q(x) \in V$, then there exists $n \in \mathbb{N}$ such that $q(x) \in q(U_n) \subset V$ (by the continuity of q and the fact that $\{U_n\}_{n=1}^{\infty}$ is a basis for X). Thus, $\{q(U_n)\}_{n=1}^{\infty}$ is a countable basis for X/G.

Now, take $x \in X$. For each $n \in \mathbb{N}$, let

$$V_n = \begin{cases} q(U_n), & \text{if } q(x) \in q(U_n), \\ X/G \setminus q(U_n), & \text{if } q(x) \notin q(U_n). \end{cases}$$

Since X/G is a T_0 space, $\bigcap_{n=1}^{\infty} V_n = \{q(x)\}$. Hence, $\bigcap_{n=1}^{\infty} q^{-1}(V_n) = q^{-1}(q(x)) = G \cdot x$. Since each V_n is either open or closed in X/G, $q^{-1}(V_n)$

is either open or closed in X. In either case, $q^{-1}(V_n)$ is a G_δ subset of X (see Lemma 1.5.7 when $q^{-1}(V_n)$ is closed). Therefore, $G \cdot x$ is a G_δ subset of X.

Q.E.D.

We are ready to prove Effros's Theorem [4].

4.2.25 Theorem *If G is a separable complete metric group acting on a separable complete metric space X, then the following are equivalent:*

(a) *For each $x \in X$, the map $\psi_x \colon G/G_x \to G \cdot x$ of Definition 4.2.21 is a homeomorphism.*

(b) *G acts micro-transitively on each orbit.*

(c) *Each orbit is of the second category (in itself).*

(d) *Each orbit is a G_δ subset of X.*

(e) *X/G is a T_0 space.*

Proof By Lemma 4.2.23, (a) and (b) are equivalent. By Theorem 4.2.15, (b) and (c) are equivalent. By Theorem 1.5.8, (d) is equivalent to the fact that each orbit is topologically complete. This latter statement is equivalent to (b) by Theorem 4.2.15. Hence, (d) is equivalent to (b). Finally, by Lemma 4.2.24, (d) is equivalent to (e).

Q.E.D.

4.2.26 Definition Let X and Y be continua. Define

$$\mathcal{C}(X, Y) = \{f \colon X \to Y \mid f \text{ is a map}\}.$$

We topologize $\mathcal{C}(X, Y)$ with the sup *metric*, ρ, given by

$$\rho((f, g)) = \sup\{d(f(x), g(x)) \mid x \in X\},$$

for every $g, f \in \mathcal{C}(X, Y)$.

4.2.27 Definition If X is a continuum, then $\mathcal{H}(X)$ denotes the *group of homeomorphisms of X*. $\mathcal{H}(X)$ acts on X as follows: If $h \in \mathcal{H}(X)$ and $x \in X$, then $\theta((h, x)) = h(x)$.

4.2.28 Theorem *If X is a continuum, then $\mathcal{C}(X, X)$ is a separable complete metric space and $\mathcal{H}(X)$ is a G_δ subset of $\mathcal{C}(X, X)$.*

Proof By [8, Theorem 1, p. 244], $\mathcal{C}(X, X)$ is a separable space. By [9, Theorem 3, p. 90], $\mathcal{C}(X, X)$ is a complete metric space. By [9, Theorem 1, p. 91], $\mathcal{H}(X)$ is a G_δ subset of $\mathcal{C}(X, X)$.

Q.E.D.

4.2.29 Corollary *If X is a continuum, then $\mathcal{H}(X)$ has a separable complete metric, ρ'.*

Proof By Theorem 4.2.28, $\mathcal{H}(X)$ is a G_δ subset of $\mathcal{C}(X, X)$. Hence, by Theorem 1.5.8, $\mathcal{H}(X)$ has a complete metric. In fact, it is known that ρ' is given by $\rho'(f, g) = \rho(f, g) + \rho(f^{-1}, g^{-1})$ for every $f, g \in \mathcal{H}(X)$.

Q.E.D.

4.2.30 Definition A metric space X has the *Property of Effros* provided that for each $\varepsilon > 0$, there exists $\delta > 0$ such that if $x, y \in X$ and $d(x, y) < \delta$, then there exists $h \in \mathcal{H}(X)$, such that $h(x) = y$ and $d(z, h(z)) < \varepsilon$ for every $z \in X$. The number δ is called an *Effros number for the given ε*. A homeomorphism $h \in \mathcal{H}(X)$ satisfying that $d(z, h(z)) < \varepsilon$ for each $z \in X$ is called an *ε-homeomorphism.*

The following theorem is known as Effros's Theorem in the theory of homogeneous continua. A slightly different version of this result is first shown by C. L. Hagopian [5].

4.2.31 Theorem *If X is a homogeneous continuum, with metric d, then X has the property of Effros.*

Proof By Corollary 4.2.29, $\mathcal{H}(X)$ is a separable complete metric space. Since X is a homogeneous continuum, $\mathcal{H}(X)$ acts transitively on X. Let $x \in X$. Since $\mathcal{H}(X)$ acts transitively on X, the orbits are of the second category. Hence, by Theorem 4.2.25, the map

$$\psi_x : \mathcal{H}(X)/\mathcal{H}(X)_x \twoheadrightarrow \mathcal{H}(X) \cdot x$$

is a homeomorphism. By Lemma 4.2.23, $\mathcal{H}(X)$ acts micro-transitively on X. Hence, by Lemma 4.2.9, the map γ_x is open.

Let $\varepsilon > 0$ be given and let $U = V_{\frac{\varepsilon}{2}}^{\rho'}(1_X)$, where 1_X is the identity map of X. Then for each $h \in U$, h is an $\frac{\varepsilon}{2}$-homeomorphism of X. Since γ_x is open, $\gamma_x(U)$ is an open subset of X such that $x \in \gamma_x(U)$. Let $\delta_x > 0$ be such that $V_{\delta_x}^d(x) \subset \gamma_x(U)$. Thus, if $y \in V_{\delta_x}^d(x)$, then there exists $h \in U$ such that $\gamma_x(h) = y$; i.e., h is an $\frac{\varepsilon}{2}$-homeomorphism such that $h(x) = y$.

Note that $\{V_{\delta_x}^d(x) \mid x \in X\}$ is an open cover of X. Let δ be a Lebesgue number for this cover (Theorem 1.6.6). Let $x, y \in X$ be such that $d(x, y) < \delta$. Then there exists $z \in X$ such that $x, y \in V_{\delta_z}^d(z)$. By the previous paragraph, there exist $h_1, h_2 \in U$ such that $\gamma_z(h_1) = x$ and $\gamma_z(h_2) = y$; i.e., $h_1(z) = x$ and $h_2(z) = y$. Let $h = h_2 \circ h_1^{-1}$. Then $h : X \twoheadrightarrow X$ is a homeomorphism such that $h(x) = y$ ($h(x) = h_2 \circ h_1^{-1}(x) = h_2\left(h_1^{-1}(x)\right) = h_2(z) = y$).

Let $w \in X$. Since

$$d(w, h(w)) = d\left(w, h_2\left(h_1^{-1}(w)\right)\right)$$

$$\leq d\left(w, h_1^{-1}(w)\right) + d\left(h_1^{-1}(w), h_2\left(h_1^{-1}(w)\right)\right)$$

$$= d\left(h_1\left(h_1^{-1}(w)\right), h_1^{-1}(w)\right) + d\left(h_1^{-1}(w), h_2\left(h_1^{-1}(w)\right)\right)$$

$$< \frac{\varepsilon}{2} + \frac{\varepsilon}{2} = \varepsilon,$$

h is an ε-homeomorphism.

Q.E.D.

As our first application of Effros's Theorem, we show that Jones's set function $\mathcal{T}$ is idempotent on closed sets, for homogeneous continua.

4.2.32 Theorem *If X is a homogeneous continuum, with metric d, then for each closed subset A of X, $\mathcal{T}^2(A) = \mathcal{T}(A)$; i.e., $\mathcal{T}$ is idempotent on closed sets.*

Proof Let A be a closed subset of X. If $A = \emptyset$, then $\mathcal{T}^2(A) = \mathcal{T}(A)$ by Corollary 3.1.14. Suppose $A \neq \emptyset$. Note that, by Remark 3.1.5, $\mathcal{T}(A) \subset \mathcal{T}^2(A)$.

Next, let $x \in X \setminus \mathcal{T}(A)$. Then there exists a subcontinuum W of X such that $x \in Int(W) \subset W \subset X \setminus A$. Let $\varepsilon > 0$ be such that $\varepsilon < d(W, A)$ and $\mathcal{V}_\varepsilon^d(x) \subset Int(W)$. Let $\delta > 0$ be an Effros number for this ε. We assume that $\delta < \varepsilon$. Since W is compact, there exist $w_1, \ldots, w_m \in W$ such that $W \subset \bigcup_{j=1}^m \mathcal{V}_\delta^d(w_j)$.

Let $j \in \{1, \ldots, m\}$. Then for each $y \in \mathcal{V}_\delta^d(w_j)$, there exists an ε-homeomorphism $h_y \colon X \twoheadrightarrow X$ such that $h_y(w_j) = y$. When $y = w_j$, we let $h_y = 1_X$, the identity map of X. Let

$$M_j = Cl\left(\bigcup_{y \in \mathcal{V}_\delta^d(w_j)} h_y(W)\right).$$

Then M_j is a subcontinuum of X such that $w_j \in \mathcal{V}_\delta^d(w_j) \subset M_j \subset X \setminus A$. Let $M = \bigcup_{j=1}^m M_j$. Then M is a subcontinuum of X and $W \subset \bigcup_{j=1}^m \mathcal{V}_\delta^d(w_j) \subset Int(M) \subset M \subset X \setminus A$. Since $Int(M) \subset X \setminus \mathcal{T}(A)$, $x \in Int(W) \subset W \subset X \setminus \mathcal{T}(A)$. Hence, $x \in X \setminus \mathcal{T}^2(A)$. Therefore, $\mathcal{T}$ is idempotent on closed sets.

Q.E.D.

As a consequence of Corollary 3.2.2 and Theorem 4.2.32, we obtain the following:

4.2.33 Corollary *If X is a homogeneous continuum such that $\mathcal{T} \colon 2^X \twoheadrightarrow 2^X$ is surjective, then X is locally connected.*

As a second application of Effros's Theorem, we prove that almost connected im kleinen (Definition 1.7.4) homogeneous continua are locally connected.

4.2.34 Theorem *If X is a homogeneous continuum which is almost connected im kleinen at some point x, then X is locally connected.*

Proof Let X be a homogeneous continuum. Let $\varepsilon > 0$ and let $\delta > 0$ be an Effros number for $\frac{\varepsilon}{2}$. Without loss of generality we assume $\delta < \frac{\varepsilon}{2}$. Since X is almost connected im kleinen at x, there exists a subcontinuum W of X such that $Int(W) \neq \emptyset$ and $W \subset \mathcal{V}^d_\delta(x)$. Let $w \in Int(W)$. Then $d(x, w) < \delta$. Hence, there exists an $\frac{\varepsilon}{2}$-homeomorphism $h: X \twoheadrightarrow X$ such that $h(w) = x$. Thus, $x \in Int(h(W))$ and $h(W) \subset \mathcal{V}^d_\varepsilon(x)$. Therefore, X is connected im kleinen at x. Since X is homogeneous, X is connected im kleinen at every point. Therefore, by Theorem 1.7.12, X is locally connected.

Q.E.D.

As a consequence of Theorem 4.2.34 we have:

4.2.35 Corollary *If X is a homogeneous continuum, then the following are equivalent:*

(1) X is locally connected;
(2) X is locally connected at some point;
(3) X is connected im kleinen at some point of X;
(4) X is almost connected im kleinen at every point of X;
(5) X is almost connected im kleinen at some point of X.

In our third application of Effros's Theorem, we show that homogeneous continua have the property of Kelley (Definition 6.1.18).

4.2.36 Theorem *If X is a homogeneous continuum, then X has the property of Kelley*

Proof Let X be a homogeneous continuum. Let $\varepsilon > 0$, and let $\delta > 0$ be an Effros number for ε. Let x and y be two points of X such that $d(x, y) < \delta$ and let A be a subcontinuum of X such that $x \in A$. Since δ is an Effros number, there exists an ε-homeomorphism $h: X \twoheadrightarrow X$ such that $h(x) = y$. Observe that $\mathcal{H}(A, h(A)) < \varepsilon$.

Q.E.D.

4.2.37 Remark Observe that Theorem 4.2.32 is also a consequence of Theorems 4.2.36 and 3.1.70.

We finish this chapter showing that any connected metric space with the property of Effros is homogeneous.

4.2.38 Theorem *If X is a connected metric space, with metric d, satisfying the property of Effros, then X is homogeneous.*

Proof Let $x, y \in X$, and let $\varepsilon > 0$. Since X has the property of Effros, there exists an Effros number $\delta > 0$ for this ε. Since X is connected, there exist finitely many points $z_1 = x, z_2, \ldots, z_{n-1}, z_n = y$ in X such that $d(z_{j-1}, z_j) < \delta$ for each $j \in \{2, \ldots, n\}$. Hence, there exists a homeomorphism $h_j \colon X \twoheadrightarrow X$ such that $h_j(z_{j-1}) = z_j$, $j \in \{2, \ldots, n\}$. Then $h = h_n \circ \cdots \circ h_2$ is a homeomorphism of X onto itself such that $h(x) = y$. Therefore, X is homogeneous.

Q.E.D.

References

1. F. D. Ancel, An Alternative Proof and Applications of a Theorem of E. G. Effros, Michigan Math. J., 34 (1987), 39–55.
2. G. E. Bredon, *Introduction to Compact Transformation Groups*, Monographs and Textbooks in Pure and Applied Mathematics, Vol. 46, Academic Press, Inc., New York, 1972.
3. G. E. Bredon, *Topology and Geometry*, Graduate Texts in Mathematics, Vol. 139, Springer-Verlag, New York, Inc., 1993.
4. E. G. Effros, Transformation Groups and C^*-Algebras, Ann. of Math., (2) 81 (1965), 38–55.
5. C. L. Hagopian, Homogeneous Plane Continua, Houston J. Math., 1 (1975), 35–41.
6. C. Hernández, O. J. Rendón, M. Tkačenko and L. M. Villegas, *Grupos Topológicos*, Libros de Texto, Manuales de Prácticas y Antologías, Universidad Autónoma Metropolitana, Unidad Iztapalapa, 1997. (Spanish)
7. T. Homma, On the Embedding of Polyhedra in Manifolds, Yokohama Math. J., 10 (1962), 5–10.
8. K. Kuratowski, *Topology*, Vol. I, Academic Press, New York, N. Y., 1966.
9. K. Kuratowski, *Topology*, Vol. II, Academic Press, New York, N. Y., 1968.
10. S. Macías and S. B. Nadler, Jr., On Hereditarily Decomposable Homogeneous Continua, Topology Proc., 34 (2009), 131–145.
11. G. McCarty, *Topology: An Introduction with Applications to Topological Groups*, Dover Publications, Inc., New York, 1988.

Chapter 5
Decomposition Theorems

We present a proof of Jones's Aposyndetic Decomposition Theorem and Rogers's Terminal Decomposition Theorem. These theorems are proven using Jones's set function $\mathcal{T}$ (Chap. 3) and Effros's Theorem (Chap. 4). We also give a construction of the Case continuum and present a sketch of the construction of the Minc–Rogers continua. Finally, we study covering spaces of any solenoid, the Menger curve, the Case continuum and one of the Minc–Rogers examples.

5.1   Jones's Theorem

We give a proof of Jones's Aposyndetic Decomposition Theorem (Theorem 5.1.18). In order to do this, we use [5, 6, 8, 13, 14, 19, 26, 29, 31, 34].

We begin this section proving that the set function $\mathcal{T}$ commutes with homeomorphisms.

5.1.1 Lemma *Let X be a compactum. If $h\colon X \twoheadrightarrow X$ is a homeomorphism, then $h(\mathcal{T}(A)) = \mathcal{T}(h(A))$ for each subset A of X.*

Proof Let A be a closed subset of X. Let $x \in X \setminus h(\mathcal{T}(A))$. Then $h^{-1}(x) \in X \setminus \mathcal{T}(A)$. Hence, there exists a subcontinuum W of X such that $h^{-1}(x) \in Int(W) \subset W \subset X \setminus A$. This implies that $x \in Int(h(W)) \subset h(W) \subset X \setminus h(A)$. Thus, $x \in X \setminus \mathcal{T}(h(A))$.

Now, let $x \in X \setminus \mathcal{T}(h(A))$. Then there exists a subcontinuum K of X such that $x \in Int(K) \subset K \subset X \setminus h(A)$. This implies that $h^{-1}(x) \in Int(h^{-1}(K)) \subset h^{-1}(K) \subset X \setminus A$. Thus, $h^{-1}(x) \in X \setminus \mathcal{T}(A)$. Hence, $x \in X \setminus h(\mathcal{T}(A))$.

Therefore, $h(\mathcal{T}(A)) = \mathcal{T}(h(A))$.

Q.E.D.

The following theorem says that, for a homogeneous continuum, the images of all singletons under $\mathcal{T}$ form a decomposition of the continuum.

© Springer International Publishing AG, part of Springer Nature 2018
S. Macías, *Topics on Continua*, https://doi.org/10.1007/978-3-319-90902-8_5

5.1.2 Theorem *Let X be a homogeneous continuum. If $\mathcal{G} = \{\mathcal{T}(\{x\}) \mid x \in X\}$, then $\mathcal{G}$ is a decomposition of X.*

Proof Let x be a point of X. We show that if $y \in \mathcal{T}(\{x\})$, then $\mathcal{T}(\{x\}) = \mathcal{T}(\{y\})$.

Let $y \in \mathcal{T}(\{x\})$. By Proposition 3.1.7 and the fact that $\mathcal{T}$ is idempotent on closed sets (Theorem 4.2.32), $\mathcal{T}(\{y\}) \subset \mathcal{T}^2(\{x\}) = \mathcal{T}(\{x\})$.

By Theorem 3.4.9, there exists a point $x_0 \in \mathcal{T}(\{x\})$ such that $\mathcal{T}(\{x_0\})$ has property BL; i.e., for each $w \in \mathcal{T}(\{x_0\})$, $\mathcal{T}(\{x_0\}) \subset \mathcal{T}(\{w\})$.

Since X is homogeneous, there exists a homeomorphism $h\colon X \twoheadrightarrow X$ such that $h(x_0) = x$. Since $y \in \mathcal{T}(\{x\})$, by Lemma 5.1.1, $h^{-1}(y) \in \mathcal{T}(\{x_0\})$. Hence, since $\mathcal{T}(\{x_0\})$ has property BL, we have that $\mathcal{T}(\{x_0\}) \subset \mathcal{T}(\{h^{-1}(y)\})$. This implies that $\mathcal{T}(\{x\}) = h(\mathcal{T}(\{x_0\})) \subset h(\mathcal{T}(\{h^{-1}(y)\})) = \mathcal{T}(\{y\})$ (Lemma 5.1.1). Consequently, $\mathcal{T}(\{x\}) \subset \mathcal{T}(\{y\})$.

Therefore, $\mathcal{G}$ is a decomposition of X.

Q.E.D.

5.1.3 Definition Let $\mathcal{G}$ be a decomposition of the compactum X. Let $\mathcal{H}$ be a family of homeomorphisms of X. We say $\mathcal{H}$ *respects* $\mathcal{G}$ provided that for each pair $G_1, G_2 \in \mathcal{G}$ and each $h \in \mathcal{H}$, either $h(G_1) = G_2$ or $h(G_1) \cap G_2 = \emptyset$.

5.1.4 Theorem *Let X be a homogeneous continuum, with metric d, and let $\mathcal{G}$ be a decomposition of X such that the elements of $\mathcal{G}$ are continua. If the homeomorphism group, $\mathcal{H}(X)$, of X respects $\mathcal{G}$, then the following hold:*

(1) $\mathcal{G}$ is a continuous decomposition of X.
(2) The elements of $\mathcal{G}$ are homogeneous mutually homeomorphic continua.
(3) The quotient space $X/\mathcal{G}$ is a homogeneous continuum.

Proof First, we show that $\mathcal{G}$ is continuous. To this end, we prove first that $\mathcal{G}$ is upper semicontinuous.

Let $G \in \mathcal{G}$, and let U be an open subset of X such that $G \subset U$. Let $\varepsilon = d(G, X \setminus U)$. Since G and $X \setminus U$ are disjoint compacta, $\varepsilon > 0$. Let $\delta > 0$ be an Effros number for this ε (Theorem 4.2.31); without loss of generality, we assume that $\delta < \varepsilon$. Let $V = \mathcal{V}_\delta^d(G)$. Then V is an open subset of X contained in U. Let $G' \in \mathcal{G}$ be such that $G' \cap V \neq \emptyset$. Let $y \in G' \cap V$, and let $x \in G$ be such that $d(x, y) < \delta$. Since δ is an Effros number, there exists an ε-homeomorphism $h\colon X \twoheadrightarrow X$ such that $h(x) = y$. Since $\mathcal{H}(X)$ respects $\mathcal{G}$ and $h(G) \cap G' \neq \emptyset$, $h(G) = G'$. Hence, $G' \subset \mathcal{V}_\varepsilon^d(G)$ (h is an ε-homeomorphism); i.e., $G' \subset U$. Thus, $\mathcal{G}$ is upper semicontinuous.

Next, we prove that $\mathcal{G}$ is lower semicontinuous. Let $G \in \mathcal{G}$, let $p, q \in G$ and let U be an open subset of X such that $p \in U$. Let $\varepsilon > 0$ be such that $\mathcal{V}_\varepsilon^d(p) \subset U$. Let δ be an Effros number for this ε, and let $V = \mathcal{V}_\delta^d(q)$. Take $G' \in \mathcal{G}$ such that $G' \cap V \neq \emptyset$, and let $z \in G' \cap V$. Hence, $d(q, z) < \delta$. Since δ is an Effros number, there exists an ε-homeomorphism $k\colon X \twoheadrightarrow X$ such that $k(q) = z$. In particular, $d(p, k(p)) < \varepsilon$. Thus, $k(p) \in U$. Since $\mathcal{H}(X)$ respects $\mathcal{G}$, and $k(q) \in G'$, $k(G) = G'$. Hence, $G' \cap U \neq \emptyset$. Thus, $\mathcal{G}$ is lower semicontinuous.

Therefore, $\mathcal{G}$ is continuous.

Observe that since $\mathcal{H}(X)$ respects $\mathcal{G}$, all the elements of $\mathcal{G}$ are homeomorphic. Let $G \in \mathcal{G}$, and let $x, y \in G$. Since X is homogeneous, there exists a homeomorphism $\ell: X \twoheadrightarrow X$ such that $\ell(x) = y$. Note that, since $\mathcal{H}(X)$ respects $\mathcal{G}$, $\ell(G) = G$. Hence, $\ell|_G: G \twoheadrightarrow G$ is a homeomorphism sending x to y. Therefore, G is homogeneous.

Finally, we prove that the quotient space $X/\mathcal{G}$ is a homogeneous continuum. By Theorem 1.7.3, $X/\mathcal{G}$ is a continuum. Let $q: X \twoheadrightarrow X/\mathcal{G}$ be the quotient map.

Let $\chi_1, \chi_2 \in X/\mathcal{G}$ be two points. Let $x_1, x_2 \in X$ be such that $q(x_1) = \chi_1$ and $q(x_2) = \chi_2$. Since X is homogeneous, there exists a homeomorphism $h: X \twoheadrightarrow X$ such that $h(x_1) = x_2$. Since $\mathcal{H}(X)$ respects $\mathcal{G}$, $x_1 \in q^{-1}(\chi_1)$, $x_2 \in q^{-1}(\chi_2)$, and $h(x_1) = x_2$, we have that $h(q^{-1}(\chi_1)) = q^{-1}(\chi_2)$.

Define $f: X/\mathcal{G} \twoheadrightarrow X/\mathcal{G}$ by

$$ f(\chi) = q \circ h \left(q^{-1}(\chi) \right). $$

Then f is well defined (because $\mathcal{H}(X)$ respects $\mathcal{G}$) and $f(\chi_1) = \chi_2$. Since $\mathcal{G}$ is a continuous decomposition, q is an open map (Theorem 1.2.23). Hence, since q is open and h is continuous, f is continuous. Note that $f^{-1}: X/\mathcal{G} \twoheadrightarrow X/\mathcal{G}$ is given by $f^{-1}(\chi) = q \circ h^{-1} \left(q^{-1}(\chi) \right)$ and it is continuous also. Thus, f is a homeomorphism. Therefore, $X/\mathcal{G}$ is homogeneous.

Q.E.D.

5.1.5 Definition A surjective map $f: X \twoheadrightarrow Y$ between continua is *completely regular* if for each $\varepsilon > 0$ and each point $y \in Y$, there exists an open set V in Y containing y such that if $y' \in V$, then there exists a homeomorphism $h: f^{-1}(y) \twoheadrightarrow f^{-1}(y')$ such that for each $x \in f^{-1}(y)$, $d(x, h(x)) < \varepsilon$.

5.1.6 Remark Note that the fibres of completely regular maps are all homeomorphic.

5.1.7 Proposition *Let X and Y be continua. If $g: X \twoheadrightarrow Y$ is a completely regular map, then g is open.*

Proof Let U be an open subset of X. Let $y \in g(U)$. Then there exists $x \in U$ such that $g(x) = y$. Let $\varepsilon > 0$ be such that $V_\varepsilon^d(x) \subset U$. Since g is completely regular, there exists an open subset V of Y such that $y \in V$ and if $y' \in V$, then there exists a homeomorphism $h: g^{-1}(y) \twoheadrightarrow g^{-1}(y')$ such that $d(z, h(z)) < \varepsilon$ for all $z \in g^{-1}(y)$. In particular, $d(x, h(x)) < \varepsilon$. Thus, $h(x) \in U$. Hence, $V \subset g(U)$. Since y is an arbitrary point of $g(U)$, $g(U)$ is open. Therefore, g is an open map.

Q.E.D.

5.1.8 Theorem *Let X be a homogeneous continuum, with metric d, and let $\mathcal{G}$ be a decomposition of X whose elements are proper nondegenerate subcontinua of X. If the homeomorphism group, $\mathcal{H}(X)$, of X respects $\mathcal{G}$, then the elements of $\mathcal{G}$ are nowhere dense, and the quotient map $q: X \twoheadrightarrow X/\mathcal{G}$ is completely regular.*

Proof Let $G \in \mathcal{G}$ and suppose $Int(G) \neq \emptyset$. Let $g' \in Int(G)$, and let $\varepsilon > 0$ be such that $\mathcal{V}_{\varepsilon}^d(g') \subset G$. Let $\delta > 0$ be an Effros number for this ε (Theorem 4.2.31). We assume that $\delta < \varepsilon$. Let $x \in X \setminus G$ be such that $d(x, G) < \delta$. Since G is compact, there exists $g \in G$ such that $d(x, g) = d(x, G)$. Now, since $d(x, g) < \delta$, there exists an ε-homeomorphism $h \colon X \twoheadrightarrow X$ such that $h(g) = x$. Hence, since $h(g') \in \mathcal{V}_{\varepsilon}^d(g') \subset G$ and $\mathcal{H}(X)$ respects $\mathcal{G}$, $h(G) = G$. Thus, $x \in h(G) \setminus G$, a contradiction. Therefore, $Int(G) = \emptyset$.

To show the quotient map $q \colon X \twoheadrightarrow X/\mathcal{G}$ is completely regular, let $\varepsilon > 0$, and let $\delta > 0$ be an Effros number for this ε (Theorem 4.2.31). Since $\mathcal{H}(X)$ respects $\mathcal{G}$, by Theorem 5.1.4 (1), $\mathcal{G}$ is a continuous decomposition. Thus, by (Theorem 1.2.23), q is open.

Let $\chi \in X/\mathcal{G}$, and let $V = q\left(\mathcal{V}_{\delta}^d(x)\right)$, where $x \in q^{-1}(\chi)$. Then V is an open subset of $X/\mathcal{G}$ containing χ. Let $\chi' \in V$, and let $x' \in q^{-1}(\chi') \cap \mathcal{V}_{\delta}^d(x)$. Then since $d(x, x') < \delta$, there exists an ε-homeomorphism $h \colon X \twoheadrightarrow X$ such that $h(x) = x'$. Since $\mathcal{H}(X)$ respects $\mathcal{G}$ and $h(q^{-1}(\chi)) \cap q^{-1}(\chi') \neq \emptyset$, $h(q^{-1}(\chi)) = q^{-1}(\chi')$. Thus, $h|_{q^{-1}(\chi)} \colon q^{-1}(\chi) \twoheadrightarrow q^{-1}(\chi')$ is a homeomorphism such that $d(z, h(z)) < \varepsilon$ for each $z \in q^{-1}(\chi)$. Therefore, q is completely regular.

Q.E.D.

5.1.9 Definition Let X be a continuum. A subcontinuum Z of X is said to be *terminal* if each subcontinuum Y of X that intersects Z satisfies either $Y \subset Z$ or $Z \subset Y$. A decomposition $\mathcal{G}$ of X such that the elements of $\mathcal{G}$ are continua is said to be *terminal* if each element of $\mathcal{G}$ is a terminal subcontinuum of X.

5.1.10 Remark It is easy to see that in the topologist sine curve X (Example 2.4.5), $\{0\} \times [-1, 1]$ is a terminal subcontinuum of X.

5.1.11 Lemma *Let X and Y be continua. If $f \colon X \twoheadrightarrow Y$ is a monotone surjective map and W is a terminal subcontinuum of X, then $f(W)$ is a terminal subcontinuum of Y.*

Proof Let W be a terminal subcontinuum of X. Let K be a subcontinuum of Y such that $K \cap f(W) \neq \emptyset$. Since f is a monotone map, $f^{-1}(K)$ is a subcontinuum of X (Lemma 2.1.12) such that $f^{-1}(K) \cap W \neq \emptyset$. Hence, since W is a terminal subcontinuum of X, either $W \subset f^{-1}(K)$ or $f^{-1}(K) \subset W$. This implies that either $f(W) \subset K$ or $K \subset f(W)$ (f is surjective). Therefore, $f(W)$ is a terminal subcontinuum of Y.

Q.E.D.

5.1.12 Corollary *Let X be a continuum. If Z is a terminal subcontinuum of X and $h \colon X \twoheadrightarrow X$ is a homeomorphism, then $h(Z)$ is a terminal subcontinuum of X.*

5.1.13 Lemma *Let X be a continuum. If X is aposyndetic, then X does not contain nondegenerate proper terminal subcontinua.*

Proof Suppose Y is a nondegenerate proper terminal subcontinuum of X. Let $y \in Y$. Let $y' \in Y \setminus \{y\}$. Then since X is aposyndetic, there exists a subcontinuum W of X such that $y \in Int(W) \subset W \subset X \setminus \{y'\}$. Since Y is terminal and $Y \setminus W \neq \emptyset$,

$W \subset Y$. Thus, y is an interior point of Y. Since y is an arbitrary point of Y, all the points of Y are interior points. Hence, Y is a nonempty open and closed proper subset of X. This contradicts the fact that X is connected. Therefore, X does not contain nondegenerate proper terminal subcontinua.

Q.E.D.

5.1.14 Theorem *Let X be a continuum. If A is a terminal subcontinuum of X, if B is a subcontinuum of X disjoint from A, and if $f \colon A \to Y$ is a map from A into the absolute neighborhood retract Y, then there exists a map $F \colon X \to Y$ such that $F|_A = f$ and $F|_B$ is homotopic to a constant map.*

Proof Since Y is an absolute neighborhood retract, there exist an open subset U of X containing A and a map $g \colon U \to Y$ such that $g|_A = f$. Without loss of generality, we assume that $U \cap B = \emptyset$. By [13, Theorem 7.1, p. 96], Y is locally contractible. Hence, there exists an open neighborhood V of a point a of A such that $V \subset U$ and $g|_V$ is homotopic to a constant map.

Note that $A \setminus V$ and $X \setminus U$ are two closed subsets of $X \setminus V$ such that no connected subset of $X \setminus V$ intersects both $A \setminus V$ and $X \setminus U$. (If K is a connected subset of $X \setminus V$ such that $K \cap (A \setminus V) \neq \emptyset$ and $K \cap (X \setminus U) \neq \emptyset$, then $Cl(K)$ is a continuum in $X \setminus V$ intersecting A and $X \setminus A$. Since A is terminal, $A \subset Cl(K)$. This implies that $V \cap Cl(K) \neq \emptyset$, a contradiction.) By Theorem 1.6.8, there exist two disjoint closed subsets X_1 and X_2 of X such that $X \setminus V = X_1 \cup X_2$, $A \setminus V \subset X_1$ and $X \setminus U \subset X_2$.

Note that $X_1 \cup A$ and X_2 are two disjoint closed subsets of X. Then, by Urysohn's Lemma, there exists a map $h \colon X \to [0, 1]$ such that $h(X_1 \cup A) = \{0\}$ and $h(X_2) = \{1\}$. Let $M = h^{-1}\left(\left[0, \frac{1}{2}\right]\right)$ and let $N = h^{-1}\left(\left[\frac{1}{2}, 1\right]\right)$. Then $X = M \cup N$, $A \subset M$, $X \setminus U \subset N$ and $M \cap N \subset V \setminus A$. Since $g|_{M \cap N}$ is homotopic to a constant map (because $M \cap N \subset V$ and $g|_V$ is homotopic to a constant map), by [8, (15.A.1)], there exists a map $k \colon N \to Y$ such that k is homotopic to a constant map and $k|_{M \cap N} = g|_{M \cap N}$.

Let $F \colon X \to Y$ be given by

$$F(x) = \begin{cases} g(x), & \text{if } x \in M, \\ k(x), & \text{if } x \in N. \end{cases}$$

Then F is well defined and continuous. Note that $F|_A = g|_A = f$ and $F|_B = k|_B$. Since $k|_B$ is homotopic to a constant map (because $B \subset X \setminus U \subset N$ and k is homotopic to a constant map), F is the desired extension of f.

Q.E.D.

5.1.15 Definition A continuum X is *cell-like* if each map of X into a compact absolute neighborhood retract is homotopic to a constant map.

5.1.16 Definition Let X and Y be a continua. We say that a surjective map $f \colon X \twoheadrightarrow Y$ is *cell-like* if for each $y \in Y$, $f^{-1}(y)$ is a cell-like continuum.

5.1.17 Theorem *Let X and Z be nondegenerate continua. Suppose that $g: X \twoheadrightarrow Z$ is a monotone and completely regular map. If z_1 is a point of Z such that $g^{-1}(z_1)$ is a terminal subcontinuum of X, then g is a cell-like map.*

Proof Let Y be a compact absolute neighborhood retract, and let

$$f: g^{-1}(z_1) \to Y$$

be a map. Let $z_2 \in Z \setminus \{z_1\}$, and let $F: X \to Y$ be an extension of f such that $F|_{g^{-1}(z_2)}$ is homotopic to a constant map (Theorem 5.1.14).

Since Y is a compact absolute neighborhood retract, by [13, Theorem 1.1, p. 111], there exists $\varepsilon > 0$ such that for any metric space W and for any two maps $k, k': W \to Y$ such that $d(k(w), k'(w)) < \varepsilon$ for every $w \in W$, k and k' are homotopic.

Since F is uniformly continuous, for this ε, there exists $\delta > 0$ such that if $d(x, x') < \delta$, then $d(F(x), F(x')) < \varepsilon$.

Let $Z' = \{z \in Z \mid F|_{g^{-1}(z)}$ is homotopic to a constant map$\}$. We show that Z' is open in Z. Let $z \in Z'$. Since g is completely regular, there exists an open subset V of Z such that $z \in V$ and if $z' \in V$, then there exists a homeomorphism $h: g^{-1}(z) \twoheadrightarrow g^{-1}(z')$ with $d(x, h(x)) < \delta$ for every $x \in g^{-1}(z)$. Let $z' \in V$, and let h be the homeomorphism guaranteed by the complete regularity of g. Hence, $d(F|_{g^{-1}(z')}(x), (F|_{g^{-1}(z)}) \circ h^{-1}(x)) < \varepsilon$ for every $x \in g^{-1}(z')$. Thus, $F|_{g^{-1}(z')}$ is homotopic to $(F|_{g^{-1}(z)}) \circ h^{-1}$. Since $z \in Z'$, $(F|_{g^{-1}(z)}) \circ h^{-1}$ is homotopic to a constant map. Hence, $F|_{g^{-1}(z')}$ is homotopic to a constant map. Therefore, $z' \in Z'$. A similar argument shows that Z' is closed in Z.

Since Z' is a nonempty open and closed subset of Z and Z is connected, $Z' = Z$. This implies that $F|_{g^{-1}(z_1)} = f$ is homotopic to a constant map. Thus, each map from $g^{-1}(z_1)$ into a compact absolute neighborhood retract is homotopic to a constant map. Hence, $g^{-1}(z_1)$ is cell-like. Since fibres of completely regular maps defined on continua are homeomorphic, we have that g is a cell-like map.

Q.E.D.

Now, we are ready to prove Jones's Aposyndetic Decomposition Theorem.

5.1.18 Theorem *Let X be a decomposable homogeneous continuum, with metric d, which is not aposyndetic. If $\mathcal{G} = \{\mathcal{T}(\{x\}) \mid x \in X\}$, then the following hold:*

(1) $\mathcal{G}$ is a continuous, monotone and terminal decomposition of X.

(2) The elements of $\mathcal{G}$ are indecomposable, cell-like, homogeneous and mutually homeomorphic continua of the same dimension as X.

(3) The quotient map $q: X \twoheadrightarrow X/\mathcal{G}$ is completely regular.

(4) The quotient space $X/\mathcal{G}$ is a one-dimensional aposyndetic homogeneous continuum, which does not contain nondegenerate proper terminal subcontinua.

Proof Since X is decomposable, there exist two points of X such that X is aposyndetic at one of them with respect to the other. Hence, the elements of $\mathcal{G}$ are nondegenerate proper subcontinua (Theorem 3.1.21) of X. By Theorem 5.1.2, $\mathcal{G}$ is a decomposition of X. By Lemma 5.1.1, the homeomorphism group of X respects $\mathcal{G}$. Hence, by Theorem 5.1.4, $\mathcal{G}$ is a continuous decomposition, the elements of $\mathcal{G}$ are mutually homeomorphic homogeneous continua and the quotient space $X/\mathcal{G}$ is a homogeneous continuum. A proof of the fact that the elements of $\mathcal{G}$ have the same dimension as X may be found in [31, Corollary 9].

Now, we show that all the elements of $\mathcal{G}$ are terminal subcontinua. To this end, suppose there exists a point $x \in X$ such that $\mathcal{T}(\{x\})$ is not terminal. Hence, there exists a subcontinuum Y of X such that $Y \cap \mathcal{T}(\{x\}) \neq \emptyset$, $Y \setminus \mathcal{T}(\{x\}) \neq \emptyset$ and $\mathcal{T}(\{x\}) \setminus Y \neq \emptyset$. Without loss of generality, we assume that $x \in \mathcal{T}(\{x\}) \setminus Y$. Let $p \in Y \setminus \mathcal{T}(\{x\})$, and let $y \in Y \cap \mathcal{T}(\{x\})$.

Since $p \in Y \setminus \mathcal{T}(\{x\})$, there exists a subcontinuum W of X such that $p \in Int(W) \subset W \subset X \setminus \{x\}$. Let $K = Y \cup W$. Then K is a subcontinuum of X, $p \in Int(K)$ and $K \subset X \setminus \{x\}$. Let $\varepsilon > 0$ such that $V_\varepsilon^d(p) \subset K$, $V_{2\varepsilon}^d(K) \subset X \setminus \{x\}$ and $\varepsilon < d(x, y)$. Let $\delta > 0$ be an Effros number for this ε (Theorem 4.2.31). Hence, for each $y' \in V_\delta^d(y)$, there exists an ε-homeomorphism $h_{y'}\colon X \twoheadrightarrow X$ such that $h_{y'}(y) = y'$ (for $y' = y$, we take $h_y = 1_X$). Then

$$
M = Cl\left(\bigcup_{y' \in V_\delta(y)} h_{y'}(K) \right)
$$

is a subcontinuum of X such that $V_\delta^d(y) \subset M \subset X \setminus \{x\}$. This contradicts the choice of y. Therefore, all the elements of $\mathcal{G}$ are terminal. Note that, this implies that all the elements of $\mathcal{G}$ are cell-like (Theorem 5.1.17) since the quotient map is completely regular (Theorem 5.1.8). Note that, by Theorem 3.4.12, the elements of $\mathcal{G}$ are indecomposable.

To finish the proof, we show that $X/\mathcal{G}$ is aposyndetic. Let $\chi_1, \chi_2 \in X/\mathcal{G}$. We see that $X/\mathcal{G}$ is aposyndetic at χ_1 with respect to χ_2. Let $x_1 \in q^{-1}(\chi_1)$ and let $x_2 \in q^{-1}(\chi_2)$. Note that X is aposyndetic at x_1 with respect to x_2. Then there exists a subcontinuum W of X such that $x_1 \subset Int(W) \subset W \subset X \setminus \{x_2\}$. Since $\mathcal{G}$ is a terminal decomposition, $\mathcal{T}(\{x_1\}) \subset W$ and $W \cap \mathcal{T}(\{x_2\}) = \emptyset$ ($\mathcal{T}(\{x\})$ is nowhere dense for every $x \in X$ (Theorem 5.1.8)). Since $\mathcal{G}$ is a continuous decomposition, by Theorem 1.2.23, q is an open map. Then

$$
\chi_1 = q(x_1) \in Int(q(W)) \subset q(W) \subset X/\mathcal{G} \setminus \{\chi_2\}.
$$

Therefore, $X/\mathcal{G}$ is aposyndetic. The fact that $X/\mathcal{G}$ is one-dimensional may be found in [34, Theorem 3].

The fact that $X/\mathcal{G}$ does not contain nondegenerate proper terminal subcontinua follows from Lemma 5.1.13.

Q.E.D.

As a consequence of Jones's Theorem we have the following three Corollaries:

5.1.19 Corollary *Let X be a decomposable homogeneous continuum. If $x \in X$, then $\mathcal{T}(\{x\})$ is the maximal terminal proper subcontinuum of X containing x.*

Proof If X is aposyndetic, then $\mathcal{T}(\{x\}) = \{x\}$ (Theorem 3.1.28). Since aposyndetic continua do not contain nondegenerate proper terminal subcontinua (Lemma 5.1.13), $\mathcal{T}(\{x\})$ is the maximal terminal proper subcontinuum of X containing x.

Suppose X is not aposyndetic, and let $x \in X$. Then, by Theorem 5.1.18 (1), $\mathcal{T}(\{x\})$ is a terminal subcontinuum of X. Suppose K is a terminal proper subcontinuum of X such that $\mathcal{T}(\{x\}) \subsetneq K$. By Lemma 5.1.11, $q(K)$ is a terminal subcontinuum of $X/\mathcal{G}$. Since $X/\mathcal{G}$ does not contain proper nondegenerate terminal subcontinua (Theorem 5.1.18 (4)), $q(K) = \{q(x)\}$. Hence, $K = q^{-1}(q(x)) = \mathcal{T}(\{x\})$, a contradiction. Therefore, $\mathcal{T}(\{x\})$ is the maximal terminal proper subcontinuum of X containing x.

Q.E.D.

5.1.20 Corollary *If X is a hereditarily decomposable and homogeneous continuum, then X is aposyndetic.*

5.1.21 Corollary *If X is an arcwise connected homogeneous continuum, then X is aposyndetic.*

5.1.22 Theorem *Let X be a decomposable homogeneous continuum. If $\mathcal{G} = \{\mathcal{T}_X(\{x\}) \mid x \in X\}$, then $\mathcal{T}_X(Z) = q^{-1}\mathcal{T}_{X/\mathcal{G}}q(Z)$ for any nonempty closed subset Z of X, where $q\colon X \twoheadrightarrow X/\mathcal{G}$ is the quotient map.*

Proof Without loss of generality, we assume that X is not aposyndetic. Let Z be a nonempty closed subset of X. We divide the proof in six steps.

Step 1. $q^{-1}q(Z) \subset \mathcal{T}_X(Z)$.
Let $x \in q^{-1}q(Z)$. Then $q(x) \in q(Z)$. Thus, there exists $z \in Z$ such that $q(z) = q(x)$. This implies that $\mathcal{T}_X(\{z\}) = \mathcal{T}_X(\{x\})$. Hence, since $\mathcal{T}_X$ is idempotent on closed sets (Theorem 4.2.32), $\mathcal{T}_X(\{x\}) = \mathcal{T}_X(\{z\}) \subset \mathcal{T}_X(Z)$. Therefore, $x \in \mathcal{T}_X(Z)$, and $q^{-1}q(Z) \subset \mathcal{T}_X(Z)$.

Step 2. $\mathcal{T}_X(Z) = q^{-1}q\mathcal{T}_X(Z)$.
Clearly, $\mathcal{T}_X(Z) \subset q^{-1}q\mathcal{T}_X(Z)$. Let $x \in q^{-1}q\mathcal{T}_X(Z)$. Then $q(x) \in q\mathcal{T}_X(Z)$. Hence, there exists $y \in \mathcal{T}_X(Z)$ such that $q(y) = q(x)$. This implies that $\mathcal{T}_X(\{y\}) = \mathcal{T}_X(\{x\})$. Thus, since $\mathcal{T}_X$ is idempotent on closed sets (Theorem 4.2.32), we have that $\mathcal{T}_X(\{y\}) \subset \mathcal{T}_X(Z)$. Hence, $x \in \mathcal{T}_X(Z)$, and $q^{-1}q\mathcal{T}_X(Z) \subset \mathcal{T}_X(Z)$. Therefore, $\mathcal{T}_X(Z) = q^{-1}q\mathcal{T}_X(Z)$.

Step 3.   *If Z is connected and $Int_X(Z) \neq \emptyset$, then $Z = q^{-1}q(Z)$.*
Clearly, $Z \subset q^{-1}q(Z)$. Let $x \in q^{-1}q(Z)$. Then $q(x) \in q(Z)$. Thus, there exists $z \in Z$ such that $q(z) = q(x)$. This implies that $\mathcal{T}_X(\{z\}) = \mathcal{T}_X(\{x\})$. Since $\mathcal{T}_X(\{x\})$ is a nowhere dense terminal subcontinuum of X (Lemma 5.1.1, Theorem 5.1.8 and Theorem 3.1.21), $\mathcal{T}_X(\{x\}) \cap Z \neq \emptyset$ and $Int_X(Z) \neq \emptyset$, we have that $\mathcal{T}_X(\{x\}) \subset Z$. In particular, $x \in Z$. Therefore, $Z = q^{-1}q(Z)$.

Step 4. $qT_X(Z) \subset T_{X/\mathcal{G}}q(Z)$.

Let $\chi \in X/\mathcal{G} \setminus T_{X/\mathcal{G}}q(Z)$. Then there exists a subcontinuum $\mathcal{W}$ of $X/\mathcal{G}$ such that $\chi \in Int_{X/\mathcal{G}}(\mathcal{W}) \subset \mathcal{W} \subset X/\mathcal{G} \setminus q(Z)$. From these inclusions we obtain that $q^{-1}(\chi) \subset Int_X(q^{-1}(\mathcal{W})) \subset q^{-1}(\mathcal{W}) \subset X \setminus q^{-1}q(Z) \subset X \setminus Z$. Hence, since q is monotone (Theorem 5.1.18), $q^{-1}(\chi) \cap T_X(Z) = \emptyset$. Thus, $qq^{-1}(\chi) \cap qT_X(Z) = \emptyset$. Therefore, $\chi \in X/\mathcal{G} \setminus qT_X(Z)$, and $qT_X(Z) \subset T_{X/\mathcal{G}}q(Z)$.

Step 5. $T_{X/\mathcal{G}}q(Z) \subset qT_X(Z)$.

Let $\chi \in X/\mathcal{G} \setminus qT_X(Z)$. Then $\{\chi\} \cap qT_X(Z) = \emptyset$. This implies that $q^{-1}(\chi) \cap q^{-1}qT_X(Z) = \emptyset$. Hence, by Step 2, $q^{-1}(\chi) \cap T_X(Z) = \emptyset$. Since T_X is idempotent on closed sets (Theorem 4.2.32), $q^{-1}(\chi) \cap T_X^2(Z) = \emptyset$. Thus, there exists a subcontinuum W of X such that $q^{-1}(\chi) \subset Int_X(W) \subset W \subset X \setminus T_X(Z) \subset X \setminus Z$. From these inclusions, since q is an open map (Theorem 5.1.18 and Proposition 5.1.7), we obtain that $\{\chi\} = qq^{-1}(\chi) \subset Int_{X/\mathcal{G}}(q(W)) \subset q(W) \subset q(X \setminus Z)$. To finish, we need to show that $q(W) \cap q(Z) = \emptyset$. Suppose there exists $\chi' \in q(W) \cap q(Z)$. Then, by Steps 3 and 1, $q^{-1}(\chi') \subset q^{-1}q(W) \cap q^{-1}q(Z) = W \cap q^{-1}q(Z) \subset W \cap T_X(Z)$, a contradiction to the election of W. Hence, $q(W) \cap q(Z) = \emptyset$, and $\chi \in X/\mathcal{G} \setminus T_{X/\mathcal{G}}q(Z)$. Therefore, $T_{X/\mathcal{G}}q(Z) \subset qT_X(Z)$.

Step 6. $qT_X(Z) = T_{X/\mathcal{G}}q(Z)$.

The equality follows from Steps 4 and 5.

From Steps 2 and 6, we have that

$$T_X(Z) = q^{-1}q'T_X(Z) = q^{-1}T_{X/\mathcal{G}}q(Z).$$

Therefore, $T_X(Z) = q^{-1}T_{X/\mathcal{G}}q(Z)$.

                                                                        Q.E.D.

5.1.23 Corollary *Let X be a decomposable homogeneous continuum and let $\mathcal{G} = \{T_X([x]) \mid x \in X\}$. Then T_X is continuous if and only if $X/\mathcal{G}$ is locally connected.*

Proof Let $q \colon X \twoheadrightarrow X/\mathcal{G}$ be the quotient map. Suppose T_X is continuous for X. Note that, by Step 4 of Theorem 5.1.22, q is $T_{XX/\mathcal{G}}$-continuous (Definition 3.3.3). Thus, $T_{X/\mathcal{G}}$ is continuous for $X/\mathcal{G}$ (Theorem 3.3.4). Since $X/\mathcal{G}$ is an aposyndetic continuum (Theorem 5.1.18) for which $T_{X/\mathcal{G}}$ is continuous, $X/\mathcal{G}$ is locally connected (Corollary 3.3.16).

Next, suppose $X/\mathcal{G}$ is locally connected. Recall that $q \colon X \twoheadrightarrow X/\mathcal{G}$ is a monotone open map (Theorem 5.1.18). Let W be a proper subcontinuum of X, and let $x \in X \setminus W$. Since $T_X(\{x\})$ is a terminal subcontinuum of X (Theorem 5.1.18), either $W \subset T_X(\{x\})$ or $T_X(\{x\}) \cap W = \emptyset$. In either case, $q(W)$ is a proper subcontinuum of $X/\mathcal{G}$. Therefore, by Theorem 3.3.2, T_X is continuous.

                                                                        Q.E.D.

5.1.24 Remark Note that Corollary 5.1.23 gives a partial answer to Question 9.2.2.

As a consequence of Theorem 3.1.39, Corollary 3.3.16 and Corollary 5.1.23, we have the following characterization of homogeneous continua for which the set function $\mathcal{T}$ is continuous:

5.1.25 Theorem *Let X be a homogeneous continuum. Then $\mathcal{T}$ is continuous for X if and only if one of the following conditions holds:*

(1) X is indecomposable.
(2) X is not aposyndetic and $X/\mathcal{G}$ is homeomorphic to the simple closed curve, $\mathcal{S}^1$, or the Menger universal curve, $\mathbb{M}$; where $\mathcal{G} = \{\mathcal{T}(\{x\}) \mid x \in X\}$.
(3) X is locally connected.

Proof Let X be a homogeneous continuum. Suppose $\mathcal{T}$ is continuous for X. If X is indecomposable, then (1) holds. Thus, suppose X is decomposable. If X is not aposyndetic, then $X/\mathcal{G}$ is locally connected, by Corollary 5.1.23. Hence, since $X/\mathcal{G}$ is also one-dimensional (Theorem 5.1.18), $X/\mathcal{G}$ is homeomorphic to the simple closed curve, $\mathcal{S}^1$, or to the Menger universal curve $\mathbb{M}$, by [5, Theorem XIII]. Therefore, (2) holds. If X is aposyndetic, then X is locally connected (Corollary 3.3.16), and (3) holds.

Conversely, if (1) holds, then $\mathcal{T}$ is a constant map by Theorem 3.1.39. Hence, $\mathcal{T}$ is continuous. Suppose (2) holds. Since $\mathcal{S}^1$ and $\mathbb{M}$ are locally connected, $\mathcal{T}$ is continuous, by Corollary 5.1.23. Finally, if (3) holds, by Theorem 3.1.31, $\mathcal{T}$ is the identity map on 2^X. Thus, $\mathcal{T}$ is continuous.

$$\text{Q.E.D.}$$

5.1.26 Theorem *Let X be a homogeneous continuum, let $\mathcal{G} = \{\mathcal{T}_X(\{x\}) \mid x \in X\}$ and let $q \colon X \to X/\mathcal{G}$ be the quotient map. If $\Im(q) \colon 2^{X/\mathcal{G}} \to 2^X$ is given by $\Im(q)(\Gamma) = q^{-1}(\Gamma)$, then $\mathcal{T}_X\left(2^X\right) \subset \Im(q)\left(2^{X/\mathcal{G}}\right)$. Moreover, $\mathcal{T}_X\left(2^X\right) = \Im(q)\left(2^{X/\mathcal{G}}\right)$ if and only if $\mathcal{T}_X$ is continuous for X.*

Proof By Theorem 5.1.18, $\mathcal{G}$ is a continuous decomposition of X. Hence, the quotient map is monotone and open (Theorem 1.2.23). Thus, $\Im(q)$ is continuous (Theorem 8.5.5). To show that $\mathcal{T}_X\left(2^X\right) \subset \Im(q)\left(2^{X/\mathcal{G}}\right)$, let $K \in \mathcal{T}_X\left(2^X\right)$. Then there exists $Z \in 2^X$ such that $\mathcal{T}_X(Z) = K$. By Theorem 5.1.22, $\mathcal{T}_X(Z) = q^{-1}\mathcal{T}_{X/\mathcal{G}}q(Z) = \Im(q)(\mathcal{T}_{X/\mathcal{G}}q(Z))$; i.e., $K = \Im(q)(\mathcal{T}_{X/\mathcal{G}}q(Z))$. Therefore, $\mathcal{T}_X\left(2^X\right) \subset \Im(q)\left(2^{X/\mathcal{G}}\right)$.

Now we prove that $\mathcal{T}_X\left(2^X\right) = \Im(q)\left(2^{X/\mathcal{G}}\right)$ if and only if $\mathcal{T}_X$ is continuous. If $\mathcal{T}_X$ is continuous for X, then $\mathcal{T}_X\left(2^X\right) = \Im(q)\left(2^{X/\mathcal{G}}\right)$, by the proof of Theorem 3.4.13. Suppose $\mathcal{T}_X\left(2^X\right) = \Im(q)\left(2^{X/\mathcal{G}}\right)$. Let $\Gamma \in 2^{X/\mathcal{G}}$. Then, by Theorem 3.1.80 (c), $\mathcal{T}_{X/\mathcal{G}}(\Gamma) = q\mathcal{T}_Xq^{-1}(\Gamma) = 2^q \circ \mathcal{T}_X \circ \Im(q)(\Gamma)$. Since $\Im(q)(\Gamma) \in \mathcal{T}_X\left(2^X\right)$, there exists $Z \in 2^X$ such that $\mathcal{T}_X(Z) = \Im(q)(\Gamma)$. Hence, $\mathcal{T}_X\Im(q)(\Gamma) = \mathcal{T}_X\mathcal{T}_X(Z) = \mathcal{T}_X(Z) = \Im(q)(\Gamma)$, the second last equality is true by Theorem 4.2.32. Thus, $\mathcal{T}_{X/\mathcal{G}}(\Gamma) = 2^q \circ \Im(q)(\Gamma) = \Gamma$. Hence, $X/\mathcal{G}$ is locally connected (Theorem 3.1.31) and, by Corollary 5.1.23, $\mathcal{T}_X$ is continuous for X.

$$\text{Q.E.D.}$$

5.1.27 Theorem *If X is a decomposable homogeneous continuum such that* $\dim(\mathcal{C}(X)) < \infty$, *then $\mathcal{T}$ is continuous for X.*

Proof If X is aposyndetic, then the theorem follows from Corollary 3.1.35 and Theorem 3.1.32. Suppose X is not aposyndetic. Then, by Theorem 5.1.18, $\mathcal{G} = \{\mathcal{T}_X(\{x\}) \mid x \in X\}$ is a monotone, terminal and continuous decomposition of X such that $X/\mathcal{G}$ is an aposyndetic continuum. Therefore, by Theorem 3.5.20, $\mathcal{T}$ is continuous for X.

Q.E.D.

Next, we prove an alternate version of Theorem 5.1.18. The awkward collection of hypotheses in this theorem is precisely the situation encountered in Theorem 5.3.23.

5.1.28 Theorem *Let Y be a continuum such that it is the union of two disjoint, nonempty sets K and E with the following properties:*

(1) Y is aposyndetic at each point of E,
(2) Y is not aposyndetic at any point of K,
(3) Y is aposyndetic at each point of K with respect to each point of E,
(4) K has a metric d such that (K, d) has the property of Effros,
(5) Each homeomorphism h of (K, d) can be extended to a homeomorphism $\hat{h}$ of Y, by defining $\hat{h}(z) = z$ for each $z \in E$,
(6) K is connected and open, and
(7) $\dim(E) = 0$ (i.e., E is totally disconnected).
  If $\mathcal{G} = \{\mathcal{T}(\{y\}) \mid y \in Y\}$, then $\mathcal{G}$ is a continuous, terminal decomposition of Y with the following properties:
(8) The degenerate elements of $\mathcal{G}$ are precisely the points of E,
(9) The nondegenerate elements of $\mathcal{G}$ are mutually homeomorphic, cell-like, homogeneous, indecomposable continua of the same dimension as Y,
(10) The quotient space $Y/\mathcal{G}$ is aposyndetic, and
(11) The quotient space $K/\mathcal{G}'$ is homogeneous, where $\mathcal{G}' = \{\mathcal{T}(\{y\}) \mid y \in K\}$.

Proof Since Y is aposyndetic at every point with respect to at least one other point (by (1) and (3)), the elements of $\mathcal{G}$ are proper subcontinua of Y (Theorem 3.1.21). Moreover, if $y \in K$, then $\mathcal{T}(\{y\}) \subset K$ since Y is aposyndetic at each point of E, by (1). By (2), the elements of $\mathcal{G}$ that are contained in K are nondegenerate. Furthermore, if $y \in E$, then $\mathcal{T}(\{y\}) = \{y\}$, since Y is aposyndetic at every point of K with respect to y, by (3).

Note that, by (4) and by Theorem 4.2.38, K is a homogeneous space. By (5) and Lemma 5.1.1, the elements of $\mathcal{G}$ contained in K are mutually homeomorphic and homogeneous.

Thus, the elements of $\mathcal{G}$ are the points of E and some nondegenerate subcontinua of K; i.e., $\mathcal{G} = \{\{y\} \mid y \in E\} \cup \{\mathcal{T}(\{y\}) \mid y \in K\}$.

The proof of the fact that $\mathcal{G}$ is a decomposition of Y is similar to the one given in Theorem 5.1.2 (using (5) to extend the homeomorphisms defined on K to homeomorphisms defined on Y).

Now, we see that $\mathcal{G}$ is continuous. First, we show that $\mathcal{G}$ is upper semicontinuous. Let $y \in E$. Then $\mathcal{T}(\{y\}) = \{y\}$. Suppose $\mathcal{G}$ is not upper semicontinuous at $\{y\}$. Then there exists an open subset U of Y such that $\{y\} \subset U$ and for each $n \in \mathbb{N}$, there exists $y_n \in \mathcal{V}_{\frac{1}{n}}^d(y)$ such that $\mathcal{T}(\{y_n\}) \setminus U \neq \emptyset$. Note that $\lim_{n\to\infty} y_n = y$. Since the hyperspace of subcontinua of Y, $\mathcal{C}(Y)$, is compact (Theorem 1.8.5), without loss of generality, we assume that the sequence $\{\mathcal{T}(\{y_n\})\}_{n=1}^{\infty}$ converges to a subcontinuum Z of Y. Note that $Z \setminus U \neq \emptyset$. Let $z \in Z \setminus U$. Since $z \in Y \setminus \mathcal{T}(\{y\})$, there exists a subcontinuum W of Y such that $z \in Int(W) \subset W \subset Y \setminus \{y\}$. Since $\lim_{n\to\infty} \mathcal{T}(\{y_n\}) = Z$, for each $n \in \mathbb{N}$, there exists $z_n \in \mathcal{T}(\{y_n\})$ such that $\lim_{n\to\infty} z_n = z$. Hence, there exists $N' \in \mathbb{N}$ such that $z_n \in Int(W)$ for every $n \geq N'$. Since $y \in Y \setminus W$, and $\lim_{n\to\infty} y_n = y$, there exists $N'' \in \mathbb{N}$ such that $y_n \in Y \setminus W$ for every $n \geq N''$. Let $n = \max\{N', N''\}$. Then $z_n \in Int(W)$ and $y_n \in Y \setminus W$. This implies that $z_n \in Y \setminus \mathcal{T}(\{y_n\})$, a contradiction to the choice of z_n. Hence, $\mathcal{G}$ is upper semicontinuous at $\{y\}$.

Let $y \in K$, and let U be an open subset of Y such that $\mathcal{T}(\{y\}) \subset U$. Without loss of generality, we assume that $U \subset K$. By a similar argument to the one given in Theorem 5.1.4, we can find an open subset V of Y such that $\mathcal{T}(\{y\}) \subset V \subset U$ and if $y' \in Y$ is such that $\mathcal{T}(\{y'\}) \cap V \neq \emptyset$, then $\mathcal{T}(\{y'\}) \subset U$. Hence, $\mathcal{G}$ is upper semicontinuous at $\mathcal{T}(\{y\})$.

Therefore, $\mathcal{G}$ is upper semicontinuous.

Next, we prove that $\mathcal{G}$ is lower semicontinuous. Let $y \in E$. Then, since $\mathcal{T}(\{y\}) = \{y\}$, clearly, $\mathcal{G}$ is lower semicontinuous at $\{y\}$.

Let $y \in K$. Let $p, q \in \mathcal{T}(\{y\})$, and let U be an open subset of Y such that $p \in U$. Without loss of generality, we assume that $U \subset K$. By a similar argument to the one given in Theorem 5.1.4, we can find an open subset V of Y such that $q \in V$ and if $y' \in Y$ is such that $\mathcal{T}(\{y'\}) \cap V \neq \emptyset$, then $\mathcal{T}(\{y'\}) \cap U \neq \emptyset$. Hence, $\mathcal{G}$ is lower semicontinuous at $\mathcal{T}(\{y\})$.

Therefore, $\mathcal{G}$ is a continuous decomposition.

Each degenerate element of $\mathcal{G}$ is obviously a terminal subcontinuum of Y. A proof of the fact that the nondegenerate elements of $\mathcal{G}$ are terminal subcontinua is similar to the one given in Theorem 5.1.18 (only the property of Effros is used).

The proof of the fact that the nondegenerate elements of $\mathcal{G}$ are indecomposable is similar to the proof given in Theorem 3.4.12.

By Theorem 1.7.3, $Y/\mathcal{G}$ is a continuum. Let $q: Y \twoheadrightarrow Y/\mathcal{G}$ be the quotient map. Then q is a monotone map. Since K has the property of Effros, $q|_K: K \twoheadrightarrow K/\mathcal{G}'$ is completely regular. Since the nondegenerate elements of $\mathcal{G}$ are terminal, by Theorem 5.1.17, each such element of $\mathcal{G}$ is cell-like. A proof of the fact that the elements of $\mathcal{G}$ have the same dimension as Y may be found in [31, Corollary 9].

Since the elements of $\mathcal{G}$ are terminal subcontinua, the proof of the fact that $Y/\mathcal{G}$ is aposyndetic is similar to the one given in Theorem 5.1.18.

Since $\mathcal{G}$ is a continuous decomposition, the proof of the fact that $K/\mathcal{G}'$ is homogeneous is similar to the one given in Theorem 5.1.4.

Q.E.D.

5.2 Detour to Covering Spaces

We present the necessary definitions and results of the theory of covering spaces to state and prove Rogers's Terminal Decomposition Theorem (Theorem 5.3.28). For this purpose, we follow [12, 13, 15, 16, 25].

5.2.1 Definition Let X and Y be metric spaces. A map $f: X \to Y$ is called a *local homeomorphism* provided that for each point $x \in X$, there exists an open subset U of X such that $x \in U$, $f(U)$ is an open subset of Y and $f|_U: U \twoheadrightarrow f(U)$ is a homeomorphism.

5.2.2 Proposition *Let X and Y be metric spaces. If $f: X \twoheadrightarrow Y$ is a surjective local homeomorphism, then $f^{-1}(y)$ is a discrete subset of X for every $y \in Y$.*

Proof Let $y \in Y$. Then, by definition, for each point $x \in f^{-1}(y)$, there exists an open subset U of X such that $x \in U$ and $f|_U: U \twoheadrightarrow f(U)$ is a homeomorphism. Then $U \cap f^{-1}(y) = \{x\}$. Hence, each point of $f^{-1}(y)$ is an isolated point in $f^{-1}(y)$. Therefore, $f^{-1}(y)$ is a discrete subset of X.

Q.E.D.

5.2.3 Corollary *Let X and Y be metric spaces. If X is compact and $f: X \twoheadrightarrow Y$ is a surjective local homeomorphism, then $f^{-1}(y)$ is finite for every $y \in Y$.*

5.2.4 Definition Let X, Y and Z be metric spaces. If $f: X \to Y$ and $g: Z \to Y$ are maps, then a *lifting of g relative to f* is a map $\tilde{g}: Z \to X$ such that $f \circ \tilde{g} = g$.

5.2.5 Proposition *Let X, Y and Z be metric spaces. Let $f: X \to Y$ be a local homeomorphism. If Z is connected and $g: Z \to Y$ is a map, then two liftings $\tilde{g}, \hat{g}: Z \to X$ of g relative to f such that $\tilde{g}(z_0) = \hat{g}(z_0)$, for some point $z_0 \in Z$, are equal.*

Proof Let $A = \{z \in Z \mid \tilde{g}(z) = \hat{g}(z)\}$. Then, since $\tilde{g}(z_0) = \hat{g}(z_0)$, $A \neq \emptyset$. Since X is a metric space, A is a closed subset of Z. To conclude that $\tilde{g} = \hat{g}$, it is enough to show that A is an open subset of Z (because Z is connected). Let $z \in A$. Then $\tilde{g}(z) = \hat{g}(z)$. Since f is a local homeomorphism, there exists an open subset U of X such that $\tilde{g}(z) = \hat{g}(z) \in U$ and $f|_U: U \twoheadrightarrow f(U)$ is a homeomorphism. Since $\tilde{g}$ and $\hat{g}$ are continuous, there exists an open subset V of Z such that $z \in V$ and $\tilde{g}(V) \cup \hat{g}(V) \subset U$. Then for each $z' \in V$, $(f \circ \tilde{g})(z') = g(z') = (f \circ \hat{g})(z')$. Hence, for each $z' \in V$, $\tilde{g}(z') = \hat{g}(z')$, since $f|_U$ is a homeomorphism. Thus, $V \subset A$. Therefore, $\tilde{g} = \hat{g}$.

Q.E.D.

5.2.6 Definition Let X be a metric space. A *covering space* of X is a pair consisting of a metric space $\widetilde{X}$ and a map $\sigma: \widetilde{X} \twoheadrightarrow X$ such that the following condition holds: for each point $x \in X$, there exists an open subset U of X such that $x \in U$, $\sigma^{-1}(U) = \bigcup_{\lambda \in \Lambda} V_\lambda$, where $\{V_\lambda\}_{\lambda \in \Lambda}$ is a family of pairwise disjoint open subsets of $\widetilde{X}$, and $\sigma|_{V_\lambda}: V_\lambda \twoheadrightarrow U$ is a homeomorphism for every $\lambda \in \Lambda$. Any such open subset U of X is called an *evenly covered* subset of X. The map σ is called a *covering map*.

5.2.7 Remark Observe that, by definition, every covering map is a local homeomorphism. The converse is not necessarily true as can be easily seen using the interval $(0, 1)$ and the map $f : (0, 1) \twoheadrightarrow \mathcal{S}^1$ given by $f(t) = \exp(4\pi t)$.

The next lemma says that the Cartesian product of covering spaces is a covering space.

5.2.8 Lemma *Let X_1 and X_2 be metric spaces. If $\sigma_1 : \widetilde{X}_1 \twoheadrightarrow X_1$ and $\sigma_2 : \widetilde{X}_2 \twoheadrightarrow X_2$ are covering maps, then $(\sigma_1 \times \sigma_2) : \widetilde{X}_1 \times \widetilde{X}_2 \twoheadrightarrow X_1 \times X_2$ is a covering map.*

Proof Let $(x_1, x_2) \in X_1 \times X_2$. Let U_j be an open subset of X_j such that $x_j \in U_j$ and U_j is evenly covered by σ_j, $j \in \{1, 2\}$. Suppose $\sigma_1^{-1}(U_1) = \bigcup_{\lambda \in \Lambda} V_\lambda^1$ and $\sigma_2^{-1}(U_2) = \bigcup_{\gamma \in \Gamma} V_\gamma^2$. Then

$$(\sigma_1 \times \sigma_2)^{-1}(U_1 \times U_2) = \left(\bigcup_{\lambda \in \Lambda} V_\lambda^1 \right) \times \left(\bigcup_{\gamma \in \Gamma} V_\gamma^2 \right).$$

Note that $(\sigma_1 \times \sigma_2)|_{V_\lambda^1 \times V_\gamma^2} : V_\lambda^1 \times V_\gamma^2 \twoheadrightarrow U_1 \times U_2$ is a homeomorphism for every $\lambda \in \Lambda$ and $\gamma \in \Gamma$. Therefore, $\sigma_1 \times \sigma_2$ is a covering map.

Q.E.D.

5.2.9 Definition Let X be a metric space. A covering space $\widetilde{X}$ of X is said to be *universal* if $\widetilde{X}$ is arcwise connected and its fundamental group is trivial.

5.2.10 Definition Let $\sigma : \widetilde{X} \twoheadrightarrow X$ be a covering map. A homeomorphism

$$\varphi : \widetilde{X} \twoheadrightarrow \widetilde{X}$$

is *covering homeomorphism* of X if $\sigma \circ \varphi = \sigma$.

5.2.11 Definition Let X and Y be metric spaces, with metrics d and d', respectively. A map $f : X \to Y$ is a *local isometry* provided that for each point $x \in X$, there exists an open subset U of X such that $x \in U$ and $d(x', x'') = d'(f(x'), f(x''))$ for every $x', x'' \in U$. The map f is an *isometry* if $d(x', x'') = d'(f(x'), f(x''))$ for every $x', x'' \in X$.

The following proposition tells us that a connected covering space of a locally connected continuum has a metric such that the covering map is a local isometry.

5.2.12 Proposition *Let X be a locally connected continuum with metric d. If $\sigma : \widetilde{X} \twoheadrightarrow X$ is a covering map, where $\widetilde{X}$ is connected, then there exists a metric $\tilde{d}$ for $\widetilde{X}$ such that σ is a local isometry and every covering homeomorphism of X is an isometry.*

Proof First, we show the existence of a metric $\tilde{d}$ for $\widetilde{X}$ such that σ is a local isometry.

Since $\widetilde{X}$ is a covering space, for each $x \in X$, there exists an evenly covered open subset U_x of X. Thus, each $\sigma^{-1}(U_x)$ may be written as

$$\sigma^{-1}(U_x) = \bigcup_{\lambda \in \Lambda_x} V_\lambda(x),$$

where each $V_\lambda(x)$ is an open subset of $\widetilde{X}$ and $\sigma|_{V_\lambda(x)} \colon V_\lambda(x) \twoheadrightarrow U_x$ is a homeomorphism. Since X is locally connected, without loss of generality, we assume that each U_x is connected.

Let $\widetilde{U}_x = \bigcup_{\lambda \in \Lambda_x} V_\lambda(x)$. Then $\widetilde{U}_x$ is an open subset of $\widetilde{X}$. Note that $\mathcal{U} = \{\widetilde{U}_x \mid x \in X\}$ is an open cover of $\widetilde{X}$. Hence, since $\widetilde{X}$ is connected, by [12, Theorem 3–4], for any two points $\tilde{x}, \tilde{y} \in \widetilde{X}$ there exists a finite subfamily $\{\widetilde{U}_1, \ldots, \widetilde{U}_n\}$ of $\mathcal{U}$ such that $\tilde{x} \in \widetilde{U}_1$, $\tilde{y} \in \widetilde{U}_n$ and $\widetilde{U}_j \cap \widetilde{U}_k \neq \emptyset$ if and only if $|j - k| \leq 1$, $j, k \in \{1, \ldots, n\}$.

Let $\tilde{z}_1 = \tilde{x}$, $\tilde{z}_n = \tilde{y}$ and let $\tilde{z}_j \in \widetilde{U}_{j-1} \cap \widetilde{U}_j$ for each $j \in \{2, \ldots, n\}$. Then $\tilde{z}_1, \ldots, \tilde{z}_n$ is a *chain of points from $\tilde{x}$ to $\tilde{y}$*. We define the length of this chain of points as

$$\sum_{j=2}^{n} d(z_{j-1}, z_j), \text{ where } z_j = \sigma(\tilde{z}_j), \ j \in \{1, \ldots, n\}.$$

Define $\tilde{d}(\tilde{x}, \tilde{y})$ to be the infimum of the lengths of chains of points from $\tilde{x}$ to $\tilde{y}$. We assert that $\tilde{d}$ is a metric. Clearly, $\tilde{d}(\tilde{x}, \tilde{y}) \geq 0$ and $\tilde{d}(\tilde{x}, \tilde{y}) = \tilde{d}(\tilde{y}, \tilde{x})$ for every $\tilde{x}, \tilde{y} \in \widetilde{X}$.

We show the triangle inequality. Let $\tilde{x}, \tilde{y}, \tilde{z} \in \widetilde{X}$. Note that if $\tilde{a}_1, \ldots, \tilde{a}_n$ is a chain of points from $\tilde{x}$ to $\tilde{y}$ and $\tilde{c}_1 \ldots, \tilde{c}_m$ is a chain of points from $\tilde{y}$ to $\tilde{z}$, then $\tilde{a}_1, \ldots, \tilde{a}_n, \tilde{c}_1 \ldots, c_m$ is a chain of points from $\tilde{x}$ to $\tilde{z}$. Hence,

$$\sum_{j=2}^{n} d(a_{j-1}, a_j) + \sum_{j=2}^{m} d(c_{j-1}, c_j) \in$$

$$\left\{ \sum_{j=2}^{k} d(z_{j-1}, z_j) \ \middle|\ \tilde{z}_1, \ldots, \tilde{z}_k \text{ is chain of points from } \tilde{x} \text{ to } \tilde{z} \right\}.$$

Since

$$\inf \left\{ \sum_{j=2}^{n} d(a_{j-1}, a_j) + \sum_{j=2}^{m} d(c_{j-1}, c_j) \ \middle|\ \tilde{a}_1, \ldots, \tilde{a}_n, \tilde{c}_1 \ldots, \tilde{c}_m \text{ is a} \right.$$

$$\left. \text{chain of points from } \tilde{x} \text{ to } \tilde{z} \right\} \leq$$

$$\inf\left\{\sum_{j=2}^{n} d(a_{j-1}, a_j) \,\middle|\, \tilde{a}_1, \ldots, \tilde{a}_n \text{ is a chain of points from } \tilde{x} \text{ to } \tilde{y}\right\}+$$

$$\inf\left\{\sum_{j=2}^{m} d(c_{j-1}, c_j) \,\middle|\, \tilde{c}_1 \ldots, \tilde{c}_m \text{ is a chain of points from } \tilde{y} \text{ to } \tilde{z}\right\},$$

$\tilde{d}(\tilde{x}, \tilde{z}) \le \tilde{d}(\tilde{x}, \tilde{y}) + \tilde{d}(\tilde{y}, \tilde{z})$. Hence, $\tilde{d}$ satisfies the triangle inequality. Therefore, $\tilde{d}$ is a metric for $\tilde{X}$.

Next, we see that σ is a local isometry. Let $\tilde{x} \in \tilde{X}$, and let $x = \sigma(\tilde{x})$. Let U_x be an open subset of X such that $x \in U_x$ and U_x is evenly covered by σ. Let $\tilde{V}_{\tilde{x}}$ be an open subset of $\sigma^{-1}(U_x)$ such that $\tilde{x} \in \tilde{V}_{\tilde{x}}$ and $\sigma|_{\tilde{V}_{\tilde{x}}}: \tilde{V}_{\tilde{x}} \twoheadrightarrow U_x$ is a homeomorphism.

Take $\tilde{w}, \tilde{z} \in \tilde{V}_{\tilde{x}}$, let $w = \sigma(\tilde{w})$ and let $z = \sigma(\tilde{z})$. Observe that $\tilde{w}, \tilde{z}$ is a chain of points from $\tilde{w}$ to $\tilde{z}$. Hence, $d(w, z)$ is the smallest length of chains of points from $\tilde{w}$ to $\tilde{z}$. Thus, $\tilde{d}(\tilde{w}, \tilde{z}) = d(w, z)$. Therefore, σ is a local isometry.

To finish the proof, we show that each covering homeomorphism is an isometry. Let $\varphi: \tilde{X} \twoheadrightarrow \tilde{X}$ be a covering homeomorphism. Let $\tilde{x}, \tilde{y} \in \tilde{X}$. Let $\tilde{z}_1, \ldots, \tilde{z}_n$ be a chain of points from $\tilde{x}$ to $\tilde{y}$. Then $\varphi(\tilde{z}_1), \ldots, \varphi(\tilde{z}_n)$ is a chain of points from $\varphi(\tilde{x})$ to $\varphi(\tilde{y})$. Since $\sigma \circ \varphi = \sigma$,

$$d(\sigma(\tilde{z}_{j-1}), \sigma(\tilde{z}_j)) = d(\sigma \circ \varphi(\tilde{z}_{j-1}), \sigma \circ \varphi(\tilde{z}_j)).$$

Hence,

$$\left\{\sum_{j=2}^{n} d(\sigma(\tilde{z}_{j-1}), \sigma(\tilde{z}_j)) \,\middle|\, \tilde{z}_1, \ldots, \tilde{z}_n \text{ is a chain of points from } \tilde{x} \text{ to } \tilde{y}\right\}$$

$$\subset \left\{\sum_{j=2}^{n} d(\sigma(\tilde{w}_{j-1}), \sigma(\tilde{w}_j)) \,\middle|\, \tilde{w}_1, \ldots, \tilde{w}_n \text{ is a chain of points from } \right.$$

$$\left. \varphi(\tilde{x}) \text{ to } \varphi(\tilde{y})\right\}.$$

Since φ is a covering homeomorphism, it is easy to see that φ^{-1} is also a covering homeomorphism. Thus, the reverse inclusion is also true. Consequently, $\tilde{d}(\tilde{x}, \tilde{y}) = \tilde{d}(\varphi(\tilde{x}), \varphi(\tilde{y}))$. Therefore, φ is an isometry.

Q.E.D.

The next theorem says that, in certain cases, the property of Effros can be lifted to a covering space.

5.2.13 Theorem *Let X be a compact and connected absolute neighborhood retract, and let $\sigma: \widetilde{X} \twoheadrightarrow X$ be a covering map, where $\widetilde{X}$ is connected. If M is a homogeneous subcontinuum of X, then $\widetilde{M} = \sigma^{-1}(M)$ has the property of Effros.*

Proof Since X is an absolute neighborhood retract, X is locally connected (by [15, (iii), p. 339]). Hence, by Proposition 5.2.12, we assume that $\widetilde{X}$ has a metric $\widetilde{d}$ such that σ is a local isometry and every covering homeomorphism is an isometry. Let $\mathcal{U}$ be a finite open cover of X by connected and evenly covered subsets of X. Let $\varepsilon > 0$ be such that 2ε is a Lebesgue number for $\mathcal{U}$ (Theorem 1.6.6).

Since X is a compact absolute neighborhood retract, by [13, Theorem 1.1, p. 111], there exists $\beta > 0$ such that for any two maps $k, k': \widetilde{X} \to X$ such that $d(k(\widetilde{x}), k'(\widetilde{x})) < \beta$, for every $\widetilde{x} \in \widetilde{X}$, there exists a homotopy $F: \widetilde{X} \times [0, 1] \to X$ such that $F((\widetilde{x}, 0)) = k(\widetilde{x})$, $F((\widetilde{x}, 1)) = k'(\widetilde{x})$, and $d(F((\widetilde{x}, t)), F((\widetilde{x}, s))) < \varepsilon$ for every $\widetilde{x} \in \widetilde{X}$ and every $s, t \in [0, 1]$ (such a homotopy is called an ε-*homotopy*). Without loss of generality, we assume that $\beta \leq \varepsilon$.

Let M be a homogeneous subcontinuum of X. Then, by Theorem 4.2.31, there exists an Effros number $\delta > 0$ for β. We assume that $\delta \leq \beta$. Let $\widetilde{p}, \widetilde{q} \in \widetilde{M}$ such that $\widetilde{d}(\widetilde{p}, \widetilde{q}) < \delta$. We show there exists an ε-homeomorphism $\widetilde{h}: \widetilde{M} \twoheadrightarrow \widetilde{M}$ such that $\widetilde{h}(\widetilde{p}) = \widetilde{q}$.

Let $p = \sigma(\widetilde{p})$ and let $q = \sigma(\widetilde{q})$. Since σ is a local isometry, $\widetilde{d}(\widetilde{p}, \widetilde{q}) = d(p, q)$ ($\delta < \varepsilon$ and 2ε is a Lebesgue number for $\mathcal{U}$). Hence, there exists a β-homeomorphism $h: M \twoheadrightarrow M$ such that $h(p) = q$. Since h is a β-homeomorphism, there exists an ε-homotopy F' between h and 1_M. Since 1_M can be extended to 1_X, there exist a map $f: X \to X$ such that $f|_M = h$ and an ε-homotopy F between f and 1_X such that $F|_{M \times [0,1]} = F'$ [13, Theorem 2.2, p. 117].

Let $G: \widetilde{X} \times [0, 1] \twoheadrightarrow X$ be given by $G = F \circ (\sigma \times 1_{[0,1]})$. Then G is a homotopy between $f \circ \sigma$ and σ because:

$$G((\widetilde{x}, 0)) = F((\sigma(\widetilde{x}), 0)) = f(\sigma(\widetilde{x})) = f \circ \sigma(\widetilde{x})$$

and

$$G((\widetilde{x}, 1)) = F((\sigma(\widetilde{x}), 1)) = \sigma(\widetilde{x}).$$

Note that $1_{\widetilde{x}}$ is a lifting of σ relative to σ. By [16, Proposição 9, p. 132], there exists a homotopy $\widetilde{G}: \widetilde{X} \times [0, 1] \to \widetilde{X}$ between $1_{\widetilde{x}}$ and a map $\widetilde{f}: \widetilde{X} \to \widetilde{X}$ with $\sigma \circ \widetilde{f} = f \circ \sigma$ such that $\sigma \circ \widetilde{G} = G$.

Let $\widetilde{h} = \widetilde{f}|_{\widetilde{M}}$. We show that $\widetilde{h}$ is desired homeomorphism. Note that $\sigma \circ \widetilde{h}(\widetilde{M}) = \sigma \circ \widetilde{f}(\widetilde{M}) = f \circ \sigma(\widetilde{M}) = f \circ \sigma(\sigma^{-1}(M)) = f(M) = h(M) = M$. Hence, $\widetilde{h}(\widetilde{M}) \subset \widetilde{M}$.

Now, we see that $\widetilde{h}$ moves no point of $\widetilde{M}$ more than ε. Let $\widetilde{x} \in \widetilde{M}$. Observe that $G|_{\widetilde{M} \times [0,1]}$ is an ε-homotopy. Hence, we have that $d(G((\widetilde{x}, t)), G((\widetilde{x}, s))) < \varepsilon$ for every $s, t \in [0, 1]$. Thus, $G(\{\widetilde{x}\} \times [0, 1])$ is contained in an element of $\mathcal{U}$. Since $\sigma(\widetilde{G}(\{\widetilde{x}\} \times [0, 1])) = G(\{\widetilde{x}\} \times [0, 1])$ and σ is a local isometry, $\widetilde{d}(\widetilde{G}((\widetilde{x}, t)), \widetilde{G}((\widetilde{x}, s))) < \varepsilon$ for every $s, t \in [0, 1]$. Since $\widetilde{G}((\widetilde{x}, 0)) = \widetilde{x} = 1_{\widetilde{M}}(\widetilde{x})$

and $\widetilde{G}((\tilde{x}, 1)) = \tilde{f}(\tilde{x}) = \tilde{h}(\tilde{x})$, $\tilde{d}(\tilde{x}, \tilde{h}(\tilde{x})) < \varepsilon$. Therefore, $\tilde{h}$ moves no point of $\widetilde{M}$ more than ε.

To see that $\tilde{h}$ is a homeomorphism, follow the procedure of the above paragraphs to construct a lift $k \colon \widetilde{M} \to \widetilde{M}$ of the homeomorphism $h^{-1} \colon M \twoheadrightarrow M$ satisfying $\tilde{d}(\tilde{z}, k(\tilde{z})) < \varepsilon$ for all $\tilde{z} \in \widetilde{M}$. Let $\tilde{x} \in \widetilde{M}$. Then $\sigma \circ k \circ \tilde{h}(\tilde{x}) = h^{-1} \circ \sigma \circ \tilde{h}(\tilde{x}) = h^{-1} \circ h \circ \sigma(\tilde{x}) = \sigma(\tilde{x})$, and $\sigma \circ \tilde{h} \circ k(\tilde{x}) = h \circ \sigma \circ k(\tilde{s}) = h \circ h^{-1} \circ \sigma(\tilde{x}) = \sigma(\tilde{x})$. Hence, $k \circ \tilde{h}(\tilde{x}), \tilde{h} \circ k(\tilde{x}) \in \sigma^{-1}(\sigma(\tilde{x}))$. Since neither $k \circ \tilde{h}$ nor $\tilde{h} \circ k$ moves a point more than 2ε, it follows that $k \circ \tilde{h}(\tilde{x}) = \tilde{x}$ and $\tilde{h} \circ k(\tilde{x}) = \tilde{x}$. Therefore, $\tilde{h}$ is an ε-homeomorphism.

Since $h(p) = q$, f is an extension of h and $\tilde{f}$ is a lift of $f \circ \sigma$, it follows that $\tilde{h}(p) \in \sigma^{-1}(q)$. Since $\tilde{d}(\tilde{p}, \tilde{h}(\tilde{p})) < \varepsilon$,

$$\tilde{d}(\tilde{q}, \tilde{h}(\tilde{p})) \le \tilde{d}(\tilde{q}, \tilde{p}) + \tilde{d}(\tilde{p}, \tilde{h}(\tilde{p})) < \delta + \varepsilon \le 2\varepsilon.$$

Hence, $\tilde{h}(\tilde{p}) = \tilde{q}$. Therefore, $\widetilde{M}$ has the property of Effros.

Q.E.D.

5.3   Rogers's Theorem

We use covering spaces techniques due to James T. Rogers, Jr. to study homogeneous continua. The material of this section is based on [1, 3, 9, 10, 16, 17, 20, 27, 31–35].

We begin with a discussion of the Poincaré model of the hyperbolic plane $\mathbb{H}$ ([3, 9] and [35]).

Let $\mathbb{H}$ be the interior of the closed unit disk D in $\mathbb{R}^2$, and let $\mathcal{S}^1$ be its boundary.

5.3.1 Definition A *geodesic* in $\mathbb{H}$ is the intersection of $\mathbb{H}$ and a circle C in $\mathbb{R}^2$ that intersects $\mathcal{S}^1$ orthogonally (straight lines through the origin are considered circles centered at ∞).

5.3.2 Definition A *reflection* in a geodesic $C \cap \mathbb{H}$ is a Euclidean inversion in the circle. A *hyperbolic isometry* of $\mathbb{H}$ is a composition of reflections in geodesics.

5.3.3 Definition The set $\mathbb{H}$ with the family of isometries is the *Poincaré model of the hyperbolic plane*. The boundary $\mathcal{S}^1$ of $\mathbb{H}$, which is not in $\mathbb{H}$, is called the *circle at ∞*.

Let $\mathbb{F}$ be a double torus. The universal covering space of $\mathbb{F}$ may be chosen to be $\mathbb{H}$ ([3] and [35]), and the group of covering homeomorphisms to be a subgroup of the orientation-preserving isometries of $\mathbb{H}$. Each covering homeomorphism (except the identity map $1_{\mathbb{H}}$) is a hyperbolic isometry. The pertinent property for us is that a hyperbolic isometry, when extended to the circle at ∞, has exactly two fixed points in $\mathcal{S}^1$ and none in $\mathbb{H}$ ([3, Theorem 9–3, p. 132] and [35, p. 410]). The universal covering map $\sigma \colon \mathbb{H} \twoheadrightarrow \mathbb{F}$ may be chosen to be a local isometry.

5.3.4 Definition A *geodesic in* $\mathbb{F}$ is the image under σ of a geodesic in $\mathbb{H}$. A geodesic is *simple* if it has no transverse intersection.

5.3.5 Definition A *simple closed geodesic* is a geodesic in $\mathbb{F}$ that is a simple closed curve. A simple closed curve in $\mathbb{F}$ is *essential* if it does not bound a disk.

Each essential simple closed curve in $\mathbb{F}$ is isotopic to a unique simple closed geodesic [32, p. 339].

Assume the universal covering space $(\mathbb{H}, \sigma)$ of $\mathbb{F}$ is constructed with the following properties. The geodesic $\mathbb{H} \cap x$-axis maps to a simple closed geodesic C_1 under σ. The geodesic $\mathbb{H} \cap y$-axis maps to a simple closed geodesic C_2 under σ. Furthermore, $C_1 \cup C_2$ is a figure eight W, and $C_1 \cap C_2 = \{v\}$.

A proof of the following theorem may be found in [27, Theorem 5].

5.3.6 Theorem *Let* $\sigma : \mathbb{H} \twoheadrightarrow \mathbb{F}$ *be the universal covering map of* $\mathbb{F}$. *Let* $\mathcal{U}$ *be a finite cover of* $\mathbb{F}$ *by evenly covered sets, and let* 2ε *be a Lebesgue number for* $\mathcal{U}$. *If* $\mathbb{K}$ *is a compact subset of* $\mathbb{H}$ *and if* $\varphi : \mathbb{H} \twoheadrightarrow \mathbb{H}$ *is a covering homeomorphism different from the identity map* $1_{\mathbb{H}}$, *then there exists* $\tilde{x} \in \mathbb{K}$ *such that* $\tilde{d}(\tilde{x}, \varphi(\mathbb{K})) > \varepsilon$.

5.3.7 Theorem *Let* $\sigma : \mathbb{H} \twoheadrightarrow \mathbb{F}$ *be the universal covering map of* $\mathbb{F}$. *Let* M *be a subspace of* $\mathbb{F}$, *and let* $\widetilde{M} = \sigma^{-1}(M)$. *Suppose* $\mathbb{L}$ *is a component of* $\widetilde{M}$. *If* $\mathbb{L}$ *is compact, then* $\sigma|_{\mathbb{L}}$ *is one-to-one.*

Proof Suppose there exist $x \in M$, and two points $\tilde{x}_1, \tilde{x}_2 \in \mathbb{L} \cap \sigma^{-1}(x)$. Let $\varphi : \mathbb{H} \twoheadrightarrow \mathbb{H}$ be a covering homeomorphism such that $\varphi(\tilde{x}_1) = \tilde{x}_2$ [16, Proposiçao 8, p. 162]. Since $\varphi \neq 1_{\mathbb{H}}$, by Theorem 5.3.6, there exists $\tilde{z} \in \varphi(\mathbb{L}) \setminus \mathbb{L}$. Hence, $\mathbb{L} \cup \varphi(\mathbb{L})$ is a subcontinuum of $\widetilde{M}$ which properly contains $\mathbb{L}$, a contradiction to the fact that $\mathbb{L}$ is a component of $\widetilde{M}$. Therefore, $\sigma|_{\mathbb{L}}$ is one-to-one.

Q.E.D.

5.3.8 Theorem *Let* $\sigma : \mathbb{H} \twoheadrightarrow \mathbb{F}$ *be the universal covering map of* $\mathbb{F}$. *Let* M *be a subspace of* $\mathbb{F}$, *and let* $\widetilde{M} = \sigma^{-1}(M)$. *Suppose* $\mathbb{L}$ *is a component of* $\widetilde{M}$. *If* M *is arcwise connected, then* $\sigma(\mathbb{L}) = M$.

Proof Let $z \in \sigma(\mathbb{L})$, and let $z' \in M$. Since M is arcwise connected, there exists a map $\alpha : [0, 1] \to M$ such that $\alpha(0) = z$ and $\alpha(1) = z'$. Let $\tilde{z} \in \mathbb{L}$ be such that $\sigma(\tilde{z}) = z$. By a proof similar to the one given in Theorem 1.3.31, it is shown that there exists a map $\tilde{\alpha} : [0, 1] \to \widetilde{M}$ such that $\tilde{\alpha}(0) = \tilde{z}$ and $\sigma \circ \tilde{\alpha} = \alpha$. Since $\tilde{\alpha}([0, 1]) \cap \mathbb{L} \neq \emptyset$ and $\mathbb{L}$ is a component of $\widetilde{M}$, $\tilde{\alpha}([0, 1]) \subset \mathbb{L}$. Hence, $\sigma(\tilde{\alpha}(1)) = z'$. Therefore, $\sigma(\mathbb{L}) = M$.

Q.E.D.

5.3.9 Definition Let $\mathcal{Q}$ be the Hilbert cube, and let $\sigma \times 1_{\mathcal{Q}} : \mathbb{H} \times \mathcal{Q} \twoheadrightarrow \mathbb{F} \times \mathcal{Q}$ be the universal covering map of $\mathbb{F} \times \mathcal{Q}$. Let X be a continuum essentially embedded in $W \times \mathcal{Q}$ (this means that the embedding is not homotopic to a constant map; note that this eliminates some continua from consideration). Let $f : X \twoheadrightarrow W$

be the projection map. Let $\widetilde{X} = (\sigma \times 1_Q)^{-1}(X)$, and let $\widetilde{f}\colon \widetilde{X} \twoheadrightarrow \widetilde{W}$ be the projection map. If $\mathbb{K}$ is a component of $\widetilde{X}$, then the set $\mathbb{E}(\mathbb{K}) = \{z \in \mathcal{S}^1 \mid z$ is a (Euclidean) limit point of $\widetilde{f}(\mathbb{K})\}$ is called the *set of ends of* $\mathbb{K}$.

Let $\widetilde{W} = \sigma^{-1}(W)$. Then $\widetilde{W}$ is the universal covering space of W, the "infinite snowflake" pictured on the next page. The set $Cl(\widetilde{W}) \cap \mathcal{S}^1$ is a Cantor set; call it $\mathfrak{Z}$ [32, p. 341].

5.3.10 Lemma *Let X be a continuum. If $\ell\colon X \to W$ is a map which is not homotopic to a constant map, then X can be essentially embedded in $W \times Q$.*

Proof By Theorem 1.1.16, there exists an embedding $g\colon X \to Q$. Let $h\colon X \to W \times Q$ be given by $h(x) = (\ell(x), g(x))$. Then h is an embedding of X into $W \times Q$ which is not homotopic to a constant map.

Q.E.D.

A proof of the following theorem may be modeled from the proof of [27, Theorem 8], using the results in [1] to obtain X as the remainder of a compactification of $[0, 1)$.

5.3.11 Theorem *Let $\sigma \times 1_Q\colon \mathbb{H} \times Q \twoheadrightarrow \mathbb{F} \times Q$ be the universal covering map of $\mathbb{F} \times Q$. Let X be a continuum essentially embedded in $W \times Q$. If $\widetilde{X} = (\sigma \times 1_Q)^{-1}(X)$, then no component of $\widetilde{X}$ is compact.*

5.3.12 Theorem *Let $\sigma \times 1_Q\colon \mathbb{H} \times Q \twoheadrightarrow \mathbb{F} \times Q$ be the universal covering map of $\mathbb{F} \times Q$. Let X be a homogeneous continuum essentially embedded in $W \times Q$, and let $\widetilde{X} = (\sigma \times 1_Q)^{-1}(X)$. Then there exists $\delta > 0$ such that if $\widetilde{x} \in \mathbb{K}$, $\widetilde{x}' \in \mathbb{K}'$ and $\widetilde{d}(\widetilde{x}, \widetilde{x}') < \delta$ (where $\mathbb{K}$ and $\mathbb{K}'$ are components of $\widetilde{X}$), then $\mathbb{E}(\mathbb{K}) = \mathbb{E}(\mathbb{K}')$.*

Proof This is an immediate consequence of Theorem 5.2.13 and the fact that each bounded homeomorphism of $\widetilde{X}$ preserves ends.

Q.E.D.

5.3.13 Notation Let $\sigma \times 1_Q\colon \mathbb{H} \times Q \twoheadrightarrow \mathbb{F} \times Q$ be the universal covering map of $\mathbb{F} \times Q$. Let X be a homogeneous continuum essentially embedded in $W \times Q$. Let $\widetilde{X} = (\sigma \times 1_Q)^{-1}(X)$, and let $\mathbb{K}$ be a component of $\widetilde{X}$. Assume each of the geodesics C_1 and C_2 has length greater than one and that 2δ is an Effros number for $\varepsilon = \frac{1}{2}$. Cover $\{v\} \times Q$ with a finite collection $\mathcal{B}$ of open δ-balls. Since $\mathbb{K}$ is unbounded (Theorem 5.3.11) and locally compact, there exist a ball $B \in \mathcal{B}$, two liftings $\widetilde{B}_0$ and $\widetilde{B}_m$ of B and a subcontinuum M of $\widetilde{W}$ meeting both $\widetilde{B}_0$ and $\widetilde{B}_m$. Let $\varphi \times 1_Q\colon \mathbb{H} \times Q \twoheadrightarrow \mathbb{H} \times Q$ be a covering homeomorphism such that $(\varphi \times 1_Q)(\widetilde{B}_0) = \widetilde{B}_m$.

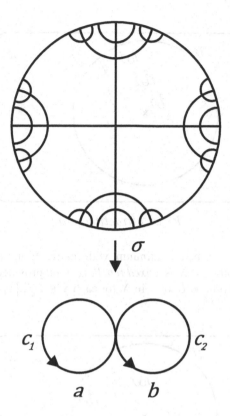

5.3.14 Theorem *If $\mathbb{K}$ and $\varphi \times 1_Q$ are as in Notation 5.3.13, then $\mathbb{E}((\varphi \times 1_Q)(\mathbb{K})) = \mathbb{E}(\mathbb{K})$.*

Proof Let $\tilde{x}_1 \in \widetilde{B}_0 \cap M$ and let $\tilde{x}_2 \in \widetilde{B}_m \cap M$. Hence,

$$\tilde{d}(\tilde{x}_2, (\varphi \times 1_Q)(\tilde{x}_1)) < 2\delta.$$

By Theorem 5.3.12, $\mathbb{E}(\mathbb{K}) = \mathbb{E}((\varphi \times 1_Q)(\mathbb{K}))$.

Q.E.D.

Compactify $\mathbb{H} \times Q$ with $\mathcal{S}^1 \times Q$. Shrink each set of the form $\{z\} \times Q$, where $z \in \mathcal{S}^1$, to a point to obtain another compactification of $\mathbb{H} \times Q$. This time the remainder is $\mathcal{S}^1$. Let $\pi \colon (\mathbb{H} \times Q) \cup \mathcal{S}^1 \twoheadrightarrow \mathbb{H} \cup \mathcal{S}^1$ be the map of this latter compactification onto the disk D obtained by naturally extending the projection map. Note that the restriction of π to $\widetilde{X}$ is just $\tilde{f}$. (If $\tilde{x} \in \widetilde{X}$, then $\sigma \circ \tilde{f}(\tilde{x}) = f \circ (\sigma \times 1_Q)(\tilde{x}) \in W$. Hence, $\tilde{f}(\tilde{x}) \in \widetilde{W}$.)

5.3.15 Definition Let X be a continuum, with metric d, and let $f \colon X \twoheadrightarrow X$ be a surjective map. A point $x \in X$ is a *fixed attracting point* provided that $f(x) = x$ and there exists a neighborhood U of x in X such that $\lim_{n \to \infty} d(x, f^n(y)) = 0$ for every $y \in U$.

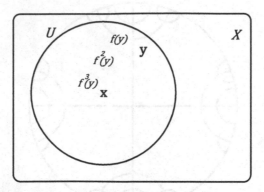

Fixed attracting point

5.3.16 Definition Let X be a continuum, with metric d, and let $f \colon X \twoheadrightarrow X$ be a surjective map. A point $x \in X$ is a *fixed repelling point* provided that $f(x) = x$ and there exists a neighborhood U of x in X for each $y \in U \setminus \{x\}$, there exists $n \in \mathbb{N}$ such that $f^n(y) \in X \setminus U$.

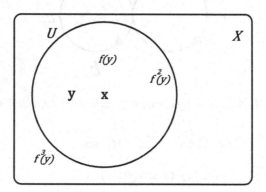

Fixed repelling point

5.3.17 Remark It is known that a hyperbolic isometry of $\mathbb{H}$ has an attracting fixed point and a repelling fixed point on $\mathcal{S}^1$.

5.3.18 Theorem *If $\mathbb{K}$ and $\varphi \times 1_{\mathcal{Q}}$ are as in Notation 5.3.13, then the attracting point of φ belongs to $\mathbb{E}(\mathbb{K})$.*

Proof Let $\mathbb{K}_n = Cl_{(\mathbb{H} \times \mathcal{Q}) \cup \mathcal{S}^1}((\varphi^n \times 1_{\mathcal{Q}})(\mathbb{K}))$. Since the hyperspace of subcontinua of $(\mathbb{H} \times \mathcal{Q}) \cup \mathcal{S}^1$, $\mathcal{C}((\mathbb{H} \times \mathcal{Q}) \cup \mathcal{S}^1)$, is compact (Theorem 1.8.5), without loss of generality, we assume that the sequence $\{\mathbb{K}_n\}_{n=1}^{\infty}$ converges to $\mathbb{P} \in \mathcal{C}((\mathbb{H} \times \mathcal{Q}) \cup \mathcal{S}^1)$. This implies that $\{\pi(\mathbb{K}_n)\}_{n=1}^{\infty}$ converges to $\pi(\mathbb{P})$.

Since $\pi(\mathbb{K}_n) \cap \mathcal{S}^1 = \mathbb{E}((\varphi^n \times 1_{\mathcal{Q}})(\mathbb{K})) = \mathbb{E}(\mathbb{K})$ for all $n \in \mathbb{N}$, by Theorem 5.3.14, and since $\{\pi(\mathbb{K}_n) \cap \mathcal{S}^1\}_{n=1}^{\infty}$ converges to $\pi(\mathbb{P}) \cap \mathcal{S}^1$, it follows that $\pi(\mathbb{P}) \cap \mathcal{S}^1 = \mathbb{E}(\mathbb{K})$. If $\tilde{y} \in \pi(\mathbb{K})$ and $\tilde{z}$ is the attracting point of φ, then $\{\varphi^n(\tilde{y})\}_{n=1}^{\infty}$ converges to $\tilde{z}$ in $\mathbb{H} \cup \mathcal{S}^1$. Since $\tilde{y} \in \pi(\mathbb{K})$, it follows that $\varphi^n(\tilde{y}) \in \pi\left((\varphi^n \times 1_Q)(\mathbb{K})\right)$, and so $\tilde{z} \in \pi(\mathbb{P})$. Hence, $\tilde{z} \in \mathbb{E}(\mathbb{K})$.

Q.E.D.

5.3.19 Theorem *If $\mathbb{K}$ and $\varphi \times 1_Q$ are as in Notation 5.3.13, then the repelling point of φ belongs to $\mathbb{E}(\mathbb{K})$.*

Proof The repelling point of φ is the attracting point of φ^{-1}. Hence, the theorem follows from Theorem 5.3.18.

Q.E.D.

5.3.20 Theorem *Let $\sigma \times 1_Q \colon \mathbb{H} \times Q \twoheadrightarrow \mathbb{F} \times Q$ be the universal covering map of $\mathbb{F} \times Q$. Let X be a homogeneous continuum essentially embedded in $W \times Q$. If $\widetilde{X} = (\sigma \times 1_Q)^{-1}(X)$ and $\mathbb{K}$ is a component of $\widetilde{X}$, then $\mathbb{E}(\mathbb{K})$ is either a two-point set or a Cantor set. Furthermore, $\mathbb{E}(\mathbb{K})$ contains a dense subset each point of which is a fixed point of a hyperbolic isometry φ such that $\varphi \times 1_Q$ is a covering homeomorphism.*

Proof Suppose $\mathbb{E}(\mathbb{K})$ has more than two points. Since $\mathbb{E}((\varphi^n \times 1_Q)(\mathbb{K})) = \mathbb{E}(\mathbb{K})$ for each $n \in \mathbb{N}$ (Theorem 5.3.14), and φ only fixes two points of $\mathcal{S}^1$ [35, p. 419], it follows that $\mathbb{E}(\mathbb{K})$ is infinite. In fact, there exists a sequence of points $\{\tilde{z}_m\}_{m=1}^{\infty}$ in $\mathbb{E}(\mathbb{K})$ that converges to the attracting point, $\tilde{z}$, of φ.

Furthermore, if $\tilde{z}' \in \mathbb{E}(\mathbb{K})$, then arbitrarily close to $\tilde{z}'$ there exists a point $\tilde{z}'' \in \mathbb{E}(\mathbb{K})$ such that $\tilde{z}''$ is the attracting fixed point of a hyperbolic isometry $\varphi'' \colon \mathbb{H} \twoheadrightarrow \mathbb{H}$ such that $\varphi'' \times 1_Q$ is a covering homeomorphism. Such a hyperbolic isometry is constructed in the same way as φ is (Notation 5.3.13). In particular, φ'' has the same properties as φ. It follows that $\mathbb{E}(\mathbb{K})$ is perfect. Therefore, $\mathbb{E}(\mathbb{K})$ is a Cantor set.

Q.E.D.

A proof of the following theorem may be found in [33, Theorem 4.1].

5.3.21 Theorem *Let $\sigma \times 1_Q \colon \mathbb{H} \times Q \twoheadrightarrow \mathbb{F} \times Q$ be the universal covering map of $\mathbb{F} \times Q$. Let X be a nonaposyndetic homogeneous continuum essentially embedded in $W \times Q$. Let $\widetilde{X} = (\sigma \times 1_Q)^{-1}(X)$ and let $\mathbb{K}$ be a component of $\widetilde{X}$. Let $\mathbb{Y} = Cl_{(\mathbb{H} \times Q) \cup \mathcal{S}^1}(\mathbb{K})$. Then $\mathbb{Y} \setminus \mathbb{K} = \mathbb{E}(\mathbb{K})$, and $\mathbb{Y}$ is connected im kleinen at each point of $\mathbb{E}(\mathbb{K})$.*

5.3.22 Corollary *Let $\sigma \times 1_Q \colon \mathbb{H} \times Q \twoheadrightarrow \mathbb{F} \times Q$ be the universal covering map of $\mathbb{F} \times Q$. Let X be a nonaposyndetic homogeneous continuum essentially embedded in $W \times Q$. Let $\widetilde{X} = (\sigma \times 1_Q)^{-1}(X)$ and let $\mathbb{K}$ be a component of $\widetilde{X}$. Let $\mathbb{Y} = Cl_{(\mathbb{H} \times Q) \cup \mathcal{S}^1}(\mathbb{K})$. Then $\mathbb{Y}$ is aposyndetic at each point of $\mathbb{E}(\mathbb{K})$.*

5.3.23 Theorem *Let $\sigma \times 1_Q \colon \mathbb{H} \times Q \twoheadrightarrow \mathbb{F} \times Q$ be the universal covering map of $\mathbb{F} \times Q$. Let X be a nonaposyndetic homogeneous continuum essentially embedded in $W \times Q$. Let $\widetilde{X} = (\sigma \times 1_Q)^{-1}(X)$ and let $\mathbb{K}$ be a component of $\widetilde{X}$. Let $\mathbb{Y} = Cl_{(\mathbb{H} \times Q) \cup \mathcal{S}^1}(\mathbb{K})$. Then the collection $\mathcal{G} = \{\mathcal{T}_{\mathbb{Y}}(\{\tilde{x}\}) \mid \tilde{x} \in \mathbb{Y}\}$ is a continuous,* *terminal decomposition of* $\mathbb{Y}$ *with the following properties:*

(1) The quotient space $\mathbb{Y}/\mathcal{G}$ *is aposyndetic.*

(2) $\mathcal{T}_{\mathbb{Y}}(\{\tilde{x}\})$ is degenerate if $\tilde{x} \in \mathbb{Y} \setminus \mathbb{K}$.

(3) The nondegenerate elements of $\mathcal{G}$ are mutually homeomorphic, cell-like, inde-
     composable homogeneous continua of the same dimension as $\mathbb{Y}$.

(4) $\mathbb{K}/\mathcal{G}'$ is homogeneous, where $\mathcal{G}' = \{\mathcal{T}_{\mathbb{Y}}(\{\tilde{x}\}) \mid \tilde{x} \in \mathbb{K}\}$.

Proof Note that $\mathbb{Y} = \mathbb{K} \cup \mathbb{E}(\mathbb{K})$, where $\mathbb{K}$ is open and connected. By Theorem 5.3.20, $\dim(\mathbb{E}(\mathbb{K})) = 0$ and each homeomorphism h of $\mathbb{K}$ can be extended
to a homeomorphism $\hat{h} \colon \mathbb{Y} \twoheadrightarrow \mathbb{Y}$ such that $\hat{h}(\tilde{z}) = \tilde{z}$ for each $\tilde{z} \in \mathbb{E}(\mathbb{K})$. By
Corollary 5.3.22, $\mathbb{Y}$ is aposyndetic at each point of $\mathbb{E}(\mathbb{K})$.

By Proposition 5.2.12, $(\mathbb{H} \times \mathcal{Q}) \cup \mathcal{S}^1$ has a metric $\tilde{d}$ such that $\sigma \times 1_{\mathcal{Q}}$ is a local
isometry. Thus, by Theorem 5.2.13, $\tilde{X}$ has a metric $\tilde{d}$ such that $\tilde{X}$ has the property
of Effros. Hence, $\mathbb{K}$ has a metric $\tilde{d}$ such that $\mathbb{K}$ has the property of Effros.

It can be shown that each point of $\mathbb{E}(\mathbb{K})$ has a local basis of open sets of $\mathbb{Y}$ whose
complements are connected. Hence, $\mathbb{Y}$ is aposyndetic at each point of $\mathbb{K}$ with respect
to each point of $\mathbb{E}(\mathbb{K})$.

Since X is not aposyndetic and homogeneous, X is not aposyndetic at any point
with respect to any other. Let $\tilde{x} \in \mathbb{K}$ and suppose $\mathbb{Y}$ is aposyndetic at $\tilde{x}$ with respect
to $\tilde{x}' \in \mathbb{K}$. Thus, there exists a subcontinuum $\mathbb{W}$ of $\mathbb{Y}$ such that $\tilde{x} \in Int_{\mathbb{Y}}(\mathbb{W}) \subset$
$\mathbb{W} \subset \mathbb{Y} \setminus \{\tilde{x}'\}$; without loss of generality, we assume that $\mathbb{W} \subset \mathbb{K}$ and $(\sigma \times 1_{\mathcal{Q}})(\tilde{x}) \neq$
$(\sigma \times 1_{\mathcal{Q}})(\tilde{x}')$. Hence, since $\sigma \times 1_{\mathcal{Q}}$ is a local isometry, $(\sigma \times 1_{\mathcal{Q}})(\mathbb{W})$ satisfies that
$(\sigma \times 1_{\mathcal{Q}})(\tilde{x}) \in Int_X((\sigma \times 1_{\mathcal{Q}})(\mathbb{W})) \subset (\sigma \times 1_{\mathcal{Q}})(\mathbb{W}) \subset X \setminus \{(\sigma \times 1_{\mathcal{Q}})(\tilde{x}')\}$, a
contradiction. Therefore, $\mathbb{Y}$ is not aposyndetic at any point of $\mathbb{K}$.

Therefore, the theorem follows from Theorem 5.1.28.

Q.E.D.

5.3.24 Theorem *Let $\sigma \times 1_{\mathcal{Q}} \colon \mathbb{H} \times \mathcal{Q} \twoheadrightarrow \mathbb{F} \times \mathcal{Q}$ be the universal covering map of*
$\mathbb{F} \times \mathcal{Q}$. *Let X be a nonaposyndetic homogeneous continuum essentially embedded*
in $W \times \mathcal{Q}$. Let $\tilde{X} = (\sigma \times 1_{\mathcal{Q}})^{-1}(X)$ and let $\mathbb{K}$ be a component of $\tilde{X}$. Let $\mathbb{Y} =$
$Cl_{(\mathbb{H} \times \mathcal{Q}) \cup \mathcal{S}^1}(\mathbb{K})$. *If $\tilde{x} \in \mathbb{K}$, then $(\sigma \times 1_{\mathcal{Q}})|_{\mathcal{T}_{\mathbb{Y}}(\{\tilde{x}\})}$ is one-to-one.*

Proof Let $\tilde{x}_1$ and $\tilde{x}_2$ be distinct points of $\mathcal{T}_{\mathbb{Y}}(\{\tilde{x}\})$ such that $(\sigma \times 1_{\mathcal{Q}})(\tilde{x}_1) = (\sigma \times$
$1_{\mathcal{Q}})(\tilde{x}_2)$. By [16, Proposição 9, p. 132], there exists a covering homeomorphism
$\varphi \colon \mathbb{H} \times \mathcal{Q} \twoheadrightarrow \mathbb{H} \times \mathcal{Q}$ such that $\varphi(\tilde{x}_1) = \tilde{x}_2$. Since φ does not fix (setwise) any
compact set (Theorem 5.3.6), it follows that $\varphi(\mathcal{T}_{\mathbb{Y}}(\{\tilde{x}\})) \cap (\mathbb{Y} \setminus \mathcal{T}_{\mathbb{Y}}(\{\tilde{x}\})) \neq \emptyset$ and
$\mathcal{T}_{\mathbb{Y}}(\{\tilde{x}\}) \cap (\mathbb{Y} \setminus \varphi(\mathcal{T}_{\mathbb{Y}}(\{\tilde{x}\}))) \neq \emptyset$. Since $\varphi(\mathcal{T}_{\mathbb{Y}}(\{\tilde{x}\})) \cap \mathcal{T}_{\mathbb{Y}}(\{\tilde{x}\}) \neq \emptyset$, this contradicts
the fact that the decomposition, $\mathcal{G}$, of $\mathbb{Y}$ is terminal (Theorem 5.3.23).

Q.E.D.

5.3.25 Theorem *Let $\sigma \times 1_{\mathcal{Q}} \colon \mathbb{H} \times \mathcal{Q} \twoheadrightarrow \mathbb{F} \times \mathcal{Q}$ be the universal covering map of*
$\mathbb{F} \times \mathcal{Q}$. *Let X be a nonaposyndetic homogeneous continuum essentially embedded*
in $W \times \mathcal{Q}$. Let $\tilde{X} = (\sigma \times 1_{\mathcal{Q}})^{-1}(X)$ and let $\mathbb{K}_1$ and $\mathbb{K}_2$ be components of $\tilde{X}$.
Let $\mathbb{Y}_j = Cl_{(\mathbb{H} \times \mathcal{Q}) \cup \mathcal{S}^1}(\mathbb{K}_j)$ $j \in \{1, 2\}$. If $\tilde{x}_1 \in \mathbb{K}_1$ and $\tilde{x}_2 \in \mathbb{K}_2$ are such that
$(\sigma \times 1_{\mathcal{Q}})(\tilde{x}_1) = (\sigma \times 1_{\mathcal{Q}})(\tilde{x}_2)$, *then $\mathcal{T}_{\mathbb{Y}_1}(\{\tilde{x}_1\})$ is homeomorphic to $\mathcal{T}_{\mathbb{Y}_2}(\{\tilde{x}_2\})$, and*
$(\sigma \times 1_{\mathcal{Q}})(\mathcal{T}_{\mathbb{Y}_1}(\{\tilde{x}_1\})) = (\sigma \times 1_{\mathcal{Q}})(\mathcal{T}_{\mathbb{Y}_2}(\{\tilde{x}_2\}))$.

Proof Let $\varphi \colon \mathbb{H} \times Q \twoheadrightarrow \mathbb{H} \times Q$ be a covering homeomorphism such that $\varphi(\tilde{x}_1) = \tilde{x}_2$ [16, Proposição 9, p. 132]. By Lemma 5.1.1,

$$\varphi(\mathcal{T}_{\mathbb{Y}_1}(\{\tilde{x}_1\})) = \mathcal{T}_{\mathbb{Y}_2}(\{\varphi(\tilde{x}_1)\}) = \mathcal{T}_{\mathbb{Y}_2}(\{\tilde{x}_2\}).$$

Hence, $\mathcal{T}_{\mathbb{Y}_1}(\{\tilde{x}_1\})$ is homeomorphic to $\mathcal{T}_{\mathbb{Y}_2}(\{\tilde{x}_2\})$, and

$$(\sigma \times 1_Q)(\mathcal{T}_{\mathbb{Y}_1}(\{\tilde{x}_1\})) = (\sigma \times 1_Q)(\mathcal{T}_{\mathbb{Y}_2}(\{\tilde{x}_2\})).$$

Q.E.D.

5.3.26 Lemma *Let $\sigma \times 1_Q \colon \mathbb{H} \times Q \twoheadrightarrow \mathbb{F} \times Q$ be the universal covering map of $\mathbb{F} \times Q$. Let X be a nonaposyndetic homogeneous continuum essentially embedded in $W \times Q$. Let $\tilde{X} = (\sigma \times 1_Q)^{-1}(X)$. If $\tilde{Z}$ is a subcontinuum of $\tilde{X}$ such that $(\sigma \times 1_Q)|_{\tilde{Z}}$ is one-to-one, then there exists an open subset $\tilde{U}$ of $\mathbb{H} \times Q$ such that $\tilde{Z} \subset \tilde{U}$ and $(\sigma \times 1_Q)|_{\tilde{U}}$ is one-to-one.*

Proof Suppose the lemma is not true. Then for each $n \in \mathbb{N}$, there exist $\tilde{x}_n, \tilde{y}_n \in \mathcal{V}^d_{\frac{1}{n}}(\tilde{Z})$ such that $\tilde{x}_n \neq \tilde{y}_n$ and $(\sigma \times 1_Q)(\tilde{x}_n) = (\sigma \times 1_Q)(\tilde{y}_n)$. Since $\mathbb{F} \times Q$ is compact and $\sigma \times 1_Q$ is a covering map, without loss of generality, we assume that the sequences $\{\tilde{x}_n\}_{n=1}^{\infty}$ and $\{\tilde{y}_n\}_{n=1}^{\infty}$ converge to $\tilde{x}_0$ and $\tilde{y}_0$, respectively. Note that $\tilde{x}_0, \tilde{y}_0 \in \tilde{Z}$.

Since $\sigma \times 1_Q$ is continuous, the sequences $\{(\sigma \times 1_Q)(\tilde{x}_n)\}_{n=1}^{\infty}$ and $\{(\sigma \times 1_Q)(\tilde{y}_n)\}_{n=1}^{\infty}$ converge to $(\sigma \times 1_Q)(\tilde{x}_0)$ and $(\sigma \times 1_Q)(\tilde{y}_0)$, respectively. Since for each $n \in \mathbb{N}$, $(\sigma \times 1_Q)(\tilde{x}_n) = (\sigma \times 1_Q)(\tilde{y}_n)$, we have that $(\sigma \times 1_Q)(\tilde{x}_0) = (\sigma \times 1_Q)(\tilde{y}_0)$. Hence, $\tilde{x}_0 = \tilde{y}_0$ ($(\sigma \times 1_Q)|_{\tilde{Z}}$ is one-to-one).

Let V be an evenly covered open subset of $\mathbb{F} \times Q$ such that $(\sigma \times 1_Q)(\tilde{x}_0) \in V$. Let $\tilde{V}$ be the open subset of $(\sigma \times 1_Q)^{-1}(V)$ such that $\tilde{x}_0 \in \tilde{V}$ and $(\sigma \times 1_Q)|_{\tilde{V}} \colon \tilde{V} \twoheadrightarrow V$ is a homeomorphism. Then $\tilde{V}$ is an open subset of $\mathbb{H} \times Q$ containing $\tilde{x}_0$. Since $\{\tilde{x}_n\}_{n=1}^{\infty}$ and $\{\tilde{y}_n\}_{n=1}^{\infty}$ converge to $\tilde{x}_0$, there exists $N \in \mathbb{N}$ such that $\tilde{x}_n, \tilde{y}_n \in \tilde{V}$ for every $n \geq N$. Hence, $(\sigma \times 1_Q)(\tilde{x}_n) \neq (\sigma \times 1_Q)(\tilde{y}_n)$ for each $n \geq N$ ($(\sigma \times 1_Q)|_{\tilde{V}} \colon \tilde{V} \twoheadrightarrow V$ is a homeomorphism), a contradiction to the choices of $\tilde{x}_n$ and $\tilde{y}_n$.

Therefore, there exists an open subset $\tilde{U}$ of $\mathbb{H} \times Q$ such that $\tilde{Z} \subset \tilde{U}$ and $(\sigma \times 1_Q)|_{\tilde{U}}$ is one-to-one.

Q.E.D.

5.3.27 Theorem *Let $\sigma \times 1_Q \colon \mathbb{H} \times Q \twoheadrightarrow \mathbb{F} \times Q$ be the universal covering map of $\mathbb{F} \times Q$. Let X be a nonaposyndetic homogeneous continuum essentially embedded in $W \times Q$. Let $\tilde{X} = (\sigma \times 1_Q)^{-1}(X)$, and let $\mathbb{K}$ be a component of $\tilde{X}$. Let $\mathbb{Y} = Cl_{(\mathbb{H} \times Q) \cup S^1}(\mathbb{K})$. If $\tilde{x} \in \mathbb{K}$, then $(\sigma \times 1_Q)(\mathcal{T}_{\mathbb{Y}}(\{\tilde{x}\}))$ is a maximal terminal proper, cell-like subcontinuum of X.*

Proof Let $\tilde{x} \in \mathbb{K}$. Recall that $\mathcal{T}_{\mathbb{Y}}(\{\tilde{x}\})$ is cell-like ((3) of Theorem 5.3.23). It follows, from Theorem 5.3.24, that $(\sigma \times 1_Q)(\mathcal{T}_{\mathbb{Y}}(\{\tilde{x}\}))$ is cell-like. Hence, $(\sigma \times 1_Q)(\mathcal{T}_{\mathbb{Y}}(\{\tilde{x}\}))$ is a proper subcontinuum of X (there exists a map $f \colon X \twoheadrightarrow W$ that is not homotopic to a constant map).

Since $(\sigma \times 1_Q)|_{\mathcal{T}_{\mathbb{Y}}(\{\tilde{x}\})}$ is one-to-one (Theorem 5.3.24), there exists an open set $\tilde{U}$ such that $\mathcal{T}_{\mathbb{Y}}(\{\tilde{x}\}) \subset \tilde{U}$ and $(\sigma \times 1_Q)|_{\tilde{U}}$ is one-to-one, by Lemma 5.3.26. Thus, $(\sigma \times 1_Q)|_{\tilde{U}} \colon \tilde{U} \twoheadrightarrow (\sigma \times 1_Q)(\tilde{U})$ is a homeomorphism.

If M is a subcontinuum of X such that $M \cap (\sigma \times 1_Q)(\mathcal{T}_{\mathbb{Y}}(\{\tilde{x}\})) \neq \emptyset$ and $M \cap (X \setminus (\sigma \times 1_Q)(\mathcal{T}_{\mathbb{Y}}(\{\tilde{x}\}))) \neq \emptyset$, then we may assume, by taking a subcontinuum if necessary, that $M \subset (\sigma \times 1_Q)(\tilde{U})$. Hence, M has a lift $\tilde{M}$ such that $\tilde{M} \cap \mathcal{T}_{\mathbb{Y}}(\{\tilde{x}\}) \neq \emptyset$ and $\tilde{M} \cap (\mathbb{Y} \setminus \mathcal{T}_{\mathbb{Y}}(\{\tilde{x}\})) \neq \emptyset$. Since $\mathcal{T}_{\mathbb{Y}}(\{\tilde{x}\})$ is terminal, $\mathcal{T}_{\mathbb{Y}}(\{\tilde{x}\}) \subset \tilde{M}$. Thus, $(\sigma \times 1_Q)(\mathcal{T}_{\mathbb{Y}}(\{\tilde{x}\})) \subset M$. Hence, $(\sigma \times 1_Q)(\mathcal{T}_{\mathbb{Y}}(\{\tilde{x}\}))$ is a terminal subcontinuum of X.

Now, we show that if T is a proper, terminal, cell-like subcontinuum of X such that $(\sigma \times 1_Q)(\mathcal{T}_{\mathbb{Y}}(\{\tilde{x}\})) \subset T$, then $T = (\sigma \times 1_Q)(\mathcal{T}_{\mathbb{Y}}(\{\tilde{x}\}))$. Since T is cell-like, the inclusion map $i \colon T \to \mathbb{F} \times Q$ lifts to a one-to-one map $\tilde{i} \colon T \to \mathbb{H} \times Q$ such that $\tilde{i}((\sigma \times 1_Q)(\tilde{x})) = \tilde{x}$. Hence, $\tilde{i}(T) \cap \mathcal{T}_{\mathbb{Y}}(\{\tilde{x}\}) \neq \emptyset$. Let $\tilde{T} = \tilde{i}(T)$.

Since $\tilde{i}$ is one-to-one, $(\sigma \times 1_Q)|_{\tilde{T}}$ is one-to-one. Thus, there exists an open set $\tilde{V}$ such that $\tilde{T} \subset \tilde{V}$ and $(\sigma \times 1_Q)|_{\tilde{V}}$ is one-to-one, by Lemma 5.3.26. Since $\mathcal{T}_{\mathbb{Y}}(\{\tilde{x}\})$ is a maximal terminal subcontinuum of $\mathbb{Y}$, either $\tilde{T}$ is not terminal or $\tilde{T} = \mathcal{T}_{\mathbb{Y}}(\{\tilde{x}\})$. Let $\tilde{N}$ be a proper subcontinuum of $\mathbb{Y}$ such that $\tilde{N} \cap \tilde{T} \neq \emptyset$ and $\tilde{N} \cap \mathbb{Y} \setminus \tilde{T} \neq \emptyset$. Again, we assume that $\tilde{N} \subset \tilde{V}$. Hence, $(\sigma \times 1_Q)(\tilde{N}) \cap T \neq \emptyset$ and $(\sigma \times 1_Q)(\tilde{N}) \cap (X \setminus T) \neq \emptyset$. Since T is terminal, $T \subset (\sigma \times 1_Q)(\tilde{N})$. Thus, $\tilde{T} \subset \tilde{N}$, and so $\tilde{T}$ is terminal. Hence, $\tilde{T} = \mathcal{T}_{\mathbb{Y}}(\{\tilde{x}\})$, and $T = (\sigma \times 1_Q)(\mathcal{T}_{\mathbb{Y}}(\{\tilde{x}\}))$.

Q.E.D.

We are ready to state and prove Rogers's Terminal Decomposition Theorem.

5.3.28 Theorem *Let* $\sigma \times 1_Q \colon \mathbb{H} \times Q \twoheadrightarrow \mathbb{F} \times Q$ *be the universal covering map of* $\mathbb{F} \times Q$. *Let* X *be a homogeneous continuum that admits a map into the figure eight* W *that is not homotopic to a constant map. Hence, we may consider* X *essentially embedded in* $\mathbb{F} \times Q$. *Let* $\tilde{X} = (\sigma \times 1_Q)^{-1}(X)$. *If* $\mathcal{G} = \{(\sigma \times 1_Q)(\mathcal{T}_{\mathbb{Y}}(\{\tilde{x}\})) \mid \tilde{x} \in \mathbb{K}$, *where* $\mathbb{K}$ *is a component of* $\tilde{X}$, *and* $\mathbb{Y} = Cl_{(\mathbb{H} \times Q) \cup \mathcal{S}^1}(\mathbb{K})\}$, *then* $\mathcal{G}$ *is a continuous decomposition such that the following hold:*

(1) $\mathcal{G}$ *is a monotone and terminal decomposition of* X.
(2) *The elements of* $\mathcal{G}$ *are mutually homeomorphic, indecomposable, cell-like terminal, homogeneous continua.*
(3) *The quotient space,* $X/\mathcal{G}$, *is a homogeneous continuum.*
(4) $X/\mathcal{G}$ *does not contain any proper, nondegenerate terminal subcontinuum.*
(5) *If* X *is decomposable, then* $X/\mathcal{G}$ *is an aposyndetic continuum; in fact, the decomposition* $\mathcal{G}$ *is Jones's decomposition.*
(6) *If the elements of* $\mathcal{G}$ *are nondegenerate, then they have the same dimension as* X *and* $X/\mathcal{G}$ *is one-dimensional.*

Proof Note that, by Theorem 5.3.27, $\mathcal{G}$ is a collection of maximal proper terminal cell-like subcontinua of X.

We show that $\mathcal{G}$ is a decomposition of X. Let $\tilde{x}_1, \tilde{x}_2 \in \tilde{X}$ be such that $(\sigma \times 1_Q)(\mathcal{T}_{\mathbb{Y}_1}(\{\tilde{x}_1\})) \cap (\sigma \times 1_Q)(\mathcal{T}_{\mathbb{Y}_2}(\{\tilde{x}_2\})) \neq \emptyset$ ($\mathbb{K}_j$ is the component of $\tilde{X}$ such that $\tilde{x}_j \in \mathbb{K}_j$ and $\mathbb{Y}_j = Cl_{(\mathbb{H} \times Q) \cup \mathcal{S}^1}(\mathbb{K}_j)$, $j \in \{1, 2\}$). Let $z \in (\sigma \times 1_Q)(\mathcal{T}_{\mathbb{Y}_1}(\{\tilde{x}_1\})) \cap$

$(\sigma \times 1_{Q})(\mathcal{T}_{\mathbb{Y}_{2}}(\{\tilde{x}_{2}\}))$. Then there exists $\tilde{z}_{j} \in \mathcal{T}_{\mathbb{Y}_{j}}(\{\tilde{x}_{j}\})$ such that $(\sigma \times 1_{Q})(\tilde{z}_{j}) = z$, $j \in \{1, 2\}$. Hence, by Theorem 5.3.25, $\mathcal{T}_{\mathbb{Y}_{1}}(\{\tilde{x}_{1}\})$ is homeomorphic to $\mathcal{T}_{\mathbb{Y}_{2}}(\{\tilde{x}_{2}\})$ and $(\sigma \times 1_{Q})(\mathcal{T}_{\mathbb{Y}_{1}}(\{\tilde{x}_{1}\})) = (\sigma \times 1_{Q})(\mathcal{T}_{\mathbb{Y}_{2}}(\{\tilde{x}_{2}\}))$. Since $\sigma \times 1_{Q}$ is a covering map,

$$X = \bigcup \{(\sigma \times 1_{Q})(\mathcal{T}_{\mathbb{Y}}(\{\tilde{x}\})) \mid \tilde{x} \in \mathbb{K}, \text{ where } \mathbb{K} \text{ is a component of}$$

$$\widetilde{X}, \text{ and } \mathbb{Y} = Cl_{(\mathbb{H} \times Q) \cup \mathcal{S}^{1}}(\mathbb{K})\}.$$

Therefore, $\mathcal{G}$ is a decomposition of X.

Now, we prove that the homeomorphism group of X, $\mathcal{H}(X)$, respects $\mathcal{G}$, and then, apply Theorem 5.1.4 to conclude that $\mathcal{G}$ is continuous.

Let $h \in \mathcal{H}(X)$, and let $\tilde{x}_{1}, \tilde{x}_{2} \in \widetilde{X}$ be such that

$$h((\sigma \times 1_{Q})(\mathcal{T}_{\mathbb{Y}_{1}}(\{\tilde{x}_{1}\}))) \cap (\sigma \times 1_{Q})(\mathcal{T}_{\mathbb{Y}_{2}}(\{\tilde{x}_{2}\})) \neq \emptyset.$$

Since $(\sigma \times 1_{Q})(\mathcal{T}_{\mathbb{Y}_{1}}(\{\tilde{x}_{1}\}))$ is terminal, $h((\sigma \times 1_{Q})(\mathcal{T}_{\mathbb{Y}_{1}}(\{\tilde{x}_{1}\})))$ is terminal (Corollary 5.1.12). Thus, since $(\sigma \times 1_{Q})(\mathcal{T}_{\mathbb{Y}_{2}}(\{\tilde{x}_{2}\}))$ is a maximal terminal subcontinuum,

$$h((\sigma \times 1_{Q})(\mathcal{T}_{\mathbb{Y}_{1}}(\{\tilde{x}_{1}\}))) \subset (\sigma \times 1_{Q})(\mathcal{T}_{\mathbb{Y}_{2}}(\{\tilde{x}_{2}\})).$$

This implies that $(\sigma \times 1_{Q})(\mathcal{T}_{\mathbb{Y}_{1}}(\{\tilde{x}_{1}\})) \subset h^{-1}((\sigma \times 1_{Q})(\mathcal{T}_{\mathbb{Y}_{2}}(\{\tilde{x}_{2}\})))$. Hence, $(\sigma \times 1_{Q})(\mathcal{T}_{\mathbb{Y}_{1}}(\{\tilde{x}_{1}\})) = h^{-1}((\sigma \times 1_{Q})(\mathcal{T}_{\mathbb{Y}_{2}}(\{\tilde{x}_{2}\})))$. Consequently, $h((\sigma \times 1_{Q})(\mathcal{T}_{\mathbb{Y}_{1}}(\{\tilde{x}_{1}\}))) = (\sigma \times 1_{Q})(\mathcal{T}_{\mathbb{Y}_{2}}(\{\tilde{x}_{2}\}))$. Therefore, $\mathcal{H}(X)$ respects $\mathcal{G}$.

Since $\mathcal{H}(X)$ respects $\mathcal{G}$, by Theorem 5.1.4, the elements of $\mathcal{G}$ are mutually homeomorphic homogeneous continua, and the quotient space $X/\mathcal{G}$ is homogeneous. The fact that the nondegenerate elements of $\mathcal{G}$ have the same dimension as X follows from [31, Theorem 8]. The proof of the indecomposability of the elements of $\mathcal{G}$ is similar to the one given in Theorem 3.4.12. The fact that $X/\mathcal{G}$ is one-dimensional may be found in [34, Theorem 6].

Next, we show that $X/\mathcal{G}$ does not contain nondegenerate proper terminal subcontinua. To this end, let Γ be a nondegenerate terminal proper subcontinuum of $X/\mathcal{G}$. Let $q \colon X \twoheadrightarrow X/\mathcal{G}$ be the quotient map. Then q is a monotone map. Thus, $q^{-1}(\Gamma)$ is a subcontinuum of X (Lemma 2.1.12). Note that

$$q^{-1}(\Gamma) = \bigcup \{(\sigma \times 1_{Q})(\mathcal{T}_{\mathbb{Y}}(\{\tilde{x}\})), \mid \tilde{x} \in \mathbb{K}, \text{ where } \mathbb{K} \text{ is a component}$$

$$\text{of } \widetilde{X}, \mathbb{Y} = Cl_{(\mathbb{H} \times Q) \cup \mathcal{S}^{1}}(\mathbb{K}) \text{ and } q((\sigma \times 1_{Q})(\mathcal{T}_{\mathbb{Y}}(\{\tilde{x}\}))) \in \Gamma\}.$$

Since Γ is a nondegenerate proper subcontinuum of $X/\mathcal{G}$, $q^{-1}(\Gamma)$ is a proper subcontinuum of X and if $q((\sigma \times 1_{Q})(\mathcal{T}_{\mathbb{Y}}(\{\tilde{x}\}))) \in \Gamma$, $(\sigma \times 1_{Q})(\mathcal{T}_{\mathbb{Y}}(\{\tilde{x}\})) \subsetneq q^{-1}(\Gamma)$. Hence, $q^{-1}(\Gamma)$ is not a terminal subcontinuum of X (each element $(\sigma \times 1_{Q})(\mathcal{T}_{\mathbb{Y}}(\{\tilde{x}\}))$ of $\mathcal{G}$ is a maximal terminal proper subcontinuum of X). Thus, there exists a subcontinuum Z of X such that $Z \cap q^{-1}(\Gamma) \neq \emptyset$, $Z \setminus q^{-1}(\Gamma) \neq \emptyset$ and $q^{-1}(\Gamma) \setminus Z \neq \emptyset$. Since $q(Z \cap q^{-1}(\Gamma)) = q(Z) \cap \Gamma$, $q(Z)$ is a subcontinuum of $X/\mathcal{G}$ such that $q(Z) \cap \Gamma \neq \emptyset$. Since Γ is a terminal subcontinuum of $X/\mathcal{G}$, either $q(Z) \subset \Gamma$ or $\Gamma \subset q(Z)$.

If $q(Z) \subset \Gamma$, then $q^{-1}(q(Z)) \subset q^{-1}(\Gamma)$. Hence, $Z \subset q^{-1}(\Gamma)$, a contradiction. Suppose, then, that $\Gamma \subset q(Z)$. This implies that $q^{-1}(\Gamma) \subset q^{-1}(q(Z))$. Thus, for each $x \in q^{-1}(\Gamma)$, there exists $z \in Z$ such that $q(z) = q(x)$. Hence, $q^{-1}(q(x)) \cap Z \neq \emptyset$. Since $q^{-1}(q(x))$ is a terminal subcontinuum of X, either $Z \subset q^{-1}(q(x))$ or $q^{-1}(q(x)) \subset Z$. If $Z \subset q^{-1}(q(x))$, then $Z \subset q^{-1}(\Gamma)$, a contradiction. Thus, $q^{-1}(q(x)) \subset Z$. Since x is an arbitrary point of $q^{-1}(\Gamma)$, $q^{-1}(\Gamma) \subset Z$, a contradiction. Therefore, $X/\mathcal{G}$ does not contain nondegenerate terminal proper subcontinua.

Finally, suppose X is decomposable. Let $\mathcal{G}' = \{\mathcal{T}_X(\{x\}) \mid x \in X\}$ be Jones's decomposition (Theorem 5.1.18). We show that $\mathcal{G} = \mathcal{G}'$. Thus, by Theorem 5.1.18, $X/\mathcal{G}$ is aposyndetic.

Let $(\sigma \times 1_Q)(\mathcal{T}_{\tilde{X}}(\{\tilde{x}\})) \in \mathcal{G}$. Let $x \in (\sigma \times 1_Q)(\mathcal{T}_{\tilde{X}}(\{\tilde{x}\}))$. Then $(\sigma \times 1_Q)(\mathcal{T}_{\tilde{X}}(\{\tilde{x}\})) \cap \mathcal{T}_X(\{x\}) \neq \emptyset$. Since both of these continua are maximal terminal subcontinua of X (Corollary 5.1.19), $(\sigma \times 1_Q)(\mathcal{T}_{\tilde{X}}(\{\tilde{x}\})) = \mathcal{T}_X(\{x\})$. Thus, $(\sigma \times 1_Q)(\mathcal{T}_{\tilde{X}}(\{\tilde{x}\})) \in \mathcal{G}'$. Hence, $\mathcal{G} \subset \mathcal{G}'$.

Let $\mathcal{T}_X(\{x\}) \in \mathcal{G}'$. Since $\sigma \times 1_Q$ is a covering map, there exists $\tilde{x} \in \tilde{X}$ such that $(\sigma \times 1_Q)(\tilde{x}) = x$. Let $\mathbb{K}$ be the component of $\tilde{X}$ such that $\tilde{x} \in \mathbb{K}$. Then $(\sigma \times 1_Q)(\mathcal{T}_{\tilde{X}}(\{\tilde{x}\})) \cap \mathcal{T}_X(\{x\}) \neq \emptyset$. Since both of these continua are maximal terminal subcontinua of X, $(\sigma \times 1_Q)(\mathcal{T}_{\tilde{X}}(\{\tilde{x}\})) = \mathcal{T}_X(\{x\})$. Thus, $\mathcal{T}_X(\{x\}) \in \mathcal{G}$. Hence, $\mathcal{G}' \subset \mathcal{G}$.

Therefore, $\mathcal{G} = \mathcal{G}'$.

Q.E.D.

5.4 Case and Minc–Rogers Continua

Aposyndetic homogeneous continua of dimension greater than one are very easy to construct, since the product of homogeneous continua is homogeneous and aposyndetic (Corollary 3.7.9). It is difficult to find one-dimensional aposyndetic homogeneous continua. In fact, R. D. Anderson showed that the simple closed curve and the Menger universal curve are the only one-dimensional locally connected homogeneous continua [5, Theorem XIII]. Hence, if the only one-dimensional aposyndetic homogeneous continua were locally connected, then Theorem 5.1.18 would imply that it would be enough to study one-dimensional indecomposable homogeneous continua.

We present a collection of one-dimensional aposyndetic homogeneous continua which are not locally connected. We use inverse limits to construct such continua. First, we present the construction of J. T. Rogers, Jr. [28] of a continuum originally constructed by J. H. Case [7]. Afterwards, we give a sketch of the generalization of Rogers's construction made by P. Minc and J. T. Rogers, Jr. [22]. Additional material for this section may be found in [4, 21, 24, 36].

The *Sierpiński universal plane curve*, $\mathbb{S}$, can be constructed by taking the unit square with boundary B, deleting the interior of the middle-ninth of that square (leaving the boundary of such a square, *C*), then deleting the interiors of the middle-ninths of each of the eight squares remaining. Let us continue in this manner step by step. The points which have not been removed constitute the required curve.

The next theorem gives a characterization of $\mathbb{S}$; a proof of it may be found in [36, Theorem 4]:

5.4.1 Theorem *In order that a plane one-dimensional locally connected continuum Z be the Sierpiński universal plane curve $\mathbb{S}$ it is necessary and sufficient that for each open connected subset U of Z and each point $z \in U$, $U \setminus \{z\}$ is connected.*

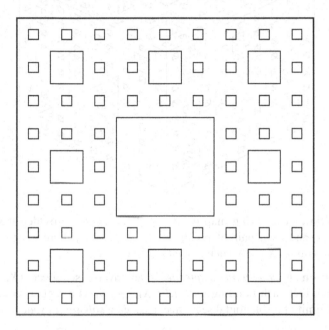

The *Menger universal curve*, $\mathbb{M}$, can be described as the set of all points of the unit cube, $[0, 1]^3$, that project, in each of the *x*, *y* and *z* directions, onto Sierpiński universal plane curves on the faces of the cube.

5.4.2 Lemma *Each point of the Menger universal curve has arbitrarily small neighborhoods with connected boundary.*

Proof Note that the point $(1, 0, 0) \in \mathbb{M}$ has arbitrarily small neighborhoods homeomorphic to $\mathbb{M}$. Since $\mathbb{M}$ is homogeneous [4, Theorem III], this is true for all points of $\mathbb{M}$.

Q.E.D.

The next theorem presents a characterization of $\mathbb{M}$; a proof of it may be found in [5, Theorem XII]:

5.4.3 Theorem *In order that a one-dimensional locally connected continuum Z be the Menger universal curve $\mathbb{M}$ it is necessary and sufficient that for each open* *connected subset U of Z, U is not embeddable in the plane, and for each point $z \in U$, $U \setminus \{z\}$ is connected.*

5.4.4 Definition A covering map $\sigma : \widetilde{X} \twoheadrightarrow X$ is *regular* provided that for each point $x \in X$ and each pair of points $\tilde{x}_1, \tilde{x}_2 \in \sigma^{-1}(x)$, there exists a covering homeomorphism $\varphi : \widetilde{X} \twoheadrightarrow \widetilde{X}$ such that $\varphi(\tilde{x}_1) = \tilde{x}_2$.

5.4.5 Definition A *solenoidal sequence* is an inverse sequence $\{X_n, f_n^{n+1}\}$ of continua such that each bonding map $f_n^m : X_m \twoheadrightarrow X_n$ is a regular covering map. The inverse limit of a solenoidal sequence is called a *solenoidal space*.

5.4.6 Theorem *If the points $p = (p_n)_{n=1}^{\infty}$ and $q = (q_n)_{n=1}^{\infty}$ of the solenoidal space $X_\infty = \varprojlim\{X_n, f_n^{n+1}\}$ have the same first coordinate, then there exists a homeomorphism of X_∞ onto itself mapping p to q.*

Proof For each $n \in \mathbb{N}$, we construct a homeomorphism $h_n : X_n \twoheadrightarrow X_n$ such that $h_n(p_n) = q_n$ and $f_n^{n+1} \circ h_{n+1} = h_n \circ f_n^{n+1}$. Then Theorems 2.1.48, 2.1.49 and 2.1.50 imply that $h = \varprojlim\{h_n\}$ is a homeomorphism of X_∞ onto itself such that $h(p) = q$.

Since $p_1 = q_1$, we take $h_1 = 1_{X_1}$. Note that $p_2, q_2 \in (f_1^2)^{-1}(p_1)$. Since f_1^2 is a regular covering map, there exists a covering homeomorphism $h_2 : X_2 \twoheadrightarrow X_2$ such that $h_2(p_2) = q_2$ and $f_1^2 \circ h_2 = f_1^2 = h_1 \circ f_1^2$.

Note that $p_3, q_3 \in (f_1^3)^{-1}(p_1)$. Since f_1^3 is a regular covering map, there exists a covering homeomorphism $h_3 : X_3 \twoheadrightarrow X_3$ such that $h_3(p_3) = q_3$ and $f_1^3 \circ h_3 = f_1^3 = h_1 \circ f_1^3$. Observe that both $h_2 \circ f_2^3$ and $f_2^3 \circ h_3$ are liftings of f_1^3 such that $h_2 \circ f_2^3(p_3) = h_2(p_2) = q_2$ and $f_2^3 \circ h_3(p_3) = f_2^3(q_3) = q_2$. Hence, $h_2 \circ f_2^3 = f_2^3 \circ h_3$, by Proposition 5.2.5.

Suppose we have defined a covering homeomorphism $h_k \colon X_k \twoheadrightarrow X_k$ for $f_1^k \colon X_k \twoheadrightarrow X_1$ such that $h_k(p_k) = q_k$, $f_1^k \circ h_k = f_1^k$ and $h_{k-1} \circ f_{k-1}^k = f_{k-1}^k \circ h_k$ for each $k \in \{1, \ldots, n-1\}$.

Since $p_n, q_n \in (f_1^n)^{-1}(p_1)$ and f_1^n is a regular covering map, there exists a covering homeomorphism $h_n \colon X_n \twoheadrightarrow X_n$ such that $h_n(p_n) = q_n$ and $f_1^n \circ h_n = f_1^n$. Again, both $h_{n-1} \circ f_{n-1}^n$ and $f_{n-1}^n \circ h_n$ are liftings of f_1^n such that $h_{n-1} \circ f_{n-1}^n(p_n) = h_{n-1}(p_{n-1}) = q_{n-1}$ and $f_{n-1}^n \circ h_n(p_n) = f_{n-1}^n(q_n) = q_{n-1}$. Hence, $h_{n-1} \circ f_{n-1}^n = f_{n-1}^n \circ h_n$ (Proposition 5.2.5).

Q.E.D.

5.4.7 Definition Let V be an open subset of a metric space Y. Call Y *stably homogeneous over V* if for each pair of points $p, q \in V$, there exists a homeomorphism $h \colon Y \twoheadrightarrow Y$ such that $h(p) = q$ and $h|_{(Y \setminus V)} = 1_{(Y \setminus V)}$.

5.4.8 Definition A metric space Y is *strongly locally homogeneous* if for each point $y \in Y$ and each open subset U of Y such that $y \in U$, there exists an open subset V of Y such that $y \in V \subset U$ and such that Y is stably homogeneous over V.

5.4.9 Definition Let $\{X_n, f_n^{n+1}\}$ be an inverse sequence of continua. If $X_\infty = \varprojlim\{X_n, f_n^{n+1}\}$, we say that X_∞ is *locally a product of an open subset of X_1 and the Cantor set, C*, provided that for each point $x \in X_\infty$, there exists an open subset U_1 of X_1 such that $x \in f_1^{-1}(U_1)$ and $f_1^{-1}(U_1)$ is homeomorphic to $U_1 \times C$.

5.4.10 Theorem *Let $\{X_n, f_n^{n+1}\}$ be a solenoidal sequence such that X_1 is strongly locally homogeneous. If $X_\infty = \varprojlim\{X_n, f_n^{n+1}\}$ and X_∞ is locally a product of an open subset of X_1 and the Cantor set, C, then X_∞ is homogeneous.*

Proof Let $p = (p_n)_{n=1}^\infty \in X_\infty$. Let $\mathbb{H}$ be the set of points $q_1 \in X_1$ such that there exists a homeomorphism of X_∞ onto itself such that takes p to a point whose first coordinate is q_1. We show that $\mathbb{H}$ is both open and closed.

Let $q_1 \in \mathbb{H}$. Then there exists a homeomorphism $g \colon X_\infty \twoheadrightarrow X_\infty$ such that $f_1(g(p)) = q_1$ (recall that $f_1 \colon X_\infty \to X_1$ is the projection map).

Since X_∞ is locally a product of an open subset of X_1 and the Cantor set, there exists an open set U_1 of X_1 containing q_1 such that $f_1^{-1}(U_1)$ is homeomorphic to $U_1 \times C$.

Since X_1 is strongly locally homogeneous, there exists an open set V of X_1 such that $q_1 \in V \subset U_1$ and X_1 is stably homogeneous over V. If $r_1 \in V$, then there exists a homeomorphism $h_1 \colon X_1 \twoheadrightarrow X_1$ such that $h_1(q_1) = r_1$ and $h_1|_{(X_1 \setminus V)} = 1_{(X_1 \setminus V)}$.

Define the map $h \colon X_\infty \twoheadrightarrow X_\infty$ as follows:

$$h|_{(X_\infty \setminus f_1^{-1}(V))} = 1_{(X_\infty \setminus f_1^{-1}(V))}$$

and

$$h|_{f_1^{-1}(V)} = (h_1|_V) \times 1_C.$$

Then h is a homeomorphism, and $f_1(h(g(p))) = r_1$. Hence, $r_1 \in \mathbb{H}$. Thus, $\mathbb{H}$ is open.

A similar argument shows that $\mathbb{H}$ is closed. Hence, $\mathbb{H} = X_1$. Then, by Theorem 5.4.6, X_∞ is homogeneous.

Q.E.D.

5.4.11 Lemma *View the Sierpiński universal plane curve $\mathbb{S}$ as a subset of an annulus $\mathbb{D}$ with boundary $B \cup C$. If $g \colon \mathbb{D} \twoheadrightarrow \mathbb{D}$ is a double-covering map; i.e., g is a covering map such that its fibres have exactly two points, such that $g(B) = B$ and $g(C) = C$, then $g^{-1}(\mathbb{S})$ is homeomorphic to $\mathbb{S}$.*

Proof First, we show that $g^{-1}(\mathbb{S})$ is connected. Let $p_2 \in g^{-1}(\mathbb{S})$, and let $p_1 \in \mathbb{S}$ be such that $g(p_2) = p_1$. Let A be an arc in $\mathbb{S}$ from p_1 to a point of B. The arc A lifts to an arc $\tilde{A}$ in $g^{-1}(\mathbb{S})$ from p_2 to a point of B, by a proof similar to the one given in Theorem 1.3.31. Hence, $g^{-1}(\mathbb{S})$ is arcwise connected.

Since g is a covering map, g preserves local properties. Hence, by Theorem 5.4.1, $g^{-1}(\mathbb{S})$ is homeomorphic to $\mathbb{S}$.

Q.E.D.

5.4.12 Lemma *View the Menger universal curve $\mathbb{M}$ as a subset of $\mathbb{D} \times [0, 1]$, where $\mathbb{D}$ is an annulus with boundary $B \cup C$. If $f \colon \mathbb{D} \times [0, 1] \twoheadrightarrow \mathbb{D} \times [0, 1]$ is given by $f = g \times 1_{[0,1]}$, where $g \colon \mathbb{D} \twoheadrightarrow \mathbb{D}$ is a double-covering map such that $g(B) = B$ and $g(C) = C$, then $f^{-1}(\mathbb{M})$ is homeomorphic to $\mathbb{M}$.*

Proof A proof similar to that of Lemma 5.4.11 shows that $f^{-1}(\mathbb{M})$ is arcwise connected.

Since f is a covering map, f preserves local properties. Hence, by Theorem 5.4.3, $f^{-1}(\mathbb{M})$ is homeomorphic to $\mathbb{M}$.

Q.E.D.

The following theorem is a consequence of [21, Theorem 5.6]:

5.4.13 Theorem *If $\{X_n, f_n^{n+1}\}$ is a solenoidal sequence of manifolds, then $X_\infty = \varprojlim\{X_n, f_n^{n+1}\}$ is locally a product of an open subset of X_1 and the Cantor set C.*

5.4.14 Definition A continuum X is said to be *colocally connected* at $x \in X$ provided that for each open subset U of X such that $x \in U$, there exists an open subset V of X such that $x \in V \subset U$ and $X \setminus V$ is connected. The continuum X is *colocally connected* if it is colocally connected at each of its points.

5.4.15 Remark Note that each colocally connected continuum is aposyndetic.

The continuum, $\mathbb{C}$, constructed in the following theorem is known as the *Case continuum.*

5.4.16 Theorem *Let $f \colon \mathbb{D} \times [0, 1] \twoheadrightarrow \mathbb{D} \times [0, 1]$ be the covering map of Lemma 5.4.12. If $X_1 = \mathbb{M}$, $X_2 = f^{-1}(\mathbb{M}), \ldots, X_n = (f^{n-1})^{-1}(\mathbb{M})$, and if for each $n \in \mathbb{N}$, $f_n^{n+1} = f|_{X_{n+1}}$, then $\{X_n, f_n^{n+1}\}$ is a solenoidal sequence of Menger* *curves and* $\mathbb{C} = \varprojlim\{X_n, f_n^{n+1}\}$ *is an aposyndetic, homogeneous, one-dimensional solenoidal space that is not locally connected.*

Proof Since the solenoidal sequence $\{X_n, f_n^{n+1}\}$ is obtained as the restriction of a solenoidal sequence of manifolds, $\mathbb{C}$ is locally homeomorphic to the product of an open subset of $X_1 = \mathbb{M}$ and the Cantor set (Theorem 5.4.13). Hence, $\mathbb{C}$ is not locally connected. Since $\mathbb{M}$ is one-dimensional, $\mathbb{C}$ is one-dimensional [24, 15.6].

By [5, Theorem XVI], $\mathbb{M}$ is strongly locally homogeneous. Hence, $\mathbb{C}$ is homogeneous, by Theorem 5.4.10.

To see $\mathbb{C}$ is aposyndetic, we show that $\mathbb{C}$ is colocally connected.

Since $\mathbb{C}$ is locally homeomorphic to the product of an open subset of $X_1 = \mathbb{M}$ and the Cantor set, by Proposition 2.1.9 and Lemma 5.4.2, each point of $\mathbb{C}$ has arbitrarily small neighborhoods of the form $U = f_n^{-1}(U_n)$, where U_n is an open subset of X_n with connected boundary B_n, and $Cl_{\mathbb{C}}(U)$ is homeomorphic to $Cl_{X_n}(U_n) \times C$. Let $U = f_n^{-1}(U_n)$ be one of such neighborhoods, and suppose $\mathbb{C} \setminus U$ is not connected. Then there exist two disjoint closed subsets C and D of $\mathbb{C}$ such that $\mathbb{C} \setminus U = C \cup D$. Let B_n be the boundary of U_n. Note that no component of $f_n^{-1}(B_n)$ intersects both C and D. Define C' to be the union of all the components K of U such that $Cl_{\mathbb{C}}(K) \cap C \neq \emptyset$. Define D' in a similar way. Note that $C \cup C'$ and $D \cup D'$ are disjoint closed subsets of $\mathbb{C}$ whose union is $\mathbb{C}$. Since $\mathbb{C}$ is connected, either $C \cup C'$ or $D \cup D'$ is empty. Hence, $\mathbb{C} \setminus U$ is connected.

Q.E.D.

To finish this section, we sketch the construction of the Minc–Rogers continua.

Fix $n \in \mathbb{N}$ such that $n \geq 3$. Let $\mathbb{T}^n = \prod_{k=1}^n \mathcal{S}^1$ be the n-dimensional torus. Let $\mathbb{Z}_\ell = \prod_{k=1}^{\ell-1}\{e\} \times \mathcal{S}^1 \times \prod_{k=\ell+1}^n\{e\}$, for $\ell \in \{1, \ldots, n\}$, where $e = (1, 0)$. One can assume that the Menger universal curve $\mathbb{M}$ is embedded in $\mathbb{T}^n$ in such a way that $\mathbb{Z}_\ell \subset \mathbb{M}$ for every $\ell \in \{1, \ldots, n\}$.

For each $\boldsymbol{\alpha} \in \mathbb{Z}^n$, $\boldsymbol{\alpha} = (\alpha_1, \ldots, \alpha_n)$, define $f_{\boldsymbol{\alpha}} \colon \mathbb{T}^n \to \mathbb{T}^n$ by

$$f_{\boldsymbol{\alpha}}((z_1, \ldots, z_n)) = (z_1^{\alpha_1}, \ldots, z_n^{\alpha_n}).$$

The proof of the next lemma is similar to the proof given for Lemma 5.4.12.

5.4.17 Lemma *With the above notation,* $f_{\boldsymbol{\alpha}}^{-1}(\mathbb{M})$ *is homeomorphic to* $\mathbb{M}$ *containing* $\mathbb{Z}_\ell$ *for every* $\ell \in \{1, \ldots, n\}$.

The proof of the following theorem is similar to the one given in Theorem 5.4.16. The spaces obtained from Theorem 5.4.18 are known as *Minc–Rogers continua.*

5.4.18 Theorem *With the above notation. Let* $\Lambda = \{\boldsymbol{\alpha}^k\}_{k=1}^\infty$ *be a sequence of elements of* $\mathbb{Z}^n$. *Assume that* $\boldsymbol{\alpha}^k = (\alpha_1^k, \ldots, \alpha_n^k)$. *Let* $\mathbb{M}_1^\Lambda = \mathbb{M}$, *and* $\mathbb{M}_k^\Lambda = f_{\boldsymbol{\alpha}^k}^{-1}(\mathbb{M}_{k-1}^\Lambda)$. *If* $\mathbb{M}^\Lambda = \varprojlim\{\mathbb{M}_k^\Lambda, f_{\boldsymbol{\alpha}^k}|_{\mathbb{M}_{k+1}^\Lambda}\}$, *then* $\mathbb{M}^\Lambda$ *is a one-dimensional colocally connected homogeneous continuum.*

5.5   Covering Spaces of Some Homogeneous Continua

We study covering spaces of certain homogeneous continua. To this end, we use
[2, 10, 11, 18, 23, 30, 33].

Let us note first that, by Corollary 1.3.30, we have that $\mathbb{R}$ is a covering space of
$\mathcal{S}^1$ since the exponential map is a covering map. It is easy to see that, for a given
$n \in \mathbb{N}$, the map $p \colon \mathcal{S}^1 \to \mathcal{S}^1$ given by $p(z) = z^n$, where $\mathcal{S}^1$ is considered as a
subset of the set of complex numbers $\mathbb{C}$, is a covering map. Hence, $\mathcal{S}^1$ is a covering
space of itself in a nontrivial way.

Let X be a continuum for which we can define a map $f_{\mathbb{H}} \colon X \to W$, where W is
the figure eight $C_1 \cup C_2$ that is not homotopic to a constant map. Let $g \colon X \to \mathcal{Q}$ be
an embedding of X into the Hilbert cube (Theorem 1.1.16). Then the map

$$(f_{\mathbb{H}}, g) \colon X \to W \times \mathcal{Q}$$

given by

$$(f_{\mathbb{H}}, g)(x) = (f_{\mathbb{H}}(x), g(x))$$

is an embedding that is not homotopic to a constant map. Let $\sigma \times 1_{\mathcal{Q}} \colon \widetilde{W} \times \mathcal{Q} \to$
$W \times \mathcal{Q}$ be the universal covering map of $W \times \mathcal{Q}$, and define $\widetilde{X}_{\mathbb{H}} = (\sigma \times 1_{\mathcal{Q}})^{-1}(X)$.
We are going to study $\widetilde{X}_{\mathbb{H}}$ for several continua.

5.5.1 Definition   A *simple triod* is a continuum which is the union of three arcs
having only one end point in common.

5.5.2 Theorem   *Let $n = \{n_k\}_{k=1}^{\infty}$ be a sequence of positive integers. Let*

$$f_{\mathbb{H}} \colon \Sigma \to W$$

*be a map that is not homotopic to a constant map, where Σ is the n-solenoid. Then
$\widetilde{\Sigma}_{\mathbb{H}}$ is homeomorphic to $\mathbb{R} \times C^*$, where C^* is the Cantor set minus one point.*

Proof Let $f_{\mathbb{H}} \colon \Sigma \to W$ be an essential map. By [2, Theorem], we know that $\widetilde{\Sigma}_{\mathbb{H}}$
is homeomorphic to $\mathbb{K}_{\mathbb{H}} \times \widetilde{\mathbb{B}}_{\mathbb{H}}$, where $\mathbb{K}_{\mathbb{H}}$ is a component of $\widetilde{\Sigma}_{\mathbb{H}}$ and $\widetilde{\mathbb{B}}_{\mathbb{H}}$ is a zero-
dimensional locally compact homogeneous space. Hence, $\mathbb{B}_{\mathbb{H}}$ is homeomorphic to
C^*.

We divide the proof in eight steps.

Step 1. $\widetilde{\Sigma}_{\mathbb{H}}$ *does not contain simple triods.*

It is known that each proper subcontinuum of Σ is an arc [11, Theorem 2]. Hence,
Σ does not contain simple triods. Since $\sigma \times 1_{\mathcal{Q}}$ is a local homeomorphism, $\widetilde{\Sigma}_{\mathbb{H}}$
does not contain simple triods.

Step 2. $\widetilde{\Sigma}_{\mathbb{H}}$ *does not contain simple closed curves.*

Suppose $\widetilde{\Sigma}_{\mathbb{H}}$ contains a simple closed curve $\widetilde{C}$. Then $(\sigma \times 1_{\mathcal{Q}})(\widetilde{C})$ is a
subcontinuum of Σ. Hence, $(\sigma \times 1_{\mathcal{Q}})(\widetilde{C})$ is an arc A [11, Theorem 2]. Let
a be one of the end points of A. Let U be an evenly covered open set of Σ

containing a. Let $x \in U \setminus A$ be in the same arc component of U containing a; so there exists an arc xa from x to a contained in U, such that $xa \cap A = \{a\}$. Let $\tilde{a} \in \tilde{C}$ be such that $(\sigma \times 1_Q)(\tilde{a}) = a$, and let $\tilde{U}$ be an open set of $\tilde{\Sigma}_{\mathbb{H}}$ containing $\tilde{a}$ so that $(\sigma \times 1_Q)|_{\tilde{U}} : \tilde{U} \twoheadrightarrow (\sigma \times 1_Q)(\tilde{U}) = U$ is a homeomorphism. Let $\tilde{x} \in \tilde{U} \cap (\sigma \times 1_Q)^{-1}(x)$. Then the arc xa can be lifted to an arc $\tilde{x}\tilde{a}$, and $\tilde{x}\tilde{a} \cap \tilde{C} = \{\tilde{a}\}$. This implies that $\tilde{\Sigma}_{\mathbb{H}}$ contains a simple triod, a contradiction to Step 1. Therefore, $\tilde{\Sigma}_{\mathbb{H}}$ does not contain a simple closed curve.

Given two points $\tilde{p}$ and $\tilde{q}$ in the same arc component of $\tilde{\Sigma}_{\mathbb{H}}$, the union of all the arcs in $\tilde{\Sigma}_{\mathbb{H}}$ that have $\tilde{p}$ as an end point and contain $\tilde{q}$ is called a *ray* starting at $\tilde{p}$.

Step 3. *A ray $\tilde{R}$ in $\tilde{\Sigma}_{\mathbb{H}}$ is the union of a countable number of arcs.*

Let $\tilde{p}$ be the starting point of $\tilde{R}$, and let $\{\tilde{p}_n\}_{n=1}^{\infty}$ be a countable dense subset of $\tilde{R}$. We assert that $\tilde{R}$ is the union of the arcs $\{\tilde{p}\tilde{p}_n\}_{n=1}^{\infty}$. Suppose there exists a point $\tilde{r}$ in $\tilde{R} \setminus \bigcup_{n=1}^{\infty} \tilde{p}\tilde{p}_n$. Consider the arc $\tilde{p}\tilde{r}$. Now, $\tilde{r}$ must be in the interior of an arc. Thus, $\tilde{p}\tilde{r}$ may be extended to an arc $\tilde{p}\tilde{s}$ so that $\tilde{r}$ is contained in the relative interior of $\tilde{p}\tilde{s}$. Since $\tilde{\Sigma}_{\mathbb{H}}$ contains no simple triods, each $\tilde{p}_n$ belongs to the arc $\tilde{p}\tilde{r}$. This implies that $\{\tilde{p}_n\}_{n=1}^{\infty}$ is not dense in $\tilde{R}$, since no $\tilde{p}_n$ is near $\tilde{s}$, a contradiction. Therefore, $\tilde{R} = \bigcup_{n=1}^{\infty} \tilde{p}\tilde{p}_n$.

Step 4. *For each point $\tilde{p}$ of an arc component $\tilde{A}$ of $\tilde{\Sigma}_{\mathbb{H}}$, $\tilde{A}$ is the union of two rays $\tilde{R}_1$ and $\tilde{R}_2$ starting at $\tilde{p}$ such that $\tilde{R}_1 \cap \tilde{R}_2 = \{\tilde{p}\}$.*

We know that $\tilde{p}$ is in the interior of an arc $\tilde{a}\tilde{b}$. Since $\tilde{\Sigma}_{\mathbb{H}}$ does not contain simple triods, we have that $\tilde{A}$ is the union of two rays starting at $\tilde{p}$ going through $\tilde{a}$ and $\tilde{b}$, respectively. Since $\tilde{\Sigma}_{\mathbb{H}}$ does not contain simple closed curves, these rays intersect only at $\tilde{p}$.

Step 5. *$\tilde{\Sigma}_{\mathbb{H}}$ has uncountably many arc components.*

This follows from the fact that $\tilde{\Sigma}_{\mathbb{H}}$ has uncountably many components.

Step 6. *The arc components of $\tilde{\Sigma}_{\mathbb{H}}$ are unbounded.*

Let $\tilde{A}$ be an arc component of $\tilde{\Sigma}_{\mathbb{H}}$, and let $\tilde{p} \in Cl_{\tilde{\Sigma}_{\mathbb{H}}}(\tilde{A})$. We show that $Cl_{\tilde{\Sigma}_{\mathbb{H}}}(\tilde{A}) = Cl_{\tilde{\Sigma}_{\mathbb{H}}}(\tilde{A}_{\tilde{p}})$, where $\tilde{A}_{\tilde{p}}$ is the arc component of $\tilde{\Sigma}_{\mathbb{H}}$ containing $\tilde{p}$. Suppose there exists a point $\tilde{q} \in Cl_{\tilde{\Sigma}_{\mathbb{H}}}(\tilde{A}_{\tilde{p}}) \setminus Cl_{\tilde{\Sigma}_{\mathbb{H}}}(\tilde{A})$. Let $\varepsilon \leq \tilde{d}(q, Cl_{\tilde{\Sigma}_{\mathbb{H}}}(\tilde{A}))$ be such that each ε-homeomorphism of Σ may be lifted to $\tilde{\Sigma}_{\mathbb{H}}$ (Theorem 5.2.13). Hence, by Theorem 4.2.31, there exists an Effros number $\delta > 0$ for this ε. Let $\tilde{a} \in \tilde{A}$ be such that $\tilde{d}(\tilde{a}, \tilde{p}) < \delta$. Hence, there exists an ε-homeomorphism $\tilde{h} : \tilde{\Sigma}_{\mathbb{H}} \twoheadrightarrow \tilde{\Sigma}_{\mathbb{H}}$ such that $\tilde{h}(\tilde{p}) = \tilde{a}$. Note that $\tilde{h}(\tilde{q}) \in Cl_{\tilde{\Sigma}_{\mathbb{H}}}(\tilde{A})$; this is a contradiction. Thus, $Cl_{\tilde{\Sigma}_{\mathbb{H}}}(\tilde{A}_{\tilde{p}}) \subset Cl_{\tilde{\Sigma}_{\mathbb{H}}}(\tilde{A})$.

A symmetric argument shows that $Cl_{\tilde{\Sigma}_{\mathbb{H}}}(\tilde{A}) \subset Cl_{\tilde{\Sigma}_{\mathbb{H}}}(\tilde{A}_{\tilde{p}})$.

Therefore, $Cl_{\tilde{\Sigma}_{\mathbb{H}}}(\tilde{A}) = Cl_{\tilde{\Sigma}_{\mathbb{H}}}(\tilde{A}_{\tilde{p}})$.

Let

$$\mathcal{G} = \{Cl_{\tilde{\Sigma}_{\mathbb{H}}}(\tilde{A}) \mid \tilde{A} \text{ is an arc component of } \tilde{\Sigma}_{\mathbb{H}}\}.$$

The above argument shows that the elements of $\mathcal{G}$ are pairwise disjoint.

Suppose that one arc component of $\tilde{\Sigma}_{\mathbb{H}}$ is bounded. Then all of them are. Hence, $\mathcal{G}$ is an uncountable collection of continua (Step 5). Since each arc component of

Σ is dense, $\sigma \times 1_Q$ sends each element of $\mathcal{G}$ onto Σ, and this contradicts the fact that the fibres of $\sigma \times 1_Q$ are countable.

Step 7. *Each arc component of $\widetilde{\Sigma}_\mathbb{H}$ is a closed subset of $\widetilde{\Sigma}_\mathbb{H}$ which is homeomorphic to $\mathbb{R}$.*

Since Σ does not contain simple closed curves, by [30, Corollary 8], there exists a one-to-one map $j \colon \mathbb{R} \to \Sigma$ such that $j(\mathbb{R})$ is a dense arc component of Σ. If $\mathbb{K}_\mathbb{H}$ is a component of $\widetilde{\Sigma}_\mathbb{H}$, then let $\tilde{j} \colon \mathbb{R} \to \mathbb{K}_\mathbb{H}$ be a lift of j so that $\tilde{j}(\mathbb{R}) \subset \mathbb{K}_\mathbb{H}$ [23, Lemma 79.1].

Let $\varepsilon > 0$ be a positive number such that each ε-homeomorphism of Σ lifts to an ε-homeomorphism of $\widetilde{\Sigma}_\mathbb{H}$ (Theorem 5.2.13), and let $2\delta > 0$ be an Effros number for this ε (Theorem 4.2.31).

Let $v \in W$ be the common point of C_1 and C_2. Cover $\{v\} \times Q$ with a finite collection $\mathcal{B}$ of δ-balls. Since the arc components of $\widetilde{\Sigma}_\mathbb{H}$ are unbounded (Step 6) and $\mathcal{B}$ is finite, we can find an arc $\widetilde{A}_0$, contained in $\tilde{j}(\mathbb{R})$, whose end points are in two different liftings, $\widetilde{B}_0$ and $\widetilde{B}_1$, of the same element B of $\mathcal{B}$.

Let $\widetilde{A}_0 = \tilde{a}_0 \tilde{a}_1$, where $\tilde{a}_k \in \widetilde{B}_k$, $k \in \{0, 1\}$. Let $(\varphi \times 1_Q) \colon \mathbb{H} \times Q \to \mathbb{H} \times Q$ be the covering homeomorphism that maps $\widetilde{B}_0$ onto $\widetilde{B}_1$. Since $(\varphi \times 1_Q)(\tilde{a}_0) \in \widetilde{B}_1$, there exists an ε-homeomorphism $\tilde{h} \colon \widetilde{\Sigma}_\mathbb{H} \to \widetilde{\Sigma}_\mathbb{H}$ such that $\tilde{h} \circ (\varphi \times 1_Q)(\tilde{a}_0) = \tilde{a}_1$. Then $\widetilde{A}_1 = \widetilde{A}_0 \cup \tilde{h} \circ (\varphi \times 1)(\widetilde{A}_0)$ is an arc, because $\tilde{a}_1 \in \widetilde{A}_0 \cap \tilde{h} \circ (\varphi \times 1)(\widetilde{A}_0)$ and $\widetilde{\Sigma}_\mathbb{H}$ does not contain either simple triods or simple closed curves. If we call $\tilde{a}_2 = \tilde{h} \circ (\varphi \times 1_Q)(\tilde{a}_1)$, we may write $\widetilde{A}_1 = \tilde{a}_0 \tilde{a}_1 \tilde{a}_2$.

Now observe that $\tilde{h} \circ (\varphi \times 1_Q)(\widetilde{A}_1)$ is an arc, and $\widetilde{A}_1 \cap \tilde{h} \circ (\varphi \times 1_Q)(\widetilde{A}_1) = \tilde{a}_1 \tilde{a}_2$. Thus, we have that $\widetilde{A}_2 = \widetilde{A}_1 \cup \tilde{h} \circ (\varphi \times 1_Q)(\widetilde{A}_1)$ is an arc. If $\tilde{a}_3 = \tilde{h} \circ (\varphi \times 1_Q)(\tilde{a}_2)$ we may write $\widetilde{A}_2 = \tilde{a}_0 \tilde{a}_1 \tilde{a}_2 \tilde{a}_3$.

If we continue in this way, we obtain a sequence $\{\widetilde{A}_n\}_{n=1}^\infty$ of arcs so that the "right" end points of the arcs tend to the attracting point z^+ of $\varphi \times 1_Q$.

Similarly, using $\left(\tilde{h} \circ (\varphi \times 1_Q) \right)^{-1}$, we can construct a sequence $\{\widetilde{A}_{-n}\}_{n=1}^\infty$ of arcs having their "left" end points tending to the repelling point z^- of $\varphi \times 1_Q$.

Thus, $\tilde{j}(\mathbb{R})$ is closed, and each point of $\tilde{j}(\mathbb{R})$ has a neighborhood homeomorphic to an open interval. Therefore, $\tilde{j}(\mathbb{R})$ is homeomorphic to $\mathbb{R}$.

Step 8. *The components of $\widetilde{\Sigma}_\mathbb{H}$ are homeomorphic to $\mathbb{R}$.*

We prove that each arc component of $\widetilde{\Sigma}_\mathbb{H}$ is a component of $\widetilde{\Sigma}_\mathbb{H}$. By [2, Corollary], we only need to show that each arc component is a quasicomponent, and for this, it is enough to prove that if $\widetilde{A}_1$ and $\widetilde{A}_2$ are two distinct arc components of $\widetilde{\Sigma}_\mathbb{H}$, then there exists a closed and open subset of $\widetilde{\Sigma}_\mathbb{H}$ containing $\widetilde{A}_1$ which is disjoint from $\widetilde{A}_2$.

Let $\widetilde{A}_1$ and $\widetilde{A}_2$ be two different arc components of $\widetilde{\Sigma}_\mathbb{H}$. Let $\tilde{a}_1 \in \widetilde{A}_1$. Since $\widetilde{\Sigma}_\mathbb{H}$ is locally homeomorphic to Σ, we can find a compact set of $\widetilde{\Sigma}_\mathbb{H}$ of the form $[r, s] \times \mathcal{C}$, such that $(r, s) \times \mathcal{C}$ is an open subset of $\widetilde{\Sigma}_\mathbb{H}$ containing $\tilde{a}_1$. Let $\widetilde{U}$ be a closed and open subset of $(r, s) \times \mathcal{C}$ containing $\tilde{a}_1$ such that $\widetilde{U} \cap \widetilde{A}_2 = \emptyset$.

Let $\widetilde{V}$ be the union of all the arc components $\widetilde{A}$ of $\widetilde{\Sigma}_\mathbb{H}$ such that $\widetilde{A} \cap \widetilde{U} \neq \emptyset$. Clearly $\widetilde{A}_1 \subset \widetilde{V}$ and $\widetilde{A}_2 \cap \widetilde{V} = \emptyset$. We show that $\widetilde{V}$ is a closed and open subset of $\widetilde{\Sigma}_\mathbb{H}$.

First, we show that $\widetilde{V}$ is open. Let $\tilde{x} \in \widetilde{V}$ and let $\widetilde{\mathcal{A}}_{\tilde{x}}$ be the arc component of $\widetilde{\Sigma}_{\mathbb{H}}$ containing $\tilde{x}$. Let $\tilde{x}_1 \in \widetilde{\mathcal{A}}_{\tilde{x}} \cap \widetilde{U}$. Take $\varepsilon > 0$ such that each ε-homeomorphism of Σ may be lifted to $\widetilde{\Sigma}_{\mathbb{H}}$ (Theorem 5.2.13), and $\mathcal{V}_\varepsilon^{\tilde{d}}(\tilde{x}_1) \subset \widetilde{U}$. Let $\delta > 0$ be an Effros number for this ε (Theorem 4.2.31). Let $\tilde{x}_2 \in \mathcal{V}_\delta^{\tilde{d}}(\tilde{x})$. Then there exists an ε-homeomorphism $\tilde{h} \colon \widetilde{\Sigma}_{\mathbb{H}} \twoheadrightarrow \widetilde{\Sigma}_{\mathbb{H}}$ such that $\tilde{h}(\tilde{x}) = \tilde{x}_2$. Hence, $\tilde{h}(\tilde{x}) \in \widetilde{\mathcal{A}}_{\tilde{x}_2}$, $\widetilde{\mathcal{A}}_{\tilde{x}_2}$ being the arc component of $\widetilde{\Sigma}_{\mathbb{H}}$ containing $\tilde{x}_2$. Therefore, $\tilde{d}\left(\tilde{h}(\tilde{x}_1), \tilde{x}_1\right) < \varepsilon$. Thus, $\widetilde{\mathcal{A}}_{\tilde{x}_2} \cap \widetilde{U} \neq \emptyset$. Consequently, $\tilde{x}_2 \in \widetilde{V}$. Therefore, $\widetilde{V}$ is open.

Now, we prove that $\widetilde{V}$ is closed. Let $\tilde{x} \in Cl_{\widetilde{\Sigma}_{\mathbb{H}}}(\widetilde{V})$, and let $\widetilde{\mathcal{A}}_{\tilde{x}}$ be the arc component of $\widetilde{\Sigma}_{\mathbb{H}}$ containing $\tilde{x}$. For every $n \in \mathbb{N}$, let $\delta_n > 0$ be an Effros number for $\frac{1}{n}$. Let $\tilde{x}_n \in \widetilde{V} \cap \mathcal{V}_{\delta_n}^{\tilde{d}}(\tilde{x})$. For each $n \in \mathbb{N}$, there exists a $\frac{1}{n}$-homeomorphism $\tilde{h}_n \colon \widetilde{\Sigma}_{\mathbb{H}} \twoheadrightarrow \widetilde{\Sigma}_{\mathbb{H}}$ so that $\tilde{h}_n(\tilde{x}_n) = \tilde{x}$.

Let $\widetilde{\mathcal{A}}_{\tilde{x}_n}$ be the arc component of $\widetilde{\Sigma}_{\mathbb{H}}$ containing $\tilde{x}_n$. Let $\tilde{x}'_n \in \widetilde{\mathcal{A}}_{\tilde{x}_n} \cap \widetilde{U}$. Then for each $n \in \mathbb{N}$, $\tilde{h}_n(\tilde{x}'_n) \in \widetilde{\mathcal{A}}_{\tilde{x}}$, and $\tilde{d}\left(\tilde{x}'_n, \tilde{h}_n(\tilde{x}'_n)\right) < \frac{1}{n}$.

Since $\widetilde{U}$ is compact, without loss of generality, we assume that the sequence, $\{\tilde{x}'_n\}_{n=1}^\infty$, converges to $\tilde{x}'$, where $\tilde{x}' \in \widetilde{U}$. Let $\varepsilon > 0$ be given, and let $n \in \mathbb{N}$ be such that $\frac{1}{n} < \frac{\varepsilon}{2}$ and $\tilde{d}(\tilde{x}'_n, \tilde{x}') < \frac{\varepsilon}{2}$. Then

$$\tilde{d}\left(\tilde{x}', \tilde{h}_n(\tilde{x}'_n)\right) \leq \tilde{d}(\tilde{x}', \tilde{x}'_n) + \tilde{d}\left(\tilde{x}'_n, \tilde{h}_n(\tilde{x}'_n)\right) < \frac{\varepsilon}{2} + \frac{1}{n} < \varepsilon.$$

Hence, for every $\varepsilon > 0$, $\mathcal{V}_\varepsilon^{\tilde{d}}(\tilde{x}') \cap \widetilde{\mathcal{A}}_{\tilde{x}} \neq \emptyset$. Therefore, $\tilde{x}' \in Cl_{\widetilde{\Sigma}_{\mathbb{H}}}(\widetilde{\mathcal{A}}_{\tilde{x}}) = \widetilde{\mathcal{A}}_{\tilde{x}}$ (Step 7), and $\widetilde{\mathcal{A}}_{\tilde{x}} \cap \widetilde{U} \neq \emptyset$. Thus, $\tilde{x} \in \widetilde{V}$, and $\widetilde{V}$ is closed.

Therefore, $\widetilde{\Sigma}_{\mathbb{H}}$ is homeomorphic to $\mathbb{R} \times C^*$.

Q.E.D.

5.5.3 Theorem *Let $\mathbb{M}$ be the Menger universal curve. If $f_{\mathbb{H}} \colon \mathbb{M} \to W$ is a map that is not homotopic to a constant map, then $\widetilde{\mathbb{M}}_{\mathbb{H}}$ has countably many components, and for each component $\mathbb{K}_{\mathbb{H}}$ of $\widetilde{\mathbb{M}}_{\mathbb{H}}$, $\mathbb{Y}_{\mathbb{K}_{\mathbb{H}}} = Cl_{(\mathbb{H} \times \mathcal{Q}) \cup \mathcal{S}^1}(\mathbb{K}_{\mathbb{H}})$ is homeomorphic to $\mathbb{M}$.*

Proof By [2, Theorem], $\widetilde{\mathbb{M}}_{\mathbb{H}}$ is homeomorphic to $\mathbb{K}_{\mathbb{H}} \times \widetilde{\mathbb{B}}_{\mathbb{H}}$. Since $\widetilde{\mathbb{M}}_{\mathbb{H}}$ is locally homeomorphic to $\mathbb{M}$, and $\mathbb{M}$ is locally connected, $\mathbb{B}_{\mathbb{H}}$ is an at most countable and discrete space. Therefore, $\mathbb{K}_{\mathbb{H}}$ is locally homeomorphic to $\mathbb{M}$. Thus, $\mathbb{K}_{\mathbb{H}}$ is connected, locally arcwise connected (hence, arcwise connected), and locally compact. Note that for each connected subset $\widetilde{U}$ of $\mathbb{K}_{\mathbb{H}}$ and each point $\tilde{z} \in \widetilde{U}$, $\widetilde{U} \setminus \{\tilde{z}\}$ is connected, and no open set of $\mathbb{K}_{\mathbb{H}}$ can be embedded in the plane.

By Theorem 5.3.21, $\mathbb{Y}_{\mathbb{K}_{\mathbb{H}}}$ is connected im kleinen at each point of $\mathbb{E}(\mathbb{K}_{\mathbb{H}}) = \mathbb{Y}_{\mathbb{K}_{\mathbb{H}}} \setminus \mathbb{K}_{\mathbb{H}}$. Hence, $\mathbb{Y}_{\mathbb{K}_{\mathbb{H}}}$ is locally connected.

Now, we show that for each point $\tilde{z}$ of $\mathbb{E}(\mathbb{K}_{\mathbb{H}})$ and each connected subset $\widetilde{U}$ of $\mathbb{Y}_{\mathbb{K}_{\mathbb{H}}}$, such that $\tilde{z} \in \widetilde{U}$, $\widetilde{U} \setminus \{\tilde{z}\}$ is connected.

Let $\tilde{z} \in \mathbb{E}(\mathbb{K}_{\mathbb{H}})$ and let $\widetilde{U}$ be a connected open set of $\mathbb{Y}_{\mathbb{K}_{\mathbb{H}}}$ containing $\tilde{z}$. Observe that $\widetilde{U}$ is arcwise connected. Let $\widetilde{U}'$ be an open connected subset of $\mathbb{Y}_{\mathbb{K}_{\mathbb{H}}}$ containing $\tilde{z}$ such that $Cl_{\mathbb{Y}_{\mathbb{K}_{\mathbb{H}}}}(\widetilde{U}') \subset \widetilde{U}$. The projection of $\widetilde{U}' \setminus \mathbb{E}(\mathbb{K}_{\mathbb{H}})$ into $\widetilde{W}$ contains an infinite number of vertices of $\widetilde{W}$. Let $\widetilde{V}'$ be the set of all vertices $\tilde{v}$ of $\widetilde{W}$ contained in the projection of $\widetilde{U}' \backslash \mathbb{E}(\mathbb{K}_{\mathbb{H}})$ into $\widetilde{W}$ for which $\mathcal{Q}_{\tilde{v}} \cap \mathbb{K}_{\mathbb{H}} \subset \widetilde{U}'$, where $\mathcal{Q}_{\tilde{v}} = \{\tilde{v}\} \times \mathcal{Q}$.

Let $v \in W$ be the common point of C_1 and C_2. Cover $(\{v\} \times \mathcal{Q}) \cap \mathbb{M}$ with a finite open cover $\mathcal{B}$, where each element of $\mathcal{B}$ is evenly covered and arcwise connected. We take the elements of $\mathcal{B}$ small enough such that for each $\tilde{v} \in \widetilde{V}'$, the lifting $\widetilde{\mathcal{B}}_{\tilde{v}}$ of $\mathcal{B}$, covering $\mathcal{Q}_{\tilde{v}} \cap \mathbb{K}_{\mathbb{H}}$, is such that $\bigcup \widetilde{\mathcal{B}}_{\tilde{v}} \subset \widetilde{U}$.

Suppose that $\widetilde{U} \backslash \{\tilde{z}\}$ is not connected. Then $\widetilde{U} \backslash \{\tilde{z}\}$ can be written as $\widetilde{U}_1 \cup \widetilde{U}_2$, where $\widetilde{U}_1$ and $\widetilde{U}_2$ are disjoint open sets of $\widetilde{U}$.

We assert that there exists an infinite subset $\widetilde{V}$ of $\widetilde{V}'$ with the property that if $\tilde{v} \in \widetilde{V}$, then $\bigcup \widetilde{\mathcal{B}}_{\tilde{v}}$ is totally contained in $\widetilde{U}_1$ or in $\widetilde{U}_2$. To show this, let $\tilde{v}' \in \widetilde{V}'$, and suppose that $\widetilde{\mathcal{B}}_{\tilde{v}'} = \{\widetilde{B}_{\tilde{v}',1}, \dots, \widetilde{B}_{\tilde{v}',\ell}\}$. Let $\tilde{x}_{\tilde{v}',j} \in \widetilde{B}_{\tilde{v}',j}$. Since $\mathbb{K}_{\mathbb{H}}$ is arcwise connected, for each $j \in \{1, \dots, \ell - 1\}$, there exists an arc $\alpha_j \colon [0,1] \to \mathbb{K}_{\mathbb{H}}$ such that $\alpha_j(0) = \tilde{x}_{\tilde{v}',1}$ and $\alpha_j(1) = \tilde{x}_{\tilde{v}',j+1}$. Since $\bigcup_{j=1}^{\ell-1} \alpha_j([0,1])$ is compact, there exists a covering homeomorphism $(\varphi \times 1_{\mathcal{Q}}) \colon \mathbb{K}_{\mathbb{H}} \to \mathbb{K}_{\mathbb{H}}$ such that $(\varphi \times 1_{\mathcal{Q}})\left(\bigcup_{j=1}^{\ell-1} \alpha_j([0,1])\right) \subset \widetilde{U}$.

Let $(\varphi \times 1_{\mathcal{Q}})(\tilde{v}') = \tilde{v}$. Since $\bigcup \widetilde{\mathcal{B}}_{\tilde{v}} \cup (\varphi \times 1_{\mathcal{Q}})\left(\bigcup_{j=1}^{\ell-1} \alpha_j([0,1])\right)$ is connected, it is contained in $\widetilde{U}_1$ or in $\widetilde{U}_2$. Hence, $\bigcup \widetilde{\mathcal{B}}_{\tilde{v}}$ is totally contained in $\widetilde{U}_1$ or in $\widetilde{U}_2$. Since this is true for every $\tilde{v}' \in \widetilde{V}'$, such a subset $\widetilde{V}$ of $\widetilde{V}'$ exists.

Now, we assert that $\bigcup_{\tilde{v} \in \widetilde{V}} (\bigcup \widetilde{\mathcal{B}}_{\tilde{v}})$ is contained in $\widetilde{U}_1$ or in $\widetilde{U}_2$. To see this, let $\tilde{v}_1$ and $\tilde{v}_2$ in $\widetilde{V}$ and suppose that $\bigcup \widetilde{\mathcal{B}}_{\tilde{v}_1} \subset \widetilde{U}_1$ and $\bigcup \widetilde{\mathcal{B}}_{\tilde{v}_2} \subset \widetilde{U}_2$. Let $\tilde{x}_k \in \bigcup \widetilde{\mathcal{B}}_{\tilde{v}_k}$, $k \in \{1,2\}$. Then $\tilde{x}_k = (\tilde{w}_k, q_k)$, where $\tilde{w}_k \in \widetilde{W}$ and $q_k \in \mathcal{Q}$. Now, $\tilde{w}_k$ and $\tilde{z}$ determine a unique arc $[\tilde{w}_k, \tilde{z}]$ in $Cl_{\mathbb{H}}(\widetilde{W})$, from $\tilde{w}_k$ to $\tilde{z}$ [33, p. 284]. Now $[\tilde{w}_1, \tilde{z}] \cap [\tilde{w}_2, \tilde{z}]$ contains an element $\tilde{v}$ of $\widetilde{V}$. Without loss of generality, we assume that $\bigcup \widetilde{\mathcal{B}}_{\tilde{v}}$ is contained in $\widetilde{U}_2$. Since $\widetilde{U}$ is arcwise connected, there exists an $\alpha \colon [0,1] \to \widetilde{U}$ such that $\alpha(0) = \tilde{x}_1$ and $\alpha(1) = \tilde{x}_2$. Since $\tilde{x}_k \in \widetilde{U}_k$, there exists a $t \in [0,1]$ such that $\alpha(t) = \tilde{z}$. On the other hand, any arc from $\tilde{x}_1$ to $\tilde{z}$ must intersect $\mathcal{Q}_{\tilde{v}} \cap \mathbb{K}_{\mathbb{H}}$, which is a contradiction. Therefore, $\bigcup_{\tilde{v} \in \widetilde{V}} (\bigcup \widetilde{\mathcal{B}}_{\tilde{v}})$ is contained in $\widetilde{U}_1$ or in $\widetilde{U}_2$, say $\widetilde{U}_1$.

A similar argument shows that $\widetilde{U}_2$ is empty. Hence, $\widetilde{U} \backslash \{\tilde{z}\}$ is connected. Therefore, by Theorem 5.4.3, $\mathbb{Y}_{\mathbb{K}_{\mathbb{H}}}$ is homeomorphic to $\mathbb{M}$.

Q.E.D.

In the following theorem we construct a covering space of the Case continuum whose components are locally connected.

5.5.4 Theorem *The Case continuum $\mathbb{C}$ has a covering space with locally connected components.*

Proof Note that, by Theorem 5.4.16, $\mathbb{C} \subset \Sigma \times [0,1]^2$, where Σ is the dyadic solenoid.

First, we use inverse limits to construct a covering space of $\Sigma \times [0,1]^2$.

Let $Y_n = \mathcal{S}^1 \times [0,1]^2$, and let $\widetilde{Y}_n = \mathbb{R} \times [0,1]^2 \times \{2^{n-1}\text{th roots of unity}\}$. Define the bonding maps $f_n^{n+1} \colon Y_{n+1} \twoheadrightarrow Y_n$ and $\tilde{f}_n^{n+1} \colon \widetilde{Y}_{n+1} \twoheadrightarrow \widetilde{Y}_n$ by $f_n^{n+1}((z,x)) = (z^2, x)$, and $\tilde{f}_n^{n+1}((r,x,t)) = (2r, x, t^2)$, respectively. Let $q_n \colon \widetilde{Y}_n \to Y_n$ be the covering map given by $q_n((r,x,t)) = (t\exp(2\pi r), x)$. Thus, we can consider the following infinite ladder:

$$\cdots \longleftarrow \widetilde{Y}_{n-1} \overset{\tilde{f}_{n-1}^n}{\longleftarrow} \widetilde{Y}_n \overset{\tilde{f}_n^{n+1}}{\longleftarrow} \widetilde{Y}_{n+1} \longleftarrow \cdots : \widetilde{Y}_\infty$$

$$\Big\downarrow q_{n-1} \qquad \Big\downarrow q_n \qquad \Big\downarrow q_{n+1} \qquad \Big\downarrow q_\infty$$

$$\cdots \longleftarrow Y_{n-1} \underset{\tilde{f}_{n-1}^n}{\longleftarrow} Y_n \underset{\tilde{f}_n^{n+1}}{\longleftarrow} Y_{n+1} \longleftarrow \cdots : Y_\infty$$

where $\widetilde{Y}_\infty = \varprojlim\{\widetilde{Y}_n, \tilde{f}_n^{n+1}\}$, $Y_\infty = \varprojlim\{Y_n, f_n^{n+1}\}$ and $q_\infty = \varprojlim\{q_n\}$. Note that $\widetilde{Y}_\infty$ is homeomorphic to $\mathbb{R} \times [0, 1]^2 \times \mathcal{C}$ and Y_∞ is homeomorphic to $\Sigma \times [0, 1]^2$.

Let $(\mathbb{R} \times [0, 1]^2 \times \{t\})_{n+1}$ be a component of $\widetilde{Y}_{n+1}$. Note that the restriction of the map $\tilde{f}_n^{n+1}$ to this component, $\tilde{f}_n^{n+1}|_{(\mathbb{R} \times [0,1]^2 \times \{t\})_{n+1}} : (\mathbb{R} \times [0, 1]^2 \times \{t\})_{n+1} \subset \widetilde{Y}_{n+1} \twoheadrightarrow (\mathbb{R} \times [0, 1]^2 \times \{t^2\})_n \subset \widetilde{Y}_n$, is a homeomorphism, since the map $g_{n+1}^n : (\mathbb{R} \times [0, 1]^2 \times \{t^2\})_n \subset \widetilde{Y}_n \twoheadrightarrow (\mathbb{R} \times [0, 1]^2 \times \{t\})_{n+1} \subset \widetilde{Y}_{n+1}$ given by $g_{n+1}^n((r, x, t^2)) = (\frac{1}{2}r, x, t)$ is its inverse.

Let $\tilde{t} = (t_n)_{n=1}^\infty \in \mathcal{C}$, and let $(\mathbb{R} \times [0, 1]^2 \times \{\tilde{t}\})_\infty$ be a component of $\widetilde{Y}_\infty$. Consider the map $\tilde{f}_n|_{(\mathbb{R} \times [0,1]^2 \times \{\tilde{t}\})_\infty} : (\mathbb{R} \times [0, 1]^2 \times \{\tilde{t}\})_\infty \subset \widetilde{Y}_\infty \twoheadrightarrow (\mathbb{R} \times [0, 1]^2 \times \{t_n\})_n \subset \widetilde{Y}_n$. We claim that this map is a homeomorphism. To see this, let us define the map $g^n : (\mathbb{R} \times [0, 1]^2 \times \{t_n\})_n \subset \widetilde{Y}_n \twoheadrightarrow (\mathbb{R} \times [0, 1]^2 \times \{\tilde{t}\})_\infty \subset \widetilde{Y}_\infty$ by

$$g^n((r, x, t_n)) = (f_1^n((r, x, t_n)), \ldots, f_{n-1}^n((r, x, t_n)), (r, x, t_n),$$

$$g_{n+1}^n((r, x, t_n)), \ldots, g_{n+m}^n((r, x, t_n)), \ldots)$$

where $g_{n+m}^n = g_{n+m}^{n+m-1} \circ \cdots \circ g_{n+1}^n$. Since $g^n \circ \tilde{f}_n|_{(\mathbb{R} \times [0,1]^2 \times \{\tilde{t}\})_\infty} = 1_{(\mathbb{R} \times [0,1]^2 \times \{\tilde{t}\})_\infty}$, and $\tilde{f}_n|_{(\mathbb{R} \times [0,1]^2 \times \{\tilde{t}\})_\infty} \circ g^n = 1_{(\mathbb{R} \times [0,1]^2 \times \{t_n\})_n}$, we have that $\tilde{f}_n|_{(\mathbb{R} \times [0,1]^2 \times \{\tilde{t}\})_\infty}$ is a homeomorphism.

Now, assume that the Menger curve $\mathbb{M}$ is embedded in Y_1, as described in Lemma 5.4.12. Since $(f_1^2)^{-1}(\mathbb{M})$ is homeomorphic to $\mathbb{M}$, setting $X_1 = \mathbb{M}$, $X_2 = (f_1^2)^{-1}(\mathbb{M}), \ldots, X_n = (f_1^n)^{-1}(\mathbb{M})$, we obtain that $\mathbb{C} = \varprojlim\{X_n, f_n^{n+1}\}$.

Let $\widetilde{X}_n = q_n^{-1}(X_n)$. Note that $\widetilde{X}_n$ is a locally compact, metric space, with 2^{n-1} components, and each component is locally homeomorphic to $\mathbb{M}$.

Let $\widetilde{\mathbb{C}} = \varprojlim\{\widetilde{X}_n, \tilde{f}_n^{n+1}\}$. Then $\widetilde{\mathbb{C}}$ is a locally compact, metric space. Let $\mathbb{K}$ be a component of $\widetilde{\mathbb{C}}$. To see that $\mathbb{K}$ is locally connected, note that $\mathbb{K}$ is contained in one of the components of $\widetilde{Y}_\infty$. Let $(\mathbb{R} \times [0, 1]^2 \times \{\tilde{t}\})_\infty$ be the component of $\widetilde{Y}_\infty$ containing $\mathbb{K}$. Since $\tilde{f}_1|_{(\mathbb{R} \times [0,1]^2 \times \{\tilde{t}\})_\infty}$ is a homeomorphism, we have that $\tilde{f}_1|_{\mathbb{K}}$ is also a homeomorphism onto $\tilde{f}_1(\mathbb{K})$, but $\tilde{f}_1(\mathbb{K}) = \widetilde{X}_1$. Since $\widetilde{X}_1$ is locally homeomorphic to $\mathbb{M}$, $\mathbb{K}$ is locally homeomorphic to $\mathbb{M}$. Thus, $\mathbb{K}$ is locally connected, because $\mathbb{M}$ is locally connected.

Q.E.D.

To finish this chapter, we show two different covering spaces of one of the Minc–Rogers continua (Theorem 5.4.18).

5.5.5 Theorem *Let Λ be the constant sequence $\{(2, 2, 1)\}$. If $\mathbb{MR} = \mathbb{M}^\Lambda$, then $\mathbb{MR}$ has a covering space whose components are not locally connected.*

Proof Note that, by Theorem 5.4.18, $\mathbb{MR} \subset \Sigma \times \Sigma \times S^1$, where Σ is the dyadic solenoid.

First, we construct a covering space for $\Sigma \times \Sigma \times S^1$ using inverse limits.

Let $Y_n = S^1 \times S^1 \times S^1$, and $\widetilde{Y}_n = \mathbb{R} \times S^1 \times S^1 \times \{2^{n-1}\text{th roots of unity}\}$. Define the bonding maps

$$f_n^{n+1} \colon Y_{n+1} \twoheadrightarrow Y_n \text{ and } \tilde{f}_n^{n+1} \colon \widetilde{Y}_{n+1} \twoheadrightarrow \widetilde{Y}_n$$

by

$$f_n^{n+1}((z_1, z_2, z_3)) = (z_1^2, z_2^2, z_3) \text{ and } \tilde{f}_n^{n+1}((r, z_2, z_3, t)) = (2r, z_2^2, z_3, t^2),$$

respectively. Let $q_n \colon \widetilde{Y}_n \twoheadrightarrow Y_n$ be the map given by $q_n((r, z_2, z_3, t)) = (t \exp(2\pi r), z_2, z_3)$. Note that q_n is a covering map. Hence, we have the following infinite ladder:

$$
\begin{array}{ccccccccc}
\cdots & \xleftarrow{} & \widetilde{Y}_{n-1} & \xleftarrow{f_{n-1}^n} & \widetilde{Y}_n & \xleftarrow{f_n^{n+1}} & \widetilde{Y}_{n+1} & \xleftarrow{} & \cdots : \widetilde{Y}_\infty \\
 & & \downarrow{q_{n-1}} & & \downarrow{q_n} & & \downarrow{q_{n+1}} & & \downarrow{q_\infty} \\
\cdots & \xleftarrow{} & Y_{n-1} & \xleftarrow{\tilde{f}_{n-1}^n} & Y_n & \xleftarrow{\tilde{f}_n^{n+1}} & Y_{n+1} & \xleftarrow{} & \cdots : Y_\infty
\end{array}
$$

where $\widetilde{Y}_\infty = \varprojlim\{\widetilde{Y}_n, \tilde{f}_n^{n+1}\}$, $Y_\infty = \varprojlim\{Y_n, f_n^{n+1}\}$ and $q_\infty = \varprojlim\{q_n\}$. Note that $\widetilde{Y}_\infty$ is homeomorphic to $\mathbb{R} \times \Sigma \times S^1 \times C$ (Theorem 2.2.10) and Y_∞ is homeomorphic to $\Sigma \times \Sigma \times S^1$.

Let $\tilde{t} = (t_n)_{n=1}^\infty \in C$, and let $(\mathbb{R} \times \Sigma \times S^1 \times \{\tilde{t}\})_\infty$ be a component of $\widetilde{Y}_\infty$. We assert that $q_\infty|_{(\mathbb{R}\times\Sigma\times S^1\times\{\tilde{t}\})_\infty}$ is one-to-one. To see this, let $(\tilde{r}, \tilde{s}, \tilde{x}, \tilde{t}) = ((r_n, s_n, x_n, t_n))_{n=1}^\infty$ and $(\tilde{r}', \tilde{s}', \tilde{x}', \tilde{t}) = ((r_n', s_n', x_n', t_n))_{n=1}^\infty$ be two points of $(\mathbb{R} \times \Sigma \times S^1 \times \{\tilde{t}\})_\infty$ such that $q_\infty((\tilde{r}, \tilde{s}, \tilde{x}, \tilde{t})) = q_\infty((\tilde{r}', \tilde{s}', \tilde{x}', \tilde{t}))$. This equality implies that, for every $n \in \mathbb{N}$, $q_n((r_n, s_n, x_n, t_n)) = q_n((r_n', s_n', x_n', t_n))$. Then $(t_n \exp(2\pi r_n), s_n, x_n) = (t_n \exp(2\pi r_n'), s_n', x_n')$. Thus, for each $n \in \mathbb{N}$, $s_n = s_n'$, $x_n = x_n'$ and $t_n \exp(2\pi r_n) = t_n \exp(2\pi r_n')$. From this last equality, we obtain that $\exp(2\pi r_n) = \exp(2\pi r_n')$. Hence, for every $n \in \mathbb{N}$, there exists an $m_n \in \mathbb{Z}$ such that $2\pi r_n + 2\pi m_n = 2\pi r_n'$, from here, $r_n + m_n = r_n'$. For $n = 1$, we have that $r_1' = r_1 + m_1$. Then $r_2' = \frac{1}{2}r_1' = \frac{1}{2}r_1 + \frac{1}{2}m_1$, $r_3' = \frac{1}{2}r_2' = \frac{1}{4}r_1 + \frac{1}{4}m_1$, etc. Now, if $m_1 \neq 0$, there exists a $k \in \mathbb{Z}$ such that $\frac{1}{2^k}m_1 \notin \mathbb{Z}$, and, on the other hand, $r_{k+1}' = \frac{1}{2^k}r_1 + \frac{1}{2^k}m_1$, a contradiction. Hence, $m_1 = 0$. Similarly, $m_n = 0$ for all $n \in \mathbb{N}$. Thus, $r_n = r_n'$, and $(\tilde{r}, \tilde{s}, \tilde{x}, \tilde{t}) = (\tilde{r}', \tilde{s}', \tilde{x}', \tilde{t})$.

Note that if $(\tilde{r}, \tilde{s}, \tilde{x}, \tilde{t}) \in (\mathbb{R} \times \Sigma \times S^1 \times C)_\infty$, then $q_\infty((\tilde{r}, \tilde{s}, \tilde{x}, \tilde{t})) = (s', \tilde{s}, \tilde{x})$; i.e., q_∞ sends the second and third coordinates of a point in $\mathbb{R} \times \Sigma \times S^1 \times C$ identically to the second and third coordinates of its image in $\Sigma \times \Sigma \times S^1$.

Let $\widetilde{U} = (\tilde{a}, \tilde{b}) \times ((\tilde{c}, \tilde{d}) \times C') \times (\tilde{e}, \tilde{f}) \times C''$ be a basic open set of $\widetilde{Y}_\infty$. Then $q_\infty(\widetilde{U}) = ((a, b) \times C''') \times ((\tilde{c}, \tilde{d}) \times C') \times (\tilde{e}, \tilde{f})$. Hence, if $\widetilde{U}_c = (\tilde{a}, \tilde{b}) \times ((\tilde{c}, \tilde{d}) \times C') \times (\tilde{e}, \tilde{f}) \times \{\tilde{t}\}$ is an open set of a component of $\widetilde{Y}_\infty$, then $q_\infty(\widetilde{U}_c) = ((a, b) \times \{t\}) \times ((\tilde{c}, \tilde{d}) \times C') \times (\tilde{e}, \tilde{f})$.

Assume that the Menger curve $\mathbb{M}$ is embedded in Y_1 in such a way that $\{e\} \times S^1 \times \{e\} \subset \mathbb{M}$ and $S^1 \times \{e\} \times \{e\} \subset \mathbb{M}$ (Lemma 5.4.17). Since $(f_1^2)^{-1}(\mathbb{M})$ is homeomorphic to $\mathbb{M}$, if we let $X_1 = \mathbb{M}$, $X_2 = (f_1^2)^{-1}(\mathbb{M}), \ldots, X_n = (f_1^n)^{-1}(\mathbb{M})$, then we obtain that $\mathbb{MR} = \lim_{\leftarrow}\{X_n, f_n^{n+1}\}$.

Let $\widetilde{X}_n = q_n^{-1}(X_n)$. Observe that $\widetilde{X}_n$ is a locally compact, metric space, with 2^{n-1} components, each of which is locally homeomorphic to $\mathbb{M}$. Let $\widetilde{\mathbb{MR}} = \lim_{\leftarrow}\{\widetilde{X}_n, \tilde{f}_n^{n+1}\}$. Let $\widetilde{U} = (\tilde{a}, \tilde{b}) \times ((\tilde{c}, \tilde{d}) \times C') \times (\tilde{e}, \tilde{f}) \times C''$ be a basic open set of $\widetilde{Y}_\infty$, then $\widetilde{U} \cap \widetilde{\mathbb{MR}}$ is a basic open set of $\widetilde{\mathbb{MR}}$. We can take $\widetilde{U}$ such that $q_\infty|_{\widetilde{U}}$ is a homeomorphism. Since $\widetilde{\mathbb{MR}} = q_\infty^{-1}(\mathbb{MR})$, we have that $q_\infty(\widetilde{U} \cap \widetilde{\mathbb{MR}}) = q_\infty(\widetilde{U}) \cap \mathbb{MR} = \left[((a, b) \times C''') \times ((\tilde{c}, \tilde{d}) \times C') \times (\tilde{e}, \tilde{f})\right] \cap \mathbb{MR}$. But this is homeomorphic to $\left[(a, b) \times (\tilde{c}, \tilde{d}) \times (\tilde{e}, \tilde{f}) \times C' \times C'''\right] \cap \mathbb{MR}$. Since $\mathbb{MR}$ is locally homeomorphic to the product of an open set of $\mathbb{M}$ and a Cantor set, there exists an open set V of $\mathbb{MR}$ so that $\left[(a, b) \times (\tilde{c}, \tilde{d}) \times (\tilde{e}, \tilde{f}) \times C' \times C'''\right] \cap \mathbb{MR}$ is homeomorphic to $V \times C' \times C'''$. Hence, $q_\infty([(\tilde{a}, \tilde{b}) \times ((\tilde{c}, \tilde{d}) \times C') \times (\tilde{e}, \tilde{f}) \times \{\tilde{t}\}] \cap \widetilde{\mathbb{MR}})$ is homeomorphic to $V \times C' \times \{t\}$. Then the components of $\widetilde{\mathbb{MR}}$ are locally homeomorphic to $V \times C'$. In particular they are not locally connected.

Q.E.D.

5.5.6 Theorem *Let Λ be the constant sequence $\{(2, 2, 1)\}$. If $\mathbb{MR} = \mathbb{M}^\Lambda$, then $\mathbb{MR}$ has a covering space whose components are locally connected.*

Proof Let the inverse sequences $\{Y_n, f_n^{n+1}\}$ and $\{X_n, f_n^{n+1}\}$ be as in the proof of Theorem 5.5.5. Hence, $\mathbb{MR} = \lim_{\leftarrow}\{X_n, f_n^{n+1}\}$.

Let $\widetilde{Y}_n = \mathbb{R} \times \mathbb{R} \times S^1 \times \{2^{n-1}\text{th roots of unity}\}$, and let $\tilde{f}_n^{n+1} \colon \widetilde{Y}_{n+1} \to \widetilde{Y}_n$ be given by $\tilde{f}_n^{n+1}((r_1, r_2, z_3, t)) = (2r_1, 2r_2, z_3, t^2)$. Define the covering map $q_n \colon \widetilde{Y}_n \to Y_n$ by

$$q_n((r_1, r_2, z_3, t)) = (t\exp(2\pi r_1), t\exp(2\pi r_2), z_3, t).$$

Thus, we may construct the following infinite and commutative ladder:

$$
\begin{array}{ccccccccc}
\cdots & \longleftarrow & \widetilde{Y}_{n-1} & \xleftarrow{\tilde{f}_{n-1}^n} & \widetilde{Y}_n & \xleftarrow{\tilde{f}_n^{n+1}} & \widetilde{Y}_{n+1} & \longleftarrow \cdots & : \widetilde{Y}_\infty \\
& & \downarrow{q_{n-1}} & & \downarrow{q_n} & & \downarrow{q_{n+1}} & & \downarrow{q_\infty} \\
\cdots & \longleftarrow & Y_{n-1} & \xleftarrow[\tilde{f}_{n-1}^n]{} & Y_n & \xleftarrow[\tilde{f}_n^{n+1}]{} & Y_{n+1} & \longleftarrow \cdots & : Y_\infty
\end{array}
$$

where $\widetilde{Y}_\infty = \lim_{\leftarrow}\{\widetilde{Y}_n, \tilde{f}_n^{n+1}\}$ and $q_\infty = \lim_{\leftarrow}\{q_n\}$. Note that $\widetilde{Y}_\infty$ is homeomorphic to $\mathbb{R} \times \mathbb{R} \times S^1 \times C$ (Theorem 2.2.10).

Let $(\mathbb{R} \times \mathbb{R} \times \mathcal{S}^1 \times \{t\})_{n+1}$ be a component of $\widetilde{Y}_{n+1}$. Note that the restriction of bonding maps $\tilde{f}_n^{n+1}|_{(\mathbb{R}\times\mathbb{R}\times\mathcal{S}^1\times\{t\})_{n+1}} : (\mathbb{R}\times\mathbb{R}\times\mathcal{S}^1 \times \{t\})_{n+1} \subset \widetilde{Y}_{n+1} \to (\mathbb{R}\times\mathbb{R}\times \mathcal{S}^1 \times \{t^2\})_n \subset \widetilde{Y}_n$ is a homeomorphism, since the following map $g_{n+1}^n : (\mathbb{R} \times \mathbb{R} \times \mathcal{S}^1\times\{t^2\})_n \subset \widetilde{Y}_n \to (\mathbb{R}\times\mathbb{R}\times\mathcal{S}^1\times\{t\})_{n+1} \subset \widetilde{Y}_{n+1}$ given by $g_{n+1}^n((r_1, r_2, z_3, t^2)) = (\frac{1}{2}r_1, \frac{1}{2}r_2, z_3, t)$ is its inverse.

Define $g_{n+m}^n = g_{n+m}^{n+m-1} \circ \cdots \circ g_{n+1}^n$. Let $\tilde{t} = (t_n)_{n=1}^\infty \in \mathcal{C}$, and let $(\mathbb{R} \times \mathbb{R} \times \mathcal{S}^1 \times \{\tilde{t}\})_\infty$ be a component of $\widetilde{Y}_\infty$. Consider the map $\tilde{f}_n|_{(\mathbb{R}\times\mathbb{R}\times\mathcal{S}^1\times\{\tilde{t}\})_\infty} : (\mathbb{R} \times \mathbb{R} \times \mathcal{S}^1 \times \{\tilde{t}\})_\infty \subset \widetilde{Y}_\infty \to (\mathbb{R} \times \mathbb{R} \times \mathcal{S}^1 \times \{t_n\})_n \subset \widetilde{Y}_n$. We claim that this map is a homeomorphism. To this end, let $g^n : (\mathbb{R} \times \mathbb{R} \times \mathcal{S}^1 \times \{t_n\})_n \subset \widetilde{Y}_n \to (\mathbb{R} \times \mathbb{R} \times \mathcal{S}^1 \times \{\tilde{t}\})_\infty \subset \widetilde{Y}_\infty$ be given by

$$g^n((r_1, r_2, z_3, t_n)) = (f_1^n((r_1, r_2, z_3, t_n)), \ldots, f_{n-1}^n((r_1, r_2, z_3, t_n)),$$

$$(r_1, r_2, z_3, t_n), g_{n+1}^n((r_1, r_2, z_3, t_n)), \ldots, g_{n+m}^n((r_1, r_2, z_3, t_n)), \ldots).$$

Since it is easy to see that $g^n \circ \tilde{f}_n|_{(\mathbb{R}\times\mathbb{R}\times\mathcal{S}^1\times\{\tilde{t}\})_\infty} = 1_{(\mathbb{R}\times\mathbb{R}\times\mathcal{S}^1\times\{\tilde{t}\})_\infty}$ and $\tilde{f}_n|_{(\mathbb{R}\times\mathbb{R}\times\mathcal{S}^1\times\{\tilde{t}\})_\infty} \circ g^n = 1_{(\mathbb{R}\times\mathbb{R}\times\mathcal{S}^1\times\{t_n\})_n}$, $\tilde{f}_n|_{(\mathbb{R}\times\mathbb{R}\times\mathcal{S}^1\times\{\tilde{t}\})_\infty}$ is a homeomorphism.

Let $\widetilde{X}_n = q_n^{-1}(X_n)$. Then $\widetilde{X}_n$ is a locally compact, metric space with 2^{n-1} components, each of which is locally homeomorphic to $\mathbb{M}$. Let $\widetilde{\mathbb{MR}} = \varprojlim\{\widetilde{X}_n, \tilde{f}_n^{n+1}\}$, and let $\mathbb{K}$ be a component of $\widetilde{\mathbb{MR}}$. $\mathbb{K}$ is contained in a component of $\widetilde{Y}_\infty$. Let $(\mathbb{R} \times \mathbb{R} \times \mathcal{S}^1 \times \{\tilde{t}\})_\infty$ be such a component. Since $\tilde{f}_1|_{(\mathbb{R}\times\mathbb{R}\times\mathcal{S}^1\times\{\tilde{t}\})_\infty}$ is a homeomorphism, we have that $\tilde{f}_1|_\mathbb{K}$ is a homeomorphism onto $\tilde{f}_1(\mathbb{K}) = \widetilde{X}_1$. Since $\widetilde{X}_1$ is locally homeomorphic to $\mathbb{M}$, $\mathbb{K}$ is locally homeomorphic to $\mathbb{M}$. Therefore, $\mathbb{K}$ is locally connected.

Q.E.D.

References

1. J. M. Aarts and P. Van Emde Boas, Continua as Remainders in Compact Extensions, Nieuw Archief voor Wiskunde 3 (1967), 34–37.
2. J. M. Aarts and L. G. Oversteegen, The Product Structure of Homogeneous Spaces, Indag. Math., 1 (1990), 1–5.
3. L. V. Ahlfors, *Conformal Invariants, Topics in Geometric Function Theory*, Series in Higher Mathematics, McGraw-Hill Book Co., New York, 1973.
4. R. D. Anderson, A Characterization of the Universal Curve and a Proof of its Homogeneity, Ann. Math., 67 (1958), 313–324.
5. R. D. Anderson, One-dimensional Continuous Curves and a Homogeneity Theorem, Ann. Math., 68 (1958), 1–16.
6. D. P. Bellamy and L. Lum, The Cyclic Connectivity of Homogeneous Arcwise Connected Continua, Trans. Amer. Math. Soc., 266 (1981), 389–396.
7. J. H. Case, Another 1-dimensional Homogeneous Continuum Which Contains an Arc, Pacific J. Math., 11(1961), 455–469.

8. C. O. Christenson and W. L. Voxman, *Aspects of Topology*, Monographs and Textbooks in Pure and Applied Math., Vol. 39, Marcel Dekker, Inc., New York, Basel, 1977.
9. R. Fenn, What is the Geometry of a Surface?, The Amer. Math. Monthly, 90 (1983), 87–98.
10. J. Grispolakis and E. D. Tymchatyn, On Confluent Mappings and Essential Mappings—A Survey, Rocky Mountain J. Math., 11 (1981), 131–153.
11. C. L. Hagopian, A Characterization of Solenoids, Pacific J. Math., 68 (1977), 425–435.
12. J. G. Hocking and G. S. Young, *Topology*, Dover Publications, Inc., New York, 1988.
13. S. T. Hu, *Theory of Retracts*, Wayne State University Press, Detroit, 1965.
14. J. Krasinkiewicz, On One-point Union of Two Circles, Houston J. Math., 2 (1976), 91–95.
15. K. Kuratowski, *Topology*, Vol. II, Academic Press, New York, N. Y., 1968.
16. E. L. Lima, *Grupo Fundamental e Espaços de Recobrimento*, Instituto de Matemática Pura e Aplicada, CNPq, (Projeto Euclides), 1993. (Portuguese)
17. J. C. Macías, *El Teorema de Descomposición Terminal de Rogers*, Tesis de Maestría, Facultad de Ciencias Físico Matemáticas, B. U. A. P., 1999. (Spanish)
18. S. Macías, Covering Spaces of Homogeneous Continua, Topology Appl., (1994), 157–177.
19. S. Macías, Homogeneous Continua for Which the Set Function T is Continuous, Topology Appl., 153 (2006), 3397–3401.
20. T. Maćkowiak and E. D. Tymchatyn, Continuous Mappings on Continua II, Dissertationes Math., 225 (1984), 1–57.
21. M. C. McCord, Inverse Limit Sequences with Covering Maps, Trans. Amer. Math. Soc., 114 (1965), 197–209.
22. P. Minc and J. T. Rogers, Jr., Some New Examples of Homogeneous Curves, Topology Proc., 10 (1985), 347–356.
23. J. Munkres, *Topology*, second edition, Prentice Hall, Upper Saddle River, NJ, 2000.
24. S. B. Nadler, Jr., *Dimension Theory: An Introduction with Exercises*, Aportaciones Matemáticas, Serie Textos # 18, Sociedad Matemática Mexicana, 2002.
25. J. T. Rogers, Jr., Homogeneous Separating Plane Continua are Decomposable, Michigan J. Math., 28 (1981), 317–321.
26. J. T. Rogers, Jr., Decompositions of Homogeneous Continua, Pacific J. Math., 99 (1982), 137–144.
27. J. T. Rogers, Jr., Homogeneous Hereditarily Indecomposable Continua Are Tree-like, Houston J. Math., 8, (1982), 421–428.
28. J. T. Rogers, Jr., An Aposyndetic Homogeneous Curve That is not Locally Connected, Houston J. Math., 9 (1983), 433–440.
29. J. T. Rogers, Jr., Cell-like Decompositions of Homogeneous Continua, Proc. Amer. Math. Soc., 87 (1983), 375–377.
30. J. T. Rogers, Jr., Homogeneous Curves That Contain Arcs, Topology Appl., 21 (1985), 95–101.
31. J. T. Rogers, Jr., Orbits of Higher-dimensional Hereditarily Indecomposable Continua, Proc. Amer. Math. Soc., 95 (1985), 483–486.
32. J. T. Rogers, Jr., Hyperbolic Ends And Continua, Michigan Math. J., 34 (1987), 337–347.
33. J. T. Rogers, Jr., Decompositions of Continua Over the Hyperbolic Plane, Trans. Amer. Math. Soc., 310 (1988), 277–291.
34. J. T. Rogers, Jr., Higher Dimensional Aposyndetic Decompositions, Proc. Amer. Math. Soc., 131 (2003), 3285–3288.
35. P. Scott, The Geometries of 3-manifolds, Bull. London Math. Soc., 15 (1983), 401–487.
36. G. T. Whyburn, Topological Characterization of the Sierpiński Curve, Fund. Math., 45 (1958), 320–324.

Chapter 6
n-Fold Hyperspaces

We give a brief overview of n-fold hyperspaces. Other hyperspaces have been studied extensively in [84] and [48]. Throughout this chapter, σ denotes the union map and $\mathcal{H}$ denotes the Hausdorff metric.

We prove that n-fold hyperspaces are unicoherent (in fact, they have trivial shape), and aposyndetic. We study arcwise accessibility of points of the n-fold symmetric products form points of the n-fold hyperspaces. We give properties of the points that arcwise disconnect the n-fold hyperspaces. We study C_n^*-smoothness, Z-sets in n-fold hyperspaces and retractions between n and m fold hyperspaces. We present properties of the n-fold hyperspaces of graphs. We give conditions in order to have an n-fold hyperspace homeomorphic to a cone, suspension or product of continua. We end the chapter with a study of strong size maps.

6.1 General Properties

We start with a result which says that the union of a connected subset of the n-fold hyperspace of a continuum has at most n components. To prepare this section, we use [1, 12–14, 17, 21, 28, 37, 39, 40, 48, 52, 56, 60, 61, 66, 67, 69, 73, 80, 84, 87, 91].

6.1.1 Lemma *Let* $n \in \mathbb{N}$*, and let* X *be a continuum. If* $\mathcal{A}$ *is a connected subset of* 2^X *such that* $\mathcal{A} \cap C_n(X) \neq \emptyset$*, then* $\sigma(\mathcal{A}) = \bigcup\{A \mid A \in \mathcal{A}\}$ *has at most* n *components.*

Proof Suppose the result is not true. Then there exists a connected subset $\mathcal{A}$ of 2^X such that $\mathcal{A} \cap C_n(X) \neq \emptyset$ and $\sigma(\mathcal{A})$ has at least $n+1$ components. Thus, we can find $n + 1$ pairwise separated subsets, $C_1, \ldots, C_{n+1}$, of X such that $\sigma(\mathcal{A}) = \bigcup_{j=1}^{n+1} C_j$. Let $A_0 \in \mathcal{A} \cap C_n(X)$. Then $A_0 \subset \bigcup_{j=1}^{n+1} C_j$. Suppose that $A_0 \cap C_j \neq \emptyset$ for each

© Springer International Publishing AG, part of Springer Nature 2018   247
S. Macías, *Topics on Continua*, https://doi.org/10.1007/978-3-319-90902-8_6

$j \in \{1, \ldots, \ell\}$ and $A_0 \cap C_j = \emptyset$ for all $j \in \{\ell+1, \ldots, n+1\}$. Note that $\ell \leq n$. Let

$$\mathcal{A}' = \left\{ A \in \mathcal{A} \,\middle|\, A \subset \bigcup_{j=1}^{\ell} C_j \right\}$$

and

$$\mathcal{A}'' = \left\{ A \in \mathcal{A} \,\middle|\, A \cap \bigcup_{j=\ell+1}^{n+1} C_j \neq \emptyset \right\}.$$

Since $A_0 \in \mathcal{A}'$, $\mathcal{A}' \neq \emptyset$. Now, let $x \in C_{n+1} \subset \sigma(\mathcal{A})$. Then there exists $A'' \in \mathcal{A}$ such that $x \in A''$. Hence, $A'' \in \mathcal{A}''$ and $\mathcal{A}'' \neq \emptyset$. Observe that $\mathcal{A} = \mathcal{A}' \cup \mathcal{A}''$.

We show that $\mathcal{A}'$ and $\mathcal{A}''$ are separated. Let $A'' \in Cl_{2^X}(\mathcal{A}') \cap \mathcal{A}''$ and let $\mathcal{E} = \left\{ E \in 2^X \mid E \subset Cl_X \left(\bigcup_{j=1}^{\ell} C_j \right) \right\}$. Then $\mathcal{E}$ is a closed subset of 2^X and $Cl_{2^X}(\mathcal{A}') \subset \mathcal{E}$. Hence, $A'' \in \mathcal{E}$. Now, since $A'' \in \mathcal{A}''$, $A'' \cap \bigcup_{j=\ell+1}^{n+1} C_j \neq \emptyset$. Thus, $Cl_X \left(\bigcup_{j=\ell+1}^{n+1} C_j \right) \cap \left(\bigcup_{j=1}^{\ell} C_j \right) \neq \emptyset$, a contradiction. Therefore, $Cl_{2^X}(\mathcal{A}') \cap \mathcal{A}'' = \emptyset$.

Let $A' \in \mathcal{A}' \cap Cl_{2^X}(\mathcal{A}'')$ and let

$$\mathcal{G} = \left\{ G \in 2^X \,\middle|\, G \cap Cl_X \left(\bigcup_{j=\ell+1}^{n+1} C_j \right) \neq \emptyset \right\}.$$

Then $\mathcal{G}$ is a closed subset of 2^X and $Cl_{2^X}(\mathcal{A}'') \subset \mathcal{G}$. Hence, $A' \in \mathcal{G}$. Now, since $A' \in \mathcal{A}'$, $A' \subset \bigcup_{j=1}^{\ell} C_j$. Thus, $Cl_X \left(\bigcup_{j=1}^{\ell} C_j \right) \cap \left(\bigcup_{j=\ell+1}^{n+1} C_j \right) \neq \emptyset$, a contradiction. Hence, $\mathcal{A}' \cap Cl_{2^X}(\mathcal{A}'') = \emptyset$.

Then $\mathcal{A}'$ and $\mathcal{A}''$ are separated subsets of 2^X and $\mathcal{A} = \mathcal{A}' \cup \mathcal{A}''$, a contradiction. Therefore, $\sigma(\mathcal{A})$ has at most n components.

$$\textbf{Q.E.D.}$$

6.1.2 Corollary *Let $n \in \mathbb{N}$. If X is a continuum and $\mathcal{A}$ is a subcontinuum of 2^X such that $\mathcal{A} \cap C_n(X) \neq \emptyset$, then $\sigma(\mathcal{A}) \in C_n(X)$.*

Proof By Lemma 1.8.11, $\sigma(\mathcal{A})$ is a closed subset of X. By Lemma 6.1.1, $\sigma(\mathcal{A})$ has at most n components. Therefore, $\sigma(\mathcal{A}) \in C_n(X)$.

$$\textbf{Q.E.D.}$$

6.1.3 Notation Let X be a continuum. To simplify notation, we write: $\langle U_1, \ldots, U_m \rangle_n$, to denote the intersection of the open set $\langle U_1, \ldots, U_m \rangle$, of the Vietoris Topology, with $C_n(X)$.

6.1.4 Lemma *Let X be a continuum. If $\mathcal{A}$ is a subcontinuum of $C_n(X)$ and $A \in \mathcal{A}$, then A intersects each component of $\sigma(\mathcal{A})$.*

Proof Let $\mathcal{A}$ be a subcontinuum of $C_n(X)$. By Lemma 1.8.11, $\sigma(\mathcal{A})$ is a compact subset of X. Suppose there exists a component C of $\sigma(\mathcal{A})$ such that $C \cap A = \emptyset$. Then, by Theorem 1.6.8, there exist two nonempty closed and disjoint subsets K and L of X such that $\sigma(\mathcal{A}) = K \cup L$, $A \subset K$ and $C \subset L$. Let

$$\mathcal{K} = \langle K \rangle_n \cap \mathcal{A} \quad \text{and} \quad \mathcal{L} = \langle X, L \rangle_n \cap \mathcal{A}.$$

Then $\mathcal{K}$ and $\mathcal{L}$ are nonempty closed and disjoint subsets of $C_n(X)$ such that $\mathcal{K} \cup \mathcal{L} = \mathcal{A}$, a contradiction to the fact that $\mathcal{A}$ is connected.

Q.E.D.

6.1.5 Proposition *Let X be a continuum and let $A, A_1, \ldots, A_k$ be subsets of X. Then:*

(1) If $\langle A \rangle_n$ is a subcontinuum of $C_n(X)$, then A is a subcontinuum of X;
(2) If $A_1, \ldots, A_k$ are subcontinua of X, then $\langle A_1, \ldots, A_k \rangle_n$ is a subcontinuum of $C_n(X)$, whenever we have that $\langle A_1, \ldots, A_k \rangle_n \neq \emptyset$.

Proof Note that (1) follows from continuity of the union map σ (Lemma 1.8.11) and Lemma 6.1.4.

To see (2), suppose that $\langle A_1, \ldots, A_k \rangle_n \neq \emptyset$. Observe that $\langle A_1, \ldots, A_k \rangle_n$ is a closed subset of $C_n(X)$. Also note that $\bigcup_{j=1}^{k} A_j \in \langle A_1, \ldots, A_k \rangle_n$. Let $B \in \langle A_1, \ldots, A_k \rangle_n$. Then there exists an order arc α from B to $\bigcup_{j=1}^{k} A_j$, by Theorem 1.8.20. Note that $\alpha \subset \langle A_1, \ldots, A_k \rangle_n$. Since B is an arbitrary point of $\langle A_1, \ldots, A_k \rangle_n$, we obtain, in fact, that $\langle A_1, \ldots, A_k \rangle_n$ is arcwise connected. Therefore, $\langle A_1, \ldots, A_k \rangle_n$ is a subcontinuum of $C_n(X)$.

Q.E.D.

6.1.6 Theorem *Let $n \in \mathbb{N}$. Then the continuum X is locally connected if and only if $C_n(X)$ is locally connected.*

Proof Suppose that X is locally connected. We see first that $\mathcal{C}(X)$ is locally connected. Let $A \in \mathcal{C}(X)$ and let $\langle U_1, \ldots, U_k \rangle_1$ be a basic subset of $\mathcal{C}(X)$ such that $A \in \langle U_1, \ldots, U_k \rangle_1$. Recall that, by definition, $A \subset \bigcup_{j=1}^{k} U_j$. Since X is locally connected, for each $a \in A$, there exists a connected open subset V_a of X such that $a \in V_a \subset Cl(V_a) \subset \bigcup_{j=1}^{k} U_j$. Since A is compact, there exist $a_1, \ldots, a_r \in A$ such that $A \subset \bigcup_{j=1}^{r} V_{a_j}$. Without loss of generality, we assume that for each $j \in \{1, \ldots, k\}$, there exists $\ell \in \{1, \ldots, r\}$ such that $a_\ell \in A \cap U_j$ and $Cl(V_{a_\ell}) \subset U_l$. Then $A \in \langle V_{a_1}, \ldots, V_{a_r} \rangle_1 \subset \langle U_1, \ldots, U_k \rangle_1$. Let $D = \bigcup_{j=1}^{r} Cl(V_{a_j})$. Note that $D \in \mathcal{C}(X)$ and $D \in \langle U_1, \ldots, U_k \rangle_1$. Since $A \subset D$, by Theorem 1.8.20, there exists an order arc $\alpha \colon [0, 1] \to \mathcal{C}(X)$ such that $\alpha(0) = A$ and $\alpha(1) = D$. Observe that for each $t \in [0, 1]$, $\alpha(t) \in \langle U_1, \ldots, U_k \rangle_1$. Let $B \in \langle V_{a_1}, \ldots, V_{a_r} \rangle_1$. Then $B \subset D$. By Theorem 1.8.20, there exists an order arc $\beta \colon [0, 1] \to \mathcal{C}(X)$ such that $\beta(0) = B$ and $\beta(1) = D$. Hence, $\alpha([0, 1]) \cup \beta([0, 1])$ is a connected subset of $\langle U_1, \ldots, U_k \rangle_1$ containing A and B. By Theorem 1.7.9, $\mathcal{C}(X)$ is connected im kleinen at A. Therefore, since A is an arbitrary point of $\mathcal{C}(X)$, by Theorem 1.7.12, $\mathcal{C}(X)$ is locally connected.

Suppose that $n \geq 2$. Since $\mathcal{C}(X)$ is locally connected, we have that $\mathcal{C}(X)^n$ is locally connected. By Corollary 1.8.7, the function $f_n \colon \mathcal{C}(X)^n \twoheadrightarrow \mathcal{F}_n(\mathcal{C}(X))$ is continuous. Thus, $\mathcal{F}_n(\mathcal{C}(X))$ is locally connected. Since the union map, σ, is continuous (Lemma 1.8.11) and $\sigma((\mathcal{F}_n(\mathcal{C}(X)))) = \mathcal{C}_n(X)$, we obtain that $\mathcal{C}_n(X)$ is locally connected.

Now, assume that $\mathcal{C}_n(X)$ is locally connected. Let x be a point in X and let U be an open subset of X containing x. Since $\mathcal{C}_n(X)$ is locally connected, there exists a connected open subset $\mathcal{V}$ of $\mathcal{C}_n(X)$ such that $\{x\} \in \mathcal{V} \subset Cl(\mathcal{V}) \subset \langle U \rangle_n$. Let $\langle V_1, \ldots, V_k \rangle_n$ be a basic open set of $\mathcal{C}_n(X)$ such that $\{x\} \in \langle V_1, \ldots, V_k \rangle_n \subset \mathcal{V}$. Let $V = \bigcap_{j=1}^k V_j$. Now, if $y \in V$, then $y \in \sigma(Cl(\mathcal{V}))$. Since $\{x\} \in Cl(\mathcal{V})$, we have that $\sigma(Cl(\mathcal{V})) \in \mathcal{C}(X)$, by Corollary 6.1.2. Thus, both x and y belong to $\sigma(Cl(\mathcal{V})) \subset U$. Hence, X is connected im kleinen at x. Since x is an arbitrary point of X, we have that X is locally connected (Theorem 1.7.12).

Q.E.D.

6.1.7 Definition A *free arc* in a metric space X is an arc α such that $\alpha \setminus \{$end points$\}$ is an open subset of X.

A proof of the following theorem may be found in [84, (1.98)], [60, Theorem 7.1] and [66, Theorem 3.4].

6.1.8 Theorem *Let X be a continuum and let n be a positive integer. Then $\mathcal{C}_n(X)$ is homeomorphic to the Hilbert cube if and only if X is locally connected and does not contain free arcs.*

6.1.9 Theorem *Let X be a continuum. If $n \in \mathbb{N}$, then $\mathcal{C}_n(X)$ is nowhere dense in $\mathcal{C}_{n+1}(X)$. In particular, $\mathcal{C}_n(X)$ is nowhere dense in 2^X.*

Proof Let $n \in \mathbb{N}$. Suppose $Int_{\mathcal{C}_{n+1}(X)}(\mathcal{C}_n(X)) \neq \emptyset$. Recall that $\mathcal{F}(X)$ is dense in 2^X (see the proof of Corollary 1.8.9). Hence, X does not belong to $Int_{\mathcal{C}_{n+1}(X)}(\mathcal{C}_n(X))$. Let $A \in Int_{\mathcal{C}_{n+1}(X)}(\mathcal{C}_n(X) \setminus \{X\})$, and suppose that $A_1, \ldots, A_k$ are the components of A. Thus, $k \leq n$. Then there exists $\varepsilon > 0$ such that $\mathcal{V}_\varepsilon^{\mathcal{H}}(A) \cap \mathcal{C}_{n+1}(X) \subset Int_{\mathcal{C}_{n+1}(X)}(\mathcal{C}_n(X))$. Without loss of generality, we assume that ε is small enough that $\mathcal{V}_\varepsilon^d(A_j) \cap \mathcal{V}_\varepsilon^d(A_\ell) = \emptyset$ if and only if $j \neq \ell$ and $j, \ell \in \{1, \ldots, k\}$. Suppose that $n - k = m$ and let $x_1, \ldots, x_{m+1} \in \mathcal{V}_\varepsilon^d(A_1) \setminus A_1$ be $m + 1$ distinct points. Let $B = A \cup \{x_1, \ldots, x_{m+1}\}$. Then $B \in \mathcal{C}_{n+1}(X) \setminus \mathcal{C}_n(X)$ and $\mathcal{H}(A, B) < \varepsilon$. Therefore $Int_{\mathcal{C}_{n+1}(X)}(\mathcal{C}_n(X)) = \emptyset$.

Similarly, $Int_{2^X}(\mathcal{C}_n(X)) = \emptyset$.

Q.E.D.

6.1.10 Lemma *Let X be a continuum, with metric d. If $n \in \mathbb{N}$, then there exist n pairwise disjoint nondegenerate subcontinua of X.*

Proof Let $n \in \mathbb{N}$, and let $x_1, \ldots, x_n$ be n distinct points of X. Let $\eta = \frac{1}{2} \min\{d(x_j, x_k) \mid j, k \in \{1, \ldots, n\}$ and $j \neq k\}$. Then $\eta > 0$ and $\mathcal{V}_\eta^d(x_j) \cap \mathcal{V}_\eta^d(x_k) =$

$\emptyset$ for each $j, k \in \{1, \ldots, n\}$ and $j \neq k$. By Corollary 1.7.28, there exists a subcontinuum A_j of X such that $\{x_j\} \subsetneq A_j \subset \mathcal{V}_n^d(x_j)$ for every $j \in \{1, \ldots, n\}$. Therefore, $A_1, \ldots, A_n$ are n pairwise disjoint nondegenerate subcontinua of X.

Q.E.D.

6.1.11 Theorem *Let X be a continuum. If $n \in \mathbb{N}$, then $\mathcal{C}_n(X)$ contains an n-cell.*

Proof Let $A_1, \ldots, A_n$ be n pairwise disjoint nondegenerate subcontinua of X (Lemma 6.1.10). For each $j \in \{1, \ldots, n\}$, let $a_j \in A_j$, and let $\alpha_j : [0, 1] \to \mathcal{C}(X)$ be an order arc such that $\alpha_j(0) = \{a_j\}$ and $\alpha_j(1) = A_j$ (Theorem 1.8.20). Then the map $\xi : [0, 1]^n \to \mathcal{C}_n(X)$ given by $\xi((t_1, \ldots, t_n)) = \alpha_1(t_1) \cup \cdots \cup \alpha_n(t_n)$ is an embedding of $[0, 1]^n$ in $\mathcal{C}_n(X)$.

Q.E.D.

If the continuum X contains decomposable continua, more can be said.

6.1.12 Theorem *Let X be a continuum and let $n \in \mathbb{N}$. If X contains k pairwise disjoint decomposable subcontinua ($k \leq n$), then $\mathcal{C}_n(X)$ contains a $(k + n)$-cell.*

Proof First suppose $k < n$. Let $M_1, \ldots, M_k$ be k pairwise disjoint decomposable subcontinua of X. Suppose that $M_j = A_j \cup B_j$, where A_j and B_j are continua, for each $j \in \{1, \ldots, k\}$. By the proof of [84, (1.145)], we may assume that for each $j \in \{1, \ldots, k\}$, $A_j \cap B_j$ is connected, $A_j \setminus (A_j \cap B_j) \neq \emptyset$, $B_j \setminus (A_j \cap B_j) \neq \emptyset$, and $[A_j \setminus (A_j \cap B_j)] \cap [B_j \setminus (A_j \cap B_j)] = \emptyset$. Let $C_{k+1}, \ldots, C_n$ be $n - k$ pairwise disjoint nondegenerate subcontinua of X such that $M_j \cap C_\ell = \emptyset$ for every $j \in \{1, \ldots, k\}$ and every $\ell \in \{k + 1, \ldots, n\}$. For each $j \in \{1, \ldots, k\}$, let $\alpha_j : [0, 1] \to \mathcal{C}(A_j)$ and $\beta_j : [0, 1] \to \mathcal{C}(B_j)$ be order arcs such that $\alpha_j(0) = A_j \cap B_j, \alpha_j(1) = A_j, \beta_j(0) = A_j \cap B_j$, and $\beta_j(1) = B_j$ (Theorem 1.8.20). For each $\ell \in \{k + 1, \ldots, n\}$, let $x_\ell \in C_\ell$. Let $\gamma_\ell : [0, 1] \to \mathcal{C}(C_\ell)$ be an order arc such that $\gamma_\ell(0) = \{x_\ell\}$ and $\gamma_\ell(1) = C_\ell, \ell \in \{k + 1, \ldots, n\}$. Since $[0, 1]^{k+n}$ is homeomorphic to $[0, 1]^{2k} \times [0, 1]^{n-k}$, we need an embedding of $[0, 1]^{2k} \times [0, 1]^{n-k}$ into $\mathcal{C}_n(X)$. The map $\xi : [0, 1]^{2k} \times [0, 1]^{n-k} \to \mathcal{C}_n(X)$ given by

$$\xi((t_1, \ldots, t_{2k}), (t_1, \ldots, t_{n-k})) =$$

$$\left(\bigcup_{j=1}^{k} (\alpha_j(t_{2j-1}) \cup \beta_j(t_{2j})) \right) \cup \left(\bigcup_{\ell=1}^{n-k} \gamma_{k+\ell}(t_\ell) \right)$$

is an embedding of $[0, 1]^{k+n}$ in $\mathcal{C}_n(X)$. When $k = n$, repeat the argument without using the γ's.

Q.E.D.

As a consequence of Theorem 6.1.12, we have the following corollaries:

6.1.13 Corollary *If X is a continuum which is not hereditarily indecomposable, then $\dim(\mathcal{C}_n(X)) \geq n + 1$.*

6.1.14 Corollary *If X is a graph topologically different from an arc and a simple closed curve, and n is a positive integer, then $\dim(C_n(X)) \geq 2n + 1$.*

6.1.15 Corollary *Let X be a continuum and let $n \in \mathbb{N}$. If X contains n pairwise disjoint decomposable subcontinua, then $C_n(X)$ contains a $2n$-cell.*

6.1.16 Corollary *If X is a continuum containing an arc, then $C_n(X)$ contains a $2n$-cell for each $n \in \mathbb{N}$.*

6.1.17 Theorem *Let X be a continuum. If $n \in \mathbb{N}$, then the following are equivalent:*

(1) 2^X is contractible;
(2) $C_n(X)$ is contractible;
(3) $C(X)$ is contractible.

Proof Suppose 2^X is contractible. Then there exists a map $H' \colon 2^X \times [0, 1] \to 2^X$ such that for each $A \in 2^X$, $H'((A, 0)) = A$ and $H'((A, 1)) = X$. Let $H \colon 2^X \times [0, 1] \to 2^X$ be the *segment homotopy associated* with H' defined by

$$H((A, t)) = \sigma \left(\{H'((A, s)) \mid 0 \leq s \leq t\} \right).$$

Then H is continuous [84, (16.3)]. Observe that for each $A \in 2^X$, $H((A, 0)) = A$, $H((A, 1)) = X$, and $H(\{A\} \times [0, 1])$ is an order arc from A to X. Note that if $A \in C_n(X)$ and $B \in H(\{A\} \times [0, 1])$, then $B \in C_n(X)$ (Theorem 1.8.20). Therefore, $G = H|_{C_n(X) \times [0,1]} \colon C_n(X) \times [0, 1] \to C_n(X)$ is a continuous function such that for each $A \in C_n(X)$, $G((A, 0)) = A$ and $G((A, 1)) = X$. Hence, $C_n(X)$ is contractible. A similar argument shows that if $C_n(X)$ is contractible, then $C(X)$ is contractible. The other implication is contained in the proof of [84, (16.7)].

Q.E.D.

6.1.18 Definition A continuum X is said to have the *property of Kelley* provided that given any $\varepsilon > 0$, there exists $\delta > 0$ such that if $a, b \in X$, $d(a, b) < \delta$, and $a \in A \in C(X)$, then there exists $B \in C(X)$ such that $b \in B$ and $\mathcal{H}(A, B) < \varepsilon$. This number δ is called a *Kelley number* for the given ε.

6.1.19 Corollary *If X is a continuum having the property of Kelley, then $C_n(X)$ is contractible for each $n \in \mathbb{N}$.*

Proof If X is a continuum having the property of Kelley, then 2^X and $C(X)$ are contractible [84, (16.5)]. Hence, the result follows from Theorem 6.1.17.

Q.E.D.

6.1.20 Lemma *Let $n \in \mathbb{N}$, and let X be a continuum having the property of Kelley. If $\mathcal{W}$ is a subcontinuum of $C_n(X)$ having nonempty interior, then $\sigma(\mathcal{W}) \in C_n(X)$ and $Int_X(\sigma(\mathcal{W})) \neq \emptyset$.*

Proof By Corollary 6.1.2, $\sigma(\mathcal{W}) \in C_n(X)$. Let $A \in Int_{C_n(X)}(\mathcal{W})$. Let $A_1, \ldots, A_k$ be the components of A. Then $k \leq n$. Let $\varepsilon > 0$ be given such that $\mathcal{V}_\varepsilon^{\mathcal{H}}(A) \cap C_n(X) \subset$

$\mathcal{W}$, and such that $\mathcal{V}_\varepsilon^d(A_j) \cap \mathcal{V}_\varepsilon^d(A_\ell) = \emptyset$ when $j \neq \ell$. Let $\delta > 0$ be a Kelley number for the given ε. Clearly, $A \subset \sigma(\mathcal{W})$. For each $j \in \{1, \ldots, k\}$, let $a_j \in A_j$. For every $j \in \{1, \ldots, k\}$, let $x_j \in \mathcal{V}_\delta^d(a_j)$. Since X has the property of Kelley, there exists a subcontinuum B_j of X such that $x_j \in B_j$ and $\mathcal{H}(A_j, B_j) < \varepsilon$ for each $j \in \{1, \ldots, k\}$. Let $B = \bigcup_{j=1}^k B_j$. Then $B \in \mathcal{C}_n(X)$. Note that $\mathcal{V}_\varepsilon^d(A) = \bigcup_{j=1}^k \mathcal{V}_\varepsilon^d(A_j)$, $\mathcal{V}_\varepsilon^d(B) = \bigcup_{j=1}^k \mathcal{V}_\varepsilon^d(B_j)$, and $\mathcal{H}(A_j, B_j) < \varepsilon$ for each $j \in \{1, \ldots, k\}$; hence, $\mathcal{H}(A, B) < \varepsilon$. Thus, $B \in \mathcal{W}$, and $B \subset \sigma(\mathcal{W})$. Therefore, $\bigcup_{j=1}^k \mathcal{V}_\delta^d(a_j) \subset \sigma(\mathcal{W})$, and $Int_X(\sigma(\mathcal{W})) \neq \emptyset$.

Q.E.D.

6.1.21 Theorem *Let $n \in \mathbb{N}$. If X is an indecomposable continuum having the property of Kelley, then X is the only point at which $\mathcal{C}_n(X)$ is locally connected.*

Proof By [73, Lemma 2.3], $\mathcal{C}_n(X)$ is locally connected at X. Suppose that $\mathcal{C}_n(X)$ is locally connected at the point $A \neq X$. Since $\mathcal{C}_n(X)$ is locally connected at A, there exists a subcontinuum $\mathcal{W}$ of $\mathcal{C}_n(X)$ such that $A \in Int_{\mathcal{C}_n(X)}(\mathcal{W})$ and $\sigma(\mathcal{W}) \neq X$. By Lemma 6.1.20, $\sigma(\mathcal{W}) \in \mathcal{C}_n(X)$ and $Int_X(\sigma(\mathcal{W})) \neq \emptyset$. Since $\sigma(\mathcal{W})$ has finitely many components and $Int_X(\sigma(\mathcal{W})) \neq \emptyset$, it follows that at least one of the components of $\sigma(\mathcal{W})$ has nonempty interior, which is impossible because $\sigma(\mathcal{W}) \neq X$ and X is an indecomposable continuum (Corollary 1.7.26).

Q.E.D.

As a consequence of Theorem 6.1.21, we have the following result:

6.1.22 Theorem *Let $n \in \mathbb{N}$. If X is a hereditarily indecomposable continuum, then X is the only point at which $\mathcal{C}_n(X)$ is locally connected.*

Proof Since hereditarily indecomposable continua have the property of Kelley [84, (16.27)], we have that the result follows from Theorem 6.1.21.

Q.E.D.

The following result is easy to establish.

6.1.23 Lemma *Let $n \in \mathbb{N}$, $n \geq 2$, and let X be a continuum. Let A be a point of $\mathcal{C}_n(X) \setminus \mathcal{C}_{n-1}(X)$, and suppose $A_1, \ldots, A_n$ are the components of A. Let $\varepsilon > 0$ be such that $\mathcal{V}_{2\varepsilon}^d(A_j) \cap \mathcal{V}_{2\varepsilon}^d(A_k) = \emptyset$ if and only if $j \neq k$ and $j, k \in \{1, \ldots, n\}$. Let B be a point of $\mathcal{C}_n(X) \setminus \mathcal{C}_{n-1}(X)$, and suppose $B_1, \ldots, B_n$ are the components of B. Then $\mathcal{H}(A, B) < \varepsilon$ if and only if $\mathcal{H}^2(\{A_1, \ldots, A_n\}, \{B_1, \ldots, B_n\}) < \varepsilon$, where $\mathcal{H}^2$ is the Hausdorff metric on $\mathcal{F}_n(\mathcal{C}(X))$ induced by $\mathcal{H}$.*

As a consequence of Lemma 6.1.23, we have the following result.

6.1.24 Theorem *Let $n \in \mathbb{N}$, $n \geq 2$. If X is a continuum, then the function $f_n: \mathcal{C}_n(X) \setminus \mathcal{C}_{n-1}(X) \to \mathcal{F}_n(\mathcal{C}(X))$ given by*

$$f_n(A) = \{K \mid K \text{ is a component of } A\}$$

is an embedding.

The next corollary says that given a continuum X, if $C(X)$ is of finite dimension, then all the n-fold hyperspaces of X are of finite dimension too.

6.1.25 Corollary *If X is a continuum such that $C(X)$ is of finite dimension then for each $n \geq 2$, $\dim(C_n(X)) \leq n(\dim(C(X)))$.*

Proof Let $n \geq 2$. Then $\dim(\mathcal{F}_n(C(X))) \leq n(\dim(C(X)))$ [21, Lemma 3.1]. Also, $\dim(C_n(X)) \leq \dim(\mathcal{F}_n(C(X)))$ (Theorem 6.1.24 and [87, 7.3]). Therefore, $\dim(C_n(X)) \leq n(\dim(C(X)))$.

Q.E.D.

6.1.26 Corollary *If X is a continuum such that all its nondegenerate proper subcontinua are arcs, then $dim(C_n(X)) = 2n$ for every $n \in \mathbb{N}$.*

Proof Since the hyperspace of subcontinua of an arc is a 2-cell [48, 5.1], $\dim(C(Y)) < 3$ for every proper subcontinuum Y of X. Hence, by [91, Theorem 2.2], $\dim(C(X)) < 3$. Also, by [48, 22.18], $\dim(C(X)) \geq 2$. Thus, $\dim(C(X)) = 2$. Therefore, by Corollaries 6.1.16 and 6.1.25, $\dim(C_n(X)) = 2n$ for every $n \in \mathbb{N}$.

Q.E.D.

6.1.27 Corollary *If X is a hereditarily indecomposable tree-like continuum (Definition 6.8.27) and $n \in \mathbb{N}$, then $\dim(C_n(X)) \leq 2n$.*

Proof By [52, 4.1], $\dim(C(X)) = 2$. Therefore, by Corollary 6.1.25, $\dim(C_n(X)) \leq 2n$.

Q.E.D.

We use the following lemma to characterize continua for which its n-fold hyperspace is homogeneous.

6.1.28 Lemma *Let n be a positive integer. Then neither $C_n([0, 1])$ nor $C_n(S^1)$ is homogeneous.*

Proof The lemma follows from the facts that there exist points of $C_n([0, 1])$ and of $C_n(S^1)$ which have open $2n$-cell neighborhoods in $C_n([0, 1])$ and $C_n(S^1)$ [69, Lemma 4.2 and Corollary 4.3], respectively, and points, like any element of $C_n(X)$ with less that n components, which do not have that property, apply the Brouwer Invariance of Domain Theorem [40, Theorem VI 9, p. 95].

Q.E.D.

The following theorem extends [84, (17.2)] to n-fold hyperspaces.

6.1.29 Theorem *If X is a continuum, $\mathcal{Q}$ is the Hilbert cube and n is a positive integer, then the following are equivalent:*

(1) $C_n(X)$ is homogeneous;
(2) X is locally connected and does not contain free arcs;
(3) $C_n(X)$ is homeomorphic to $\mathcal{Q}$.

Proof Suppose $C_n(X)$ is homogeneous. Since $C_n(X)$ is homogeneous and locally connected at X [73, Lemma 2.3], $C_n(X)$ is locally connected. Hence, X is locally connected, by Theorem 6.1.6. Suppose X contains a free arc α. Let $a \in Int(\alpha)$. Then $\{a\}$ has arbitrary small neighborhoods in $C_n(X)$ homeomorphic to $C_n([0, 1])$. Hence, since $C_n(X)$ is homogeneous, $\dim(C_n(X)) = 2n$. Thus, X is a graph, by Theorem 6.9.3. Then, by Corollary 6.1.14, X must be an arc or a simple closed curve. But, by Lemma 6.1.28, neither $C_n([0, 1])$ nor $C_n(\mathcal{S}^1)$ is homogeneous, a contradiction. Therefore, X does not contain a free arc.

Now, if X is a locally connected continuum without free arcs, then $C_n(X)$ is homeomorphic to Q, by Theorem 6.1.8.

Finally, since Q is homogeneous [80, Theorem 6.1.6], if $C_n(X)$ is homeomorphic to Q, then $C_n(X)$ is homogeneous.

Q.E.D.

6.1.30 Theorem *If X is a finite-dimensional continuum, then $C_n(X)$ is not homeomorphic to $\mathcal{F}_n(X)$ for any positive integer n.*

Proof Let n be a positive integer. Suppose $C_n(X)$ is homeomorphic to $\mathcal{F}_n(X)$. Since X is finite dimensional, $\dim(\mathcal{F}_n(X)) = \dim(X^n) < \infty$ [87, 22.12]. Hence, $\dim(C_n(X)) < \infty$ and $\dim(X) = 1$ [56, Theorem 2.1].

Since $\dim(\mathcal{C}(X)) \geq 2$ [48, 22.18] and $\dim(X) = 1$, $\mathcal{C}(X)$ is not homeomorphic to $\mathcal{F}_1(X)$ ($\mathcal{F}_1(X)$ is homeomorphic to X).

Suppose that $n \geq 2$. Note that, by [40, Theorem III 4, p. 33], $\dim(X^n) \leq n$. On the other hand, since $\dim(X) = 1$, by [39, (a), p. 197], $\dim(X^n) \geq n$. Therefore, $\dim(X^n) = n$. Thus, $\dim(C_n(X)) = n$. Hence, X is hereditarily indecomposable, by Corollary 6.1.13. Since $C_n(X)$ is arcwise connected (Corollary 1.8.12), $\mathcal{F}_n(X)$ is arcwise connected. Thus, X is arcwise connected [21, Lemma 2.2]. A contradiction to the fact that X is hereditarily indecomposable.

Therefore, $C_n(X)$ is not homeomorphic to $\mathcal{F}_n(X)$.

Q.E.D.

6.1.31 Theorem *If X is an arc-smooth continuum and n is a positive integer, then $C_n(X)$ is arc-smooth.*

Proof By [28, Theorem II-3-B], X is freely contractible; i.e., there exist a point p and a homotopy $R: X \times [0, 1] \to X$ such that for each x in X: (1) $R((x, 0)) = p$, (2) $R((x, 1)) = x$ and (3) $R((R((x, s)), t)) = R((x, \min\{s, t\}))$ for all $s, t \in [0, 1]$.

Define $G: C_n(X) \times [0, 1] \to C_n(X)$ by

$$G((A, t)) = \{R((a, t)) \mid a \in A\}.$$

Note that G is continuous. Also observe that for each $A \in C_n(X)$, $G((A, 0)) = \{p\}$ and $G((A, 1)) = A$. It is easy to verify that $G((G((A, s)), t)) = G((A, \min\{s, t\}))$ for all $s, t \in [0, 1]$ and each $A \in C_n(X)$. Hence, $C_n(X)$ is freely contractible. Therefore, $C_n(X)$ is arc-smooth [28, Theorem II-3-B].

Q.E.D.

6.2 Unicoherence

We show that $C_n(X)$ has trivial shape and it is unicoherent for every $n \in \mathbb{N}$. To this end, we follow [5, 9, 22, 23, 41, 48, 51, 53, 60, 67, 80, 91, 98].

6.2.1 Definition A compactum X has *trivial shape* if X is homeomorphic to $\varprojlim\{X_n, f_n^{n+1}\}$, where each X_n is an absolute neighborhood retract, and for each $n \in \mathbb{N}$, there exists $m \geq n$ such that f_n^m is homotopic to a constant map.

A proof of the following theorem may be found in [53, 2.1].

6.2.2 Theorem *A continuum X has trivial shape if and only if each map from X into an absolute neighborhood retract is homotopic to a constant map.*

6.2.3 Theorem *If X is a continuum, then $C_n(X)$ has trivial shape for each $n \in \mathbb{N}$.*

Proof Let $n \in \mathbb{N}$. Note that by Theorem 2.1.56, X is homeomorphic to $\varprojlim\{X_m, f_m^{m+1}\}$, where each X_m is a polyhedron. Hence, by [98, Théorème II$_m$], $C_n(X_m)$ is an absolute retract. Also, by Theorem 2.3.4, $C_n(X)$ is homeomorphic to $\varprojlim\{C_n(X_m), C_n(f_m^{m+1})\}$. Since absolute retracts are contractible [80, 1.6.7], by Theorem 1.3.12, all the maps $C_n(f_k^m)$ are homotopic to a constant map. Therefore, $C_n(X)$ has trivial shape.

$$\text{Q.E.D.}$$

6.2.4 Remark Given a continuum X, $\check{H}^1(X; \mathbb{Z})$ denotes the first Čech cohomology group of X with integer coefficients. It is known that $\check{H}^1(X; \mathbb{Z}) = \{0\}$ if and only if each map from X into $\mathcal{S}^1$ is homotopic to a constant map [22, 8.1]. This implies, by [48, 19.7], that if $\check{H}^1(X; \mathbb{Z}) = \{0\}$, then X is unicoherent.

6.2.5 Theorem *If X is a continuum, then for every $n \in \mathbb{N}$, each map from $C_n(X)$ to the unit circle, $\mathcal{S}^1$, is homotopic to a constant map. In particular, we have that $C_n(X)$ is unicoherent.*

Proof By Theorem 2.1.56, X is homeomorphic to the inverse limit of an inverse sequence of compact and connected polyhedra, $\{X_m, f_m^{m+1}\}$. Hence, $C_n(X_m)$ is a continuum which is an absolute retract [98, Théorème II$_m$] for each $n \in \mathbb{N}$.

By [5, 2.4, p. 86] and [22, 8.1], each map from $C_n(X_m)$ into $\mathcal{S}^1$ is homotopic to a constant map. Then we have that $\check{H}^1(C_n(X_m); \mathbb{Z}) = \{0\}$ [22, 8.1]. By the continuity theorem for Čech cohomology [92, Theorem 7–7], $\check{H}^1(C_n(X), \mathbb{Z}) = \{0\}$. Thus, every map from $C_n(X)$ into $\mathcal{S}^1$ is homotopic to a constant map [22, 8.1]. Therefore, by Remark 6.2.4, we have that $C_n(X)$ is unicoherent for each $n \in \mathbb{N}$.

$$\text{Q.E.D.}$$

Next, we show that not all n-fold symmetric products of continua have trivial shape.

From the proof of [41, 1.1], we obtain the following lemma. It is worth mentioning that this result does not extend to n-fold symmetric products for $n \geq 3$.

It is known that $\mathcal{F}_3(\mathcal{S}^1)$ is homeomorphic to the 3-sphere $\mathcal{S}^3$ [9] and any map from $\mathcal{S}^3$ into $\mathcal{S}^1$ is homotopic to a constant map [23, p. 343].

6.2.6 Lemma *Let X be a continuum such that there exists a map from X onto $\mathcal{S}^1$ that is not homotopic to a constant map. Then there exists a map from $\mathcal{F}_2(X)$ onto $\mathcal{S}^1$ that is not homotopic to a constant map.*

6.2.7 Theorem *If X is a circle-like continuum, which is not arc-like, then $\mathcal{F}_2(X)$ does not have trivial shape.*

Proof By [51, 3.1], there exists a map from X onto $\mathcal{S}^1$ that is not homotopic to a constant map. Then, by Lemma 6.2.6, there exists a map from $\mathcal{F}_2(X)$ onto $\mathcal{S}^1$ that is not homotopic to a constant map. Therefore, $\mathcal{F}_2(X)$ does not have trivial shape (Theorem 6.2.2).

Q.E.D.

6.3 Aposyndesis

We prove that for every $n \in \mathbb{N}$, $\mathcal{C}_n(X)$ is colocally connected and finitely aposyndetic. We also prove that $\mathcal{C}_n(X)$ is zero-dimensional aposyndetic. The material of this section is based on [3, 48, 59, 60, 73, 77, 79, 80, 87, 98].

6.3.1 Theorem *Let X be a continuum, with metric d, and let $n \subset \mathbb{N}$ be given. Then $\mathcal{C}_n(X)$ is colocally connected.*

Proof Let A be a point of $\mathcal{C}_n(X)$. We consider two cases.

First assume that $A \in \mathcal{C}_n(X) \setminus \mathcal{F}_n(X)$. Let $\varepsilon > 0$ be given such that $(\mathcal{V}_\varepsilon^{\mathcal{H}}(A) \cap \mathcal{C}_n(X)) \cap \mathcal{F}_n(X) = \emptyset$. To see that $\mathcal{C}_n(X) \setminus (\mathcal{V}_\varepsilon^{\mathcal{H}}(A) \cap \mathcal{C}_n(X))$ is connected, let $B \subseteq \mathcal{C}_n(X) \setminus (\mathcal{V}_\varepsilon^{\mathcal{H}}(A) \cap \mathcal{C}_n(X))$, and let $B_1, \ldots, B_\ell$ be the components of B. Then $\ell \leq n$. Since $B \in \mathcal{C}_n(X) \setminus (\mathcal{V}_\varepsilon^{\mathcal{H}}(A) \cap \mathcal{C}_n(X))$, we have that $\mathcal{H}(A, B) \geq \varepsilon$. Hence, either $A \not\subset \mathcal{V}_\varepsilon^d(B)$ or $B \not\subset \mathcal{V}_\varepsilon^d(A)$. If $A \not\subset \mathcal{V}_\varepsilon^d(B)$, then there exists a point a in A such that $a \notin \mathcal{V}_\varepsilon^d(B)$. Thus, for every $b \in B$, $d(a, b) \geq \varepsilon$. For each $j \in \{1, \ldots, \ell\}$, let $b_j \in B_j$. Let $\alpha \colon [0, 1] \to \mathcal{C}_n(X)$ be an order arc from $\{b_1, \ldots, b_\ell\}$ to B (Theorem 1.8.20). Since for each $t \in [0, 1]$, $\alpha(t) \subset B$, and for every $b \in B$, $d(a, b) \geq \varepsilon$, we have that $\alpha(t) \notin \mathcal{V}_\varepsilon^{\mathcal{H}}(A) \cap \mathcal{C}_n(X)$ for any $t \in [0, 1]$. Thus, $\alpha([0, 1]) \cup \mathcal{F}_n(X) \subset \mathcal{C}_n(X) \setminus (\mathcal{V}_\varepsilon^{\mathcal{H}}(A) \cap \mathcal{C}_n(X))$.

If $B \not\subset \mathcal{V}_\varepsilon^d(A)$, then there exists a point b in B such that $b \notin \mathcal{V}_\varepsilon^d(A)$. Thus, for every point a of A, $d(b, a) \geq \varepsilon$. Without loss of generality, we assume that $b \in B_1$. For each $j \in \{2, \ldots, \ell\}$, let b_j be any point of B_j. Let $\beta \colon [0, 1] \to \mathcal{C}_n(X)$ be an order arc from $\{b, b_2, \ldots, b_\ell\}$ to B. Since for each $t \in [0, 1]$, $b \in \beta(t) \subset B$ and $d(b, a) \geq \varepsilon$, for each $a \in A$, we have that $\beta(t) \notin \mathcal{V}_\varepsilon^{\mathcal{H}}(A) \cap \mathcal{C}_n(X)$ for any $t \in [0, 1]$. Thus, $\beta([0, 1]) \cup \mathcal{F}_n(X) \subset \mathcal{C}_n(X) \setminus (\mathcal{V}_\varepsilon^{\mathcal{H}}(A) \cap \mathcal{C}_n(X))$.

Therefore, if $A \in \mathcal{C}_n(X) \setminus \mathcal{F}_n(X)$ and $\varepsilon > 0$ is given such that $(\mathcal{V}_\varepsilon^{\mathcal{H}}(A) \cap \mathcal{C}_n(X)) \cap \mathcal{F}_n(X) = \emptyset$, then each element B of $\mathcal{C}_n(X) \setminus (\mathcal{V}_\varepsilon^{\mathcal{H}}(A) \cap \mathcal{C}_n(X))$ can be joined with $\mathcal{F}_n(X)$ by an order arc completely contained in $\mathcal{C}_n(X) \setminus (\mathcal{V}_\varepsilon^{\mathcal{H}}(A) \cap \mathcal{C}_n(X))$.

Now, assume that $A \in \mathcal{F}_n(X)$. Suppose $A = \{a_1, \ldots, a_k\}$ and let $\varepsilon > 0$ be given such that $\mathcal{V}_\varepsilon^d(a_j) \cap \mathcal{V}_\varepsilon^d(a_m) = \emptyset$ if and only if $j \neq m$ and $j, m \in \{1, \ldots, k\}$. Let $B \in \mathcal{C}_n(X) \setminus (\mathcal{V}_\varepsilon^{\mathcal{H}}(A) \cap \mathcal{C}_n(X))$ and let $B_1, \ldots, B_\ell$ be the components of B. Then $\ell \leq n$. Hence, $\mathcal{H}(A, B) \geq \varepsilon$. Thus, we have that either $A \not\subset \mathcal{V}_\varepsilon^d(B)$ or $B \not\subset \mathcal{V}_\varepsilon^d(A)$.

If $A \not\subset \mathcal{V}_\varepsilon^d(B)$, then there exists a point a in A such that for each point b of B, $d(a, b) \geq \varepsilon$. Without loss of generality, we assume that $a = a_1$. Hence, $B \cap \mathcal{V}_\varepsilon^d(a_1) = \emptyset$. Let $\alpha: [0, 1] \to \mathcal{C}_n(X)$ be an order arc from B to X. We claim that for each $t \in [0, 1]$, $\mathcal{H}(\alpha(t), A) \geq \varepsilon$. To show this, suppose it is not true. Then there exists a point t_0 in $[0, 1]$ such that $\mathcal{H}(\alpha(t_0), A) < \varepsilon$. Thus, we have that $A \subset \mathcal{V}_\varepsilon^d(\alpha(t_0))$ and $\alpha(t_0) \subset \mathcal{V}_\varepsilon^d(A)$. Since $A \subset \mathcal{V}_\varepsilon^d(\alpha(t_0))$, we obtain that for each $j \in \{1, \ldots, k\}$, $\alpha(t_0) \cap \mathcal{V}_\varepsilon^d(a_j) \neq \emptyset$. On the other hand, since $\alpha(t_0) \subset \mathcal{V}_\varepsilon^d(A) = \bigcup_{j=1}^k \mathcal{V}_\varepsilon^d(a_j)$, and these balls are pairwise disjoint, we have that each component of $\alpha(t_0)$ is contained in one such ball. In particular, there exists a component of $\alpha(t_0)$ contained in $\mathcal{V}_\varepsilon^d(a_1)$. Hence, $B \cap \mathcal{V}_\varepsilon^d(a_1) \neq \emptyset$, a contradiction. Therefore, $\alpha([0, 1]) \subset \mathcal{C}_n(X) \setminus (\mathcal{V}_\varepsilon^{\mathcal{H}}(A) \cap \mathcal{C}_n(X))$.

If $B \not\subset \mathcal{V}_\varepsilon^d(A)$, then given an order arc $\beta: [0, 1] \to \mathcal{C}_n(X)$ from B to X, we have that $\beta([0, 1]) \subset \mathcal{C}_n(X) \setminus (\mathcal{V}_\varepsilon^{\mathcal{H}}(A) \cap \mathcal{C}_n(X))$.

Therefore, if $A \in \mathcal{F}_n(X)$, where $A = \{a_1, \ldots, a_k\}$, and $\varepsilon > 0$ is given such that $\mathcal{V}_\varepsilon^d(a_j) \cap \mathcal{V}_\varepsilon^d(a_m) = \emptyset$ if and only if $j \neq m$ and $j, m \in \{1, \ldots, k\}$, then each element $B \in \mathcal{C}_n(X) \setminus (\mathcal{V}_\varepsilon^{\mathcal{H}}(A) \cap \mathcal{C}_n(X))$ can be joined with X by an order arc contained in $\mathcal{C}_n(X) \setminus (\mathcal{V}_\varepsilon^{\mathcal{H}}(A) \cap \mathcal{C}_n(X))$.

Therefore $\mathcal{C}_n(X)$ is colocally connected.

Q.E.D.

6.3.2 Definition A continuum X is said to be *finitely aposyndetic* provided that for each finite subset K of X, $\mathcal{T}(K) = K$.

6.3.3 Corollary *If X is a continuum and $n \in \mathbb{N}$, then $\mathcal{C}_n(X)$ is aposyndetic and finitely aposyndetic.*

Proof Clearly, any colocally connected continuum is aposyndetic. Given a continuum X and a positive integer n, since $\mathcal{C}_n(X)$ is unicoherent (Theorem 6.2.5) and aposyndetic, we have that $\mathcal{C}_n(X)$ is finitely aposyndetic [3, Corollary 1].

Q.E.D.

Our goal is to prove that the n-fold hyperspaces of a continuum are zero-dimensional aposyndetic.

6.3.4 Definition A continuum X is *zero-dimensional aposyndetic* provided that for each totally disconnected closed subset A of X and for each $p \in X \setminus A$, there exists a subcontinuum W of X such that $p \in Int(W) \subset W \subset X \setminus A$.

6.3.5 Remark Let us observe that for compact metric spaces, the concepts of total disconnectedness and zero-dimensionality coincide [87, 4.7]

6.3.6 Definition Let K and L be closed disjoint subsets of the connected metric space X. A closed set B *cuts weakly between K and L in X* provided that whenever C is a closed connected set in X that intersects each of K and L, then C intersects B.

6.3.7 Definition A metric space X is *s-connected between the closed sets K and L* provided that whenever B is a closed set in X that cuts weakly between K and L, then some component E of B cuts weakly between K and L. A connected space X is said to be *s-connected* provided that whenever K and L are disjoint closed connected subsets of X, then X is *s*-connected between K and L.

A proof of the next theorem may be found in [77, Theorem 3]:

6.3.8 Theorem *If X is an inverse limit of absolute retracts, then X is s-connected.*

6.3.9 Theorem *If X is a continuum and n is positive integer, then $C_n(X)$ is s-connected.*

Proof Let X be a continuum and let $n \in \mathbb{N}$. By Theorem 2.1.56, there exists an inverse sequence $\{P_m^{m+1}, f_m^{m+1}\}$ of compact connected polyhedra with surjective bonding maps such that $X = \varprojlim\{P_m^{m+1}, f_m^{m+1}\}$. Since compact connected polyhedra are locally connected continua (by Theorem 2.6.15 and [80, p. 50]), $C_n\left(P_m^{m+1}\right)$ is an absolute retract [98, Théorème II$_m$]. By Theorem 2.3.4, we have that $C_n(X) = \varprojlim\{C_n\left(P_m^{m+1}\right), C_n\left(f_m^{m+1}\right)\}$. Therefore, by Theorem 6.3.8, $C_n(X)$ is *s*-connected.

Q.E.D.

The following lemma is a special case of [48, 17.3].

6.3.10 Lemma *Let X be a continuum, let n be a positive integer, and let $\mu \colon C_n(X) \twoheadrightarrow [0, 1]$ be a Whitney map for $C_n(X)$. Then, for any given $\varepsilon > 0$, there exists $\delta > 0$ such that if $A, B \in C_n(X)$, $A \subset B$ and $\mu(B) - \mu(A) < \delta$, then $\mathcal{H}(A, B) < \varepsilon$.*

The next result is well known and it is called *Avoidance Lemma*. It is often used in dimension theory [87, 8.1].

6.3.11 Lemma *Let X be a separable metric space. Let Z be a subset of X such that $dim(Z) \leq 0$. If K and L are two nonempty disjoint closed subsets of X, then there exists a closed subset B of X such that B separates K and L in X and $B \cap Z = \emptyset$.*

Now we are ready to prove that the n-fold hyperspaces of a continuum are zero-dimensional aposyndetic.

6.3.12 Theorem *If X is a continuum and $n \in \mathbb{N}$, then $C_n(X)$ is zero-dimensional aposyndetic.*

Proof Let $\mu \colon C_n(X) \twoheadrightarrow [0, 1]$ be a Whitney map for $C_n(X)$. Let Z be a closed totally disconnected subset of $C_n(X)$ and let $A \in C_n(X) \setminus Z$. If $A = X$, then, by [73, Lemma 2.3], $C_n(X)$ is locally connected at A; hence, $C_n(X)$ is zero-dimensional aposyndetic at A. So we assume that $A \neq X$. Since Z is closed and $A \in C_n(X) \setminus Z$, we have that there exists an $\varepsilon > 0$ such that $Cl_{C_n(X)}\left(\mathcal{V}_\varepsilon^{\mathcal{H}}(A)\right) \cap Z = \emptyset$. Let $\delta_1 > 0$ be as in Lemma 6.3.10 for $\frac{\varepsilon}{4}$ and let $\delta = \min\left\{\delta_1, \frac{1-\mu(A)}{2}\right\}$.

Let $\mathcal{K} = \mu^{-1}\left(\mu(A) + \frac{3\delta}{8}\right)$ and $\mathcal{L} = \mu^{-1}\left(\mu(A) + \frac{\delta}{8}\right)$. Since $\mathcal{Z}$ is totally disconnected, $\dim(\mathcal{Z}) = 0$, by Remark 6.3.5. Hence, by Lemma 6.3.11, there exists a closed subset $\mathcal{B}$ of $\mathcal{C}_n(X)$ such that $\mathcal{B}$ separates $\mathcal{K}$ and $\mathcal{L}$ in $\mathcal{C}_n(X)$ and $\mathcal{B} \cap \mathcal{Z} = \emptyset$. Note that, since $\mathcal{B}$ separates $\mathcal{K}$ and $\mathcal{L}$ in $\mathcal{C}_n(X)$, $\mathcal{B}$ cuts weakly between $\mathcal{K}$ and $\mathcal{L}$ in $\mathcal{C}_n(X)$. Thus, by Theorem 6.3.9, there exists a component $\mathcal{E}$ of $\mathcal{B}$ such that $\mathcal{E}$ cuts weakly between $\mathcal{K}$ and $\mathcal{L}$ in $\mathcal{C}_n(X)$. Observe that $\mathcal{E} \cap \mathcal{Z} = \emptyset$.

Let $\mathcal{U} = \mathcal{V}_{\frac{\varepsilon}{4}}^{\mathcal{H}}(A) \cap \mu^{-1}\left(\left(\mu(A) - \frac{\delta}{8}, \mu(A) + \frac{\delta}{8}\right)\right)$. For each $D \in Cl_{\mathcal{C}_n(X)}(\mathcal{U})$, consider an order arc from D to X, let G_D be the element of this order arc such that $\mu(G_D) = \mu(A) + \frac{3\delta}{8}$, we can reparametrize linearly the order arc in such a way that we obtain an order arc $\alpha_D \colon [0, 1] \to \mathcal{C}_n(X)$ such that $\alpha_D(0) = D$ and $\alpha_D(1) = G_D$. Note that, since $\alpha_D([0, 1]) \cap \mathcal{K} \neq \emptyset$ and $\alpha_D([0, 1]) \cap \mathcal{L} \neq \emptyset$, $\alpha_D([0, 1]) \cap \mathcal{E} \neq \emptyset$. Also, for each $t \in [0, 1]$, we have that

$$\mu\left(\alpha_D(t)\right) - \mu(D) \leq \mu(A) + \frac{3\delta}{8} - \mu(D) \leq$$

$$\frac{3\delta}{8} + |\mu(A) - \mu(D)| \leq \frac{3\delta}{8} + \frac{\delta}{8} < \delta.$$

Hence, for each $t \in [0, 1]$, $\mathcal{H}(D, \alpha_D(t)) < \frac{\varepsilon}{4}$.

Let $\mathcal{M}' = \left(\bigcup_{D \in Cl_{\mathcal{C}_n(X)}(\mathcal{U})} \alpha_D([0, 1])\right) \cup \mathcal{E}$. Thus, $\mathcal{M}'$ is the union of a connected set and arcs that intersect it. Therefore, $\mathcal{M}'$ is connected. Also, for each $D \in Cl_{\mathcal{C}_n(X)}(\mathcal{U})$ and for each $t \in [0, 1]$, we have that

$$\mathcal{H}(A, \alpha_D(t)) \leq \mathcal{H}(A, D) + \mathcal{H}(D, \alpha_D(t)) \leq \frac{\varepsilon}{4} + \frac{\varepsilon}{4} < \varepsilon.$$

As a consequence of this, $\left(\bigcup_{D \in Cl_{\mathcal{C}_n(X)}(\mathcal{U})} \alpha_D([0, 1])\right) \subset \mathcal{V}_\varepsilon^{\mathcal{H}}(A)$. It follows that $Cl_{\mathcal{C}_n(X)}\left(\bigcup_{D \in Cl_{\mathcal{C}_n(X)}(\mathcal{U})} \alpha_D([0, 1])\right) \subset Cl_{\mathcal{C}_n(X)}\left(\mathcal{V}_\varepsilon^{\mathcal{H}}(A)\right)$.

Let $\mathcal{M} = Cl_{\mathcal{C}_n(X)}\left(\bigcup_{D \in Cl_{\mathcal{C}_n(X)}(\mathcal{U})} \alpha_D([0, 1])\right) \cup \mathcal{E}$. We have that $\mathcal{M}$ is connected, that $A \in \mathcal{U} \subset \mathcal{M}$ and that $\mathcal{M} \cap \mathcal{Z} = \emptyset$. Hence, $\mathcal{M}$ is a subcontinuum of $\mathcal{C}_n(X)$ such that $A \in Int_{\mathcal{C}_n(X)}(\mathcal{M})$ and $\mathcal{M} \cap \mathcal{Z} = \emptyset$. Therefore, $\mathcal{C}_n(X)$ is zero-dimensional aposyndetic.

Q.E.D.

6.4 Arcwise Accessibility

We present results concerning arcwise accessibility of points of the n-fold symmetric product from the n-fold hyperspace of a given continuum. Here, we only use [62, 83].

6.4.1 Definition Let X be a continuum. Let Σ_1 and Σ_2 be two arcwise connected closed subsets of 2^X, such that $\Sigma_2 \subset \Sigma_1$. A member A of Σ_2 is said to be *arcwise accessible from $\Sigma_1 \setminus \Sigma_2$ beginning with K* if and only if there exists an arc $\alpha : [0, 1] \to \Sigma_1$ such that $\alpha(0) = K$, $\alpha(1) = A$ and $\alpha(t) \in \Sigma_1 \setminus \Sigma_2$ for all $t < 1$.

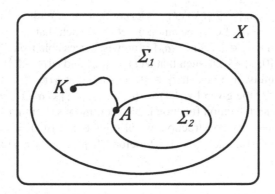

We begin observing that in the proof of [83, (2.2)], Sam B. Nadler, Jr. actually showed the following:

6.4.2 Theorem *Let X be a continuum. If A is a nondegenerate subcontinuum of X and q is any point of A, then there exist a point $p \in A \setminus \{q\}$ and an order arc $\alpha : [0, 1] \to \mathcal{C}(A)$ such that $\alpha(0) = \{q\}$, $\alpha(1) = A$ and $p \notin \alpha(t)$ for any $t < 1$.*

6.4.3 Theorem *Let $n \in \mathbb{N}$, $n \geq 2$. Let X be a continuum and let A be an element of $\mathcal{C}_n(X)$ having exactly n components and at least one of them is nondegenerate. Then A is arcwise accessible from $\mathcal{C}_{n+1}(X) \setminus \mathcal{C}_n(X)$, beginning with an element in $\mathcal{F}_{n+1}(X) \setminus \mathcal{F}_n(X)$.*

Proof Suppose $A_1, \ldots, A_n$ are the components of A and that A_n is not degenerate. For each $j \in \{1, \ldots, n\}$, let $q_j \in A_j$. By Theorem 6.4.2, there exist $q_{n+1} \in A_n$ and an order arc $\alpha_n : [0, 1] \to \mathcal{C}(A_n)$ such that $\alpha_n(0) = \{q_n\}$, $\alpha_n(1) = A_n$ and $q_{n+1} \notin \alpha_n(t)$ for any $t < 1$. For each $j \in \{1, \ldots, n - 1\}$, let $\alpha_j : [0, 1] \to \mathcal{C}(A_j)$ be an order arc such that $\alpha_j(0) = \{q_j\}$, $\alpha_j(1) = A_j$ (Theorem 1.8.20), or a constant map if $A_j = \{q_j\}$.

Let $\gamma : [0, 1] \to \mathcal{C}_{n+1}(X)$ be given by $\gamma(t) = \alpha_1(t) \cup \ldots \cup \alpha_n(t) \cup \{q_{n+1}\}$. Since for each $j \in \{1, \ldots, n\}$, α_j is continuous and the union is also continuous (Lemma 1.8.11), then γ is continuous. Also, we have that $\gamma(0) = \{q_1, \ldots, q_{n+1}\}$, $\gamma(1) = A$, and $\gamma(t) \in \mathcal{C}_{n+1}(X) \setminus \mathcal{C}_n(X)$ for each $t < 1$. Therefore, A is arcwise accessible from $\mathcal{C}_{n+1}(X) \setminus \mathcal{C}_n(X)$, beginning with an element in $\mathcal{F}_{n+1}(X) \setminus \mathcal{F}_n(X)$.
                                                                        Q.E.D.

6.4.4 Corollary *Let $n \in \mathbb{N}$, $n \geq 3$. Let X be a continuum. If x is a point of X such that $\{x\}$ is arcwise accessible from $\mathcal{C}_2(X) \setminus \mathcal{C}(X)$ with an arc $\alpha : [0, 1] \to \mathcal{C}_2(X)$*

such that $\alpha([0, 1]) \cap (\mathcal{C}_2(X) \setminus \mathcal{F}_2(X)) \neq \emptyset$, then $\{x\}$ is arcwise accessible from $\mathcal{C}_n(X) \setminus \mathcal{C}(X)$.

6.4.5 Theorem *Let $n \in \mathbb{N}$, $n \geq 3$. Let X be a continuum, and let a be a point of X such that $\{a\}$ is arcwise accessible from $2^X \setminus \mathcal{C}(X)$. Then each element A of $\mathcal{F}_n(X) \setminus \mathcal{F}_{n-1}(X)$ containing a is arcwise accessible from $2^X \setminus \mathcal{C}_n(X)$.*

Proof Let $\{a_1, \ldots, a_n\} \in \mathcal{F}_n(X) \setminus \mathcal{F}_{n-1}(X)$ containing a. Without loss of generality, we assume that $a = a_n$. Let U be an open set of X such that $a_n \in U$, and for each $j \in \{1, \ldots, n-1\}$, $a_j \notin U$. Since $\{a_n\}$ is arcwise accessible from $2^X \setminus \mathcal{C}(X)$, there exists an arc $\alpha \colon [0, 1] \to 2^X$ such that $\alpha(1) = \{a_n\}$ and $\alpha(t) \in 2^X \setminus \mathcal{C}(X)$ for each $t < 1$. By continuity, there exists $t_1 \in [0, 1)$ such that $\alpha(t) \in U$ for each $t \geq t_1$. Let $\beta \colon [0, 1] \to 2^X$ be given by $\beta(s) = \{a_1, \ldots, a_{n-1}\} \cup \alpha((1-s)t_1 + s)$. Since α is continuous and the union function is also continuous (Lemma 1.8.11), we have that β is continuous. By construction, we also have that $\beta(1) = \{a_1, \ldots, a_n\}$, and $\beta(t) \in 2^X \setminus \mathcal{C}_n(X)$ for each $t < 1$. Therefore, $\{a_1, \ldots, a_n\}$ is arcwise accessible from $2^X \setminus \mathcal{C}_n(X)$.

Q.E.D.

From the proof of Theorem 6.4.5, we have the following:

6.4.6 Corollary *Let $n, m \in \mathbb{N}$, $n \geq 3$. Let X be a continuum and let a be a point of X such that $\{a\}$ is arcwise accessible from $\mathcal{C}_m(X) \setminus \mathcal{C}(X)$. Then each element A of $\mathcal{F}_n(X) \setminus \mathcal{F}_{n-1}(X)$ such that $a \in A$ is arcwise accessible from $\mathcal{C}_{m+n-1}(X) \setminus \mathcal{C}_n(X)$.*

6.4.7 Theorem *Let $n \in \mathbb{N}$, $n \geq 2$. Let X be a continuum with metric d. Suppose that for each $j \in \{1, \ldots, n\}$, a_j is a point of X such that $\{a_j\}$ is not arcwise accessible from $2^X \setminus \mathcal{C}(X)$. Then $\{a_1, \ldots, a_n\}$ is not arcwise accessible from $2^X \setminus \mathcal{C}_n(X)$.*

Proof Suppose that $\{a_1, \ldots, a_n\}$ is arcwise accessible from $2^X \setminus \mathcal{C}_n(X)$. Then there exists an arc $\alpha \colon [0, 1] \to 2^X$ such that $\alpha(1) = \{a_1, \ldots, a_n\}$ and $\alpha(t) \in 2^X \setminus \mathcal{C}_n(X)$ for each $t < 1$. Let $U_1, \ldots, U_n$ be open sets of X such that for each $j \in \{1, \ldots, n\}$, $a_j \in U_j$, and $Cl(U_j) \cap Cl(U_k) = \emptyset$ if and only if $j, k \in \{1, \ldots, n\}$ and $j \neq k$. By continuity, there exists $t_1 \in [0, 1)$ such that $\alpha(t) \in \langle U_1, \ldots, U_n \rangle$ for each $t \geq t_1$.

For each $j \in \{1, \ldots, n\}$, let $\beta_j \colon [0, 1] \to 2^X$ be given by $\beta_j(s) = \alpha((1-s)t_1 + s) \cap U_j$. Since the family $\{U_1, \ldots, U_n\}$ of open sets has pairwise disjoint closures, β_j is well defined for each $j \in \{1, \ldots, n\}$. To see that each β_j is continuous, let $j \in \{1, \ldots, n\}$ and $\varepsilon > 0$ be such that $d(Cl(U_k), Cl(U_\ell)) > \varepsilon$, for each $k, \ell \in \{1, \ldots, n\}$ and $k \neq \ell$. Let $s_0 \in [0, 1]$. By continuity, there exists $\delta > 0$ such that if $|s_0 - s_1| < \delta$, then $\mathcal{H}(\alpha((1-s_0)t_1 + s_0), \alpha((1-s_1)t_1 + s_1)) < \varepsilon$. Let $x \in \alpha((1-s_0)t_1 + s_0) \cap U_j$. Since $\mathcal{H}(\alpha((1-s_0)t_1 + s_0), \alpha((1-s_1)t_1 + s_1)) < \varepsilon$, there exists $y \in \alpha((1-s_1)t_1 + s_1)$ such that $d(x, y) < \varepsilon$. Since $y \in \bigcup_{k=1}^{n} U_k$, there exists $\ell \in \{1, \ldots, n\}$ such that $y \in U_\ell$. Since $d(x, y) < \varepsilon$ and $d(Cl(U_k), Cl(U_\ell)) > \varepsilon$ for each $k \in \{1, \ldots, n\}$ and $k \neq \ell$, we have that $\ell = j$ and $y \in U_j$. Therefore, $y \in \alpha((1-s_1)t_1 + s_1) \cap U_j$, and $\alpha((1-s_0)t_1 + s_0) \cap U_j \subset \mathcal{V}_\varepsilon^d(\alpha((1-s_1)t_1 + s_1) \cap U_j)$. Similarly $\alpha((1-s_1)t_1 + s_1) \cap U_j \subset \mathcal{V}_\varepsilon^d(\alpha((1-s_0)t_1 + s_0) \cap U_j)$. Thus,

$\mathcal{H}(\alpha((1 - s_0)t_1 + s_0) \cap U_j, \alpha((1 - s_1)t_1 + s_1) \cap U_j) = \mathcal{H}(\beta_j(s_0), \beta_j(s_1)) < \varepsilon.$
Hence, β_j is continuous.

Note that for each $j \in \{1, \ldots, n\}$, $\beta_j(1) = \{a_j\}$. Since, for each $j \in \{1, \ldots, n\}$, $\{a_j\}$ is not arcwise accessible from $2^X \setminus \mathcal{C}(X)$, we have that, for each $j \in \{1, \ldots, n\}$, there exists $s_j \in [0, 1)$ such that $\beta_j(s) \in \mathcal{C}(X)$ for every $s \geq s_j$. Let $s_* = \max\{s_1, \ldots, s_n\}$. Then for each $j \in \{1, \ldots, n\}$, $\beta_j(s) \in \mathcal{C}(X)$ for every $s \geq s_*$.

Let $s \geq s_*$. Then $\bigcup_{j=1}^{n} \beta_j(s) = \bigcup_{j=1}^{n} \alpha((1 - s)t_1 + s) \cap U_j = \alpha((1 - s)t_1 + s) \cap \bigcup_{j=1}^{n} U_j = \alpha((1 - s)t_1 + s)$. Also, if $s \geq s_*$, then $\bigcup_{j=1}^{n} \beta_j(s) \in \mathcal{C}_n(X)$. Thus, for each $s \geq s_*$, $\alpha((1 - s)t_1 + s) \in \mathcal{C}_n(X)$, a contradiction. Therefore, $\{a_1, \ldots, a_n\}$ is not arcwise accessible from $2^X \setminus \mathcal{C}_n(X)$.

Q.E.D.

6.4.8 Corollary *Let $n \in \mathbb{N}$, $n \geq 2$. If X is a hereditarily indecomposable continuum, then $\{a_1, \ldots, a_n\}$ is not arcwise accessible from $2^X \setminus \mathcal{C}_n(X)$ for any $\{a_1, \ldots, a_n\} \in \mathcal{F}_n(X)$.*

Proof Since X is hereditarily indecomposable, no singleton is arcwise accessible from $2^X \setminus \mathcal{C}(X)$ [83, (3.4)]. Thus, we have the result follows from Theorem 6.4.7.

Q.E.D.

6.5 Points That Arcwise Disconnect

We consider when a point arcwise disconnects $\mathcal{C}_n(X)$. We also study the arc components of $\mathcal{C}_n(X) \setminus \{X\}$, where X is an indecomposable continuum. To this end, we follow [25, 36, 45, 57, 60, 61, 64, 66, 73, 84, 86].

6.5.1 Lemma *Let X be a continuum and let $n \in \mathbb{N}$ be given. If A is a proper subcontinuum of X, then $\mathcal{C}_n(X) \setminus \mathcal{C}_n(A)$ is arcwise connected.*

Proof Each element of $\mathcal{C}_n(X) \setminus \mathcal{C}_n(A)$ can be joined with X with an order arc contained in $\mathcal{C}_n(X) \setminus \mathcal{C}_n(A)$ (Theorem 1.8.20).

Q.E.D.

The proof of the following theorem is similar to the one given in [84, (11.3)].

6.5.2 Theorem *Let X be a continuum and let $n \in \mathbb{N}$ be given. If $A \in \mathcal{C}_n(X)$ is such that $\mathcal{C}_n(X) \setminus \{A\}$ is not arcwise connected, then A is connected.*

The next theorem characterizes indecomposable continua.

6.5.3 Theorem *A nondegenerate continuum X is indecomposable if and only if for each $n \in \mathbb{N}$, $\mathcal{C}_n(X) \setminus \{X\}$ is not arcwise connected.*

Proof Suppose X is decomposable. We show that for each $n \in \mathbb{N}$, $\mathcal{C}_n(X) \setminus \{X\}$ is arcwise connected. The proof is done by induction. The result is known for $n = 1$ [84, (1.51)]. Let $n = 2$, and let A and B be two elements of $\mathcal{C}_2(X) \setminus \{X\}$. If both A and B are connected, then there exists an arc in $\mathcal{C}(X) \setminus \{X\}$ joining A and B [84,

(1.51)]. Suppose A has two components, say A_1 and A_2. Since X is decomposable, there exist two proper subcontinua H and K of X such that $X = K \cup H$. We show there exists an arc in $C_2(X) \setminus \{X\}$ joining A and an element of $C(X)$. We have to consider several cases.

If either $A \subset H$ or $A \subset K$, then there exists an order arc joining A with H or K (Theorem 1.8.20), and we are done.

If $A_2 = K$, then let $x \in H \cap K$. Let $\alpha \colon [0, 1] \to C_2(X)$ be an order arc joining $A_1 \cup \{x\}$ and $A_1 \cup K = A$. Let $\beta \colon [0, 1] \to C_2(X)$ be an order arc joining $A_1 \cup \{x\}$ and H. Then $\alpha([0, 1]) \cup \beta([0, 1])$ contains an arc having A and H as its end points, contained in $C_2(X) \setminus \{X\}$.

If $A_1 \subset H$, $A_2 \cap (H \cap K) \neq \emptyset$, and $A_2 \neq K$, then $H \cup A_2$ is a proper subcontinuum of X, and we can take an order arc joining A to $H \cup A_2$.

If $A_1 \cap (H \cap K) \neq \emptyset$ and $A_2 \cap (H \cap K) \neq \emptyset$, then let $a_j \in A_j \cap H$ for $j \in \{1, 2\}$. Let $\alpha \colon [0, 1] \to C_2(X)$ be an order arc joining $\{a_1, a_2\}$ and A, and let $\beta \colon [0, 1] \to C_2(X)$ be an order arc having $\{a_1, a_2\}$ and H as its end points. Then $\alpha([0, 1]) \cup \beta([0, 1])$ is contained in $C_2(X) \setminus \{X\}$ and contains an arc joining A and H.

If $A_1 \subset H \setminus K$ and $A_2 \subset K \setminus H$, then let $x \in H \cap K$. Let $\alpha \colon [0, 1] \to C_2(X)$ be an order arc from A to $H \cup A_2$. Let $\beta[0, 1] \to C_2(X)$ be an order arc from $\{x\} \cup A_2$ to $H \cup A_2$. Let $\gamma \colon [0, 1] \to C_2(X)$ be an order arc from $\{x\} \cup A_2$ to K. Then $\alpha([0, 1]) \cup \beta([0, 1]) \cup \gamma([0, 1])$ contains an arc joining A and K contained in $C_2(X) \setminus \{X\}$. The rest of the cases are similar to the ones treated.

Now, let $n \geq 3$ and suppose $C_n(X) \setminus \{X\}$ is arcwise connected. We show that $C_{n+1}(X) \setminus \{X\}$ is arcwise connected. Let A and B be two points in $C_{n+1}(X) \setminus \{X\}$. If both A and B belong to $C_n(X)$, by induction hypothesis we can find an arc in $C_n(X) \setminus \{X\} \subset C_{n+1}(X) \setminus \{X\}$ having A and B as its end points. Hence, assume that A has $n + 1$ components. Let $A_1, \ldots, A_{n+1}$ be the components of A. Since X is decomposable, there exist two proper subcontinua H and K of it such that $X = H \cup K$. Since $n \geq 3$, at least two components of A intersect either H or K; suppose that two components of A intersect H. Without loss of generality, we assume that $A_n \cap H \neq \emptyset$ and $A_{n+1} \cap H \neq \emptyset$. For each $j \in \{1, \ldots, n - 1\}$, let $a_j \in A_j$. Take $a_n \in A_n \cap H$ and $a_{n+1} \in A_{n+1} \cap H$. Let $\alpha \colon [0, 1] \to C_{n+1}(X)$ be an order arc from $\{a_1, \ldots, a_{n+1}\}$ to A. Let $\beta \colon [0, 1] \to C_2(X)$ be an order arc from $\{a_n, a_{n+1}\}$ to H. Let $\gamma \colon [0, 1] \to C_{n+1}(X)$ be given by $\gamma(t) = \{a_1, \ldots, a_{n-1}\} \cup \beta(t)$. Then γ is continuous, $\gamma(0) = \{a_1, \ldots, a_{n-1}\} \cup \beta(0) = \{a_1 \ldots, a_{n+1}\}$ and $\gamma(1) = \{a_1, \ldots, a_{n-1}\} \cup \beta(1) = \{a_1, \ldots, a_{n-1}\} \cup H \in C_n(X) \setminus \{X\}$. Hence, $\alpha([0, 1]) \cup \gamma([0, 1])$ is contained in $C_{n+1}(X) \setminus \{X\}$ and contains an arc having A and $\gamma(1)$ as its end points. Similarly, if B has $n + 1$ components, we can find an arc in $C_{n+1}(X) \setminus \{X\}$ having B and an element of $C_n(X)$ as its end points. Thus, by induction hypothesis, we are done.

The proof of the reverse implication is similar to the one given in [84, (11.4)].

Q.E.D.

The next theorem tells us more about what type of subcontinua may arcwise disconnect the hyperspaces.

6.5.4 Theorem *Let X be a continuum, and let E be a nondegenerate proper subcontinuum of X. Consider the following statements:*

(1) E is a terminal subcontinuum of X.
(2) $2^X \setminus \{E\}$ is not arcwise connected.
(3) For each $n \in \mathbb{N}$, $C_n(X) \setminus \{E\}$ is not arcwise connected.
(4) $C(X) \setminus \{E\}$ is not arcwise connected.

> *Then (1) implies (2), (3) and (4). Furthermore, if E is decomposable then all four statements are equivalent.*

Proof The proofs of (1) implies (2) and (4) are given in [84, (11.5)]. The proof of (1) implies (3) is similar to the one given in [84, (11.5)].

Now suppose E is a decomposable nondegenerate subcontinuum of X. The equivalence between (1), (2) and (4) is given in [84, (11.5)]. Suppose E is not terminal, then $E \neq X$. We show that for each $n \in \mathbb{N}$, $C_n(X) \setminus \{E\}$ is arcwise connected. This is done by induction.

For $n = 1$ the result is known. Suppose $C_n(X) \setminus \{E\}$ is arcwise connected. To show that $C_{n+1}(X) \setminus \{E\}$ is arcwise connected, let A and B be two points in $C_{n+1}(X) \setminus \{E\}$. If both A and B belong to $C_{n+1}(X) \setminus C_{n+1}(E)$, then there exists an arc having A and B as its end points and contained in $C_{n+1}(X) \setminus \{E\}$ (Lemma 6.5.1). If both A and B belong to $C_{n+1}(E)$, then by Theorem 6.5.3, there exists an arc joining A and B and contained in $C_{n+1}(E) \setminus \{E\} \subset C_{n+1}(X) \setminus \{E\}$. Thus, suppose, without loss of generality, that $A \in C_{n+1}(E) \setminus \{E\}$ and $B \in C_{n+1}(X) \setminus C_n(E)$.

If A has less than $n + 1$ components then, by induction hypothesis, there exists an arc in $C_n(X) \setminus \{E\}$ joining A and X. So, assume A has exactly $n + 1$ components. Since E is decomposable, by Theorem 6.5.3, there exists an arc joining A and an element A' of $C_n(E) \setminus \{E\}$. By induction hypothesis, there exists an arc in $C_n(X) \setminus \{E\}$ joining A' and X. Hence, there exists an arc in $C_{n+1}(X) \setminus \{E\}$ joining A and X. Since $B \in C_{n+1}(X) \setminus C_{n+1}(E)$, by Lemma 6.5.1, there exits an arc having B and X as its end points.

Therefore, there exists an arc in $C_{n+1}(X) \setminus \{E\}$ joining A and B.

Q.E.D.

6.5.5 Theorem *If X is a continuum, then for any $E \in 2^X$, the following are equivalent:*

(1) $2^X \setminus \{E\}$ is not arcwise connected.
(2) $C(X) \setminus \{E\}$ is not arcwise connected.
(3) For each $n \in \mathbb{N}$, $C_n(X) \setminus \{E\}$ is not arcwise connected.

Proof The proof of the equivalence of (1) and (2) is given in [84, (11.8)]. Clearly (3) implies (2). We show that (2) implies (3).

Let $E \in 2^X$ and suppose $C(X) \setminus \{E\}$ is not arcwise connected. Then $E \in C(X)$ [84, (11.3)]. Let $n \geq 2$. If $E = X$, then, by Theorem 6.5.3, X is indecomposable and $C_n(X) \setminus \{E\}$ is not arcwise connected.

Suppose $E \neq X$. If E is decomposable, then, by Theorem 6.5.4, E is a terminal subcontinuum of X and $C_n(X) \setminus \{E\}$ is not arcwise connected.

Thus, assume E is indecomposable. Let $B \in \mathcal{C}(E) \subset \mathcal{C}_n(X)$ and let $A \in \mathcal{C}_n(X) \setminus \mathcal{C}_n(E)$. Let $\alpha \colon [0, 1] \to \mathcal{C}_n(X)$ be an arc such that $\alpha(0) = B$ and $\alpha(1) = A$. Let $\beta \colon [0, 1] \to \mathcal{C}_n(X)$ be given by $\beta(t) = \sigma(\alpha([0, t]))$. Then β is an order arc from B to $\sigma(\alpha([0, 1]))$. Hence, $\beta([0, 1]) \subset \mathcal{C}(X)$ [84, (11.1)]. In particular, $\sigma(\alpha([0, 1]))$ is connected. Since $\beta(0) = B \subset E$, $\beta(1) = \sigma(\alpha([0, 1]))$, $(\sigma(\alpha([0, 1]))) \cap (\mathcal{C}(X) \setminus \mathcal{C}(E)) \neq \emptyset$, and $\mathcal{C}(X) \setminus \{E\}$ is not arcwise connected, we have that there exists $t_0 \in [0, 1]$ such that $\beta(t_0) = E$. Let $t_1 = \min\{t \in [0, 1] \mid \beta(t) = E\}$. Then $t_1 > 0$ and $\beta(t_1) = E$. Since for each $t < t_1$, $\beta(t)$ is a proper subcontinuum of E and E is indecomposable, we have that $\beta(t)$ is nowhere dense in E (Corollary 1.7.26). Note that, for every $t < t_1$, $E = \beta(t) \cup (\sigma(\alpha([t, t_1])))$, and $\sigma(\alpha([t, t_1]))$ is compact. Hence, $E = \sigma(\alpha([t, t_1]))$. By continuity, we have that $E = \alpha(t_1)$.

Therefore, $\mathcal{C}_n(X) \setminus \{E\}$ is not arcwise connected.

Q.E.D.

6.5.6 Theorem *If E, A and B are subcontinua of the continuum X and $n \in \mathbb{N}$, then the following are equivalent:*

(1) If Γ is an arc in $\mathcal{C}(X)$ such that $A, B \in \Gamma$, then $E \in \Gamma$.
(2) If Γ is an arc in $\mathcal{C}_n(X)$ such that $A, B \in \Gamma$, then $E \in \Gamma$.
(3) If Γ is an arc in 2^X such that $A, B \in \Gamma$, then $E \in \Gamma$.

Proof Clearly (3) implies (2) and (2) implies (1). The proof of the implication from (1) to (3) is in [84, (11.13)].

Q.E.D.

6.5.7 Corollary *Let E be a subcontinuum of the continuum X. If Λ is an arc component of $2^X \setminus \{E\}$ and $\Lambda \cap \mathcal{C}_n(X) \neq \emptyset$, for some $n \in \mathbb{N}$, then $\Lambda \cap \mathcal{C}_n(X)$ is an arc component of $\mathcal{C}_n(X) \setminus \{E\}$.*

6.5.8 Theorem *If X is a continuum, then the following are equivalent:*

(1) X is hereditarily indecomposable.
(2) For each nondegenerate subcontinuum E of X, $2^X \setminus \{E\}$ is not arcwise connected.
(3) For each nondegenerate subcontinuum E of X, $\mathcal{C}_n(X) \setminus \{E\}$ is not arcwise connected, for each $n \in \mathbb{N}$.
(4) For each nondegenerate subcontinuum E of X, $\mathcal{C}(X) \setminus \{E\}$ is not arcwise connected.

Proof By Theorem 6.5.5, we have that (2), (3) and (4) are equivalent. The proof of the equivalence between (4) and (1) is given in [84, (11.15)].

Q.E.D.

The following lemma is easy to establish.

6.5.9 Lemma *Let X be a continuum, let n be a positive integer and let $A \in \mathcal{C}_n(X)$. Then $\mathcal{C}_n(X) \setminus \{A\}$ is not arcwise connected if and only if $\mathcal{C}_n(X) \setminus (\{A\} \cup \mathcal{F}_n(X))$ is not arcwise connected.*

6.5.10 Lemma *Let A be a proper decomposable subcontinuum of a continuum X and let n be a positive integer. If $C_n(X) \setminus \{A\}$ is not arcwise connected, then $C_n(X) \setminus \{A\}$ has exactly two arc components.*

Proof Observe that $C_n(X) \setminus C_n(A)$ is arcwise connected by Lemma 6.5.1. Since A is a decomposable continuum, $C_n(A) \setminus \{A\}$ is arcwise connected by Theorem 6.5.3. Hence, since $C_n(X) \setminus \{A\} = (C_n(X) \setminus C_n(A)) \cup C_n(A) \setminus \{A\}$, we have that $C_n(X) \setminus \{A\}$ has exactly two arc components.

Q.E.D.

6.5.11 Definition Let X be a continuum, and let $p \in X$. The *composant of p in X* is the union of all proper subcontinua of X containing p.

6.5.12 Notation Given a continuum X, a positive integer n and a composant κ of X, let $C_n(\kappa)$ denote the set $C_n(\kappa) = \{A \in C_n(X) \mid A \subset \kappa\}$.

In the next two theorems we describe the arc components of $C_n(X) \setminus \{X\}$, where X is an indecomposable continuum.

6.5.13 Theorem *Let n be an integer greater than one. If κ is a composant of an indecomposable continuum X, then $C_n(\kappa)$ is an arc component of $C_n(X) \setminus \{X\}$.*

Proof By Theorem 6.5.3, $C_n(X) \setminus \{X\}$ is not arcwise connected. First, observe that $C_n(\kappa)$ is arcwise connected. To see this, let $A \in C_n(\kappa)$. Then, since A only has finitely many components and X is indecomposable, it is easy to show that there exists a proper subcontinuum B of X containing A. Thus, there exists an order arc from A to B (Theorem 1.8.20). Since $C(\kappa)$ is arcwise connected [84, (1.52.1)], we have that $C_n(\kappa)$ is arcwise connected.

Let $\mathcal{A}$ be the arc component of $C_n(X) \setminus \{X\}$ containing $C_n(\kappa)$, and suppose there exists $B \in \mathcal{A} \setminus C_n(\kappa)$. Let $A \in C(\kappa)$. Since A and B belong to $\mathcal{A}$, there exists an arc $\alpha \colon [0, 1] \to \mathcal{A}$ such that $\alpha(0) = A$ and $\alpha(1) = B$. Let $\beta \colon [0, 1] \to C_n(X)$ be given by $\beta(t) = \sigma(\alpha([0, t]))$. Then β is well defined (Corollary 6.1.2), β is an order arc and $\beta(0) = \alpha(0) = A$. Hence, $\beta(t) \in C(X)$ for each $t \in [0, 1]$ [84, (1.11)], and $B \subset \beta(1)$. Since B is not contained in κ and $B \subset \beta(1)$, we have that $\beta(1)$ is a subcontinuum of X intersecting two different composants of it. Thus, $\beta(1) = X$. Let $t_0 = \min\{t \in [0, 1] \mid \beta(t) = X\}$. Then $\beta(t_0) = X$ and $t_0 > 0$. Observe that if $0 \le t < t_0$, then $\beta(t)$ is a nowhere dense subset of X (Corollary 1.7.26). Note that for each $0 < t < t_0$, $X = \beta(t_0) = \beta(t) \cup (\sigma(\alpha([t, t_0])))$. Since $\beta(t)$ is nowhere dense in X, we have that $X = \sigma(\alpha([t, t_0]))$ for each $t < t_0$. By continuity, $\alpha(t_0) = X$, a contradiction. Therefore, $C_n(\kappa) = \mathcal{A}$.

Q.E.D.

6.5.14 Theorem *Let n be an integer greater than one, and let X be an indecomposable continuum. If $\mathcal{A}$ is an arc component of $C_n(X) \setminus \{X\}$, which is not of the form $C_n(\kappa)$, where κ is a composant of X, then there exist finitely many composants*

$\kappa_1, \ldots, \kappa_\ell$ of X and there exists a one-to-one map from

$$\mathcal{R} = \bigcup \left\{ \prod_{j=1}^{\ell} \mathcal{C}_{r_j}(\kappa_j) \; \Big| \; \{r_1, \ldots, r_\ell\} \in \mathcal{N} \right\} \subset \prod_{j=1}^{\ell} \mathcal{C}_{n-\ell+1}(\kappa_j)$$

(with the "max" metric ρ_1) onto $\mathcal{A}$, where

$$\mathcal{N} = \left\{ \{r_1, \ldots, r_\ell\} \; \Big| \; 1 \le r_1, \ldots, r_\ell \le n - \ell + 1 \text{ and } \sum_{j=1}^{\ell} r_j = n \right\}.$$

Proof Let A_0 be a point of $\mathcal{A}$, and let $\kappa_1, \ldots, \kappa_\ell$ be the composants of X which intersect A_0. Since $\mathcal{A}$ is not of the form $\mathcal{C}_n(\kappa)$, $\ell \ge 2$.

First, we show that every element of $\mathcal{A}$ intersects each κ_j for each $j \in \{1, \ldots, \ell\}$. To see this, suppose that there exists a point B of $\mathcal{A}$ such that $B \cap \kappa_j = \emptyset$, for some $j \in \{1, \ldots, \ell\}$. Since A_0 and B belong to $\mathcal{A}$, there exists an arc $\alpha \colon [0, 1] \to \mathcal{A}$ such that $\alpha(0) = A_0$ and $\alpha(1) = B$. Let $\beta \colon [0, 1] \to \mathcal{C}_n(X)$ be given by $\beta(t) = \sigma(\alpha([0, t]))$. Then β is well defined (Corollary 6.1.2), β is an order arc, $\beta(0) = \alpha(0) = A_0$, and $\beta(1) = \sigma(\alpha([0, 1]))$ is an element of $\mathcal{C}_n(X)$ intersecting κ_j and containing B. On the other hand, the map $\gamma \colon [0, 1] \to \mathcal{C}_n(X)$ given by $\gamma(t) = \sigma(\alpha([1 - t, 1]))$ is also well defined, γ is an order arc, $\gamma(0) = \alpha(1) = B$ and $\gamma(1) = \sigma(\alpha([0, 1])) = \beta(1)$. Thus, there exists an order arc in $\mathcal{C}_n(X)$ from B to $\beta(1)$, $B \cap \kappa_j = \emptyset$ and $\beta(1) \cap \kappa_j \ne \emptyset$; this contradicts Theorem 1.8.20 if $\sigma(\alpha([0, 1]))$ is a proper subset of X. Otherwise, a similar argument to the one given in Theorem 6.5.13 shows that there exists $t_0 \in [0, 1]$ such that $\alpha(t_0) = X$, which is also a contradiction. Therefore, every element of $\mathcal{A}$ intersects each κ_j, $j \in \{1, \ldots, \ell\}$. A similar argument proves that if $A \in \mathcal{A}$ and $A \cap \kappa \ne \emptyset$, for some composant κ of X, then $\kappa \in \{\kappa_1, \ldots, \kappa_\ell\}$.

Let $f \colon \mathcal{R} \to \mathcal{A}$ be given by

$$f((A_1, \ldots, A_\ell)) = \bigcup_{j=1}^{\ell} A_j.$$

Let us see first that f is well defined. Clearly, $f((A_1, \ldots, A_\ell)) \in \mathcal{C}_n(X)$. On the other hand, for each $j \in \{1, \ldots, \ell\}$, A_j and $A_0 \cap \kappa_j$ both belong to $\mathcal{C}_{r_j}(\kappa_j)$, where $\{r_1, \ldots, r_\ell\} \in \mathcal{N}$. Since $\mathcal{C}_{r_j}(\kappa_j)$ is arcwise connected, there exists an arc $\alpha_j \colon [0, 1] \to \mathcal{C}_{r_j}(\kappa_j)$ such that $\alpha_j(0) = A_0 \cap \kappa_j$ and $\alpha_j(1) = A_j$. Hence, $\alpha \colon [0, 1] \to \mathcal{C}_n(X) \setminus \{X\}$ given by $\alpha(t) = \bigcup_{j=1}^{\ell} \alpha_j(t)$ is a path joining A_0 and $\bigcup_{j=1}^{\ell} A_j$. Therefore, $\bigcup_{j=1}^{\ell} A_j \in \mathcal{A}$.

Let $A \in \mathcal{A}$. For each $j \in \{1, \ldots, \ell - 1\}$, let r_j be the number of components of A contained in κ_j. Let $r_\ell = n - \sum_{j=1}^{\ell-1} r_j$. Then $\sum_{j=1}^{\ell} r_j = n$. Hence, $\{r_1, \ldots, r_\ell\} \in$

$\mathcal{N}$, $(A\cap\kappa_1, \ldots, A\cap\kappa_\ell) \in \prod_{j=1}^{\ell} C_{r_j}(\kappa_j)$ and $f(A\cap\kappa_1, \ldots, A\cap\kappa_\ell) = A$. Therefore, f is surjective.

Let $(A_1, \ldots, A_\ell)$ and $(B_1, \ldots, B_\ell)$ be two different points of $\mathcal{R}$. Then $A_{j_0} \neq B_{j_0}$ for some $j_0 \in \{1, \ldots, \ell\}$. Hence, $\bigcup_{j=1}^{\ell} A_j \neq \bigcup_{j=1}^{\ell} B_j$, being a disjoint union. Therefore, f is one-to-one.

To see that f is continuous, let $\varepsilon > 0$ be given. Let $(A_1, \ldots, A_\ell)$ and $(B_1, \ldots, B_\ell)$ be two points of $\mathcal{R}$ such that

$$\rho_1\left((A_1, \ldots, A_\ell), (B_1, \ldots, B_\ell)\right) < \frac{\varepsilon}{2}.$$

Then for each $j \in \{1, \ldots, \ell\}$, $\mathcal{H}(A_j, B_j) < \frac{\varepsilon}{2}$. Hence, for every $j \in \{1, \ldots, \ell\}$, $A_j \subset \mathcal{V}_{\frac{\varepsilon}{2}}^d(B_j) \subset \mathcal{V}_{\frac{\varepsilon}{2}}^d\left(\bigcup_{r=1}^{\ell} B_r\right)$ and $B_j \subset \mathcal{V}_{\frac{\varepsilon}{2}}^d(A_j) \subset \mathcal{V}_{\frac{\varepsilon}{2}}^d\left(\bigcup_{r=1}^{\ell} A_r\right)$. Thus, we have that $\bigcup_{j=1}^{\ell} A_j \subset \mathcal{V}_{\frac{\varepsilon}{2}}^d\left(\bigcup_{j=1}^{\ell} B_j\right)$ and $\bigcup_{j=1}^{\ell} B_j \subset \mathcal{V}_{\frac{\varepsilon}{2}}^d\left(\bigcup_{j=1}^{\ell} A_j\right)$. This implies that

$$\mathcal{H}\left(\bigcup_{j=1}^{\ell} A_j, \bigcup_{j=1}^{\ell} B_j\right) \leq \frac{\varepsilon}{2} < \varepsilon.$$

Therefore, f is continuous.

Q.E.D.

The next theorem gives us an intrinsic characterization of the arc components of $\mathcal{C}_n(X) \setminus \{X\}$.

6.5.15 Theorem *Let n be an integer greater than one, and let X be an indecomposable continuum. Then $\mathcal{A}$ is an arc component of $\mathcal{C}_n(X) \setminus \{X\}$ if and only if there exists a finite number of composants $\kappa_1, \ldots, \kappa_m$ of X such that $\mathcal{A} = \langle \kappa_1, \ldots, \kappa_m \rangle_n$.*

Proof Suppose $\mathcal{A}$ is an arc component of $\mathcal{C}_n(X)\setminus\{X\}$. Let A_0 be a point of $\mathcal{A}$ and let $\kappa_1, \ldots, \kappa_m$ be the composants of X which intersect A_0. Let $f\colon \mathcal{R} \to \mathcal{C}_n(X)$ be the map given in Theorem 6.5.14. It is easy to see that $f(\mathcal{R}) = \langle \kappa_1, \ldots, \kappa_m \rangle_n$. Hence, $\mathcal{A} = \langle \kappa_1, \ldots, \kappa_m \rangle_n$.

Next, let $\kappa_1, \ldots, \kappa_m$ be m distinct composants of X. Again, using the map $f\colon \mathcal{R} \to \mathcal{C}_n(X)$ given in Theorem 6.5.14 and the fact that $f(\mathcal{R}) = \langle \kappa_1, \ldots, \kappa_m \rangle_n$, we have that $\langle \kappa_1, \ldots, \kappa_m \rangle_n$ is arcwise connected.

Now, suppose Ω is an arcwise connected subset of $\mathcal{C}_n(X) \setminus \{X\}$ containing $\langle \kappa_1, \ldots, \kappa_m \rangle_n$. We show that $\Omega = \langle \kappa_1, \ldots, \kappa_m \rangle_n$. Let W be a point of Ω and let $A \in \langle \kappa_1, \ldots, \kappa_m \rangle_n$. Repeating the arguments given in the second paragraph of the proof of Theorem 6.5.14, we have that A and W intersect the same composants of X, namely, $\kappa_1, \ldots, \kappa_m$. Thus, $\Omega \subset \langle \kappa_1, \ldots, \kappa_m \rangle_n$. Therefore, $\langle \kappa_1, \ldots, \kappa_m \rangle_n$ is an arc component of $\mathcal{C}_n(X) \setminus \{X\}$.

Q.E.D.

6.5.16 Theorem *Let X be an indecomposable continuum and let n be a positive integer. If $\mathcal{A}$ is an arc component of $C_n(X) \setminus \{X\}$, then $\mathcal{A} \setminus \mathcal{F}_n(X)$ is an arc component of $C_n(X) \setminus (\{X\} \cup \mathcal{F}_n(X))$.*

Proof By Theorem 6.5.2, no element of $\mathcal{F}_n(X)$ arcwise disconnects $C_n(X)$. Thus, $\mathcal{A} \setminus \mathcal{F}_n(X)$ is arcwise connected. Let $\mathcal{B}$ be the arc component of $C_n(X) \setminus (\{X\} \cup \mathcal{F}_n(X))$ containing $\mathcal{A} \setminus \mathcal{F}_n(X)$. Note that $\mathcal{B}$ is an arcwise connected subset of $C_n(X) \setminus \{X\}$ and $\mathcal{B} \cap \mathcal{A} \neq \emptyset$. Hence, $\mathcal{B} \subset \mathcal{A}$, and $\mathcal{B} = \mathcal{A} \setminus \mathcal{F}_n(X)$.

Q.E.D.

6.5.17 Theorem *Let X be a continuum such that $C(X)$ is finite dimensional. If A is a nondegenerate indecomposable proper subcontinuum of X, then at most a finite number of composants of A have the property that some subcontinuum of X contains a point of $X \setminus A$ and a point of the composant but does not contain A; also, $C_n(X) \setminus \{A\}$ has uncountably many arc components.*

Proof The first part follows by the proof of [84, (∗), p. 312]. By the proof of [84, (v), p. 312], $C(X) \setminus \{A\}$ has uncountably many arc components. Hence, $C_n(X) \setminus \{A\}$ is not arcwise connected, by Theorem 6.5.5, and has uncountably many arc components by [86, 11.15] and Theorem 6.5.13.

Q.E.D.

6.5.18 Theorem *Let n be an integer greater than one. Let X be an indecomposable continuum. If $\mathcal{A}$ is an arc component of $C_n(X) \setminus \{X\}$, which is not of the form $C_n(\kappa)$, where κ is a composant of X, then for any arc $\alpha \colon [0, 1] \to \mathcal{A}$, $\sigma(\alpha([0, 1]))$ is not connected.*

Proof Let $\alpha \colon [0, 1] \to \mathcal{A}$ be an arc and suppose $\sigma(\alpha([0, 1]))$ is connected. Observe that $\alpha(0)$ is a nonconnected subset of $\sigma(\alpha([0, 1]))$ intersecting at least two composants of X. Hence, $\sigma(\alpha([0, 1])) = X$. An argument similar to the one given in the proof of Theorem 6.5.13 shows that there exists a point t_0 in $[0, 1]$ such that $\alpha(t_0) = X$, which is not possible. Therefore, $\sigma(\alpha([0, 1]))$ is not connected.

Q.E.D.

6.5.19 Theorem *Let X and Y be continua, where X is indecomposable with the property of Kelley, and let n and m be positive integers. If $C_n(X)$ is homeomorphic to $C_m(Y)$, then Y is indecomposable.*

Proof Let $h \colon C_n(X) \to C_m(Y)$ be a homeomorphism. Since X is indecomposable and has the property of Kelley, X is the only point at which $C_n(X)$ is locally connected, by Theorem 6.1.21. Thus, $h(X)$ is the only point at which $C_m(Y)$ is locally connected. Since $C_m(Y)$ is always locally connected at Y [73, Lemma 2.3], we have that $h(X) = Y$.

Now, since X is indecomposable, $C_n(X) \setminus \{X\}$ is not arcwise connected, by Theorem 6.5.3. Hence, $C_m(Y) \setminus \{Y\}$ is not arcwise connected. Therefore, Y is indecomposable, by Theorem 6.5.3.

Q.E.D.

The following theorem is similar to Theorem 6.5.19, but the property of Kelley is not necessary.

6.5.20 Theorem *Let X and Y be continua, where X is indecomposable. If 2^X is homeomorphic to 2^Y, then Y is indecomposable.*

Proof Let $h: 2^X \to 2^Y$ be a homeomorphism. Since X is indecomposable, X is the only point at which 2^X is locally connected [84, (1.139)]. Thus, $h(X)$ is the only point at which 2^Y is locally connected. Since 2^Y is always locally connected at Y [84, (1.136)], we have that $h(X) = Y$.

Now, since X is indecomposable, $2^X \setminus \{X\}$ is not arcwise connected [84, (11.4)]. Hence, $2^Y \setminus \{Y\}$ is not arcwise connected. Therefore, Y is indecomposable [84, (11.4)].

Q.E.D.

6.5.21 Remark In [25, Example 4.5] the authors present two continua X_1 and $Y = X_1 \cup X_2$ such that X_1 is indecomposable, Y is decomposable and $\mathcal{C}(X_1)$ is homeomorphic to $\mathcal{C}(Y)$. Theorem 6.5.19 shows that this cannot happen when X_1 has the property of Kelley. Also, observe that even though $\mathcal{C}(X_1)$ and $\mathcal{C}(Y)$ are homeomorphic, by Theorem 6.5.20, 2^{X_1} is not homeomorphic to 2^Y.

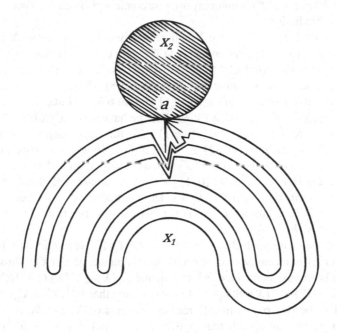

6.5.22 Theorem *Let X be an indecomposable continuum and let n and m be positive integers. If Y is a hereditarily decomposable continuum, then $\mathcal{C}_m(Y)$ is not homeomorphic to $\mathcal{C}_n(X)$.*

Proof Let us suppose that $\mathcal{C}_m(Y)$ is homeomorphic to $\mathcal{C}_n(X)$. Let $h: \mathcal{C}_n(X) \to \mathcal{C}_m(Y)$ be a homeomorphism. Since X is indecomposable, $\mathcal{C}_n(X) \setminus \{X\}$ is not arcwise connected. Recall that since X is indecomposable, X has uncountably many composants [36, Theorem 3–46]. Also, for each composant κ of X, $C_n(\kappa)$ is an arc component of $C_n(X) \setminus \{X\}$ by Theorem 6.5.13. Hence, we have that $C_n(X) \setminus \{X\}$ has uncountably many arc components. Then $C_m(Y) \setminus \{h(X)\}$ has uncountably many arc components. Since $C_m(Y) \setminus \{h(X)\}$ is not arcwise connected, $h(X) \in C(Y)$ by Theorem 6.5.2. Since Y is a hereditarily decomposable continuum, $h(X)$ is a decomposable subcontinuum of Y. Hence, by Lemma 6.5.10, $C_m(Y) \setminus \{h(X)\}$ has exactly two arc components, a contradiction. Therefore, $C_m(Y)$ is not homeomorphic to $C_n(X)$.

Q.E.D.

We end this section showing that hereditarily indecomposable continua have unique *n*-fold hyperspaces.

6.5.23 Theorem *Let $n, m \in \mathbb{N}$, $m \geq 2$. Let X be a hereditarily indecomposable continuum. If Y is a continuum such that $C_m(Y)$ is homeomorphic to $C_n(X)$, then Y is homeomorphic to X.*

Proof Let $h \colon C_n(X) \twoheadrightarrow C_m(Y)$ be a homeomorphism. We consider first the case $n = 1$. Since X is hereditarily indecomposable, $C(X)$ is uniquely arcwise connected [84, (1.61)]. Hence, $C_m(Y)$ is uniquely arcwise connected. Then, by Theorem 6.1.11, $m = 1$. A contradiction.

Suppose now that $n \geq 2$. Since X is hereditarily indecomposable, X is the only point at which $C_n(X)$ is locally connected (Theorem 6.1.22). Hence, $h(X) = Y$.

Now, we show that $h(C(X)) \subset C(Y)$. To see this, let $A \in C(X) \setminus \mathcal{F}_n(X)$, then $C_n(X) \setminus \{A\}$ is not arcwise connected (Theorem 6.5.8). Thus, $C_m(Y) \setminus \{h(A)\}$ is not arcwise connected. Hence, $h(A) \in C(Y)$ (Theorem 6.5.2). Since $C(Y)$ is closed in $C_m(Y)$ and $\mathcal{F}_1(X) \subset Cl_{C_n(X)}(C(X) \setminus \mathcal{F}_1(X))$, we have that $h(C(X)) \subset C(Y)$.

Next, we prove that $h(\mathcal{F}_1(X)) \subset \mathcal{F}_1(Y)$. To this end, suppose there exists a point $\{x\}$ in $\mathcal{F}_1(X)$ such that $h(\{x\}) \in C(Y) \setminus \mathcal{F}_1(Y)$. Then there exists an order arc $\alpha \colon [0, 1] \to C_2(Y)$ such that $\alpha(0) \in \mathcal{F}_2(Y)$, $\alpha(1) = \{x\}$, and $\alpha([0, 1)) \subset C_2(Y) \setminus C(Y)$ [83, (2.2)]. Hence, $h^{-1} \circ \alpha \colon [0, 1] \to C_n(X)$ is an arc such that $h^{-1} \circ \alpha(1) = \{x\}$ and $h^{-1} \circ \alpha([0, 1)) \subset C_n(X) \setminus C(X)$. This contradicts the fact that in 2^X, such arcs do not exist (i.e., singletons are not arcwise accessible from $2^X \setminus C(X)$) [83, (3.4)]. Therefore, $h(\mathcal{F}_1(X)) \subset \mathcal{F}_1(Y)$.

Let $Y' \in C(Y)$ be such that $\mathcal{F}_1(Y') = h(\mathcal{F}_1(X))$. Note that $C(Y') \subset C(Y)$ and $h^{-1}(C(Y'))$ is an arcwise connected subcontinuum of $C_n(X)$. Thus, $C(X) \cap h^{-1}(C(Y'))$ is arcwise connected [83, (5.2)], and $\mathcal{F}_1(X) \subset C(X) \cap h^{-1}(C(Y'))$. We claim that $h^{-1}(Y') \in C(X)$. Suppose, to the contrary, that $h^{-1}(Y') \in C_n(X) \setminus C(X)$. Let x_1 and x_2 be two points in different composants of X. Let $\beta_1, \beta_2 \colon [0, 1] \to h^{-1}(C(Y'))$ be two arcs such that $\beta_j(0) = \{x_j\}$ and $\beta_j(1) = h^{-1}(Y')$, $j \in \{1, 2\}$. By [83, (3.4)], we have that $\beta_j([0, 1)) \cap C(X) \neq \{\{x_j\}\}$, $j \in \{1, 2\}$. Hence, $(\beta_1([0, 1)) \cup \beta_2([0, 1))) \cap C(X)$ contains an arc from $\{x_1\}$ to $\{x_2\}$ [83, (5.2)]. Since X is indecomposable and x_1 and x_2 are in different composants of X, we have that $X = \sigma([(\beta_1([0, 1)) \cup \beta_2([0, 1))) \cap C(X)])$ [84, (1.51)]. Then

$X \in (\beta_1([0, 1]) \cup \beta_2([0, 1])) \cap C(X)$ [84, (1.50)]. Thus, $X \in h^{-1}(C(Y'))$. It follows that $Y = h(X) \in C(Y')$ and $Y = Y'$, a contradiction. Therefore, $h^{-1}(Y') \in C(X)$.

We assert that $h^{-1}(Y') = X$. To see this, let x_1' and x_2' be two points in different composants of X. Since $C(X) \cap h^{-1}(C(Y'))$ is arcwise connected, there exist two arcs $\alpha_1, \alpha_2 : [0, 1] \to C(X) \cap h^{-1}(C(Y'))$ such that $\alpha_j(0) = \{x_j'\}$ and $\alpha_j(1) = h^{-1}(Y')$, $j \in \{1, 2\}$. Then $\alpha_1([0, 1]) \cup \alpha_2([0, 1])$ contains an arc from $\{x_1'\}$ to $\{x_2'\}$. Since X is indecomposable, we have that $X = \sigma(\alpha_1([0, 1]) \cap \alpha_2([0, 1]))$ [84, (1.51)]. Thus, $X \in \alpha_1([0, 1]) \cup \alpha_2([0, 1])$ [84, (1.50)]. Then $h(X) \in C(Y')$. Hence, since $h(X) = Y$, $Y \in C(Y')$. Thus, $Y' = Y$. Then, by definition of Y', $h(\mathcal{F}_1(X)) = \mathcal{F}_1(Y)$. Therefore, Y is homeomorphic to X.

Q.E.D.

6.5.24 Corollary *Let* $n \in \mathbb{N}$, *and let* X *be a hereditarily indecomposable continuum. If* Y *is a continuum such that* $C_n(Y)$ *is homeomorphic to* $C_n(X)$, *then* X *and* Y *are homeomorphic.*

Proof By Theorem 6.5.23, we only have to consider the case when $n = 1$. Since X is hereditarily indecomposable, $C(X)$ is uniquely arcwise connected [84, (1.61)]. Hence, $C(Y)$ is uniquely arcwise connected. Thus, by [84, (1.61)], Y is hereditarily indecomposable. Therefore, Y is homeomorphic to X [84, (0.60)].

Q.E.D.

6.6 C_n^*-Smoothness

We study the continuity of taking n-fold hyperspaces. For this, we use [33–35, 49, 71, 85, 87].

6.6.1 Definition Let $n \in \mathbb{N}$. A continuum X is C_n^*-*smooth at* $A \in C_n(X)$, provided that for any sequence $\{A_k\}_{k=1}^\infty$ of elements of $C_n(X)$ converging to A, the sequence $\{C_n(A_k)\}_{k=1}^\infty$ of hyperspaces converges to $C_n(A)$; i.e., the map $C_n^* : C_n(X) \to 2^{2^X}$ given by $C_n^*(A) = C_n(A)$ is continuous at A. A continuum X is C_n^*-*smooth* if it is C_n^*-smooth at each element of $C_n(X)$; i.e., C_n^* is continuous.

6.6.2 Remark For $n = 1$, C_n^*-smoothness is called C^*-*smoothness* instead of C_1^*-smoothness. C^*-smoothness is introduced in [84, Chapter XV].

We start with C^*-smoothness of homogeneous continua.

6.6.3 Theorem *If* X *is a* C^*-*smooth homogeneous continuum, then* X *is indecomposable. Moreover, if* X *is a* C^*-*smooth homogeneous plane continuum, then* X *is hereditarily indecomposable.*

Proof Let X be a C^*-smooth homogeneous continuum. Then X is hereditarily unicoherent [33, (3.4)]. By [49, Theorem 1], X is indecomposable.

If X is a C^*-smooth homogeneous plane continuum, we have that X is indecomposable. Since any indecomposable homogeneous plane continuum is hereditarily indecomposable [35, Theorem 1], X is hereditarily indecomposable.

Q.E.D.

The following result is easy to prove.

6.6.4 Lemma *Let* $n \in \mathbb{N}$, *let* X *be a continuum and let* $\{A_k\}_{k=1}^{\infty}$ *be a sequence in* $C_n(X)$ *converging to* A. *If* $\lim_{k \to \infty} C_n(A_k)$ *exists, then* $\lim_{k \to \infty} C_n(A_k) \subset C_n(A)$.

6.6.5 Theorem *Let* X *be a continuum. If* A *is a subcontinuum of* X, *then the following are equivalent:*

(1) X *is* C^*-*smooth at* A;
(2) $C_n^* |_{C(X)}$ *is continuous at* A *for all* $n \in \mathbb{N}$;
(3) $C_n^* |_{C(X)}$ *is continuous at* A *for some* $n \in \mathbb{N}$.

Proof Assume X is C^*-smooth at A, and let $n \geq 2$. We prove $C_n^* |_{C(X)}$ is continuous at A. Let $\{A_k\}_{k=1}^{\infty}$ be a sequence of subcontinua of X converging to A. Let B be any element of $C_n(A)$. Let $B_1, \ldots, B_\ell$ ($\ell \leq n$) be the components of B. Hence, each B_j is a subcontinuum of A, $j \in \{1, \ldots, \ell\}$. Since X is C^*-smooth at A, there exist subcontinua $B_k^1, \ldots, B_k^\ell$ of A_k, for each $k \in \mathbb{N}$, such that $\lim_{k \to \infty} B_k^j = B_j$ for each $j \in \{1, \ldots, \ell\}$. Hence, $B_k = \bigcup_{j=1}^{\ell} B_k^j$ is an element of $C_n(A_k)$, for each $k \in \mathbb{N}$, and $\lim_{k \to \infty} B_k = B$ (Lemma 1.8.11). Therefore, $C_n(A) \subset \lim_{k \to \infty} C_n(A_k)$. By Lemma 6.6.4, we may conclude that $\lim_{k \to \infty} C_n(A_k) = C_n(A)$. Therefore, $C_n^* |_{C(X)}$ is continuous at A.

Assume $C_n^* |_{C(X)}$ is continuous at A for some $n \geq 1$ and some element A of $C(X)$. We prove X is C^*-smooth at A. Let $\{A_k\}_{k=1}^{\infty}$ be a sequence of subcontinua of X converging to A. Let B be a nondegenerate proper subcontinuum of A. Let $x_1, \ldots, x_{n-1}$ be $n - 1$ distinct points in $A \setminus B$. Let $D = B \cup \{x_1, \ldots, x_{n-1}\}$. Since $C_n^* |_{C(X)}$ is continuous at A, there exists $D_k \in C_n(A_k)$, for each $k \in \mathbb{N}$, such that the sequence $\{D_k\}_{k=1}^{\infty}$ converges to D. Since D has n components, we assume without loss of generality that D_k also has n components for any $k \in \mathbb{N}$. Since n is the maximum number of components we allow, there exists a component D_k^1 of D_k such that $\{D_k^1\}_{k=1}^{\infty}$ converges to B. Therefore, $C(A) \subset \lim_{k \to \infty} C(A_k)$. By Lemma 6.6.4, we conclude that $\lim_{k \to \infty} C(A_k) = C(A)$.

The fact that (2) implies (3) is obvious.

Q.E.D.

6.6.6 Definition A continuum X is *absolutely* C^*-*smooth*, provided that for any continuum Z in which X can be embedded and for each sequence $\{A_k\}_{k=1}^{\infty}$ of elements of $C(Z)$ converging to X, the sequence $\{C(A_k)\}_{k=1}^{\infty}$ of hyperspaces converges to $C(X)$.

With a proof similar to the one given for Theorem 6.6.5, we have the following result:

6.6.7 Theorem *Let X be a continuum. Then the following statements are equivalent:*

(1) X is absolutely C^*-smooth;
(2) for any continuum Z in which X is embedded, $C_n^*|_{C(Z)}$ is continuous at X for all $n \in \mathbb{N}$;
(3) for any continuum Z in which X is embedded, $C_n^*|_{C(Z)}$ is continuous at X for some $n \in \mathbb{N}$.

6.6.8 Lemma *Let $n \in \mathbb{N}$, let X be a continuum, with metric d, let A be an indecomposable subcontinuum of X and let $\{B_m\}_{m=1}^{\infty}$ be a sequence of elements of $C_n(X)$ converging to A. Then there exists a subsequence $\{B_{m_k}\}_{k=1}^{\infty}$ of $\{B_m\}_{m=1}^{\infty}$ such that for each k, there exists a component D_k of B_{m_k} such that the sequence $\{D_k\}_{k=1}^{\infty}$ of continua converges to A.*

Proof Since A is an indecomposable continuum, A has uncountably many mutually disjoint composants [86, 11.15 and 11.17]. Let $a_1, \dots, a_{n+1}$ be $n+1$ points in $n+1$ distinct composants of A. We may assume that $\mathcal{V}_{\frac{1}{\ell}}^d(a_i) \cap \mathcal{V}_{\frac{1}{\ell}}^d(a_j) = \emptyset$ if $i \neq j$ for each positive integer ℓ.

Since $\{B_m\}_{m=1}^{\infty}$ converges to A, for each ℓ, there exists an integer m_ℓ such that $\mathcal{H}(A, B_{m_\ell}) < \frac{1}{\ell}$. Thus, $B_{m_\ell} \cap \mathcal{V}_{\frac{1}{\ell}}^d(a_j) \neq \emptyset$ for each $j \in \{1, \dots, n+1\}$. Since B_{m_ℓ} has at most n components, we have that at least one of the components of B_{m_ℓ} intersects two of the balls $\mathcal{V}_{\frac{1}{\ell}}^d(a_j)$, $j \in \{1, \dots, n+1\}$.

Since we only have $n + 1$ balls, there exist $j_0, j_1 \in \{1, \dots, n+1\}$ such that for infinitely many indices k, B_{m_k} has a component D_k such that $D_k \cap \mathcal{V}_{\frac{1}{k}}^d(a_{j_0}) \neq \emptyset$ and $D_k \cap \mathcal{V}_{\frac{1}{k}}^d(a_{j_1}) \neq \emptyset$ for each k. Since $C(X)$ is compact (Theorem 1.8.5), we assume without loss of generality that the sequence $\{D_k\}_{k=1}^{\infty}$ converges to a subcontinuum D of A. Since a_{j_0} and a_{j_1} belong to D and they are in different composants of A, we conclude that $D = A$.

Q.E.D.

6.6.9 Remark The converse of Lemma 6.6.8 is false, as can be seen from the argument of [71, Example 3.4].

6.6.10 Lemma *Let X be a decomposable continuum, with metric d, and let A and B be nondegenerate proper subcontinua of X such that $X = A \cup B$. Assume that there exist two order arcs $\alpha, \beta : [0, 1] \to C(X)$ with the following properties: $\alpha(0) \in \mathcal{F}_1(A), \alpha(1) = A, \beta(0) \in \mathcal{F}_1(B), \beta(1) = B$ and $(A \cap B) \cap (\alpha(t) \cup \beta(t)) = \emptyset$ for each $t \in [0, 1)$. Then X is not C_n^*-smooth at X for any $n \geq 2$.*

Proof Suppose X is C_n^*-smooth at X. Let $\{t_m\}_{m=1}^{\infty}$ be an increasing sequence of numbers in $[0, 1)$ converging to 1. For each $m \in \mathbb{N}$, let $D_m = \alpha(t_m) \cup \beta(t_m)$. For each $m \in \mathbb{N}$, $(A \cap B) \cap (\alpha(t_m) \cup \beta(t_m)) = \emptyset$, hence, $D_m \in C_2(X) \setminus C(X)$.

Let R be a component of $A \cap B$. Let H and K be proper subcontinua of A and B, respectively, such that they properly contain R (Corollary 1.7.28). Let $x_1, \ldots, x_{n-1}$ be $n - 1$ distinct points of $X \setminus (H \cup K)$. Let $L = \{x_1, \ldots, x_{n-1}\} \cup (H \cup K)$. Let $\varepsilon > 0$ be such that the following hold:

$$\mathcal{V}_{2\varepsilon}^d(x_i) \cap \mathcal{V}_{2\varepsilon}^d(x_j) = \emptyset \text{ if and only if } i \neq j,$$

$$\{x_1, \ldots, x_{n-1}\} \cap \mathcal{V}_{2\varepsilon}^d(H \cup K) = \emptyset,$$

$$\bigcup_{j=1}^{n-1} \mathcal{V}_{2\varepsilon}^d(x_j) \cap (H \cup K) = \emptyset,$$

$$H \setminus \mathcal{V}_{2\varepsilon}^d(K) \neq \emptyset, \text{ and } K \setminus \mathcal{V}_{2\varepsilon}^d(H) \neq \emptyset.$$

Since X is $\mathcal{C}_n^*$-smooth at X, there exists $m_0 \in \mathbb{N}$ such that if $m \geq m_0$, then there exists $E_m \in \mathcal{C}_n(D_m)$ such that $\mathcal{H}(E_m, L) < \varepsilon$. Let $m' \geq m_0$. Then $E_{m'} \subset \mathcal{V}_{\varepsilon}^d(L) = \left(\bigcup_{j=1}^{n-1} \mathcal{V}_{\varepsilon}^d(x_j)\right) \cup \mathcal{V}_{\varepsilon}^d(H \cup K)$, $E_{m'} \cap \mathcal{V}_{\varepsilon}^d(x_j) \neq \emptyset$ for each $j \in \{1, \ldots, n-1\}$, and $E_{m'} \cap \mathcal{V}_{\varepsilon}^d(H \cup K) \neq \emptyset$. Hence, $E_{m'}$ has exactly n components. Let $G_1, \ldots, G_n$ be the components of $E_{m'}$. Since the ε-balls about each $x_1, \ldots, x_{n-1}$ and $H \cup K$ are pairwise disjoint we assume, without loss of generality, that $G_j \subset \mathcal{V}_{\varepsilon}^d(x_j)$ for each $j \in \{1, \ldots, n-1\}$ and $G_n \subset \mathcal{V}_{\varepsilon}^d(H \cup K)$. Since G_n is a subcontinuum of $D_{m'}$, G_n is contained either in $\alpha(t_{m'})$ or in $\beta(t_{m'})$. Suppose that G_n is contained in $\alpha(t_{m'})$. Let $x \in K \setminus \mathcal{V}_{2\varepsilon}^d(H)$. Then for each point z of $E_{m'}$, $d(y, z) \geq \varepsilon$. This is a contradiction; therefore, X is not $\mathcal{C}_n^*$-smooth at X.

Q.E.D.

The following result characterizes the class of continua for which the map $\mathcal{C}_n^*$ is continuous for $n \geq 2$.

6.6.11 Theorem *A continuum X is $\mathcal{C}_n^*$-smooth for some $n \geq 2$ if and only if X is hereditarily indecomposable.*

Proof Suppose X is hereditarily indecomposable. Then X is $\mathcal{C}_n^*$-smooth by Lemma 6.6.8 and [84, (1.207.8)].

Suppose that X is $\mathcal{C}_n^*$-smooth for some integer $n \geq 2$. Then condition (3) of Theorem 6.6.7 is satisfied. Hence, X is $\mathcal{C}^*$-smooth by Theorem 6.6.7. Since X is $\mathcal{C}^*$-smooth, X is hereditarily unicoherent [33, (3.4)].

Suppose X is decomposable. Then there exist two proper subcontinua A and B of X such that $X = A \cup B$.

Let $a \in A \setminus B$ and $b \in B \setminus A$. Let $\alpha, \beta : [0, 1] \to \mathcal{C}(X)$ be order arcs such that $\alpha(0) = \{a\}$, $\alpha(1) = A$, $\beta(0) = \{b\}$ and $\beta(1) = B$ (Theorem 1.8.20). Let t_0 and s_0 be points of $[0, 1]$ such that $\alpha(t_0) \cap \beta(s_0) \neq \emptyset$ and such that for each $t < t_0$ and each $s < s_0$, $\alpha(t) \cap \beta(s) = \emptyset$. Note that $t_0 > 0$ and $s_0 > 0$. Let $\{t_k\}_{k=1}^{\infty}$ and $\{s_k\}_{k=1}^{\infty}$ be increasing sequences in $[0, 1]$ converging to t_0 and s_0, respectively.

Let $Y = \alpha(t_0) \cup \beta(s_0)$. Then Y is a subcontinuum of X. Then, by Lemma 6.6.10, X is not C_n^*-smooth at Y, a contradiction. Therefore, X is indecomposable.

A similar argument shows that each subcontinuum of X is indecomposable.

Q.E.D.

We now present some results about the points at which a continuum X is C_n^*-smooth.

6.6.12 Theorem *Let X be a continuum and let A be an element of $C_n(X)$ for some $n \geq 2$. If X is C_n^*-smooth at A, then X is C^*-smooth at each component of A.*

Proof Let A be an element of $C_n(X)$ and suppose X is C_n^*-smooth at A. Observe that if A is connected, then X is C^*-smooth at A by Theorem 6.6.5.

Suppose A has at least two components. Let $A_1, \ldots, A_k$ be the components of A. We show that X is C^*-smooth at A_1. Let $\{K_m\}_{m=1}^\infty$ be a sequence of subcontinua of X converging to A_1. Without loss of generality, we assume that $K_m \cap \left(\bigcup_{j=2}^k A_j \right) = \emptyset$ for each $m \in \mathbb{N}$. Let L be a subcontinuum of A_1.

Let $\alpha \colon [0, 1] \to C(X)$ be an order arc such that $\alpha(0) \in \mathcal{F}_1(A_2)$ and $\alpha(1) = A_2$ (Theorem 1.8.20). Let $\{t_m\}_{m=1}^\infty$ be an increasing sequence of numbers in $[0, 1)$ converging to 1. For each $m \in \mathbb{N}$, let $p_m^{(1)}, \ldots, p_m^{(n-k)}$ be $n - k$ distinct points in $A_2 \setminus \alpha(t_m)$.

For each $m \in \mathbb{N}$, let

$$F_m = K_m \cup \alpha(t_m) \cup \left(\bigcup_{j=3}^k A_j \right) \cup \{p_m^{(1)}, \ldots, p_m^{(n-k)}\}.$$

Then $\lim_{m \to \infty} F_m = A$. Since X is C_n^*-smooth at A, for every $m \in \mathbb{N}$, there exists an element D_m of $C_n(F_m)$ such that $\lim_{m \to \infty} D_m = L \cup \alpha(t_1) \cup \left(\bigcup_{j=3}^k A_j \right) \cup \{p_1^{(1)}, \ldots, p_1^{(n-k)}\}$.

For each $m \in \mathbb{N}$, let $L_m = D_m \cap K_m$. Then L_m is a subcontinuum of K_m and $\lim_{m \to \infty} L_m = L$. Therefore, X is C^*-smooth at A_1. Similarly, X is C^*-smooth at the other components of A.

Q.E.D.

6.6.13 Lemma *Let C be a closed subset of a space Z. Let $A = Cl(Z \setminus C)$ and let $B = Cl(Z \setminus A)$. Then $A = Cl(Z \setminus B)$.*

Proof Since A is closed in Z, $Cl(Int(A)) \subset A$; thus, since

$$A = Cl(Z \setminus C) = Cl(Int(Z \setminus C)) \subset Cl(Int(Cl(Z \setminus C))) = Cl(Int(A)),$$

we have that $A = Cl(Int(A))$. Therefore, since $Int(A) = Z \setminus Cl(Z \setminus A) = Z \setminus B$, $A = Cl(Z \setminus B)$.

Q.E.D.

6.6.14 Theorem *If X is an irreducible continuum such that X is C_n^*-smooth at X for some $n \geq 2$, then X is indecomposable.*

Proof Assume that a and b are points about which X is irreducible. Suppose X is decomposable. Let C be a nondegenerate proper subcontinuum of X, with nonempty interior, containing b. Let $A = Cl(X \setminus C)$ and $B = Cl(X \setminus A)$.

Then A and B are subcontinua of X (Theorem 1.7.31) containing a and b, respectively. Note that $A = Cl(X \setminus B)$, by Lemma 6.6.13. Since $A \cap B = Bd(A) = Bd(B)$ and since $B = Cl(X \setminus A)$ (and $A = Cl(X \setminus B)$), A (and B, respectively) is irreducible between a (b, respectively) and any point of $A \cap B$ [86, 11.42].

Let $\alpha, \beta \colon [0, 1] \to C(X)$ be order arcs such that $\alpha(0) = \{a\}$, $\alpha(1) = A$, $\beta(0) = \{b\}$ and $\beta(1) = B$ (Theorem 1.8.20).

Notice that for any $t \in [0, 1)$, $(A \cap B) \cap \alpha(t) = \emptyset$ and $(A \cap B) \cap \beta(t) = \emptyset$. By Lemma 6.6.10, X is not C_n^*-smooth at X, a contradiction. Therefore, X is indecomposable.

$$\text{Q.E.D.}$$

6.6.15 Theorem *Let X be a continuum and let A be an element of $C_n(X)$ with exactly n components, $n \geq 2$. Then X is C_n^*-smooth at A if and only if X is C^*-smooth at each component of A.*

Proof Let $A \in C_n(X) \setminus C_{n-1}(X)$. If X is C_n^*-smooth at A, then X is C^*-smooth at each component of A by Theorem 6.6.12.

Let A be an element of $C_n(X)$ with n components $A_1, \ldots, A_n$. Suppose X is C^*-smooth at each A_j for each $j \in \{1, \ldots, n\}$.

Let $\{B_k\}_{k=1}^{\infty}$ be a sequence of elements of $C_n(X)$ converging to A. Since A has n components, without loss of generality, we assume that B_k has n components, $B_k^1, \ldots, B_k^n$, for each $k \in \mathbb{N}$. In fact, we may suppose that $\lim_{k \to \infty} B_k^j = A_j$ for each $j \in \{1, \ldots, n\}$.

Let C be an element of $C_n(A)$. Let $A_{j_1}, \ldots, A_{j_\ell}$ be the components of A intersecting C; i.e., $C = \bigcup_{i=1}^{\ell}(A_{j_i} \cap C)$. Let $C_{j_i} = A_{j_i} \cap C$ for each $i \in \{1, \ldots, \ell\}$. Since X is C^*-smooth at A_{j_i}, there exists a subcontinuum $D_k^{j_i}$ of $B_k^{j_i}$ for each $i \in \{1, \ldots, \ell\}$ such that $\lim_{k \to \infty} D_k^{j_i} = C_{j_i}$. For $k \in \mathbb{N}$, let $D_k = \bigcup_{i=1}^{\ell} D_k^{j_i}$. Hence, $D_k \in C_n(B_k)$ and $\lim_{k \to \infty} D_k = C$. Therefore, X is C_n^*-smooth at A by Lemma 6.6.4.

$$\text{Q.E.D.}$$

6.6.16 Theorem *Let X be a continuum. If A is an element of $C_n(X)$, for some $n \geq 2$, such that all the components of A are indecomposable and X is C^*-smooth at each component of A, then X is C_n^*-smooth at A.*

Proof Let A be an element of $C_n(X)$. Let $A_1, \ldots, A_\ell$ ($\ell \leq n$) be the components of A. Suppose A_j is an indecomposable continuum and X is C^*-smooth at A_j for each $j \in \{1, \ldots, \ell\}$.

Let $\{B_k\}_{k=1}^{\infty}$ be a sequence of elements of $C_n(X)$ converging to A. Let C be an element of $C_n(A)$. Let $A_{j_1}, \ldots, A_{j_s}$ be the components of A intersecting C; i.e., $C = \bigcup_{i=1}^{s}(A_{j_i} \cap C)$. Let $C_{j_1}^1, \ldots, C_{j_i}^{\ell_{j_i}}$ be the components of $A_{j_i} \cap C$ for each $i \in \{1, \ldots, s\}$.

In what follows, $k \in \mathbb{N}$ and $i \in \{1, \dots, s\}$. Since each A_{j_i} is indecomposable, by a similar argument to the one given in Lemma 6.6.8, there exist components $B_k^{j_i}$ of B_k such that $\lim\limits_{k \to \infty} B_k^{j_i} = A_{j_i}$. Since X is $\mathcal{C}^*$-smooth at each A_{j_i}, there exist subcontinua $D_{k,1}^{j_i}, \dots, D_{k,\ell_{j_i}}^{j_i}$ of $B_k^{j_i}$ such that $\lim\limits_{k \to \infty} D_{k,m}^{j_i} = C_{j_i}^m$ for each $m \in \{1, \dots, \ell_{j_i}\}$. Let $D_k^{j_i} = \bigcup_{m=1}^{\ell_{j_i}} D_{k,m}^{j_i}$ and let $D_k = \bigcup_{i=1}^{s} D_k^{j_i}$. Then $D_k \in \mathcal{C}_n(B_k)$ and $\lim\limits_{k \to \infty} D_k = C$. Therefore, X is $\mathcal{C}_n^*$-smooth at A by Lemma 6.6.4.

Q.E.D.

6.6.17 Corollary *Let X be a continuum and let $n \geq 2$. If A is an element of $\mathcal{C}_n(X)$ such that all the components of A are hereditarily indecomposable, then X is $\mathcal{C}_n^*$-smooth at A.*

Proof The corollary follows from Theorem 6.6.16 and the fact that hereditarily indecomposable continua are absolutely $\mathcal{C}^*$-smooth continua (by [84, (14.14.1)] and [34, 3.2]).

Q.E.D.

6.7 Z-Sets

We present results showing conditions that imply that the n-fold symmetric products of a continuum X are Z-sets in the n-fold hyperspaces of X and in the hyperspace of closed subsets of X. The material of this section is based on [8, 16, 20, 26–28, 30, 31, 48, 70, 84].

6.7.1 Definition A closed subset A of a metric space Y, with metric d, is a *Z-set* in Y provided that for every $\varepsilon > 0$, there exists a map $f_\varepsilon \colon Y \to Y \setminus A$ such that $d(y, f_\varepsilon(y)) < \varepsilon$ for all $y \in Y$.

We begin with a useful result that gives necessary and sufficient conditions for the hyperspace of singletons to be a Z-set in the hyperspace of subcontinua and in the hyperspace of closed subsets.

6.7.2 Theorem *Let X be a continuum and let $\mathcal{A}$ be either $\mathcal{C}(X)$ or 2^X. Then $\mathcal{F}_1(X)$ is a Z-set in $\mathcal{A}$ if and only if for each $\varepsilon > 0$, there exists a map $f_\varepsilon \colon \mathcal{F}_1(X) \to \mathcal{A} \setminus \mathcal{F}_1(X)$ such that $\mathcal{H}(\{x\}, f_\varepsilon(\{x\})) < \varepsilon$ for each $x \in X$.*

Proof We present the proof when $\mathcal{A}$ is $\mathcal{C}(X)$. The proof when $\mathcal{A}$ is 2^X is similar.

If $\mathcal{F}_1(X)$ is a Z-set in $\mathcal{C}(X)$, then the existence of the maps f_ε is evident.

Conversely, fix $\varepsilon > 0$. By assumption there exists a map $f_\varepsilon \colon \mathcal{F}_1(X) \to \mathcal{C}(X) \setminus \mathcal{F}_1(X)$ such that $\mathcal{H}(\{x\}, f_\varepsilon(\{x\})) < \varepsilon$ for every $x \in X$.

For each $A \in \mathcal{C}(X)$, let

$$g_\varepsilon(A) = \bigcup \{f_\varepsilon(\{a\}) \mid a \in A\}.$$

Then g_ε maps $\mathcal{C}(X)$ into $\mathcal{C}(X) \setminus \mathcal{F}_1(X)$, by Corollary 6.1.2; also, g_ε is continuous (Lemma 1.8.11).

We only need to show that $\mathcal{H}(A, g_\varepsilon(A)) < \varepsilon$ for each $A \in \mathcal{C}(X)$.

Let $A \in \mathcal{C}(X)$ and let $a \in A$. Since $\mathcal{H}(\{a\}, f_\varepsilon(\{a\})) < \varepsilon$, $a \in \mathcal{V}_\varepsilon(f_\varepsilon(\{a\})) \subset \mathcal{V}_\varepsilon(g_\varepsilon(A))$. Thus, $A \subset \mathcal{V}_\varepsilon(g_\varepsilon(A))$. Next, let $x \in g_\varepsilon(A)$. Then there exists $a \in A$ such that $x \in f_\varepsilon(\{a\})$. Hence, since $\mathcal{H}(\{a\}, f_\varepsilon(\{a\})) < \varepsilon$, $x \in \mathcal{V}_\varepsilon(\{a\}) \subset \mathcal{V}_\varepsilon(A)$. Thus, $g_\varepsilon(A) \subset \mathcal{V}_\varepsilon(A)$.

Therefore, we have proved that $\mathcal{H}(A, g_\varepsilon(A)) < \varepsilon$.

Q.E.D.

6.7.3 Theorem *Let X be a continuum. If $\mathcal{F}_1(X)$ is a Z-set in $\mathcal{C}(X)$, then $\mathcal{F}_n(X)$ is a Z-set in 2^X and in $\mathcal{C}_n(X)$ for any n.*

Proof Let $\varepsilon > 0$ be given. By assumption, there exists a map $f_\varepsilon \colon \mathcal{C}(X) \to \mathcal{C}(X) \setminus \mathcal{F}_1(X)$ such that $\mathcal{H}(K, f_\varepsilon(K)) < \varepsilon$ for all $K \in \mathcal{C}(X)$. For each $A \in \mathcal{C}_n(X)$, let

$$g_\varepsilon(A) = \bigcup \{f_\varepsilon(K) \mid K \text{ is a component of } A\}.$$

Then g_ε maps $\mathcal{C}_n(X)$ into $\mathcal{C}_n(X) \setminus \mathcal{F}_n(X)$, by Corollary 6.1.2; also, g_ε is continuous (Lemma 1.8.11), and $\mathcal{H}(A, g_\varepsilon(A)) < \varepsilon$ for each $A \in \mathcal{C}_n(X)$ (since $\mathcal{H}(K, f_\varepsilon(K)) < \varepsilon$ for each component K of A). This proves the theorem for $\mathcal{C}_n(X)$. The proof for 2^X is similar.

Q.E.D.

6.7.4 Corollary *Let X be a continuum. Assume that $\{\{p\}\}$ is a Z-set in $\mathcal{C}(X)$. Then for all $A \in \mathcal{F}_n(X)$ such that $p \in A$, $\{A\}$ is a Z-set in $\mathcal{C}_n(X)$ for each n, and $\{A\}$ is a Z-set in 2^X.*

As a consequence of Theorems 6.7.2 and 6.7.3, we have the following result:

6.7.5 Corollary *Let Y be a continuum. If $\mathcal{F}_1(Y)$ is a Z-set in $\mathcal{C}(Y)$, then for any continuum X, $\mathcal{F}_n(X \times Y)$ is a Z-set in $2^{X \times Y}$ and in $\mathcal{C}_n(X \times Y)$ for any n.*

Proof Let $\varepsilon > 0$ be given. Since $\mathcal{F}_1(Y)$ is a Z-set in $\mathcal{C}(Y)$, there exits a map $f_\varepsilon \colon \mathcal{F}_1(Y) \to \mathcal{C}(Y) \setminus \mathcal{F}_1(Y)$ such that $\mathcal{H}_Y(\{y\}, f_\varepsilon(\{y\})) < \varepsilon$.

Define $g_\varepsilon \colon \mathcal{F}_1(X \times Y) \to \mathcal{C}(X \times Y)$ by $g_\varepsilon(\{(x, y)\}) = \{\{x\}\} \times f_\varepsilon(\{y\})$. Then g_ε is a map such that $g_\varepsilon(\mathcal{C}(X \times Y)) \cap \mathcal{F}_1(X \times Y) = \emptyset$.

Clearly, $\mathcal{H}_{X \times Y}(\{(x, y)\}, g_\varepsilon(\{(x, y)\})) < \varepsilon$ for each $(x, y) \in X \times Y$.

Therefore, $\mathcal{F}_1(X \times Y)$ is a Z-set in $\mathcal{C}(X \times Y)$ by Theorem 6.7.2. Hence, $\mathcal{F}_n(X \times Y)$ is a Z-set in $2^{X \times Y}$ and in $\mathcal{C}_n(X \times Y)$ by Theorem 6.7.3.

Q.E.D.

6.7.6 Definition Given a continuum X and a closed subset A of X, $\mathcal{C}(A, X)$ denotes the family of all subcontinua of X containing A.

6.7.7 Theorem *If X is a continuum with the property of Kelley, then $\mathcal{F}_n(X)$ is a Z-set in 2^X and in $\mathcal{C}_n(X)$ for every n.*

Proof Let $\varepsilon > 0$ be given. Let $\mu \colon \mathcal{C}(X) \to [0, 1]$ be a Whitney map. Let $F_\mu^* \colon \mathcal{C}(X) \times [0, 1] \to \mathcal{C}(\mathcal{C}(X))$ be given by

$$F_\mu^*((A, t)) = \begin{cases} \{K \in \mathcal{C}(A, X) \mid K \in \mu^{-1}(t)\}, & \text{if } \mu(A) \leq t; \\ \{A\}, & \text{if } \mu(A) \geq t. \end{cases}$$

Since X has the property of Kelley, F_μ^* is continuous [20, Lemma 2]. Let $G_\mu \colon \mathcal{C}(X) \times [0, 1] \to \mathcal{C}(X)$ be given by $G_\mu((A, t)) = \bigcup F_\mu^*((A, t))$. Then G_μ is a well defined and continuous function (Corollary 6.1.2 and Lemma 1.8.11). Observe that for each $A \in \mathcal{C}(X)$, $G_\mu((A, 0)) = A$ and A is properly contained in $G_\mu((A, t))$ for each $t > \mu(A)$.

Since G_μ is uniformly continuous, for the given ε, there exists $t_0 \in (0, 1)$ such that $\mathcal{H}(G_\mu((A, 0)), G_\mu((A, t_0)) < \varepsilon$.

Let $K \colon \mathcal{C}(X) \to \mathcal{C}(X)$ be given by $K(A) = G_\mu((A, t_0))$. Then K is continuous. Also, $K(\mathcal{C}(X)) \cap \mathcal{F}_1(X) = \emptyset$ and $\mathcal{H}(A, K(A)) = \mathcal{H}(G_\mu((A, 0)), G_\mu((A, t_0))) < \varepsilon$ for each $A \in \mathcal{C}(X)$. Therefore, $\mathcal{F}_1(X)$ is a Z-set of $\mathcal{C}(X)$. Hence, $\mathcal{F}_n(X)$ is a Z-set in 2^X and in $\mathcal{C}_n(X)$ by Theorem 6.7.3.

Q.E.D.

6.7.8 Corollary *Let X be a continuum. If for each positive ε, there exist a subcontinuum Y_ε of X with the property of Kelley and a map $f_\varepsilon \colon X \to Y$ such that $d(x, f_\varepsilon(x)) < \varepsilon$ for each $x \in X$, then $\mathcal{F}_n(X)$ is a Z-set in 2^X and in $\mathcal{C}_n(X)$.*

For the next result, we need the following definitions.

6.7.9 Definition A *dendroid* is an arcwise connected and hereditarily unicoherent continuum. A *fan* is a dendroid with exactly one ramification point (i.e., with only one point which is the common part of three otherwise disjoint arcs) [16]. The unique ramification point of a fan F is called the *top of F*; τ always denotes the top of a fan.

6.7.10 Definition A dendroid D is said to be *smooth at a point p of D* provided that whenever $\{x_j\}_{j=1}^\infty$ is a sequence in D converging to a point x of D, then the sequence of arcs $\{px_j\}_{j=1}^\infty$ converges to the arc px. A fan F is *smooth* if it is smooth at the top τ.

6.7.11 Corollary *For the following classes of continua, $\mathcal{F}_n(X)$ is a Z-set in 2^X and in $\mathcal{C}_n(X)$, for any n:*

(1) homogeneous continua;
(2) hereditarily indecomposable continua;
(3) fans;
(4) smooth dendroids.

Proof Parts (1) and (2) follow from Theorem 6.7.7 and the fact that, in either case, the continuum has the property of Kelley [84, (16.26) and (16.27)].

Now, (3) and (4) follow from Corollary 6.7.8 and the fact that if X is a fan or a smooth dendroid, then for any $\varepsilon > 0$, there exist a tree T contained in X and a map $f_\varepsilon \colon X \to T$ such that $d(x, f_\varepsilon(x)) < \varepsilon$ (see [26, Theorem 1] and [27, Theorem 2]).

Q.E.D.

We do not know if $\mathcal{F}_n(X)$ is a Z-set in 2^X or in $C_n(X)$ for every dendroid X. This result seems to depend on the still unanswered problem of whether every dendroid can be ε-retracted onto a tree. The problem is implicit in [27].

Related to Theorem 6.7.7 (by Corollary 6.1.19, if X is a continuum with the property of Kelley, then $C_n(X)$ is contractible for each $n \in \mathbb{N}$), we have the following theorem:

6.7.12 Theorem *If X is continuum such that $C(X)$ is contractible, then $\mathcal{F}_n(X)$ is a Z-set in 2^X and in $C_n(X)$ for any n.*

Proof By Theorem 6.7.3, it is enough to prove that $\mathcal{F}_1(X)$ is a Z-set in $C(X)$. Since $C(X)$ is contractible, there exists a homotopy $H' \colon C(X) \times [0, 1] \to C(X)$ such that for each $A \in C(X)$, $H'((A, 0)) = A$ and $H'((A, 1)) = X$. Let $H \colon C(X) \times [0, 1] \to C(X)$ be the segment homotopy associated to H' defined by

$$H((A, t)) = \sigma\left(\{H'((A, s)) \mid s \in [0, t]\}\right).$$

Then H is continuous [84, (16.3)]. Note that for every $A \in C(X)$, $H((A, 0)) = A$, $H((A, 1)) = X$ and $H(\{A\} \times [0, 1])$ is an order arc from A to X. Let $\mu \colon C(X) \to [0, 1]$ be a Whitney map. Define $G \colon C(X) \times [0, 1] \to C(X)$ by letting $G((A, t))$ be the unique element of $\{H((A, s)) \mid s \in [0, 1]\}$ such that $\mu(G((A, t))) = t + (1 - t)\mu(A)$. Then G is continuous.

If $t > 0$, then the map $G_t \colon C(X) \to C(X)$ given by $G_t(A) = G((A, t))$ maps $C(X)$ into $C(X) \setminus \mathcal{F}_1(X)$, and if t is close to zero, then G_t is near the identity map on $C(X)$. Therefore, $\mathcal{F}_1(X)$ is a Z-set in $C(X)$.

Q.E.D.

6.7.13 Corollary *If X is a contractible continuum, then $\mathcal{F}_n(X)$ is a Z-set in 2^X and in $C_n(X)$ for every n.*

Proof Since X is contractible, by [84, (16.8)], $C(X)$ is contractible. The corollary now follows from Theorem 6.7.12.

Q.E.D.

Now we focus on a special class of continua, namely, the class of arc-smooth continua.

6.7.14 Theorem *Let X be a continuum such that $C(X)$ is arc-smooth at some point $B \in C(X) \setminus \mathcal{F}_1(X)$, then $\mathcal{F}_n(X)$ is a Z-set in 2^X and in $C_n(X)$ for each $n \in \mathbb{N}$.*

Proof Let $\varepsilon > 0$ be given. By [28, Theorem II-2-A], there exists a map $H \colon C(X) \times [0, 1] \to C(X)$ such that $H((A, 0)) = B$ and $H((A, 1)) = A$ for each $A \in C(X)$, and $H((H((A, s)), t)) = H((A, st))$ for all $A \in C(X)$ and all $s, t \in [0, 1]$. As seen from the proof of [28, Theorem II-4-G], H has the following property: $H((A, t)) \neq A$ for $A \neq B$ and $t \neq 1$.

Let $G: C(X) \times [0, 1] \to C(X)$ be given by

$$G((A, t)) = \sigma \left(\{ H((A, 1 - s)) \mid 0 \leq s \leq t \leq 1 \} \right).$$

Then G is a continuous function. Since for each $A \in \mathcal{F}_1(X)$ and each $t > 0$, $H((A, 1 - t)) \neq A$, we have that $G((A, t)) \neq A$. In fact, $G((A, t)) \notin \mathcal{F}_1(X)$ [84, (16.5)].

By the continuity of G, for the given ε, there exists $t_0 \in [0, 1]$ such that $\mathcal{H}(G((A, 0)), G((A, t_0))) < \varepsilon$.

Let $F: C(X) \to C(X)$ be given by $F(A) = G((A, t_0))$. Then F is continuous. Also, $F(C(X)) \cap \mathcal{F}_1(X) = \emptyset$ and $\mathcal{H}(A, F(A)) = \mathcal{H}(G((A, 0)), G((A, t_0))) < \varepsilon$ for each $A \in C(X)$. Therefore, $\mathcal{F}_1(X)$ is a Z-set of $C(X)$. Hence, $\mathcal{F}_n(X)$ is a Z-set in 2^X and in $C_n(X)$ for each n by Theorem 6.7.3.

Q.E.D.

Note that there exist continua which are not arc-smooth, but which have arc-smooth hyperspaces of subcontinua [31, 6.8]. However, we have the following result:

6.7.15 Corollary *If X is an arc-smooth continuum, then $\mathcal{F}_n(X)$ is a Z-set in 2^X and in $C_n(X)$ for every n.*

Proof Since X is arc-smooth, $C(X)$ is arc-smooth at X [30, Corollary 2]. Hence, $\mathcal{F}_1(X)$ is a Z-set in $C(X)$ by Theorem 6.7.14. Therefore, $\mathcal{F}_n(X)$ is a Z-set in 2^X and in $C_n(X)$ for each n, by Theorem 6.7.3.

Q.E.D.

Let us observe that a very similar proof to the one for Theorem 6.7.14 gives the following two results:

6.7.16 Theorem *Let n be a positive integer and let X be a continuum such that $C_n(X)$ is arc-smooth at some point $B \in C_n(X) \setminus \mathcal{F}_n(X)$. Then $\mathcal{F}_n(X)$ is a Z-set in $C_n(X)$.*

6.7.17 Theorem *Let n be a positive integer and let X be a continuum such that 2^X is arc-smooth at some point $B \in 2^X \setminus \mathcal{F}_n(X)$. Then $\mathcal{F}_n(X)$ is a Z-set in 2^X.*

6.7.18 Theorem *Let X be a continuum. If there exists a map $h: X \times [0, 1] \to C(X)$ such that $h((x, 0)) = \{x\}$ and $h((x, t)) \neq \{x\}$ for all $x \in X$ and all $t > 0$, then $\mathcal{F}_n(X)$ is a Z-set in 2^X and in $C_n(X)$.*

Proof Let $\varepsilon > 0$ be given. Since the union map (Lemma 1.8.11) and h are uniformly continuous, there exists $t_0 \in (0, 1)$ such that diam $\left(\bigcup h(\{x\} \times [0, t_0]) \right) < \varepsilon$.

Let $f: \mathcal{F}_1(X) \to C(X)$ be given by $f(\{x\}) = \bigcup h(\{x\} \times [0, t_0])$. Then f is well defined and continuous by Corollary 6.1.2 and Lemma 1.8.11. Also, $f(\mathcal{F}_1(X)) \cap \mathcal{F}_1(X) = \emptyset$ and $\mathcal{H}(\{x\}, f(\{x\})) < \varepsilon$ for each $x \in X$. Therefore, $\mathcal{F}_1(X)$ is a Z-set in

$C(X)$, by Theorem 6.7.2. Hence, $\mathcal{F}_n(X)$ is a Z-set in 2^X and in $C_n(X)$ for all n by Theorem 6.7.3.

Q.E.D.

Next, we present a number of examples related to the previous results.

6.7.19 Example There exists a compactification X of a ray with an arc as remainder for which $\mathcal{F}_1(X)$ is not a Z-set in $C(X)$.

For any two points x and y in $\mathbb{R}^2$, $\overline{xy}$ denotes the convex arc in $\mathbb{R}^2$ joining x and y (Notation 3.6.32). For each positive integer m, let

$$a_m = \left(\frac{1}{m}, -1 \right), \ b_m = \left(\frac{6m+5}{6m(m+1)}, \frac{1}{2} \right), \ c_m = \left(\frac{3m+2}{3m(m+1)}, 0 \right),$$

$$d_m = \left(\frac{2m+1}{2m(m+1)}, 1 \right), \ e_m = \left(\frac{3m+1}{3m(m+1)}, -\frac{1}{2} \right),$$

$$f_m = \left(\frac{6m+1}{6m(m+1)}, 0 \right),$$

$$A_m = \overline{a_m b_m}, \ B_m = \overline{b_m c_m}, \ C_m = \overline{c_m d_m},$$

$$D_m = \overline{d_m e_m}, \ E_m = \overline{e_m f_m}, \ F_m = \overline{f_m a_{m+1}}.$$

Let $X_0 = \{0\} \times [-1, 1]$ and let

$$X_1 = \bigcup_{m=1}^{\infty} (A_m \cup B_m \cup C_m \cup D_m \cup E_m \cup F_m)$$

Let $X = X_0 \cup X_1$. We claim that $\mathcal{F}_1(X)$ is not a Z-set in $C(X)$. Suppose, to the contrary, that $\mathcal{F}_1(X)$ is a Z-set in $C(X)$. Let $\varepsilon > 0$ be given such that $\varepsilon < \frac{1}{8}$. Then, by Theorem 6.7.2, there exists a map $h_\varepsilon: \mathcal{F}_1(X) \to C(X) \setminus \mathcal{F}_1(X)$ such that $\mathcal{H}(\{x\}, h_\varepsilon(\{x\})) < \varepsilon$ for each $x \in X$.

Since X_1 is arcwise connected and $d(a_1, x) > \varepsilon$ for each $x \in X_0$, we have that for each $x \in X_1$, $h_\varepsilon(\{x\}) \cap X_0 = \emptyset$.

We claim that $h_\varepsilon(\{(0, 0)\}) \cap X_0 = \emptyset$. Suppose, to the contrary, that $h_\varepsilon(\{(0, 0)\}) \subset X_0$. Observe that the sequences $\{c_m\}_{m=1}^{\infty}$ and $\{f_m\}_{m=1}^{\infty}$ converge to $(0, 0)$. Also note that for each positive integer m, $\inf\{\pi_2(h_\varepsilon(\{c_m\}))\} \geq 0$ and $\sup\{\pi_2(h_\varepsilon(\{f_m\}))\} \leq 0$, where $\pi_2: X \twoheadrightarrow [-1, 1]$ is the natural projection. Thus, by continuity, on one hand $\inf\{\pi_2(h_\varepsilon(\{(0, 0)\}))\} \geq 0$ and on the other hand, $\sup\{\pi_2(h_\varepsilon(\{(0, 0)\}))\} \leq 0$. Hence, $h_\varepsilon(\{(0, 0)\}) = \{(0, 0)\}$, a contradiction. Therefore, $h_\varepsilon(\{(0, 0)\}) \cap X_0 = \emptyset$.

Since $h_\varepsilon(\{(0, 0)\}) \cap X_0 = \emptyset$ and since X_0 is arcwise connected, we have that $h_\varepsilon(\{x\}) \cap X_0 = \emptyset$ for each $x \in X_0$.

Let M be a positive integer such that a_M is the closest element of $\{a_m\}_{m=1}^{\infty}$ to $h_{\varepsilon}(\{(0, -1)\})$. Since $\mathcal{H}(h_{\varepsilon}(\{(0, 1)\}), \{(0, 1)\}) < \varepsilon$, $\bigcup\{h_{\varepsilon}(\{(0, t)\}) \mid t \in [-1, 1]\}$ contains either f_{M-1} or b_M. Without loss of generality, we assume that b_M belongs to $\bigcup\{h_{\varepsilon}(\{(0, t)\}) \mid t \in [-1, 1]\}$. Hence, c_M also belongs to $\bigcup\{h_{\varepsilon}(\{(0, t)\}) \mid t \in [-1, 1]\}$.

Let $t_1 = \sup\{s \in [-1, 1] \mid b_M \notin h_{\varepsilon}(\{(0, s)\})\}$. Note that $\pi_2(h_{\varepsilon}(\{(0, t_1)\})) \subset \left[\frac{1}{2} - 2\varepsilon, \frac{1}{2}\right]$. Since $\mathcal{H}(\{(0, t_1)\}, h_{\varepsilon}(\{(0, t_1)\})) < \varepsilon$, $t_1 \in \left[\frac{1}{2} - 2\varepsilon, \frac{1}{2}\right]$.

Let $t_2 = \sup\{s \in [t_1, 1] \mid c_M \notin h_{\varepsilon}(\{(0, s)\})\}$. Let us note that $\pi_2(h_{\varepsilon}(\{(0, t_2)\})) \subset [0, 2\varepsilon]$. Since $\mathcal{H}(\{(0, t_2)\}, h_{\varepsilon}(\{(0, t_2)\})) < \varepsilon$, $t_2 \in [0, 2\varepsilon]$; also, since $\varepsilon < \frac{1}{8}$, $2\varepsilon < \frac{1}{2} - 2\varepsilon$. Hence, $t_2 < t_1$. This contradicts the definition of t_2.

Therefore $\mathcal{F}_1(X)$ is not a Z-set in $\mathcal{C}(X)$.

6.7.20 Example There exists a compactification X of a ray with an arc as remainder such that $\mathcal{F}_1(X)$ is a Z-set in $\mathcal{C}(X)$ and yet $\mathcal{C}(X)$ contains a positive Whitney level $\mu^{-1}(t_0)$, that is not a Z-set in $\mu^{-1}([t_0, 1])$, $t_0 < 1$.

We use the notation for convex arcs in Notation 3.6.32. The construction of this continuum is similar to the continuum constructed in Example 6.7.19, we lower the sequence of points $\{c_m\}_{m=1}^{\infty}$ and rise the sequence of points $\{f_m\}_{m=1}^{\infty}$.

For each m, let $c'_m = \left(\frac{3m+2}{3m(m+1)}, -\frac{1}{16}\right)$ and $f'_m = \left(\frac{6m+1}{6m(m+1)}, \frac{1}{16}\right)$. Let a_m, b_m, d_m and e_m be as in Example 6.7.19. Let $A_m = \overline{a_m b_m}$, $B'_m = \overline{b_m c'_m}$, $C'_m = \overline{c'_m d_m}$, $D_m = \overline{d_m e_m}$, $E'_m = \overline{e_m f'_m}$ and $F'_m = \overline{f'_m a_{m+1}}$. Let X_0 be as in Example 6.7.19 and let $X'_1 = \bigcup_{m=1}^{\infty}(A_m \cup B'_m \cup C'_m \cup D_m \cup E'_m \cup F'_m)$, and let $X = X_0 \cup X'_1$. Then, X is a chainable continuum. Let $\mu \colon \mathcal{C}(X) \to [0, 1]$ be a Whitney map. Let $t_0 = \mu\left(\{0\} \times \left[-\frac{1}{16}, \frac{1}{16}\right]\right)$. Then, it is not difficult to see that $\mu^{-1}(t_0)$ is homeomorphic to the continuum in Example 6.7.19; also $\mu^{-1}([t_0, 1])$ is homeomorphic to $\mathcal{C}\left(\mu^{-1}(t_0)\right)$ by [84, (14.73.12) and (14.76.3)]. Therefore, $\mu^{-1}(t_0)$ is not a Z-set in $\mu^{-1}([t_0, 1])$ by Example 6.7.19.

6.7.21 Example There exists a chainable continuum Y such that $\mathcal{C}(Y)$ contains a Whitney level $\mu^{-1}(t_0)$, with $t_0 > 0$, that is not a Z-set in $\mu^{-1}([0, t_0])$.

Let $Y = (\{0\} \times [-1, 1]) \cup \left\{\left(x, \sin\left(\frac{1}{x}\right)\right) \in \mathbb{R}^2 \mid x \in [-1, 1] \setminus \{0\}\right\}$. Then, Y is a chainable continuum. Let $\mu \colon \mathcal{C}(Y) \to [0, 1]$ be a Whitney map. Let $t_0 = \mu(\{0\} \times [-1, 1])$. Then, it is not difficult to see that $\mu^{-1}(t_0)$ is an arc. Suppose $\mu^{-1}(t_0)$ is a Z-set in $\mu^{-1}([0, t_0])$. Then for each $\varepsilon > 0$, there exists a map

$$f_{\varepsilon} \colon \mu^{-1}([0, t_0]) \to (\mu^{-1}([0, t_0]) \setminus \mu^{-1}(t_0)),$$

such that $\mathcal{H}(A, f_{\varepsilon}(A)) < \varepsilon$ for all $A \in \mu^{-1}([0, t_0])$.

Let A_- be the element of $\mu^{-1}(t_0)$ containing $(-1, \sin(-1))$ and let A_+ be the element of $\mu^{-1}(t_0)$ containing $(1, \sin(1))$. Since $\mu^{-1}(t_0)$ is an arc, there must exist an arc in $\mu^{-1}([0, t_0]) \setminus \mu^{-1}(t_0)$ joining $f_{\varepsilon}(A_-)$ with $f_{\varepsilon}(A_+)$ in $\mu^{-1}([0, t_0]) \setminus \mu^{-1}(t_0)$, but it is not difficult to see that there does not exist arc joining $f_{\varepsilon}(A_-)$

with $f_\varepsilon(A_+)$ in $\mu^{-1}([0, t_0]) \setminus \mu^{-1}(t_0)$, a contradiction. Therefore, $\mu^{-1}(t_0)$ is not a Z-set in $\mu^{-1}([0, t_0])$.

6.7.22 Example There exists a chainable continuum X that does not have the property of Kelley, but for which $\mathcal{F}_n(X)$ is a Z-set in $\mathcal{C}_n(X)$ for each n.

Let $X = (\{0\} \times [-1, 1]) \cup \left\{\left(x, \sin\left(\frac{1}{x}\right)\right) \in \mathbb{R}^2 \,\middle|\, x \in (0, 1]\right\}$, and let $\mathcal{G}$ be the decomposition of X whose nondegenerate elements are $\{(0, y), (0, 1 - y)\}$ for each $y \in \left[0, \frac{1}{2}\right]$. Clearly, $\mathcal{G}$ is an upper semicontinuous decomposition of X. Hence, by Theorem 1.7.3, $Z = X/\mathcal{G}$ is a continuum called M-*continuum* (note that $X/\mathcal{G}$ does not have the property of Kelley). It is easy to see that for each $\varepsilon > 0$, there exists a map $f_\varepsilon: Z \to f_\varepsilon(Z)$ such that $f_\varepsilon(Z)$ is an arc and $d(z, f_\varepsilon(z)) < \varepsilon$. Since arcs have the property of Kelley, $\mathcal{F}_n(Z)$ is a Z-set in $\mathcal{C}_n(Z)$ for each n, by Corollary 6.7.8.

6.7.23 Example There exists a chainable continuum $X = X_1 \cup X_2$, where X_1 and X_2 are chainable continua such that $\mathcal{F}_1(X_j)$ is a Z-set in $\mathcal{C}(X_j)$ for $j \in \{1, 2\}$, $X_1 \cap X_2$ is a single point, and $\mathcal{F}_1(X)$ is not a Z-set in $\mathcal{C}(X)$.

Let

$$X_1 = (\{0\} \times [-3, -1]) \cup \left\{\left(x, -2 + \sin\left(\frac{1}{x}\right)\right) \in \mathbb{R}^2 \,\middle|\, x \in [-1, 0)\right\},$$

let

$$X_2 = (\{0\} \times [-1, 1]) \cup \left\{\left(x, \sin\left(\frac{1}{x}\right)\right) \in \mathbb{R}^2 \,\middle|\, x \in (0, 1]\right\},$$

and let $X = X_1 \cup X_2$. By Corollary 6.7.8, $\mathcal{F}(X_j)$ is a Z-set in $\mathcal{C}(X_j)$, $j \in \{1, 2\}$. It is not difficult to see that $\{(0, -1)\}$ is not a Z-set in $\mathcal{C}(X)$.

6.7.24 Example There exists a chainable continuum X such that $\mathcal{F}_1(X \times X)$ is not a Z-set in $\mathcal{C}(X \times X)$.

Let X be the continuum constructed in Example 6.7.23 and let $p = (0, -1)$.

We use the following easy-to-prove fact: If $G: \mathcal{C}(X) \to \mathcal{C}(X)$ is a map near enough to the identity, then $G(\{p\}) = \{p\}$.

If $\mathcal{F}_1(X \times X)$ is a Z-set in $\mathcal{C}(X \times X)$, there exists a map $f: \mathcal{C}(X \times X) \to \mathcal{C}(X \times X) \setminus \mathcal{F}_1(X \times X)$ as near to the identity as we please. Let $\pi_j: X \times X \twoheadrightarrow X$ be the projection onto the j-th factor for $j \in \{1, 2\}$. Since $f(\{(p, p)\})$ is nondegenerate, either $\pi_1(f(\{(p, p)\}))$ or $\pi_2(f(\{(p, p)\}))$ is nondegenerate, say $\pi_1(f(\{(p, p)\}))$ is nondegenerate. Let $h: X \to X \times X$ be the map defined by $h(x) = (x, p)$ for each $x \in X$. Let $G = \mathcal{C}(\pi_1) \circ f \circ \mathcal{C}(h)$. Note that with the usual metric on $X \times X$, h is an isometry and π_1 is nonexpansive. Thus, $G: \mathcal{C}(X) \to \mathcal{C}(X)$ is a map as near the identity as f is; also, $G(\{p\}) \neq \{p\}$, This contradicts the fact stated above. Therefore, $\mathcal{F}_1(X \times X)$ is not a Z-set in $\mathcal{C}(X \times X)$.

6.7.25 Example There exists a dendroid X such that $C(X)$ is not contractible but $\mathcal{F}_n(X)$ is a Z-set in 2^X and in $C_n(X)$ for every n.

Let A be the convex arc in $\mathbb{R}^2$ from $(-1, 0)$ to $(1, 0)$. For each positive integer m, let B_m be the convex arc in $\mathbb{R}^2$ from $(-1, 0)$ to $\left(0, \frac{1}{m}\right)$, and let C_m be the convex arc in $\mathbb{R}^2$ from $(1, 0)$ to $\left(0, -\frac{1}{m}\right)$. Let

$$X = A \bigcup \left(\bigcup_{m=1}^{\infty} B_m \right) \bigcup \left(\bigcup_{m=1}^{\infty} C_m \right).$$

Then X is a dendroid. It is easy to see that $C(X)$ is not contractible. It is also easy to show that for every $\varepsilon > 0$, there exist a tree T_ε and a map $f_\varepsilon: X \twoheadrightarrow T_\varepsilon$ such that $d(x, f_\varepsilon(x)) < \varepsilon$ for each $x \in X$. Hence, by Corollary 6.7.8, $\mathcal{F}_n(X)$ is a Z-set in 2^X and in $C_n(X)$.

6.7.26 Remark Regarding Example 6.7.25, we note that $\{(0, 0)\}$ is an R^3-set in X [48, 24.12]. This is stronger than the fact that $C(X)$ is not contractible [48, 78.15].

6.7.27 Definition A map $r: X \twoheadrightarrow Y$ between continua is called an *r-map* provided that there exists a map $g: Y \to X$ such that $r \circ g = 1_Y$.

6.7.28 Example There exists a monotone and open *r*-map $f: Y \twoheadrightarrow X$ between continua such that $\mathcal{F}_1(Y)$ is a Z-set in $C(Y)$ but $\mathcal{F}_1(X)$ is not a Z-set in $C(X)$.

Let X be the continuum in Example 6.7.19, let $Y = X \times [0, 1]$ and let $f: Y \twoheadrightarrow X$ be the projection map. Then f is a monotone open *r*-map, $\mathcal{F}_1(Y)$ is a Z-set in $C(Y)$ by Corollary 6.7.5, and $\mathcal{F}_1(X)$ is not a Z-set in $C(X)$ by Example 6.7.19.

6.7.29 Definition A map $f: X \to Y$ between continua is said to be *simple* provided that for each $y \in Y$, $f^{-1}(y)$ has at most two points [8].

6.7.30 Example There exists a continuum X such that $\mathcal{F}_n(X)$ is not a Z-set in $C_n(X)$ for any n; furthermore, X is a simple image of a chainable continuum Y such that $\mathcal{F}_n(Y)$ is a Z-set in $C_n(Y)$ for each n.

Let Y be the space X_2 of Example 6.7.23. By Corollary 6.7.8, $\mathcal{F}_n(Y)$ is a Z-set in $C_n(Y)$.

Let X be the quotient space of Y obtained by identifying the point $(0, -1)$ with the point $(0, 1)$. Hence, X is a compactification of a ray with a simple closed curve as remainder.

Suppose $\mathcal{F}_n(X)$ is a Z-set in $C_n(X)$. Let $\varepsilon > 0$ be given. Hence, there exists a map $h_\varepsilon: C_n(X) \to C_n(X) \backslash \mathcal{F}_n(X)$ such that $\mathcal{H}(A, h_\varepsilon(A)) < \varepsilon$ for every $A \in C_n(X)$.

Let $q: Y \twoheadrightarrow X$ be the quotient map. Clearly, q is a simple map. Let $S = q(\{0\} \times [-1, 1])$; thus, S is a simple closed curve.

Let $Y_0 = \left\{ \left(z, \sin\left(\frac{1}{z}\right)\right) \in \mathbb{R}^2 \mid z \in (0, 1] \right\}$. Since $q(Y_0)$ is arcwise connected and $d(q((1, \sin(1))), x) > \varepsilon$ for each $x \in S$, we have that for each $x \in q(Y_0)$, $h_\varepsilon(\{x\}) \cap S = \emptyset$.

A similar argument to the one given for $\{(0, 0)\}$ in Example 6.7.19 shows $h_\varepsilon(\{q((0, 1))\}) \cap S = \emptyset$. As a consequence of this, since S is arcwise connected, we have that $h_\varepsilon(\{x\}) \cap S = \emptyset$ for all $x \in S$.

A similar argument to the one given for $\{0\} \times [-1, 1]$ in Example 6.7.19 shows that there exists a point $x \in S$ such that $\mathcal{H}(\{x\}, h_\varepsilon(\{x\})) \geq \varepsilon$. Therefore $\mathcal{F}_n(X)$ is not a Z-set in $\mathcal{C}_n(X)$.

6.8 Retractions

We present results about retractions between the hyperspaces of locally connected continua. For this purpose, we use [2, 4, 6, 7, 10, 15, 18, 19, 29, 32, 40, 48, 50, 53–55, 61, 65, 67, 81, 84, 86, 96–98].

6.8.1 Lemma *Let* $n \in \mathbb{N}$. *If* X *is a continuum containing an open set with uncountably many components, then* $\mathcal{C}_n(X)$ *contains an open set with uncountably many components.*

Proof Let U be an open subset of X with uncountably many components. Let $\Gamma = \langle U \rangle_n$. Then Γ is an open subset of $\mathcal{C}_n(X)$, and $\bigcup \Gamma \subset U$. Since for each $x \in U$, $\{x\} \in \Gamma$, we have that $\bigcup \Gamma = U$. By Lemma 6.1.1, if Λ is a component of Γ, then $\bigcup \Lambda$ has at most n components. Since U has uncountably many components, $\bigcup \Gamma = U$ and for each component Λ of Γ, $\bigcup \Lambda$ has at most n components, we have that Γ has uncountably many components.

                                                                              Q.E.D.

We begin considering maps between some of the hyperspaces.

6.8.2 Theorem *Let* $n \in \mathbb{N}$. *Then for any continuum* X, *there exists a map of* 2^X *onto* $\mathcal{C}_n(X)$.

Proof If X is locally connected, then 2^X and $\mathcal{C}_n(X)$ are locally connected continua [98, Théorème II and Théorème II$_m$]. Hence, the result follows from [86, 8.19].

Suppose that X is not locally connected. Then there exists a map f of 2^X onto the cone over the Cantor set [84, (1.39)]. Also, there exists a map g of the cone over the Cantor set onto $\mathcal{C}_n(X)$ [50, Remark on p. 29 and Theorem 2.7]. Thus, $g \circ f$ is a map of 2^X onto $\mathcal{C}_n(X)$.

                                                                              Q.E.D.

6.8.3 Theorem *Let* $n, m \in \mathbb{N}$. *If* X *is a continuum containing an open set with uncountably many components, then there exists a map from* $\mathcal{C}_n(X)$ *onto* $\mathcal{C}_m(X)$.

Proof Since X contains an open subset with uncountably many components, $\mathcal{C}_n(X)$ contains an open subset with uncountably many components (Lemma 6.8.1). Thus, there exists a map f from $\mathcal{C}_n(X)$ onto the cone over the Cantor set [2, Theorem II]. Also, there exists a map g from the cone over the Cantor set onto $\mathcal{C}_m(X)$ [50, Remark on p. 29 and Theorem 2.7]. Hence, $g \circ f$ is a map of $\mathcal{C}_n(X)$ onto $\mathcal{C}_m(X)$.

                                                                              Q.E.D.

6.8.4 Definition Let Z be a metric space. By a *deformation* we mean a map $H: Z \times [0, 1] \to Z$ such that for each $z \in Z$, $H((z, 1)) = z$. Let $A = \{H((z, 0)) \mid z \in Z\}$. If the map $h: Z \to A$ given by $h(z) = H((z, 0))$ is a retraction from Z onto A, then H is a *deformation retraction from Z onto A*. If H is a deformation retraction from Z onto A such that for each $a \in A$ and each $t \in [0, 1]$, $H((a, t)) = a$, then H is a *strong deformation retraction from Z onto A*. The set A is called *deformation retract* (*strong deformation retract*, respectively).

The next lemma provides a sufficient condition for a continuum X for the nonexistence of a deformation of the m-fold hyperspace of X onto a subset of the n-fold hyperspace of X, $m > n$.

6.8.5 Lemma *Let $n, m \in \mathbb{N}$ and let X be a nonlocally connected continuum, with metric d. Let p be a point of X at which X is not connected im kleinen, and let $A \subset C_n(X)$ be such that $\{p\} \in A$. If $m > n$, then there does not exist a deformation, H, from $C_m(X)$ onto A such that $H(((\{p\}, t)) = \{p\}$ for each $t \in [0, 1]$.*

Proof Since X is not connected im kleinen at p, there exist a neighborhood U of p and a sequence $\{K_j\}_{j=1}^{\infty}$ of components of $Cl_X(U)$ converging to a continuum $K \subset Cl_X(U)$ such that $p \in K$ and such that for each $j \in \mathbb{N}$, $K_j \cap K = \emptyset$ [96, (12.1), p. 18]. Let $\{p_j\}_{j=1}^{\infty}$ be a sequence in $Cl_X(U)$ converging to p and such that $p_j \in K_j$ for each $j \in \mathbb{N}$.

Suppose there exists a deformation $H: C_m(X) \times [0, 1] \to C_m(X)$ from $C_m(X)$ onto A such that for each $t \in [0, 1]$, $H(((\{p\}, t)) = \{p\}$. Let H' be the segment homotopy associated with H [84, (16.3)], given by $H'((A, t)) = \sigma(\{H((A, s)) \mid 0 \leq s \leq t\})$. Observe that for each $t \in [0, 1]$, $H'(((\{p\}, t)) = \{p\}$ and for each $A \in C_m(X)$, $H'((A, 0)) = H((A, 0)) \in C_n(X)$. Let $\varepsilon > 0$ be such that $V_\varepsilon^d(p) \subset U$. Then there exists $\delta > 0$ such that if $A \in C_m(X)$ and $\mathcal{H}(A, \{p\}) < \delta$, then for each $t \in [0, 1]$, $\mathcal{H}(H'((A, t)), H'(((\{p\}, t))) = \mathcal{H}(H'((A, t)), \{p\}) < \varepsilon$. For each $j \in \mathbb{N}$, let $A_j = \{p\} \cup \{p_{j+\ell}\}_{\ell=1}^{m-1}$. Hence, there exists $j_0 \in \mathbb{N}$ such that if $j \geq j_0$, then $\mathcal{H}(A_j, \{p\}) < \delta$. Choose $j \geq j_0$. Then $\{H'((A_j, t)) \mid t \in [0, 1]\}$ is a connected subset of $V_\varepsilon^{\mathcal{H}}(\{p\})$ such that $H'((A_j, 0)) \in C_n(X)$. Thus, $\sigma(\{H'((A_j, t)) \mid t \in [0, 1]\})$ is a closed subset of X having at most n components (by Corollary 6.1.2). Since $A_j \subset \sigma(\{H'((A_j, t)) \mid t \in [0, 1]\}) \subset V_\varepsilon^d(p) \subset U$ and since $A_j \cap K \neq \emptyset$ and $A_j \cap K_{j+\ell} \neq \emptyset$ for each $\ell \in \{1, \ldots, m - 1\}$, we obtain a contradiction. We conclude that there does not exist a deformation H of $C_m(X)$ onto A such that for each $t \in [0, 1]$, $H(((\{p\}, t)) = \{p\}$.

Q.E.D.

6.8.6 Corollary *Let $m, n \in \mathbb{N}$. If X is a nonlocally connected continuum and $m > n$, then there does not exist a strong deformation retraction from $C_m(X)$ onto $\mathcal{F}_n(X)$.*

6.8.7 Theorem *Let $n \in \mathbb{N}$, $n \geq 2$, and let X be a continuum. If $\mathcal{F}_1(X)$ is a deformation retract from 2^X, then $\mathcal{F}_1(X)$ is a deformation retract from $C_n(X)$.*

Proof Let $H: 2^X \times [0, 1] \to 2^X$ be a deformation retraction from 2^X onto $\mathcal{F}_1(X)$. Let $r(A) = H((A, 0))$, for each $A \in 2^X$.

Define $G: C_n(X) \times [0, 1] \to C_n(X)$ by

$$
G((A, t)) = \begin{cases} \sigma(\{H((A, s)) \mid 0 \le s \le 2t\}), & \text{if } t \in \left[0, \dfrac{1}{2}\right]; \\[2mm] \sigma(\{H((A, s)) \mid 2t - 1 \le s \le 1\}), & \text{if } t \in \left[\dfrac{1}{2}, 1\right]. \end{cases}
$$

Note that if $t = \frac{1}{2}$, then both definitions of G give the value $\sigma(\{H((A, s)) \mid 0 \le s \le 1\})$. Let $t \in \left[0, \frac{1}{2}\right]$. Since H is continuous, $\{H((A, s)) \mid 0 \le s \le 2t\}$ is a closed connected subset of 2^X that contains the element $H((A, 0)) = r(A)$ and $r(A) \in \mathcal{F}_1(X)$. By Corollary 6.1.2, $G((A, t)) \in C_n(X)$ for each $A \in C_n(X)$. Similarly, since $H((A, 1)) = A$, $G((A, t)) \in C_n(X)$ for each $t \in \left[\frac{1}{2}, 1\right]$ and each $A \in C_n(X)$. Hence, G is well defined. Since H and σ (Lemma 1.8.11) are continuous, it follows that G is continuous also.

Note that $G((A, 0)) = r(A)$ and $G((A, 1)) = A$. Therefore, $\mathcal{F}_1(X)$ is a deformation retract of $C_n(X)$.

Q.E.D.

The proof of the following theorem is similar to the one given in Theorem 6.8.7.

6.8.8 Theorem *Let* $n \in \mathbb{N}$, $n \ge 2$, *and let* X *be a continuum. If* $\mathcal{F}_1(X)$ *is a deformation retract of* $C_n(X)$, *then* $\mathcal{F}_1(X)$ *is a deformation retract of* $\mathcal{C}(X)$.

6.8.9 Definition A metric ρ for a continuum X is said to be *convex* provided that given two points x and y of X there exists a point z in X such that $\rho(x, z) = \frac{\rho(x,y)}{2} = \rho(z, y)$.

The following result is proved by R H Bing [4] and E. E. Moise [81], independently.

6.8.10 Theorem *Every locally connected continuum admits a convex metric.*

6.8.11 Definition If X is a locally connected continuum with a convex metric ρ, let $K_\rho : [0, \infty) \times 2^X \to 2^X$ be given by

$$
K_\rho((t, A)) = \{x \in X \mid \rho(x, y) \le t \text{ for some } y \in A\}.
$$

6.8.12 Remark If X is a locally connected continuum with a convex metric ρ, then given two points x and y of X, there exists an arc $\gamma : [0, \rho(x, y)] \to X$ such that γ is an isometry [84, (0.65.3) (a)]. This implies that the set $K_\rho((t, A))$ has at most as many components as A for each $t \in [0, \infty)$. Also, observe that if $t \ge \mathrm{diam}(X)$, then $K_\rho((t, A)) = X$ for each $A \in 2^X$.

6.8.13 Theorem *Let* $n \in \mathbb{N}$ *and let* X *be a locally connected continuum with a convex metric* ρ. *Then the function* $\alpha_\rho^n : 2^X \to \mathbb{R}$ *given by*

$$
\alpha_\rho^n(A) = \inf\{t \ge 0 \mid K_\rho((t, A)) \in C_n(X)\}
$$

satisfies the following:

(a) $K_\rho((\alpha_\rho^n(A), A))$ *belongs to* $C_n(X)$ *for each* $A \in 2^X$.

(b) *If* A *and* B *belong to* 2^X, $t \geq 0$, $K_\rho((t, A)) \in C_n(X)$ *and* $\mathcal{H}_\rho(A, B) \leq \eta$, *then* $K_\rho((t + \eta, B)) \in C_n(X)$.

(c) α_ρ^n *is continuous.*

Proof Since X is locally connected with a convex metric, K_ρ is continuous [84, (0.65.3) (f)]. To see $K_\rho((\alpha_\rho^n(A), A))$ belongs to $C_n(X)$ for each $A \in 2^X$, observe that, by the definition of α_ρ^n, there exists a decreasing sequence, $\{t_j\}_{j=1}^\infty$, of real numbers converging to $\alpha_\rho^n(A)$ such that $K_\rho((t_j, A))$ belongs to $C_n(X)$ for each $j \in \mathbb{N}$. By the continuity of K_ρ, we have that $\lim_{j\to\infty} K_\rho((t_j, A)) = K_\rho((\alpha_\rho^n(A), A))$. Since $C_n(X)$ is compact (Theorem 1.8.5), we have that $K_\rho((\alpha_\rho^n(A), A)) \in C_n(X)$.

Now, suppose A and B belong to 2^X, $t \geq 0$, $K_\rho((t, A)) \in C_n(X)$ and $\mathcal{H}_\rho(A, B) \leq \eta$. Observe first that $A \subset K_\rho((t, A)) \subset K_\rho((t + \eta, B))$. Let $x \in K_\rho((t + \eta, B))$, then there exists $b \in B$ such that $\rho(x, b) \leq t + \eta$. Since $b \in B$ and $\mathcal{H}_\rho(A, B) \leq \eta$, there exists $a \in A$ such that $\rho(a, b) \leq \eta$. Since ρ is a convex metric, there exist arcs $\gamma_1 \colon [0, \rho(x, b)] \to X$ and $\gamma_2 \colon [0, \rho(b, a)] \to X$ such that $\gamma_1(0) = x$, $\gamma_1(\rho(x, b)) = b$, $\gamma_2(0) = b$, $\gamma_2(\rho(b, a)) = a$, and both γ_1 and γ_2 are isometries [84, (0.65.3) (a)]. Let $G_x = \gamma_1([0, \rho(x, b)]) \cup \gamma_2([0, \rho(b, a)])$. Then G_x is a connected subset of $K_\rho((t + \eta, B))$ containing x and intersecting $K_\rho((t, A))$. Hence, $K_\rho((t + \eta, B))$ has at most as many components as $K_\rho((t, A))$. Therefore, $K_\rho((t + \eta, B)) \in C_n(X)$.

To show α_ρ^n is continuous, let $\eta > 0$. Let A and B be two elements of 2^X such that $\mathcal{H}_\rho(A, B) \leq \eta$. By (a), $K_\rho((\alpha_\rho^n(A), A))$ belongs to $C_n(X)$. Hence, by (b), $K_\rho((\alpha_\rho^n(A) + \eta, B))$ also belongs to $C_n(X)$. By definition of $\alpha_\rho^n(B)$, we have that $\alpha_\rho^n(B) \leq \alpha_\rho^n(A) + \eta$. Interchanging the roles of A and B, in the above argument, we obtain $\alpha_\rho^n(A) \leq \alpha_\rho^n(B) + \eta$. Hence, $|\alpha_\rho^n(A) - \alpha_\rho^n(B)| \leq \eta$. Therefore, α_ρ^n is continuous.

Q.E.D.

6.8.14 Theorem *Let* X *be a locally connected continuum with a convex metric* ρ *and let* n *and* m *be positive integers such that* $m \geq n$. *Then*

(1) *If* $A \in C_m(X)$ *and* $s \in [0, 1]$, *then* $\alpha_\rho^n(K_\rho((s \cdot \alpha_\rho^n(A), A))) = (1 - s) \cdot \alpha_\rho^n(A)$.

(2) *Let* $R \colon C_m(X) \twoheadrightarrow C_n(X)$ *be given by* $R(A) = K_\rho((\alpha_\rho^n(A), A))$. *If* $A \in R^{-1}(B)$ *and* $s \in [0, 1]$, *then* $K_\rho((s \cdot \alpha_\rho^n(A), A)) \in R^{-1}(B)$.

Proof We show (1). Note that, by [84, (0.65.3) (c)], we have that:

$$K_\rho(((1 - s) \cdot \alpha_\rho^n(A), K_\rho((s \cdot \alpha_\rho^n(A), A)))) =$$

$$K_\rho(((1 - s) \cdot \alpha_\rho^n(A) + s \cdot \alpha_\rho^n(A), A)) =$$

$$K_\rho((\alpha_\rho^n(A), A)).$$

Now, it follows that $\alpha_\rho^n(K_\rho((s \cdot \alpha_\rho^n(A), A))) = (1 - s) \cdot \alpha_\rho^n(A)$.

To prove (2), let $B \in C_n(X)$, let $A \in R^{-1}(B)$ and let $s \in [0, 1]$. Then it follows from the definition of R, (1) and [84, (0.65.3) (c)], that

$$R(K_\rho((s \cdot \alpha_\rho^n(A), A))) =$$

$$K_\rho((\alpha_\rho^n(K_\rho((s \cdot \alpha_\rho^n(A), A))), K_\rho((s \cdot \alpha_\rho^n(A), A)))) =$$

$$K_\rho(((1 - s) \cdot \alpha_\rho^n(A), K_\rho((s \cdot \alpha_\rho^n(A), A)))) =$$

$$K_\rho((\alpha_\rho^n(A), A)) = B.$$

Q.E.D.

The following results give conditions for the existence of retractions between the hyperspaces of locally connected continua.

6.8.15 Theorem *Let $n, m \in \mathbb{N}$, $m > n$, and let X be a continuum. Then $C_n(X)$ is a strong deformation retract of $C_m(X)$ if and only if X is locally connected.*

Proof If X is not locally connected, then there does not exist a strong deformation retraction from $C_m(X)$ onto $C_n(X)$ by Lemma 6.8.5.

If X is a locally connected continuum, then the map $H: C_m(X) \times [0, 1] \twoheadrightarrow C_m(X)$ given by

$$H((A, t)) = K_\rho \left(((1 - t)\alpha_\rho^n(A), A) \right)$$

is a strong deformation retraction from $C_m(X)$ onto $C_n(X)$ (Remark 6.8.12 and Theorem 6.8.13).

Q.E.D.

The proof of the following theorem is similar to the one given for Theorem 6.8.15.

6.8.16 Theorem *Let $n \in \mathbb{N}$ and let X be a continuum. Then $C_n(X)$ is a strong deformation retract of 2^X if and only if X is locally connected.*

6.8.17 Theorem *Let X be a locally connected continuum and let n and m be positive integers such that $m \geq n$. Then $\mathcal{F}_n(X)$ is a retract of $C_m(X)$ if and only if $\mathcal{F}_n(X)$ is an absolute retract.*

Proof If $\mathcal{F}_n(X)$ is an absolute retract, then, clearly, $\mathcal{F}_n(X)$ is a retract of $C_m(X)$. Suppose $\mathcal{F}_n(X)$ is a retract of $C_m(X)$. Since X is locally connected, $C_m(X)$ is an absolute retract [98, Théorème II_m]. Hence, $\mathcal{F}_n(X)$ is an absolute retract [55, Theorem 6, p. 341].

Q.E.D.

6.8.18 Corollary *Let $n \in \mathbb{N}$. If X is a locally connected continuum, then $\mathcal{F}_1(X)$ is a retract of $C_n(X)$ if and only if X is an absolute retract.*

6.8.19 Theorem *Let* $n \in \mathbb{N}$. *If* X *is a locally connected continuum, then the following are equivalent:*

(1) X *is an absolute retract.*
(2) $\mathcal{F}_1(X)$ *is a retract of* $C_n(X)$.
(3) $\mathcal{F}_1(X)$ *is a deformation retract of* $C_n(X)$.
(4) $\mathcal{F}_1(X)$ *is a strong deformation retract of* $C_n(X)$.

Proof Clearly (4) implies (3), and (3) implies (2). Note that (2) implies (1) by Corollary 6.8.18. Suppose now, X is an absolute retract. Then $C_n(X)$ is an absolute retract [98, Théorème II$_m$]. Hence, there exists a retraction $r_n : 2^X \to C_n(X)$. By [32, 2.5], there exists a strong deformation retraction $G : 2^X \times [0, 1] \to 2^X$, from 2^X onto $\mathcal{F}_1(X)$. Thus, $r_n \circ (G|_{C_n(X) \times [0,1]})$ is a strong deformation retraction from $C_n(X)$ onto $\mathcal{F}_1(X)$. Therefore, (1) implies (4).

Q.E.D.

6.8.20 Theorem *Let* $n, m \in \mathbb{N}$, *with* $m \geq n$. *If* X *is a locally connected continuum, then the following hold:*

(1) $C_n(X)$ *is a retract of* 2^X.
(2) $C_n(X)$ *is a deformation retract of* 2^X.
(3) $C_n(X)$ *is a strong deformation retract of* 2^X.
(4) $C_n(X)$ *is a retract of* $C_m(X)$.
(5) $C_n(X)$ *is a deformation retract of* $C_m(X)$.
(6) $C_n(X)$ *is a strong deformation retract of* $C_m(X)$.

Proof By [98, Théorème II$_m$], $C_n(X)$ is an absolute retract. Hence, (1) and (4) hold. Since X is locally connected, X has the property of Kelley. Thus, $C_n(X)$ is contractible (Corollary 6.1.19) and 2^X is contractible too (Theorem 6.1.17). This implies, by [97, 32E/4], the equivalence of (1) and (2) and of (4) and (5). By Theorems 6.8.16 and 6.8.15, (3) and (6) are equivalent to the fact that X is locally connected.

Q.E.D.

The proof of the following theorem is similar to the proof of [32, 2.5].

6.8.21 Theorem *Let* X *be a continuum and let* $n \in \mathbb{N}$. *If* $\mathcal{F}_n(X)$ *is an absolute retract, then* $\mathcal{F}_n(X)$ *is a strong deformation retract of* 2^X.

Proof Since $\mathcal{F}_n(X)$ is an absolute retract, $\mathcal{F}_n(X) \times \mathcal{Q}$ is homeomorphic to $\mathcal{Q}$ [15, 44.1 and 22.1]. Let $q_0 \in \mathcal{Q}$. Since $\{q_0\}$ is a Z-set in $\mathcal{Q}$ [15, (ii), p. 2], it follows that $\mathcal{F}_n(X) \times \{q_0\}$ is a Z-set in $\mathcal{F}_n(X) \times \mathcal{Q}$, and since $\{q_0\}$ is a strong deformation retract of $\mathcal{Q}$, it follows that $\mathcal{F}_n(X) \times \{q_0\}$ is a strong deformation retract of $\mathcal{F}_n(X) \times \mathcal{Q}$. Let $h : (\mathcal{F}_n(X) \times \mathcal{Q}) \times [0, 1] \to \mathcal{F}_n(X) \times \mathcal{Q}$ be a strong deformation retraction from $\mathcal{F}_n(X) \times \mathcal{Q}$ onto $\mathcal{F}_n(X) \times \{q_0\}$. Since X is locally connected, X has the property of Kelley [84, (16.11)]. Hence, by Theorem 6.7.7, $\mathcal{F}_n(X)$ is a Z-set in 2^X. By [84, (1.97)], 2^X is homeomorphic to $\mathcal{Q}$. Let $f : \mathcal{F}_n(X) \to \mathcal{F}_n(X) \times \{q_0\}$ be the homeomorphism given by $f(A) = (A, q_0)$. By Anderson's homeomorphism extension theorem [48, 11.9.1], f can be extended to a homeomorphism

$F : 2^X \twoheadrightarrow \mathcal{F}_n(X) \times \mathcal{Q}$. It follows that the map

$$H : 2^X \times [0, 1] \twoheadrightarrow 2^X$$

given by

$$H((A, t)) = F^{-1}\left(h\left((F(A), t)\right)\right)$$

is a strong deformation retraction from 2^X onto $\mathcal{F}_n(X)$.

Q.E.D.

As a consequence of Theorem 6.8.21, we have:

6.8.22 Theorem *Let X be a continuum and let $n, m \in \mathbb{N}$ with $m \geq n$. If $\mathcal{F}_n(X)$ is an absolute retract, then $\mathcal{F}_n(X)$ is a strong deformation retraction of $\mathcal{C}_m(X)$.*

Proof Since $\mathcal{F}_n(X)$ is an absolute retract, there exists a strong deformation retraction $H : 2^X \times [0, 1] \twoheadrightarrow 2^X$ from 2^X onto $\mathcal{F}_n(X)$ by Theorem 6.8.21.

Since X is locally connected, $\mathcal{C}_m(X)$ is an absolute retract [98, Théorème II$_m$]. Hence, there exists a retraction $r : 2^X \twoheadrightarrow \mathcal{C}_m(X)$. Thus, the map $r \circ \left(H|_{\mathcal{C}_m(X) \times [0,1]}\right) : \mathcal{C}_m(X) \times [0, 1] \twoheadrightarrow \mathcal{C}_m(X)$ is a strong deformation retraction from $\mathcal{C}_m(X)$ onto $\mathcal{F}_n(X)$.

Q.E.D.

6.8.23 Corollary *Let X be a continuum and let $n, m \in \mathbb{N}$ with $m \geq n$. If X is an absolute retract, then $\mathcal{F}_n(X)$ is a strong deformation retract of both 2^X and $\mathcal{C}_m(X)$.*

Proof Since X is an absolute retract, $\mathcal{F}_n(X)$ is an absolute retract [29, p. 316]. Now the corollary follows from Theorems 6.8.22 and 6.8.21.

Q.E.D.

Theorem 6.8.19 may be strengthened as follows:

6.8.24 Theorem *Let X be a locally connected continuum and let n and m be positive integers such that $m \geq n$. Then the following are equivalent:*

(1) $\mathcal{F}_n(X)$ is an absolute retract;
(2) $\mathcal{F}_n(X)$ is a retract of $\mathcal{C}_m(X)$;
(3) $\mathcal{F}_n(X)$ is a deformation retract of $\mathcal{C}_m(X)$;
(4) $\mathcal{F}_n(X)$ is a strong deformation retract of $\mathcal{C}_m(X)$.

Proof Clearly, (4) implies (3) and (3) implies (2). By Theorem 6.8.17, (2) implies (1). By Theorem 6.8.22, (1) implies (4).

Q.E.D.

6.8.25 Definition A continuum X is *uniformly pathwise connected* provided that X is a continuous image of the cone over the Cantor set.

6.8.26 Remark Let us note that the definition of uniformly pathwise connected continuum given in Definition 6.8.25 is not the original, but it is equivalent [54, Theorem 3.5].

6.8.27 Definition A *tree* is a graph (Definition 6.9.1) without simple closed curves. A continuum X is *tree-like* if it is the inverse limit of trees with surjective bonding maps.

6.8.28 Theorem *If X is a continuum with trivial shape and Y is a subcontinuum of X which is a retract of X, then Y has trivial shape.*

Proof Let $r: X \twoheadrightarrow Y$ be a retraction. Let Z be an absolute neighborhood retract and let $f: Y \to Z$ be a map. Since X has trivial shape, the map $f \circ r: X \to Z$ is homotopic to a constant map [53, Theorem 2.1 (A)]. Let $H: X \times [0, 1] \to Z$ be a homotopy such that $H((x, 0)) = f \circ r(x)$ and $H((x, 1)) = z_0$ for all $x \in X$ and some $z_0 \in Z$. Let $G = H|_{Y \times [0,1]}$. Then $G: Y \times [0, 1] \to Z$ is a homotopy such that $G((y, 0)) = f \circ r(y) = f(y)$ and $H((y, 1)) = z_0$ for each $y \in Y$. Hence, f is homotopic to a constant map. Therefore, Y has trivial shape [53, Theorem 2.1 (A)].

Q.E.D.

6.8.29 Theorem *Let X be a one-dimensional continuum and let $n \in \mathbb{N}$. If there exists a retraction from $C_n(X)$ onto $\mathcal{F}_1(X)$, then X is a uniformly pathwise connected dendroid.*

Proof By Theorem 6.2.3, $C_n(X)$ has trivial shape. Trivial shape is preserved by retractions (Theorem 6.8.28). Hence, since $\mathcal{F}_1(X)$ is homeomorphic to X, X has trivial shape. Since every one-dimensional continuum with trivial shape is tree-like [53, Theorem 2.1 (B)], we have that X is tree-like. Since trees are hereditarily unicoherent, by Corollary 2.1.28, tree-like continua are hereditarily unicoherent. Thus, X is hereditarily unicoherent. Furthermore, for each $n \in \mathbb{N}$, there exists a map from the cone over the Cantor set onto $C_n(X)$ [50, Remark on p. 29 and Theorem 2.7]. Thus, $C_n(X)$ is uniformly pathwise connected and X is also uniformly pathwise connected. Therefore, X is a uniformly pathwise connected dendroid.

Q.E.D.

6.8.30 Definition A continuum X is *weakly chainable* if it is a continuous image of an arc-like continuum.

6.8.31 Definition A continuum Z lying in the Hilbert cube Q is *movable* provided that for any neighborhood V of Z in Q, there exists a neighborhood U of Z in Q such that for any neighborhood W of Z in Q, there exists a homotopy $\xi_W: U \times [0, 1] \to V$ such that $\xi_W((z, 0)) = z$ and $\xi_W((z, 1)) \in W$ for each $z \in U$.

6.8.32 Theorem *Let X be a continuum and let n and m be positive integers such that $m \geq n$. If $\mathcal{F}_n(X)$ is a retract of $C_m(X)$, then:*

(1) X is arcwise connected;
(2) $\mathcal{F}_n(X)$ is arcwise decomposable;
(3) $\mathcal{F}_n(X)$ is uniformly pathwise connected;

(4) $\mathcal{F}_n(X)$ *is weakly chainable;*
(5) $\mathcal{F}_n(X)$ *has trivial shape;*
(6) $\mathcal{F}_n(X)$ *is movable.*

Proof Suppose $\mathcal{F}_n(X)$ is a retract of $\mathcal{C}_m(X)$. Since *n*-fold hyperspaces are arcwise connected continua (Theorem 1.8.10 and Corollary 1.8.12), we have that $\mathcal{F}_n(X)$ is arcwise connected. Thus, X is arcwise connected [19, 2.7]. By [50, Remark on p. 29 and Theorem 2.7], there exists a map from the cone over the Cantor set $\mathfrak{F}_{\mathfrak{C}}$ onto $\mathcal{C}_m(X)$. Hence, there exists a map from $\mathfrak{F}_{\mathfrak{C}}$ onto $\mathcal{F}_n(X)$. This implies that $\mathcal{F}_n(X)$ is arcwise decomposable, by Theorem 1.7.40, and uniformly pathwise connected (Definition 6.8.25). By [19, 3.3], $\mathcal{C}_m(X)$ is weakly chainable. Thus, $\mathcal{F}_n(X)$ is weakly chainable. Since $\mathcal{C}_m(X)$ has trivial shape (Theorem 6.2.3), $\mathcal{F}_n(X)$ has trivial shape by Theorem 6.8.28. Since $\mathcal{F}_n(X)$ has trivial shape, $\mathcal{F}_n(X)$ is movable [6, (3.2)].

Q.E.D.

Recall that if $\mathcal{S}^1$ is the unit circle, then $\mathcal{F}_2(\mathcal{S}^1)$ is homeomorphic to the Möbius strip [7, p. 877], which is not unicoherent. Note the following:

6.8.33 Corollary *Let X be a continuum and let $m \geq 2$. If $\mathcal{F}_2(X)$ is a retract of $\mathcal{C}_m(X)$, then $\mathcal{F}_2(X)$ is unicoherent.*

Proof By Theorem 6.8.32, $\mathcal{F}_2(X)$ has trivial shape. Hence, by Theorem 6.2.2, every map from $\mathcal{F}_2(X)$ into any absolute neighborhood retract is homotopic to a constant map. In particular, every map from $\mathcal{F}_2(X)$ into $\mathcal{S}^1$ is homotopic to a constant map. Hence, by [55, Theorem 1, p. 434], $\mathcal{F}_2(X)$ is unicoherent.

Q.E.D.

6.8.34 Corollary *An arc-like continuum X is an arc if and only if $\mathcal{F}_n(X)$ is a retract of $\mathcal{C}_m(X)$ for some positive integers m and n with $m \geq n$.*

Proof Let X be an arc-like continuum such that there $\mathcal{F}_n(X)$ is a retract of $\mathcal{C}_m(X)$ for some positive integers m and n such that $m \geq n$. By Theorem 6.8.32, X is arcwise connected. Since the arc is the only arcwise connected arc-like continuum, we have that X is an arc.

The reverse implication follows from Corollary 6.8.23

Q.E.D.

6.8.35 Corollary *If X is a circle-like continuum and $m \geq 2$ is an integer, then there does not exist a retraction from $\mathcal{C}_m(X)$ onto $\mathcal{F}_2(X)$.*

Proof If X is also an arc-like continuum, it follows from [10, Theorem 3] that X is either indecomposable or the union of two indecomposable continua. Hence, by Corollary 6.8.34, there does not exist a retraction from $\mathcal{C}_m(X)$ onto $\mathcal{F}_2(X)$. Suppose X is a circle-like continuum which is not arc-like. Then, by Theorem 6.2.7, $\mathcal{F}_2(X)$ does not have trivial shape. Therefore, by Theorem 6.8.32, there does not exist a retraction from $\mathcal{C}_m(X)$ onto $\mathcal{F}_2(X)$.

Q.E.D.

6.8.36 Theorem *If X is a 1-dimensional continuum containing a simple closed curve and n is an integer such that $n \geq 2$, then there does not exist a retraction from $C_n(X)$ onto $\mathcal{F}_2(X)$.*

Proof By definition, a simple closed curve does not have trivial shape. Since X is a 1-dimensional continuum containing a simple closed curve S, there exists a retraction from X onto S [40, Theorem VI 4, p. 83]. Hence, there exists a retraction from $\mathcal{F}_2(X)$ onto $\mathcal{F}_2(S)$. Since $\mathcal{F}_2(S)$ does not have trivial shape (Theorem 6.2.7) by Theorem 6.8.28, $\mathcal{F}_2(X)$ does not have trivial shape. Therefore, by Theorem 6.8.32, there does not exist a retraction from $C_n(X)$ onto $\mathcal{F}_2(X)$.

Q.E.D.

6.9 Graphs

We present results about the n-fold hyperspaces of graphs. To this end, we follow [11, 40, 44, 46, 47, 50, 61, 66, 69, 78, 84, 87].

6.9.1 Definition A *graph* is a continuum which can be written as the union of finitely many arcs, any two of which are either disjoint or intersect only in one or both of their end points.

6.9.2 Definition Let X be a graph and let A be a subset of X. Let β be a cardinal number. We say that *A is of order less than or equal to β in X,* written ord$(A, X) \leq \beta$, provided that for each open subset U of X containing A, there exists an open subset V of X such that $A \subset V \subset U$ and $Bd_X(V)$ has cardinality less than or equal to β. We say that *A is of order β in X,* written ord$(A, X) = \beta$, provided that ord$(A, X) \leq \beta$ and ord$(A, X) \not\leq \alpha$ for any cardinal number $\alpha < \beta$. A point x of a graph X is *a ramification point of X* if and only if ord$(\{x\}, X) \geq 3$. A point a of a graph X is an *end point of X* if and only if ord$(\{a\}, X) = 1$.

The next theorem characterizes graphs as the class of locally connected continua X having finite dimensional hyperspaces $C_n(X)$.

6.9.3 Theorem *A locally connected continuum X is a graph if and only if for each $n \in \mathbb{N}$, $C_n(X)$ is of finite dimension.*

Proof Suppose X is a graph. Then, by [50, Lemma 5.2], dim$(C(X)) < \infty$. Hence, given $n \in \mathbb{N}$, by Corollary 6.1.25, we have that dim$(C_n(X)) \leq n(\dim(C(X)) < \infty$.

Suppose X is a locally connected continuum such that for each $n \in \mathbb{N}$, dim$(C_n(X)) < \infty$. Then we have that dim$(C(X)) < \infty$. Thus, X is a graph [50, Lemma 5.2].

Q.E.D.

6.9.4 Theorem *Let $n \in \mathbb{N}$, $n \geq 2$. Then $C_n([0, 1]) \setminus C_{n-1}([0, 1])$ is embeddable in $\mathbb{R}^{2n}$.*

Proof Given $A \in C_n([0, 1]) \setminus C_{n-1}([0, 1])$, without loss of generality, we assume that $[a_1, a_1'], \ldots, [a_n, a_n']$ are the components of A and $a_1 \leq a_1' < a_2 \leq a_2' < \cdots < a_n \leq a_n'$. Define $\xi: C_n([0, 1]) \setminus C_{n-1}([0, 1]) \to \mathbb{R}^{2n}$ by $\xi(A) = (a_1, a_1', \ldots, a_n, a_n')$. Clearly, ξ is a one-to-one function. To see its continuity, let $\varepsilon > 0$ and let $B = \bigcup_{j=1}^n [b_j, b_j']$ be a point of $C_n([0, 1]) \setminus C_{n-1}([0, 1])$ such that $\mathcal{H}(A, B) < \varepsilon$. Then for each $j \in \{1, \ldots, n\}$, $|a_j - b_j| < \varepsilon$ and $|a_j' - b_j'| < \varepsilon$. Hence, $D(\xi(A), \xi(B)) = \max\{|a_j - b_j|, |a_j' - b_j'| \mid j \in \{1, \ldots, n\}\} < \varepsilon$. Thus, ξ is continuous. Note that $\xi\left(\mathcal{V}_\varepsilon^{\mathcal{H}}(A) \cap C_n([0, 1]) \setminus C_{n-1}([0, 1])\right) = \mathcal{V}_\varepsilon^{\mathbb{R}^{2n}}(\xi(A)) \cap \xi(C_n([0, 1]) \setminus C_{n-1}([0, 1]))$. Therefore, ξ is an embedding.

Q.E.D.

6.9.5 Theorem *Let $n \in \mathbb{N}$, and let $\mathcal{E}_n = \{(x_1, \ldots, x_{2n}) \in \mathbb{R}^{2n} \mid 0 \leq x_1 \leq \cdots \leq x_{2n} \leq 1\}$. Then the function $f: \mathcal{E}_n \to C_n([0, 1])$ given by*

$$f((x_1, \ldots, x_{2n})) = \bigcup_{j=1}^n [x_{2j-1}, x_{2j}]$$

is continuous and surjective. Furthermore, $\mathcal{E}_n / \mathcal{G}_f$ is homeomorphic to $C_n([0, 1])$.

Proof Let $\varepsilon > 0$ be given. If $(x_1, \ldots, x_{2n})$ and $(x_1', \ldots, x_{2n}')$ belong to $\mathcal{E}_n$ and $D((x_1, \ldots, x_{2n}), (x_1', \ldots, x_{2n}')) = \max\{|x_j - x_j'| \mid j \in \{1, \ldots, 2n\}\} < \varepsilon$, then we have that

$$\mathcal{H}(f((x_1, \ldots, x_{2n})), f((x_1', \ldots, x_{2n}'))) \leq$$

$$\max\{|x_j - x_j'| \mid j \in \{1, \ldots, 2n\}\} < \varepsilon.$$

Thus, f is continuous. If $A = \bigcup_{j=1}^k [x_j, y_j]$, where $k \leq n$, then $(x_1, y_1, \ldots, x_k, y_k, \underbrace{y_k, \ldots, y_k}_{2n-2k \text{ times}})$ belongs to $\mathcal{E}_n$ and

$$f((x_1, y_1, \ldots, x_k, y_k, y_k, \ldots, y_k)) = A.$$

Hence, f is surjective. The fact that $\mathcal{E}_n / \mathcal{G}_f$ is homeomorphic to $C_n([0, 1])$ follows from Theorem 1.2.10.

Q.E.D.

6.9.6 Notation Let $\mathcal{S}^1$ denote the unit circle in the plane. If $\theta \in \mathcal{S}^1$ and $t \in [0, 1)$, then let $A(\theta, t)$ be the arc (possibly degenerate) contained in $\mathcal{S}^1$ having θ as mid point and length $2\pi t$. For $t = 1$, $A(\theta, 1)$ denotes $\mathcal{S}^1$.

The following lemma is easily established.

6.9.7 Lemma *Let $\varepsilon > 0$ be given. If θ and φ belong to $\mathcal{S}^1$, t and s belong to $[0, 1]$ and satisfy that $\|\theta - \varphi\| < \varepsilon$ and $|t - s| < \frac{\varepsilon}{2\pi}$, then $\mathcal{H}(A(\theta, t), A(\varphi, s)) < \varepsilon$.*

6.9.8 Notation Let $n \in \mathbb{N}$. If $n > 1$, let $\mathbb{T}^n = \underbrace{\mathcal{S}^1 \times \cdots \times \mathcal{S}^1}_{n \text{ times}}$. If $n = 1$, then
$\mathbb{T}^1 = \mathcal{S}^1$.

6.9.9 Theorem *Let $n \in \mathbb{N}$. If $f : \mathbb{T}^n \times [0, 1]^n \to \mathcal{C}_n(\mathcal{S}^1)$ is a function given by*

$$f((\theta_1, \ldots, \theta_n, t_1, \ldots, t_n)) = \bigcup_{j=1}^{n} A(\theta_j, t_j),$$

then f is continuous and surjective. Furthermore, $(\mathbb{T}^n \times [0, 1]^n)/\mathcal{G}_f$ is homeomorphic to $\mathcal{C}_n(\mathcal{S}^1)$.

Proof Let $\varepsilon > 0$ be given. If $(\theta_1, \ldots, \theta_n, t_1, \ldots, t_n), (\varphi_1, \ldots, \varphi_n, s_1, \ldots, s_n)$ $\in \mathbb{T}^n \times [0, 1]^n$ and $D((\theta_1, \ldots, \theta_n, t_1, \ldots, t_n), (\varphi_1, \ldots, \varphi_n, s_1, \ldots, s_n)) < \frac{\varepsilon}{2\pi}$, then $|\theta_j - \varphi_j| < \frac{\varepsilon}{2\pi} < \varepsilon$ and $|t_j - s_j| < \frac{\varepsilon}{2\pi}$ for each $j \in \{1, \ldots, n\}$. By Lemma 6.9.7, it follows that $\bigcup_{j=1}^n A(\theta_j, t_j) \subset \mathcal{V}_\varepsilon^d \left(\bigcup_{j=1}^n A(\varphi_j, s_j) \right)$ and $\bigcup_{j=1}^n A(\varphi_j, s_j) \subset \mathcal{V}_\varepsilon^d \left(\bigcup_{j=1}^n A(\theta_j, t_j) \right)$. Thus, $\mathcal{H} \left(\bigcup_{j=1}^n A(\theta_j, t_j), \bigcup_{j=1}^n A(\varphi_j, s_j) \right) < \varepsilon$. Therefore, f is continuous.

Let $A \in \mathcal{C}_n(\mathcal{S}^1)$, and suppose $A_1, \ldots, A_k$ are the components of A, with $k \leq n$. For each $j \in \{1, \ldots, k\}$, let θ_j be the midpoint of A_j and let t_j be the length of A_j. Then $\left(\theta_1, \ldots, \theta_k, \underbrace{\theta_k, \ldots, \theta_k}_{n-k \text{ times}}, \frac{t_1}{2\pi}, \ldots, \frac{t_k}{2\pi}, \underbrace{\frac{t_k}{2\pi}, \ldots, \frac{t_k}{2\pi}}_{n-k \text{ times}} \right)$ is an element of $\mathbb{T}^n \times [0, 1]^n$ whose image under f is A. Thus, f is surjective.

The fact that $(\mathbb{T}^n \times [0, 1]^n)/\mathcal{G}_f$ is homeomorphic to $\mathcal{C}_n(\mathcal{S}^1)$ follows from Theorem 1.2.10.

Q.E.D.

As a consequence of Corollary 6.1.26 we have the following Theorem.

6.9.10 Theorem $\dim(\mathcal{C}_n([0, 1])) = \dim(\mathcal{C}_n(\mathcal{S}^1)) = 2n$ *for every $n \in \mathbb{N}$.*

The following theorem is a nice consequence of Theorem 6.9.10.

6.9.11 Theorem *If X is either an arc-like or a circle-like continuum and n is a positive integer, then $\dim(\mathcal{C}_n(X)) \leq 2n$.*

Proof Let n be a positive integer, let X be an arc-like continuum and let $\varepsilon > 0$ be given. Then there exists an ε-map $f : X \twoheadrightarrow [0, 1]$ (Theorem 2.4.22). Hence, by Lemma 8.2.11, the induced map $\mathcal{C}_n(f) : \mathcal{C}_n(X) \twoheadrightarrow \mathcal{C}_n([0, 1])$ is an ε-map with respect to the Hausdorff metric . Thus, since $\dim(\mathcal{C}_n([0, 1])) = 2n$ (Theorem 6.9.10), we have that $\dim(\mathcal{C}_n(X)) \leq 2n$ [87, 15.5].

The proof for the case when X is circle-like is similar.

Q.E.D.

The proof of the following theorem is due to R. Schori:

6.9.12 Theorem $C_2([0, 1])$ *is homeomorphic to* $[0, 1]^4$.

Proof Let $D^1 = \{A \in C_2([0, 1]) \mid 1 \in A\}$ and $D_0^1 = \{A \in C_2([0, 1]) \mid \{0, 1\} \subset A\}$. We divide the proof in three steps.

Step 1. $C_2([0, 1])$ *is homeomorphic to* $K(D^1)$ (the cone over D^1).

Let $f : K(D^1) \twoheadrightarrow C_2([0, 1])$ be given by $f((A, t)) = (1 - t)A = \{(1 - t)a \mid a \in A\}$. Since $f((A, 1)) = \{0\}$ for every $A \in C_2([0, 1])$, f is a well defined map. Let $A, B \in D^1$ and let $s, t \in [0, 1]$ be such that $f((A, t)) = f((B, s))$. Since $1 \in A \cap B$, $1 - t = \max(1 - t) A$ and $1 - s = \max(1 - s) B$. This implies that $t = s$. If $t = s = 1$, then $(A, 1)$ and $(B, 1)$ both represent the vertex of $K(D^1)$. Hence, we assume that $t < 1$. Since $(1 - t)A = (1 - t)B$, $A = B$. Therefore, f is one-to-one. Now, let $A \in C_2([0, 1]) \setminus \{\{0\}\}$. Take $t = 1 - \max A$. Then $\max\left(\frac{1}{1-t}\right) A = 1$. Thus, $\left(\frac{1}{1-t}\right) A \in D^1$ and $f\left(\left(\left(\frac{1}{1-t}\right) A, t\right)\right) = A$. Hence, f is surjective. Therefore, $C_2([0, 1])$ is homeomorphic to $K(D^1)$.

Step 2. D^1 *is homeomorphic to* $K(D_0^1)$.

Let $g : K(D_0^1) \twoheadrightarrow D^1$ be given by $g((A, t)) = t + (1 - t)A = \{t + (1 - t)a \mid a \in A\}$. Proceeding as in Step 1, it can be seen that g is a homeomorphism.

Step 3. D_0^1 *is homeomorphic to* $[0, 1]^2$.

Let $T = \{(a, b) \in \mathbb{R}^2 \mid 0 \le a \le b \le 1\}$, and let $S = T/\Delta$, where $\Delta = \{(a, b) \in T \mid a = b\}$. Note that S is homeomorphic to $[0, 1]^2$. Let $h : T \twoheadrightarrow D_0^1$ be given by $h((a, b)) = [0, a] \cup [b, 1]$. Then h is continuous. Note that $h((a, b)) = h((c, d))$ if and only if $a = c$ and $b = d$. Hence, by Theorem 1.2.10, S and D_0^1 are homeomorphic.

Q.E.D.

Recall that for any continuum X and any $n \in \mathbb{N}$, $C_n(X)$ is unicoherent (Theorem 6.2.5). The next theorem, whose proof is contained in [44, Lemma 2.3], says that unicoherence of these hyperspaces may be destroyed by removing a point.

6.9.13 Theorem $C_2(\mathcal{S}^1) \setminus \{\mathcal{S}^1\}$ *is not unicoherent.*

The proof of the following theorem is the content of [47].

6.9.14 Theorem $C_2(\mathcal{S}^1)$ *is homeomorphic to the cone over the solid torus.*

The proof of the following theorem is the content of [44] and [46].

6.9.15 Theorem *Let X be a graph and let Y be a continuum such that $C_n(Y)$ is homeomorphic to $C_n(X)$. Then:*

(1) if $n = 1$ and X is neither an arc nor a simple closed curve, then Y is homeomorphic to X,
(2) if $n \ge 2$, then Y is homeomorphic to X.

6.9.16 Definition A finite dimensional separable metric space Z is a *Cantor manifold* if for any subset B of Z such that $\dim(B) \le \dim(Z) - 2$, $Z \setminus B$ is connected.

A proof of the next theorem may be found in [69, Theorem 4.6].

6.9.17 Theorem *If $n \in \mathbb{N}$, then $C_n([0, 1])$ and $C_n(S^1)$ are $2n$-dimensional Cantor manifolds.*

6.9.18 Notation If X is a graph, then

$$R(X) = \{p \in X \mid ord(\{p\}, X) \geq 3\}.$$

The set $R(X)$ is called the *set of ramification points*.

A proof of the next result may be found in [78, Theorem 2.4].

6.9.19 Theorem *Let X be a graph and let A be an element of $C_n(X)$. Then*

(1) $\dim_A(C_n(X)) = 2n + \sum_{p \in R(X) \cap A}(ord(\{p\}, X) - 2)$;
(2) $\dim(C_n(X)) = \dim_X(C_n(X))$.
(3) $\dim(C_n(X)) = 2n + \dim(C(X))$.

6.9.20 Remark Note that part (3) of Theorem 6.9.19 is not true for either [0, 1] or S^1 by Theorem 6.9.10.

The next theorem characterizes the graphs whose n-fold hyperspaces are Cantor manifolds.

6.9.21 Theorem *Let X be a graph and let n be a positive integer. Then $C_n(X)$ is a Cantor manifold if and only if X is either an arc or a simple closed curve.*

Proof Suppose X is neither an arc nor a simple closed curve. Then X has a ramification point p. Hence, if A is an element of $C_n(X)$ with exactly n components and such that $p \in A$, we have that $\dim_A(C_n(X)) \geq (2n - 1) + ord(\{p\}, X) \geq (2n - 1) + 3 = 2n + 1$. Also, if x an element of X that is not a ramification point, then there exists a subarc L of X such that $x \in Int_X(L)$ and L does not contain a ramification point of X. Thus, $C_n(L)$ is a closed $2n$-dimensional neighborhood of $\{x\}$, by Theorem 6.9.10, that is a Cantor manifold, by Theorem 6.9.17. This implies that $\dim_{\{x\}}(C_n(X)) = 2n$. Since Cantor manifolds have the same dimension at each of its points [40, A), pp. 93 and 94], we obtain that $C_n(X)$ is not a Cantor manifold.

If X is either an arc or a simple closed curve, by Theorem 6.9.17, $C_n(X)$ is a $2n$-dimensional Cantor manifold.

Q.E.D.

The next theorem characterizes the n-fold hyperspaces that are cells.

6.9.22 Theorem *Let X be a continuum and let n and k be positive integers. Then $C_n(X)$ is homeomoprhic to $[0, 1]^k$ if and only if X is an arc or a simple closed curve, $n \in \{1, 2\}$ and $k \in \{2, 4\}$. Moreover, if $n = 2$, then X is an arc and $k = 4$.*

Proof Suppose $C_n(X)$ is homeomorphic to $[0, 1]^k$. Then, by Theorem 6.1.6, X is locally connected. Since $\dim(C_n(X)) = k < \infty$, by Theorem 6.9.3, X is a graph. Since $[0, 1]^k$ is a Cantor manifold [40, Example VI 11, p. 93], by Theorem 6.9.21, X is an arc or a simple closed curve. Hence, by Theorem 6.9.10, k is an even number. Suppose $n \geq 3$. Note that each point of $[0, 1]^k$ has a local basis of closed neighborhoods homemorphic to $[0, 1]^k$. Hence, by [46, Lemma 3.4], $C_n(X)$ cannot be homeomorphic to $[0, 1]^k$. Thus, $n \in \{1, 2\}$ and $k \in \{2, 4\}$.

If $n = 1$, then $C(X)$ is homeomorphic to $[0, 1]^2$ [84, (0.54) and (0.55)]. Now, if $n = 2$, then $C_2([0, 1])$ is homeomorphic to $[0, 1]^4$, by Theorem 6.9.12, and $C_2(S^1)$ is homemorphic to the cone over a solid torus, by Theorem 6.9.14.

$$\text{Q.E.D.}$$

6.9.23 Remark Observe that Theorem 6.9.22 gives a negative answer to Question 9.4.2.

6.10   Cones, Suspensions and Products

The material of this section is based on [16, 22, 24, 39, 55, 56, 63, 66, 69, 72, 84–87]. Recall that given a space Z, $K(Z)$ denotes the cone over Z.

6.10.1 Definition Let F be a fan. By an *end point of a fan F* we mean an end point in the classical sense, that is, a point e of F which is a nonseparating point of any arc in F that contains e; $E(F)$ denotes the set of all end points of a fan F. A *leg of a fan F* is the unique arc in F from the top τ to some end point of F. Given two points x and y of a fan F, xy denotes the unique arc in F joining x and y. Given an $m \in \mathbb{N}$, an *m-od* is a fan for which $E(F)$ has exactly m elements.

Theorem 6.9.12 gives an example of a continuum whose 2-fold hyperspace is a cone. Corollary 6.10.3 extends this example. Note that Theorem 6.9.14 also gives such an example.

Given a fan F, let $\mathcal{G}(F)$ denote either of the hyperspaces 2^F, $C_n(F)$, or $\mathcal{F}_n(F)$, for $n \in \mathbb{N}$.

6.10.2 Theorem *If F is a fan, with metric d, which is homeomorphic to the cone over a compact metric space, then $\mathcal{G}(F)$ is homeomorphic to the cone over a continuum.*

Proof Let F be a fan which is a cone. Then F is smooth. We assume by [24, Corollary 4, p. 90] and [16, Theorem 9, p. 27], that F is embedded in $\mathbb{R}^2$, $\tau = (0, 0)$ is the top of F and the legs of F are convex arcs of length one [72, 4.2]. Given two points a and b of $\mathbb{R}^2$, $\overline{ab}$ denotes the convex arc in $\mathbb{R}^2$ whose end points are a and b (Notation 3.6.32), and $||a||$ denotes the norm of a in $\mathbb{R}^2$. Given an element A of $\mathcal{G}(F)$ and $r \geq 0$, $rA = \{ra \mid a \in A\}$. Note that for $r = 0$, $rA = \{(0, 0)\} = \{\tau\}$.

Let $E(F) = \{e_\lambda\}_{\lambda \in \Lambda}$. Then $F = K(E(F))$ by [72, 4.2]. Note that this equality implies that $E(F)$ is closed in F. Hence, $E(F)$ is a compactum.

Let $\mathcal{B} = \bigcup \{\{A \in \mathcal{G}(F) \mid e_\lambda \in A\} \mid \lambda \in \Lambda\}$.

Let $\varphi \colon \mathcal{B} \times I \to \mathcal{G}(F)$ be given by

$$\varphi((A, t)) = (1 - t)A.$$

Clearly, φ is well defined. Observe that if $t \in [0, 1)$ and $A \in \mathcal{B}$, then $\{\tau\} \in \varphi((A, t))$ if and only if $\tau \in A$. We show that φ is continuous. Let $\varepsilon > 0$ be given and let $\delta = \frac{\varepsilon}{2}$. Let $A, B \in \mathcal{G}(F)$ and let $t, s \in [0, 1]$ be such that $\mathcal{H}(A, B) < \delta$ and $|t - s| < \delta$. Let $a \in A$. Then there exists $b \in B$ such that $||a - b|| < \delta$. Note that

$$||(1 - t)a - (1 - s)b|| \leq ||(1 - t)a - (1 - t)b|| + ||(1 - t)b - (1 - s)b|| \leq$$

$$(1 - t)||a - b|| + |s - t|||b|| \leq ||a - b|| + |s - t| < 2\delta = \varepsilon.$$

Thus, $\varphi((A, t)) \subset \mathcal{V}_\varepsilon^d(\varphi((B, s)))$. Similarly, we have that $\varphi((B, s)) \subset \mathcal{V}_\varepsilon^d(\varphi((A, t)))$. Therefore, $\mathcal{H}(\varphi((A, t)), \varphi((B, s))) < \varepsilon$ and φ is continuous.

We show that φ is one-to-one on $\mathcal{B} \times [0, 1)$. Let $t, s \in [0, 1)$, and let $A, B \in \mathcal{B}$. Suppose that $\varphi((A, t)) = \varphi((B, s))$. Since $A, B \in \mathcal{B}$, there exist $e_\lambda, e_{\lambda'} \in E(F)$ such that $e_\lambda \in A$ and $e_{\lambda'} \in B$.

Case (1) $\tau \notin A$.

Then $\tau \notin B$. Let $\overline{ae_\lambda}$ be the component of A containing e_λ. Let $\overline{be_{\lambda'}}$ be the component of B containing $e_{\lambda'}$. Since $\varphi((A, t)) = \varphi((B, s))$, there exist $\overline{b_{\lambda'}c_{\lambda'}} \subset A$ and $\overline{b_\lambda c_\lambda} \subset B$ such that

$$\overline{(1 - t)a(1 - t)e_\lambda} = \overline{(1 - s)b_\lambda(1 - s)c_\lambda}$$

and

$$\overline{(1 - s)b(1 - s)e_{\lambda'}} = \overline{(1 - t)b_{\lambda'}(1 - t)c_{\lambda'}}.$$

From the first equality we obtain that $(1 - t)e_\lambda = (1 - s)c_\lambda$, which implies that $1 - t = (1 - s)||c_\lambda|| \leq 1 - s$. From the second inequality we obtain that $(1 - s)e_{\lambda'} = (1 - t)c_{\lambda'}$, which implies that $1 - s = (1 - t)||c_{\lambda'}|| \leq 1 - t$. Therefore, $t = s$. Hence, $A = B$.

Case (2) $\tau \in A$.

Then $\tau \in B$. Let us observe that either $\overline{\tau e_\lambda} \subset A$ or there exists $a \in A$ such that $\overline{ae_\lambda} \subset A$. In either case, as in Case (1), we conclude that $(1 - t)e_\lambda = (1 - s)c_\lambda$, for some $c_\lambda \in E(B)$, and that $(1 - s)e_{\lambda'} = (1 - t)c_{\lambda'}$, for some $c_{\lambda'} \in E(A)$. These two equalities imply that $t = s$. Hence, $A = B$.

We show that φ is surjective. Let $B \in \mathcal{G}(F)$. If $B = \{\tau\}$, then $\varphi((A, 1)) = \{\tau\}$ for any $A \in \mathcal{B}$. Thus, assume $B \neq \{\tau\}$. If $B \cap E(F) \neq \emptyset$, then $\varphi((B, 0)) = B$.

Suppose $B \cap E(F) = \emptyset$. Let $t = \inf\{||b - e_\lambda|| \mid b \in B \text{ and } e_\lambda \in E(F)\}$. Since $B \neq \{\tau\}$, $t \neq 1$. Then there exists $\lambda_0 \in \Lambda$ such that $||b_{\lambda_0} - e_{\lambda_0}|| = t$, where $b_{\lambda_0} \in B \cap \overline{\tau e_{\lambda_0}}$.

Let $A = \frac{1}{1-t}B$. Note that for λ_0, $\frac{1}{1-t}b_{\lambda_0} \in A \cap \overline{\tau, e_{\lambda_0}}$. Since $\frac{1}{1-t}b_{\lambda_0} = e_{\lambda_0}$, we have that $e_{\lambda_0} \in A$. Hence, $A \in \mathcal{B}$, and $\varphi((A, t)) = \frac{1-t}{1-t}B = B$.

By Theorem 1.2.10, the hyperspace $\mathcal{G}(F)$ is homeomorphic to $K(\mathcal{B})$. Since no point of $\mathcal{G}(F)$ arcwise disconnects $\mathcal{G}(F)$ [84, (11.5)], we have that $\mathcal{B}$ is a continuum.

Q.E.D.

Note that a similar proof to the one given for Theorem 6.10.2 shows:

6.10.3 Corollary *If* $\mathcal{G}([0, 1]) \in \{2^{[0,1]}, \mathcal{C}_n([0, 1]), \mathcal{F}_n([0, 1])\}$, *then* $\mathcal{G}([0, 1])$ *is homeomorphic to the cone over a continuum.*

Since, clearly, a simple m-od is a fan homeomorphic to the cone over a finite set, we have the following:

6.10.4 Corollary *Let* m *and* n *be positive integers. If* F *is a simple* m-od, *and* $\mathcal{G}(F) \in \{\mathcal{F}_n(F), \mathcal{C}_n(F)\}$, *then* $\mathcal{G}(F)$ *is homeomorphic to the cone over a finite-dimensional continuum.*

The next theorem shows that for $n \geq 2$, no n-fold hyperspace of a finite-dimensional continuum is homeomorphic to its cone.

6.10.5 Theorem *Let* X *be a finite-dimensional continuum. Then for each integer* $n \geq 2$, $\mathcal{C}_n(X)$ *is not homeomorphic to* $K(X)$.

Proof Let $n \geq 2$ and suppose $\mathcal{C}_n(X)$ is homeomorphic to $K(X)$. Since X is of finite dimension, $K(X)$ is of finite dimension too. In fact $\dim(K(X)) = \dim(X) + 1$ [84, (8.0)]. Since $\mathcal{C}(X) \subset \mathcal{C}_n(X)$ and $\mathcal{C}_n(X)$ is homeomorphic to $K(X)$, we have that $\dim(\mathcal{C}(X)) < \infty$. Hence, by the dimension theorem, $\dim(X) = 1$ [56, Theorem 2.1]. Thus, $\dim(K(X)) = 2$, and $\dim(\mathcal{C}_n(X)) = 2$. Therefore, since $\mathcal{C}_n(X)$ contains an n-cell (Theorem 6.1.11), $n = 2$. We consider two cases.

Case (1) *X contains a proper decomposable subcontinuum.*
Then, by Theorem 6.1.12, $\mathcal{C}_2(X)$ contains a 3-cell, a contradiction to the fact that $\dim(\mathcal{C}_2(X)) = 2$.

Case (2) *All proper subcontinua of X are indecomposable.*
Then, by Lemma 1.7.29, X is hereditarily indecomposable. Hence, $K(X)$ is uniquely arcwise connected. On the other hand, since $\mathcal{C}_2(X)$ contains 2-cells (Theorem 6.1.11), $\mathcal{C}_2(X)$ is not uniquely arcwise connected.
Therefore, $\mathcal{C}_n(X)$ is not homeomorphic to $K(X)$.

$$\text{Q.E.D.}$$

6.10.6 Lemma *Let* X *be a continuum such that* $C(X)$ *is finite-dimensional. If* A *is a nondegenerate indecomposable proper subcontinuum of* X, *then at most a finite number of composants of* A *have the property that some subcontinuum of* X *contains a point of* $X \setminus A$ *and a point of the composant but does not contain* A. *Also,* $\mathcal{C}_n(X) \setminus \{A\}$ *has uncountably many arc components.*

Proof The first part follows by the proof of [84, (*), p. 312]. By the proof [84, (v), p. 312], $C(X) \setminus \{A\}$ has uncountably many arc components. Hence, $\mathcal{C}_n(X) \setminus \{A\}$ is not arcwise connected (Theorem 6.5.5) and has uncountably many arc components by [86, 11.15] and Theorem 6.5.13.

$$\text{Q.E.D.}$$

6.10.7 Theorem *Let* X *be a continuum and let* $n \geq 2$ *be an integer. If* Z *is a finite-dimensional continuum such that* $K(Z)$ *is homeomorphic to* $\mathcal{C}_n(X)$, *then* $\dim(X) =$

1 *and X contains at most one nondegenerate indecomposable continuum. Hence, X is not hereditarily indecomposable.*

Proof Let $n \geq 2$. Let $h: C_n(X) \to K(Z)$ be a homeomorphism. The proof of the fact that $\dim(X) = 1$ is similar to the one given in Theorem 6.10.5.

Suppose X contains two nondegenerate indecomposable continua, A and B. Then $C_n(X) \setminus \{A\}$ and $C_n(X) \setminus \{B\}$ have infinitely many arc components (Lemma 6.10.6). Hence, $K(Z) \setminus \{h(A)\}$ and $K(Z) \setminus \{h(B)\}$ both have infinitely many arc components. On the other hand, for each $p \in K(Z) \setminus \{v_Z\}$, $K(Z) \setminus \{p\}$ has at most two arc components. Therefore, X contains at most one nondegenerate indecomposable subcontinuum.

Q.E.D.

6.10.8 Theorem *Let X be a continuum and let $n \geq 2$. Then every 2-cell in $C_n(X)$ is nowhere dense.*

Proof First, suppose $n \geq 3$. Let $\mathcal{U}$ be any nonempty open set in $C_n(X)$. Then there exists $A \in \mathcal{U}$ such that A has exactly n components, $A_1, \ldots, A_n$ (Theorem 6.1.9). By Corollary 1.7.28, for each $j \in \{1, \ldots, n\}$, there exists a subcontinuum B_j of X such that B_j contains A_j properly, $B_j \cap B_\ell = \emptyset$ if $j \neq \ell$ and $B = \bigcup_{j=1}^{n} B_j \in \mathcal{U}$.

For each $j \in \{1, \ldots, n\}$, let $\alpha_j: [0, 1] \to \mathcal{C}(X)$ be an order arc such that $\alpha_j(0) = A_j$ and $\alpha_j(1) = B_j$ (Theorem 1.8.20). Define $\gamma((t_1, \ldots, t_n)) = \bigcup_{j=1}^{n} \alpha_j(t_j)$. Then $\gamma([0, 1]^n)$ is an n-cell contained in $\mathcal{U}$. Since $n \geq 3$, $\mathcal{U}$ cannot be a 2-cell.

Next, suppose $n = 2$. Assume that $\mathcal{D}$ is a 2-cell in $C_2(X)$ with nonempty interior in $C_2(X)$. Then there exists $D \in Int_{C_2(X)}(\mathcal{D})$ such that D has two nondegenerate components, D_1 and D_2 (Theorem 6.1.9 and Corollary 1.7.28). Let $\varepsilon > 0$ be such that if $E \in C_2(X)$ and $\mathcal{H}(E, D) < \varepsilon$, then $E \in Int_{C_2(X)}(\mathcal{D})$.

Let $\mu: \mathcal{C}(D_2) \to [0, 1]$ be a Whitney map. Let $\varphi: \mathcal{C}(D_2) \to C_2(X)$ be given by $\varphi(B) = D_1 \cup B$. Then φ is an embedding of $\mathcal{C}(D_2)$ into $\{G \in C_2(X) \mid G \subset D\}$. Let $t_0 \in (0, 1)$ be such that if $\mu(B) \geq t_0$, then $\mathcal{H}(\varphi(B), D) < \varepsilon$. Let $\mathcal{B} = \{\varphi(B) \mid t_0 < \mu(B) < 1\}$. Then $\mathcal{B}$ is a 2-dimensional subset of $Int_{C_2(X)}(\mathcal{D})$, $\dim(\mathcal{B}) = 2$ is seen using the fact that no zero-dimensional set separates $\mathcal{C}(D_2)$ by [84, (2.15)]. Hence, $Int_{C_2(X)}(\mathcal{B}) \neq \emptyset$ [87, 10.2]. Thus, letting $B_0 \in \mathcal{C}(D_2)$ be such that $\varphi(B_0) \in Int_{C_2(X)}(\mathcal{B})$, we have that $\varphi(B_0)$ is not arcwise accessible from $C_2(X) \setminus \mathcal{B}$. However, let $\beta: [0, 1] \to \mathcal{C}(D_1)$ be an order arc such that $\beta(0) \in \mathcal{F}_1(D_1)$ and $\beta(1) = D_1$ (Theorem 1.8.20). Then $\beta(s) \cup B_0 \notin \mathcal{B}$ for any $s \in [0, 1)$ and $\beta(1) \cup B_0 = D_1 \cup B_0 = \varphi(B_0) \in \mathcal{B}$, a contradiction.

Therefore, every 2-cell in $C_n(X)$ is nowhere dense.

Q.E.D.

The following theorem gives conditions on an indecomposable continuum X in order to have its n-fold hyperspaces homeomorphic to a cone over a finite-dimensional continuum.

6.10.9 Theorem *Let X be a continuum containing a nondegenerate indecomposable subcontinuum A. Let $n \geq 2$ be an integer, and let Z be a finite-dimensional* *continuum such that $K(Z)$ is homeomorphic to $C_n(X)$. If $h: C_n(X) \to K(Z)$ is a homeomorphism, then*

(1) $h(A) = v_Z$;
(2) Z has uncountably many arc components. In particular, Z is not locally connected;
(3) $\dim(C_n(X)) \geq 2n$ and $\dim(Z) \geq 2n - 1$;
(4) Each point z of Z is contained in an arc in Z and some points of Z belong to locally connected subcontinua of Z whose dimension is at least $2n - 1$;
(5) No point of $K(Z) \setminus \{v_Z\}$ arcwise disconnects $K(Z)$;
(6) If $A = X$, then X does not contain a nondegenerate proper terminal subcontinuum;
(7) Z is not irreducible. In particular, Z is decomposable.

Proof

(1) The proof is similar to the proof of Theorem 6.10.7.
(2) By Lemma 6.10.6, $C_n(X) \setminus \{A\}$ has uncountably many arc components. Since $h(A) = v_Z$ (by (1)), $K(Z) \setminus \{v_Z\}$ has uncountably many arc components. Thus, since $K(Z) \setminus \{v_Z\}$ is homeomorphic to $Z \times [0, 1)$, we conclude that Z has uncountably many arc components.
(3) By Theorem 6.10.7, each subcontinuum of X, distinct from A, is decomposable. Thus, by Corollary 6.1.15, $C_n(X)$ contains a $2n$-cell. Hence, $\dim(C_n(X)) \geq 2n$. Since $K(Z)$ is homeomorphic to $C_n(X)$ and $\dim(K(Z)) = \dim(Z) + 1$ [84, (8.0)], we have that $\dim(Z) \geq 2n - 1$.
(4) Let z be any point of Z. Let $\pi: K(Z) \setminus \{v_Z\} \twoheadrightarrow Z$ be the projection map. We consider two cases.

First, suppose there exists $t_0 \in [0, 1)$ such that $h^{-1}((z, t_0)) \in C_n(X) \setminus C(X)$. Let $B = h^{-1}((z, t_0))$ and let $B_1, \ldots, B_k$ be the components of B, where $k \in \{2, \ldots, n\}$. By Corollary 1.7.28, for each $j \in \{1, \ldots, k\}$, there exists a subcontinuum C_j of X containing B_j properly. We assume, without loss of generality, that $C_j \cap C_\ell = \emptyset$ if $j \neq \ell$.

For each $j \in \{1, \ldots, k\}$, let $\alpha_j: [0, 1] \to C(X)$ be an order arc (Theorem 1.8.20) such that $\alpha_j(0) = B_j$ and $\alpha_j(1) = C_j$. Let $\alpha: [0, 1]^k \to C_n(X)$ be given by $\alpha((t_1, \ldots, t_k)) = \bigcup_{j=1}^k \alpha_j(t_j)$. Let $\mathcal{D} = \alpha([0, 1]^k)$. Then $\mathcal{D}$ is a k-cell such that $B \in \mathcal{D}$ and $A \notin \mathcal{D}$. Thus, $h(\mathcal{D})$ is a k-cell containing the point (z, t_0) and not containing v_Z. Hence, $\pi(h(\mathcal{D}))$ is a locally connected subcontinuum of Z containing z. Since $k \geq 2$, $\pi(h(\mathcal{D}))$ is nondegenerate. Thus, z is contained in an arc by [86, 8.23].

Next, suppose that $h^{-1}((z, t)) \in C(X)$ for each $t \in [0, 1)$. Since $h(A) = v_Z$, there exists $t' \in [0, 1)$ such that $h^{-1}((z, t')) \neq A$ and $h^{-1}((z, t')) \notin \mathcal{F}_1(X)$. Let $E = h^{-1}((z, t'))$. Since $E \neq A$ and E is nondegenerate, E is a decomposable continuum (by Theorem 6.10.7). Hence, there exist two proper subcontinua K and H of E such that $E = H \cup K$.

Suppose, first, that A is not contained in E. Take $x_1 \in H \setminus K$ and $x_2 \in K \setminus H$. Let $\beta_j: [0, 1] \to C(X)$ be an order arc such that $\beta_j(0) = \{x_j\}$, $j \in \{1, 2\}$,

$\beta_1(1) = H$ and $\beta_2(1) = K$. Let $\beta \colon [0, 1]^2 \to C_n(X)$ be given by $\beta((t_1, t_2)) = \beta_1(t_1) \cup \beta_2(t_2)$. Let $\mathcal{G} = \beta\left([0, 1]^2\right)$. Then $\mathcal{G}$ is a locally connected subcontinuum of $C_n(X)$ such that $\mathcal{G}$ contains a 2-cell and such that $E \in \mathcal{G}$ and $A \notin \mathcal{G}$. Thus, $h(\mathcal{G})$ is a locally connected subcontinuum of $K(Z)$ containing a 2-cell, such that $(z, t') \in h(\mathcal{G})$ and $v_Z \notin h(\mathcal{G})$. Hence, $\pi(h(\mathcal{G}))$ is a nondegenerate locally connected subcontinuum of Z containing z. Thus, z is in an arc by [86, 8.23].

Suppose next A is contained in E. Since A is indecomposable, $E \neq A$. Hence, there exists a point $x_1 \in E \setminus A$. Suppose that $x_1 \in H$. Choose a point $x_2 \in K \setminus \{x_1\}$. Then we just repeat the argument in the preceding paragraph to construct a nondegenerate locally connected subcontinuum of Z containing z.

This completes the proof of the first part of (4). We prove the second part of (4) as follows:

By Corollary 6.1.15, there exists a $2n$-cell $\mathcal{E}$ in $C_n(X)$. We may choose $\mathcal{E}$ such that $A \notin \mathcal{E}$. Let $B \in \mathcal{E}$. Then $h(\mathcal{E})$ is a $2n$-cell such that $h(B) \in \mathcal{E}$. Hence, $\pi(h(\mathcal{E}))$ is a locally connected subcontinuum of Z containing $\pi(h(B))$, and $\dim(\pi(h(\mathcal{E}))) \geq 2n - 1$ (by [87, 20.10] since $\pi(h(\mathcal{E})) \times [0, 1)$ contains $h(\mathcal{E})$ and, thus, has dimension at least $2n$).

(5) By (4), each point of $K(Z)$ lies in the cone over an arc. Hence, (5) follows easily.

(6) This is a consequence of Theorem 6.10.7, Theorem 6.5.4 and part (5) of this theorem.

(7) Suppose there exist two points z_1 and z_2 of Z such that Z is irreducible between them.

First, we prove that both z_1 and z_2 belong to $\pi(h(\mathcal{F}_n(X)))$. Suppose this is not true. Note that $\mathcal{F}_n(X)$ intersects all the arc components of $C_n(X) \setminus \{A\}$. Hence, there exist arcs α_1 and α_2 in Z such that one end point of α_j is z_j and the other point of α_j is in $\pi(h(\mathcal{F}_n(X)))$, for $j \in \{1, 2\}$. Hence, by irreducibility, $Z = \alpha_1 \cup \alpha_2 \cup \pi(h(\mathcal{F}_n(X)))$. On the other hand, Z cannot contain free arcs (otherwise, $C_n(X)$ contains 2-cells with nonempty interior, which contradicts Theorem 6.10.8). Thus, we have proved that z_1 and z_2 belong to $\pi(h(\mathcal{F}_n(X)))$.

Let t_1 and t_2 be points of $[0, 1)$ such that (z_1, t_1) and (z_2, t_2) belong to $h(\mathcal{F}_n(X))$. Let B_1 and B_2 be the elements of $\mathcal{F}_n(X)$ such that $h(B_1) = (z_1, t_1)$ and $h(B_2) = (z_2, t_2)$. Take $x_1 \in B_1$ and $x_2 \in B_2$. Let

$$\mathcal{B}_1 = \{\{x_1\} \cup B \mid B \in \mathcal{F}_{n-1}(X)\}$$

and

$$\mathcal{B}_2 = \{\{x_2\} \cup B \mid B \in \mathcal{F}_{n-1}(X)\}.$$

Then $\mathcal{B}_1$ and $\mathcal{B}_2$ are subcontinua of $C_n(X)$ containing B_1 and B_2, respectively; also, $\mathcal{B}_1 \cap \mathcal{F}_{n-1}(X) \neq \emptyset$ and $\mathcal{B}_2 \cap \mathcal{F}_{n-1}(X) \neq \emptyset$. Hence, $\mathcal{B}_1 \cup \mathcal{B}_2 \cup \mathcal{F}_{n-1}(X)$ is a subcontinuum of $C_n(X)$ which does not intersect all the arc components of $C_n(X) \setminus \{A\}$, which we prove as follows: By Lemma 6.10.6, $C_n(X) \setminus \{A\}$ has uncountably many arc components. Let $a_1, \ldots, a_n$ be n points of A in n distinct composants, $\kappa_1, \ldots, \kappa_n$, of A such that $\{x_1, x_2\} \cap \bigcup_{j=1}^{n} \kappa_j = \emptyset$; if $A \neq X$, by Lemma 6.10.6, we may take n composants that are not accessible from $X \setminus A$. Let $\mathcal{G}$ be the arc component of $C_n(X) \setminus \{A\}$ containing $\{a_1, \ldots, a_n\}$. Then $\mathcal{G} \cap (\mathcal{B}_1 \cup \mathcal{B}_2 \cup \mathcal{F}_{n-1}(X)) = \emptyset$.

Since $\mathcal{B}_1 \cup \mathcal{B}_2 \cup \mathcal{F}_{n-1}(X)$ does not intersect all the arc components of $C_n(X) \setminus \{A\}$, we have that $h(\mathcal{B}_1 \cup \mathcal{B}_2 \cup \mathcal{F}_{n-1}(X))$ is a subcontinuum of $K(Z)$ containing both (z_1, t_1) and (z_2, t_2), which does not intersect all the arc components of $K(Z) \setminus \{v_Z\}$. Then $\pi(h(\mathcal{B}_1 \cup \mathcal{B}_2 \cup \mathcal{F}_{n-1}(X)))$ is a proper subcontinuum of Z containing z_1 and z_2, a contradiction. Therefore, Z is not irreducible.

Q.E.D.

Now we turn our attention to suspensions.

6.10.10 Theorem *If X is a finite-dimensional continuum, then $C_n(X)$ is not homeomorphic to $\Sigma(X)$, for any integer $n \geq 2$.*

Proof Let $n \geq 2$. Suppose $C_n(X)$ is homeomorphic to $\Sigma(X)$. With an argument similar to the one given for the proof of Theorem 6.10.5, we obtain that $n = 2$ and no nondegenerate proper subcontinuum of X is decomposable.

Assume that each proper subcontinuum of X is indecomposable. Then, by Lemma 1.7.29, X is hereditarily indecomposable. It is well known that $\Sigma(X)$ is not arcwise disconnected by any of its points. On the other hand, since X is hereditarily indecomposable, for each $A \in \mathcal{C}(X) \setminus \mathcal{F}_1(X)$, $C_2(X) \setminus \{A\}$ is not arcwise connected, by Theorem 6.5.8, a contradiction. Therefore, $C_n(X)$ is not homeomorphic to $\Sigma(X)$.

Q.E.D.

Let us recall that R. Schori proved that $C_2([0, 1])$ is homeomorphic to $[0, 1]^4$ (Theorem 6.9.12). Note that $[0, 1]^4$ is homeomorphic to $\Sigma([0, 1]^3)$. In connection with this we have the following:

6.10.11 Theorem *Let X be a continuum and let $n \geq 2$ be an integer. If Z is a finite-dimensional continuum such that $\Sigma(Z)$ is homeomorphic to $C_n(X)$, then X is hereditarily decomposable, and X does not contain nondegenerate proper terminal subcontinua. Also, Z is arcwise connected.*

Proof Let $h \colon C_n(X) \twoheadrightarrow \Sigma(Z)$ be a homeomorphism. Suppose X contains a nondegenerate indecomposable subcontinuum A. Since Z is finite-dimensional, with a similar argument to the one given in the proof of Theorem 6.10.5, we have that $\mathcal{C}(X)$ is finite-dimensional. Then $C_n(X) \setminus \{A\}$ is not arcwise connected, by Theorem 6.5.17. Hence, $\Sigma(Z) \setminus \{h(A)\}$ is not arcwise connected. A contradiction to the fact that $\Sigma(Z)$ is not arcwise disconnected by any of its points. Therefore, X is hereditarily decomposable.

Now, suppose X contains a nondegenerate proper terminal subcontinuum B. Then $C_n(X) \setminus \{B\}$ is not arcwise connected, by Theorem 6.5.4. Hence, like in the previous paragraph, we obtain a contradiction. Therefore, X does not contain nondegenerate proper terminal subcontinua.

To show that Z is arcwise connected, it is enough to prove that $\Sigma(Z) \setminus \{v^+, v^-\}$ is arcwise connected, where v^+ and v^- are the vertexes of $\Sigma(Z)$.

Let $A_1 = h^{-1}(v^+)$ and let $A_2 = h^{-1}(v^-)$. We prove that $C_n(X) \setminus \{A_1, A_2\}$ is arcwise connected. Since X does not contain nondegenerate proper terminal subcontinua, $C_n(X) \setminus \{A_j\}$, $j \in \{1, 2\}$, is arcwise connected, by Theorems 6.5.2 and 6.5.4. We consider three cases:

Case (1) $A_1, A_2 \in \mathcal{C}(X)$.
Observe that, by Theorem 6.5.4, $\mathcal{C}(X) \setminus \{A_j\}$ is arcwise connected, $j \in \{1, 2\}$. Hence, $\mathcal{C}(X) \setminus \{A_1, A_2\}$ is arcwise connected [84, (9.2)]. Therefore, $C_n(X) \setminus \{A_1, A_2\}$ is arcwise connected.

Case (2) $A_1, A_2 \in C_n(X) \setminus \mathcal{C}(X)$.
Given $B \in C_n(X) \setminus \{A_1, A_2\}$, it is easy to construct an arc from B to X in $C_n(X) \setminus \{A_1, A_2\}$. Therefore, $C_n(X) \setminus \{A_1, A_1\}$ is arcwise connected.

Case (3) $A_1 \in C_n(X) \setminus \mathcal{C}(X)$ and $A_2 \in \mathcal{C}(X)$.
Suppose $A_2 \neq X$. Let $B \in C_n(X) \setminus \{A_1, A_2\}$. Since X is decomposable and does not contain nondegenerate proper terminal subcontinua, $\mathcal{C}(X) \setminus \{A_2\}$ is arcwise connected, by Theorems 6.5.3 and 6.5.4. Now, it is easy to construct an arc from B to X. Hence, in this case, $C_n(X) \setminus \{A_1, A_2\}$ is arcwise connected.

Assume that $A_2 = X$. Let $B_0, B_1 \in C_n(X) \setminus \{A_1, X\}$. Since X is decomposable, $C_n(X) \setminus \{X\}$ is arcwise connected, by Theorem 6.5.3. Thus, there exists an arc $\alpha \colon [0, 1] \to C_n(X) \setminus \{X\}$ such that $\alpha(0) = B_0$ and $\alpha(1) = B_1$. Suppose that $A_1 \in \alpha([0, 1])$. Let k be the number of components of A_1. Note that $k \geq 2$. Then there exists a k-cell, $\mathcal{K}$, in $C_n(X) \setminus \mathcal{C}(X)$ such that $A_1 \in \mathcal{K}$, (see the proof of Theorem 6.1.11). Now it is easy to find an arc $\beta \colon [0, 1] \to C_n(X) \setminus \{A_1, X\}$ such that $\beta(0) = B_0$ and $\beta(1) = B_1$. Therefore, $C_n(X) \setminus \{A_1, X\}$ is arcwise connected. Therefore, Z is arcwise connected.

Q.E.D.

Next, we compare n-fold hyperspaces with products of continua. We start with a couple of definitions.

6.10.12 Definition A continuum X is *acyclic* if $\check{H}^1(X, \mathbb{Z}) = 0$; i.e., the first Čech cohomology group with integer coefficients is trivial.

6.10.13 Definition A continuum X has *property* (b) provided that each map $f \colon X \to S^1$ is homotopic to a constant map, where S^1 is the unit circle in the plane.

6.10.14 Theorem *Let X be an acyclic continuum. If X is homeomorphic to the product of two nondegenerate continua Y and Z, then Y and Z are acyclic.*

Proof Since X is acyclic, X has property (b) [22, Theorem 8.1]. Also, since the projection maps π_Y and π_Z are monotone, Y and Z both have property (b) [55, Theorem 2, p. 434]. Therefore, Y and Z are both acyclic [22, Theorem 8.1].

Q.E.D.

6.10.15 Theorem *Let X be a continuum and let $n \geq 2$. If $C_n(X)$ is homeomorphic to $Y \times Z$, where Y and Z are nondegenerate finite-dimensional continua, then X is hereditarily decomposable and X has no nondegenerate proper terminal continua. Also, Y and Z are arcwise connected and acyclic.*

Proof Suppose that $C_n(X)$ is homeomorphic to $Y \times Z$, where Y and Z are finite-dimensional continua. Hence, by [39, Theorem III 4, p. 33], $C_n(X)$ is finite-dimensional. Since $C_n(X)$ is arcwise connected (Corollary 1.8.12), we have that both Y and Z are arcwise connected continua. By the proof of Theorem 6.2.5, $C_n(X)$ is acyclic. Thus, by Theorem 6.10.14, Y and Z are acyclic.

Suppose X contains an indecomposable subcontinuum A. Note that $C_n(X) \setminus \{A\}$ is not arcwise connected, by Theorem 6.5.17. On the other hand, it is easy to show that no point arcwise disconnects the product of two arcwise connected continua, a contradiction.

The fact that X does not contain terminal subcontinua follows from Theorem 6.5.4 and the fact that no point arcwise disconnects the products of two arcwise connected continua.

Q.E.D.

6.10.16 Lemma *Let X be a dendroid and let n be an integer greater than one. Then $C_n(X)$ is finite-dimensional if and only if X is a graph.*

Proof If $C_n(X)$ is finite-dimensional, then $C(X)$ is finite-dimensional. Hence, X is a graph [85, (2.6)]. If X is a graph, then $C_n(X)$ is finite-dimensional by Theorem 6.9.3.

Q.E.D.

6.10.17 Theorem *Let X and Z be finite-dimensional continua and let n be an integer greater than one. If $C_n(X)$ is homeomorphic to $X \times Z$, then X is a tree.*

Proof Since $C_n(X)$ is arcwise connected (Corollary 1.8.12), X and Z are arcwise connected. By the proof of Theorem 6.2.5, $C_n(X)$ is acyclic. Hence, X and Z are acyclic, by Theorem 6.10.14. Also, X is hereditarily decomposable, by Theorem 6.10.15. Since X is finite-dimensional, X is a dendroid [85, (1.2)]. Since $X \times Z$ is finite-dimensional, $C_n(X)$ is finite-dimensional. Thus, X is a graph, by Lemma 6.10.16. Therefore, since X is acyclic, X is a tree.

Q.E.D.

A proof of the next theorem may be found in [69, Theorem 4.9].

6.10.18 Theorem $C_2(\mathcal{S}^1)$ *is not homeomorphic to a product $Y \times D$ for any one-dimensional continuum D and any continuum Y.*

6.10.19 Corollary $C_2(\mathcal{S}^1)$ *is not homeomorphic to a product $Z \times [0, 1]^k$ for any continuum Z and any positive integer k.*

Proof Apply Theorem 6.10.18 with $Y = Z \times [0, 1]^{k-1}$.

Q.E.D.

6.11 Strong Size Maps

We study strong size maps. These maps are a nice generalization of Whitney maps to n-fold hyperspaces. Strong size maps, like Whitney maps restricted to the hyperspace of subcontinua, are open (Theorem 6.11.16) and monotone (Theorem 6.11.18). We follow [22, 29, 38, 42, 43, 58, 68, 74–76, 82, 88–90, 92–96].

6.11.1 Definition Let X be a continuum and let $n \in \mathbb{N}$. A map $\mu \colon C_n(X) \to [0, 1]$ is a *strong size map* provided that

(1) $\mu(A) = 0$ if $A \in \mathcal{F}_n(X)$;
(2) if $A \subset B$, $A \neq B$ and $B \notin \mathcal{F}_n(X)$, then $\mu(A) < \mu(B)$
and
(3) $\mu(X) = 1$.

First, we show that strong size maps exist. To this end, we need the following:

6.11.2 Definition A *partially ordered space* is a topological space Z endowed with a partial order $\leq$ whose graph is a closed subset of $Z \times Z$.

6.11.3 Notation If Z is a partially ordered space and $z \in Z$, we write $L(z) = \{p \in Z \mid p \leq z\}$ and $M(z) = \{p \in Z \mid z \leq p\}$, and if $A \subset Z$, then $L(A) = \bigcup\{L(a) \mid a \in A\}$ and $M(A) = \bigcup\{M(a) \mid a \in A\}$. An element m of the partially ordered space Z is *minimal (maximal)* if, whenever $z \in Z$ and $z \leq m$ ($m \leq z$), it follows that $m = z$. The set of minimal elements of Z is denoted by $Min(Z)$ and the set of maximal elements of Z is denoted by $Max(Z)$.

The following three theorems are Theorems 2.2, 2.3 and Lemma 3.2 of [94]:

6.11.4 Theorem *If K is a compact subset of a partially ordered space, then $L(K)$ and $M(K)$ are closed sets.*

6.11.5 Theorem *If x and y are elements of a compact partially ordered space and if $M(x) \cap L(y) = \emptyset$, then there exist disjoint open sets U and V such that $x \in U = M(U)$ and $y \in V = L(V)$.*

6.11.6 Theorem *Suppose Z is a compact partially ordered space such that $Min(Z)$ and $Max(Z)$ are disjoint closed sets, Q is a closed subset containing $(Min(P)) \cup (Max(P))$, and suppose A and B are disjoint nonempty closed subsets such that $A = M(A)$ and $B = L(B)$. If $f \colon Q \to [0, 1]$ is a continuous order-preserving function such that $f(Min(Z)) = \{0\}$ and $f(Max(P)) = \{1\}$, then f admits a continuous order-preserving extension $\hat{f} \colon Z \to [0, 1]$ such that $\hat{f}(a) \geq \inf f(A \cap Q)$ for each $a \in A$ and $\hat{f}(b) \leq \sup f(B \cap Q)$ for each $b \in B$.*

The proof of the following result may be found in [93, Theorem].

6.11.7 Theorem *If Z is a compact metric partially ordered space such that Min(Z) and Max(Z) are disjoint closed sets, then there exists a map $\mu \colon Z \to [0, 1]$ such that $\mu(x) = 0$ for $x \in Min(Z)$, $\mu(x) = 1$ for $x \in Max(Z)$ and $\mu(x) < \mu(y)$ if $x < y$.*

6.11.8 Remark For $A, B \in C_n(X)$, define $A < B$ if $A \subset B$, $A \neq B$ and $B \notin \mathcal{F}_n(X)$. We denote $A \leq B$ if $A < B$ or $A = B$. Then $C_n(X)$ is a partially ordered space with respect to this order. Note that $Min(C_n(X)) = \mathcal{F}_n(X)$ and $Max(C_n(X)) = \{X\}$. Also these sets are closed and disjoint.

As a consequence of Theorem 6.11.7, we have:

6.11.9 Theorem *Let X be a continuum and let $n \in \mathbb{N}$. Then there exists a strong size map for $C_n(X)$.*

We can do better, we present two constructions of strong size maps. The first one is a modification of [95, pp. 275–276].

6.11.10 Example Let X be a continuum with a bounded metric d, bounded by 1, let $n \in \mathbb{N}$ and let $A \in C_n(X)$. For each integer $m \geq 2$ and $K \in \mathcal{F}_m(A)$, where $K = \{k_1, \ldots, k_m\}$, it is possible that $k_j = k_l$ for $j \neq l$. Let $\omega_m(K) = \min\{d(k_j, k_l) \mid j \neq l\}$. Define $\mu_m(A) = \sup\{\omega_m(K) \mid K \in \mathcal{F}_m(A)\}$. Then the function $\mu \colon C_n(X) \to [0, 1]$ given by $\mu(A) = \Sigma_{j=n+1}^{\infty} \frac{1}{2^j} \mu_j(A)$ is a strong size map. To see this, note that μ is continuous. Observe that if $A \in \mathcal{F}_n(X)$, then, clearly, $\mu_m(A) = 0$ for all $m > n$. This implies that $\mu(A) = 0$. Now, if $A \in C_n(X) \setminus \mathcal{F}_n(X)$, then A has more than $n + 1$ elements. Hence, $\mu_{n+1}(A) > 0$, and $\mu(A) > 0$. Now, suppose A and B are two elements of $C_n(X)$ such that $A \subsetneq B$ and $B \in C_n(X) \setminus \mathcal{F}_n(X)$. The proof of the fact that $\mu(A) < \mu(B)$ is similar to the argument given in [95, pp. 275–276].

The next example is modification of an example that appears in [42].

6.11.11 Example Let X be a continuum and let $n \in \mathbb{N}$. For each $A \in C_n(X)$, let

$$\mu_m(A) = \inf\left\{\varepsilon > 0 \,\middle|\, \text{there exist } p_1, \ldots, p_m \in X\right.$$

$$\left.\text{such that } A \subset \bigcup_{j=1}^{m} \mathcal{V}_\varepsilon(p_j)\right\}.$$

Then $\mu \colon C_n(X) \to [0, 1]$ given by $\mu(A) = \sum_{m=n}^{\infty} \frac{1}{2^m} \mu_m(A)$ is a strong size map. To show this, note μ is continuous. Observe that if $A \in \mathcal{F}_n(X)$, then, clearly, $\mu_m(A) = 0$ for all $m > n$. This implies that $\mu(A) = 0$. If $A \in C_n(X) \setminus \mathcal{F}_n(X)$, then A has more than n elements. Thus, $\mu_n(A) > 0$, and $\mu(A) > 0$. Now, suppose A and B are two elements of $C_n(X)$ such that $A \subsetneq B$ and $B \in C_n(X) \setminus \mathcal{F}_n(X)$. Without loss of generality, we assume that $A \in C_n(X) \setminus \mathcal{F}_n(X)$ also. Let $b \in B \setminus A$ and let $\varepsilon > 0$ be such that $d(b, A) > 2\varepsilon$ and we assume ε is small enough that we need more than n

points of X to cover A with open balls of radius ε. Let

$$\ell = \min \left\{ m \geq n \, \Big| \, \text{there exist } x_1, \ldots, x_m \in X \right.$$

$$\left. \text{such that } A \subset \bigcup_{j=1}^{m} V_\varepsilon(x_j) \right\}.$$

Since A is compact, there exists $\delta > 0$ such that $\delta < \varepsilon$ and $A \subset \bigcup_{j=1}^{m} V_\delta(x_j)$. This implies that $\mu_\ell(A) < \varepsilon$. We claim that $\mu_\ell(B) \geq \varepsilon$. Suppose this is not true and assume that $\mu_\ell(B) < \varepsilon$. Then there exist $b_1, \ldots, b_\ell \in X$ such that $B \subset \bigcup_{j=1}^{\ell} V_\varepsilon(b_j)$. Note that $b \in \bigcup_{j=1}^{\ell} V_\varepsilon(b_j)$. Without loss of generality, we assume that $b \in V_\varepsilon(b_\ell)$. Then $A \subset \bigcup_{j=1}^{\ell-1} V_\varepsilon(b_j)$, a contradiction to the election of ℓ. Thus, $\mu_\ell(B) \geq \varepsilon$. Therefore, $\mu_\ell(A) < \mu_\ell(B)$.

In the next theorem we show if $\mathfrak{C}$ is a nonempty closed subset of the n-fold hyperspace of a continuum X and $\mu \colon \mathfrak{C} \to [0, 1]$ is a strong size map, then μ can be extended to a strong size map defined on $C_n(X)$. This implies, in particular, that Whitney maps defined on the hyperspace of subcontinua of a continuum can be extended to a strong size map on the n-fold hyperspace of that continuum.

6.11.12 Theorem *Let X be a continuum and let $n \in \mathbb{N}$. If $\mathfrak{C}$ is a nonempty closed subset of $C_n(X)$ and $\mu \colon \mathfrak{C} \to [0, 1]$ is a strong size map, then μ can be extended to a strong size map μ_n defined on $C_n(X)$.*

Proof Let $\mathfrak{C}$ be a nonempty closed subset of $C_n(X)$ and let $\mu \colon \mathfrak{C} \to [0, 1]$ be a strong size map. Without loss of generality we assume that $\mathcal{F}_n(X) \cup \{X\} \subset \mathfrak{C}$ (If this is not true, let $\mathfrak{K} = \mathfrak{C} \cup \mathcal{F}_n(X) \cup \{X\}$ and note that μ can be extended to a strong size map μ' on $\mathfrak{K}$ by defining $\mu'(X) = 1$ and $\mu'(A) = 0$ for each $A \in \mathcal{F}_n(X)$).

Let $\mathfrak{U}$ be a countable base for $C_n(X)$ and let

$$\mathfrak{B} = \{(\mathcal{U}, \mathcal{V}) \mid M(Cl(\mathcal{U})) \cap L(Cl(\mathcal{V})) = \emptyset \text{ and } \mathcal{U}, \mathcal{V} \in \mathfrak{U}\}.$$

Then $\mathfrak{B}$ is countable and we may enumerate its elements $\mathfrak{B} = \{(\mathcal{U}_k, \mathcal{V}_k) \mid k$ is a positive integer$\}$. By Theorem 6.11.4 the sets $M(Cl(\mathcal{U}))$ and $L(Cl(\mathcal{V}))$ are closed. Hence, by Theorem 6.11.6, for each positive integer k, there exists a continuous order-preserving function $\omega_k \colon C_n(X) \to [0, 1]$ such that $\omega_k|_{\mathfrak{C}} = \mu$ and:

$$\omega_k(A) \geq \inf \mu(M(Cl(\mathcal{U}_k)) \cap \mathfrak{C}) \text{ if } A \in (M(Cl(\mathcal{U}_k))),$$

$$\omega_k(B) \leq \max \mu(L(Cl(\mathcal{V}_k)) \cap \mathfrak{C}) \text{ if } B \in (L(Cl(\mathcal{V}_k))).$$

Define $\mu_n \colon C_n(X) \to [0, 1]$ by $\mu_n(A) = \sum_{k=1}^{\infty} \frac{1}{2^k} \omega_k(A)$ for all $A \in C_n(X)$. Observe that μ_n is a continuous extension of μ. Since each ω_k is order-preserving, μ_n is also order preserving.

We need to show that if A and B are elements of $C_n(X)$ and $A < B$ (Rermark 6.11.8), then $\mu_n(A) < \mu_n(B)$. It suffices to prove that there exists a positive integer k such that $\omega_k(A) < \omega_k(B)$.

Let $t_A = \sup \mu(L(A) \cap \mathfrak{C})$ and let $t_B = \inf \mu(M(B) \cap \mathfrak{C})$. Since μ is a strong size map, $t_A < t_B$. Let $\varepsilon > 0$ be such that $\varepsilon < \frac{1}{2}(t_B - t_A)$. By Theorem 6.11.5, there exist two disjoint open subsets $\mathcal{U}$ and $\mathcal{V}$ of $C_n(X)$ such that $A \in \mathcal{V} = L(\mathcal{V})$ and $B \in \mathcal{U} = M(\mathcal{U})$ and, by compactness, we may assume that $\mu(\mathcal{V} \cap \mathfrak{C}) \subset [0, t_A + \varepsilon)$ and $\mu(\mathcal{U} \cap \mathfrak{C}) \subset (t_B - \varepsilon, 1]$. It follows that there is a positive integer k such that $A \in \mathcal{V}_k \subset Cl(\mathcal{V}_k) \subset \mathcal{V}$ and $B \in \mathcal{U}_k \subset Cl(\mathcal{U}_k) \subset \mathcal{U}$, from here we obtain:

$$\omega_k(A) \le t_A + \varepsilon < t_B - \varepsilon \le \omega_k(B).$$

Q.E.D.

6.11.13 Corollary *Let X be a continuum. If $\mu \colon C(X) \to [0, 1]$ is a Whitney map, μ can be extended to a strong size map, μ_n, defined on $C_n(X)$.*

Some consequences of the definition of strong size map are:

6.11.14 Lemma *Let X be a continuum and let $n \in \mathbb{N}$. If $\alpha \colon [0, 1] \to C_n(X)$ is an order arc and $\mu \colon C_n(X) \twoheadrightarrow [0, 1]$ is a strong size map, then $\mu(\alpha(s)) < \mu(\alpha(t))$ for each $0 \le s < t \le 1$.*

6.11.15 Lemma *Let X be a continuum and let $n \in \mathbb{N}$. If $\alpha \colon [0, 1] \to C_n(X)$ is an order arc and $\mu \colon C_n(X) \twoheadrightarrow [0, 1]$ is a strong size map, then the map $\varphi \colon [0, 1] \twoheadrightarrow [0, 1]$ given by $\varphi(t) = \mu(\alpha(t))$ is a homeomorphism.*

6.11.16 Theorem *Let X be a continuum and let $n \in \mathbb{N}$. If $\mu \colon C_n(X) \twoheadrightarrow [0, 1]$ is a strong size map, then μ is open.*

Proof Let $\mathcal{U}$ be an open subset of $C_n(X)$, let $A \in \mathcal{U}$ and let $\alpha \colon [0, 1] \to C_n(X)$ be an order arc from an element of $\mathcal{F}_n(X)$ to X passing through A, Theorem 1.8.20. By Lemma 6.11.15, the map $\varphi \colon [0, 1] \twoheadrightarrow [0, 1]$ given by $\varphi(t) = \mu(\alpha(t))$ is a homeomorphism. Since α is continuous, $\alpha^{-1}(\mathcal{U})$ is an open subset of $[0, 1]$ containing $\alpha^{-1}(A)$. Hence, $\varphi(\alpha^{-1}(\mathcal{U})) \subset \mu(\mathcal{U})$ is an open subset of $[0, 1]$ containing $\mu(A)$. Therefore, μ is an open map.

Q.E.D.

6.11.17 Remark Let X be a continuum and let $n \in \mathbb{N}$. It is known that Whitney levels for Whitney maps defined on $C_n(X)$ are not necessarily connected.

6.11.18 Theorem *Let X be a continuum and let $n \in \mathbb{N}$. If $\mu \colon C_n(X) \twoheadrightarrow [0, 1]$ is a strong size map, then μ is monotone.*

Proof Note that, by definition, $\mu^{-1}(0) = \mathcal{F}_n(X)$. By Corollary 1.8.8, $\mathcal{F}_n(X)$ is a continuum. Hence, $\mu^{-1}(0)$ is connected. Using order arcs, we see that if $0 <$

$t < 1$, $\mu^{-1}([0, t])$ and $\mu^{-1}([t, 1])$ are connected. Thus, since $C_n(X)$ is unicoherent, Theorem 6.2.5, $\mu^{-1}(t) = \mu^{-1}([0, t]) \cap \mu^{-1}([t, 1])$ is connected. Therefore, μ is monotone.

Q.E.D.

6.11.19 Definition Let X be a continuum and let $n \in \mathbb{N}$. If $\mu \colon C_n(X) \twoheadrightarrow [0, 1]$ is a strong size map, then each set of the form $\mu^{-1}(t)$, for $t \in [0, 1)$, is a *strong size level* for $C_n(X)$.

6.11.20 Definition Let X be a continuum and let $\mathcal{S}$ be a subset of $C_n(X) \setminus \mathcal{F}_n(X)$. Then $\mathcal{S}$ is an *antichain* if given two elements S and S' of $\mathcal{S}$ such that $S \subset S'$, then $S = S'$.

The following theorem gives us a characterization of strong size levels.

6.11.21 Theorem *Let X be a nondegenerate continuum. Then a nonempty closed subset $\mathcal{S}$ of $C_n(X)$ is a strong size level if and only if either $\mathcal{S} = \mathcal{F}_n(X)$ or $\mathcal{S} \cap \mathcal{F}_n(X) = \emptyset$, $\mathcal{S}$ is an antichain and every order arc from an element of $\mathcal{F}_n(X)$ to X intersects $\mathcal{S}$.*

Proof If $\mathcal{S}$ is a strong size level, it is clear that either $\mathcal{S} = \mathcal{F}_n(X)$ or $\mathcal{S} \cap \mathcal{F}_n(X) = \emptyset$, $\mathcal{S}$ is an antichain and every order arc from an element of $\mathcal{F}_n(X)$ to X intersects $\mathcal{S}$.

Suppose that $\mathcal{S} \cap \mathcal{F}_n(X) = \emptyset$, $\mathcal{S}$ is an antichain and every order arc from an element of $\mathcal{F}_n(X)$ to X intersects $\mathcal{S}$. Define $\mu' \colon \mathcal{S} \to [0, 1]$, by $\mu'(S) = \frac{1}{2}$ for each $S \in \mathcal{S}$. Observe that μ' is a strong size map for $\mathcal{S}$. Hence, by Theorem 6.11.12, there exists a strong size map $\mu \colon C_n(X) \twoheadrightarrow [0, 1]$ such that $\mu|_{\mathcal{S}} = \mu'$. By construction, $\mathcal{S} \subset \mu^{-1}(\frac{1}{2})$. To show that $\mu^{-1}(\frac{1}{2}) \subset \mathcal{S}$, let $A \in \mu^{-1}(\frac{1}{2})$ and let $\alpha_A \colon [0, 1] \to C_n(X)$ be an order arc from some point of $\mathcal{F}_n(X)$ to X passing through A (Theorem 1.8.20), by hypothesis, $\alpha_A([0, 1]) \cap \mathcal{S} \neq \emptyset$, let $A' \in \alpha_A([0, 1]) \cap \mathcal{S}$. Since μ is a strong size map, $A' \in \mu^{-1}(\frac{1}{2})$, and either $A' \subset A$ or $A \subset A'$, we have that $A = A'$. Therefore, $\mathcal{S} = \mu^{-1}(\frac{1}{2})$.

If $\mathcal{S} = \mathcal{F}_n(X)$, then $\mathcal{S} = \mu^{-1}(0)$ for each strong size map μ.

Q.E.D.

Recall that in Theorem 6.1.11 we prove that n-fold hyperspaces always contain n-cells. In the next theorem, we prove that a strong size level for the n-fold hyperspace always contains an $(n - 1)$-cell.

6.11.22 Theorem *Let X be a continuum, let $n \geq 2$, let $\mu \colon C_n(X) \twoheadrightarrow [0, 1]$ be a strong size map and let $t \in (0, 1)$. If $A \in (C_n(X) \setminus C_{n-1}(X)) \cap \mu^{-1}(t)$, then there exist an $(n - 1)$-cell $\mathcal{A}$ contained in $\mu^{-1}(t)$ such that $A \in \mathcal{A}$.*

Proof Let $A \in (C_n(X) \setminus C_{n-1}(X)) \cap \mu^{-1}(t)$. Let $A_1, \ldots, A_n$ be the components of A. For each $j \in \{1, \ldots, n\}$, let a_j be a point of A_j, and let $\alpha_j \colon [0, 1] \to C(X)$ be an order arc such that $\alpha_j(0) = \{a_j\}$, $\alpha_j(1) = X$ and $\alpha_j(\frac{1}{2}) = A_j$. Define the function $\xi \colon [0, 1]^n \to C_n(X)$ by $\xi((r_1, \ldots, r_n)) = \bigcup_{j=1}^{n} \alpha_j(r_j)$. Then ξ is well defined and continuous. Since $A \in C_n(X) \setminus C_{n-1}(X)$ and ξ is continuous, there exists an $\varepsilon > 0$

such that $\xi\left(\prod_{j=1}^{n}\left[\frac{1}{2}-\varepsilon,\frac{1}{2}+\varepsilon\right]_{j}\right) \subset C_n(X) \setminus C_{n-1}(X)$, where $\left[\frac{1}{2}-\varepsilon,\frac{1}{2}+\varepsilon\right]_{j}$ is a copy of $\left[\frac{1}{2}-\varepsilon,\frac{1}{2}+\varepsilon\right]$.

Note that $A = \xi((\frac{1}{2},\ldots,\frac{1}{2}))$ and $\mu\left(\xi((\frac{1}{2},\ldots,\frac{1}{2}+\varepsilon))\right) > t$. By the continuity of α and μ, there exists a $\delta > 0$ such that $\delta < \varepsilon$ and

$$\mu\left(\xi\left(\left(s_1,\ldots,s_{n-1},\frac{1}{2}+\varepsilon\right)\right)\right) > t$$

for all $(s_1,\ldots,s_{n-1}) \in \prod_{j=1}^{n-1}\left[\frac{1}{2}-\delta,\frac{1}{2}+\delta\right]_{j}$.

Define $\theta\colon \prod_{j=1}^{n-1}\left[\frac{1}{2}-\delta,\frac{1}{2}\right]_{j} \to \left[0,\frac{1}{2}+\varepsilon\right]$ as follows: For each point $(s_1,\ldots,s_{n-1})$ of $\prod_{j=1}^{n-1}\left[\frac{1}{2}-\delta,\frac{1}{2}\right]_{j}$, let $\theta((s_1,\ldots,s_{n-1}))$ be the unique element of $\left[0,\frac{1}{2}+\varepsilon\right]$ such that

$$\mu(\xi((s_1,\ldots,s_{n-1}),\theta((s_1,\ldots,s_{n-1})))) = t.$$

To prove that θ is a continuous function, let $\{(s_1^k,\ldots,s_{n-1}^k)\}_{k=1}^{\infty}$ be a sequence of elements of $\prod_{j=1}^{n-1}\left[\frac{1}{2}-\delta,\frac{1}{2}\right]_{j}$ converging to $(s_1,\ldots,s_{n-1}) \in \prod_{j=1}^{n-1}\left[\frac{1}{2}-\delta,\frac{1}{2}\right]_{j}$. Consider the sequence $\{\theta((s_1^k,\ldots,s_{n-1}^k))\}_{k=1}^{\infty}$ of elements of $\left[0,\frac{1}{2}+\varepsilon\right]$. Since this set is compact, $\{\theta((s_1^k,\ldots,s_{n-1}^k))\}_{k=1}^{\infty}$ has a convergent subsequence $\{\theta((s_1^{k_\ell},\ldots,s_{n-1}^{k_\ell}))\}_{\ell=1}^{\infty}$ converging to $s \in \left[0,\frac{1}{2}+\varepsilon\right]$. Then, since μ and ξ are continuous, we have:

$$\lim_{\ell\to\infty} \mu\left(\xi\left((s_1^{k_\ell},\ldots,s_{n-1}^{k_\ell}),\theta((s_1^{k_\ell},\ldots,s_{n-1}^{k_\ell}))\right)\right) =$$

$$\mu\left(\xi\left(\lim_{\ell\to\infty}(s_1^{k_\ell},\ldots,s_{n-1}^{k_\ell}), \lim_{\ell\to\infty}\theta((s_1^{k_\ell},\ldots,s_{n-1}^{k_\ell}))\right)\right) =$$

$$\mu(\xi((s_1,\ldots,s_{n-1},s))).$$

Since for each positive integer ℓ,

$$\mu\left(\xi\left((s_1^{k_\ell},\ldots,s_{n-1}^{k_\ell}),\theta((s_1^{k_\ell},\ldots,s_{n-1}^{k_\ell}))\right)\right) = t,$$

we obtain that $\mu(\xi((s_1,\ldots,s_{n-1},s))) = t$ and $\theta((s_1,\ldots,s_{n-1})) = s$. Hence, θ is continuous.

Let $\chi: \prod_{j=1}^{n-1} \left[\frac{1}{2} - \delta, \frac{1}{2}\right]_j \to \mu^{-1}(t)$ be given by

$$\chi((s_1, \ldots, s_{n-1})) = \xi((s_1, \ldots, s_{n-1}), \theta((s_1, \ldots, s_{n-1}))).$$

Then χ is an embedding of the $(n - 1)$-cell $\prod_{j=1}^{n-1} \left[\frac{1}{2} - \delta, \frac{1}{2}\right]_j$ into $\mu^{-1}(t)$.

Q.E.D.

6.11.23 Corollary *Let X be a continuum, let $n \geq 2$, let $\mu: C_n(X) \twoheadrightarrow [0, 1]$ be a strong size map and let $t \in (0, 1)$. Then $\mu^{-1}(t)$ contains an $(n - 1)$-cell.*

Proof Since $C_{n-1}(X)$ has empty interior in $C_n(X)$, Theorem 6.1.9, we have that $\mu^{-1}(t) \cap (C_n(X) \setminus C_{n-1}(X)) \neq \emptyset$. Now, the corollary follows from Theorem 6.11.22.

Q.E.D.

6.11.24 Definition A topological property P is called a *strong size property* if whenever a continuum X has property P, so does every strong size level of $C_n(X)$ for each $n \in \mathbb{N}$.

6.11.25 Corollary *The property of being a hereditarily indecomposable continuum is not a strong size property.*

6.11.26 Corollary *The property of being an irreducible continuum is not a strong size property.*

6.11.27 Lemma *Let X be a continuum, let $n \in \mathbb{N}$, let $\mu: C_n(X) \twoheadrightarrow [0, 1]$ be a strong size map, and let $S = \mu^{-1}(t)$ be a strong size level for $C_n(X)$. Let $A, B \in S$, let $A_1, \ldots, A_\ell$ be the components of A and let $B_1, \ldots, B_m$ be the components of B. If $P \in \mathcal{F}_n(X)$ is such that $P \subset A \cap B$, $P \cap A_j \neq \emptyset$ for each $j \in \{1, \ldots, \ell\}$ and $P \cap B_k \neq \emptyset$ for all $k \in \{1, \ldots, m\}$, then there exists an arc in S joining A and B.*

Proof Let $\alpha, \beta: [0, 1] \to C_n(X)$ be two order arcs such that $\alpha(0) = P$, $\alpha(1) = A$, $\beta(0) = P$ and $\beta(1) = B$, Theorem 1.8.20. Given $s \in [0, 1]$, define $f_s: [0, 1] \to C_n(X)$ by $f_s(r) = \alpha(s) \cup \beta(r)$. Then f_s is well defined and continuous. Since $\mu(f_s(0)) = \mu(\alpha(s) \cup \beta(0)) = \mu(\alpha(s)) \leq t$ and $\mu(f_s(1)) = \mu(\alpha(s) \cup \beta(1)) = \mu(\alpha(s) \cup B) \geq t$, there exists $r_s \in [0, 1]$ such that $\mu(f_s(r_s)) = t$.

Let $\gamma: [0, 1] \to S$ be given by $\gamma(s) = \alpha(s) \cup \beta(r_s)$. We show γ is well defined. To this end, let $s \in [0, 1]$ and suppose there exists $r \in [0, 1]$ such that $\alpha(s) \cup \beta(r) \in S$. Since β is an order arc, we have that either $\beta(r) \subset \beta(r_s)$ or $\beta(r_s) \subset \beta(r)$. Without loss of generality we assume that $\beta(r) \subset \beta(r_s)$. Then $\alpha(s) \cup \beta(r) \subset \alpha(s) \cup \beta(r_s)$. Since μ is a strong size map, we obtain that $\alpha(s) \cup \beta(r) = \alpha(s) \cup \beta(r_s)$. Thus, γ is well defined.

To see that γ is continuous, let $\{s_m\}_{m=1}^\infty$ be sequence of elements of $[0, 1]$ converging to an element s of $[0, 1]$. Then the corresponding sequence $\{r_{s_m}\}_{m=1}^\infty$ has a convergent subsequence $\{r_{s_{m_k}}\}_{k=1}^\infty$. Let r be the limit of the sequence $\{r_{s_{m_k}}\}_{k=1}^\infty$. Since α and β are continuous, we have that $\lim_{k \to \infty} \gamma(s_{m_k}) = \lim_{k \to \infty} (\alpha(s_{m_k}) \cup \beta(r_{s_{m_k}})) = \alpha(s) \cup \beta(r)$. By definition of γ, $\gamma(s) = \alpha(s) \cup \beta(r_s)$. Since both

$\alpha(s) \cup \beta(r)$ and $\alpha(s) \cup \beta(r_s)$ belong to $\mathcal{S}$ and either $\alpha(s) \cup \beta(r) \subset \alpha(s) \cup \beta(r_s)$ or $\alpha(s) \cup \beta(r_s) \subset \alpha(s) \cup \beta(r)$, we have that $\alpha(s) \cup \beta(r) = \alpha(s) \cup \beta(r_s)$. Therefore, γ is continuous.

Q.E.D.

Our next goal is to prove that for an integer $n \geq 3$, the strong size levels of $\mathcal{C}_n(X)$ are acyclic (Corollary 6.11.55). To this end, we follow James T. Rogers, Jr. [90]. We include most of the details.

6.11.28 Remark Note that acyclicity is not a Whitney property [89, Example 2] and it is for 1-dimensional continua [90, Corollary 7]. Since we do not require any additional properties to the continuum X, Theorem 6.11.54 says that, for $n \geq 3$, the levels of strong size maps are much nicer than the Whitney levels. In particular, Corollary 6.11.55, tells us the acyclicity is a strong size property.

6.11.29 Lemma *If X is a continuum, then $\check{H}^0(X)$ is trivial.*

Proof The result follows form three the facts: (1) each continuum is an inverse limit of connected polyhedra [76, Theorem 2], (2) the 0th reduced cohomology group of a connected polyhedron is trivial [82, 42.2], and (3) the continuity theorem for Čech cohomology [92, Theorem 7–7].

Q.E.D.

6.11.30 Definition Let X be a continuum, let $n \in \mathbb{N}$ and let B and A be two elements of $\mathcal{C}_n(X)$. We say that the pair (B, A) satisfies *property* (OA) provided that $B \subset A$ and each component of A intersects B.

6.11.31 Remark Observe that the condition in Definition 6.11.30 guaratees the existence of an order arc, in $\mathcal{C}_n(X)$, from B to A when $B \subset A$ and $B \neq A$, Theorem 1.8.20.

6.11.32 Definition Let X be a continuum and let $n \in \mathbb{N}$. If $B \in \mathcal{C}_n(X)$, define:

$$\mathcal{C}_n(B, X) = \{A \in \mathcal{C}_n(X) \mid B \subset A\};$$

$$\mathcal{OA}_n(B, X) = \{A \in \mathcal{C}_n(X) \mid (B, A) \text{ satisfies property } (OA)\}.$$

If $A \in \mathcal{OA}_n(B, X)$, then

$$\mathcal{OA}_n(B, A) = \{D \in \mathcal{OA}_n(B, X) \mid D \subset A\}.$$

6.11.33 Definition Let X be a continuum. A nonempty collection Σ of closed subsets of X is called a *structure* if Σ is closed with respect to finite unions, finite intersections, and intersections of (set theoretic) chains ordered by inclusion.

6.11.34 Definition Let X be a continuum. If Σ is a structure on X, then an element P of Σ is called an *indecomposable set* provided that whenever $P = A \cup B$, for some elements A and B of Σ, we have that $P = A$ or $P = B$.

6.11.35 Notation Let X be a continuum and let $n \in \mathbb{N}$. We consider two structures in $\mathcal{C}_n(X)$. If $\mathcal{B}$ is a closed subset of $\mathcal{C}_n(X)$, let

$$\mathcal{M}(\mathcal{B}) = \sigma \left(\{ \mathcal{O}\mathcal{A}_n(B, X) \mid B \in \mathcal{B} \} \right).$$

6.11.36 Remark Note that if $\mathcal{B}$ and $\mathcal{D}$ are two closed subsets of $\mathcal{C}_n(X)$, then $\mathcal{M}(\mathcal{B} \cup \mathcal{D}) = \mathcal{M}(\mathcal{B}) \cup \mathcal{M}(\mathcal{D})$ and $\mathcal{M}(\mathcal{B}) \cap \mathcal{M}(\mathcal{D}) = \mathcal{M}(\mathcal{M}(\mathcal{B}) \cap \mathcal{M}(\mathcal{D}))$. Also, if $\{\mathcal{B}_\lambda\}_{\lambda \in \Lambda}$ is a (set theoretic) chain ordered by inclusion, then $\mathcal{M} \left(\bigcap_{\lambda \in \Lambda} \mathcal{M}(\mathcal{B}_\lambda) \right) = \bigcap_{\lambda \in \Lambda} \mathcal{M}(\mathcal{B}_\lambda)$.

6.11.37 Lemma *Let X be a continuum and let $n \in \mathbb{N}$. If $\mathcal{B}$ is a closed subset of $\mathcal{C}_n(X)$, then $\mathcal{M}(\mathcal{B})$ is closed in $\mathcal{C}_n(X)$.*

Proof Let $D \in Cl_{\mathcal{C}_n(X)}(\mathcal{M}(\mathcal{B}))$. Then there exists a sequence $\{D_m\}_{m=1}^\infty$ of elements of $\mathcal{M}(\mathcal{B})$ converging to D. For each $m \in \mathbb{N}$, there exists $B_m \in \mathcal{B}$ such that $D_m \in \mathcal{O}\mathcal{A}_n(B_m, X)$. Since $\mathcal{B}$ is compact, there exists a subsequence $\{B_{m_k}\}_{k=1}^\infty$ of the sequence $\{B_m\}_{m=1}^\infty$ that converges to an element B of $\mathcal{B}$. Since $D_{m_k} \in \mathcal{O}\mathcal{A}_n(B_{m_k}, X)$, we have that $D \in \mathcal{O}\mathcal{A}_n(B, X)$. Therefore, $\mathcal{M}(\mathcal{B})$ is closed in $\mathcal{C}_n(X)$.
                                                                    Q.E.D.

6.11.38 Definition Let X be a continuum and let $n \in \mathbb{N}$. We define:

$$\Sigma_1 = \{\mathcal{M}(\mathcal{B}) \mid \mathcal{B} \text{ is a closed subset of } \mathcal{C}_n(X)\}.$$

By Remark 6.11.36 and Lemma 6.11.37, we have the following:

6.11.39 Lemma *If X is a continuum and $n \in \mathbb{N}$, then Σ_1 is a structure.*

6.11.40 Remark Note that the indecomposable sets of Σ_1 are the sets of the form $\mathcal{M}(\{B\})$ where $B \in \mathcal{C}_n(X)$. Since $\mathcal{M}(\{B\})$ is homeomorphic to $\mathcal{O}\mathcal{A}_n(B, X)$ and this set is an absolute retract [68, 4.3], the indecomposable sets of Σ_1 have all its reduced Čech cohomology groups trivial. Hence, all the reduced Čech cohomology groups of each member of Σ_1 are trivial [90, Theorem 2].

The second structure is found in $\mathcal{O}\mathcal{A}_n(Z, X)$, where Z is an arbitrary point of $\mathcal{C}_n(X)$.

6.11.41 Notation Let X be a continuum and let $n \in \mathbb{N}$. If $\mathcal{B}$ is a closed subset of $\mathcal{O}\mathcal{A}_n(Z, X)$, let

$$\mathcal{L}(\mathcal{B}) = \sigma \left(\{ \mathcal{O}\mathcal{A}_n(Z, B) \mid B \in \mathcal{B} \} \right).$$

6.11.42 Remark Observe that if $\mathcal{B}$ and $\mathcal{D}$ are two closed subsets of $\mathcal{C}_n(X)$, then $\mathcal{L}(\mathcal{B} \cup \mathcal{D}) = \mathcal{L}(\mathcal{B}) \cup \mathcal{L}(\mathcal{D})$ and $\mathcal{L}(\mathcal{B}) \cap \mathcal{L}(\mathcal{D}) = \mathcal{L}(\mathcal{L}(\mathcal{B}) \cap \mathcal{L}(\mathcal{D}))$. Also, if $\{\mathcal{B}_\lambda\}_{\lambda \in \Lambda}$ is a (set theoretic) chain ordered by inclusion, then $\mathcal{L} \left(\bigcap_{\lambda \in \Lambda} \mathcal{L}(\mathcal{B}_\lambda) \right) = \bigcap_{\lambda \in \Lambda} \mathcal{L}(\mathcal{B}_\lambda)$.

6.11.43 Lemma *Let X be a continuum, let $n \in \mathbb{N}$ and let $Z \in \mathcal{C}_n(X)$. If $\mathcal{B}$ is a closed subset of $\mathcal{O}\mathcal{A}_n(Z, X)$, then $\mathcal{L}(\mathcal{B})$ is closed in $\mathcal{C}_n(X)$.*

Proof Let $D \in Cl_{C_n(X)}(\mathcal{L}(\mathcal{B}))$. Then there exists a sequence $\{D_m\}_{m=1}^{\infty}$ of elements of $\mathcal{L}(\mathcal{B})$ converging to D. For each $m \in \mathbb{N}$, there exists $B_m \in \mathcal{B}$ such that $D_m \in \mathcal{OA}_n(Z, B_m)$. Since $\mathcal{OA}_n(Z, X)$ is a continuum [68, 4.3], $\mathcal{B}$ is compact. Then there exists a subsequence $\{B_{m_k}\}_{k=1}^{\infty}$ of the sequence $\{B_m\}_{m=1}^{\infty}$ that converges to an element B of $\mathcal{B}$. Since $D_{m_k} \in \mathcal{OA}_n(Z, B_{m_k})$, we obtain that $D \in \mathcal{OA}_n(Z, B)$. Therefore, $\mathcal{L}(\mathcal{B})$ is closed in $C_n(X)$.

Q.E.D.

6.11.44 Definition Let X be a continuum and let $n \in \mathbb{N}$. For an element Z of $C_n(X)$, let

$$\Sigma_2 = \{\mathcal{L}(\mathcal{B}) \mid \mathcal{B} \text{ is a closed subset of } \mathcal{OA}_n(Z, X)\}.$$

By Remark 6.11.42 and Lemma 6.11.43, we obtain:

6.11.45 Lemma *If X is a continuum and $n \in \mathbb{N}$, then Σ_2 is a structure.*

6.11.46 Remark Note that the indecomposable sets of Σ_2 are the sets of the form $\mathcal{L}(\{B\})$ where $B \in \mathcal{OA}_n(Z, X)$. Since $\mathcal{L}(\{B\})$ is homeomorphic to $\mathcal{OA}_n(Z, B)$ and this set is an absolute retract [68, 4.3], all the reduced Čech cohomology groups of the indecomposable sets of Σ_2 are trivial. Thus, all the reduced Čech cohomology groups of each member of Σ_2 are trivial [90, Theorem 2].

6.11.47 Definition Let X be a continuum, let $n \in \mathbb{N}$ and let $\mu\colon C_n(X) \twoheadrightarrow [0, 1]$ be a strong size map. For an element Z of $C_n(X)$ and an element $t \in [\mu(Z), 1]$, let

$$\mathcal{D}_n(Z, t) = \mathcal{M}(\{Z\}) \cap \mu^{-1}(t).$$

As a consequence of Lemma 6.11.27, we have:

6.11.48 Lemma *Let X be a continuum, let $n \in \mathbb{N}$ and let $\mu\colon C_n(X) \twoheadrightarrow [0, 1]$ be a strong size map. If $Z \in C_n(X)$, then $\mathcal{D}_n(Z, t)$ is an arcwise connected continuum.*

The proof of the following theorem is similar to the one given in [90, Theorem 4].

6.11.49 Theorem *Let X be a continuum, let $n \in \mathbb{N}$, let $\mu\colon C_n(X) \twoheadrightarrow [0, 1]$ be a strong size map and let $Z \in C_n(X)$. If $t \in [\mu(Z), 1]$, then all the reduced Čech cohomology groups of $\mathcal{D}_n(Z, t)$ are trivial.*

Proof Consider the pair $\{\mathcal{M}(\mathcal{D}_n(Z, t)), \mathcal{L}(\mathcal{D}_n(Z, t))\}$ of subsets of $\mathcal{OA}_n(Z, X)$. For an integer $m \geq 0$, consider the following part of the reduced Mayer–Vietoris sequence:

$$\check{H}^m(\mathcal{M}(\mathcal{D}_n(Z, t))) \oplus \check{H}^m(\mathcal{L}(\mathcal{D}_n(Z, t))) \to$$

$$\check{H}^m(\mathcal{D}_n(Z, t)) \to \check{H}^{m+1}(\mathcal{OA}_n(Z, X))$$

for this pair. By Remarks 6.11.40 and 6.11.46, we have that $\check{H}^m(\mathcal{M}(\mathcal{D}_n(Z, t)))$ and $\check{H}^m(\mathcal{L}(\mathcal{D}_n(Z, t)))$ are both trivial. Since $\mathcal{OA}_n(Z, X)$ is an absolute retract [68, 4.3],

$\check{H}^{m+1}(\mathcal{O}\mathcal{A}_n(Z, X))$ is trivial too. Hence, $\check{H}^m(\mathcal{D}_n(Z, t))$ is trivial. Therefore, all the reduced Čech cohomology groups of $\mathcal{D}_n(Z, t)$ are trivial.

Q.E.D.

6.11.50 Definition Let X be a continuum, let $n \in \mathbb{N}$ and let $\mu\colon C_n(X) \twoheadrightarrow [0, 1]$ be a strong size map. Let $s, t \in [0, 1]$ be such that $s \le t$. Define $_n\gamma_s^t\colon \mu^{-1}(s) \to \mu^{-1}(t)$ by $_n\gamma_s^t(Z) = \mathcal{D}_n(Z, t)$.

The next Lemma shows that $_n\gamma_s^t$ is an upper semicontinuous function.

6.11.51 Lemma *Let X be a continuum, let $n \in \mathbb{N}$ and let $\mu\colon C_n(X) \twoheadrightarrow [0, 1]$ be a strong size map. If $s, t \in [0, 1]$ are such that $s \le t$, then $_n\gamma_s^t$ is upper semicontinuous.*

Proof Let $\{Z_m\}_{m=1}^{\infty}$ be a sequence of elements of $\mu^{-1}(s)$ that converges to an element Z of $\mu^{-1}(s)$. Let $Y \in \limsup_n \gamma_s^t(Z_m)$. Then there exists a subsequence $\{m_k\}_{k=1}^{\infty}$ of the natural sequence such that for each $k \in \mathbb{N}$, there exists $Y_{m_k} \in {_n\gamma_s^t(Z_{m_k})}$ such that the sequence $\{Y_{m_k}\}_{k=1}^{\infty}$ converges to Y. Since for all $k \in \mathbb{N}$, $Z_{m_k} \subset Y_{m_k}$ and $\{Z_{m_k}\}_{k=1}^{\infty}$ converges to Z, we have that $Z \subset Y$. It is easy to see that $Y \in \mathcal{O}\mathcal{A}_n(Z, X)$. Hence, $Y \in {_n\gamma_s^t(Z)}$. Therefore, $_n\gamma_s^t$ is upper semicontinuous.

Q.E.D.

6.11.52 Theorem *Let X be a continuum, let $n \in \mathbb{N}$ and let $\mu\colon C_n(X) \twoheadrightarrow [0, 1]$ be a strong size map. If $s, t \in [0, 1]$ are such that $s \le t$, then $_n\gamma_s^t$ induces a monomorphism $(_n\gamma_s^t)^*\colon \check{H}^1(\mu^{-1}(t)) \to \check{H}^1(\mu^{-1}(s))$.*

Proof By Theorem 6.11.49, all the reduced Čech cohomology groups of $_n\gamma_s^t(Z)$ are trivial. Suppose $t \ne 1$, the result is clear for $t = 1$. Let $B \in \mu^{-1}(t)$ and suppose that $B_1, \dots, B_m$ are the components of B. Note that $(_n\gamma_s^t)^{-1}(B) = \langle B_1, \dots, B_m \rangle_n \cap \mu^{-1}(s)$ and this set is a proper continuum of $\mu^{-1}(s)$ by [38, Theorem 2.14]. The result now follows from Lemmas 6.11.51, 6.11.29 and [90, Theorem 3].

Q.E.D.

6.11.53 Corollary *Let X be a continuum, let $n \in \mathbb{N}$ and let $\mu\colon C_n(X) \twoheadrightarrow [0, 1]$ be a strong size map. If $t \in [0, 1]$, then $_n\gamma_0^t$ induces a monomorphism $(_n\gamma_0^t)^*\colon \check{H}^1(\mu^{-1}(t)) \to \check{H}^1(\mu^{-1}(0))$.*

6.11.54 Theorem *Let X be a continuum, let $n \ge 3$ be an integer and let $\mu\colon C_n(X) \twoheadrightarrow [0, 1]$ be a strong size map. If $\mathcal{S} = \mu^{-1}(t)$ is a strong size level, then $\mathcal{S}$ is acyclic.*

Proof By [58, Theorem 8], each map from $\mathcal{F}_n(X)$ into the unit circle in the plane is homotopic to a constant map. This implies, by [22, 8.1], that $\check{H}^1(\mathcal{F}_n(X))$ is trivial; i.e., $\mathcal{F}_n(X)$ is acyclic. The theorem now follows from the fact that $\mu^{-1}(0) = \mathcal{F}_n(X)$ and Corollary 6.11.53.

Q.E.D.

6.11.55 Corollary *The property of being acyclic is a strong size property for each integer $n \ge 3$.*

6.11.56 Corollary *The property of being acyclic is a strong size property for locally connected continua.*

Proof Let X be a locally connected continuum. For $n \geq 3$, the corollary follows from Corollary 6.11.55. Suppose $n = 2$. By [29, Satz 1], [96, (7.4)] and [22, 8.1], we have that $\mathcal{F}_2(X)$ is acyclic. Hence, since $\mu^{-1}(0) = \mathcal{F}_2(X)$, by Corollary 6.11.53, $\mu^{-1}(t)$ is acyclic for all $t \in (0, 1]$. If $n = 1$, the corollary follows from [43, p. 253], [96, (7.4)] and [22, 8.1]. Therefore, the property of being acyclic is a strong size property for locally connected continua.

Q.E.D.

We end the section with the following list of strong size properties, the proofs may be found in [38] and [74].

6.11.57 Theorem *The following are strong size properties:*

(1) *Local connectedness.*
(2) *Arc connectedness.*
(3) *Aposyndesis.*
(4) *Countable aposyndesis.*
(5) *continuum chainability.*

Several versions of reversible strong size properties have been studied also in [75] and [88].

References

1. G. Acosta, Continua with Unique Hyperspace, in *Continuum Theory: Proceedings of the Special Session in Honor of Professor Sam B. Nadler, Jr.'s 60th Birthday.* Lecture Notes in Pure and Applied Mathematics Series, Vol. 230, Marcel Dekker, Inc., New York, Basel, 2002, 33–49. (eds.: Alejandro Illanes, Ira Wayne Lewis and Sergio Macías.)
2. D. P. Bellamy, The Cone Over the Cantor Set-continuous Maps From Both Directions, Proc. Topology Conference Emory University, Atlanta, Ga., 1970, 8–25. (ed. J. W. Rogers, Jr.)
3. D. E. Bennett, Aposyndetic Properties of Unicoherent Continua, Pacific J. Math., 37 (1971), 585–589.
4. R H Bing, Partitioning a Set, Bull. Amer. Math. Soc., 55 (1949), 1101–1110.
5. K. Borsuk, *Theory of Retracts,* Monografie Mat. Vol. 44, PWN (Polish Scientific Publishers), Warszawa, 1967.
6. K. Borsuk, Some Remarks on Shape Properties of Compacta, Fund. Math., 85 (1974), 185–195.
7. K. Borsuk and S. Ulam, On Symmetric Products of Topological Spaces, Bull. Amer. Math. Soc., 37 (1931), 875–882.
8. K. Borsuk and R. Molski, On a Class of Continuous Mappings, Fund. Math., 45 (1957), 84–98.
9. R. Bott, On the Third Symmetric Potency of S_1, Fund. Math., 39 (1952), 364–368.
10. C. E. Burgess, Chainable Continua and Indecomposability, Pacific J. Math., 9 (1959), 653–659.
11. J. Camargo, D. Herrera and S. Macías, Cells and n-fold Hyperspaces, to appeat in Colloquium Mathematicum.
12. J. Camargo and S. Macías, On Stongly Freely Decomposable and Induced Maps, Glasnik Math., 48(68), (2013), 429–442.

13. E. Castañeda, A Unicoherent Continuum for Which its Second Symmetric Product is not Unicoherent, Topology Proc., 23 (1998), 61–67.

14. E. Castañeda, *Productos Simétricos*, Tesis Doctoral, Facultad de Ciencias, U. N. A. M., 2003. (Spanish)

15. T. A. Chapman, *Lecture Notes on Hilbert Cube Manifolds*, C. B. M. S. Regional Conf. Series in Math., 28 (Amer. Math. Soc., Providence, RI, 1975).

16. J. J. Charatonik, On Fans, Dissertationes Math. (Rozprawy Mat.), 54 (1967), 1–37.

17. J. J. Charatonik and A. Illanes, Local Connectedness in Hyperspaces, Rocky J. Math., 36 (2006), 811–856.

18. J. J. Charatonik, A. Illanes, S. Macías, Induced Mappings on the Hyperspace $C_n(X)$ of a Continuum X, Houston J. Math., 28 (2002), 781–805.

19. J. J. Charatonik and S. Macías, Mappings of Some Hyperspaces, JP Jour. Geometry & Topology, 4(1) (2004), 53–80.

20. W. J. Charatonik, On the Property of Kelley in Hyperspaces, Topology, Proceedings of the International Topological Conference held in Leningrad, 1982, Lecture Notes in Math. 1060, Springer Verlag, 1984, 7–10.

21. D. Curtis and N. T. Nhu, Hyperspaces of Finite Subsets Which are Homeomorphic to $\aleph_0$-dimensional Linear Metric Spaces, Topology Appl., 19 (1985), 251–260.

22. C. H. Dowker, Mapping Theorems for Non-compact Spaces, Amer. J. Math., 69 (1947), 200–242.

23. J. Dugundji, *Topology*, Allyn and Bacon, Inc., Boston, London, Sydney, Toronto, 1966.

24. C. A. Eberhart, A Note on Smooth Fans, Colloq. Math., 20 (1969), 89–90.

25. C. Eberhart and S. B. Nadler, Jr., Hyperspaces of Cones and Fans, Proc. Amer. Math. Soc., 77 (1979), 279–288.

26. J. B. Fugate, Retracting Fans onto Finite Fans, Fund. Math., 71 (1971), 113–125.

27. J. B. Fugate, Small Retractions of Smooth Dendroids onto Trees, Fund. Math., 71 (1971), 255–262.

28. J. B. Fugate, G. R. Gordh and L. Lum, Arc-smooth Continua, Trans. Amer. Math. Soc., 265 (1981), 545–561.

29. T. Ganea, Symmetische Potenz Topologischer Räume, Math. Nachrichten, 11 (1954), 305–316.

30. J. T. Goodykoontz, Jr., Hyperspaces of Arc-smooth Continua, Houston J. Math., 7 (1981), 33–41.

31. J. T. Goodykoontz, Jr., Arc Smoothness in Hyperspaces, Topology Appl., 15 (1983), 131–150.

32. J. T. Goodykoontz, Some Retractions and Deformation Retractions on 2^X and $C(X)$, Topology Appl., 21 (1985), 121–133.

33. J. Grispolakis, S. B. Nadler, Jr. and E. D. Tymchatyn, Some Properties of Hyperspaces with Applications to Continua Theory, Can. J. Math., 31 (1979), 197–210.

34. J. Grispolakis and E. D. Tymchatyn, Weakly Confluent Mappings and the Covering Property of Hyperspaces, Proc. Amer. Math. Soc., 74 (1979), 177–182.

35. C. L. Hagopian, Indecomposable Homogeneous Plane Continua are Hereditarily Indecomposable, Trans. Amer. Math. Soc., 224 (1976), 339–350.

36. J. G. Hocking and G. S. Young, *Topology*, Dover Publications, Inc., New York, 1988.

37. H. Hosokawa, Induced Mappings on Hyperspaces, Tsukuba J. Math., 21 (1997), 239–250.

38. H. Hosokawa, Strong Size Levels of $C_n(X)$, Houston J. Math., 37 (2011), 955–965.

39. W. Hurewicz, Sur la Dimension des Produits Cartésiens, Annals of Math., 36 (1935), 194–197.

40. W. Hurewicz and H. Wallman, *Dimension Theory*, Princeton Univ. Press, Princeton, N. J., 1948.

41. A. Illanes, Multicoherence of Symmetric Products, An. Inst. Mat. Univ. Nac. Autónoma México, 25 (1985), 11–24.

42. A. Illanes, Monotone and Open Whitney maps, Proc. Amer. Math. Soc., 98 (1986), 516–518.

43. A. Illanes, Multicoherence of Whitney Levels, Topology Appl., 68 (1996), 251–265.

44. A. Illanes, The Hyperspace $C_2(X)$ for a Finite Graph X is Unique, Glansnik Mat., 37(57) (2002), 347–363.

45. A. Illanes, Comparing n-fold and m-fold Hyperspaces, Topology Appl., 133 (2003), 179–198.

46. A. Illanes, Finite Graphs Have Unique Hyperspaces $C_n(X)$, Topology Proc., 27 (2003), 179–188.

47. A. Illanes, A Model for the Hyperspace $C_2(S^1)$, Q. & A. in General Topology, 22 (2004), 117–130.

48. A. Illanes and S. B. Nadler, Jr., *Hyperspaces: Fundamentals and Recent Advances,* Monographs and Textbooks in Pure and Applied Math., Vol. 216, Marcel Dekker, New York, Basel, 1999.

49. F. B. Jones, Certain Homogeneous Unicoherent Indecomposable Continua, Proc. Amer. Math. Soc., 2 (1951), 855–859.

50. J. L. Kelley, Hyperspaces of a Continuum, Trans. Amer. Math. Soc., 52 (1942), 22–36.

51. J. Krasinkiewicz, On the Hyperspaces of Snake-like and Circle-like Continua, Fund. Math., 84 (1974), 155–164.

52. J. Krasinkiewicz, On the Hyperspaces of Hereditarily Indecomposable Continua, Fund. Math., 84 (1974), 175–186.

53. J. Krasinkiewicz, Curves Which Are Continuous Images of Tree-like Continua Are Movable, Fun. Math., 89 (1975), 233–260.

54. W. Kuperberg, Uniformly Pathwise Connected Continua, in *Studies in Topology*, Proc. Conf. Univ. North Carolina, Charlotte, NC, 1974, Academic Press, New York, 1975, 315–324. (eds. N. M. Stavrakas and K. R. Allen.)

55. K. Kuratowski, *Topology*, Vol. II, Academic Press, New York, N. Y., 1968.

56. M. Levin and Y. Sternfeld, The Space of Subcontinua of a 2-dimensional Continuum is Infinitely Dimensional, Proc. Amer. Math. Soc., 125 (1997), 2771–2775.

57. J. C. Macías, *El n-ésimo Pseudohiperespacio Suspensión de Continuos*, Tesis de Doctorado, Facultad de Ciencias Físico Matemáticas, B. U. A. P., 2008. (Spanish)

58. S. Macías, On Symmetric Products of Continua, Topology Appl., 92 (1999), 173–182.

59. S. Macías, Aposyndetic Properties of Symmetric Products of Continua, Topology Proc., 22 (1997), 281–296.

60. S. Macías, On the Hyperspaces $C_n(X)$ of a Continuum X, Topology Appl., 109 (2001), 237–256.

61. S. Macías, On the Hyperspaces $C_n(X)$ of a Continuum X, II, Topology Proc., 25 (2000), 255–276.

62. S. Macías, On Arcwise Accessibility in Hyperspaces, Topology Proc., 26 (2001–2002), 247–254.

63. S. Macías, Fans Whose Hyperspaces Are Cones, Topology Proc., 27 (2003), 217–222.

64. S. Macías, Correction to the paper "On the Hyperspaces $C_n(X)$ of a continuum X, II", Topology Proc., 30 (2006), 335–340.

65. S. Macías, On the n-fold Hyperspace Suspension of Continua, II, Glasnik Mat., 41(61) (2006), 335–343.

66. S. Macías, On n-fold Hyperspaces, Glasnik Mat., 44(64) (2009), 479–492.

67. S. Macías, Retractions and Hyperspaces, Glasnik Mat., 46(66) (2011), 473–483.

68. S. Macías, Deformation Retracts and Hilbert cubes in n-fold hyperspaces, Topology Proc., 40 (2012), 215–226.

69. S. Macías and S. B. Nadler, Jr., n-fold Hyperspaces, Cones and Products, Topology Proc., 26 (2001–2002), 255–270.

70. S. Macías and S. B. Nadler, Jr., Z-sets in Hyperspaces, Q. & A. in General Topology, 19 (2001), 227–241.

71. S. Macías and S. B. Nadler, Jr., Smoothness in n-fold Hyperspaces, Glasnik Mat., 37(57) (2002), 365–373.

72. S. Macías and S. B. Nadler, Jr., Fans Whose Hyperspace of Subcontinua are Cones, Topology Appl., 126 (2002), 29–36.

73. S. Macías and S. B. Nadler, Jr., Various Types of Local Connectedness in n-fold Hyperspaces, Topology Appl., 54 (2007), 39–53.

74. S. Macías and César Piceno, Strong Size Properties, Glasnik Mat., 48(68) (2013), 103–114.

75. S. Macías and César Piceno, More on Strong Size Properties, Glasnik Mat., 50(70) (2015), 467–488.
76. S. Mardešić and J. Segal, ε-mappings onto Polyhedra, Trans. Amer. Math. Soc., 109 (1963), 146–164.
77. M. M. Marsh, s-connected Spaces and the Fixed Point Property, Topology Proc., 8 (1983), 85–97.
78. V. Martínez-de-la-Vega, Dimension of n-fold Hyperspaces of Graphs, Houston J. Math., 32 (2006), 783–799.
79. J. M. Martínez-Montejano, Zero-dimensional Closed Set Aposyndesis and Hyperspaces, Houston J. Math., 32 (2006), 1101–1105.
80. J. van Mill, *Infinite-Dimensional Topology*, North Holland, Amsterdam, 1989.
81. E. E. Moise, Grille Decomposition and Convexification Theorems for Compact Locally Connected Continua, Bull. Amer. Math. Soc., 55 (1949), 1111–1121.
82. J. R. Munkres, *Elements of Algebraic Topology*, Addison-Wesley, Reading, MA, 1984.
83. S. B. Nadler, Jr., Arcwise Accessibility in Hyperspaces, Dissertationes Math. (Rozprawy Mat.), 138 (1976), 1–29.
84. S. B. Nadler, Jr., *Hyperspaces of Sets,* Monographs and Textbooks in Pure and Applied Math., Vol. 49, Marcel Dekker, New York, Basel, 1978. Reprinted in: Aportaciones Matemáticas de la Sociedad Matemática Mexicana, Serie Textos # 33, 2006.
85. S. B. Nadler, Jr., Continua Whose Hyperspace is a Product, Fund. Math., 108 (1980), 49–66.
86. S. B. Nadler, Jr., *Continuum Theory: An Introduction,* Monographs and Textbooks in Pure and Applied Math., Vol. 158, Marcel Dekker, New York, Basel, Hong Kong, 1992.
87. S. B. Nadler, Jr., *Dimension Theory: An Introduction with Exercises*, Aportaciones Matemáticas, Serie Textos # 18, Sociedad Matemática Mexicana, 2002.
88. L. Paredes-Rivas and P. Pellicer-Covarrubias, On Strong Size Levels, Topology Appl., 160 (2013), 1816–1828.
89. A. Petrus, Contractibility of Whitney Continua in $C(X)$, General Topology Appl., 9 (1978), 275–288.
90. J. T. Rogers, Jr., Applications of a Vietoris–Begle Theorem for Multi-valued Maps to the Cohomology of Hyperspaces, Michigan Math. J., 22 (1975), 315–319.
91. J. T. Rogers, Jr., Dimension and the Whitney Subcontinua of $C(X)$, Gen. Top. and its Applications, 6 (1976), 91–100.
92. A. H. Wallace, *Algebraic Topology, Homology and Cohomology,* W. A. Benjamin Inc., 1970.
93. L. E. Ward, Jr. A Note on Whitney Maps, Canad. Math.Bull., 23 (1980), 373–374.
94. L. E. Ward, Extending Whitney Maps, Pacific J. Math., 93 (1981), 465–469.
95. H. Whitney, On Regular Families of Curves I, Proc. Nat. Acad. Sci., 18 (1932), 275–278.
96. G. T. Whyburn, *Analytic Topology,* Amer. Math. Soc. Colloq. Publ., Vol. 28, Amer. Math. Soc., Providence, R. I., 1942.
97. S. Willard, *General Topology*, Addison-Wesley Publishing Co., 1970.
98. M. Wojdisławski, Rétractes Absolus et Hyperespaces des Continus, Fund. Math., 32 (1939), 184–192.

Chapter 7
n-Fold Hyperspace Suspensions

In 1979 Sam B. Nadler, Jr. introduced the hyperspace suspension of a continuum
[41] to present examples of disk-like continua with the fixed point property. A
thorough investigation of these spaces is presented in [12]. Then an extension to
n-fold hyperspace suspensions is done in [30].

We show that the *n*-fold hyperspace suspension of a continuum has the same
dimension as the *n*-fold hyperspace of the continuum. We prove that *n*-fold hyperspace suspensions are zero-dimensional aposyndetic. We give sufficient conditions
to have the *n*-fold hyperspace suspensions are contractible. We present results of
local connectedness. In particular, we characterize the arc as the only continuum
for which its *n*-fold hyperspace suspensions are cells. We give properties of
points that arcwise disconnect these spaces. We present necessary or sufficient
conditions in order to have that *n*-fold hyperspace suspensions are homeomorphic
to a cone, suspension or product of continua. We present sufficient conditions
to obtain that *n*-fold hyperspace suspensions have the fixed point property. We
study absolute *n*-fold hyperspace suspensions. We end the chapter proving that
hereditarily indecomposable continua have unique *n*-fold hyerpsace suspensions.

7.1 General Properties

The material for this section comes from [1, 5, 8, 10–13, 15, 18–20, 25, 27–32, 34–
37, 40, 42–46].

7.1.1 Definition Let X be a continuum and let $n \in \mathbb{N}$. The *n-fold hyperspace suspension of* X, denoted by $HS_n(X)$, is the quotient space $HS_n(X) = \mathcal{C}_n(X)/\mathcal{F}_n(X)$,
with the quotient topology.

7.1.2 Remark Let X be a continuum and let $n \in \mathbb{N}$. Note that, in fact, $HS_n(X)$ is
not a hyperspace, since it is not a subset of the power set of X. However, we keep
the name, following Sam B. Nadler, Jr.

© Springer International Publishing AG, part of Springer Nature 2018
S. Macías, *Topics on Continua*, https://doi.org/10.1007/978-3-319-90902-8_7

As a consequence of Theorem 1.7.3, we have:

7.1.3 Theorem *If X is a continuum and $n \in \mathbb{N}$, then $HS_n(X)$ is an arcwise connected continuum.*

7.1.4 Notation Let X be a continuum and let $n \in \mathbb{N}$. Then $q_X^n \colon C_n(X) \twoheadrightarrow HS_n(X)$ denotes the quotient map. Also, T_X^n and F_X^n denote the points $q_X^n(X)$ and $q_X^n(\mathcal{F}_n(X))$, respectively. We write $HS(X)$, T_X and F_X instead of $HS_1(X)$ T_X^1 and F_X^1, respectively.

7.1.5 Remark Let X be a continuum and let $n \in \mathbb{N}$. Note that $HS_n(X) \setminus \{F_X^n\}$ and $HS_n(X) \setminus \{T_X^n, F_X^n\}$ are homeomorphic to $C_n(X) \setminus \mathcal{F}_n(X)$ and $C_n(X) \setminus (\{X\} \cup \mathcal{F}_n(X))$, respectively, using the appropriate restriction of q_X^n.

7.1.6 Example Consider $[0, 1]$. Then, by [40, (0.54)], $\mathcal{C}([0, 1])$ is a 2-cell and $\mathcal{F}_1(X)$ is properly contained in the manifold boundary of $\mathcal{C}([0, 1])$. Hence, $HS([0, 1])$ is a 2-cell.

7.1.7 Example Let $X = \mathcal{S}^1$. Then, by [40, (0.55)], $\mathcal{C}(X)$ is a 2-cell and $\mathcal{F}_1(X)$ is equal to the manifold boundary of $\mathcal{C}(X)$. Thus, $HS(X)$ is a 2-sphere.

Before we present a model for $HS_2([0, 1])$ (Theorem 7.1.9), we need the following:

7.1.8 Lemma *There exists an embedding φ of $\mathcal{C}_2([0, 1])$ in $\mathbb{R}^4$ such that $\varphi(\mathcal{F}_2([0, 1]))$ is convex.*

Proof The proof of the lemma is a matter of making several observations about the proof that $\mathcal{C}_2([0, 1])$ is a 4-cell as done in Theorem 6.9.12.

Let $K(Y)$ be the quotient space $(Y \times [0, 1])/(Y \times \{1\})$; i.e., the cone over Y. We use notation in the proof of Theorem 6.9.12:

- $D^1 = \{A \in \mathcal{C}_2([0, 1]) \mid 1 \in A\}$, $D_0^1 = \{A \in \mathcal{C}_2([0, 1]) \mid 0, 1 \in A\}$.
- $f \colon K(D^1) \to \mathcal{C}_2(I)$ is the homeomorphism given by

$$f((A, t)) = \begin{cases} (1 - t)A, & \text{if } t < 1; \\ \{0\}, & \text{if } t = 1. \end{cases}$$

- $g \colon K(D_0^1) \to D^1$ is the homeomorphism given by

$$g((A, t)) = \begin{cases} t + (1 - t)A, & \text{if } t < 1; \\ \{1\}, & \text{if } t = 1. \end{cases}$$

In Step 3 of the proof of Theorem 6.9.12, it is shown that D_0^1 is a 2-cell S. The only element of $\mathcal{F}_2([0, 1])$ in D_0^1 is $\{0, 1\}$, which corresponds to the point $(0, 1)$ under the homeomorphism in the proof of Theorem 6.9.12. We consider $\{0, 1\}$ to correspond to the point $(0, 1, 0, 0)$ in $\mathbb{R}^4$.

The elements of $\mathcal{F}_2([0, 1])$ in D^1 are those of the form $\{a, 1\}$, $a \in [0, 1]$. From the formula for g, we have that

$$g^{-1}(\{a, 1\}) = (\{0, 1\}, a) \text{ for all } \{a, 1\}.$$

Thus, letting $(0, 1, 1, 0)$ be the vertex of $K(D_0^1)$, we can consider the elements of $\mathcal{F}_2([0, 1])$ in D^1 as corresponding under g^{-1} to the line segment L in $\mathbb{R}^4$ from the point $(0, 1, 0, 0)$ to the point $(0, 1, 1, 0)$; specifically, $g^{-1}(\{a, 1\}) = (0, 1, a, 0)$ for all $\{a, 1\}, a \in [0, 1]$.

From the formula of f, we have that

$$f^{-1}(\{a, b\}) = \begin{cases} \left(\frac{1}{\max\{a,b\}}\{a, b\}, 1 - \max\{a, b\}\right), & \text{if } \{a, b\} \neq \{0\}; \\ \text{vertex } v \text{ of } K(D^1), & \text{if } \{a, b\} = \{0\}. \end{cases}$$

Hence, letting $(0, 1, 0, 1)$ denote the vertex of $K(D^1)$, we can consider $\mathcal{F}_2([0, 1])$ to correspond to the line segments in $\mathbb{R}^4$ joining all points of L to $(0, 1, 0, 1)$; specifically, $f^{-1}(\{a, b\}) = (0, 1, \min\{a, b\}, 1 - \max\{a, b\})$ for all $\{a, b\} \in \mathcal{F}_2([0, 1])$. Note that $f^{-1}(\{a, 1\}) = g^{-1}(\{a, 1\})$ for all $a \in [0, 1]$.

Therefore, $\varphi = f^{-1}$ satisfies our lemma.

Q.E.D.

7.1.9 Theorem $HS_2([0, 1])$ *is homeomorphic to* $[0, 1]^4$.

Proof The hyperspace $C_2([0, 1])$ is a 4-cell (Theorem 6.9.12). Let β denote the manifold boundary of $C_2([0, 1])$.

Next, $\mathcal{F}_2([0, 1]) \subset \beta$. This can be seen from the proof of Lemma 7.1.8 or by considering the maps $f_\varepsilon : C_2([0, 1]) \to C_2([0, 1]) \setminus \mathcal{F}_2([0, 1])$, $\varepsilon > 0$, given by

$$f_\varepsilon(A) = Cl(V_\varepsilon(A)) \text{ for each } A \subset C_2([0, 1]);$$

the maps f_ε show that each element of $\mathcal{F}_2([0, 1])$ is an unstable value of the identity map on $C_2([0, 1])$ (as defined in [19, VI1, p. 74]). Thus, since $C_2([0, 1])$ is a 4-cell, $\mathcal{F}_2([0, 1]) \subset \beta$ [19, Example VI2, p. 75].

By Lemma 7.1.8, we can assume that $C_2([0, 1])$ is in $\mathbb{R}^4$ with $\mathcal{F}_2([0, 1])$ being convex. Thus, since $\mathcal{F}_2([0, 1]) \subset \beta$ (as just proved) and since β is a 3-sphere, it follows that the quotient space $\beta/\mathcal{F}_2([0, 1])$ is a 3-sphere by [1, Theorem, p. 21].

Finally, we prove that $HS_2([0, 1])$ is a 4-cell. Let $q : \beta \twoheadrightarrow \beta/\mathcal{F}_2([0, 1])$ denote the quotient map. Since β and $\beta/\mathcal{F}_2([0, 1])$ are 3-spheres and $\mathcal{F}_2([0, 1])$ is compact, q is a cellular map [46, 1.8.1, p. 44]. Hence, q can be extended to a map f from the 4-cell $C_2([0, 1])$ onto a 4-cell I^4 such that f is a homeomorphism on the manifold interior of $C_2([0, 1])$ onto the manifold interior of $[0, 1]^4$ [15, Theorem, p. 1279]. Note that the only nondegenerate fibre of f is $q^{-1}(\{\mathcal{F}_2([0, 1])\}) = \mathcal{F}_2([0, 1])$, which is also the only nondegenerate fibre of the quotient map

$$q_I^2 : C_2([0, 1]) \to C_2([0, 1])/\mathcal{F}_2([0, 1]) = HS_2([0, 1]).$$

Therefore, $q_I^2 \circ f^{-1}$ is a homeomorphism of $[0, 1]^4$ onto $HS_2([0, 1])$ (continuity is by [42, 3.22, p. 45]). Therefore, $HS_2([0, 1])$ is homeomorphic to $[0, 1]^4$.

Q.E.D.

7.1.10 Theorem *If X is a continuum and $n \in \mathbb{N}$, then $HS_n(X)$ is locally arcwise connected at T_X^n and F_X^n. Moreover, any neighborhood of F_X^n contains simple closed curves passing through F_X^n.*

Proof By [37, Lemma 2.3], $C_n(X)$ is locally arcwise connected at X. Hence, by Remark 7.1.5, $HS_n(X)$ is locally arcwise connected at T_X^n . Let $\mathfrak{U}$ be an open set of $HS_n(X)$ containing F_X^n. Then $(q_X^n)^{-1}(\mathfrak{U})$ is an open set of $C_n(X)$ containing $\mathcal{F}_n(X)$. Now, let $\mu \colon C_n(X) \twoheadrightarrow [0, 1]$ be a strong size map. Since $(q_X^n)^{-1}(\mathfrak{U})$ is open, $\mu(\mathcal{F}_n(X)) = \{0\}$ and μ is open (Theorem 6.11.16), there exists $t_0 < 1$ such that $[0, t_0] \subsetneqq \mu\left((q_X^n)^{-1}(\mathfrak{U})\right)$. We claim that there exists $t \in \mu\left((q_X^n)^{-1}(\mathfrak{U})\right) \setminus [0, t_0]$ such that $\mu^{-1}([0, t)) \subset (q_X^n)^{-1}(\mathfrak{U})$. Suppose this is not true. Then for each $n \in \mathbb{N}$, there exists $B_n \in \mu^{-1}(t_0 + \frac{1}{n}) \setminus (q_X^n)^{-1}(\mathfrak{U})$. Since, by Corollary 1.8.12, $C_n(X)$ is a continuum, without loss of generality, we assume that the sequence $\{B_n\}_{n=1}^{\infty}$ converges to a point B of $C_n(X)$. In fact, $B \in C_n(X) \setminus (q_X^n)^{-1}(\mathfrak{U})$. Since μ is continuous, $\mu(B) \leq t_0$, a contradiction. Therefore, there exists $t \in \mu\left((q_X^n)^{-1}(\mathfrak{U})\right) \setminus [0, t_0]$ such that $\mu^{-1}([0, t)) \subset (q_X^n)^{-1}(\mathfrak{U})$. Since μ is monotone (Theorem 6.11.18), $\mu^{-1}([0, t))$ is connected (Lemma 2.1.12). Clearly, $q_X^n(\mu^{-1}([0, t)))$ is an open arcwise connected subset of $HS_n(X)$ and $q_X^n(\mu^{-1}([0, t))) \subset \mathfrak{U}$. Therefore, $HS_n(X)$ is locally arcwise connected at F_X^n.

The fact that any neighborhood of F_X^n contains simple closed curves passing through F_X^n, follows easily from the arguments given in the previous paragraph.

Q.E.D.

7.1.11 Theorem *If X is an indecomposable continuum with the property of Kelley and $n \in \mathbb{N}$, then T_X^n and F_X^n are the only points at which $HS_n(X)$ is locally connected.*

Proof By Theorem 7.1.10, $HS_n(X)$ is locally connected at T_X^n and F_X^n. By Theorem 6.1.21, X is the only point at which $C_n(X)$ is locally connected. Therefore, by Remark 7.1.5, T_X^n and F_X^n are the only points at which $HS_n(X)$ is locally connected.

Q.E.D.

As a consequence of Theorem 7.1.11, we have:

7.1.12 Corollary *If X is a hereditarily indecomposable continuum and $n \in \mathbb{N}$, then T_X^n and F_X^n are the only points at which $HS_n(X)$ is locally connected.*

Proof Since hereditarily indecomposable continua have the property of Kelley [40, (16.27)], the corollary now follows from Theorem 7.1.11.

Q.E.D.

7.1.13 Theorem *If X is a continuum and $n \in \mathbb{N}$, then $HS_n(X)$ has property (b). In particular, $HS_n(X)$ is unicoherent.*

Proof Note that, by Theorem 6.2.5, $C_n(X)$ has property (b). Since the quotient map, q_X^n, is monotone, by [27, Theorem 2 (ii), p. 434], $HS_n(X)$ has property (b). The fact that $HS_n(X)$ is unicoherent, follows from Remark 6.2.4.

Q.E.D.

7.1.14 Lemma *If X is a continuum with a free arc, then there exists $\chi \in HS_n(X)$ such that $\dim_\chi(HS_n(X)) = 2n$.*

Proof Let α be a free arc contained in X. Then, by Theorem 6.9.10, $C_n(\alpha)$ is a $2n$-dimensional subspace of $C_n(X)$, and there exists a point $A \in C_n(\alpha) \setminus \mathcal{F}_n(\alpha)$ such that $\dim_A(C_n(X)) = 2n$. Hence, by Remark 7.1.5, $\dim_{q_X^n(A)}(HS_n(X)) = 2n$.

Q.E.D.

The next result is a technical lemma, which gives us a way to define a metric for certain quotient spaces.

7.1.15 Lemma *Let (Y, d) and Z be continua. If A is a subcontinuum of Y and B is a subcontinuum of Z, then there exists a metric ρ for Y/A such that $(Y/A, \rho)$ is homeomorphic to Y/A with the quotient topology. Moreover, if $\varepsilon > 0$ and $f : Y \to Z$ is a map such that $f(A) = B$ and $diam(f^{-1}(f(y))) < \varepsilon$ for each $y \in Y$, then the map $f^* : Y/A \to Z/B$ given by $f^*([y]) = [f(y)]$, where $[y]$ denotes the equivalence class of y, is such that $diam((f^*)^{-1}(f^*([y]))) < \varepsilon$ for all $[y] \in Y/A$.*

Proof Let $\Gamma_A = \{A \cup \{y\} \mid y \in Y\}$, as a subspace of 2^Y, be with the Hausdorff metric $\mathcal{H}$. Define $\xi : Y/A \to \Gamma_A$ by

$$\xi([y]) = A \cup \{y\},$$

for each $[y] \in Y/A$. It is not difficult to see that ξ is a homeomorphism from Y/A, with the quotient topology, onto Γ_A. Define $\rho : Y/A \times Y/A \to [0, \infty)$ by

$$\rho([y_1], [y_2]) = \mathcal{H}(\xi([y_1]), \xi([y_2])),$$

for each $[y_1], [y_2] \in Y/A$. Since ξ is a homeomorphism, ρ is a metric for Y/A such that $(Y/A, \rho)$ is homeomorphic to Y/A with the quotient topology.

Let $\varepsilon > 0$ and let $f : Y \to Z$ be a map such that $diam(f^{-1}(f(y))) < \varepsilon$ for each $y \in Y$ and $f(A) = B$. Note that f^* is well defined and continuous [10, Theorem 7.7, p. 17, and Theorem 4.3, p. 126]. Now, let $[y_1] \in Y/A$ and let $[y_2] \in (f^*)^{-1}(f^*([y_1]))$. We show that $\rho([y_1], [y_2]) < \varepsilon$. To this end, it is enough to prove that $A \cup \{y_1\} \subset \mathcal{V}_\varepsilon^d(A \cup \{y_2\})$ and $A \cup \{y_2\} \subset \mathcal{V}_\varepsilon^d(A \cup \{y_1\})$. First, we prove that $A \cup \{y_1\} \subset \mathcal{V}_\varepsilon^d(A \cup \{y_2\})$. If $f^*([y_1]) = B$, then $f(y_1) \in B$. Since $f(A) = B$, there exists $a_1 \in A$ such that $f(a_1) = y_1$. This implies that $A \cup \{y_1\} \subset \mathcal{V}_\varepsilon^d(A \cup \{y_2\})$. Suppose that $f^*([y_1]) \neq B$. Then $f(y_1) \in Z \setminus B$. Hence, by the assumption on f and the fact that $[y_2] \in (f^*)^{-1}(f^*([y_1]))$, we have that $d(y_1, y_2) < \varepsilon$. Thus, $A \cup \{y_1\} \subset \mathcal{V}_\varepsilon^d(A \cup \{y_2\})$. By symmetry, we have that $A \cup \{y_2\} \subset \mathcal{V}_\varepsilon^d(A \cup \{y_1\})$. Therefore, $\rho([y_1], [y_2]) < \varepsilon$.

Q.E.D.

7.1.16 Remark One can construct examples to show that Lemma 7.1.15 would not necessarily remain valid if the condition $f(A) = B$ were replaced by $f(A) \subset B$; e.g.: Let $Y = Z$, let $f: Y \twoheadrightarrow Z$ be the identity map, let A be a proper subcontinuum of Y and let $B = Z$.

7.1.17 Remark Let X be a continuum and let $n \in \mathbb{N}$. Then Lemma 7.1.15 can be used to define a metric on $HS_n(X)$ in the following way: Let

$$\Im_n(X) = \{\mathcal{F}_n(X) \cup \{A\} \mid A \in \mathcal{C}_n(X)\}.$$

Note that $\Im_n(X) \subset \mathcal{C}_2(\mathcal{C}_n(X))$. Define $\xi_n: HS_n(X) \twoheadrightarrow \Im_n(X)$ by

$$\xi_n(\chi) = \mathcal{F}_n(X) \cup (q_X^n)^{-1}(\chi).$$

Then ξ_n is a homeomorphism. Next, define

$$\rho_X^n: HS_n(X) \times HS_n(X) \to [0, \infty)$$

by

$$\rho_X^n(\chi_1, \chi_2) = \mathcal{H}^2(\xi_n(\chi_1), \xi_n(\chi_2)),$$

where $\mathcal{H}^2$ is the Hausdorff metric on $\mathcal{C}_2(\mathcal{C}_n(X))$, induced by the Hausdorff metric on $\mathcal{C}_n(X)$. Then ρ_X^n is the required metric.

7.1.18 Definition Let F be a fan with top τ. Define

$$\mathcal{T}[\mathcal{C}(F)] = \bigcup_{e \in E(F)} \mathcal{C}(\tau e),$$

where τe denotes the unique arc in F from τ to e. $\mathcal{T}[\mathcal{C}(F)]$ is called the *two dimensional part of* $\mathcal{C}(F)$ [35, Lemma 3.1].

The next theorem gives us a model for the hyperspace suspension of a smooth fan, namely, its hyperspace of subcontinua.

7.1.19 Theorem *If F is a smooth fan, then $HS(F)$ is homeomorphic to $\mathcal{C}(F)$.*

Proof Note that if e_1 and e_2 are two different elements of $E(F)$, then $\mathcal{C}(\tau e_1) \cap \mathcal{C}(\tau e_2) = \{\{\tau\}\}$. Also observe that each $\mathcal{C}(\tau e)$, $e \in E(F)$, is a 2-cell [40, (0.54)]. The symbol $\partial(\mathcal{C}(\tau e))$ denotes the manifold boundary of $\mathcal{C}(\tau e)$. By [11, Theorem 3.1], $\mathcal{C}(F) = \mathcal{C}_\tau(F) \cup \mathcal{T}[\mathcal{C}(F)]$, where $\mathcal{C}_\tau(F) = \{A \in \mathcal{C}(F) \mid \tau \in A\}$ is homeomorphic to the Hilbert cube or $[0, 1]^n$ for some $n \in \mathbb{N}$. Note that

$$\mathcal{F}_1(F) = \bigcup_{e \in E(F)} \mathcal{F}_1(\tau e) \subset \bigcup_{e \in E(F)} \partial(\mathcal{C}(\tau e)) \subset \mathcal{T}[\mathcal{C}(F)].$$

Since F is a smooth fan, by [40, (0.54)] and the fact that when we shrink an arc in the manifold boundary of a two-cell we obtain a two-cell, we have that $\mathcal{T}[\mathcal{C}(F)]/\mathcal{F}_1(F)$ is homeomorphic to $\mathcal{T}[\mathcal{C}(F)]$. Since $\mathcal{F}_1(F) \cap \mathcal{C}_\tau(F) = \{\{\tau\}\}$, we have that $HS(F)$ is homeomorphic to $\mathcal{C}_\tau(F) \cup [\mathcal{T}[\mathcal{C}(F)]/\mathcal{F}_1(F)]$. Hence, $HS(F)$ is homeomorphic to $\mathcal{C}(F)$.

Q.E.D.

7.1.20 Theorem *If X is a continuum and $n \in \mathbb{N}$, then $HS_n(X)$ is uniformly pathwise connected.*

Proof It is known that for each $n \in \mathbb{N}$, $\mathcal{C}_n(X)$ is a continuous image of the cone over the Cantor set [25, Remark on p. 29 and Theorem 2.7]. Hence, $HS_n(X)$ is a continuous image of the cone over the Cantor set. Therefore, $HS_n(X)$ is uniformly pathwise connected.

Q.E.D.

7.1.21 Theorem *Let X be a finite-dimensional continuum. Then $\dim(\mathcal{C}_n(X)) < \infty$ if and only if $\dim(HS_n(X)) < \infty$. Moreover, $\dim(\mathcal{C}_n(X)) = \dim(HS_n(X))$.*

Proof Suppose that $\dim(\mathcal{C}_n(X)) < \infty$. Hence, $\dim(X) = 1$ [28, Theorem 2.1]. Thus, by [8, Lemma 3.1], $\dim(\mathcal{F}_n(X)) \leq n$. Since, by Theorem 6.1.11, $\mathcal{C}_n(X)$ contains n-cells, $\dim(\mathcal{C}_n(X)) \geq n$. Then there exists $A \in \mathcal{C}_n(X) \setminus \mathcal{F}_n(X)$ such that $\dim_A(\mathcal{C}_n(X)) = \dim(\mathcal{C}_n(X))$. Therefore, $\dim(\mathcal{C}_n(X) \setminus \mathcal{F}_n(X)) = \dim(\mathcal{C}_n(X))$. Since, by Remark 7.1.5, $\mathcal{C}_n(X) \setminus \mathcal{F}_n(X)$ is homeomorphic to $HS_n(X) \setminus \{F_X^n\}$ and also $\dim(HS_n(X) \setminus \{F_X^n\}) = \dim(HS_n(X))$ [19, Corollary 2, p. 32], we have that $\dim(\mathcal{C}_n(X) \setminus \mathcal{F}_n(X)) = \dim(HS_n(X))$. Therefore, $\dim(\mathcal{C}_n(X)) = \dim(HS_n(X))$.

Now, assume that $\dim(HS_n(X)) < \infty$. Since $\dim(HS_n(X)) = \dim(HS_n(X) \setminus \{F_X^n\})$ [19, Corollary 2, p. 32] and $HS_n(X) \setminus \{F_X^n\}$ is homeomorphic to $\mathcal{C}_n(X) \setminus \mathcal{F}_n(X)$ (Remark 7.1.5), we obtain that $\dim(HS_n(X)) = \dim(\mathcal{C}_n(X) \setminus \mathcal{F}_n(X))$. Suppose $\dim(\mathcal{C}_n(X)) = \infty$. Since $\mathcal{C}_n(X) = (\mathcal{C}_n(X) \setminus \mathcal{F}_n(X)) \cup \mathcal{F}_n(X))$ and $\dim(\mathcal{F}_n(X)) \leq n \cdot \dim(X) < \infty$ [8, Lemma 3.1], we have that $\dim(\mathcal{C}_n(X) \setminus \mathcal{F}_n(X)) = \infty$, a contradiction. Hence, $\dim(\mathcal{C}_n(X)) < \infty$. Since $\mathcal{F}_n(X)$ is now where dense in $\mathcal{C}_n(X)$, $\dim(\mathcal{C}_n(X)) = \dim(\mathcal{C}_n(X) \setminus \mathcal{F}_n(X))$. Therefore, $\dim(\mathcal{C}_n(X)) = \dim(HS_n(X))$.

From what we have done, now it is clear that $\dim(\mathcal{C}_n(X)) = \infty$ if and only if $\dim(HS_n(X)) = \infty$. Therefore, $\dim(\mathcal{C}_n(X)) = \dim(HS_n(X))$.

Q.E.D.

As a consequence of Theorem 7.1.21 and [28, Theorem 2.1], we have:

7.1.22 Corollary *Let X be a continuum and let $n \in \mathbb{N}$. If $\dim(HS_n(X)) < \infty$, then $\dim(X) = 1$.*

7.1.23 Theorem *If X is a continuum and $n \in \mathbb{N}$, then $HS_n(X)$ contains an n-cell.*

Proof By Theorem 6.1.11, $\mathcal{C}_n(X)$ contains an n-cell. Note that by the construction made in the proof of Theorem 6.1.11, $\mathcal{C}_n(X) \setminus \mathcal{F}_n(X)$ also contains an n-cell. Therefore, by Remark 7.1.5, $HS_n(X)$ contains an n-cell.

Q.E.D.

7.1.24 Theorem *Let X be a continuum and let $n \in \mathbb{N}$. If X contains n pairwise disjoint decomposable continua, then $HS_n(X)$ contains a $2n$-cell.*

Proof By Theorem 6.1.12, $C_n(X)$ contains a $2n$-cell. Observe that by the construction made in Theorem 6.1.12, $C_n(X) \setminus \mathcal{F}_n(X)$ contains a $2n$-cell too. Therefore, by Remark 7.1.5, $HS_n(X)$ contains a $2n$-cell.

Q.E.D.

7.1.25 Theorem *Let X be a continuum and let $n \in \mathbb{N}$. If $C_n(X)$ is a finite-dimensional Cantor manifold such that $\dim(C_n(X)) \geq n + 2$, then $HS_n(X)$ is a finite-dimensional Cantor manifold.*

Proof By Theorem 7.1.21, $\dim(HS_n(X)) < \infty$. Let $\dim(HS_n(X)) = k$. Suppose $HS_n(X)$ is not a Cantor manifold. Then there exists a subset $\mathcal{A}$ of $HS_n(X)$ such that $\dim(\mathcal{A}) \leq k - 2$ and $HS_n(X) \setminus \mathcal{A}$ is not connected. We assume, without loss of generality that $\mathcal{A}$ is closed in $HS_n(X)$ [19, B], p. 94]. Hence, there exist two disjoint open subsets $\mathcal{D}$ and $\mathcal{E}$ of $HS_n(X)$ such that $HS_n(X) \setminus \mathcal{A} = \mathcal{D} \cup \mathcal{E}$. Then $C_n(X) \setminus (q_X^n)^{-1}(\mathcal{A}) = (q_X^n)^{-1}(\mathcal{D}) \cup (q_X^n)^{-1}(\mathcal{E})$, where $(q_X^n)^{-1}(\mathcal{D})$ and $(q_X^n)^{-1}(\mathcal{E})$ are disjoint open subsets of $C_n(X)$. To obtain a contradiction to the fact that $C_n(X)$ is a Cantor manifold, we show that $\dim((q_X^n)^{-1}(\mathcal{A})) \leq k-2$ (since, by Theorem 7.1.21, $\dim(C_n(X)) = \dim(HS_n(X))$).

Suppose that $F_X^n \in HS_n(X) \setminus \mathcal{A}$. Then $(q_X^n)^{-1}(\mathcal{A})$ is homeomorphic to $\mathcal{A}$ [19, A), p. 24]. Hence, $\dim((q_X^n)^{-1}(\mathcal{A})) \leq k - 2$.

Observe that, since $\mathcal{C}(X) \subset C_n(X)$ and $\dim(C_n(X)) < \infty$, we have that $\dim(X) = 1$ [28, Theorem 2.1]. Hence, $\dim(\mathcal{F}_n(X)) \leq n$ [8, Lemma 3.1].

Suppose $F_X^n \in \mathcal{A}$. Then $(q_X^n)^{-1}(\mathcal{A}) = (q_X^n)^{-1}(\mathcal{A} \setminus \{F_X^n\}) \cup (q_X^n)^{-1}(\{F_X^n\}) = (q_X^n)^{-1}(\mathcal{A} \setminus \{F_X^n\}) \cup \mathcal{F}_n(X)$. Since, by Remark 7.1.5 and [19, Theorem III1, p. 26], $\dim((q_X^n)^{-1}(\mathcal{A} \setminus \{F_X^n\})) \leq \dim(\mathcal{A}) \leq k - 2$. Also, since $\dim(\mathcal{F}_n(X)) \leq n \leq k - 2$ and $\dim((q_X^n)^{-1}(\mathcal{A} \setminus \{F_X^n\})) \leq k - 2$, $\dim((q_X^n)^{-1}(\mathcal{A})) \leq k - 2$ [19, Corollary 1, p. 32].

Q.E.D.

7.1.26 Corollary *For each $n \in \mathbb{N}$, $HS_n([0, 1])$ and $HS_n(\mathcal{S}^1)$ are $2n$-dimensional Cantor manifolds.*

Proof We consider two cases. If $n = 1$, then by Example 7.1.6 $HS([0, 1])$ is a 2-cell and by Example 7.1.7, $HS(\mathcal{S}^1)$ is a 2-sphere. Since 2-cells and 2-spheres are Cantor manifolds [19, Example VI11, p. 93], we are done in this case.

Now, suppose $n \geq 2$. By Theorem 6.9.17, $C_n([0, 1])$ and $C_n(\mathcal{S}^1)$ are $2n$-dimensional Cantor manifolds. Since $n \geq 2$, $2n \geq n+2$; i.e., $\dim(C_n([0, 1])) \geq n+2$ and $\dim(C_n(\mathcal{S}^1)) \geq n + 2$. Thus, the corollary follows from Theorems 7.1.25 and 7.1.21.

Q.E.D.

In fact, next theorem characterizes the graphs whose n-fold hyperspace suspensions are Cantor manifolds.

7.1.27 Theorem *Let X be a graph and let $n \in \mathbb{N}$. Then $HS_n(X)$ is a Cantor manifold if and only if X is either an arc or a simple closed curve.*

Proof Suppose X is neither an arc nor a simple closed curve. Then X has a ramification point p. Hence, by the proof of Theorem 6.9.21, there exists an element A of $C_n(X)$ with exactly n components such that $\dim_A(C_n(X)) \geq 2n + 1$. Since $q_X^n|_{C_n(X) \setminus \mathcal{F}_n(X)}$ is a homeomorphism (Remark 7.1.5), we have that $\dim_{q_X^n(A)}(HS_n(X)) \geq 2n + 1$. Now, by Lemma 7.1.14, there exists an element χ in $HS_n(X)$ such that $\dim_\chi(HS_n(X)) = 2n$. Since Cantor manifolds have the same dimension at each of its points [19, A), pp. 93 and 94], we obtain that $HS_n(X)$ is not a Cantor manifold.

If X is either an arc or a simple closed curve, by Corollary 7.1.26, $HS_n(X)$ is a $2n$-dimensional Cantor manifold.

Q.E.D.

7.1.28 Theorem *If X is a continuum and $n \geq 2$, then any 2-cell in $HS_n(X)$ is nowhere dense.*

Proof The result follows from the fact that any 2-cell in $C_n(X)$ is nowhere dense, Theorem 6.10.8.

Q.E.D.

We continue our study of n-fold hyperspace suspensions by noting that if $m < n$, then the m-fold hyperspace suspension of a continuum may be embedded in its n-fold hyperspace suspension.

7.1.29 Theorem *Let X be a continuum. Let n and m be positive integers such that $n > m$. Then $HS_m(X)$ may be embedded in $HS_n(X)$.*

Proof Let $i_{m,n} \colon C_m(X) \to C_n(X)$ be the inclusion map. Observe that $i_{m,n}$ is a relation preserving map [10, p. 16]. Now let $h_{m,n} \colon HS_m(X) \to HS_n(X)$ be given by

$$h_{m,n}(\chi) = \begin{cases} F_X^n, & \text{if } \chi = F_X^m; \\ q_X^n\left(i_{m,n}\left(\left(q_X^m\right)^{-1}(\chi)\right)\right), & \text{if } \chi \neq F_X^m. \end{cases}$$

Note that $h_{m,n}$ is continuous by [10, 4.3, p. 126]. It is clear that $h_{m,n}$ is one-to-one. Since the spaces are compacta, $h_{m,n}$ is an embedding. Therefore, $HS_m(X)$ may be embedded in $HS_n(X)$.

Q.E.D.

The next two theorems are used in the proof of Theorem 7.9.6.

7.1.30 Theorem *Let X be a continuum and let $n \in \mathbb{N}$. If $\chi \in HS_n(X) \setminus \{T_X^n, F_X^n\}$, then $HS_n(X) \setminus \{T_X^n, \chi\}$ is arcwise connected.*

Proof Let $\chi' \in HS_n(X) \setminus \{T_X^n, F_X^n, \chi\}$. We show that there exists an arc joining χ' and F_X^n contained in $HS_n(X) \setminus \{T_X^n, \chi\}$. Let $A_1, \ldots, A_k$ be the components of $\left(q_X^n\right)^{-1}(\chi')$.

Suppose that $\left(q_X^n\right)^{-1}(\chi') \setminus \left(q_X^n\right)^{-1}(\chi) \neq \emptyset$. Let $a_1 \in \left(q_X^n\right)^{-1}(\chi') \setminus \left(q_X^n\right)^{-1}(\chi)$. Without loss of generality, we assume that $a_1 \in A_1$. For each $j \in \{2, \ldots, k\}$, let $a_j \in A_j$. By Theorem 1.8.20, there exists an order arc $\alpha: [0, 1] \to C_n(X)$ such that $\alpha(0) = \{a_1, \ldots, a_k\}$ and $\alpha(1) = \left(q_X^n\right)^{-1}(\chi')$. Note that, by construction, $\{X, \left(q_X^n\right)^{-1}(\chi)\} \cap \alpha([0, 1]) = \emptyset$. Hence, $q_X^n \circ \alpha: [0, 1] \to HS_n(X)$ is an arc such that $q_X^n \circ \alpha(0) = F_X^n$ and $q_X^n \circ \alpha(1) = \chi'$ and $\{T_X^n, \chi\} \cap \left(q_X^n \circ \alpha([0, 1])\right) = \emptyset$.

Next, suppose that $\left(q_X^n\right)^{-1}(\chi') \subset \left(q_X^n\right)^{-1}(\chi)$. For each $j \in \{1, \ldots, k\}$, let $a_j \in A_j$. By Theorem 1.8.20, there exists an order arc $\beta: [0, 1] \to C_n(X)$ such that $\beta(0) = \{a_1, \ldots, a_k\}$ and $\beta(1) = \left(q_X^n\right)^{-1}(\chi')$. Again, by construction, $\{X, \left(q_X^n\right)^{-1}(\chi)\} \cap \beta([0, 1]) = \emptyset$. Thus, $q_X^n \circ \beta: [0, 1] \to HS_n(X)$ is an arc such that $q_X^n \circ \beta(0) = F_X^n, q_X^n \circ \beta(1) = \chi'$ and $\{T_X^n, \chi\} \cap \left(q_X^n \circ \beta([0, 1])\right) = \emptyset$. Therefore, $HS_n(X) \setminus \{T_X^n, \chi\}$ is arcwise connected.

Q.E.D.

7.1.31 Theorem *Let X be a continuum and let $n \in \mathbb{N}$. If $\chi \in HS_n(X) \setminus \{F_X^n\}$ is such that $HS_n(X) \setminus \{F_X^n, \chi\}$ is not arcwise connected, then $\left(q_X^n\right)^{-1}(\chi) \in C(X)$.*

Proof Suppose there exists $\chi \in HS_n(X) \setminus \{F_X^n\}$ such that $HS_n(X) \setminus \{F_X^n, \chi\}$ is not arcwise connected. Hence, $C_n(X) \setminus (\{\left(q_X^n\right)^{-1}(\chi)\} \cup \mathcal{F}_n(X)$ is not arcwise connected. Since $C_n(X) \setminus \mathcal{F}_n(X)$ is arcwise connected (by Theorem 6.5.2 and the fact that singletons do not arcwise disconnect $C_n(X)$), we have that $C_n(X) \setminus \{\left(q_X^n\right)^{-1}(\chi)\}$ is not arcwise connected. Thus, $\left(q_X^n\right)^{-1}(\chi) \in C(X)$, by Theorem 6.5.2.

Q.E.D.

The next theorem shows that for indecomposable continua n-fold hyperspaces and n-fold hyperspace suspensions are topologically different.

7.1.32 Theorem *If X is an indecomposable continuum and $n \in \mathbb{N}$, then $C_n(X)$ is not homeomorphic to $HS_n(X)$.*

Proof Note that if X is an indecomposable continuum, then $C_n(X) \setminus \{X\}$ is not arcwise connected, by Theorem 6.5.3. Meanwhile, it is easy to see that no point arcwise disconnects $HS_n(X)$.

Q.E.D.

Next, we compare n-fold hyperspace suspensions with n-fold symmetric products.

7.1.33 Theorem *If X is a finite-dimensional continuum, then $HS_n(X)$ is not homeomorphic to $\mathcal{F}_n(X)$ for any integer n greater than one.*

Proof The proof is similar to the proof given for Theorem 6.1.30. We just need to mention that $\dim(HS_n(X)) = \dim(\mathcal{C}_n(X))$, by Theorem 7.1.21, and that $HS_n(X)$ is arcwise connected, by Theorem 7.1.3.

Q.E.D.

The proofs of the following two theorems are similar to the ones given in [29, Theorems 9 and 11], respectively.

7.1.34 Theorem *Let X be a finite-dimensional continuum. If $HS(X)$ is homeomorphic to $\mathcal{F}_2(X)$, then X is homeomorphic to $[0, 1]$.*

Proof Suppose $HS(X)$ is homeomorphic to $\mathcal{F}_2(X)$. Since X is a finite-dimensional continuum, by the proof of [8, 3.1], we have that $\dim(\mathcal{F}_2(X)) \leq 2 \cdot \dim(X)$. Since $HS(X)$ is homeomorphic to $\mathcal{F}_2(X)$, $\dim(HS(X)) \leq 2 \cdot \dim(X)$. Thus, $\dim(X) = 1$, by Corollary 7.1.22. Hence, $2 \leq \dim(\mathcal{C}(X)) = \dim(HS(X)) = \dim(\mathcal{F}_2(X)) \leq 2$, the first inequality follows from [11, Theorem 1]. Therefore, $\dim(\mathcal{C}(X)) = \dim(HS(X)) = 2$. Now, by [45, Theorem 1], X is atriodic. Since, by Theorem 7.1.3, $HS(X)$ is arcwise connected, $\mathcal{F}_2(X)$ is arcwise connected. Thus, X is arcwise connected [8, 2.2].

Now, suppose X is not unicoherent. Then by the proof of [20, 1.1], there exists a map $f: \mathcal{F}_2(X) \to \mathcal{S}^1$ such that f is not homotopic to a constant map. Since $HS(X)$ has property (b), Theorem 7.1.13, $\mathcal{F}_2(X)$ has property (b), a contradiction. Therefore, X is unicoherent. Thus, by Sorgenfrey's Theorem [42, 11.34], X is an irreducible continuum. Since X is arcwise connected, X is an arc.

Q.E.D.

7.1.35 Theorem *Let X be a finite-dimensional continuum. If $n \geq 3$, then $HS(X)$ is not homeomorphic to $\mathcal{F}_n(X)$.*

Proof Let n be an integer greater than two. Suppose $HS(X)$ is homeomorphic to $\mathcal{F}_n(X)$. Since X is finite-dimensional, by the proof of [8, 3.1], we have that $\dim(\mathcal{F}_n(X)) \leq n \cdot \dim(X)$. Since $HS(X)$ is homeomorphic to $\mathcal{F}_n(X)$, $\dim(HS(X)) < \infty$. Hence, $\dim(X) = 1$, by Corollary 7.1.22. Therefore, $\dim(\mathcal{F}_n(X)) \leq n$. Since X is finite-dimensional, $\dim(\mathcal{F}_n(X)) = \dim(X^n)$ [43, 22.12]. Also, since $\dim(X) = 1$, $\dim(X^n) \geq n$ [18, (a), p. 197]. Thus, $n = \dim(\mathcal{F}_n(X)) = \dim(HS(X)) = \dim(\mathcal{C}(X))$. By [45, Theorem 1], X does not contain $(n+1)$-ods. Hence, since $HS_n(X)$ is arcwise connected (Theorem 7.1.3), $\mathcal{F}_n(X)$ is arcwise connected. Thus, X is arcwise connected [8, 2.2]. As a consequence of this, X contains a free arc [29, Theorem 11]. This implies that $\mathcal{C}(X)$ and $HS(X)$ contain a 2-dimensional subset with nonempty interior. But, since $n \geq 3$, $\mathcal{F}_n(X)$ does not contain 2-dimensional subsets with nonempty interior. Therefore, $HS(X)$ is not homeomorphic to $\mathcal{F}_n(X)$.

Q.E.D.

As a consequence of Theorems 6.9.11 and 7.1.21, we have:

7.1.36 Theorem *If X is either an arc-like or a circle-like continuum and $n \in \mathbb{N}$, then $\dim(HS_n(X)) \leq 2n$.*

Now, we consider arc-smoothness on n-fold hyperspace suspensions.

7.1.37 Remark Note that if X is the unit circle, then, since $\mathcal{C}(X)$ is a 2-cell and $\mathcal{F}_1(X)$ is the manifold boundary of the cell [40, (0.55)], $\mathcal{C}(X)$ is arc-smooth [13, p. 545], but $HS(X)$ is not arc-smooth, because $HS(X)$ is a 2-sphere, which is not contractible (arc-smooth continua are contractible [13, II-3-A]).

7.1.38 Theorem *If X is an arc-smooth continuum and $n \in \mathbb{N}$, then $HS_n(X)$ is arc-smooth.*

Proof Let $G \colon \mathcal{C}_n(X) \times [0, 1] \to \mathcal{C}_n(X)$ be the map defined in the proof of Theorem 6.1.31. Let $K \colon HS_n(X) \times [0, 1] \to HS_n(X)$ be given by

$$K((\chi, t)) = \begin{cases} F_X^n, & \text{if } \chi = F_X^n; \\ q_X^n\left(G\left((q_X^n)^{-1}(\chi), t\right)\right), & \text{if } \chi \neq F_X^n. \end{cases}$$

Then K is continuous by [10, 4.3, p. 126]. Observe that for each $\chi \in HS_n(X)$, $K((\chi, 0)) = F_X^n$, $K((\chi, 1)) = \chi$. It is also easy to see that $K((K((\chi, s)), t)) = K((\chi, \min\{s, t\}))$ for all $s, t \in [0, 1]$ and each $\chi \in HS_n(X)$. Hence, $HS_n(X)$ is freely contractible. Therefore, by [13, Theorem II-3-B], $HS_n(X)$ is arc-smooth.

Q.E.D.

7.2   Contractibility

We give sufficient conditions for a continuum to have its n-fold hyperspace suspensions contractible. To this end, we use [3, 4, 10, 32, 39, 40].

7.2.1 Theorem *Let X be a continuum and let $n \in \mathbb{N}$. If X is contractible, then $HS_n(X)$ is contractible.*

Proof Let $R \colon X \times [0, 1] \to X$ be such that $R((x, 0)) = x$ and $R((x, 1)) = q$ for every $x \in X$ and some $q \in X$. Define $G \colon \mathcal{C}_n(X) \times [0, 1] \to \mathcal{C}_n(X)$ by $G((A, t)) = R(A \times \{t\})$. Note that $G((A, 0)) = A$, $G((A, 1)) = \{q\}$ for every $A \in \mathcal{C}_n(X)$ and if $A \in \mathcal{F}_n(X)$, then $G((A, t)) \in \mathcal{F}_n(X)$ for each $t \in [0, 1]$.

Let

$$K \colon HS_n(X) \times [0, 1] \to HS_n(X)$$

be given by

$$K((\chi, t)) = \begin{cases} F_X^n, & \text{if } \chi = F_X^n; \\ q_X^n\left(G\left((q_X^n)^{-1}(\chi), t\right)\right), & \text{if } \chi \neq F_X^n. \end{cases}$$

Note that K is continuous by [10, 4.3, p. 126]. Observe also that

$$K((\chi, 0)) = \chi$$

and

$$K((\chi, 1)) = F_X^n.$$

for each $\chi \in H S_n(X)$.
   Therefore, $H S_n(X)$ is contractible.

Q.E.D.

   We omit the proof of the following theorem because it is similar to the one given in Theorem 7.2.1.

7.2.2 Theorem *Let X be a continuum and let $n \in \mathbb{N}$. If $\mathcal{F}_n(X)$ is a strong deformation retract of $C_n(X)$, then $H S_n(X)$ is contractible.*

7.2.3 Theorem *If X is a locally connected continuum and $n \in \mathbb{N}$, then $H S_n(X) \setminus \{F_X^n\}$ is contractible.*

Proof Since X is locally connected, we assume that X has a convex metric ρ ([4] and [39]). Without loss of generality, we assume that $\text{diam}(X) \leq 1$. Let $K_\rho \colon [0, 1] \times C_n(X) \to C_n(X)$ be given by

$$K_\rho((t, A)) = \{x \in X \mid \rho(x, y) \leq t \text{ for some } y \in A\}.$$

K_ρ is continuous [40, (0.65.3)(f)], $K_\rho((0, A)) = A$ and $K_\rho((1, A)) = X$.
   Note that $K_\rho([0, 1] \times (C_n(X) \setminus \mathcal{F}_n(X))) \subset C_n(X) \setminus \mathcal{F}_n(X)$. Hence, $C_n(X) \setminus \mathcal{F}_n(X)$ is contractible. Therefore, $H S_n(X) \setminus \{F_X^n\}$ is contractible by Remark 7.1.5.

Q.E.D.

7.3 Aposyndesis

We prove that for every $n \in \mathbb{N}$, $H S_n(X)$ is colocally connected and finitely aposyndetic. We also show that the $H S_n(X)$ is zero-dimensional aposyndetic. For this purpose, we follow [2, 30, 32].

7.3.1 Theorem *If X is a continuum and $n \in \mathbb{N}$, then $H S_n(X)$ is colocally connected.*

Proof Let $\chi \in H S_n(X)$. We consider three cases.

Case (1) $\chi = F_X^n$.
   For each $\varepsilon > 0$, let $\mathcal{U}_\varepsilon = \mathcal{V}_\varepsilon^{\mathcal{H}}(\mathcal{F}_n(X))$. Then $\{q_X^n(\mathcal{U}_\varepsilon) \mid \varepsilon > 0\}$ forms a base of open sets about F_X^n.

Fix $\varepsilon > 0$. Let $\chi_1 \in HS_n(X) \setminus q_X^n(\mathcal{U}_\varepsilon)$. Then $(q_X^n)^{-1}(\chi_1) \in C_n(X) \setminus \mathcal{U}_\varepsilon$. Let $\beta: [0, 1] \to C_n(X)$ be an order arc such that $\beta(0) = (q_X^n)^{-1}(\chi_1)$ and $\beta(1) = X$ (Theorem 1.8.20). Then $\beta([0, 1]) \subset C_n(X) \setminus \mathcal{U}_\varepsilon$. Hence, $q_X^n \circ \beta: [0, 1] \to HS_n(X)$ is an arc such that $q_X^n \circ \beta(0) = \chi_1, q_X^n \circ \beta(1) = T_X^n$ and $q_X^n \circ \beta([0, 1]) \subset HS_n(X) \setminus q_X^n(\mathcal{U}_\varepsilon)$.

Case (2) $\chi = T_X^n$.

For each $\varepsilon > 0$, let $\mathcal{W}_\varepsilon = \mathcal{V}_\varepsilon^{\mathcal{H}}(X)$. Then $\{q_X^n(\mathcal{W}_\varepsilon) \mid \varepsilon > 0\}$ forms a base of open sets about T_X^n.

Fix $\varepsilon > 0$. Let $\chi_2 \in HS_n(X) \setminus q_X^n(\mathcal{W}_\varepsilon)$. Then $(q_X^n)^{-1}(\chi_2) \in C_n(X) \setminus \mathcal{W}_\varepsilon$. Let $G \in \mathcal{F}_n((q_X^n)^{-1}(\chi_2))$ be such that there exists an order arc $\gamma: [0, 1] \to C_n(X)$ such that $\gamma(0) = G$ and $\gamma(1) = (q_X^n)^{-1}(\chi_2)$ (Theorem 1.8.20). Then $\gamma([0, 1]) \subset C_n(X) \setminus \mathcal{W}_\varepsilon$. Hence, $q_X^n \circ \gamma: [0, 1] \to HS_n(X)$ is an arc such that $q_X^n \circ \gamma(0) = F_X^n$, $q_X^n \circ \gamma(1) = \chi_2$ and $q_X^n \circ \gamma([0, 1]) \subset HS_n(X) \setminus q_X^n(\mathcal{W}_\varepsilon)$.

Case (3) $\chi \in HS_n(X) \setminus \{T_X^n, F_X^n\}$.

For each $\varepsilon > 0$, let $\mathcal{L}_\varepsilon = \mathcal{V}_\varepsilon^{\mathcal{H}}((q_X^n)^{-1}(\chi))$. Then $\{q_X^n(\mathcal{L}_\varepsilon) \mid \varepsilon > 0\}$ forms a base of open sets about χ.

Fix $\varepsilon > 0$ such that $q_X^n(\mathcal{L}_\varepsilon) \cap \{T_X^n, F_X^n\} = \emptyset$. Let $\chi_3 \in HS_n(X) \setminus q_X^n(\mathcal{L}_\varepsilon)$. If $(q_X^n)^{-1}(\chi_3) \not\subset (q_X^n)^{-1}(\chi)$, then let $\alpha: [0, 1] \to C_n(X)$ be an order arc such that $\alpha(0) = (q_X^n)^{-1}(\chi_3)$ and $\alpha(1) = X$. Then $\alpha([0, 1]) \subset C_n(X) \setminus \mathcal{V}_\varepsilon^{\mathcal{H}}((q_X^n)^{-1}(\chi))$. Hence, $q_X^n \circ \alpha: [0, 1] \to HS_n(X)$ is an arc such that $q_X^n \circ \alpha(0) = \chi_3, q_X^n \circ \alpha(1) = T_X^n$ and $q_X^n \circ \alpha([0, 1]) \subset HS(X) \setminus q_X(\mathcal{L}_\varepsilon)$.

Suppose $(q_X^n)^{-1}(\chi_3) \subset (q_X^n)^{-1}(\chi)$. Let $J \in \mathcal{F}_n((q_X^n)^{-1}(\chi_3))$ be such that there exists an order arc $\kappa: [0, 1] \to C_n(X)$ such that $\kappa(0) = J$ and $\kappa(1) = (q_X^n)^{-1}(\chi_3)$ (Theorem 1.8.20). Then $\kappa([0, 1]) \subset C_n(X) \setminus \mathcal{V}_\varepsilon^{\mathcal{H}}((q_X^n)^{-1}(\chi))$. Hence, $q_X^n \circ \kappa: [0, 1] \to HS_n(X)$ is an arc such that $q_X^n \circ \kappa(0) = F_X^n$, $q_X^n \circ \kappa(1) = \chi_3$ and $q_X^n \circ \kappa([0, 1]) \subset HS_n(X) \setminus q_X(\mathcal{L}_\varepsilon)$.

Q.E.D.

Clearly, colocal connectedness implies aposyndesis. Hence, we have the following result as a consequence of Theorem 7.3.1:

7.3.2 Corollary *If X is a continuum and $n \in \mathbb{N}$, then $HS_n(X)$ is aposyndetic.*

7.3.3 Corollary *If X is a continuum and $n \in \mathbb{N}$, then $HS_n(X)$ is finitely aposyndetic.*

Proof By Theorem 7.1.13, $HS_n(X)$ is unicoherent. Any aposyndetic unicoherent continuum is finitely aposyndetic [2, Corollary 1].

Q.E.D.

The next theorem extends Corollary 7.3.3 to zero-dimensional aposyndesis.

7.3.4 Theorem *Let X be a continuum and let $n \in \mathbb{N}$. Then $HS_n(X)$ is zero-dimensional aposyndetic.*

Proof Since $HS_n(X)$ is locally connected at T_X^n and at F_X^n, $HS_n(X)$ is zero-dimensional aposyndetic at T_X^n and at F_X^n.

Let $\chi \in HS_n(X) \setminus \{T_X^n, F_X^n\}$, and let $\mathcal{Z}$ be a zero-dimensional closed subset of $HS_n(X)$ such that $\chi \notin \mathcal{Z}$. Since $\mathcal{Z}$ is closed and $\chi \notin \mathcal{Z}$, there exists $\varepsilon > 0$ such that

$$Cl_{\mathcal{C}_n(X)} \left(\mathcal{V}_{\varepsilon}^{\mathcal{H}} \left(\left(q_X^n \right)^{-1} (\chi) \right) \right) \cap \mathcal{F}_n(X) = \emptyset$$

and

$$Cl_{HS_n(X)} \left(q_X^n \left(\mathcal{V}_{\varepsilon}^{\mathcal{H}} \left(\left(q_X^n \right)^{-1} (\chi) \right) \right) \right) \cap \mathcal{Z} = \emptyset.$$

Let $\mathcal{Z}' = \mathcal{Z} \setminus \{F_X^n\}$. Then $\dim(\mathcal{Z}') \leq 0$ and $\mathcal{V}_{\varepsilon}^{\mathcal{H}} \left(\left(q_X^n \right)^{-1} (\chi) \right) \cap \left(q_X^n \right)^{-1} (\mathcal{Z}') = \emptyset$. By Theorem 6.3.12, there exists a subcontinuum $\mathcal{M}$ of $\mathcal{C}_n(X)$ such that $\left(q_X^n \right)^{-1} (\chi) \in Int_{\mathcal{C}_n(X)}(\mathcal{M})$ and $\mathcal{M} \cap \left(q_X^n \right)^{-1} (\mathcal{Z}') = \emptyset$.[1] Note that $\mathcal{M}$ may be constructed in such a way that $\mathcal{F}_n(X) \cap \mathcal{M} = \emptyset$. Hence, $q_X^n(\mathcal{M})$ is a subcontinuum of $HS_n(X)$ such that $\chi \in Int_{HS_n(X)} \left(q_X^n(\mathcal{M}) \right)$ and $q_X^n(\mathcal{M}) \cap \mathcal{Z}' = \emptyset$. Since $\mathcal{M} \cap \mathcal{F}_n(X) = \emptyset$, $F_X^n \notin q_X^n(\mathcal{M})$. Thus, $q_X^n(\mathcal{M}) \cap \mathcal{Z} = \emptyset$. Therefore, $HS_n(X)$ is zero-dimensional aposyndetic.

Q.E.D.

7.4 Local Connectedness

We present results about n-fold hyperspace suspensions of locally connected continua. In order to do this, we use [5–7, 10, 12, 16, 19, 21, 22, 26, 27, 30, 34, 38, 40, 42, 47].

7.4.1 Lemma *Let X be a continuum and let n be a positive integer. If $\mathcal{C}_n(X) \setminus \mathcal{F}_n(X)$ is locally connected, then X is locally connected.*

Proof First we prove the case $n = 1$. Suppose X is not locally connected. Then there exists a point p in X such that X is not connected im kleinen at p (Theorem 1.7.12). Hence, there exists $\delta > 0$ such that if V is a neighborhood of p in X and $V \subset \mathcal{V}_{\delta}^d(p)$, then V is not connected.

Let V be a neighborhood of p in X such that $Cl(V) \subset \mathcal{V}_{\delta}^d(p)$ and let A' be the component of $Cl(V)$ containing p. Then, by [47, (12.1), p. 18], we have that $p \in A'$, $A' \subset \mathcal{V}_{\delta}^d(p)$ and there exist a subcontinuum A of A' and a sequence $\{A_m\}_{m=1}^{\infty}$ of subcontinua of X such that $p \in A$, $\{A_m\}_{m=1}^{\infty}$ converges to A, $A_{\ell} \cap A_m = \emptyset$ if $\ell \neq m$ and $A_m \cap A = \emptyset$ for all $m \in \mathbb{N}$. In fact, the component of $\mathcal{V}_{\delta}^d(p)$ that contains p is disjoint from A_m for each $m \in \mathbb{N}$.

---

[1] The only place in the proof of Theorem 6.3.12 at which the hypothesis of the zero-dimensional subset $\mathcal{Z}$ being closed is to construct an open set about the point whose closure misses $\mathcal{Z}$. To construct $\mathcal{M}$ only the fact that $\dim(\mathcal{Z}) \leq 0$ is used.

Since $A \in \langle \mathcal{V}_\delta^d(p) \rangle \cap (\mathcal{C}(X) \setminus \mathcal{F}_1(X))$ and $\mathcal{C}(X) \setminus \mathcal{F}_1(X)$ is locally connected, there exists a connected open subset $\mathcal{W}$ of $\mathcal{C}(X) \setminus \mathcal{F}_1(X)$ such that $A \in \mathcal{W} \subset Cl_{\mathcal{C}(X)}(\mathcal{W}) \subset \langle \mathcal{V}_\delta^d(p) \rangle \cap (\mathcal{C}(X) \setminus \mathcal{F}_1(X))$. Since $\{A_m\}_{m=1}^\infty$ converges to A, there exists $M \in \mathbb{N}$ such that $A_m \in \mathcal{W}$ if $m \geq M$. Let $W = \bigcup Cl_{\mathcal{C}(X)}(\mathcal{W})$. Note that W is a subcontinuum of X (Corollary 6.1.2), $W \subset \mathcal{V}_\delta^d(p)$ and $A_m \subset W$ for each $m \geq M$. As a consequence of this, we have that $A \subset W$. This contradicts the fact that every element of the sequence $\{A_m\}_{m=1}^\infty$ is disjoint from the component of $\mathcal{V}_\delta^d(p)$ that contains p. Therefore, X is locally connected.

Now, assume $n \geq 2$. Let x be a point in X and let U be an open set of X such that $x \in U$. Thus, by Corollary 1.7.28, there exists a nondegenerate subcontinuum A of X such that $x \in A \subset U$. Note that $A \in \langle U \rangle_n$. In fact, $A \in \langle U \rangle_n \setminus \mathcal{F}_n(X)$. Since $\mathcal{C}_n(X) \setminus \mathcal{F}_n(X)$ is locally connected, there exists a connected open set $\mathcal{W}$ of $\mathcal{C}_n(X) \setminus \mathcal{F}_n(X)$ such that $A \in \mathcal{W} \subset Cl_{\mathcal{C}_n(X)}(\mathcal{W}) \subset \langle U \rangle_n \setminus \mathcal{F}_n(X)$ (it is not difficult to see that the closure may be taken in $\mathcal{C}_n(X)$). Let $\varepsilon > 0$ be such that $\mathcal{V}_\varepsilon^{\mathcal{H}}(A) \cap \mathcal{C}_n(X) \subset \mathcal{W}$. Let $K = \bigcup Cl_{\mathcal{C}_n(X)}(\mathcal{W})$. Then, K is a subcontinuum of X, by Lemma 6.1.4. Let $y \in \mathcal{V}_\varepsilon^d(x)$. Note that $B = A \cup \{y\} \in \mathcal{V}_\varepsilon^{\mathcal{H}}(A) \cap \mathcal{C}_n(X) \subset \mathcal{W}$ and $y \in K$. Then, $x \in Int_X(K) \subset U$. Therefore, X is connected im kleinen at x. Thus, since x is an arbitrary point of X, X is connected im kleinen at each of its points. Therefore, by Theorem 1.7.12, X is locally connected.

Q.E.D.

7.4.2 Theorem *A continuum X is locally connected if and only if $HS_n(X)$ is locally connected for each $n \in \mathbb{N}$.*

Proof If X is locally connected, then, by Theorem 6.1.6, $\mathcal{C}_n(X)$ is locally connected. Thus, since $HS_n(X) = q_X^n(\mathcal{C}_n(X))$, $HS_n(X)$ is locally connected [27, Theorem 5, p. 257].

If $HS_n(X)$ is locally connected, then $HS_n(X) \setminus \{F_X^n\}$ is locally connected [27, Theorem 3, p. 230]. Since $q_X^n|_{\mathcal{C}_n(X) \setminus \mathcal{F}_n(X)} : \mathcal{C}_n(X) \setminus \mathcal{F}_n(X) \twoheadrightarrow HS_n(X) \setminus \{F_X^n\}$ is a homeomorphism, we have that $\mathcal{C}_n(X) \setminus \mathcal{F}_n(X)$ is locally connected. Therefore, by Lemma 7.4.1, X is locally connected.

Q.E.D.

By Theorems 6.9.3 and 7.1.21, we have the following characterization of graphs:

7.4.3 Theorem *A locally connected continuum X is a graph if and only if for each $n \in \mathbb{N}$, $HS_n(X)$ is finite-dimensional.*

It is well known that the hyperspace of subcontinua of a locally connected continuum without free arcs is homeomorphic to the Hilbert cube [40, (1.98)]. The following example shows that the corresponding result for the hyperspace suspension of such a continuum does not hold.

7.4.4 Example There exists a locally connected continuum X without free arcs such that $HS(X)$ is not homeomorphic to the Hilbert cube. Let $X = \mathcal{S}^1 \times [0, 1]$. Let $r : X \to \mathcal{S}^1 \times \{0\}$ be the projection map. Then r is a retraction of X onto $\mathcal{S}^1 \times \{0\}$.

Hence, the induced map $HS(r): HS(X) \to HS(\mathcal{S}^1 \times \{0\})$ given by

$$HS(r)(\chi) = \begin{cases} q_Y^n(\mathcal{C}(r)((q_X^n)^{-1}(\chi))), & \text{if } \chi \neq F_X; \\ F_Y, & \text{if } \chi = F_X \end{cases}$$

is a retraction. Note that $HS(r)$ is continuous by [10, 4.3, p. 126]. Thus, since $HS(\mathcal{S}^1 \times \{0\})$ is homeomorphic to $\mathcal{S}^2$, by Example 7.1.7, $HS(X)$ does not have the fixed point property. Therefore, $HS(X)$ is not homeomorphic to the Hilbert cube, since the Hilbert cube has the fixed point property [38, Corollary 3.5.3]. In this case, $HS(X)$ is not homeomorphic to $\mathcal{C}(X)$ (Theorem 6.1.8).

The next theorem says that if we have a contractible locally connected continuum without free arcs, then its n-fold hyperspace suspension is homeomorphic to the Hilbert cube.

7.4.5 Theorem *If X is a contractible locally connected continuum without free arcs, then $HS_n(X)$ is homeomorphic to the Hilbert cube $\mathcal{Q}$. In particular, $HS_n(X)$ is homeomorphic to $\mathcal{C}_n(X)$.*

Proof Let n be a positive integer. Since X is a locally connected continuum without free arcs, by Theorem 6.1.8, $\mathcal{C}_n(X)$ is homeomorphic to the Hilbert cube $\mathcal{Q}$. Since X is contractible, $\mathcal{F}_n(X)$ is contractible [22, Theorem 2.3]. Hence, $\mathcal{F}_n(X)$ has the shape of a point (in the sense of Borsuk) [3, 5.5, p. 28]. Thus, $\mathcal{C}_n(X) \setminus \mathcal{F}_n(X)$ is homeomorphic to $\mathcal{Q} \setminus \{p\}$ for some point $p \in \mathcal{Q}$ [7, 25.2]. Since $\mathcal{C}_n(X) \setminus \mathcal{F}_n(X)$ is homeomorphic to $HS_n(X) \setminus \{F_X^n\}$, we have that $\mathcal{Q} \setminus \{p\}$ is homeomorphic to $HS_n(X) \setminus \{F_X^n\}$. Therefore, $HS_n(X)$ is homeomorphic to $\mathcal{Q}$ [6, 2.23].

Q.E.D.

7.4.6 Corollary *The n-fold hyperspace suspension of the Hilbert cube is homeomorphic to the Hilbert cube for each $n \in \mathbb{N}$.*

7.4.7 Theorem *If X is a continuum, then $HS(X)$ is embeddable in $\mathbb{R}^2$ if and only if X is an arc.*

Proof Suppose $HS(X)$ is embedded in $\mathbb{R}^2$. Since, by Theorem 7.1.13, $HS(X)$ has property (b), $HS(X)$ does not separate $\mathbb{R}^2$ [19, VI 13, p. 100]. Since $HS(X)$ is aposyndetic (Corollary 7.3.2) and does not separate $\mathbb{R}^2$, $HS(X)$ is locally connected [24, Theorem 1, p. 139]. Hence, X is locally connected (by Theorem 7.4.2). Thus, by Theorem 7.4.3, X is a graph. Using [40, (1.109) and (1.100)], it is easy to see that given a graph Z, $\mathcal{C}(Z)$ is not embeddable in $\mathbb{R}^2$ if and only if Z is neither an arc nor a simple closed curve. Hence, given a graph Z, $HS(Z)$ is not embeddable in $\mathbb{R}^2$ if and only if Z is neither an arc nor a simple closed curve. Since the hyperspace suspension of a simple closed curve is a 2-sphere (Example 7.1.7), which is not embeddable in $\mathbb{R}^2$ (as can be easily seen from the 2-dimensional version of Borsuk–Ulam Theorem [27, p. 477]), we conclude that X is an arc.

On the other hand, if X is an arc, then, by Example 7.1.6, $HS(X)$ is a 2-cell. Therefore, $HS(X)$ is embeddable in $\mathbb{R}^2$.

Q.E.D.

As a consequence of Theorem 7.4.7, we have that $[0, 1]$ has unique hyperspace suspension.

7.4.8 Theorem *If X is a continuum such that $HS(X)$ is homeomorphic to $HS([0, 1])$, then X is homeomorphic to $[0, 1]$.*

Proof Since $HS(X)$ is homeomorphic to $HS([0, 1])$, we have that $HS(X)$ is embeddable in $\mathbb{R}^2$. Hence, by Theorem 7.4.7, X is homeomorphic to $[0, 1]$.

Q.E.D.

7.4.9 Theorem *If X is a continuum such that $HS(X)$ is homogeneous, then X is either a locally connected continuum without free arcs or a simple closed curve.*

Proof By Theorem 7.1.10, $HS(X)$ is locally connected at T_X. Hence, $HS(X)$ is locally connected, since $HS(X)$ is homogeneous. By Theorem 7.4.2, X is locally connected. Suppose X contains a free arc. Hence, $\dim(HS(X)) = 2$, since $HS(X)$ is homogeneous. Thus, by Theorem 7.4.3, X is a graph. Using [40, (1.109) and (1.100)], it is easy to see that X is either an arc or a simple closed curve. Since the hyperspace suspension of an arc is a 2-cell (Example 7.1.6), which is not homogeneous, we have that X is a simple closed curve.

Q.E.D.

7.4.10 Corollary *If X is a continuum such that $HS(X)$ is homogeneous and finite-dimensional, then $HS(X)$ is homeomorphic to the 2-sphere.*

Proof Suppose $HS(X)$ is homogeneous. Then, by Theorem 7.4.9, X is either a locally connected continuum without free arcs or a simple closed curve. Suppose X is a locally connected continuum without free arcs. Then $\mathcal{C}(X)$ is homeomorphic to the Hilbert cube [40, (1.98)]. Hence, $\dim(\mathcal{C}(X)) = \infty$. Thus, $\dim(HS(X)) = \infty$ (by Theorem 7.1.21), a contradiction to our hypothesis. Therefore, X is a simple closed curve, and $HS(X)$ is a 2-sphere (Example 7.1.7).

Q.E.D.

As a consequence of Corollary 7.4.10 we obtain that $\mathcal{S}^1$ has unique n-fold hyperspace suspension.

7.4.11 Theorem *If X is a continuum such that $HS(X)$ is homeomorphic to $HS(\mathcal{S}^1)$, then X is homeomorphic to $\mathcal{S}^1$.*

Proof Since $HS(X)$ is homeomorphic to $HS(\mathcal{S}^1)$, we have that $HS(X)$ is homogeneous and finite-dimensional. Thus, by Corollary 7.4.10, X is homeomorphic to $\mathcal{S}^1$.

Q.E.D.

Before we extend Theorems 7.4.8 and 7.4.11 to the 2-fold hyperspace suspension, we need the following two results.

7.4.12 Lemma *Each $A \in C_2(S^1) \setminus \{S^1\}$ has a 4-cell neighborhood in $C_2(S^1)$ and the point S^1 does not have a 4-cell neighborhood in $C_2(S^1)$. Hence, the point $T^2_{S^1}$ of $HS_2(S^1)$ does not have a 4-cell neighborhood in $HS_2(S^1)$.*

Proof Each $A \in C_2(S^1) \setminus \{S^1\}$ is contained in the interior of an arc $B \subset S^1$; clearly, $C_2(B)$ is a neighborhood of A in $C_2(S^1)$, and $C_2(B)$ is a 4-cell by Theorem 6.9.12.

We now prove that S^1 does not have a 4-cell neighborhood in $C_2(S^1)$.

$C_2(S^1)$ is the cone $\mathcal{K}$ over a solid (3-dimensional) torus (Theorem 6.9.14). Clearly, the vertex v of $\mathcal{K}$ is the only point of $\mathcal{K}$ that could fail to have a 4-cell neighborhood in $\mathcal{K}$. In fact, v does not have such a neighborhood, which we show as follows: If $\mathcal{N}$ were a 4-cell neighborhood of v in $\mathcal{K}$, then $\mathcal{N} \setminus \{v\}$ would retract onto a solid torus that is a level of $\mathcal{K}$ in $\mathcal{N}$. Thus, $\mathcal{N} \setminus \{v\}$ would not be simply connected. However, $\mathcal{N} \setminus \{v\}$ is simply connected since a 4-cell minus any point is simply connected. Therefore, v does not have a 4-cell neighborhood in $\mathcal{K}$.

Hence, from the first part of our lemma, any homeomorphism of $C_2(S^1)$ onto $\mathcal{K}$ must take S^1 to v. Therefore, S^1 does not have a 4-cell neighborhood in $C_2(S^1)$.

The last part of our lemma now follows from Remark 7.1.5 since, by the definition of hyperspace suspension, $HS_2(S^1)$ is locally the same at the point $T^2_{S^1}$ as $C_2(S^1)$ is at the point S^1.

Q.E.D.

7.4.13 Theorem *Let X be a graph, and let $n \in \mathbb{N}$. If Y is a continuum such that $HS_n(X)$ is homeomorphic to $HS_n(X)$, then Y is a graph.*

Proof Since X is a graph, X is locally connected. Then, by Theorem 7.4.2, $HS_n(X)$ is locally connected. Thus, $HS_n(Y)$ is locally connected, and hence, so is Y (by Theorem 7.4.2).

Since X is a graph, by Theorem 7.4.3, $\dim(HS_n(X)) < \infty$. Thus, $\dim(HS_n(Y)) < \infty$ and, by Theorem 7.4.3, Y is a graph.

Q.E.D.

7.4.14 Theorem *Let X be a continuum. If $HS_2(X)$ is homeomorphic to $HS_2([0, 1])$, then X is homeomorphic to $[0, 1]$; if $HS_2(X)$ is homeomorphic to $HS_2(S^1)$, then X is homeomorphic to S^1.*

Proof Assume that $HS_2(X)$ is homeomorphic to $HS_2([0, 1])$. Then, by Theorem 7.1.9, $HS_2(X)$ is a 4-cell. By Theorem 7.4.13, X is a graph. Thus, it follows easily that $\text{ord}(\{x\}, X) \le 2$ for all $x \in X$ (otherwise, $\dim_\chi(HS_2(X)) = 5$ when $(q^2_X)^{-1}(\chi)$ is the disjoint union of a simple triod and an arc contained in the manifold interior of a free arc in X). Therefore, X is homeomorphic to $[0, 1]$ or X is homeomorphic to S^1. However, $HS_2(S^1)$ is not a 4-cell, by Lemma 7.4.12. Therefore, X is homeomorphic to $[0, 1]$. This proves the first part of the theorem.

To prove the second part, assume that $HS_2(X)$ is homeomorphic to $HS_2(S^1)$. Then $HS_2(X)$ is a 4-dimensional Cantor manifold, by Corollary 7.1.26. Thus, $\dim_\chi(HS_2(X)) = 4$ for all $\chi \in HS_2(X)$ [19, A], p. 93]. Also, X is a graph by Theorem 7.4.13. Hence, $\text{ord}(\{x\}, X) \le 2$ for all $x \in X$ (as at the beginning of the proof). Thus, X is homeomorphic to $[0, 1]$ or X is homeomorphic to S^1. Since, by

Lemma 7.4.12, $HS_2(S^1)$ is not a 4-cell, we know from Theorem 7.1.9 that X is not homeomorphic to $[0, 1]$. Therefore, X is homeomorphic to S^1.

Q.E.D.

As a kind of extension of Theorem 7.4.14, we have the following:

7.4.15 Theorem *Let* n *an integer greater than two. If* X *is a continuum such that* $HS_n(X)$ *is homeomorphic to either* $HS_n([0, 1])$ *or* $HS_n(S^1)$, *then* X *is homeomorphic to either* $[0, 1]$ *or* S^1.

Proof Suppose $HS_n(X)$ is homeomorphic to $HS_n([0, 1])$, the other case is similar. Since $HS_n(X)$ is homeomorphic to $HS_n([0, 1])$, by Theorem 7.4.2, X is locally connected. By Corollary 7.1.26, we have that $\dim(HS_n(X)) = 2n$. By Theorem 7.1.21 and Corollary 6.1.14, X is an arc or a simple closed curve.

Q.E.D.

The following result is proved in [16, Theorem 3.2]:

7.4.16 Theorem *Let* X *be a graph, let* Y *be a continuum and let* $n \in \mathbb{N}$. *If* $HS_n(Y)$ *is homeomorphic to* $HS_n(X)$, *then* Y *is homeomorphic to* X.

The proof of the following lemma is similar to the one given for Lemma 6.1.28.

7.4.17 Lemma *If* n *is an integer greater than one, then neither* $HS_n([0, 1])$ *nor* $HS_n(S^1)$ *is homogeneous.*

7.4.18 Remark With respect to Lemma 7.4.17, observe that for S^1 the result is not true when $n = 1$, since $HS(S^1)$ is a 2-sphere (Example 7.1.7), which is homogeneous. The lemma is true for $HS([0, 1])$, since this is homeomorphic to a 2-cell (Example 7.1.6).

Theorem 7.4.9 may be extended to n-fold hyperspace suspensions.

7.4.19 Theorem *Let* X *be a continuum and let* n *be a positive integer greater than one. If* $HS_n(X)$ *is homogeneous, then* X *is a locally connected continuum without free arcs.*

Proof Let n be an integer greater than one. By Theorem 7.1.10, $HS_n(X)$ is locally connected at T_X^n. Hence, since $HS_n(X)$ is homogeneous, $HS_n(X)$ is locally connected. Thus, by Theorem 7.4.2, X is locally connected. Suppose X contains a free arc. Then, since $HS_n(X)$ is homogeneous, $\dim(HS_n(X)) = 2n$, by Lemma 7.1.14. Since X is locally connected and $\dim(HS_n(X)) = 2n$, X is a graph, by Theorem 7.4.3. Since $2n = \dim(HS_n(X)) = \dim(C_n(X))$ (the second equality follows by Theorem 7.1.21), by Corollary 6.1.14, X is an arc or a simple closed curve. A contradiction, because, by Lemma 7.4.17, neither $HS_n([0, 1])$ nor $HS_n(S^1)$ is homogeneous. Therefore, X does not contain free arcs.

Q.E.D.

7.4.20 Corollary *Let* X *be a continuum and let* n *be a positive integer greater than one. If* $HS_n(X)$ *is homogeneous, then* $\dim(HS_n(X)) = \infty$.

Proof The corollary follows from the Theorems 7.4.19, 6.1.8, and 7.1.21

Q.E.D.

7.4.21 Theorem *If X is a graph, then* $\dim_{F_X^n}(HS_n(X)) \leq \dim_{T_X^n}(HS_n(X))$.

Proof First note that if for each point $x \in X$, $\mathrm{ord}(\{x\}, X) \leq 2$, then X is either an arc or a simple closed curve [42, 8.40 (b)]. Since, by Corollary 7.1.26, $HS_n([0, 1])$ and $HS_n(\mathcal{S}^1)$ are $2n$-dimensional Cantor manifolds, both spaces are dimensionally homogeneous by [19, A), p. 93]. Hence, in both cases we have that $\dim_{F_X^n}(HS_n(X)) = \dim_{T_X^n}(HS_n(X))$

Suppose there exists a point $x \in X$ such that $\mathrm{ord}(\{x\}, X) \geq 3$. By the formula in Theorem 6.9.19 for $\dim_A(\mathcal{C}_n(X))$ for any $A \in \mathcal{C}_n(X)$, it follows immediately that $\dim_X(\mathcal{C}_n(X)) = \dim(\mathcal{C}_n(X))$. Thus, by Remark 7.1.5, $\dim_X(\mathcal{C}_n(X)) = \dim_{T_X^n}(HS_n(X))$. Hence, we have that $\dim_{T_X^n}(HS_n(X)) = \dim(HS_n(X))$. Therefore,

$$\dim_{F_X^n}(HS_n(X)) \leq \dim_{T_X^n}(HS_n(X)).$$

Q.E.D.

We end this section with a characterization of the n-fold hyperspace suspensions that are cells.

7.4.22 Theorem *Let X be a continuum and let n and k be positive integers. Then $IIS_n(X)$ is homeomoprhic to $[0, 1]^k$ if and only if X is an arc, $n \in \{1, 2\}$ and $k \in \{2, 4\}$.*

Proof Suppose $HS_n(X)$ is homeomorphic to $[0, 1]^k$. Then, by Theorem 7.4.2, X is locally connected. Since $\dim(HS_n(X)) = k < \infty$, by Theorem 7.4.3, X is a graph. Since $[0, 1]^k$ is a Cantor manifold [19, Example VI 11, p. 93], by Theorem 7.1.27, X is an arc or a simple closed curve. Hence, k is an even number, by Corollary 7.1.26. Suppose $n \geq 3$. Note that each point of $[0, 1]^k$ has a local basis of closed neighborhoods homemorphic to $[0, 1]^k$. Hence, since $q_X^n|_{\mathcal{C}_n(X) \setminus \mathcal{F}_n(X)}$ is a homeomorphism (Remark 7.1.5), by [21, Lemma 3.4], $HS_n(X)$ cannot be homeomorphic to $[0, 1]^k$. Thus, $n \in \{1, 2\}$ and $k \in \{2, 4\}$.

If $n = 1$, then $HS_1([0, 1])$ is homeomorphic to a 2-cell (Example 7.1.6) and $HS_1(\mathcal{S}^1)$ is homeomorphic to a 2-sphere (Example 7.1.7). Thus, if $n = 1$, then X is an arc and $k = 2$.

If $n = 2$, then $HS_2([0, 1])$ is homeomorphic to a 4-cell, by Theorem 7.1.9, and $HS_2(\mathcal{S}^1)$ cannot be homeomorphic to a 4-cell since $T_{\mathcal{S}^1}^2$ does not have 4-cell neighborhoods in $HS_2(\mathcal{S}^1)$, by Lemma 7.4.12. Hence, if $n = 2$, then X is an arc and $k = 4$.

Q.E.D.

7.5   Points That Arcwise Disconnect

In this section we consider indecomposable continua X and show that if we remove the set $\{T_X^n, F_X^n\}$ from $HS_n(X)$ it becomes arcwise disconnected and vice versa. We study the arc components of $HS_n(X) \setminus \{T_X^n, F_X^n\}$. To this end, we use [30, 32–34].

7.5.1 Theorem   *Let X be a continuum, let $n \in \mathbb{N}$ and let $A \in C_n(X)$. Then $C_n(X) \setminus \{A\}$ is not arcwise connected if and only if $HS_n(X) \setminus \{q_X^n(A), F_X^n\}$ is not arcwise connected.*

Proof   Suppose $C_n(X) \setminus \{A\}$ is not arcwise connected. Then $A \in C(X)$ by Theorem 6.5.2. Hence, $C_n(X) \setminus C_n(A)$ is arcwise connected, Lemma 6.5.1. Thus, there exists $B \in C(A) \setminus F_1(X)$ such that A belongs to each arc in $C_n(X)$ joining B and X.

Assume $HS_n(X) \setminus \{q_X^n(A), F_X^n\}$ is arcwise connected. Then there exists an arc $\alpha \colon [0, 1] \to HS_n(X) \setminus \{q_X^n(A), F_X^n\}$ such that $\alpha(0) = q_X^n(B)$ and $\alpha(1) = T_X^n$. Since q_X^n is a homeomorphism on $C_n(X) \setminus F_n(X)$ onto $HS_n(X) \setminus \{F_X^n\}$, $(q_X^n)^{-1} \circ \alpha([0, 1])$ is an arc in $C_n(X) \setminus (\{A\} \cup F_n(X))$ joining B to X, a contradiction. Therefore, $HS_n(X) \setminus \{q_X^n(A), F_X^n\}$ is not arcwise connected.

Now, suppose $HS_n(X) \setminus \{q_X^n(A), F_X^n\}$ is not arcwise connected. Then $A \in C(X)$ by Theorem 7.1.31. Assume $C_n(X) \setminus \{A\}$ is arcwise connected. Then, by Lemma 6.5.9, $C_n(X) \setminus (\{A\} \cup F_n(X))$ is arcwise connected. Thus, since $HS_n(X) \setminus \{q_X^n(A), F_X^n\} = q_X^n(C_n(X) \setminus (\{A\} \cup F_n(X)))$, we have that $HS_n(X) \setminus \{q_X^n(A), F_X^n\}$ is arcwise connected, a contradiction. Therefore, $C_n(X) \setminus \{A\}$ is not arcwise connected.

$\hfill$ **Q.E.D.**

7.5.2 Lemma   *If X is a decomposable continuum and $n \in \mathbb{N}$, then $C_n(X) \setminus (\{X\} \cup F_n(X))$ is arcwise connected.*

Proof   Since X is decomposable, there exist two nondegenerate proper subcontinua H and K of X such that $X = H \cup K$. Without loss of generality, we assume that $H \cap K$ has a nondegenerate component, say L (Corollary 1.7.28).

We first consider the case $n = 1$. Let C be a nondegenerate proper subcontinuum of X. We show that there exists an arc joining C and H in $C(X) \setminus (\{X\} \cup F_1(X))$. We consider four cases.

Case (1)   Either $C \subset H$ or $H \subset C$.

In either case, there exists an order arc joining C with H in $C(X) \setminus (\{X\} \cup F_1(X))$, by Theorem 1.8.20.

Case (2) $H \cap C = \emptyset$.

In this case $C \subset K$. Let α be an order arc from C to K. Let β an order arc from L to K. Let γ be an order arc from L to H. Then $\alpha \cup \beta \cup \gamma$ contains an arc joining C with H in $C(X) \setminus (\{X\} \cup F_1(X))$.

Case (3) $H \cap C$ has at least one nondegenerate component, say M.

If α_1 is an order arc from M to H and β_1 is an order arc from M to C, then $\alpha_1 \cup \beta_1$ contains an arc joining C with H in $C(X) \setminus (\{X\} \cup F_1(X))$.

Case (4) All the components of $H \cap C$ are degenerate.

Let $\{p\}$ be a component of $H \cap C$. Let U be an open set in C such that $p \in U$. Then, there exists a subcontinuum M_1 in C such that $p \in M_1 \subset U$, by Corollary 1.7.28. Let α_2 be an arc from H to $H \cup M_1$. Let β_2 be an order arc form M_1 to $H \cup M_1$. Let γ_2 be an order arc from M_1 to C. Then $\alpha_2 \cup \beta_2 \cup \gamma_2$ contains an arc joining C with H in $\mathcal{C}(X) \setminus (\{X\} \cup \mathcal{F}_1(X))$.

Therefore, $\mathcal{C}(X) \setminus (\{X\} \cup \mathcal{F}_1(X))$ is arcwise connected.

Next, suppose $n = 2$. Let $A \in \mathcal{C}_2(X) \setminus (\{X\} \cup \mathcal{F}_2(X))$. We show there exists an arc in $\mathcal{C}_2(X) \setminus (\{X\} \cup \mathcal{F}_2(X))$ from A to an element of $\mathcal{C}(X) \setminus (\{X\} \cup \mathcal{F}_1(X))$. Hence, we assume that A is not connected. Let A_1 and A_2 be the components of A and assume A_1 is nondegenerate.

If either $A \subset H$ or $A \subset K$, then there exists an order arc joining A with either H or K, respectively, by Theorem 1.8.20.

If $A_2 = K$, then $A_1 \subset H$. Let $\alpha_1 \colon [0, 1] \to \mathcal{C}_2(X)$ be an order arc such that $\alpha_1(0) = A_1 \cup L$ and $\alpha_1(1) = A_1 \cup K = A$. Let $\beta_1 \colon [0, 1] \to \mathcal{C}_2(X)$ be an order arc such that $\beta_1(0) = A_1 \cup L$ and $\beta_1(1) = H$. Then, $\alpha_1([0, 1]) \cup \beta_1([0, 1])$ is contained in $\mathcal{C}_2(X) \setminus (\{X\} \cup \mathcal{F}_2(X))$ and contains an arc joining A with H.

If $A_1 \subset H$, $A_2 \cap (H \cap K) \neq \emptyset$ and $A_2 \neq K$, then $H \cup A_2$ is a proper subcontinuum of X. Note that there exists an order arc from A to $H \cup A_2$.

Suppose $A_j \cap (H \cap K) \neq \emptyset$, $j \in \{1, 2\}$. Let $a_j \in A_j \cap H$. Let D_1 be a nondegenerate proper subcontinuum of A_1 containing a_1, Corollary 1.7.28. Note that $D_1 \cup H$ is a proper subcontinuum of X. Let $\alpha_2 \colon [0, 1] \to \mathcal{C}_2(X)$ be an order arc such that $\alpha_2(0) = D_1 \cup \{a_2\}$ and $\alpha_2(1) = A$. Now, let $\beta_2 \colon [0, 1] \to \mathcal{C}_2(X)$ be an order arc such that $\beta_2(0) = D_1 \cup \{a_2\}$ and $\beta_2(1) = D_1 \cup H$. Hence, $\alpha_2([0, 1]) \cup \beta_2([0, 1])$ is contained in $\mathcal{C}_2(X) \setminus (\{X\} \cup \mathcal{F}_2(X))$ and contains an arc joining A with $D_1 \cup H$.

Suppose $A_1 \subset H \setminus K$ and $A_2 \subset K \setminus H$. Let $\alpha_3 \colon [0, 1] \to \mathcal{C}_2(X)$ be an order arc such that $\alpha_3(0) = A$ and $\alpha_3(1) = H \cup A_2$. Now, let $\beta_3 \colon [0, 1] \to \mathcal{C}_2(X)$ be an order arc such that $\beta_3(0) = L \cup A_2$ and $\beta_3(1) = H \cup A_2$. Next, let $\gamma_3 \colon [0, 1] \to \mathcal{C}_2(X)$ be an order arc such that $\gamma_3(0) = L \cup A_2$ and $\gamma_3(1) = K$. Thus, $\alpha_3([0, 1]) \cup \beta_3([0, 1]) \cup \gamma_3([0, 1])$ is contained in $\mathcal{C}_2(X) \setminus (\{X\} \cup \mathcal{F}_2(X))$ and contains an arc joining A with K.

The rest of the cases are treated similarly.

Now, let $n \geq 3$ and assume that $\mathcal{C}_n(X) \setminus (\{X\} \cup \mathcal{F}_n(X))$ is arcwise connected. We show that $\mathcal{C}_{n+1}(X) \setminus (\{X\} \cup \mathcal{F}_{n+1}(X))$ is also arcwise connected. Let $A \in \mathcal{C}_{n+1}(X) \setminus (\{X\} \cup \mathcal{F}_{n+1}(X))$. We show there exists an arc in $\mathcal{C}_{n+1}(X) \setminus (\{X\} \cup \mathcal{F}_{n+1}(X))$ joining A with an element of $\mathcal{C}_n(X) \setminus (\{X\} \cup \mathcal{F}_n(X))$. Hence, suppose A has $n + 1$ components, say $A_1, \ldots, A_{n+1}$, and assume A_1 is nondegenerate. For each $j \in \{2, \ldots, n + 1\}$, let $a_j \in A_j$.

Suppose $A_1 = K$. Let $x \in A_1 \cap H$, and let D_1' be a nondegenerate proper subcontinuum of A_1 containing x, Corollary 1.7.28. Note that $D_1' \cup H$ is a proper subcontinuum of X. Let $\alpha_4 \colon [0, 1] \to \mathcal{C}_{n+1}(X)$ be an order arc such that $\alpha_4(0) = D_1' \cup \{a_2, \ldots, a_{n+1}\}$ and $\alpha_4(1) = A$. Now, let $\beta_4 \colon [0, 1] \to \mathcal{C}_{n+1}(X)$ be an order arc such that $\beta_4(0) = D_1' \cup \{a_2, \ldots, a_{n+1}\}$ and $\beta_4(1) = D_1' \cup H$. Hence, $\alpha_4([0, 1]) \cup$

$\beta_4([0, 1])$ is contained in $C_{n+1}(X) \setminus (\{X\} \cup \mathcal{F}_{n+1}(X))$ and contains an arc joining A with $D_1' \cup H$.

Suppose $A_1 \subset K$, $A_1 \neq K$ and $A \setminus A_1 \subset H$. Let $\alpha_5 \colon [0, 1] \rightarrow C_{n+1}(X)$ be an order arc such that $\alpha_5(0) = A_1 \cup \{a_2, \ldots, a_{n+1}\}$ and $\alpha_5(1) = A$. Now, let $\beta_5 \colon [0, 1] \rightarrow C_{n+1}(X)$ be an order arc such that $\beta_5(0) = A_1 \cup \{a_2, \ldots, a_{n+1}\}$ and $\beta_5(1) = A_1 \cup H$. Note that $A_1 \cup H \in C_n(X) \setminus (\{X\} \cup \mathcal{F}_n(X))$. Thus, $\alpha_5([0, 1]) \cup \beta_5([0, 1])$ is contained in $C_{n+1}(X) \setminus (\{X\} \cup \mathcal{F}_{n+1}(X))$ and contains an arc joining A with $A_1 \cup H$.

Suppose $A_1 \cap (H \cap K) \neq \emptyset$, $A_1 \neq H$ and $A_1 \neq K$. Without loss of generality, we assume that $a_2 \in H$. Let $\alpha_6 \colon [0, 1] \rightarrow C_{n+1}(X)$ be an order arc such that $\alpha_6(0) = A_1 \cup \{a_2, \ldots, a_{n+1}\}$ and $\alpha_6(1) = A$. Now, let $\beta_6 \colon [0, 1] \rightarrow C_{n+1}(X)$ be an order arc such that $\beta_6(0) = A_1 \cup \{a_2\}$ and $\beta_6(1) = A_1 \cup H$. Next, let $\gamma_6 \colon [0, 1] \rightarrow C_{n+1}(X)$ be given by $\gamma_6(t) = \beta_6(t) \cup \{a_3, \ldots, a_{n+1}\}$. Then, γ_6 is an order arc such that $\gamma_6(0) = A_1 \cup \{a_2, \ldots, a_{n+1}\}$ and $\gamma_6(1) = A_1 \cup \{a_3, \ldots, a_{n+1}\} \cup H$. Note that $A_1 \cup \{a_3, \ldots, a_{n+1}\} \cup H \in C_n(X) \setminus (\{X\} \cup \mathcal{F}_n(X))$. Hence, $\alpha_6([0, 1]) \cup \gamma_6([0, 1])$ is contained in $C_{n+1}(X) \setminus (\{X\} \cup \mathcal{F}_{n+1}(X))$ and contains an arc joining A with $A_1 \cup \{a_3, \ldots, a_{n+1}\} \cup H$.

The rest of the cases are treated similarly.

Q.E.D.

7.5.3 Lemma *Let A be a proper decomposable subcontinuum of a continuum X and let $n \in \mathbb{N}$. If $HS_n(X) \setminus \{q_X^n(A), F_X^n\}$ is not arcwise connected, then $HS_n(X) \setminus \{q_X^n(A), F_X^n\}$ has exactly two arc components.*

Proof The lemma follows from Lemma 6.5.9 and the following facts: $C_n(A) \setminus (\{A\} \cup \mathcal{F}_n(A)) = C_n(A) \setminus (\{A\} \cup \mathcal{F}_n(X))$, $q_X^n(C_n(A) \setminus (\{A\} \cup \mathcal{F}_n(X))) = q_X^n(C_n(A)) \setminus \{q_X^n(A), F_X^n\}$ and

$$HS_n(X) \setminus \{q_X^n(A), F_X^n\} =$$

$$(q_X^n(C_n(X)) \setminus q_X^n(C_n(A))) \cup (q_X^n(C_n(A)) \setminus \{q_X^n(A), F_X^n\}).$$

Q.E.D.

7.5.4 Theorem *Let $n \in \mathbb{N}$. A continuum X is indecomposable if and only if $HS_n(X) \setminus \{T_X^n, F_X^n\}$ is not arcwise connected.*

Proof Suppose X is an indecomposable continuum. Then $C_n(X) \setminus \{X\}$ is not arcwise connected, by Theorem 6.5.3. Hence, there exist two points A and B in two different arc components of $C_n(X) \setminus \{X\}$. Thus, any arc in $HS_n(X) \setminus \{T_X^n\}$ joining $q_X^n(A)$ and $q_X^n(B)$ must contain F_X^n. Therefore, $q_X^n(A)$ and $q_X^n(B)$ belong to two different arc components of $HS_n(X) \setminus \{T_X^n, F_X^n\}$.

Now, suppose X is a decomposable continuum. By Lemma 7.5.2, $C_n(X) \setminus (\{X\} \cup \mathcal{F}_n(X))$ is arcwise connected. Since, by Remark 7.1.5, $C_n(X) \setminus (\{X\} \cup \mathcal{F}_n(X))$ is homeomorphic to $HS_n(X) \setminus \{T_X^n, F_X^n\}$. $HS_n(X) \setminus \{T_X^n, F_X^n\}$ is arcwise connected.

Q.E.D.

The next result is the corresponding theorem to Theorem 6.5.14, for n-fold hyperspace suspensions.

7.5.5 Theorem *Let X be an indecomposable continuum and let $n \in \mathbb{N}$. If $\mathcal{A}$ is an arc component of $HS_n(X) \setminus \{T_X^n, F_X^n\}$, then either $\mathcal{A}$ is homeomorphic to $q_X^n(\mathcal{C}_n(\kappa) \setminus \mathcal{F}_n(\kappa))$ for some composant κ of X, or there exist finitely many composants $\kappa_1, \ldots, \kappa_\ell$ of X and there exists a one-to-one map from*

$$\mathcal{L} = \bigcup \left\{ \prod_{j=1}^{\ell} \mathcal{C}_{r_j}(\kappa_j) \setminus \prod_{j=1}^{\ell} \mathcal{F}_{r_j}(\kappa_j) \,\Big|\, \{r_1, \ldots, r_\ell\} \in \mathcal{N} \right\},$$

where $\mathcal{N}$ is defined in Theorem 6.5.14, onto $\mathcal{A}$.

Proof Let κ be a composant of X. By Theorem 6.5.13, $\mathcal{C}_n(\kappa)$ is an arc component of $\mathcal{C}_n(X) \setminus \{X\}$. By Theorem 6.5.16, $\mathcal{C}_n(\kappa) \setminus \mathcal{F}_n(\kappa) = \mathcal{C}_n(\kappa) \setminus \mathcal{F}_n(X)$ is an arc component of $\mathcal{C}_n(X) \setminus (\{X\} \cup \mathcal{F}_n(X))$. Hence, $q_X^n(\mathcal{C}_n(\kappa) \setminus \mathcal{F}_n(\kappa))$ is an arc component of $HS_n(X) \setminus \{T_X^n, F_X^n\}$.

Now, suppose $\mathcal{A}$ is an arc component of $HS_n(X) \setminus \{T_X^n, F_X^n\}$ and $\mathcal{A}$ is not homeomorphic to $q_X^n(\mathcal{C}_n(\kappa) \setminus \mathcal{F}_n(\kappa))$ for any composant κ of X. Let $\mathcal{B}$ be the arc component of $\mathcal{C}_n(X) \setminus \{X\}$ containing $(q_X^n)^{-1}(\mathcal{A})$. Note that $\mathcal{B} \setminus \mathcal{F}_n(X)) = (q_X^n)^{-1}(\mathcal{A})$. Hence, $(q_X^n)^{-1}(\mathcal{A})$ is an arc component of $\mathcal{C}_n(X) \setminus (\{X\} \cup \mathcal{F}_n(X))$, by Theorem 6.5.16. By Theorem 6.5.14, there exist finitely many composants $\kappa_1, \ldots, \kappa_\ell$ of X and a one-to-one map $f \colon \mathcal{R} \to \mathcal{B}$. Let $g = f|_{\mathcal{L}}$. Then g is continuous, one-to-one and its image is contained in $\mathcal{B} \setminus \mathcal{F}_n(X)$. To see g is onto, let $B \in \mathcal{B} \setminus \mathcal{F}_n(X)$. Note that B has at least one nondegenerate component. For each $j \in \{1, \ldots, \ell - 1\}$, let r_j be the number of components of B contained in κ_j. Let $r_\ell = n - \sum_{j=1}^{\ell-1} r_j$. Then $\sum_{j=1}^{\ell} r_j = n$ and $\{r_1, \ldots, r_\ell\} \in \mathcal{N}$. Hence, $(B \cap \kappa_1, \ldots, B \cap \kappa_\ell) \in \prod_{j=1}^{\ell} \mathcal{C}_{r_j}(\kappa_j) \setminus \prod_{j=1}^{\ell} \mathcal{F}_{r_j}(\kappa_j)$ and $g(B \cap \kappa_1, \ldots, B \cap \kappa_\ell) = B$. Therefore, $h = q_X^n \circ g$ is the desired map.

Q.E.D.

7.5.6 Lemma *If X is an indecomposable continuum, then $HS_n(X) \setminus \{T_X^n, F_X^n\}$ has uncountably many arc components.*

Proof Since X is an indecomposable continuum, by Theorem 7.5.4, $HS_n(X) \setminus \{T_X^n, F_X^n\}$ is not arcwise connected. Since indecomposable continua have uncountably many composants [36, Theorem 3–46], and for each composant κ of X, $q_X^n(\mathcal{C}_n(\kappa) \setminus \mathcal{F}_n(\kappa))$ is an arc component of $HS_n(X) \setminus \{T_X^n, F_X^n\}$ (Theorem 7.5.5), we have that $HS_n(X) \setminus \{T_X^n, F_X^n\}$ has uncountably many arc components.

Q.E.D.

7.5.7 Theorem *Let X and Y be continua, where X is indecomposable with the property of Kelley, and let n and m be positive integers. If $HS_n(X)$ is homeomorphic to $HS_m(Y)$, then Y is indecomposable.*

Proof Let $h: HS_n(X) \twoheadrightarrow HS_m(Y)$ be a homeomorphism. Since X is indecomposable and has the property of Kelley, T_X^n and F_X^n are the only two points at which $HS_n(X)$ is locally connected, Theorem 7.1.11. Thus, $h(T_X^n)$ and $h(F_X^n)$ are the only two points at which $HS_n(Y)$ is locally connected. Since $HS_n(Y)$ is always locally connected at T_Y^n and F_Y^n by Theorem 7.1.10, we have that $\{h(T_X^n), h(F_X^n)\} = \{T_Y^n, F_Y^n\}$. Since X is indecomposable, $HS_n(X) \setminus \{T_X^n, F_X^n\}$ is not arcwise connected by Theorem 7.5.4. Hence, $HS_n(Y) \setminus \{h(T_X^n), h(F_X^n)\} = HS_n(Y) \setminus \{T_Y^n, F_Y^n\}$ is not arcwise connected. Therefore, Y is indecomposable by Theorem 7.5.4.

Q.E.D.

7.5.8 Lemma *Let X be a continuum and let $n \in \mathbb{N}$. If χ_1 and χ_2 are two points of $HS_n(X)$ such that $HS_n(X) \setminus \{\chi_1, \chi_2\}$ is not arcwise connected, then $F_X^n \in \{\chi_1, \chi_2\}$.*

Proof Suppose $F_X^n \notin \{\chi_1, \chi_2\}$. By Theorem 7.1.30, we may assume that $T_X^n \notin \{\chi_1, \chi_2\}$.

First, we show that there exists an arc in $HS_n(X) \setminus \{\chi_1, \chi_2\}$ joining F_X^n and T_X^n. If $(q_X^n)^{-1}(\chi_1) \cup (q_X^n)^{-1}(\chi_2) \neq X$, then there exists a point $x \in X \setminus ((q_X^n)^{-1}(\chi_1) \cup (q_X^n)^{-1}(\chi_2))$. Let α be an order arc in $\mathcal{C}(X)$ from $\{x\}$ to X by Theorem 1.8.20. Note that

$$\alpha \cap \{(q_X^n)^{-1}(\chi_1), (q_X^n)^{-1}(\chi_2)\} = \emptyset.$$

Hence, $q_X^n(\alpha)$ is an arc in $HS_n(X) \setminus \{\chi_1, \chi_2\}$ from F_X^n to T_X^n.

Now, assume that $(q_X^n)^{-1}(\chi_1) \cup (q_X^n)^{-1}(\chi_2) = X$. Since $X \notin \{(q_X^n)^{-1}(\chi_1), (q_X^n)^{-1}(\chi_2)\}$, there exist nondegenerate components A_1 and B_1 of $(q_X^n)^{-1}(\chi_1)$ and $(q_X^n)^{-1}(\chi_2)$, respectively, such that $A_1 \cap B_1 \neq \emptyset$. Let $x \in A_1 \cap B_1$ and let $\varepsilon = \frac{1}{2} \min\{\text{diam}(A_1), \text{diam}(B_1)\}$. By Corollary 1.7.28, there exist two nondegenerate continua A and B such that $x \in A \subset A_1 \cap \mathcal{V}_\varepsilon(x)$ and $x \in B \subset B_1 \cap \mathcal{V}_\varepsilon(x)$. Let α_1 be an order arc in $\mathcal{C}(X)$ from $\{x\}$ to A (Theorem 1.8.20). Let α_2 be an order arc in $\mathcal{C}(X)$ from A to $A \cup B$. Let α_3 be an order arc in $\mathcal{C}(X)$ from $A \cup B$ to X. Then $\alpha_1 \cup \alpha_2 \cup \alpha_3$ is an arc in $\mathcal{C}_n(X) \setminus \{(q_X^n)^{-1}(\chi_1), (q_X^n)^{-1}(\chi_2)\}$ joining $\{x\}$ and X. Hence, $q_X^n(\alpha_1 \cup \alpha_2 \cup \alpha_3)$ is an arc in $HS_n(X) \setminus \{\chi_1, \chi_2\}$ joining F_X^n and T_X^n.

Let $\chi \in HS_n(X) \setminus \{\chi_1, \chi_2\}$. Now it is easy to construct an arc in $HS_n(X) \setminus \{\chi_1, \chi_2\}$ joining χ with either F_X^n or T_X^n.

Therefore, $HS_n(X) \setminus \{\chi_1, \chi_2\}$ is arcwise connected.

Q.E.D.

The following theorem shows that indecomposable continua and hereditarily decomposable continua do not share n-fold hyperspace suspensions.

7.5.9 Theorem *Let X be an indecomposable continuum and let n and m be positive integers. If Y is a hereditarily decomposable continuum, then $HS_m(Y)$ is not homeomorphic to $HS_n(X)$.*

Proof Suppose that $HS_m(Y)$ is homeomorphic to $HS_n(X)$ and let $h: HS_n(X) \twoheadrightarrow HS_m(Y)$ be a homeomorphism. Since X is indecomposable, by Lemma 7.5.6, $HS_n(X) \setminus \{T_X^n, F_X^n\}$ has uncountably many arc components. Then $HS_m(Y) \setminus$

$\{h(T_X^n), h(F_X^n)\}$ has uncountably many arc components. Since $HS_m(Y) \setminus \{h(T_X^n), h(F_X^n)\}$ is not arcwise connected, by Lemma 7.5.8, $F_Y^m \in \{h(T_X^n), h(F_X^n)\}$. Let $\chi \in HS_m(Y)$ be such that $\{\chi, F_Y^m\} = \{h(T_X^n), h(F_X^n)\}$. Since $HS_m(Y) \setminus \{\chi, F_Y^m\}$ is not arcwise connected, $(q_Y^m)^{-1}(\chi) \in \mathcal{C}(Y)$ (Theorem 7.1.31). Since Y is hereditarily decomposable, $(q_Y^m)^{-1}(\chi)$ is a decomposable subcontinuum of Y. Hence, by Lemma 7.5.3, $HS_m(Y) \setminus \{\chi, F_Y^m\}$ has exactly two arc components, a contradiction. Therefore, $HS_m(Y)$ is not homeomorphic to $HS_n(X)$.

Q.E.D.

7.6 Cones, Suspensions and Products

We begin comparing n-fold hyperspace suspensions with cones. Note that, for $n \in \{1, 2\}$, by Example 7.1.6 and Theorem 7.1.9, we have that $HS_n([0, 1])$ is a topological cone. The material for this section is based on [9, 12, 19, 30, 34, 40, 42].

7.6.1 Theorem *If X is a finite-dimensional continuum, then $HS_n(X)$ is not homeomorphic to $K(X)$ for any integer $n \geq 2$.*

Proof Let $n \geq 2$. Suppose $HS_n(X)$ is homeomorphic to $K(X)$. With an argument similar to the one given in the proof of Theorem 6.10.10, we obtain that $n = 2$ and no nondegenerate proper subcontinuum of X is decomposable.

Suppose each proper subcontinuum of X is indecomposable. Then, by Lemma 1.7.29, X is hereditarily indecomposable. Hence, $K(X)$ is uniquely arcwise connected. On the other hand, since, by Theorem 7.1.23, $HS_2(X)$ contains 2-cells, $HS_2(X)$ is not uniquely arcwise connected.

Therefore, $HS_n(X)$ is not homeomorphic to $K(X)$ for any integer n greater than one.

Q.E.D.

The following lemma is easy to prove.

7.6.2 Lemma *Let Z be an arcwise connected continuum. If $x_1, x_2 \in K(Z)$, then $K(Z) \setminus \{x_1, x_2\}$ is arcwise connected.*

Observe that $HS_2([0, 1])$ is homeomorphic to $[0, 1]^4$, by Theorem 7.1.9. Hence, $HS_2([0, 1])$ is homeomorphic to $K([0, 1]^3)$. In connection with this we have:

7.6.3 Theorem *Let X be a continuum. If Z is a finite-dimensional continuum such that $K(Z)$ is homeomorphic to $HS_n(X)$, for some integer n greater than one, then X is hereditarily decomposable, and X does not contain nondegenerate proper terminal subcontinua. Also, Z is arcwise connected.*

Proof Let n be an integer greater than one and let $h \colon HS_n(X) \twoheadrightarrow K(Z)$ be a homeomorphism. Since $HS_n(X)$ is not arcwise disconnected by any of its points, $K(Z)$ is not arcwise disconnected by any of its points. Hence, Z is arcwise connected.

Suppose X contains an indecomposable continuum A. Since Z is finite-dimensional, we have that $HS_n(X)$ is finite-dimensional. This implies that $\dim(HS_n(X)) = \dim(C_n(X))$, Theorem 7.1.21. Thus, $C_n(X) \setminus \{A\}$ is not arcwise connected, Theorem 6.5.17. Hence, $HS_n(X) \setminus \{q_X^n(A), F_X^n\}$ is not arcwise connected (compare with Theorem 7.1.30). This implies that $K(Z) \setminus \{h(q_X^n(A)), h(F_X^n)\}$ is not arcwise connected, a contradiction to Lemma 7.6.2. Therefore, X is hereditarily decomposable.

Now, assume X contains a nondegenerate proper terminal subcontinuum B. Then $C_n(X) \setminus \{B\}$ is not arcwise connected, Theorem 6.5.4. Repeating the argument of the previous paragraph, we obtain, again, a contradiction to Lemma 7.6.2. Therefore, X does not contain nondegenerate proper terminal subcontinua.

Q.E.D.

We turn our attention to the comparison of n-fold hyperspace suspensions and suspensions. Observe that, for $n \in \{1, 2\}$, by Example 7.1.6 and Theorem 7.1.9, we have that $HS_n([0, 1])$ is a topological suspension. Also, by Example 7.1.7, $HS(S^1)$ is a topological suspension.

7.6.4 Definition A continuum X is a C–H *continuum* provided that $C(X)$ is homeomorphic to $K(X)$, the cone over X.

A proof for the next theorem may be found in [12, Theorem 6.5].

7.6.5 Theorem *If X is a finite-dimensional C–H continuum, then there exists a homeomorphism $h \colon C(X) \twoheadrightarrow K(X)$ such that $h(\mathcal{F}_1(X)) = \mathcal{B}(X)$, where $\mathcal{B}(X)$ is the base of the cone.*

As a consequence of Theorem 7.6.5, we have the following result which says that, in certain cases, the hyperspace suspension is a topological suspension.

7.6.6 Theorem *If X is a finite-dimensional C–H continuum, then $HS(X)$ is homeomorphic to $\Sigma(X)$, the suspension over X.*

Next, we show that Theorem 7.6.6 is not true for $n \geq 2$.

7.6.7 Theorem *Let X be a finite-dimensional continuum. Then, for each integer $n \geq 2$, $HS_n(X)$ is not homeomorphic to $\Sigma(X)$.*

Proof Let $n \geq 2$. Suppose $HS_n(X)$ is homeomorphic to $\Sigma(X)$. With an argument similar to the one given in the proof of Theorem 6.10.10, we obtain that $n = 2$ and no nondegenerate proper subcontinuum of X is decomposable.

Suppose each proper subcontinuum of X is indecomposable. Then, by Lemma 1.7.29, X is hereditarily indecomposable. Hence, it is easy to see that $\Sigma(X)$ does not contain 2-cells. On the other hand, $HS_2(X)$ contains 2-cells (Theorem 7.1.23). Therefore, $HS_n(X)$ is not homeomorphic to $\Sigma(X)$.

Q.E.D.

The following lemma is easy to establish:

7.6.8 Lemma *Let X be a continuum and let v^+ and v^- be the vertexes of $\Sigma(X)$. If $x_1, x_2 \in \Sigma(X)$ are such that $\{x_1, x_2\} \neq \{v^+, v^-\}$, then $\Sigma(X) \setminus \{x_1, x_2\}$ has at most two arc components. Hence, $\{v^+, v^-\}$ is the only set of points such that $\Sigma(X) \setminus \{v^+, v^-\}$ may have more than two arc components.*

7.6.9 Theorem *Let X be a continuum and let $n \geq 2$ be an integer. If Z is a finite-dimensional continuum such that $\Sigma(Z)$ is homeomorphic to $HS_n(X)$, then $\dim(X) = 1$ and X contains at most one nondegenerate indecomposable continuum. Hence, X is not hereditarily indecomposable.*

Proof Let $n \geq 2$ be an integer. Let $h \colon HS_n(X) \twoheadrightarrow \Sigma(Z)$ be a homeomorphism. By Corollary 7.1.22, $\dim(X) = 1$.

Suppose X has two nondegenerate indecomposable continua, A and B. Then, by Theorems 7.1.21 and 6.5.17, $C_n(X) \setminus \{A\}$ and $C_n(X) \setminus \{B\}$ both have infinitely many arc components. Hence, $\Sigma(X) \setminus \{h(q_X^n(A)), h(F_X^n)\}$ and $\Sigma(X) \setminus \{h(q_X^n(B)), h(F_X^n)\}$ have infinitely many arc components. This contradicts Lemma 7.6.8. Therefore, X contains at most one indecomposable subcontinuum.

Q.E.D.

7.6.10 Theorem *Let X be a continuum containing an indecomposable subcontinuum A. Let $n \geq 2$ be an integer, and let Z be a finite-dimensional continuum such that $\Sigma(Z)$ is homeomorphic to $HS_n(X)$. If $h \colon HS_n(X) \twoheadrightarrow \Sigma(Z)$ is a homeomorphism, then*

(1) $h(\{q_X^n(A), F_X^n\}) = \{v^+, v^-\}$;

(2) Z has uncountably many arc components. In particular, Z is not locally connected;

(3) $\dim(HS_n(X)) \geq 2n$ and $\dim(Z) \geq 2n - 1$;

(4) Each point of Z is contained in an arc in Z and some points of Z belong to locally connected subcontinua of Z whose dimension is at least $2n - 1$;

(5) If (z, t) and (z', t') are two points of $\Sigma(Z)$ such that $\{(z, t), (z', t')\} \neq \{v^+, v^-\}$, then $\Sigma(Z) \setminus \{(z, t), (z', t')\}$ is arcwise connected;

(6) If $A = X$, then X does not contain a nondegenerate proper terminal subcontinuum.

Proof

(1) This is a consequence of the proof of Theorem 7.6.9.

(2) Since A is an indecomposable subcontinuum of X, by Theorems 7.6.9, 7.1.21 and 6.5.17, $C_n(X) \setminus \{A\}$ has uncountably many arc components. Hence, by Theorem 7.5.1, $HS_n(X) \setminus \{q_X^n(A), F_X^n\}$ is not arcwise connected. In fact, $HS_n(X) \setminus \{q_X^n(A), F_X^n\}$ has uncountably many arc components. Since $h(\{q_X^n(A), F_X^n\}) = \{v^+, v^-\}$ (by (1)), $\Sigma(Z) \setminus \{v^+, v^-\}$ has uncountably many arc components. Thus, since $\Sigma(Z) \setminus \{v^+, v^-\}$ is homeomorphic to $Z \times (0, 1)$, we conclude that Z has uncountably many arc components.

(3) By Theorem 7.6.9, each subcontinuum of X, distinct from A, is decomposable. Thus, $HS_n(X)$ contains a $2n$-cell (by Theorem 7.1.24). Hence, $\dim(HS_n(X)) \geq 2n$. Since $\Sigma(Z)$ is homeomorphic to $HS_n(X)$ and $\dim(\Sigma(Z)) = \dim(Z) + 1$ [40, (8.0)], we have that $\dim(Z) \geq 2n - 1$.

(4) Let z be a point of Z. We consider two cases.

First, we assume that there exists a point t_0 in $(0, 1)$ such that $(q_X^n)^{-1}(h^{-1}((z, t_0))) \in C_n(X) \setminus C(X)$. Let $B = (q_X^n)^{-1}(h^{-1}((z, t_0)))$, and let $B_1, \ldots, B_k$ be the components of B, where $2 \leq k \leq n$. Note that, at least one B_j is nondegenerate. By Corollary 1.7.28, for each $j \in \{1, \ldots, k\}$, there exists a subcontinuum C_j of X containing B_j properly. We assume without loss of generality that $C_j \cap C_\ell = \emptyset$ if $j \neq \ell$.

For each $j \in \{1, \ldots, k\}$, let $\alpha_j : [0, 1] \to C(X)$ be an order arc (Theorem 1.8.20) such that $\alpha_j(0) = B_j$ and $\alpha_j(1) = C_j$. Let $\alpha : [0, 1]^k \to C_n(X) \setminus \mathcal{F}_n(X)$ be given by $\alpha((t_1, \ldots, t_k)) = \bigcup_{j=1}^k \alpha_j(t_j)$. Let $\mathcal{D} = \alpha([0, 1]^k)$. Then, $\mathcal{D}$ is a k-cell such that $B \in \mathcal{D}$, $A \notin \mathcal{D}$ and $\mathcal{D} \cap \mathcal{F}_n(X) = \emptyset$. Thus, $h\left(q_X^n(\mathcal{D})\right)$ is a k-cell containing the point (z, t_0) and such that $h\left(q_X^n(\mathcal{D})\right) \cap \{v^+, v^-\} = \emptyset$. Let $\pi : \Sigma(Z) \setminus \{v^+, v^-\} \twoheadrightarrow Z$ be the projection map. Then $\pi\left(h\left(q_X^n(\mathcal{D})\right)\right)$ is a locally connected subcontinuum of Z containing z; since $k \geq 2$, $\pi\left(h\left(q_X^n(\mathcal{D})\right)\right)$ is nondegenerate. Thus, z is contained in an arc by [42, 8.23].

Next, suppose that $(q_X^n)^{-1}(h^{-1}((z, t))) \in C(X)$ for each $t \in (0, 1)$. Since $h(\{q_X^n(A), F_X^n\}) = \{v^+, v^-\}$, there exists $t' \in (0, 1)$ such that $(q_X^n)^{-1}(h^{-1}((z, t'))) \neq A$. Let $E = (q_X^n)^{-1}(h^{-1}((z, t')))$. Since $E \neq A$ and E is nondegenerate, E is a decomposable continuum (Theorem 7.6.9). Hence, there exist two proper subcontinua H and K of E such that $E = H \cup K$.

Suppose, first, that A is not contained in E. Let $x_1 \in H \setminus K$ and $x_2 \in K \setminus H$. By Corollary 1.7.28, for each $j \in \{1, 2\}$, there exists a nondegenerate subcontinuum G_j of E such that $x_1 \in G_1 \subset H \setminus K$ and $x_2 \in G_2 \subset K \setminus H$. Let $\beta_j : [0, 1] \to C(X)$ be an order arc such that $\beta_j(0) = G_j$, $\beta_1(1) = H$ and $\beta_2(1) = K$. Let $\beta : [0, 1]^2 \to C_n(X)$ be given by $\beta((t_1, t_2)) = \beta_1(t_1) \cup \beta_2(t_2)$. Let $\mathcal{G} = \beta([0, 1]^2)$. Then $\mathcal{G}$ is a locally connected subcontinuum of $C_n(X) \setminus \mathcal{F}_n(X)$ such that $\mathcal{G}$ contains a 2-cell and such that $E \in \mathcal{G}$ and $A \notin \mathcal{G}$. Thus, $h\left(q_X^n(\mathcal{G})\right)$ is a locally connected subcontinuum of $\Sigma(Z)$ containing a 2-cell, such that $(z, t') \in h\left(q_X^n(\mathcal{G})\right)$ and $\{v^+, v^-\} \cap h\left(q_X^n(\mathcal{G})\right) = \emptyset$. Hence, $\pi\left(h\left(q_X^n(\mathcal{G})\right)\right)$ is a nondegenerate locally connected subcontinuum of Z containing z. Thus, z is contained in an arc by [42, 8.23].

Assume next, A is contained in E. Since A is indecomposable, $E \neq A$. Hence, there exists a point $x_1 \in E \setminus A$. Suppose $x_1 \in H$. Choose a point $x_2 \in K \setminus \{x_1\}$. Then choose a subcontinuum G_1 such that $x_1 \in G_1 \subset H \setminus A$ (Corollary 1.7.28) and repeat the argument in the preceding paragraph with $G_2 = \{x_2\}$ to construct a nondegenerate locally connected subcontinuum of Z containing z.

This completes the proof of the first part of (4). We prove the second part of (4) as follows:

By Theorem 7.1.24, there exists a $2n$-cell $\mathcal{E}$ in $HS_n(X)$. We may choose $\mathcal{E}$ such that $q_X^n(A) \notin \mathcal{E}$. Let $B \in \mathcal{E}$. Then $h(\mathcal{E})$ is a $2n$-cell such that $h(B) \in h(\mathcal{E})$. Hence, $\pi(h(\mathcal{E}))$ is a locally connected subcontinuum of Z containing $\pi(h(B))$, and

$\dim(\pi(h(\mathcal{E}))) \geq 2n - 1$ (by [19, p. 34] since $\pi(h(\mathcal{E})) \times (0, 1)$ contains $h(\mathcal{E})$ and, thus, has dimension at least $2n$).

(5) By (4), each point of $\Sigma(Z)$ lies in the suspension over an arc. Hence, (5) follows easily.

(6) This is a consequence of Theorems 7.6.9, 6.5.4 and (5).

Q.E.D.

We turn our attention to the comparison of n-fold hyperspace suspensions and products of continua.

The following lemma is easy to prove.

7.6.11 Lemma *If Y and Z are nondegenerate arcwise connected continua, then $Y \times Z \setminus \{(y_1, z_1), (y_2, z_2)\}$ is arcwise connected for any two points (y_1, z_1) and (y_2, z_2) of $Y \times Z$.*

The proof of the following theorem is similar to the proof of Theorem 6.10.15, we need to use Lemma 7.6.11. To conclude that Y and Z are acyclic, we apply Theorem 6.10.14, [9, 8.1] and Theorem 7.1.13.

7.6.12 Theorem *Let X be a continuum, and let $n \in \mathbb{N}$. If Y and Z are nondegenerate finite-dimensional continua such that $Y \times Z$ is homeomorphic to $HS_n(X)$, then X is hereditarily decomposable and does not contain nondegenerate proper terminal subcontinua. Also, Y and Z are arcwise connected and acyclic.*

The proof of the following theorem is similar to the one given for Theorem 6.10.17, we need to use Theorem 7.6.12, Theorem 7.1.21, [9, 8.1] and Theorem 7.1.13.

7.6.13 Theorem *Let X and Z be finite-dimensional continua and let n be an integer greater than one. If $HS_n(X)$ is homeomorphic to $X \times Z$, then X is a tree.*

7.7 Fixed Points

We give three classes of continua such that their n-fold hyperspace suspensions have the fixed point property. In order to do this, we follow [12, 17, 26, 27, 34, 40, 41].

We start quoting a result of J. Krasinkiewicz [26, 1.1]:

7.7.1 Theorem *Let Z be a continuum with property (b). If for each $\varepsilon > 0$, there exist an ε-map $f_\varepsilon \colon Z \to D$, where D is a 2-cell, and a nonempty subset A_ε of $f_\varepsilon^{-1}(\partial(D))$, where $\partial(D)$ is the manifold boundary of D, such that $f_\varepsilon|_{A_\varepsilon} \colon A_\varepsilon \to \partial(D)$ is not homotopic to a constant map, then Z has the fixed point property.*

The following result is the motivation of Sam B. Nadler, Jr., to define hyperspaces suspensions. It is readily seen that the hyperspace suspension of arc-like continua are disk-like continua.

7.7.2 Theorem *If X is a chainable continuum, then $HS(X)$ has the fixed point property.*

Proof Let

$$\Gamma_0 = \{A \in \mathcal{C}([0, 1]) \mid 0 \in A\}$$

and

$$\Gamma_1 = \{A \in \mathcal{C}([0, 1]) \mid 1 \in A\}.$$

By [40, (0.54)], $\mathcal{C}([0, 1])$ is a 2-cell such that Γ_0 and Γ_1 are arcs in the manifold boundary of $\mathcal{C}([0, 1])$. In fact, the manifold boundary of $\mathcal{C}([0, 1])$ is $\Gamma_0 \cup \Gamma_1 \cup \mathcal{F}_1([0, 1])$. Hence, $HS([0, 1])$ is a 2-cell whose manifold boundary is $\mathfrak{S} = q_{[0,1]}(\Gamma_0) \cup q_{[0,1]}(\Gamma_1)$ and $q_{[0,1]}(\Gamma_0) \cap q_{[0,1]}(\Gamma_1) = \{T_{[0,1]}, F_{[0,1]}\}$.

Let $\varepsilon > 0$. Since X is a chainable continuum, there exists an ε-map $f\colon X \twoheadrightarrow [0, 1]$. Let $\mathcal{C}(f)\colon \mathcal{C}(X) \twoheadrightarrow \mathcal{C}([0, 1])$ be the induced map given by $\mathcal{C}(f)(A) = f(A)$. Note that, by Corollary 8.2.3, $\mathcal{C}(f)$ is continuous. By Lemma 8.2.11, $\mathcal{C}_n(f)$ is a ε-map.

Note that $\mathcal{C}(f)(\mathcal{F}_1(X)) = \mathcal{F}_1([0, 1])$. By Lemma 8.2.12, there exists an ε-map $HS(f)\colon HS(X) \twoheadrightarrow HS([0, 1])$. By the Kuratowski–Zorn Lemma, there exists a subcontinuum M of X such that $f(M) = [0, 1]$ and $f(L) \neq [0, 1]$ for any proper subcontinuum L of M. Let $x_0, x_1 \in M$ be such that $f(x_0) = 0$ and $f(x_1) = 1$. By Theorem 1.8.20, there exist order arcs $\alpha_0, \alpha_1\colon [0, 1] \to \mathcal{C}(X)$ such that $\alpha_j(0) = \{x_j\}$ and $\alpha_j(1) = M$, $j \in \{1, 2\}$. Let $\Lambda_j = \alpha_j([0, 1])$, $j \in \{1, 2\}$. Observe that $\Lambda_0 \cap \Lambda_1 = \{M\}$ and $\Lambda_j \cap \mathcal{F}_1(X) = \{\{x_j\}\}$, $j \in \{1, 2\}$. Also note that $\mathcal{C}(f)(\Lambda_j) = \Gamma_j$, $j \in \{1, 2\}$. Then we have that $\mathfrak{S}' = q_X(\Lambda_0) \cup q_X(\Lambda_1)$ is a simple closed curve, where $q_X(\Lambda_0)$ and $q_X(\Lambda_1)$ are arcs such that $q_X(\Lambda_0) \cap q_X(\Lambda_1) = \{T_X, F_X\}$. Since $HS(f) \circ q_X = q_{[0,1]} \circ \mathcal{C}(f)$, we have that $HS(f)(q_X(\Lambda_j)) = q_{[0,1]}(\Gamma_j)$, $j \in \{1, 2\}$, $HS(f)(T_X) = T_{[0,1]}$ and $HS(f)(F_X) = F_{[0,1]}$. Therefore, $HS(f)|_{\mathfrak{S}'}\colon \mathfrak{S}' \twoheadrightarrow \mathfrak{S}$ is not homotopic to a constant map. Hence, by Theorem 7.7.1, $HS(X)$ has the fixed point property.

Q.E.D.

7.7.3 Remark Note that Theorem 7.7.2 gives a partial positive answer to Question 9.1.5.

As a corollary of the proof of Theorem 7.7.2 and Theorem 2.5.13, we have:

7.7.4 Corollary *If X is a chainable continuum, then $HS(X)$ is a disk-like continuum with the fixed point property.*

Now, we turn our attention to continua with surjective semispan zero.

7.7.5 Definition A continuum X has *surjective semispan zero* provided that for each subcontinuum Z of $X \times X$ such that $\pi_1(Z) = X$, $Z \cap \Delta_X \neq \emptyset$, where $\pi_1\colon X \times X \twoheadrightarrow X$ is the projection to the first factor and $\Delta_X = \{(x, x) \mid x \in X\}$.

7.7.6 Theorem *Let Y be a continuum such that Y has surjective semispan zero. If $f\colon X \to Y$ is a map from a continuum X onto Y, then the induced map $HS(f)\colon HS(X) \to HS(Y)$ is universal.*

Proof Let $g\colon HS(X) \to HS(Y)$ be any map. We show that there exists $A \in \mathcal{C}(X)$ such that $g(q_X(A)) = HS(f)(q_X(A))$.

Let $\mu\colon \mathcal{C}(Y) \twoheadrightarrow [0, 1]$ be a Whitney map for $\mathcal{C}(Y)$. With the help of μ, we define $\mu'\colon HS(Y) \twoheadrightarrow [0, 1]$ by $\mu'(F_Y) = 0$ and $\mu'(\chi) = \mu(q_Y^{-1}(\chi))$ for each $\chi \in HS(Y) \setminus \{F_Y\}$. Clearly, μ' is continuous and $\mu'(\chi) = 0$ if and only if $\chi = F_Y$.
Let

$$\mathcal{R} = \{A \in \mathcal{C}(X) \mid \mu'(g(q_X(A))) = \mu'(HS(f)(q_X(A)))\}.$$

Note that $\mu' \circ HS(f) \circ q_X\colon \mathcal{C}(X) \to [0, 1]$ is continuous and onto, thus, universal (Theorem 2.6.8). Hence, there exists $A \in \mathcal{C}(X)$ such that $\mu'(g(q_X(A))) = \mu'(HS(f)(q_X(A)))$. Therefore, $\mathcal{R}$ is a nonempty closed subset of $\mathcal{C}(X)$.

Claim 1 $\mathcal{R}$ cuts weakly between $\mathcal{F}_1(X)$ and $\{X\}$ in $\mathcal{C}(X)$.

Let $\mathcal{L}$ be a subcontinuum of $\mathcal{C}(X)$ such that $X \in \mathcal{L}$ and $\mathcal{L} \cap \mathcal{F}_1(X) \neq \emptyset$. Clearly, $\mu' \circ HS(f) \circ q_X(\mathcal{L}) = [0, 1]$; i.e., $\mu' \circ HS(f) \circ q_X|_{\mathcal{L}}\colon \mathcal{L} \to [0, 1]$ is universal (Theorem 2.6.8). Thus, there exists $A \in \mathcal{L}$ such that $\mu' \circ HS(f) \circ q_X(A) = \mu' \circ g \circ q_X(A)$. Note that this implies that $A \in \mathcal{R}$. Therefore, $\mathcal{R} \cap \mathcal{L} \neq \emptyset$, and $\mathcal{R}$ cuts weakly between $\mathcal{F}_1(X)$ and $\{X\}$ in $\mathcal{C}(X)$.

Since $\mathcal{C}(X)$ is s connected (Theorem 6.3.9) and $\mathcal{R}$ cuts weakly between $\mathcal{F}_1(X)$ and $\{X\}$ in $\mathcal{C}(X)$, there exists a component $\mathcal{K}$ of $\mathcal{R}$ with the same property.

Claim 2 $X = \bigcup\{A \mid A \in \mathcal{K}\}$.

Let x be a point of X. Let γ be an order arc from $\{x\}$ to X in $\mathcal{C}(X)$ (Theorem 1.8.20). Since $\mathcal{K}$ cuts weakly between $\mathcal{F}_1(X)$ and $\{X\}$ in $\mathcal{C}(X)$, $\gamma \cap \mathcal{K} \neq \emptyset$. Let $A \in \gamma \cap \mathcal{K}$. Then, since $A \in \gamma$, we obtain that $x \in A$. Hence, $X = \bigcup\{A \mid A \in \mathcal{K}\}$.

Claim 3 If $F_Y \in g(q_X(\mathcal{K}))$, then $g(q_X(A)) = HS(f)(q_X(A))$ for some $A \in \mathcal{C}(X)$.

Suppose there exists an element A of $\mathcal{K}$ such that $g(q_X(A)) = F_Y$. Then, $\mu'(g(q_X(A))) = 0$. Since $A \in \mathcal{K} \subset \mathcal{R}$, we have that $0 = \mu'(g(q_X(A))) = \mu'(HS(f)(q_X(A)))$. Since $\mu'(HS(f)(q_X(A))) = 0$ if and only if $HS(f)(q_X(A)) = F_Y$, we have that $g(q_X(A)) = HS(f)(q_X(A))$.

From now on, we assume that $F_Y \notin g(q_X(\mathcal{K}))$. Hence,

$$q_Y^{-1}|_{g(q_X(\mathcal{K}))}\colon g(q_X(\mathcal{K})) \to \mathcal{C}(Y)$$

is an embedding.

Claim 4 There exists an element A in $\mathcal{K}$ such that either $f(A) \subset q_Y^{-1}(g(q_X(A)))$ or $q_Y^{-1}(g(q_X(A))) \subset f(A)$.

Suppose the claim is not true; i.e., suppose that for each $A \in \mathcal{C}(X)$, $f(A) \not\subset q_Y^{-1}(g(q_X(A)))$ and $q_Y^{-1}(g(q_X(A))) \not\subset f(A)$. By continuity of the maps and compactness of $\mathcal{K}$, there exists an $\varepsilon > 0$ such that

$$f(A) \not\subset \mathcal{V}_\varepsilon(q_Y^{-1}(g(q_X(A)))) \text{ and } q_Y^{-1}(g(q_X(A))) \not\subset \mathcal{V}_\varepsilon(f(A)).$$

Now, for each $A \in \mathcal{K}$, let

$$[\![A]\!] = \left[f(A) \times q_Y^{-1}(g(q_X(A))) \right] \setminus \left[\mathcal{V}_\varepsilon(q_Y^{-1}(g(q_X(A)))) \times \mathcal{V}_\varepsilon(f(A)) \right].$$

Then $[\![A]\!]$ is a subcontinuum of $Y \times Y$ such that $\pi_1([\![A]\!]) = f(A)$ and $[\![A]\!] \cap \Delta_Y = \emptyset$ [17, Lemma 3.1].

Let $M = \bigcup \{ [\![A]\!] \mid A \in \mathcal{K} \}$. It is easy to see that M is a subcontinuum of $Y \times Y$. Since $\pi_1([\![A]\!]) = f(A)$ and $X = \bigcup \{ A \mid A \in \mathcal{K} \}$ (Claim 2), we have that

$$\pi_1(M) = \bigcup \{ f(A) \mid A \in \mathcal{K} \} = f \left(\bigcup \{ A \mid A \in \mathcal{K} \} \right) = f(X) = Y.$$

Since for each $A \in \mathcal{K}$, $[\![A]\!] \cap \Delta_Y = \emptyset$, we obtain that $M \cap \Delta_Y = \emptyset$, which is a contradiction, since Y has surjective semispan zero. Therefore, there exists $A \in \mathcal{K}$ such that either $f(A) \subset q_Y^{-1}(g(q_X(A)))$ or $q_Y^{-1}(g(q_X(A))) \subset f(A)$.

To finish the proof, let A be an element of $\mathcal{K}$ such that either $f(A) \subset q_Y^{-1}(g(q_X(A)))$ or $q_Y^{-1}(g(q_X(A))) \subset f(A)$. Since $\mathcal{K} \subset \mathcal{R}$, $\mu'(g(q_X(A))) = \mu'(HS(f)(q_X(A)))$; i.e.,

$$\mu'(g(q_X(A))) = \mu'(q_Y(f(A))),$$

by definition of $HS(f)$. By definition of μ', we obtain

$$\mu(q_Y^{-1}(g(q_X(A)))) = \mu(f(A)).$$

Now, by the election of A and the properties of Whitney maps, we have that $q_Y^{-1}(g(q_X(A))) = f(A)$. Hence,

$$g(q_X(A)) = q_Y(f(A)).$$

Thus, by definition of $HS(f)$, we obtain

$$g(q_X(A)) = HS(f)(q_X(A)).$$

Therefore, $HS(f)$ is universal.

Q.E.D.

As a consequence of Theorem 7.7.6 and Proposition 2.6.4, we obtain:

7.7.7 Corollary *If X is a continuum with surjective semispan zero, then $HS(X)$ has the fixed point property.*

7.7.8 Remark   Observe that Corollary 2.6.10 implies that chainable continua have surjective semispan zero. Hence, Theorem 7.7.2 is a consequence of Corollary 7.7.7

The following theorem shows that the *n*-fold hyperspace suspension of an absolute retract is an absolute retract.

7.7.9 Theorem   *If X is an absolute retract and $n \in \mathbb{N}$, then $HS_n(X)$ is an absolute retract.*

Proof Without loss of generality, we assume that X is embedded in the Hilbert cube $\mathcal{Q}$ (Theorem 1.1.16). Since X is an absolute retract, there exists a retraction $r \colon \mathcal{Q} \twoheadrightarrow X$. Then it is easy to verify that the induced map $HS_n(r) \colon HS_n(\mathcal{Q}) \twoheadrightarrow HS_n(X)$, given by:

$$HS_n(r)(\chi) = \begin{cases} q_X^n(\mathcal{C}_n(r)((q_{\mathcal{Q}}^n)^{-1}(\chi))), & \text{if } \chi \neq F_{\mathcal{Q}}^n; \\ F_X^n, & \text{if } \chi = F_{\mathcal{Q}}^n; \end{cases}$$

is a retraction (note that, by Theorem 8.2.6, $HS(r)$ is continuous). Since $HS_n(\mathcal{Q})$ is homeomorphic to $\mathcal{Q}$ (Corollary 7.4.6), $HS_n(X)$ is an absolute retract [27, Theorem 7, p. 341].

Q.E.D.

Since absolute retracts have the fixed point property [27, Theorem 11, p. 343], we have the following:

7.7.10 Corollary   *If X is an absolute retract and $n \in \mathbb{N}$, then $HS_n(X)$ has the fixed point property.*

7.8   Absolute *n*-Fold Hyperspace Suspensions

In 1971 de Groot [14] defined the notion of an absolute suspension. The general idea of his definition is to consider the vertexes of a suspension as distinguished points of the suspension: A continuum X is an *absolute suspension* provided that for any two points p and q of X, there exists a continuum $Y(p, q)$, depending on p and q, such that (X, p, q) is homeomorphic to $(\Sigma(Y(p, q)), v^+, v^-)$, where v^+ and v^- are the vertexes of $\Sigma(Y(p, q))$. This means that there exists a homeomorphism $h \colon X \twoheadrightarrow Y$ such that $h(p) = v^+$ and $h(q) = v^-$. In analogy with this definition we define absolute *n*-fold hyperspace suspensions [36], by considering the points T_X^n and F_X^n as special points, as follows:

7.8.1 Definition   A continuum X is said to be an *absolute n-fold hyperspace suspension* provided that for each pair of different points p and q of X, there exists a continuum $Y(p, q)$ such that (X, p, q) is homeomorphic to $(HS_n(Y(p, q)), T_{Y(p,q)}^n, F_{Y(p,q)}^n)$. When $n = 1$, we call X an *absolute hyperspace suspension* (omitting 1-fold).

7.8.2 Remark Note that in the definition of absolute *n*-fold hyperspace suspension, the space $Y(p, q)$ may not be homeomorphic to the space $Y(q, p)$. On the other hand, in de Groot's definition of absolute suspension it can be assumed that the space $Y(p, q)$ is homeomorphic to the space $Y(q, p)$ since there exists a homeomorphism of any suspension onto itself that interchanges the vertexes.

7.8.3 Example The Hilbert cube, $\mathcal{Q}$, is an absolute *n*-fold hyperspace suspension. To see this, note that by Corollary 7.4.6, $HS_n(\mathcal{Q})$ is homeomorphic to $\mathcal{Q}$. Hence, for any two points p and q of $\mathcal{Q}$, $(\mathcal{Q}, p, q)$ is homeomorphic to $(HS_n(\mathcal{Q}), T^n_{\mathcal{Q}}, F^n_{\mathcal{Q}})$ by Anderson's Theorem on homogeneity [23, 11.9.1].

7.8.4 Theorem *If a continuum X is an absolute n-fold hyperspace suspension, for some $n \in \mathbb{N}$, then X is a unicoherent locally connected continuum.*

Proof By definition, for any two different points p and q of X, there exists a continuum $Y(p, q)$ such that (X, p, q) is homeomorphic to $(HS_n(Y(p, q)), T^n_{Y(p,q)}, F^n_{Y(p,q)})$. Since, by Theorem 7.1.10, $HS_n(Y(p, q))$ is locally connected at $T^n_{Y(p,q)}$, X is locally connected at p. Since p is arbitrary, we have that X is a locally connected continuum. By Theorem 7.1.13, $HS_n(Y(p, q))$ is unicoherent. Therefore, X is unicoherent.

Q.E.D.

7.8.5 Theorem *If a continuum X is an absolute n-fold hyperspace suspension, then X is dimensionally homogeneous.*

Proof Let X be an absolute *n*-fold hyperspace suspension. By Theorem 7.8.4, X is a locally connected continuum. Let p and q be two points of X. Since X is an absolute *n*-fold hyperspace suspension, there exists a continuum $Y(p, q)$ such that (X, p, q) is homeomorphic to $(HS_n(Y(p, q)), T^n_{Y(p,q)}, F^n_{Y(p,q)})$. Thus, $HS_n(Y(p, q))$ is a locally connected continuum. Hence, by Theorem 7.4.2, $Y(p, q)$ is a locally connected continuum.

If $\dim(X) < \infty$, then $Y(p, q)$ is a graph by Theorem 7.4.3. Since (X, p, q) is homeomorphic to $(HS_n(Y(p, q)), T^n_{Y(p,q)}, F^n_{Y(p,q)})$, $\dim_q(X) \leq \dim_p(X)$, by Theorem 7.4.21. Interchanging the roles of p and q, we have that $\dim_p(X) \leq \dim_q(X)$. Therefore, X is dimensionally homogeneous.

If $\dim(X) = \infty$, then $\dim(HS_n(Y(p, q))) = \infty$. Since $Y(p, q)$ is locally connected and $\dim(HS_n(Y(p, q))) = \infty$, by Theorem 7.4.3, $Y(p, q)$ is not a graph. By Theorem 7.1.21, $\dim(\mathcal{C}_n(Y(p, q))) = \infty$. Thus, by [44, 2.9], $\dim_{Y(p,q)}(\mathcal{C}_n(Y(p, q))) = \infty$. Hence, by Remark 7.1.5, $\dim_{T^n_X}(HS_n(Y(p, q)), T^n_{Y(p,q)}, F^n_{Y(p,q)}) = \infty$. Therefore, $\dim_p(X) = \infty$. Since p is an arbitrary point of X, X is dimensionally homogeneous.

Q.E.D.

7.8.6 Theorem *Let X be a finite-dimensional continuum and let $n \in \mathbb{N}$. If X is absolute n-fold hyperspace suspension, then either X is homeomorphic to $HS_n([0, 1])$ or X is homeomorphic to $HS_n(\mathcal{S}^1)$.*

Proof Suppose X is an absolute n-fold hyperspace suspension. Thus, by Theorem 7.8.4, X is a locally connected continuum. Note that X is also dimensionally homogeneous by Theorem 7.8.5. Let p and q be two points of X. Hence, since X is an absolute n-fold hyperspace suspension, there exists a continuum $Y(p, q)$ such that (X, p, q) is homeomorphic to $(HS_n(Y(p, q)), T^n_{Y(p,q)}, F^n_{Y(p,q)})$. Since X is finite-dimensional, by Theorem 7.4.3, $Y(p, q)$ is a graph. Thus, by Lemma 7.1.14, there exists $\chi \in HS_n(Y(p, q))$ such that $\dim_\chi(HS_n(Y(p, q))) = 2n$. Therefore, since X is dimensionally homogeneous, $\dim(HS_n(Y(p, q))) = 2n$. Hence, by Theorem 7.1.21 and Corollary 6.1.14, $Y(p, q)$ is either an arc or a simple closed curve. Thus, either X is homeomorphic to $HS_n([0, 1])$ or X is homeomorphic to $HS_n(\mathcal{S}^1)$.

Q.E.D.

The next theorem shows that when $n = 1$, the converse of Theorem 7.8.6 is false for $[0, 1]$ but true for $\mathcal{S}^1$. As a corollary, we obtain that the 2-sphere is the only finite-dimensional absolute hyperspace suspension. We show in Theorem 7.8.9 that when $n = 2$, the converse of Theorem 7.8.6 is false for both $[0, 1]$ and $\mathcal{S}^1$.

7.8.7 Theorem $HS([0, 1])$ *is not an absolute hyperspace suspension, but* $HS(\mathcal{S}^1)$ *is an absolute hyperspace suspension.*

Proof By Example 7.1.6, $HS([0, 1])$ is a 2-cell. Let Y be a continuum such that $HS(Y)$ is homeomorphic to $HS([0, 1])$. Then Y is an arc (Theorem 7.4.8). Furthermore, it follows from [40, (0.54)] that T_Y and F_Y are points of the manifold boundary of $HS([0, 1])$. Thus, considering two points χ_1 and χ_2 in the manifold interior of $HS([0, 1])$, we see that $HS([0, 1])$ is not an absolute hyperspace suspension.

To prove the second part of our theorem, recall from Example 7.1.7 that $HS(\mathcal{S}^1)$ is a 2-sphere. Hence, $HS(\mathcal{S}^1)$ has the following property: If $\{\chi_1, \chi_2\}$ and $\{\chi'_1, \chi'_2\}$ are two-point subsets of $HS(\mathcal{S}^1)$, then there exists a homeomorphism of $HS(\mathcal{S}^1)$ onto $HS(\mathcal{S}^1)$ taking χ_k to χ'_k for each $k \in [1, 2]$. It now follows immediately that $HS(\mathcal{S}^1)$ is an absolute hyperspace suspension.

Q.E.D.

7.8.8 Corollary *Let X be a finite-dimensional continuum. Then X is an absolute hyperspace suspension if and only if X is a 2-sphere.*

Proof Assume that X is a finite-dimensional absolute hyperspace suspension. By Theorem 7.8.6, X is homeomorphic to $HS([0, 1])$ or X is homeomorphic to $HS(\mathcal{S}^1)$. Hence, by Theorem 7.8.7, X is homeomorphic to $HS(\mathcal{S}^1)$. Therefore, X is a 2-sphere by Example 7.1.7.

Conversely, a 2-sphere is an absolute hyperspace suspension by Theorem 7.8.7 since $HS(\mathcal{S}^1)$ is a 2-sphere (Example 7.1.7).

Q.E.D.

7.8.9 Theorem $HS_2([0, 1])$ *and* $HS_2(\mathcal{S}^1)$ *are not absolute 2-fold hyperspace suspensions.*

Proof By Theorem 7.1.9, $H S_2([0, 1])$ is a 4-cell. It is actually the case that $T_{[0,1]}^2$ and $F_{[0,1]}^2$ are both points of the manifold boundary of $H S_2([0, 1])$. However, we need not verify this. Instead, we simply choose χ_1 and χ_2 as follows: Let χ_1 and χ_2 be points in the manifold interior of $H S_2([0, 1])$ if at least one of $T_{[0,1]}^2$ and $F_{[0,1]}^2$ is in the manifold boundary of $H S_2([0, 1])$; let χ_1 and χ_2 be points in the manifold boundary of $H S_2([0, 1])$ if $T_{[0,1]}^2$ and $F_{[0,1]}^2$ are both in the manifold interior of $H S_2([0, 1])$. Then we see that $(H S_2([0, 1]), \chi_1, \chi_2)$ is not homeomorphic to $(H S_2([0, 1]), T_{[0,1]}^2, F_{[0,1]}^2)$ Therefore, $H S_2([0, 1])$ is not an absolute 2-fold hyperspace suspension (since the continuum $Y(\chi_1, \chi_2)$ in the definition of absolute 2-fold hyperspace suspension must be $[0, 1]$ by Theorem 7.4.14).

We have left to prove that $H S_2(S^1)$ is not an absolute 2-fold hyperspace suspension. Let $A, B \in \mathcal{C}_2(S^1) \setminus \{S^1\}$ such that $A \notin \mathcal{F}_2(S^1)$ and $B \notin \mathcal{F}_2(S^1)$. By Lemma 7.4.12, A and B have 4-cell neighborhoods in $\mathcal{C}_2(S^1)$. Hence, $q_{S^1}^2(A)$ and $q_{S^1}^2(B)$ have 4-cell neighborhoods in $H S_2(S^1)$. By the second part of Lemma 7.4.12 and Remark 7.1.5, $T_{S^1}^2$ does not have a 4-cell neighborhood in $H S_2(S^1)$. Thus, we obtain that $(H S_2(S^1), q_{S^1}^2(A), q_{S^1}^2(B))$ is not homeomorphic to $(H S_2(S^1), T_{S^1}^2, F_{S^1}^2)$. Therefore, $H S_2(S^1)$ is not an absolute 2-fold hyperspace suspension (by Theorem 7.4.14).

Q.E.D.

As a consequence of Theorems 7.8.6 and 7.8.9, we have:

7.8.10 Corollary *No finite-dimensional continuum is an absolute 2-fold hyperspace suspension.*

The following result follows from [16, Corollary 3.4].

7.8.11 Theorem *For each $n \geq 3$, there does not exist a finite-dimensional absolute n-fold hyperspace suspension.*

7.8.12 Definition A metric space Z is said to be a *Hilbert cube manifold* (*Q-manifold*) provided that for each point $z \in Z$, there exists a neighborhood W of z in Z such that W is homeomorphic to the Hilbert cube.

7.8.13 Theorem *Let X be an infinite-dimensional continuum and let $n \in \mathbb{N}$. If X is an absolute n-fold hyperspace suspension, then X is a unicoherent Q-manifold.*

Proof Suppose X is an infinite-dimensional absolute n-fold hyperspace suspension. Thus, by Theorem 7.8.4, X is a unicoherent locally connected continuum. Let p and q be two points of X. By hypothesis, there exists a continuum $Y(p, q)$ such that (X, p, q) is homeomorphic to $(H S_n(Y(p, q)), T_{Y(p,q)}^n, F_{Y(p,q)}^n)$.

Since X is a locally connected continuum, $H S_n(Y(p, q))$ is locally connected also. Hence, by Theorem 7.4.2, $Y(p, q)$ is a locally connected continuum. Thus, $\mathcal{C}_n(Y(p, q))$ is locally connected by Theorem 6.1.6.

Since $\dim(X) = \infty$ and X is dimensionally homogeneous (Theorem 7.8.5), we see from Lemma 7.1.14 that $Y(p, q)$ does not have a free arc. Hence, by Theorem 6.1.8, $\mathcal{C}_n(Y(p, q))$ is a Hilbert cube. This implies that

$Y(p, q)$ has a Hilbert cube neighborhood $\mathcal{W}$ in $\mathcal{C}_n(Y(p, q))$ such that $\mathcal{W} \cap \mathcal{F}_n(Y(p, q)) = \emptyset$. Hence, by Remark 7.1.5 and since (X, p, q) is homeomorphic to $(HS_n(Y(p, q)), T^n_{Y(p,q)}, F^n_{Y(p,q)})$, p has a Hilbert cube neighborhood W in X. Therefore, X is a Q-manifold.

Q.E.D.

7.8.14 Remark Note that $\mathcal{S}^1 \times Q$ is a Q-manifold which is not an n-fold hyperspace suspension since it does not have property (b), and n-fold hyperspace suspensions have property (b), by Theorem 7.1.13. The material for this section is based on [16, 23, 36, 40, 43].

7.9 Hereditarily Indecomposable Continua

We show hereditarily indecomposable continua have unique n-fold hyperspace suspensions. First we establish the case for hyperspace suspensions (Theorem 7.9.5) and later the general case (Theorem 7.9.7). To this end, we use [12, 23, 32, 40].

7.9.1 Lemma *Let X be a hereditarily indecomposable continuum. Let Y be a continuum such that $HS(Y)$ is homeomorphic to $HS(X)$, and let $h\colon HS(X) \twoheadrightarrow HS(Y)$ be a homeomorphism. If B_1 and B_2 are two distinct elements of $\mathcal{C}(X) \setminus \mathcal{F}_1(X)$ such that $B_1 \subset B_2$ then $q_Y^{-1}hq_X(B_1) \subset q_Y^{-1}hq_X(B_2)$.*

Proof Since X is hereditarily indecomposable, $\mathcal{C}(X)$ is uniquely arcwise connected [40, (1.61)]. Hence, arbitrary small neighborhoods of $q_X(X)$ in $HS(X)$ do not contain simple closed curves. Since arbitrary small neighborhoods of F_X and F_Y contain simple closed curves (Theorem 7.1.10), $h(q_X(X)) = q_Y(Y)$ and $h(F_X) = F_Y$.

Let α be an order arc from $q_Y^{-1}hq_X(B_1)$ to Y in $\mathcal{C}(Y)$ (Theorem 1.8.20). Thus, $q_X^{-1}h^{-1}q_Y(\alpha)$ is an arc from B_1 to X. Since the only arc in $\mathcal{C}(X)$ from B_1 to X contains B_2, we have that $q_Y^{-1}hq_X(B_2) \subset \alpha$. Since α is an order arc, we have that $q_Y^{-1}hq_X(B_1) \subset q_Y^{-1}hq_X(B_2)$. Since $q_X|_{\mathcal{C}(X) \setminus \mathcal{F}_1(X)}$, $h|_{HS(X) \setminus \{q_X(X), F_X\}}$ and $q_Y|_{\mathcal{C}(Y) \setminus \mathcal{F}_1(Y)}$ are homeomorphisms (Remark 7.1.5) and $B_1 \neq B_2$, $q_Y^{-1}hq_X(B_1) \neq q_Y^{-1}hq_X(B_2)$.

Q.E.D.

7.9.2 Lemma *Let X be a hereditarily indecomposable continuum. If Y is a continuum such that $HS(Y)$ is homeomorphic to $HS(X)$, then Y is hereditarily indecomposable.*

Proof Let $h\colon HS(X) \twoheadrightarrow HS(Y)$ be a homeomorphism. Let $\mu\colon \mathcal{C}(Y) \to [0, 1]$ be a Whitney map and let $t \in (0, 1)$. Let $\mathcal{B} = q_X^{-1}h^{-1}q_Y\left(\mu^{-1}(t)\right)$. We show that $\mathcal{B}$ is a Whitney level of $\mathcal{C}(X)$. To this end, it is enough to show that $\mathcal{B}$ intersects each order arc in $\mathcal{C}(X)$ joining any singleton $\{x\}$ with X, and if $B_1, B_2 \in \mathcal{B}$ and $B_1 \subset B_2$, then $B_1 = B_2$ [23, 23.8].

Let $\{x\} \in \mathcal{F}_1(X)$ and let γ be an order arc in $\mathcal{C}(X)$ from $\{x\}$ to X. Suppose $\mathcal{B} \cap \gamma = \emptyset$. Thus, $q_Y^{-1}hq_X(\gamma \setminus \{\{x\}\}) \cap \mu^{-1}(t) = \emptyset$, a contradiction to [23, 23.8] by Lemma 7.9.1. Therefore, $\mathcal{B} \cap \gamma \neq \emptyset$.

Let $B_1, B_2 \in \mathcal{B}$ be such that $B_1 \subset B_2$. Suppose that $B_1 \neq B_2$. Thus, we have that $q_Y^{-1}hq_X(B_1), q_Y^{-1}hq_X(B_2) \in \mu^{-1}(t)$, $q_Y^{-1}hq_X(B_1) \subset q_Y^{-1}hq_X(B_2)$ (Lemma 7.9.1) and $q_Y^{-1}hq_X(B_1) \neq q_Y^{-1}hq_X(B_2)$, a contradiction to [23, 23.8]. Therefore, $B_1 = B_2$.

Therefore, $\mathcal{B}$ is a Whitney level in $\mathcal{C}(X)$. Since X is hereditarily indecomposable, $\mathcal{B}$ is a hereditarily indecomposable continuum [40, (14.1)]. Since $\mathcal{B}$ is homeomorphic to $\mu^{-1}(t)$, this is a hereditarily indecomposable continuum. Therefore, Y is hereditarily indecomposable [40, (14.54)].

Q.E.D.

7.9.3 Notation If X is a hereditarily indecomposable continuum and α is an order arc in $\mathcal{C}(X)$ with X as one of its end points, then the other end point of α is denoted by $ep(\alpha)$. If $A \in \mathcal{C}(X) \setminus \{X\}$, then α_A denotes the order arc from A to X.

The following lemma is easy to establish:

7.9.4 Lemma *Let X be a continuum. If $\{\alpha_n\}_{n=1}^\infty$ is a sequence of order arcs having X as an end point and converging to the order arc α, then the sequence of end points $\{ep(\alpha_n)\}_{n=1}^\infty$ converges to $ep(\alpha)$.*

We are ready to prove that hereditarily indecomposable continua have unique hyperspace suspensions.

7.9.5 Theorem *Let X be a hereditarily indecomposable continuum. If Y is a continuum such that $HS(Y)$ is homeomorphic to $HS(X)$, then Y is homeomorphic to X.*

Proof By Lemma 7.9.2, Y is hereditarily indecomposable.

Let $h: HS(X) \twoheadrightarrow HS(Y)$ be a homeomorphism. Note that $h(T_X) = T_Y$ and $h(F_X) = F_Y$ (Theorem 7.1.10). Define $\hat{h}: \mathcal{C}(X) \to \mathcal{C}(Y)$ by

$$\hat{h}(A) = \begin{cases} q_Y^{-1}hq_X(A), & \text{if } A \in \mathcal{C}(X) \setminus \mathcal{F}_1(X); \\ ep(q_Y^{-1}hq_X(\alpha_A \setminus \{A\})), & \text{if } A \in \mathcal{F}_1(X). \end{cases}$$

Since X and Y are hereditarily indecomposable, $\hat{h}$ is well defined (Lemma 7.9.1).

Let $\{x_1\}$ and $\{x_2\}$ be two distinct points of $\mathcal{F}_1(X)$. Thus, there exist two open sets $\mathcal{U}_1, \mathcal{U}_2$ of $\mathcal{C}(X)$ such that $\{x_1\} \in \mathcal{U}_1$, $\{x_2\} \in \mathcal{U}_2$ and $\mathcal{U}_1 \cap \mathcal{U}_2 = \emptyset$. Hence, $\alpha_{\{x_1\}} \cap \mathcal{U}_2 = \emptyset$ and $\alpha_{\{x_2\}} \cap \mathcal{U}_1 = \emptyset$. Thus, $q_X(\alpha_{\{x_1\}}) \neq q_X(\alpha_{\{x_2\}})$ and $ep(q_Y^{-1}hq_X(\alpha_{\{x_1\}})) \neq ep(q_Y^{-1}hq_X(\alpha_{\{x_2\}}))$. Therefore, $\hat{h}$ is one-to-one.

Let $\{y\} \in \mathcal{F}_1(Y)$. Hence, $\{x\} = ep(q_X^{-1}h^{-1}q_Y(\alpha_{\{y\}} \setminus \{\{y\}\}))$ satisfies that $\hat{h}(\{x\}) = \{y\}$. Therefore, $\hat{h}$ is onto.

To show $\hat{h}$ is continuous, let $\{A_n\}_{n=1}^\infty$ be a sequence of elements in $\mathcal{C}(X)$ converging to $\{x\}$. Note that the sequence $\{\alpha_{A_n}\}_{n=1}^\infty$ converges to $\alpha_{\{x\}}$.

Let $\mu\colon C(Y) \to [0, 1]$ be a Whitney map and let $t \in (0, 1)$. Hence, by the proof of Lemma 7.9.2, $\mathcal{B} = ep(q_X^{-1}h^{-1}q_Y(\mu^{-1}(t)))$ is a Whitney level in $C(X)$. Without loss of generality, we assume that for all $n \in \mathbb{N}$, $A_n \in \alpha_{A_n} \cap \mathcal{B}$. For each $n \in \mathbb{N}$, let α'_{A_n} and α''_{A_n} be the subarcs of α_{A_n} from $\alpha_{A_n} \cap \mathcal{B}$ to X and from $ep(\alpha_{A_n})$ to $\alpha_{A_n} \cap \mathcal{B}$, respectively. Let $\alpha'_{\{x\}}$ and $\alpha''_{\{x\}}$ be the subarcs of $\alpha_{\{x\}}$ from $\alpha_{\{x\}} \cap \mathcal{B}$ to X and from $\{x\}$ to $\alpha_{\{x\}} \cap \mathcal{B}$, respectively.

Since $C(X)$ is uniquely arcwise connected [40, (1.61)], $\{\alpha'_{A_n}\}_{n=1}^{\infty}$ converges to $\alpha'_{\{x\}}$. By continuity of $\hat{h} = q_Y^{-1}hq_X$ on $C(X) \setminus \mathcal{F}_1(X)$, $\{\hat{h}(\alpha'_{A_n})\}_{n=1}^{\infty}$ converges to $\hat{h}(\alpha'_{\{x\}})$.

Observe that for each $n \in \mathbb{N}$, $\hat{h}(\alpha'_{A_n}) \cap \hat{h}(\alpha''_{A_n}) = \hat{h}(\alpha'_{A_n} \cap \alpha''_{A_n})$. Also note that, $\hat{h}(\alpha'_{\{x\}}) \cap \hat{h}(\alpha''_{\{x\}}) = \hat{h}(\alpha'_{\{x\}} \cap \alpha''_{\{x\}})$. By continuity of $\hat{h}$, $\lim_{n\to\infty} \hat{h}(\alpha'_{A_n}) \cap \hat{h}(\alpha''_{A_n}) = \hat{h}(\alpha'_{\{x\}}) \cap \hat{h}(\alpha''_{\{x\}})$. Since $C(Y)$ is uniquely arcwise connected, we have that $\lim_{n\to\infty} \hat{h}(\alpha_{A_n}) = \hat{h}(\alpha_{\{x\}})$. Hence, $\lim_{n\to\infty} \hat{h}(A_n) = \hat{h}(\{x\})$ (Lemma 7.9.4). Thus, $\hat{h}$ is a homeomorphism. Therefore, Y is homeomorphic to X, by Corollary 6.5.24 .

Q.E.D.

7.9.6 Theorem *Let X be a hereditarily indecomposable continuum, and let n and m be positive integers greater than or equal to two. If Y is a continuum such that $HS_m(Y)$ is homeomorphic to $HS_n(X)$, then Y is homeomorphic to X.*

Proof Let $h\colon HS_n(X) \twoheadrightarrow HS_m(Y)$ be a homeomorphism. Observe that $h(\{T_X^n, F_X^n\}) = \{T_Y^m, F_Y^m\}$ (Theorem 7.1.10). Let $\chi \in HS_n(X)$ be such that $(q_X^n)^{-1}(\chi) \in C(X) \setminus (\{X\} \cup \mathcal{F}_1(X))$. Then $C_n(X) \setminus \{(q_X^n)^{-1}(\chi)\}$ is not arcwise connected (Theorem 6.5.8). Hence, $HS_n(X) \setminus \{F_X^n, \chi\}$ is not arcwise connected (Theorem 7.5.1). Thus, since h is a homeomorphism, $HS_m(Y) \setminus \{h(F_X^n), h(\chi)\}$ is not arcwise connected. Then, by Theorem 7.1.30 and the proven fact that $h(\{T_X^n, F_X^n\}) = \{T_Y^m, F_Y^m\}$, we obtain that $h(F_X^n) = F_Y^m$ and $h(T_X^n) = T_Y^m$. We also have that $(q_Y^n)^{-1}(h(\chi)) \in C(Y)$ by Theorem 7.1.31.

Hence, we have a homeomorphism

$$\ell\colon C_n(X) \setminus \mathcal{F}_n(X) \to C_m(Y) \setminus \mathcal{F}_m(Y)$$

given by

$$\ell(A) = (q_Y^m)^{-1}(h(q_X^n(A))).$$

Note that $\ell(X) = Y$ and $\ell(C(X) \setminus \mathcal{F}_1(X)) \subset C(Y) \setminus \mathcal{F}_1(Y)$. As in the proof of Theorem 7.9.5, $\ell|_{C(X)\setminus\mathcal{F}_1(X)}$ can be extended to a map $\hat{\ell}\colon C(X) \to C(Y)$ in such a way that $\hat{\ell}$ is one-to-one and $\hat{\ell}(\mathcal{F}_1(X)) \subset \mathcal{F}_1(Y)$.

Let $Z \in C(Y)$ such that $\mathcal{F}_1(Z) = h(\mathcal{F}_1(X))$. Then $\hat{\ell}(C(X)) = C(Z)$. Since $\hat{\ell}(X) = \ell(X) = Y$, $Y \in C(Z)$. Hence, $Z = Y$ and $\hat{\ell}(\mathcal{F}_1(X)) = \mathcal{F}_1(Y)$. Therefore, Y is homeomorphic to X.

Q.E.D.

As a consequence of Theorems 7.9.5 and 7.9.6 we have that hereditarily indecomposable continua have unique n-fold hyperspace suspensions.

7.9.7 Corollary *Let X be a hereditarily indecomposable continuum, and let n ∈ ℕ. If Y is a continuum such that $HS_n(Y)$ is homeomorphic to $HS_n(X)$, then Y is homeomorphic to X.*

References

1. R. J. Bean, Decompositions of E^3 with a Null Sequence of Starlike Equivalent Nondegenerate Elements are E^3, Illinois J. Math., 11 (1967), 21–23.
2. D. E. Bennett, Aposyndetic Properties of Unicoherent Continua, Pacific J. Math., 37 (1971), 585–589.
3. K. Borsuk, *Theory of Shape*, Lectures, Fall, 1970, in: Lecture Notes Series, Vol. 28, Aarthus Universitet, Matematisk Institut, 1973.
4. R H Bing, Partitioning a Set, Bull. Amer. Math. Soc., 55 (1949), 1101–1110.
5. J. Camargo, D. Herrera and S. Macías, Cells and n-fold Hyperspaces, to appear in Colloquium Mathematicum.
6. R. E. Chandler, *Hausdorff Compactifications,* Lecture Notes in Pure and Applied Math., Vol. 23, Marcel Dekker, Inc., New York, Basel, 1976.
7. T. A. Chapman, *Lectures on Hilbert cube manifolds,* Conf. Board of the Math. Sci., Regional Conference Series in Math., no. 28, Amer. Math. Soc., Providence, R. I., 1976.
8. D. Curtis and N. T. Nhu, Hyperspaces of Finite Sets which are Homeomorphic to $\aleph_0$-dimensional Linear Metric Spaces, Topology Appl., 19 (1985), 251–260.
9. C. H. Dowker, Mapping Theorems for Noncompact Spaces, Amer. J. Math., 68 (1947), 200–242.
10. J. Dugundji, *Topology,* Allyn and Bacon, Inc., Boston, 1966.
11. C. Eberhart and S. B. Nadler, Jr., Hyperspaces of Cones and Fans, Amer. Math. Soc., 77 (1979), 279–288.
12. R. Escobedo, M. de J. López, S. Macías, On the Hyperspace Suspension of a Continuum, Topology Appl., 192 (2004), 109–124.
13. J. B. Fugate, G. R. Gordh, Jr., and L. Lum, Arc-smooth continua, Trans. AMer. Math. Soc., 265 (1981), 545–561.
14. J. de Groot, On the Topological Characterization of Manifolds, in: *General Topology and its Relations to Modern Analysis and Algebra, III,* Academia, Praha, 1971, 155–158.
15. W. Haver, A Characterization Theorem for Cellular Maps, Bull. Amer. Math. Soc., 76 (1970), 1277–1280.
16. D. Herrera-Carrasco, A. Illanes, F. Macías-Romero and F. Vázquez-Juárez, Finite Graphs have Unique $HS_n(X)$, Topology Proc., 44 (2014), 75–95.
17. H. Hosokawa, The Span of Hyperspaces, Houston J. Math., 25 (1999), 35–41.
18. W. Hurewicz, Sur la Dimension des Produits Cartésiens, Annals of Math., 36 (1935), 194–197.
19. W. Hurewicz and H. Wallman, *Dimension Theory,* Princeton University Press, Princeton, NJ, 1948.
20. A. Illanes, Multicoherence of Symmetric Products, An. Inst. Mat. Univ. Nac. Autónoma de México, 25 (1985), 11–24.
21. A. Illanes, Finite graphs X have unique hyperspaces $C_n(X)$, Topology Proc., 27 (2003), 179–188.
22. A. Illanes, S. Macías and S. B. Nadler, Jr., Symmetric Products and Q-manifolds, Contemp. Math., 246 (1999), 137–141.

23. A. Illanes and S. B. Nadler, Jr., *Hyperspaces: Fundamentals and Recent Advances,* Mono graphs and Textbooks in Pure and Applied Math., Vol. 216, Marcel Dekker, New York, Basel, 1999.
24. F. B. Jones, Concerning Aposyndetic and Non-aposyndetic Continua, Bull. Amer. Math. Soc., 58 (1952), 137–151.
25. J. L. Kelley, Hyperspaces of a Continuum, Trans. Amer. Math. Soc., 52 (1942), 22–36.
26. J. Krasinkiewicz, On the Hyperspaces of Snake-like and Circle-like Continua, Fund. Math., 83 (1974), 155–164.
27. K. Kuratowski, *Topology,* Vol. II, Academic Press, New York, N. Y., 1968.
28. M. Levin and Y. Sternfeld, The Space of Subcontinua of a 2-dimensional Continuum is Infinitely Dimensional, Proc. Amer. Math. Soc., 125 (1997), 2771–2775.
29. S. Macías, On Symmetric Products of Continua, Topology Appl., 92 (1999), 173–182.
30. S. Macías, On the n-fold Hyperspace Suspension of Continua, Topology Appl., 192 (2004), 125–138.
31. S. Macías, Induced Maps on n-fold Hyperspace Suspensions, Topology Proc., 28 (2004), 143–152.
32. S. Macías, On the n-fold Hyperspace Suspension of Continua, II, Glasnik Mat., 41(61) (2006), 335–343.
33. S. Macías, Correction to the Paper "On the Hyperpspace $C_n(X)$ of a Continuum X, II", Topology Proc., 30 (2006), 335–340.
34. S. Macías, On n-fold Hyperspaces, Glasnik Mat., 44(64) (2009), 479–492.
35. S. Macías and S. B. Nadler, Jr., Fans Whose Hyperspace of Subcontinua are Cones, Topology Appl., 126 (2002), 29–36.
36. S. Macías and S. B. Nadler, Jr., Absolute n-fold Hyperspace Suspensions, Colloq. Math., 105 (2006), 221–231.
37. S. Macías and S. B. Nadler, Jr., Various Types of Local Connectedness in n-fold Hyperspaces, Topology Appl., 154 (2007), 39–53.
38. J. van Mill, *Infinite-Dimensional Topology,* North Holland, Amsterdam, 1989.
39. E. E. Moise, Grille Decomposition and Convexification Theorems for Compact Metric Locally Connected Continua, Bull. Amer. Math. Soc., 55 (1949), 1111–1121.
40. S. B. Nadler, Jr., *Hyperspaces of Sets,* Monographs and Textbooks in Pure and Applied Math., Vol. 49, Marcel Dekker, New York, Basel, 1978. Reprinted in: Aportaciones Matemáticas de la Sociedad Matemática Mexicana, Serie Textos # 33, 2006.
41. S. B. Nadler, Jr., A Fixed Point Theorem for Hyperspace Suspensions, Houston J. Math., 5 (1979), 125–132.
42. S. B. Nadler, Jr., *Continuum Theory: An Introduction,* Monographs and Textbooks in Pure and Applied Math., Vol. 158, Marcel Dekker, New York, Basel, Hong Kong, 1992.
43. S. B. Nadler, Jr., *Dimension Theory: An Introduction with Exercises,* Aportaciones Matemáticas de la Sociedad Matemática Mexicana, Serie Textos # 18, 2002.
44. S. B. Nadler, Jr., Absolute Hyperspaces and Hyperspaces that are Absolute Cones and Absolute Suspensions, Bol. Soc. Mat. Mexicana, (3) 11 (2005), 121–129.
45. J. T. Rogers, Jr., Dimension of Hyperspaces, Bull. Acad. Polon. Sci., Sér. Sci., Astronom. Phys., 20 (1972), 177–179.
46. T. B. Rushing, *Topological Embeddings,* Academic Press, New York, 1973.
47. G. T. Whyburn, *Analytic Topology,* Amer. Math. Soc. Colloq. Publ., Vol. 28, Amer. Math. Soc., Providence, R. I., 1942.

Chapter 8
Induced Maps on n-Fold Hyperspaces

We begin considering some classes of maps between continua proving some of their general properties. Then we consider the following problem: Let $\mathcal{A}$ be class of maps. What are the relationships between the following statements:

(1) $f \in \mathcal{A}$;
(2) $\mathcal{C}_n(f) \in \mathcal{A}$;
(3) $HS_n(f) \in \mathcal{A}$?

8.1 General Maps

We start by recalling the definition of the types of maps we use:

8.1.1 Definition A map $f \colon X \to Y$ between continua is said to be:

- *almost monotone* provided that it is surjective and $f^{-1}(Q)$ is connected for each subcontinuum with nonempty interior Q of Y.
- *atomic* if it is surjective and for every subcontinuum K of X such that $f(K)$ is nondegenerate, we have that $K = f^{-1}(f(K))$;
- *confluent* provided that for each subcontinuum Q of Y and every component K of $f^{-1}(Q)$, we have that $f(K) = Q$;
- *freely decomposable* if whenever A and B are proper subcontinua of Y such that $Y = A \cup B$, then there exist two proper subcontinua A' and B' of X, such that $X = A' \cup B'$, $f(A') \subset A$ and $f(B') \subset B$;
- a *homeomorphism* provided that the inverse function, f^{-1}, exists and is continuous;
- *interior at the point* $p \in X$ if it is surjective and for each open subset U of X which contains p, $f(p) \in Int(f(U))$;
- *joining* provided that for every subcontinuum Q of Y and for each two components C_1 and C_2 of $f^{-1}(Q)$, we have that $f(C_1) \cap f(C_2) \neq \emptyset$;

© Springer International Publishing AG, part of Springer Nature 2018
S. Macías, *Topics on Continua*, https://doi.org/10.1007/978-3-319-90902-8_8

- *k-to*-1 if $f^{-1}(y)$ has exactly k points for each $y \in Y$;
- *light* provided that $f^{-1}(f(x))$ is totally disconnected for every $x \in X$;
- *MO* if it is surjective and there exist a continuum Z, a monotone map $h \colon Z \to Y$ and an open map $g \colon X \to Z$ such that $f = h \circ g$;
- *monotone* provided that it is surjective and $f^{-1}(y)$ is connected for each $y \in Y$;
- *near-homeomorphism* if it is the uniform limit of homeomorphisms from X onto Y;
- *OM* provided that it is surjective and there exist a continuum Z, an open map $g \colon Z \to Y$ and a monotone map $h \colon X \to Z$ such that $f = g \circ h$;
- *open* if it is surjective and for every open set U of X, $f(U)$ is open in Y;
- *pseudo-confluent* provided that it is surjective and for each irreducible subcontinuum Q of Y, there exists a component C of $f^{-1}(Q)$ such that $f(C) = Q$;
- *quasi-interior* if for every point $y \in Y$ and each open subset U of X containing a component of $f^{-1}(y)$, we have that $y \in Int(f(U))$;
- *quasi-monotone* provided that it is surjective and for each subcontinuum Q of Y with nonempty interior, $f^{-1}(Q)$ has finitely many components and if C is a component of $f^{-1}(Q)$, then $f(C) = Q$;
- *semi-confluent* if for every subcontinuum Q of Y and for each two components C_1 and C_2 of $f^{-1}(Q)$, either $f(C_1) \subset f(C_2)$ or $f(C_2) \subset f(C_1)$;
- *semi-interior at the point* $p \in X$, provided that it is surjective and for every open subset U of X containing p, there exists a point $z \in U$ such that $f(z) \in Int(f(U))$;
- *semi-open* if it is surjective and for each open subset U of X, $Int(f(U)) \neq \emptyset$;
- *simple* provided that for every point $y \in Y$, $f^{-1}(y)$ consists of at most two points;
- *strongly freely decomposable* if whenever A and B are proper subcontinua of Y such that $Y = A \cup B$, it follows that $f^{-1}(A)$ and $f^{-1}(B)$ are connected[1];
- *weakly confluent* provided that for every subcontinuum Q of Y, there exists a subcontinuum K of X such that $f(K) = Q$;
- *weakly monotone* if it is surjective and for every subcontinuum Q of Y with nonempty interior, if C is a component of $f^{-1}(Q)$, then $f(C) = Q$.

In *Diagram I* (next page), we show the relationships between some of the classes defined in Definition 8.1.1. A proof of these facts may be found in [29]. Although the equivalence between *OM* maps an quasi-interior maps is proved below (Theorem 8.1.10). Other implications and equivalences are also proved below.

We begin with a theorem which characterizes the continuity of a function.

8.1.2 Theorem *Let X and Y be compacta. Then a surjective function $f \colon X \twoheadrightarrow Y$ is continuous if and only if for each point $y \in Y$ and each sequence $\{y_n\}_{n=1}^{\infty}$ of points of Y converging to y, we have that $\limsup f^{-1}(y_n) \subset f^{-1}(y)$.*

---

[1]Strongly freely decomposable maps are also known as feebly monotone maps in the literature.

Proof Suppose f is continuous. Let $y \in Y$ and let $\{y_n\}_{n=1}^{\infty}$ be a sequence of points of Y which converges to y. Let $x \in \limsup f^{-1}(y)$ and for every $m \in \mathbb{N}$, consider the open ball about x, $\mathcal{V}_{\frac{1}{m}}^{d}(x)$. Then, for each $m \in \mathbb{N}$, there exists $n_m \in \mathbb{N}$ such that $\mathcal{V}_{\frac{1}{m}}^{d}(x) \cap f^{-1}(y_{n_m}) \neq \emptyset$. Let $x_m \in \mathcal{V}_{\frac{1}{m}}^{d}(x) \cap f^{-1}(y_{n_m})$. Note that $\{x_m\}_{m=1}^{\infty}$ converges to x. Hence, since f is continuous, $f(x) = y$. Thus, $x \in f^{-1}(y)$. Therefore, $\limsup f^{-1}(y_n) \subset f^{-1}(y)$.

Next, suppose f is not continuous at $x \in X$. Then there exists a sequence $\{x_n\}_{n=1}^{\infty}$ of points of X converging to x such that the sequence $\{f(x_n)\}_{n=1}^{\infty}$ does not converge to $f(x)$. Since Y is compact, without loss of generality, we assume that $\{f(x_n)\}_{n=1}^{\infty}$ converges to a point $y \in Y$. Observe that $x \in \limsup f^{-1}(f(x_n))$ but $x \in X \setminus f^{-1}(y)$. Therefore, $\limsup f^{-1}(y_n) \not\subset f^{-1}(y)$.

Q.E.D.

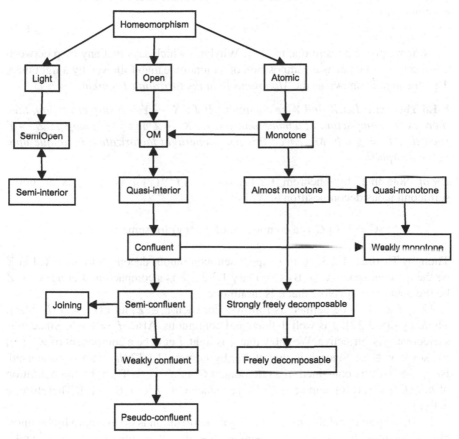

Diagram I

8.1.3 Corollary *Let X and Y be compacta. Then a surjective function $f : X \twoheadrightarrow Y$ is continuous if and only if for each $x \in X$ and each sequence of points*

$\{x_n\}_{n=1}^\infty$ *converging to* x, *there exists a subsequence* $\{x_{n_k}\}_{k=1}^\infty$ *of* $\{x_n\}_{n=1}^\infty$ *such that* $\{f(x_{n_k})\}_{n=1}^\infty$ *converges to* $f(x)$.

Proof If f is continuous, it is clear that the condition stated in the corollary is satisfied.

Suppose that the condition stated in the corollary holds. Let $y \in Y$ and let $\{y_n\}_{n=1}^\infty$ be a sequence of points which converges to y. Since f is surjective, for each $n \in \mathbb{N}$, $f^{-1}(y_n) \neq \emptyset$. Let $x \in \limsup f^{-1}(y_n)$. As we show in the first paragraph of the proof of Theorem 8.1.2, there exist a subsequence $\{y_{n_k}\}_{k=1}^\infty$ of $\{y_n\}_{n=1}^\infty$ and a point $x_{n_k} \in f^{-1}(y_{n_k})$ such that $\{x_{n_k}\}_{n=1}^\infty$ converges to x. By hypothesis, there exists a subsequence $\left\{x_{n_{k_j}}\right\}_{j=1}^\infty$ of $\{x_{n_k}\}_{n=1}^\infty$ such that $\left\{f\left(x_{n_{k_j}}\right)\right\}_{j=1}^\infty$ converges to $f(x)$. Since for each $j \in \mathbb{N}$, $f\left(x_{n_{k_j}}\right) = y_{n_{k_j}}$ and the sequence $\left\{y_{n_{k_j}}\right\}_{j=1}^\infty$ converges to y, we have that $f(x) = y$. Thus, $x \in f^{-1}(y)$. Therefore, by Theorem 8.1.2, f is continuous.

Q.E.D.

Now we prove a result due to G. T. Whyburn which says that any map between compacta may be put as a composition of a monotone map followed by a light map. This theorem is known as the *Monotone-light Factorization Theorem*.

8.1.4 Theorem *Let X and Y be compacta. If $f : X \twoheadrightarrow Y$ is a surjective map, then there exist a compactum Z, a monotone map $h : X \twoheadrightarrow Z$ and a light map $g : Z \twoheadrightarrow Y$ such that $f = g \circ h$. Moreover, this monotone-light factorization is unique upto homeomorphism.*

Proof Note that, by Theorem 1.2.17, $\mathcal{G}_f = \{f^{-1}(y) \mid y \in Y\}$ is an upper semicontinuous decomposition of X. Let

$$\mathcal{G} = \{G \mid G \text{ is a component of } f^{-1}(y) \text{ for some } y \in Y\}.$$

Then, by Theorem 1.2.31, $\mathcal{G}$ is an upper semicontinuous decomposition of X. Let Z be the quotient space $X/\mathcal{G}$. By Corollary 1.2.22, Z is a compactum. Let $h : X \twoheadrightarrow Z$ be the quotient map. Note that h is monotone.

Let $g : Z \twoheadrightarrow Y$ be defined as follows: For each $z \in Z$, let $g(z) = f(h^{-1}(z))$. Then, by [35, 3.22], g is well defined and continuous. Also, $f = g \circ h$. Since f is surjective, g is surjective. We show that g is light. Let C be a component of $g^{-1}(y)$, for some $y \in Y$. Since h is monotone, by Lemma 2.1.12, $h^{-1}(C)$ is connected. Hence, $h^{-1}(C)$ is contained in a component C' of $f^{-1}(y)$. Then, by the definition of h, $h(C') = \{z\}$, for some $z \in Z$. Hence, since $C \subset h(C')$, $C = \{z\}$. Therefore, g is light.

Next, we prove that this monotone-light factorization is unique upto homeomorphism. Suppose there exist a compactum Z', a monotone map $h' : X \twoheadrightarrow Z'$ and a light map $g' : Z' \twoheadrightarrow Y$ such that $f = g' \circ h'$. Let $y \in Y$ and let C be a component of $f^{-1}(y)$. Then $h'(C)$ is a subcontinuum of Z' such that $g' \circ h'(C) = \{y\}$. Since g' is light, there exists $z' \in Z$ such that $h'(C) = \{z'\}$. Then $C \subset (h')^{-1}(z')$.

Since h' is monotone and $f((h')^{-1}(z')) = \{y\}$, we have that $(h')^{-1}(z') \subset C$ and $C = (h')^{-1}(z')$. Now, observe that, by construction, there exists $z \in Z$ such that $C = h^{-1}(z)$. Thus, by [35, 3.22], we obtain that the functions $\ell_1 \colon Z \twoheadrightarrow Z'$ and $\ell_2 \colon Z' \twoheadrightarrow Z$ given by $\ell_1(z) = h'(h^{-1}(z))$ and $\ell_2(z') = h((h')^{-1}(z'))$ are well defined and continuous. It is clear that $\ell_2 \circ \ell_1 = 1_Z$ and $\ell_1 \circ \ell_2 = 1_{Z'}$. Therefore, Z and Z' are homeomorphic. Also note that $h = \ell_2 \circ h'$ and $h' = \ell_1 \circ h$.

We show that $g = g' \circ \ell_1$ and $g' = g \circ \ell_2$. Let $z \in Z$. Then $g' \circ \ell_1(z) = g' \circ h'(h^{-1}(z)) = f(h^{-1}(z)) = g(z)$. Similarly, if $z' \in Z'$, then $g \circ \ell_2(z') = g \circ h((h')^{-1}(z')) = f((h')^{-1}(z')) = g'(z')$. Therefore, the monotone-light factorization is unique upto homeomorphism.

Q.E.D.

We prove a lemma which helps us to show that the composition of quasi-interior maps is a quasi-interior map.

8.1.5 Lemma *Let X and Y be continua. If $f \colon X \to Y$ is a quasi-interior map, then f is confluent.*

Proof Assume f is quasi-interior but not confluent. Then there exist a subcontinuum Q of Y and a component C of $f^{-1}(Q)$ such that $f(C) \neq Q$. Let $y \in Q \setminus f(C)$. Since X is compact metric, there exists an open subset U of X such that $C \subset U$, $Cl(U) \cap f^{-1}(y) = \emptyset$ and $Bd(U) \cap f^{-1}(Q) = \emptyset$ (Theorem 1.6.8). By Theorems 1.2.31 and 1.2.19, we assume that U is of the form $U = \bigcup \{D \mid D$ is a component of $f^{-1}(w)$ and $f^{-1}(w) \cap V \neq \emptyset\}$, where V is an open subset of X. We show that $Bd(f(U)) \subset f(Bd(U))$. Suppose there exists a point $z \in Bd(f(U)) \setminus f(Bd(U))$. We assert that $z \in f(U)$. If not, there exists a sequence $\{z_n\}_{n=1}^{\infty}$ of points of $f(U)$ converging to z. For each $n \in \mathbb{N}$, let $x_n \in U \cap f^{-1}(z_n)$. Since X is compact, there exists a subsequence $\{x_{n_j}\}_{j=1}^{\infty}$ of $\{x_n\}_{n=1}^{\infty}$ converging to a point $x \in Cl(U)$. Note that $x \notin Bd(U)$ because $f(x) = z$ and $z \notin f(Bd(U))$. Hence, $x \in U$. Observe that U contains a component of $f^{-1}(z)$. Then, since f is quasi-interior, $z \in Int(f(U))$. This contradicts the fact that $z \in Bd(f(U))$. Thus, $Bd(f(U)) \subset f(Bd(U))$.

Since $Bd(U) \cap f^{-1}(Q) = \emptyset$ and $Bd(f(U)) \subset f(Bd(U))$, we have that $Q \cap Bd(f(U)) = \emptyset$. Since, also, $y \in Q \setminus f(U)$, we obtain a contradiction to the fact that Q is connected. Therefore, f is confluent.

Q.E.D.

8.1.6 Definition A class of maps $\mathcal{A}$ has the *composition property* if for each two maps $f \colon X \to Y$ and $g \colon Y \to Z$ belonging to $\mathcal{A}$ their composition $g \circ f$ belongs to $\mathcal{A}$.

8.1.7 Theorem *The following classes of maps have the composition property: almost monotone, atomic, confluent, freely decomposable, homeomorphisms, light, monotone, open, pseudo-confluent, quasi-interior, quasi-monotone, semi-open, strongly freely decomposable and weakly confluent.*

Proof Let $f: X \to Y$ and $g: Y \to Z$ be maps.

Suppose f and g are almost monotone. Let Q be a subcontinuum of Z with nonempty interior. Since g is almost monotone, $g^{-1}(Q)$ is connected. Note that $Int_Y(g^{-1}(Q)) \neq \emptyset$. Hence, since f is almost monotone, $f^{-1}(g^{-1}(Q))$ is connected. Therefore, $g \circ f$ is almost monotone.

Assume f and g are atomic. Let K be a subcontinuum of X such that $g \circ f(K)$ is nondegenerate. Then $f(K)$ is a subcontinuum of Y such that $g(f(K))$ is nondegenerate. Since g is atomic, we have that $f(K) = g^{-1}(g(f(K)))$. Since $f(K)$ is nondegenerate and f is atomic, $K = f^{-1}(f(K))$. Therefore, $K = f^{-1}(g^{-1}(g(f(K)))) = (g \circ f)^{-1}(g \circ f(K))$, and $g \circ f$ is atomic.

Suppose f and g are confluent. Let Q be a subcontinuum of Z and let C be a component of $(g \circ f)^{-1}(Q)$. Then $f(C)$ is a subcontinuum of Y and $g(f(C)) \subset Q$. Let D be the component of $g^{-1}(Q)$ containing $f(C)$. Since g is confluent, $g(D) = Q$. Let C' be the component of $f^{-1}(D)$ containing C. Since f is confluent, $f(C') = D$. Thus, $g(f(C')) = Q$. Since C is a component of $(g \circ f)^{-1}(Q)$, C' is connected and $C \subset C'$, we have that $C = C'$. Therefore, $g \circ f(C) = Q$, and $g \circ f$ is confluent.

Assume f and g are freely decomposable. Let A and B be proper subcontinua of Z such that $Z = A \cup B$. Since g is freely decomposable, there exist two proper subcontinua A' and B' of Y such that $Y = A' \cup B'$, $g(A') \subset A$ and $g(B') \subset B$. Since f is freely decomposable, there exist two proper subcontinua A'' and B'' of X such that $X = A'' \cup B''$, $f(A'') \subset A'$ and $f(B'') \subset B'$. Hence, $g \circ f(A'') \subset A$ and $g \circ f(B'') \subset B$. Therefore, $g \circ f$ is freely decomposable.

If f and g are homeomorphisms, then it is well known that $g \circ f$ is a homeomorphism.

Suppose f and g are light. Let $x \in X$ and let C be a component of $(g \circ f)^{-1}(g \circ f(x))$. Since g is light and $g \circ f(C) = \{f(x)\}$, $f(C)$ is degenerate. Hence, since f is light and $f(C)$ is degenerate, C is degenerate. Therefore, $g \circ f$ is light.

Assume f and g are monotone. Let $z \in Z$. Since g is monotone, $g^{-1}(z)$ is connected. Hence, since f is monotone, by Lemma 2.1.12, $f^{-1}(g^{-1}(z))$ is connected. Therefore, $g \circ f$ is monotone.

It is clear that if f and g are open, then $g \circ f$ is open.

Suppose f and g are pseudo-confluent. Let Q be an irreducible subcontinuum of Z. Suppose Q is irreducible about the points z_1 and z_2. Since g is pseudo-confluent, there exists a component D_0 of $g^{-1}(Q)$ such that $g(D_0) = Q$. Let $y_j \in g^{-1}(z_j) \cap D_0$, $j \in \{1, 2\}$. By [23, Theorem 2–10], there exists an irreducible subcontinuum D of D_0, irreducible about y_1 and y_2. Observe that $g(D) = Q$. Since D is an irreducible subcontinuum of Y and f is pseudo-confluent, there exists a component C' of $f^{-1}(D)$ such that $f(C') = D$. Hence, $g(f(C')) = Q$. Let C be the component of $(g \circ f)^{-1}(Q)$ which contains C'. Then $g \circ f(C) = Q$. Therefore, $g \circ f$ is pseudo-confluent.

Assume f and g are quasi-interior maps. Let $z \in Z$, let C be a component of $(g \circ f)^{-1}(z)$ and let U be an open subset of X such that $C \subset U$. Note that $f(C)$ is a subcontinuum of Y and $f(C) \subset g^{-1}(z)$. Let C' be the component of $g^{-1}(z)$ which contains $f(C)$. Thus, we have

$$C \subset f^{-1}(f(C)) \subset f^{-1}(C') \subset f^{-1}(g^{-1}(z))$$

and, consequently, C is a component of $f^{-1}(C')$. Since f is confluent (Lemma 8.1.5), it follows that $f(C) = C'$. Given a point $y \in C'$, let $x \in C$ be such that $f(x) = y$ and let C_y be the component of $f^{-1}(y)$ which contains x. Thus, $C \cap C_y \neq \emptyset$ and $C_y \subset f^{-1}(y) \subset f^{-1}(C')$. Hence, $C_y \subset C \subset U$. Since f is quasi-interior, we obtain that $y \in Int(f(U))$ for each $y \in C'$. This means that $C' \subset Int(f(U))$. Now, since g is quasi-interior, $z \in Int(g(Int(f(U))) \subset Int(g \circ f(U))$. Therefore, $g \circ f$ is quasi-interior.

Suppose f and g are quasi-monotone. Let Q be a subcontinuum of Z with nonempty interior. Since g is quasi-monotone, $g^{-1}(Q)$ has a finite number of components and g maps each of them onto Q. Hence, each component of $g^{-1}(Q)$ has nonempty interior in Y, because Q has nonempty interior. Since f is quasi-monotone, for each component C of $g^{-1}(Q)$, $f^{-1}(C)$ has a finite number of components and f maps each of them onto C. Thus, $(g \circ f)^{-1}(Q)$ has a finite number of components and $g \circ f$ maps each of them onto Q, because, as we showed above, every component of $(g \circ f)^{-1}(Q)$ is a component of $f^{-1}(C)$, for some component C of $g^{-1}(Q)$. Therefore, $g \circ f$ is quasi-monotone.

Assume f and g are semi-open. Let U be an open subset of X. Since f is semi-open, $Int(f(U)) \neq \emptyset$. Then, since g is semi-open, $Int(g(Int(f(U)))) \neq \emptyset$. Since $g(Int(f(U))) \subset g(f(U))$, we obtain that $Int(g \circ f(U)) \neq \emptyset$. Therefore, $g \circ f$ is semi-open.

Suppose f and g are strongly freely decomposable. Let A and B be subcontinua of Z such that $Z = A \cup B$. Since g is strongly freely decomposable, $g^{-1}(A)$ and $g^{-1}(B)$ are connected. Then $g^{-1}(A)$ and $g^{-1}(B)$ are proper subcontinua of Y such that $Y = g^{-1}(A) \cup g^{-1}(B)$. Since f is strongly freely decomposable, $f^{-1}(g^{-1}(A))$ and $f^{-1}(g^{-1}(B))$ are connected. Therefore, $g \circ f$ is strongly freely decomposable.

Assume f and g are weakly confluent. Let Q be a subcontinuum of Z. Since g is weakly confluent, there exists a subcontinuum W of Y such that $g(W) = Q$. Since f is weakly confluent, there exists a subcontinuum K of X such that $f(K) = W$. Therefore, $g \circ f(K) = Q$, and $g \circ f$ is weakly confluent.

Q.E.D.

8.1.8 Lemma *Let X and Y be continua, let $y \in Y$ and let $f : X \twoheadrightarrow Y$ be a surjective map. If U is an open subset of X which contains $f^{-1}(y)$, then $y \in Int(f(U))$.*

Proof Note that, by Theorem 1.2.17, $\mathcal{G}_f = \{f^{-1}(y) \mid y \in Y\}$ is an upper semicontinuous decomposition. Let $y \in Y$ and let U be an open subset of X which contains $f^{-1}(y)$. Since $\mathcal{G}_f$ is an upper semicontinuous decomposition, there exists an open set V of X such that $f^{-1}(y) \subset V$ and if $f^{-1}(y') \cap V \neq \emptyset$, then

$f^{-1}(y') \subset U$. Let $W = \bigcup\{f^{-1}(y') \mid f^{-1}(y') \cap V \neq \emptyset\}$. By Theorem 1.2.19, W is an open subset of X. Note that $f^{-1}(y) \subset W$ and $W \subset U$. Hence, $f(W)$ is an open subset of Y and $y \in f(W)$. Therefore, $y \in Int(f(U))$.

Q.E.D.

8.1.9 Corollary *All monotone maps are quasi-interior.*

8.1.10 Theorem *Let X and Y be continua and let $f \colon X \twoheadrightarrow Y$ be a map. Then f is OM if and only if f is quasi-interior.*

Proof Note that every open map is quasi-interior. By Corollary 8.1.9, every monotone map is quasi-interior. Hence, by Theorem 8.1.7, every OM map is quasi-interior.

Suppose f is quasi-interior. By Theorem 8.1.4, there exist a continuum Z, a monotone map $h \colon X \twoheadrightarrow Z$ and a light map $g \colon Z \twoheadrightarrow Y$ such that $f = g \circ h$. We show that g is an open map. Let U be an open subset of Z and let $y \in g(U)$. Then there exists $z \in U$ such that $g(z) = y$. Then $h^{-1}(z)$ is a component of $f^{-1}(y)$. Observe that $h^{-1}(U)$ is an open subset of X which contains the component $h^{-1}(z)$ of $f^{-1}(y)$. Since f is quasi-interior and $f(h^{-1}(U)) = g(U)$, $y \in Int(f(h^{-1}(U))) = Int(g(U))$. Since y is an arbitrary point of $g(U)$, $g(U)$ is open. Therefore, g is an open map, and f is an OM map.

Q.E.D.

As a consequence of the Theorems 8.1.7 and 8.1.10, we have the following:

8.1.11 Corollary *The class of OM maps has the composition property.*

8.1.12 Definition A class of maps $\mathcal{A}$ has the *composition factor property* provided that for each two maps $f \colon X \to Y$ and $g \colon Y \to Z$ if $g \circ f$ belongs to $\mathcal{A}$, then g belongs to $\mathcal{A}$.

8.1.13 Theorem *The following classes of maps have the composition factor property: confluent, joining, monotone, open, pseudo-confluent, quasi-interior, quasi-monotone, semi-confluent, semi-open, weakly confluent and weakly monotone.*

Proof Let $f \colon X \to Y$ and $g \colon Y \to Z$ be maps.

Suppose $g \circ f$ is confluent. Let Q be a subcontinuum of Z and let C be a component of $g^{-1}(Q)$. Let D' be a component of $f^{-1}(C)$. Note that $g \circ f(D') \subset Q$. Let D be the component of $(g \circ f)^{-1}(Q)$ containing D'. Since $g \circ f$ is confluent, $g \circ f(D) = Q$. Observe that $f(D)$ is a subcontinuum of $g^{-1}(Q)$ such that $C \cap f(D) \neq \emptyset$. Hence, since C is a component of $g^{-1}(Q)$, $f(D) \subset C$. Thus, $g(C) = Q$. Therefore, g is confluent.

Assume $g \circ f$ is joining. Let Q be a subcontinuum of Z and let C_1 and C_2 be two components of $g^{-1}(Q)$. Let D'_j be a component of $f^{-1}(C_j)$, $j \in \{1, 2\}$. Let D_j be the component of $(g \circ f)^{-1}(Q)$ which contains D'_j, $j \in \{1, 2\}$. Since $g \circ f$ is joining, $g \circ f(D_1) \cap g \circ f(D_2) \neq \emptyset$. Note that $f(D_j) \subset C_j$, $j \in \{1, 2\}$, by a similar argument to the one given in the previous paragraph. Hence, $g(C_1) \cap g(C_2) \neq \emptyset$. Therefore, g is joining.

Suppose $g \circ f$ is monotone. Let $z \in Z$. Since $g \circ f$ is mononote, $(g \circ f)^{-1}(z)$ is connected. Then $f((g \circ f)^{-1}(z)) = g^{-1}(z)$ is connected. Therefore, g is monotone.

Assume $g \circ f$ is open. Let U be an open subset of Y. Then $f^{-1}(U)$ is an open subset of X. Since $g \circ f$ is open, $g \circ f(f^{-1}(U)) = g(U)$ is open in Z. Therefore, g is open.

Suppose $g \circ f$ is pseudo-confluent. Let Q be an irreducible subcontinuum of Z. Since $g \circ f$ is pseudo-confluent, there exists a component D of $(g \circ f)^{-1}(Q)$ such that $g \circ f(D) = Q$. Let C be the component of $g^{-1}(Q)$ which contains $f(D)$. Then, as before, $g(C) = Q$. Therefore, g is pseudo-confluent.

Assume $g \circ f$ is quasi-interior. Let $z \in Z$, let C be a component of $g^{-1}(z)$ and let U be an open subset of Y which contains C. Let D' be a component of $f^{-1}(C)$ and let D be the component of $(g \circ f)^{-1}(z)$ containing D'. Then $f^{-1}(U)$ is an open subset of X which contains D. Since $g \circ f$ is quasi-interior and $g \circ f(f^{-1}(U)) = g(U)$, $z \in Int(g \circ f(f^{-1}(U))) \subset Int(g(U))$. Therefore, g is quasi-interior.

Suppose $g \circ f$ is quasi-monotone. Let Q be a subcontinuum of Z with nonempty interior. Since $g \circ f$ is quasi-monotone $(g \circ f)^{-1}(Q)$ has finitely many components, say $D_1, \ldots, D_m$, and $g \circ f(D_j) = Q$ for each $j \in \{1, \ldots, m\}$. Let C_j be the component of $g^{-1}(Q)$ containing $f(D_j)$, $j \in \{1, \ldots, m\}$. Hence, $g^{-1}(Q)$ has at most m components and $g(C_j) = Q$ for all $j \in \{1, \ldots, m\}$ (see the arguments above). Therefore, g is quasi-monotone.

Assume $g \circ f$ is semi-open. Let V be an open subset of Y. Then $f^{-1}(V)$ is an open subset of X. Since $g \circ f$ is semi-open, $Int(g \circ f(f^{-1}(V))) \neq \emptyset$. Since $f(f^{-1}(V)) \subset V$, we have that $Int(g(V)) \neq \emptyset$. Therefore, g is semi-open.

Assume $g \circ f$ is weakly confluent. Let Q be a subcontinuum of Z. Since $g \circ f$ is weakly confluent, there exists a subcontinuum K of X such that $g \circ f(K) = Q$. Hence, $f(K)$ is a subcontinuum of Y such that $g(f(K)) = Q$. Therefore, g is weakly confluent.

Suppose $g \circ f$ is weakly monotone. Let Q be a subcontinuum of Z with nonempty interior. Let C be a component of $g^{-1}(Q)$. Let D' be a component of $f^{-1}(C)$, and let D be the component of $(g \circ f)^{-1}(Q)$ which contains D'. Since $g \circ f$ is weakly monotone, $g \circ f(D) = Q$. Then, as we did above, $f(D) \subset C$ and $g(C) = Q$. Therefore, g is weakly monotone.

Q.E.D.

8.1.14 Remark On [19, p. 139], the authors claim that the class of (strongly) freely decomposable maps has the composition factor property. We do not know if this is true or not. We are able to prove it when the first map is surjective, namely: Assume $g \circ f$ is freely decomposable. Let A and B be subcontinua of Z such that $Z = A \cup B$. Since $g \circ f$ is freely decomposable, there exist two proper subcontinua A' and B' of X such that $X = A' \cup B'$, $g \circ f(A') \subset A$ and $g \circ f(B') \subset B$. Then $f(A')$ and $f(B')$ are proper subcontinua of Y such that $Y = f(A') \cup f(B')$, $g(f(A')) \subset A$ and $g(f(B')) \subset B$. Therefore, g is freely decomposable. (Suppose $g \circ f$ is strongly freely decomposable. Let A and B be proper subcontinua of Z such that $Z = A \cup B$. Since $g \circ f$ is strongly freely decomposable, $(g \circ f)^{-1}(A)$ and $(g \circ f)^{-1}(B)$

are connected. Note that $f\left((g \circ f)^{-1}(A)\right) = g^{-1}(A)$ and $f\left((g \circ f)^{-1}(B)\right) = g^{-1}(B)$. Hence, $g^{-1}(A)$ and $g^{-1}(B)$ are proper subcontinua of Y. Therefore, g is strongly freely decomposable.)

8.1.15 Remark Observe that in [29, (5.14)] says *The class of homeomorphisms has the composition factor property*. This is true if we assume that the first map is surjective; i.e., it is a homeomorphism itself. If we do not assume that, then we have the following example: Let $f \colon [0, 1] \to \mathcal{S}^1$ given by $f(t) = \exp(\pi t)$ and let $g \colon \mathcal{S}^1 \twoheadrightarrow [-1, 1]$ given by $g((x, y)) = x$. Then $g \circ f \colon [0, 1] \twoheadrightarrow [-1, 1]$ is a homeomorphism but g is not.

As a consequence of the Theorems 8.1.13 and 8.1.10, we have the following:

8.1.16 Corollary *The class of OM maps has the composition factor property.*

The following theorem characterizes OM maps.

8.1.17 Theorem *Let X and Y be continua and let $f \colon X \to Y$ be a map. Then f is OM if and only if for each point $y \in Y$ and every sequence of points $\{y_m\}_{m=1}^\infty$ of Y converging to y, we have that $\limsup f^{-1}(y_m)$ intersects each component of $f^{-1}(y)$.*

Proof Suppose f is OM. Then, by Theorem 8.1.10, f is quasi-interior. Let $y \in Y$, let C be a component of $f^{-1}(y)$ and let U be an open subset of X containing C. Since f is quasi-interior, $y \in Int(f(U))$. Let $\{y_m\}_{m=1}^\infty$ be a sequence of points of Y converging to y. Then there exists $N \in \mathbb{N}$ such that $y_m \in f(U)$ for each $m \geq N$. Hence, $f^{-1}(y_m) \cap U \neq \emptyset$ for all $m \geq N$. Therefore, $[\limsup f^{-1}(y_m)] \cap C \neq \emptyset$.

Next, suppose f is not OM. Then, by Theorem 8.1.10, f is not quasi-interior. Hence, there exist a point $y \in Y$, a component C of $f^{-1}(y)$ and an open subset U of X such that $f^{-1}(y) \subset U$ and $y \in Y \setminus Int(f(U))$. Then there exists a sequence $\{y_m\}_{m=1}^\infty$ of points of $Y \setminus f(U)$ converging to y. Hence, for each $m \in \mathbb{N}$, $f^{-1}(y_m) \subset X \setminus U$ and $\limsup f^{-1}(y_m) \subset X \setminus U \subset X \setminus C$. Therefore, the condition stated in the theorem is not satisfied.

Q.E.D.

The next result characterizes open maps.

8.1.18 Theorem *Let X and Y be continua and let $f \colon X \to Y$ be a map. Then f is open if and only if for each point $y \in Y$ and every sequence of points $\{y_n\}_{n=1}^\infty$ of Y converging to y, we have that $\lim f^{-1}(y_n) = f^{-1}(y)$.*

Proof Suppose f is open. Let $y \in Y$ and let $\{y_n\}_{n=1}^\infty$ be a sequence of points of Y converging to y. By the continuity of f, we have that $\limsup f^{-1}(y_n) \subset f^{-1}(y)$ (Theorem 8.1.2). Thus, we show that $f^{-1}(y) \subset \liminf f^{-1}(y_n)$.

Let $x \in f^{-1}(y)$ and let U be an open subset of X containing x. Since f is open, $f(U)$ is an open subset of Y and $y \in f(U)$. Since $\{y_n\}_{n=1}^\infty$ converges to y, there exists $N \in \mathbb{N}$ such that $y_n \in f(U)$ for all $n \geq N$. Hence, $f^{-1}(y_n) \cap U \neq \emptyset$ for each $n \geq N$. Thus, $x \in \liminf f^{-1}(y_n)$ and $\lim f^{-1}(y_n) = f^{-1}(y)$.

Now, assume f is not open. Then there exists an open subset U of X such that $f(U)$ is not open in Y. Let $y \in f(U) \setminus Int(f(U))$. Then for every $n \in \mathbb{N}$, $V_{\frac{1}{n}}^{d}(y) \cap (Y \setminus f(U)) \neq \emptyset$. For each $n \in \mathbb{N}$, let $y_n \in V_{\frac{1}{n}}^{d}(y) \cap Y \setminus f(U)$. Hence, $\{y_n\}_{n=1}^{\infty}$ is a sequence of points in $Y \setminus f(U)$ converging to y. Let $x \in f^{-1}(y) \cap U$. Then $x \in f^{-1}(y) \setminus \liminf f^{-1}(y_n)$. Therefore, $\lim f^{-1}(y_n) \neq f^{-1}(y)$.

Q.E.D.

Theorem 8.1.18 may be rewritten in the following way and it is used in the proof of Theorem 8.5.22.

8.1.19 Theorem *Let $f: X \to Y$ be a map between continua. Then f is open if and only if for each sequence $\{y_n\}_{n=1}^{\infty}$ of points of Y such that $\lim\limits_{n \to \infty} y_n = y$, for some point $y \in Y$, and for any $x \in f^{-1}(y)$, there exists a sequence $\{x_n\}_{n=1}^{\infty}$ of points of X such that $\lim\limits_{n \to \infty} x_n = x$ and $x_n \in f^{-1}(y_n)$, for each $n \in \mathbb{N}$.*

Proof Suppose that $f: X \to Y$ is an open map between continua. Let $\{y_n\}_{n=1}^{\infty}$ be a sequence of points of Y such that $\lim\limits_{n \to \infty} y_n = y$, for some point $y \in Y$. Let $x \in f^{-1}(y)$. We know that $\lim\limits_{n \to \infty} f^{-1}(y_n) = f^{-1}(y)$, by Theorem 8.1.18.

Let $\{U_n\}_{n=1}^{\infty}$ be a sequence of open subsets of X such that $x \in U_n$, $U_{n+1} \subset U_n$, for each $n \in \mathbb{N}$, and $\bigcap_{n=1}^{\infty} U_n = \{x\}$. Since $x \in \lim\limits_{n \to \infty} f^{-1}(y_n)$, for all $m \in \mathbb{N}$, there exists $k_m \in \mathbb{N}$ such that $U_m \cap f^{-1}(y_n) \neq \emptyset$ for each $n \geq k_m$. Without loss of generality, we assume that $k_j < k_\ell$, if $j < \ell$. We define the sequence $\{x_n\}_{n=1}^{\infty}$ of elements of X in the following way:

(1) $x_n \in f^{-1}(y_n)$ if $n < k_1$;
(2) $x_n \in f^{-1}(y_n) \cap U_m$ if $k_m \leq n < k_{m+1}$.

Clearly, $\lim\limits_{n \to \infty} x_n = x$ and $f(x_n) = y_n$ for all $n \in \mathbb{N}$.

The converse implication follows from Theorem 8.1.18.

Q.E.D.

8.1.20 Theorem *Any weakly monotone map onto an arcwise connected continuum is joining*

Proof Let X and Y be continua, where Y is arcwise connected, and let $f: X \twoheadrightarrow Y$ be a map. Suppose f is not joining. Then there exist a subcontinuum Q of Y and two components C_1 and C_2 of $f^{-1}(Q)$ such that $f(C_1) \cap f(C_2) = \emptyset$.

We assert that there exists a subcontinuum R of Y with nonempty interior such that $R \cap f(C_j) \neq \emptyset$ and $R \cap f(C_k) = \emptyset$, where $j, k \in \{1, 2\}$ and $j \neq k$. To see this, let $N \in \mathbb{N}$ be such that $V_{\frac{1}{N}}^{d}(f(C_1)) \cap V_{\frac{1}{N}}^{d}(f(C_2)) = \emptyset$. For each $n \geq N$, let E_n be the component of $Y \setminus V_{\frac{1}{n}}^{d}(f(C_2))$ containing $f(C_1)$ and let D_n be the component of $Y \setminus V_{\frac{1}{n}}^{d}(f(C_1))$ which contains $f(C_2)$. Since $Y \setminus V_{\frac{1}{n}}^{d}(f(C_2))$ and $Y \setminus V_{\frac{1}{n}}^{d}(f(C_1))$

are closed, each of the sets E_n and D_n is a continuum. Furthermore,

$$Y = \bigcup_{n=N}^{\infty} (E_n \cup D_n).$$

Indeed, if $y \in Y$, then there exists an irreducible arc yc between y and the set $f(C_1) \cup f(C_2)$, because Y is arcwise connected. If $c \in f(C_1)$, then $f(C_2) \cap yc = \emptyset$. Then there exists an n such that $yc \cap V_{\frac{1}{n}}^d(f(C_2)) = \emptyset$. Therefore, $y \in E_n$. Hence, by Theorem 1.5.12, there exists an n_0 such that $E_{n_0} \cup D_{n_0}$ has nonempty interior. Thus, either E_{n_0} has nonempty interior or D_{n_0} has nonempty interior. Therefore, such a continuum R exists.

Without loss of generality, we assume that $R \cap f(C_2) = \emptyset$. Note that $R \cup Q$ is a subcontinuum of Y with nonempty interior and C_2 is a component of $f^{-1}(R \cup Q)$. Therefore, since $f(C_2) \neq R \cup Q$, f is not weakly monotone

Q.E.D.

8.1.21 Theorem *Let X and Y be continua, where Y is locally connected. If $f : X \twoheadrightarrow Y$ is a weakly monotone map, then f is confluent.*

Proof Let Q be a subcontinuum of Y and let C be a component of $f^{-1}(Q)$. For each $n \in \mathbb{N}$, let Q_n be the closure of the component of $V_{\frac{1}{n}}^d(Q)$ which contains Q. Then, by Lemma 1.7.11, each Q_n is a subcontinuum of Y with nonempty interior. For each $n \in \mathbb{N}$, let C_n be the component of $f^{-1}(Q_n)$ containing C. Since f is weakly monotone, $f(C_n) = Q_n$, $n \in \mathbb{N}$. Observe that $C = \bigcap_{n=1}^{\infty} C_n$ and $f(C) = \bigcap_{n=1}^{\infty} f(C_n) = \bigcap_{n=1}^{\infty} Q_n = Q$. Therefore, f is confluent.

Q.E.D.

8.1.22 Theorem *Let X and Y be continua, where X is locally connected. If $f : X \twoheadrightarrow Y$ is a weakly monotone map, then f is quasi-monotone.*

Proof Note that, by [35, 8.16], Y is locally connected. Let Q be a subcontinuum of Y with nonempty interior. We show that $f^{-1}(Q)$ has finitely many components. Let $y \in Int(Q)$. Since Y is locally connected, there exists a connected open subset U of Y such that $y \in U \subset Q$. Then $f^{-1}(U)$ is an open subset of X. Since X is locally connected, each component of $f^{-1}(U)$ is connected. The components of $f^{-1}(U)$ form an open cover of the compact set $f^{-1}(y)$. Hence, a finite number of the components of $f^{-1}(U)$ suffice to cover $f^{-1}(y)$. Thus, $f^{-1}(U)$ has a finite number of components, say k. It follows that $f^{-1}(Q)$ can have no more than k components, since each component of $f^{-1}(Q)$ intersects $f^{-1}(y)$. Therefore, f is quasi-monotone.

Q.E.D.

As consequence of Theorems 8.1.21 and 8.1.22 (see Diagram I), we have the following result:

8.1.23 Theorem *Let X and Y be continua, where X is locally connected. If $f: X \twoheadrightarrow Y$ is a map, then the following are equivalent:*

(1) f is confluent;
(2) f is weakly monotone;
(3) f is quasi-monotone.

8.1.24 Theorem *If $f: X \twoheadrightarrow Y$ is an atomic map between continua, then f is monotone.*

Proof Let $y \in Y$ and let K and L be two closed subsets of X such that $f^{-1}(y) = K \cup L$, $K \cap L = \emptyset$ and $K \neq \emptyset$. Since X is a metric space, there exist two disjoint open subsets U and V of X such that $K \subset U$ and $L \subset V$. Let $x \in K$ and let H be the component of $Cl(U)$ which contains x. By Theorem 1.7.27, $H \cap Bd(U) \neq \emptyset$. Let $z \in H \cap Bd(U)$. Since $Cl(U) \cap V = \emptyset$ and $Bd(U) = Cl(U) \setminus U$, we have that $z \in X \setminus f^{-1}(y)$. Hence, $f(H)$ is a nondegenerate subcontinuum of Y. Since f is atomic, $H = f^{-1}(f(H))$. Since $y = f(x)$ and $f(x) \in f(H)$, we obtain that $K \cup L = f^{-1}(y) \subset H \subset Cl(U)$. Thus, $L \subset H$ and $L \cap H = \emptyset$. Hence, $L = \emptyset$. This implies that $f^{-1}(y)$ is connected. Therefore, f is monotone.

Q.E.D.

The next theorem characterizes atomic maps between continua.

8.1.25 Theorem *A map $f: X \twoheadrightarrow Y$ between continua is atomic if and only if $f^{-1}(y)$ is a terminal subcontinuum in X for each $y \in Y$.*

Proof Suppose f is an atomic map. Let $y \in Y$. By Theorem 8.1.24, f is monotone. Hence, $f^{-1}(y)$ is a subcontinuum of X. Let K be a subcontinuum of X such that $K \cap f^{-1}(y) \neq \emptyset$ and $K \cap (X \setminus f^{-1}(y)) \neq \emptyset$. Then $f(K)$ is a nondegenerate subcontinuum of Y and $y \in f(K)$. Since f is atomic, $K = f^{-1}(f(K))$. In particular, $f^{-1}(y) \subset K$. Therefore, $f^{-1}(y)$ is a terminal subcontinuum of X.

Next, suppose that $f^{-1}(y)$ is a terminal subcontinuum of X for each $y \in Y$. Let K be a subcontinuum of X such that $f(K)$ is nondegenerate. Let $y \in f(K)$. Note that $K \cap f^{-1}(y) \neq \emptyset$ and $K \cap (X \setminus f^{-1}(y)) \neq \emptyset$. Since $f^{-1}(y)$ is a terminal subcontinuum of X, $f^{-1}(y) \subset K$. Hence, $f^{-1}(f(K)) \subset K$. Since $K \subset f^{-1}(f(K))$ is always true, $K = f^{-1}(f(K))$. Therefore, f is an atomic map.

Q.E.D.

Since arcwise connected continua do not contain nondegenerate proper terminal subcontinua, as a consequence of Theorem 8.1.25, we have:

8.1.26 Corollary *Let X and Y be continua, where X is arcwise connected. If $f: X \twoheadrightarrow Y$ is an atomic map, then f is a homeomorphism.*

The proof of the following theorem is similar to the one given in [15, Proposition 2.11].

8.1.27 Theorem *If $f: X \twoheadrightarrow Y$ is a surjective map between continua which is not joining, then there exist a subcontinuum Q of Y and two components C_1 and C_2 of $f^{-1}(Q)$ such that $f(C_j)$ is nondegenerate, $j \in \{1, 2\}$, and $f(C_1) \cap f(C_2) = \emptyset$.*

Proof Since f is not joining, there exist a subcontinuum Q' of Y and two components C_1' and C_2' of $f^{-1}(Q')$ such that $f(C_1') \cap f(C_2') = \emptyset$. Suppose $f(C_1')$ is degenerate. Note that $C_1' \cap f^{-1}(f(C_2')) = \emptyset$. Hence, by Theorem 1.6.8, there exist two disjoint closed subsets F_1 and F_2 of X such that $f^{-1}(Q') = F_1 \cup F_2$, $C_1' \subset F_1$ and $f^{-1}(f(C_2')) \subset F_2$. Since X is a metric space, there exist two disjoint open subsets U_1 and U_2 of X such that $F_j \subset U_j$, $j \in \{1, 2\}$. Let $\{V_n\}_{n=1}^\infty$ be a sequence of open subsets of Y such that $Cl(V_{n+1}) \subset V_n$ for each $n \in \mathbb{N}$ and $Q' = \bigcap_{n=1}^\infty V_n$. Thus, by Lemma 1.6.7, there exists $N \in \mathbb{N}$ such that $f^{-1}(Cl(V_N)) \subset U_1 \cup U_2$. Hence, by Corollary 1.7.28, there exists a subcontinuum C_1'' of $f^{-1}(Cl(V_N))$ such that $C_1' \subsetneqq C_1''$. Note that $C_1' \subset U_1$ and $C_1'' \subset U_1$. Let $Q = Q' \cup f(C_1'')$ and let C_1 be the component of $f^{-1}(Q)$ which contains C_1''. Since $f(C_1'') \subset f(C_1)$ and $f(C_1'')$ is nondegenerate, $f(C_1)$ is nondegenerate. Since $Q \subset Cl(V_N)$, $C_1 \subset U_1$. Thus, $C_1 \cap f^{-1}(f(C_2')) = \emptyset$, and $f(C_1) \cap f(C_2') = \emptyset$.

If $f(C_2')$ is degenerate, then repeating the above argument, we find another subcontinuum of Y such that its preimage under f has two nondegenerate components such that their images under f have empty intersection.

Q.E.D.

8.1.28 Corollary *If $f: X \twoheadrightarrow Y$ is a surjective map between continua which is not semi-confluent, then there exist a subcontinuum Q of Y and two components C_1 and C_2 of $f^{-1}(Q)$ such that $f(C_j)$ is nondegenerate, $j \in \{1, 2\}$, and $f(C_j) \not\subset f(C_k)$, where $j, k \in \{1, 2\}$ and $j \neq k$.*

Proof Since f is not semi-confluent, there exist a subcontinuum Q' of Y and two components C_1' and C_2' such that $f(C_j') \not\subset f(C_k')$, where $j, k \in \{1, 2\}$ and $j \neq k$. If $f(C_j')$ is degenerate, $j \in \{1, 2\}$, then $f(C_1') \cap f(C_2') = \emptyset$. Hence, by Theorem 8.1.27, there exist a subcontinuum Q of Y and two components C_1 and C_2 of $f^{-1}(Q)$ such that $f(C_1) \cap f(C_2) = \emptyset$. In particular, $f(C_j) \not\subset f(C_k)$, where $j, k \in \{1, 2\}$ and $j \neq k$.

Q.E.D.

8.1.29 Theorem *If $f: X \twoheadrightarrow Y$ is a monotone and light map between continua, then f is a homeomorphism.*

Proof Let $y \in Y$. Since f is light and surjective, $f^{-1}(y)$ is nonempty and totally disconnected. Since f is monotone, $f^{-1}(y)$ is connected. Thus, $f^{-1}(y)$ is a singleton. Since y is an arbitrary point of Y, f is injective. Therefore, f is a homeomorphism.

Q.E.D.

The following theorem is an immediate consequence of the definition of light map.

8.1.30 Theorem *If $f\colon X \to Y$ is a light map between continua, then for any nonempty subset A of X, $f|_A\colon A \to Y$ is light.*

8.1.31 Theorem *Let $f\colon X \to Y$ be a map between continua. Suppose that there exists a point $y \in Y$ such that $f^{-1}(y)$ is not connected. If C_1 and C_2 are two different components of $f^{-1}(y)$, then there exists an open subset U of Y such that $y \in U$, and C_1 and C_2 are contained in different components of $f^{-1}(U)$.*

Proof Let $y \in Y$ be such that $f^{-1}(y)$ is not connected and let C_1 and C_2 be two components of $f^{-1}(y)$. First, we show that there exists $N \in \mathbb{N}$ such that C_1 and C_2 are contained in different components of $f^{-1}(Cl(\mathcal{V}^d_{\frac{1}{N}}(y)))$.

Suppose that for each $n \in \mathbb{N}$, C_1 and C_2 are contained in the same component, R_n, of $f^{-1}(Cl(\mathcal{V}^d_{\frac{1}{n}}(y)))$. Since, for every $n \in \mathbb{N}$, $R_{n+1} \subset R_n$, Theorem 1.7.2, $R = \bigcap^\infty_{n=1} R_n$ is a subcontinuum of X such that $C_1 \cup C_2 \subset R$. Observe that $f(R) = f\left(\bigcap^\infty_{n=1} R_n\right) = \bigcap^\infty_{n=1} f(R_n) \subset \bigcap^\infty_{n=1} Cl(\mathcal{V}^d_{\frac{1}{n}}(y)) = \{y\}$. A contradiction to the fact that C_1 and C_2 are components of $f^{-1}(y)$. Thus, there exists $N \in \mathbb{N}$ such that C_1 and C_2 are contained in different components of $f^{-1}(Cl(\mathcal{V}^d_{\frac{1}{N}}(y)))$.

Let $U = \mathcal{V}^d_{\frac{1}{N}}(y)$. Then U is an open subset of Y that contains y, and C_1 and C_2 are contained in different components of $f^{-1}(U)$.

Q.E.D.

8.1.32 Theorem *Let $f\colon X \twoheadrightarrow Y$ be a map between continua. Then f is open if an only if f is interior at each point of X.*

Proof It is clear that every open map is interior at each of the points of its domain.

Assume f is interior at each point of X. Let U be an open subset of X, and let $x \in U$. Since f is interior at x, $f(x) \in Int(f(U))$. Since x is an arbitrary point of U, $f(U)$ is open in Y. Therefore, f is an open map.

Q.E.D.

8.1.33 Theorem *Let $f\colon X \to Y$ be a light map between continua and let U be an open subset of X. If $q \in f(U) \setminus Int(f(U))$, then there exist a sequence $\{z_n\}^\infty_{n=1}$ of points of $Y \setminus f(U)$ converging to q and, for each $n \in \mathbb{N}$, a subcontinuum E_n of Y which contains z_n such that the sequence $\{E_n\}^\infty_{n=1}$ converges to a subcontinuum E_0 of $f(U)$. Moreover, if f is confluent, then E_0 may be constructed in such a way that there exists a subcontinuum D of U such that $f(D) = E_0$.*

Proof Let U be an open subset of X and let $p \in U$ be such that $f(p) \in f(U) \setminus Int(f(U))$. Let $q = f(p)$. By Corollary 1.7.28, there exists a nondegenerate subcontinuum A of X such that $p \in A \subset U$. Since f is light, $f(A)$ is a nondegenerate subcontinuum of Y and $q \in f(A)$. Let $\{y_n\}^\infty_{n=1}$ be a sequence of points of $Y \setminus f(U)$ converging to q. For each $m \in \mathbb{N}$, let B_{nm} be the component of $Cl(\mathcal{V}^d_{\frac{1}{m}}(y_n))$ that contains y_n. By Theorem 1.7.27, $\frac{1}{m} \leq \mathrm{diam}(B_{nm}) \leq \frac{2}{m}$ for all $n \in \mathbb{N}$.

Since $\mathcal{C}(Y)$ is compact (Theorem 1.8.5), without loss of generality, we assume that each sequence $\{B_{nm}\}_{n=1}^{\infty}$ converges to a subcontinuum B_m of Y. Observe that $\frac{1}{m} \leq \operatorname{diam}(B_m) \leq \frac{2}{m}$ for each $m \in \mathbb{N}$. Note that, since $q \in f(A) \cap B_m$, $f(A) \cap B_m \neq \emptyset$ for all $m \in \mathbb{N}$. To show the existence of the nondegenerate subcontinuum of $f(U)$ we consider two cases:

Case (1)   *There exists $k \in \mathbb{N}$ such that $B_m \subset f(U)$ for each $m \geq k$.*

Let $m_0 \geq k$. Then $\{B_{nm_0}\}_{n=1}^{\infty}$ is a sequence in $\mathcal{C}(Y)$ which converges to B_{m_0} and such that $y_n \in B_{nm_0}$ for each $n \in \mathbb{N}$. Then B_{m_0} is a nondegenerate subcontinuum of $f(U)$ ($\frac{1}{m_0} \leq \operatorname{diam}(B_{m_0}) \leq \frac{2}{m_0}$).

Case (2)   *There exists a subsequence $\{B_{m_k}\}_{k=1}^{\infty}$ of $\{B_m\}_{m=1}^{\infty}$ such that $B_{m_k} \cap (Y \setminus f(U)) \neq \emptyset$ for all $k \in \mathbb{N}$.*

For each $k \in \mathbb{N}$, let $z_k \in B_{m_k} \cap (Y \setminus f(U))$. Recall that $\frac{1}{m_k} \leq \operatorname{diam}(B_{m_k}) \leq \frac{2}{m_k}$ for every $k \in \mathbb{N}$. Hence, since $q \in B_{m_k}$, we have that the sequence $\{B_{m_k}\}_{k=1}^{\infty}$ converges to $\{q\} \subset f(A)$. Thus, $\{z_k\}_{k=1}^{\infty}$ converges to q. Let $E_k = B_{m_k} \cup f(A)$. Then each E_k is a subcontinuum of Y and the sequence $\{E_k\}_{k=1}^{\infty}$ converges to $f(A)$. Thus, $f(A)$ is a nondegenerate subcontinuum of $f(U)$.

Now, assume that f is confluent. Observe that in Case (2), A is already a subcontinuum of U. Suppose there exists $k \in \mathbb{N}$ such that $B_m \subset f(U)$ for all $m \geq k$. Let D_m be the component of $f^{-1}(B_m)$ which contains p. Since f is light, $\{p\}$ is a component of $f^{-1}(q)$. Hence, $\{D_m\}_{m=1}^{\infty}$ converges to $\{p\}$ since $\{B_m\}_{m=1}^{\infty}$ converges to $\{q\}$. By Lemma 1.6.7, there exists $\ell \in \mathbb{N}$ such that $D_\ell \subset U$. Since f is confluent, $f(D_\ell) = B_\ell$. Also, B_ℓ is a nondegenerate subcontinuum of $f(U)$ and $\{B_{n\ell}\}_{n=1}^{\infty}$ converges to B_ℓ, where $y_n \in B_{n\ell}$ for every $n \in \mathbb{N}$.

Q.E.D.

8.1.34 Theorem *Let $f \colon X \twoheadrightarrow Y$ be a surjective map between continua. Then f is semi-open if and only if f is semi-interior at each point of X.*

Proof Suppose f is semi-open. Let x be a point of X and let U be an open subset of X containing x. Since f is semi-open, $Int(f(U)) \neq \emptyset$. Hence, there exists $z \in U$ such that $f(z) \in Int(f(U))$. Thus, f is semi-interior at x. Therefore, since x is an arbitrary point of X, f is semi-interior at each point of X.

Now, suppose f is semi-interior at each point of X. Let U be an open subset of X, and let $x \in U$. Since f is semi-interior at x, there exists a point $z \in U$ such that $f(z) \in Int(f(U))$. In particular, $Int(f(U)) \neq \emptyset$. Therefore, f is semi-open.

Q.E.D.

We finish this section with some results about freely decomposable and strongly freely decomposable maps.

8.1.35 Example Let K be a nonconnected compactum and let $\Sigma(K)$ be its suspension (Definition 1.2.13). Then the projection map $f \colon \Sigma(K) \twoheadrightarrow [0, 1]$ is strongly freely decomposable.

8.1.36 Theorem *Let $f \colon X \to Y$ be a map between continua. If f is confluent and freely decomposable, then f is strongly freely decomposable.*

Proof Let A and B be subcontinua of Y such that $Y = A \cup B$. Since f is freely decomposable, there exist two proper subcontinua A' and B' of X such that $X = A' \cup B'$, $f(A') \subset A$ and $f(B') \subset B$. Let K be the component of $f^{-1}(A)$ that contains A'. We prove that $K = f^{-1}(A)$. Suppose this is not true. Then there exists a component K' of $f^{-1}(A)$ distinct from K. Thus, $K' \subset B'$ and, consequently, $f(K') \subset B$. This contradicts the confluence of f. Similarly, $f^{-1}(B)$ is connected. Therefore, f is a strongly freely decomposable map.

Q.E.D.

8.1.37 Theorem *Let $f : X \to Y$ be a map between continua, where Y is locally connected. If f is strongly freely decomposable, then f is confluent.*

Proof Suppose f is not confluent. Then there exist a subcontinuum Q of Y and a component K of $f^{-1}(Q)$ such that $f(K) \neq Q$. Let $a \in Q \setminus f(K)$. By Theorem 3.1.38, there exist two subcontinua A and B of Y such that $Y = A \cup B$, $a \in A \setminus B$ and $f(K) \subset B \setminus A$. Hence, since f is strongly freely decomposable, $f^{-1}(A \cup Q)$ is a subcontinuum of X containing K and $f^{-1}(A)$. Note that $K \cap f^{-1}(A) = \emptyset$. Let K' be a subcontinuum of X such that $K \subsetneq K' \subset f^{-1}(A \cup Q) \setminus f^{-1}(A)$ (Corollary 1.7.28). Thus, $f(K') \subset Q$, contradicting the fact that K is a component of $f^{-1}(Q)$.

Q.E.D.

8.1.38 Example Let X be an equilateral triangle, let Y be a simple triod inscribed in X with its end points at vertexes of X, and let $f : X \to Y$ be the map that "collapses" X onto Y. Then f is a freely decomposable map that is not confluent. Hence, by Theorem 8.1.37, f is not strongly freely decomposable.

8.1.39 Theorem *Let $f : X \to Y$ be a freely decomposable map between continua, where Y is locally connected. If y is a nonseparating point of Y, then $f^{-1}(y)$ is connected.*

Proof Let y be a nonseparating point of Y. By [37, (4.15), p. 50], there exist arbitrarily small connected open subsets of Y, containing y, whose complements are connected. Thus, there exist two sequences $\{A_n\}_{n=1}^{\infty}$ and $\{B_n\}_{n=1}^{\infty}$ of subcontinua of Y such that for every $n \in \mathbb{N}$, $Y = A_n \cup B_n$, $y \in A_{n+1} \subset A_n \setminus B_n$, and $\bigcap_{n=1}^{\infty} A_n = \{y\}$. Since f is freely decomposable, for each $n \in \mathbb{N}$, there exist two subcontinua A'_n and B'_n of X such that $X = A'_n \cup B'_n$, $f(A'_n) \subset A_n$ and $f(B'_n) \subset B_n$. Since for all $n \in \mathbb{N}$, $A'_{n+1} \subset A'_n$, we have that $f^{-1}(y) = \bigcap_{n=1}^{\infty} A'_n$. Therefore, $f^{-1}(y)$ is connected, Theorem 1.7.2.

Q.E.D.

As a consequence of Theorem 8.1.39 and Lemma 2.1.12, we obtain:

8.1.40 Theorem *Let $f : X \to Y$ be a freely decomposable map between continua, where Y is locally connected. If Y does not have separating points, then f is monotone.*

8.1.41 Theorem *Let $f: X \to Y$ be a map between continua, where Y is locally connected. If f is almost monotone, then f is monotone.*

Proof Let $f: X \to Y$ be a map. Suppose that f is not monotone. Then there exists a subcontinuum Q of Y such that $f^{-1}(Q)$ is not connected. Let D and E be two different components of $f^{-1}(Q)$. Since Y is locally connected, we have that there exists a subcontinuum K of Y such that $Q \subset Int_Y(K)$ and D and E are contained in different components of $f^{-1}(K)$. Therefore, $f^{-1}(K)$ is not connected and f is not almost monotone.

Q.E.D.

8.1.42 Theorem *Let $f: X \to Y$ be a strongly freely decomposable map between continua. If X is unicoherent, then f is almost monotone.*

Proof Let Q be a subcontinuum of Y such that $Int_Y(Q) \neq \emptyset$. We show that $f^{-1}(Q)$ is connected. If $Y \setminus Q$ is connected, then $Y = Q \cup Cl_Y(Y \setminus Q)$, where Q and $Cl_Y(Y \setminus Q)$ are proper subcontinua of Y. Since f is strongly freely decomposable, $f^{-1}(Q)$ is connected.

Now, assume that $Y \setminus Q$ is not connected. Then $Y \setminus Q = E \cup F$, where E and F are disjoint open sets. Note that $Q \cup E$ and $Q \cup F$ are proper subcontinua of Y, by Lemma 1.7.23. Clearly, $Y = (Q \cup E) \cup (Q \cup F)$. Hence, $f^{-1}(Q \cup E)$ and $f^{-1}(Q \cup F)$ are connected. Observe that $X = f^{-1}(Q \cup E) \cup f^{-1}(Q \cup F)$ and $f^{-1}(Q) = f^{-1}(Q \cup E) \cap f^{-1}(Q \cup F)$. Since X is unicoherent, $f^{-1}(Q)$ is connected. Therefore, f is almost monotone.

Q.E.D.

Theorem 8.1.42 shows that if a strongly freely decomposable map is defined on a unicoherent continuum, then f is almost monotone. The following example shows that Theorem 8.1.42 is not true for freely decomposable maps.

8.1.43 Example Let X be the irreducible V-Λ continuum defined on [25, p. 191]. Let V one of the V's of X and let p and r be the end points of V. Let $Y = X/\{p, r\}$ and let $q: X \twoheadrightarrow Y$ be the quotient map. Clearly, q is freely decomposable, X is unicoherent and q is not almost monotone.

The material for this section is taken from [3–7, 9, 15, 18, 19, 23, 25, 26, 29, 30, 35–37, 39].

8.2 Induced Maps

We recall the definition of induced maps between hyperspaces and introduce the definition of induced map between hyperspace suspensions. To this end, we use [7, 11, 17, 20, 28].

8.2.1 Definition Let $f: X \to Y$ be a map between compacta. Then $2^f: 2^X \to 2^Y$ given by $2^f(A) = f(A)$ is called the *induced map between the hyperspaces of closed subsets of X and Y*. For each $n \in \mathbb{N}$, the maps $C_n(f): C_n(X) \to C_n(Y)$

and $\mathcal{F}_n(f)\colon \mathcal{F}_n(X) \to \mathcal{F}_n(Y)$ given by $\mathcal{C}_n(f) = 2^f|_{\mathcal{C}_n(X)}$ and $\mathcal{F}_n(f) = 2^X|_{\mathcal{F}_n(X)}$ are called the *induced maps between the n-fold hyperspaces of X and Y*, and the *induced maps between the n-fold symmetric products of X and Y*, respectively.

8.2.2 Theorem *Let $f\colon X \to Y$ be a map between compacta. Then 2^f is continuous.*

Proof Let $A \in 2^X$, and let $\varepsilon > 0$. Let $\delta > 0$ given by the uniform continuity of f. Then

$$2^f \left(\mathcal{V}_\delta^{\mathcal{H}_X}(A) \right) \subset \mathcal{V}_\varepsilon^{\mathcal{H}_Y}(2^f(A)).$$

Therefore, 2^f is continuous.

Q.E.D.

8.2.3 Corollary *Let $f\colon X \to Y$ be a map between compacta. Then for each $n \in \mathbb{N}$, $\mathcal{C}_n(f)$ and $\mathcal{F}_n(f)$ are continuous.*

8.2.4 Theorem *Let $f\colon X \to Y$ be a map between continua, let $n \in \mathbb{N}$ and let K be a subcontinuum of Y. If L is a component of $f^{-1}(K)$, then $\mathcal{C}_n(L)$ is a component of $\mathcal{C}_n(f)^{-1}(\mathcal{C}_n(K))$.*

Proof If $A \in \mathcal{C}_n(L)$, then $f(A)$ has at most n components and is a subset of K. Hence, $\mathcal{C}_n(f)(\mathcal{C}_n(L)) \subset \mathcal{C}_n(K)$.

Let $\mathcal{L}$ be the component of $\mathcal{C}_n(f)^{-1}(\mathcal{C}_n(K))$ containing $\mathcal{C}_n(L)$. Note that $f(\bigcup \mathcal{L}) \subset K$. By Lemma 6.1.4, each component of $\bigcup \mathcal{L}$ intersects L. Hence, $\bigcup \mathcal{L}$ is connected. Since L is a component of $f^{-1}(K)$, $L = \bigcup \mathcal{L}$. This implies that $\mathcal{C}_n(L) = \mathcal{L}$.

Q.E.D.

8.2.5 Definition Let $f\colon X \to Y$ be a map between continua. Then the function $HS_n(f)\colon HS_n(X) \to HS_n(Y)$, given by

$$HS_n(f)(A) = \begin{cases} q_Y^n(\mathcal{C}_n(f)((q_X^n)^{-1}(A))), & \text{if } A \neq F_X^n; \\ F_Y^n, & \text{if } A = F_X^n; \end{cases}$$

is called the *induced map by f between the n-fold hyperspace suspensions of X and Y*.

8.2.6 Theorem *Let $f\colon X \to Y$ be a map between continua. Then the function $HS_n(f)\colon HS_n(X) \to HS_n(Y)$ is continuous. Also, the following diagram*

$$\begin{array}{ccc} \mathcal{C}_n(X) & \xrightarrow{\ \mathcal{C}_n(f)\ } & \mathcal{C}_n(Y) \\ {\scriptstyle q_X^n} \downarrow & & \downarrow {\scriptstyle q_Y^n} \quad (*) \\ HS_n(X) & \xrightarrow[HS_n(f)]{} & HS_n(Y) \end{array}$$

is commutative.

Proof By [17, 4.3, p. 126], $HS_n(f)$ is continuous. The commutativity of $(*)$ follows from the definition of $HS_n(f)$.

Q.E.D.

8.2.7 Proposition *If* $f: X \twoheadrightarrow Y$ *is a surjective map between continua, and* n *is a positive integer, then both* $HS_n(f)^{-1}(F_Y^n)$ *and* $HS_n(f)^{-1}(T_Y^n)$ *are pathwise connected subcontinua of* $HS_n(X)$.

Proof Let $\chi \in HS_n(f)^{-1}(F_Y^n) \setminus \{F_X^n\}$. Then, by the commutativity of $(*)$ of Theorem 8.2.6, $C_n(f)((q_X^n)^{-1}(\chi)) \in \mathcal{F}_n(Y)$. For each component C of $(q_X^n)^{-1}(\chi)$, let $x_C \in C$. Note that the set $\{x_C \mid C$ is a component of $(q_X^n)^{-1}(\chi)\}$ belongs to $\mathcal{F}_n(X)$ and, by Theorem 1.8.20, there exists an order arc $\gamma: [0, 1] \to C_n(X)$ such that $\gamma(0) = \{x_C \mid C$ is a component of $(q_X^n)^{-1}(\chi)\}$ and $\gamma(1) = (q_X^n)^{-1}(\chi)$. Hence, $C_n(f)(\gamma(t)) \in \mathcal{F}_n(Y)$ for every $t \in [0, 1]$. Let $\eta: [0, 1] \to HS_n(X)$ be given by $\eta(t) = q_X^n(\gamma(t))$. Then η is an arc such that $\eta(0) = F_X^n$, $\eta(1) = \chi$ and $HS_n(f)(\eta(t)) = F_Y^n$ for each $t \in [0, 1]$. Therefore, $HS_n(f)^{-1}(F_Y^n)$ is pathwise connected.

Let $\chi \in HS_n(f)^{-1}(T_Y^n) \setminus \{T_X^n\}$. Then, by the commutativity of $(*)$ of Theorem 8.2.6, $C_n(f)((q_X^n)^{-1}(\chi)) = Y$. As we did in the previous paragraph, we can use the image of an order arc in $C_n(X)$ from $(q_X^n)^{-1}(\chi)$ to X, under q_X^n, to see that there exists an arc from χ to T_X^n contained in $HS_n(f)^{-1}(T_Y^n)$. Therefore, $HS_n(f)^{-1}(T_Y^n)$ is pathwise connected.

Q.E.D.

The next theorem gives a necessary and sufficient condition in order to have that the induced maps are surjective.

8.2.8 Theorem *Let* $f: X \to Y$ *be a map between continua, and let* $n \in \mathbb{N}$. *Then the following are equivalent:*

(1) f *is weakly confluent;*
(2) $C_n(f)$ *is surjective;*
(3) $HS_n(f)$ *is surjective.*

Proof Suppose f is weakly confluent. We show that $C_n(f)$ is surjective. Note that, the case $n = 1$ follows directly from the definitions. So, assume $n \geq 2$. Let $B \in C_n(Y)$. Suppose $B_1, \ldots, B_k$, $k \leq n$, are the components of B. Since f is weakly confluent, for each $j \in \{1, \ldots, k\}$, there exists a subcontinuum A_j of X such that $f(A_j) = B_j$. Let $A = \bigcup_{j=1}^{k} A_j$. Then $C_n(f)(A) = B$. Therefore, $C_n(f)$ is surjective.

If $C_n(f)$ is surjective, then it follows easily, by $(*)$ of Theorem 8.2.6, that $HS_n(f)$ is surjective.

Suppose $HS_n(f)$ is surjective. Then there exists a point $\chi \in HS_n(X)$ such that $HS_n(f)(\chi) = T_Y^n$. This implies, by $(*)$ of Theorem 8.2.6, that $C_n(f)((q_X^n)^{-1}(\chi)) = Y$. Hence, $C_n(f)(X) = Y$; i.e., $f(X) = Y$. Therefore, f is surjective.

Let $B \in \mathcal{C}_n(Y)$. If $B \in \mathcal{F}_n(Y)$, then, since f is surjective, there exists $A \in \mathcal{F}_n(X)$ such that $f(A) = B$. Thus, $\mathcal{C}_n(f)(A) = B$. Suppose now that $B \in \mathcal{C}_n(Y) \setminus \mathcal{F}_n(Y)$. Then $q_Y^n(B) \in HS_n(Y) \setminus \{F_Y^n\}$ Since $HS_n(f)$ is surjective, there exists $\chi \in HS_n(X) \setminus \{F_X^n\}$ such that $HS_n(f)(\chi) = q_Y^n(B)$. This implies, by $(*)$ of Theorem 8.2.6, that $\mathcal{C}_n(f)((q_X^n)^{-1}(\chi)) = B$. Therefore, $\mathcal{C}_n(f)$ is surjective.

Suppose now that $\mathcal{C}_n(f)$ is surjective. Hence f is surjective, since $\mathcal{C}_n(f)(X) = Y$. Let A_1 be a proper subcontinuum of Y. Let $A_2, \ldots, A_n$ be $n-1$ pairwise disjoint subcontinua of Y such that $A_1 \cap A_j = \emptyset$ for each $j \in \{2, \ldots, n\}$. Since $\mathcal{C}_n(f)$ is surjective, there exists $B \in \mathcal{C}_n(X)$ such that $\mathcal{C}_n(f)(B) = \bigcup_{j=1}^{n} A_j$. Then B has n components, say $B_1, \ldots, B_n$, and $f(B_k) = A_1$ for some $k \in \{1, \ldots, n\}$. Therefore, f is weakly confluent.

Q.E.D.

The next two results are related to homeomorphisms.

8.2.9 Theorem *Let $f \colon X \to Y$ be a map between continua, and let $n \in \mathbb{N}$. Then the following are equivalent:*

(1) f is a homeomorphism;
(2) $\mathcal{C}_n(f)$ is a homeomorphism;
(3) $HS_n(f)$ is a homeomorphism.

Proof Suppose f is a homeomorphism. Then f is a weakly confluent map. Hence, by Theorem 8.2.8, $\mathcal{C}_n(f)$ is surjective. Now, let $A_1, A_1 \in \mathcal{C}_n(X)$ be such that $\mathcal{C}_n(f)(A_1) = \mathcal{C}_n(f)(A_2)$. Then $f(A_1) = f(A_2)$. Since f is injective, $A_1 = A_2$. Thus, $\mathcal{C}_n(f)$ is injective. Therefore, $\mathcal{C}_n(f)$ is a homeomorphism.

If $\mathcal{C}_n(f)$ is a homeomorphism, then, by Theorem 8.2.8, $HS_n(f)$ is surjective. Since $\mathcal{C}_n(f)$ is injective, by Remark 7.1.5, we have that $HS_n(f)|_{HS_n(X) \setminus \{F_X^n\}}$ is injective. Hence, since $HS_n(f)(F_X^n) = F_Y^n$, $HS_n(f)$ is injective. Therefore, $HS_n(f)$ is a homeomorphism.

Suppose $HS_n(f)$ is a homeomorphism. By Theorem 8.2.8, $\mathcal{C}_n(f)$ is surjective. Note that $\mathcal{C}_n(f)|_{\mathcal{C}_n(X) \setminus \mathcal{F}_n(X)}$ is injective and

$$\mathcal{C}_n(f)(\mathcal{C}_n(X) \setminus \mathcal{F}_n(X)) \subset \mathcal{C}_n(Y) \setminus \mathcal{F}_n(Y).$$

Let $B \in \mathcal{C}(Y) \setminus \mathcal{F}_1(Y)$. Since f is weakly confluent (by Theorem 8.2.8), there exists $A \in \mathcal{C}(X)$ such that $f(A) = B$. Hence, $\mathcal{C}_n(f)(A) = B$ and $\mathcal{C}_n(f)^{-1}(B) = \{A\}$. Thus, $\mathcal{C}_n(f)^{-1}(B) \cap \mathcal{C}(X) \neq \emptyset$. Then, by Corollary 6.1.2, we have that $\bigcup \mathcal{C}_n(f)^{-1}(B) \in \mathcal{C}(X)$. Since $\bigcup \mathcal{C}_n(f)^{-1}(B) = f^{-1}(B)$, we have that $f^{-1}(B)$ is connected.

Let $y \in Y$, and let $\{K_m\}_{m=1}^{\infty}$ be a decreasing sequence of nondegenerate subcontinua of Y such that $\bigcap_{m=1}^{\infty} K_m = \{y\}$. This implies that $\bigcap_{m=1}^{\infty} f^{-1}(K_m) = f^{-1}(\{y\})$. Hence, $f^{-1}(\{y\})$ is connected. If $f^{-1}(\{y\}) \in \mathcal{C}(X) \setminus \mathcal{F}_1(X)$, then $\mathcal{C}_n(f)(f^{-1}(\{y\})) \in \mathcal{C}(Y) \setminus \mathcal{F}_1(Y)$. But, $\mathcal{C}_n(f)(f^{-1}(\{y\})) = f(f^{-1}(\{y\})) = \{y\}$, a contradiction. Therefore, $f^{-1}(\{y\}) \in \mathcal{F}_n(X)$. Since $f^{-1}(\{y\}) \in \mathcal{C}(X)$, $f^{-1}(\{y\}) \in \mathcal{F}_n(X) \cap \mathcal{C}(X) = \mathcal{F}_1(X)$. Therefore, f is a homeomorphism.

Q.E.D.

As a consequence of Corollary 8.1.26 and Theorem 8.2.9, we obtain:

8.2.10 Theorem *Let $f: X \to Y$ be a map between continua, and let $n \in \mathbb{N}$. Then the following are equivalent:*

(1) f is a homeomorphism;
(2) $C_n(f)$ is an atomic map;
(3) $HS_n(f)$ is an atomic map.

Proof Suppose f is a homeomorphism. Since, it is clear that homeomorphisms are atomic maps, by Theorem 8.2.9, $C_n(f)$ and $HS_n(f)$ are atomic maps.

Suppose that $C_n(f)$ is an atomic map. Since, by Corollary 1.8.12, $C_n(X)$ is an arcwise connected continuum, $C_n(f)$ is a homeomorphism, Corollary 8.1.26. Therefore, by Theorem 8.2.9, f is a homeomorphism.

Similarly, using Theorem 7.1.3, if $HS_n(f)$ is an atomic map, then f is a homeomorphism.

Q.E.D.

We now turn our attention to ε-maps.

8.2.11 Lemma *Let $f: X \to Y$ be a weakly confluent map between continua, and let $\varepsilon > 0$ be given. If f is an ε-map, then for each $n \in \mathbb{N}$, the induced map $C_n(f)$ is an ε-map.*

Proof Note that, by Theorem 8.2.8, $C_n(f)$ is surjective. Let $B \in C_n(Y)$, and let A_1 and A_2 be two elements of $C_n(f)^{-1}(B)$. Note that $C_n(f)(A_1) = C_n(f)(A_2) = B$. Thus, for each $a_1 \in A_1$, there exists $a_2 \in A_2$ such that $f(a_1) = f(a_2)$. Since f is an ε-map, $d(a_1, a_2) < \varepsilon$. Hence, $A_1 \subset \mathcal{V}_\varepsilon^{\mathcal{H}}(A_2)$. Similarly, $A_2 \subset \mathcal{V}_\varepsilon^{\mathcal{H}}(A_1)$. Thus, $\mathcal{H}(A_1, A_2) < \varepsilon$. Therefore, $C_n(f)$ is a ε-map.

Q.E.D.

As a consequence of Lemma 7.1.15 and Remark 7.1.17, we have:

8.2.12 Theorem *Let $f: X \twoheadrightarrow Y$ be a weakly confluent map between continua, let $n \in \mathbb{N}$, and let $\varepsilon > 0$. If $C_n(f)$ is an ε-map, then $HS_n(f)$ is an ε-map.*

8.2.13 Lemma *Let $g, f: X \to Y$ be maps between continua, let $n \in \mathbb{N}$ and let $\varepsilon > 0$. If $d(f(x), g(x)) < \varepsilon$ for every $x \in X$, then:*

(1) $\mathcal{H}(C_n(f)(A), C_n(g)(A)) < \varepsilon$, for each $A \in C_n(X)$;
    and
(2) $\rho_Y^n(HS_n(f)(\chi), HS_n(g)(\chi)) < \varepsilon$ for each $\chi \in HS_n(X)$, where ρ_Y^n is defined in Remark 7.1.17.

Proof We show (1) first. Let $A \in C_n(X)$ and let $a \in A$. By hypothesis, $d(f(a), g(a)) < \varepsilon$. Hence, $C_n(f)(A) \subset \mathcal{V}_\varepsilon^{\mathcal{H}}(C_n(g)(A))$. Similarly, $C_n(g)(A) \subset \mathcal{V}_\varepsilon^{\mathcal{H}}(C_n(f)(A))$. Therefore, $\mathcal{H}(C_n(f)(A), C_n(g)(A)) < \varepsilon$.

Now, we prove (2). Note that, by (1),

$$\mathcal{H}_Y(\mathcal{C}_n(f)(A), \mathcal{C}_n(g)(A)) < \varepsilon$$

for each $A \in \mathcal{C}_n(X)$.

Observe that, since $HS_n(f)(F_X^n) = HS_n(g)(F_X^n) = F_Y^n$, we have that $\rho_Y^n(HS_n(f)(F_X^n), HS_n(g)(F_X^n)) = 0 < \varepsilon$.

Let $\chi \in HS_n(X)\backslash\{F_X^n\}$. Then there exists $A \in \mathcal{C}_n(X)\backslash\mathcal{F}_n(X)$ such that $q_X^n(A) = \chi$. Hence,

$$\rho_Y^n(HS_n(f)(\chi), HS_n(g)(\chi)) = \mathcal{H}_Y^2(\mathcal{F}_n(Y) \cup \{f(A)\}, \mathcal{F}_n(Y) \cup \{g(A)\})$$

$$= \mathcal{H}_Y(f(A), g(A)) = \mathcal{H}_Y(\mathcal{C}_n(f)(A), \mathcal{C}_n(g)(A)) < \varepsilon.$$

Q.E.D.

8.2.14 Definition A surjective map $f \colon X \twoheadrightarrow Y$ between continua is said to be *refinable* provided that for each $\varepsilon > 0$, there exists an ε-map $g \colon X \to Y$ such that $d(f(x), g(x)) < \varepsilon$ for every $x \in X$.

8.2.15 Definition A continuum Y is in $Class(W)$ provided that for each continuum X, every surjective map $f \colon X \twoheadrightarrow Y$ is weakly confluent.

8.2.16 Theorem *Let X and Y be continua and let $n \in \mathbb{N}$. If Y is in $Class(W)$ and $f \colon X \twoheadrightarrow Y$ is a refinable map, then both $\mathcal{C}_n(f)$ and $HS_n(f)$ are refinable.*

Proof Note that, by Theorem 8.2.8, $\mathcal{C}_n(f)$ and $HS_n(f)$ are surjective. Let $\varepsilon > 0$. Since f is refinable, there exists an ε-map $g \colon X \twoheadrightarrow Y$ such that $d(f(x), g(x)) < \varepsilon$ for each $x \in X$. Since g is an ε-map and Y is in $Class(W)$, by Lemma 8.2.11 and Theorem 8.2.8, $\mathcal{C}_n(g)$ is an ε-map.

By Lemma 8.2.13, $\mathcal{H}(\mathcal{C}_n(f)(A), \mathcal{C}_n(g)(A)) < \varepsilon$ for all $A \in \mathcal{C}_n(X)$. Therefore, $\mathcal{C}_n(f)$ is refinable.

By Theorem 8.2.12, $HS_n(g)$ is an ε-map. Now, by Lemma 8.2.13, $\rho_Y^n(HS_n(f)(\chi), HS_n(g)(\chi)) < \varepsilon$ for every $\chi \in HS_n(X)$. Therefore, $HS_n(f)$ is refinable.

Q.E.D.

8.2.17 Remark By [20, Section 3, Example, pp. 3–8] the assumption of Y being in $Class(W)$ is necessary in Theorem 8.2.16.

We end this section with a result about almost monotone maps.

8.2.18 Theorem *Let $f \colon X \to Y$ be a map between continua and let $n \in \mathbb{N}$. If $\mathcal{C}_n(f)$ is an almost monotone map, then f is also almost monotone.*

Proof Let $f: X \to Y$ be a map between continua such that $\mathcal{C}_n(f)$ is almost monotone. Let B be a subcontinuum of Y such that $Int(B) \neq \emptyset$. Then $\langle B \rangle_n$ is a subcontinuum $\mathcal{C}_n(Y)$, by Proposition 6.1.5. Moreover, $\langle B \rangle_n$ has nonempty interior, because $\langle Int(B) \rangle_n \subset \langle B \rangle_n$. Since $\mathcal{C}_n(f)$ is almost monotone, $\mathcal{C}_n(f)^{-1}(\langle B \rangle_n)$ is connected. Note that $\mathcal{C}_n(f)^{-1}(\langle B \rangle_n) = \langle f^{-1}(B) \rangle_n$. Hence, $f^{-1}(B)$ is connected, by Proposition 6.1.5. Therefore, f is almost monotone.

Q.E.D.

8.3 Confluent Maps

We study Confluent maps and its relationships with semi-confluent, joining maps and pseudo-confluent maps. We begin with a characterization of confluent maps. The material for this section is based on [4, 7, 11, 15, 22, 27, 32].

8.3.1 Theorem *Let $f: X \to Y$ be a map between continua, and let $n \in \mathbb{N}$. Then*

$$\bigcup \mathcal{C}_n(f)^{-1}(B) = f^{-1}(B) \text{ for each } B \in \mathcal{C}_n(Y)$$

if and only if f is confluent.

Proof Suppose the equality holds. Let Q_1 be a subcontinuum of Y, and let K be a component of $f^{-1}(Q_1)$. Let $Q_2, \ldots, Q_n \in \mathcal{C}(Y)$ be such that $Q_1 \cap Q_j = \emptyset$ and $Q_j \cap Q_\ell = \emptyset$, if $j \neq \ell$ and $j, \ell \in \{2, \ldots, n\}$. Let $Q = \bigcup_{\ell=1}^n Q_\ell$. Note that $Q \in \mathcal{C}_n(Y)$.

Let $x \in K \subset f^{-1}(Q_1) \subset f^{-1}(Q)$. Since $\bigcup \mathcal{C}_n(f)^{-1}(Q) = f^{-1}(Q)$, there exists $E \in \mathcal{C}_n(f)^{-1}(Q)$ such that $x \in E$. Let $E_1, \ldots, E_r, r \leq n$, be the components of E. Since $E \in \mathcal{C}_n(f)^{-1}(Q)$, $f(E) = \bigcup_{j=1}^r f(E_j) = \bigcup_{\ell=1}^n Q_\ell$. Since $Q_j \cap Q_\ell = \emptyset$ if $j \neq \ell$ and $j, \ell \in \{1, \ldots, n\}$, we have that $r = n$ and for each $j \in \{1, \ldots, n\}$ there exists $\ell \in \{1, \ldots, n\}$ such that $f(E_j) = Q_\ell$. Without loss of generality, we assume that $f(E_1) = Q_1$. Thus, $x \in E_1$. Since $x \in E_1 \cap K$, $E_1 \subset f^{-1}(Q_1)$ and K is a component of $f^{-1}(Q_1)$, $E_1 \subset K$. Therefore, $f(K) = Q_1$, and f is confluent.

Suppose f is confluent. Let $B \in \mathcal{C}_n(Y)$. It is clear that

$$\bigcup \mathcal{C}_n(f)^{-1}(B) \subset f^{-1}(B).$$

Let $x \in f^{-1}(B)$. Assume B has k components, say $B_1, \ldots, B_k$. For each $j \in \{1, \ldots, k\}$, let A_j be a component of $f^{-1}(B_j)$. Without loss of generality, we assume that $x \in A_1$ and $f(x) \in B_1$. Let $A = \bigcup_{j=1}^k A_j$. Since f is confluent, $f(A) = B$. Thus, $A \in \mathcal{C}_n(f)^{-1}(B)$ and $x \in \bigcup \mathcal{C}_n(f)^{-1}(B)$. Therefore, the equality holds.

Q.E.D.

8.3.2 Theorem *Let $f: X \to Y$ be a map between continua. If $\mathcal{C}_n(f)$ is confluent, then f is confluent.*

Proof Let Q be a subcontinuum of Y and let C be a component of $f^{-1}(Q)$. Let $\mathcal{K}$ be the component of $\mathcal{C}_n(f)^{-1}(\mathcal{F}_1(Q))$ which intersects $\mathcal{F}_1(C)$. Since $\mathcal{C}_n(f)$ is confluent, $\mathcal{C}_n(f)(\mathcal{K}) = \mathcal{F}_1(Q)$. Hence, $\bigcup\mathcal{K} = C$. This implies that $f(C) = Q$. Therefore, f is confluent.

Q.E.D.

The purpose of our next three results is to show that if f is a confluent map whose range space is locally connected, then $\mathcal{C}_2(f)$ is confluent, Theorem 8.3.5. We give an example showing that this is not true for $n \geq 3$ in Example 8.4.5.

The first step is to show that if $f\colon X \twoheadrightarrow Y$ is a confluent map, then the components of the preimage of an arc in $\mathcal{C}_2(Y)$ under $\mathcal{C}_2(f)$ are mapped onto the arc by $\mathcal{C}_2(f)$.

8.3.3 Theorem *Let $f\colon X \twoheadrightarrow Y$ be a confluent map. If $\mathcal{L}$ is an arc in $\mathcal{C}_2(Y)$, then for each component $\mathcal{K}$ of $\mathcal{C}_2(f)^{-1}(\mathcal{L})$, we have that $\mathcal{C}_2(f)(\mathcal{K}) = \mathcal{L}$.*

Proof Let $\mathcal{L}$ be an arc in $\mathcal{C}_2(Y)$, and let $\alpha\colon [0, 1] \twoheadrightarrow \mathcal{L}$ be a homeomorphism. Let $\mathcal{K}$ be a component of $\mathcal{C}_2(f)^{-1}(\mathcal{L})$.

First, we show that $\alpha(0) \in \mathcal{C}_2(f)(\mathcal{K})$. To this end, suppose $\alpha(0) \notin \mathcal{C}_2(f)(\mathcal{K})$. Then there exist two disjoint closed sets $\mathcal{K}_0$ and $\mathcal{K}_1$ such that $\mathcal{C}_2(f)^{-1}(\mathcal{L}) = \mathcal{K}_0 \cup \mathcal{K}_1$, where $\mathcal{C}_2(f)^{-1}(\alpha(0)) \subset \mathcal{K}_0$ and $\mathcal{K} \subset \mathcal{K}_1$ (Theorem 1.6.8). Let $t_0 = \min\{t \in [0, 1] \mid \alpha(t) \in \mathcal{C}_2(f)(\mathcal{K}_1)\}$. Note that $t_0 > 0$. Let $K \in \mathcal{K}_1$ be such that $\mathcal{C}_2(f)(K) = \alpha(t_0)$. For each $t \in [0, t_0)$, we find $K_t \in \mathcal{C}_2(X)$ such that $\mathcal{C}_2(f)(K_t) = \alpha(t)$. To this end, let $t \in [0, t_0)$. We consider two cases.

Case (1) *K is connected.*

Observe that $\alpha(t_0)$ is connected. Let M be the component of $f^{-1}(\alpha(t_0))$ containing K. Now, let M_t be the component of $f^{-1}\left(\bigcup\alpha([t, t_0])\right)$ containing M. Since $\alpha(t_0)$ is connected, $\bigcup\alpha([t, t_0])$ is connected, Corollary 6.1.2. Since f is confluent and $\alpha(t) \subset \bigcup\alpha([t, t_0])$, $f(M_t) = \bigcup\alpha([t, t_0])$ and there exists $K_t \in \mathcal{C}_2(X)$ such that $K_t \subset M_t$ and $\mathcal{C}_2(f)(K_t) = \alpha(t)$.

Case (2) *K is not connected.*

Note that $\bigcup\alpha([t, t_0]) \in \mathcal{C}_2(Y)$, by Corollary 6.1.2. Let

$$M = \bigcup\{C \mid C \text{ is a component of } f^{-1}(\alpha(t_0)) \text{ and } C \cap K \neq \emptyset\}.$$

Observe that $M \in \mathcal{K}_1$ since there exists an order arc from K to M (Theorem 1.8.20) contained in $\mathcal{C}_2(f)^{-1}(\alpha(t_0))$. Now, let

$$M_t = \bigcup\Big\{C \mid C \text{ is a component of } f^{-1}\left(\bigcup\alpha([t, t_0])\right) \text{ and}$$

$$C \cap M \neq \emptyset\Big\}.$$

Note that there exists an order arc from M to M_t (Theorem 1.8.20). Hence, $M_t \in \mathcal{C}_2(X)$.

Since f is confluent, each component of M_t is mapped onto a component of $\bigcup \alpha([t, t_0])$. Thus, $f(M_t) = \bigcup \alpha([t, t_0])$. By Lemma 6.1.4, $\alpha(t)$ intersects each component of $\bigcup \alpha([t, t_0])$. Since $\alpha(t) \subset \bigcup \alpha([t, t_0])$ and $f(M_t) = \bigcup \alpha([t, t_0])$, there exists $K_t \in \mathcal{C}_2(X)$ such that $K_t \subset M_t$, each component of M_t intersects K_t and $\mathcal{C}_2(f)(K_t) = \alpha(t)$.

Let $\{t_m\}_{m=1}^{\infty}$ be an increasing sequence in $[0, t_0)$ converging to t_0. Since $\mathcal{C}_2(X)$ is compact (Corollary 1.8.12), without loss of generality, we assume that the sequence $\{K_{t_m}\}_{m=1}^{\infty}$ converges to $K_0 \in \mathcal{C}_2(X)$. Observe that $K_0 \subset \bigcap_{m=1}^{\infty} M_{t_m} = M$ and each component of K_0 intersects M. Thus, there exists an order arc from K_0 to M (Theorem 1.8.20) in $\mathcal{C}_2(f)^{-1}(\alpha(t_0))$. Therefore, $K_0 \in \mathcal{K}_1$. On the other hand, each $K_{t_m} \in \mathcal{K}_0$. This implies that $\mathcal{K}_0 \cap \mathcal{K}_1 \neq \emptyset$, a contradiction. Therefore, $\alpha(0) \in \mathcal{C}_2(f)(\mathcal{K})$.

A similar argument shows that $\alpha(1) \in \mathcal{C}_2(f)(\mathcal{K})$. Therefore, $\mathcal{C}_2(f)(\mathcal{K}) = \mathcal{L}$.

Q.E.D.

As a consequence of Theorem 8.3.3, we have the following:

8.3.4 Corollary *Let $f: X \twoheadrightarrow Y$ be a confluent map between continua. If $\mathcal{B}$ is an arcwise connected subcontinuum of $\mathcal{C}_2(Y)$, then for each component $\mathcal{K}$ of $\mathcal{C}_2(f)^{-1}(\mathcal{B})$, we have that $\mathcal{C}_2(f)(\mathcal{K}) = \mathcal{B}$.*

Using Theorem 8.3.2 and Corollary 8.3.4, we obtain the following:

8.3.5 Theorem *Let $f: X \twoheadrightarrow Y$ be a surjective map between continua. Consider the following statements:*

(1) f is confluent;
(2) $\mathcal{C}_2(f)$ is confluent.

Then (2) implies (1). If Y is locally connected, then (1) implies (2).

The next result generalizes [22, 2.7] to maps between 2-fold hyperspaces. It gives us a sufficient condition for a confluent map f whose range is an inverse limit of locally connected continua to have its induced map $\mathcal{C}_2(f)$ confluent.

8.3.6 Theorem *Let Y be a continuum and assume that $\{Y_n, g_n^{n+1}\}$ is an inverse sequence of locally connected continua and confluent bonding maps such that $Y = \varprojlim\{Y_n, g_n^{n+1}\}$. If $f: X \twoheadrightarrow Y$ is a confluent map, then $\mathcal{C}_2(f)$ is confluent.*

Proof Note that, by Theorem 2.1.23, each g_n is confluent. Then, by Theorem 8.1.7, $g_n \circ f: X \twoheadrightarrow Y_n$ is confluent for each $n \in \mathbb{N}$. Since every Y_n is locally connected, by Theorem 8.3.5, $\mathcal{C}_2(g_n^{n+1})$ and $\mathcal{C}_2(g_n \circ f)$ are confluent, for each $n \in \mathbb{N}$. Since $\mathcal{C}_2(g_n^{n+1}) \circ \mathcal{C}_2(g_{n+1} \circ f) = \mathcal{C}_2(g_n^{n+1} \circ g_{n+1} \circ f) = \mathcal{C}_2(g_n \circ f)$, by Theorem 2.1.33, the sequence of maps $\{\mathcal{C}_2(g_n \circ f)\}_{n=1}^{\infty}$ induces a map

$$g: \mathcal{C}_2(X) \twoheadrightarrow \varprojlim\{\mathcal{C}_2(Y_n), \mathcal{C}_2(g_n^{n+1})\}.$$

By Theorem 2.1.52, g is confluent. Let $h\colon \varprojlim\{C_2(Y_n), C_2(g_n^{n+1})\} \twoheadrightarrow C_2(Y)$ be the homeomorphism given by Theorem 2.3.4. Let $A \in C_2(X)$. Then

$$h \circ g(A) = h((C_2(g_n \circ f)(A))_{n=1}^{\infty}) = \varprojlim\{g_n \circ f(A), g_n^{n+1}|_{g_{n+1} \circ f(A)}\} =$$

$$f(A) = C_2(f)(A).$$

Hence, $h \circ g = C_2(f)$. Therefore, by Theorem 8.1.7, $C_2(f)$ is confluent.

Q.E.D.

8.3.7 Theorem *Let $f\colon X \twoheadrightarrow Y$ be a surjective map between continua. If $C_n(f)$ is joining, then f is confluent.*

Proof Suppose f is not confluent. By [15, Proposition 2.11], there exist a subcontinuum L of Y and a component D of $f^{-1}(L)$ such that $f(D)$ is a nondegenerate proper subcontinuum of L. Let $p \in L \setminus f(D)$. By Theorem 1.8.20, there exists an order arc $\mathcal{L}_1$ in $C(Y)$ from $\{p\}$ to L. Also, there exists an order arc $\mathcal{L}_2$ in $C(Y)$ from $f(D)$ to L. Let $\mathcal{L} = \mathcal{L}_1 \cup \mathcal{L}_2 \cup \mathcal{F}_1(L)$ and let

$$\mathcal{K} = \{\{y_1, \ldots, y_{n-1}\} \cup A \mid A \in \mathcal{L}\},$$

where $y_1, \ldots, y_{n-1}$ are $n-1$ distinct points of $Y \setminus L$. Then $\mathcal{K}$ is a subcontinuum of $C_n(Y)$.

Let P_j be a component of $f^{-1}(y_j)$ for each $j \in \{1, \ldots, n-1\}$. Let $\mathcal{D}$ and $\mathcal{E}$ be the components of $C_n(f)^{-1}(\mathcal{K})$ such that $P_1 \cup \cdots \cup P_{n-1} \cup D \in \mathcal{D}$ and $\{P_1 \cup \cdots \cup P_{n-1} \cup P \mid P \in \mathcal{F}_1(D)\} \subset \mathcal{E}$. Note that $C_n(f)(\bigcup \mathcal{D}) \subset \{y_1, \ldots, y_{n-1}\} \cup L$. Hence, we can put $\bigcup \mathcal{D} = Q_1 \cup \cdots \cup Q_{n-1} \cup E$, where Q_j is a connected subset of $f^{-1}(y_j)$ for every $j \in \{1, \ldots, n-1\}$ and $E \subset f^{-1}(L)$.

By Lemma 6.1.4, each component of $\bigcup \mathcal{D}$ intersects $P_1 \cup \cdots \cup P_{n-1} \cup D$. Since $P_1, \ldots, P_{n-1}$ and D are components of $f^{-1}(y_1), \ldots, f^{-1}(y_{n-1})$ and $f^{-1}(L)$, respectively, we have that $Q_j \subset P_j$ for each $j \in \{1, \ldots, n-1\}$ and $E \subset D$. Thus, $\bigcup \mathcal{D} = P_1 \cup \cdots \cup P_{n-1} \cup D$.

Let $R \in \mathcal{D}$. Then $R = R_1 \cup \cdots \cup R_{n-1} \cup D'$, where $R_j \subset P_j$ for all $j \in \{1, \ldots, n-1\}$ and $D' \subset D$. Hence, $f(D') \subset L \setminus \{p\}$. By the definition of $\mathcal{K}$, we obtain that $C_n(f)(R) \in \{\{y_1, \ldots, y_{n-1}\} \cup A \mid A \in \mathcal{L}_2 \cup \mathcal{F}_1(L)\}$. Observe that $\{\{y_1, \ldots, y_{n-1}\} \cup A \mid A \in \mathcal{L}_2\}$ and $\{\{y_1, \ldots, y_{n-1}\} \cup A \mid A \in \mathcal{F}_1(L)\}$ are disjoint closed subsets of $\mathcal{K}$. Since $C_n(f)(\mathcal{D})$ is connected and the sets $\{\{y_1, \ldots, y_{n-1}\} \cup f(D) \in C_n(f)(\mathcal{D})\}$ and $\{\{y_1, \ldots, y_{n-1}\} \cup A \mid A \in \mathcal{L}_2\}$ are also disjoint, we have that $C_n(f)(R) \in \{\{y_1, \ldots, y_{n-1}\} \cup A \mid A \in \mathcal{L}_2\}$. Thus, we conclude that $C_n(f)(\mathcal{D}) \subset \{\{y_1, \ldots, y_{n-1}\} \cup A \mid A \in \mathcal{L}_2\}$.

Similarly, we obtain that

$$C_n(f)(\mathcal{E}) \subset \{\{y_1, \ldots, y_{n-1}\} \cup A \mid A \in \mathcal{F}_1(L)\}.$$

Therefore, $C_n(f)(\mathcal{D}) \cap C_n(f)(\mathcal{E}) = \emptyset$ and $C_n(f)$ is not joining.

Q.E.D.

Since each confluent map is semi-confluent and each semi-confluent map is joining (see Diagram I), from Theorem 8.3.7, we have the following:

8.3.8 Corollary *Let $f: X \twoheadrightarrow Y$ be a map between continua.*

(1) If $C_n(f)$ is semi-confluent, then f is semi-confluent.
(2) If $C_n(f)$ is joining, then f is joining.

8.3.9 Theorem *Let $f: X \twoheadrightarrow Y$ be a surjective map between continua. If $H S_n(f)$ is joining, then f is confluent.*

Proof Suppose f is not confluent. By [15, Proposition 2.11], there exist a subcontinuum L of Y and a component D of $f^{-1}(L)$ such that $f(D)$ is nondegenerate proper subcontinuum of L. We consider two cases.

Case (1) $n = 1$.
Let $p \in L \setminus f(D)$. By Theorem 1.8.20, there exists an order arc $\mathcal{L}_1$ in $C(Y)$ from $f(D)$ to L. Also, there exists an order arc $\mathcal{L}_2'$ in $C(Y)$ from $\{p\}$ to L. Let $\mu: C(L) \to [0, 1]$ be a Whitney map (Remark 1.8.18). For each $t \in [0, \mu(f(D))]$, let $\{E_t\} = \mu^{-1}(t) \cap \mathcal{L}_2'$. Let $t_0 \in (0, \mu(f(D)))$ be such that $E_{t_0} \cap f(D) = \emptyset$. Let $\mathcal{L}_2$ be an order arc in $C(Y)$ from E_{t_0} to L. Let $\mathcal{L} = \mathcal{L}_1 \cup \mathcal{L}_2 \cup \mu^{-1}(t_0)$. Since $\mu^{-1}(t_0)$ is a subcontinuum of $C(Y)$ [32, (14.2)], $\mathcal{L}_1 \cap \mu^{-1}(t_0) = \emptyset$ and $\mathcal{L}_2 \cap \mu^{-1}(t_0) = \{E_{t_0}\}$, we have that $\mathcal{L}$ is a subcontinuum of $C(Y)$. Observe that $\mathcal{L} \cap \mathcal{F}_1(Y) = \emptyset$. Since $\mu(f(D)) > t_0$, there exists a subcontinuum D_0 of D such that $\mu(f(D_0)) = t_0$. Thus, $f(D_0) \subset L \setminus \{p\}$. Let $\mathcal{D}$ and $\mathcal{E}$ be components of $C(f)^{-1}(\mathcal{L})$ such that $D \in \mathcal{D}$ and $D_0 \in \mathcal{E}$.
We show that $C(f)(\mathcal{D}) = \{f(D)\}$. First note that $f(\bigcup \mathcal{D}) \subset L$. Since $D \in \mathcal{D}$, by Lemma 6.1.4, $\bigcup \mathcal{D}$ is connected. Also, since D is a component of $f^{-1}(L)$, $D = \bigcup \mathcal{D}$. Let $E \in \mathcal{D}$. Then $E \subset D$. By the definition of $\mathcal{L}$, either $f(E) = f(D)$ or $f(E) \in \mu^{-1}(t_0)$. Assume $f(E) \in \mu^{-1}(t_0)$. Since $C(f)(\mathcal{D})$ is connected and $f(D) \in C(f)(\mathcal{D})$, we have that $E_{t_0} \in C(f)(\mathcal{D})$ ($\mathcal{L} \setminus \{E_{t_0}\}$ is not connected). This is a contradiction, because $\bigcup \mathcal{D} = D$ and $E_{t_0} \subset L \setminus f(D)$. Therefore, $f(E) = f(D)$ and $C(f)(\mathcal{D}) = \{f(D)\}$.
Similarly, $f(\bigcup \mathcal{E}) \subset L$. Since $D_0 \subset \bigcup \mathcal{E}$, we obtain that $f(\bigcup \mathcal{E}) \subset f(D) \subset L \setminus E_{t_0}$. Let $E \in \mathcal{E}$. Then either $f(E) \in \mathcal{L}_1$ or $f(E) \in \mu^{-1}(t_0)$. Assume $f(E) \in \mathcal{L}_1$. Since $D_0 \in \mathcal{E}$, $C(f)(\mathcal{E}) \cap \mu^{-1}(t_0) \neq \emptyset$. Hence, by the definition of $\mathcal{L}$, $E_{t_0} \in C(f)(\mathcal{E})$. This contradicts the fact that $f(\bigcup \mathcal{E}) \subset L \setminus E_{t_0}$. Therefore,

$f(E) \in \mu^{-1}(t_0)$. Thus, $\mathcal{C}(f)(\mathcal{E}) \subset \mu^{-1}(t_0)$. In this way, we have that $\mathcal{C}(f)(\mathcal{D}) \cap \mathcal{C}(f)(\mathcal{E}) = \emptyset$. Since $\mathcal{L} \cap \mathcal{F}_1(Y) = \emptyset$, $q_X(\mathcal{D})$ and $q_X(\mathcal{E})$ are components of $HS(f)^{-1}(q_Y(\mathcal{L}))$. Also:

$$HS(f)(q_X(\mathcal{D})) \cap HS(f)(q_X(\mathcal{E})) = \emptyset.$$

Therefore, $HS(f)$ is not joining.

Case (2) $n \geq 2$.

Let $p \in L \setminus f(D)$. By Theorem 1.8.20, there exists an order arc $\mathcal{L}_1$ in $\mathcal{C}(Y)$ from $\{p\}$ to L. Also, there exists an order arc $\mathcal{L}_2$ in $\mathcal{C}(Y)$ from $f(D)$ to L. Let $\mathcal{L} = \mathcal{L}_1 \cup \mathcal{L}_2 \cup \mathcal{F}_1(Y)$ and let $y_1, \ldots, y_{n-1}$ be $n-1$ distinct points of $Y \setminus L$. By Theorem 1.7.27, there exists a subcontinuum Q of Y such that $\{y_1\} \subsetneq Q \subset Y \setminus (L \cup \{y_2, \ldots, y_{n-1}\})$. Let

$$\mathcal{K} = \{Q \cup \{y_2, \ldots, y_{n-1}\} \cup A \mid A \in \mathcal{L}\}.$$

Then $\mathcal{K}$ is a subcontinuum of $\mathcal{C}_n(Y)$ and $\mathcal{K} \cap \mathcal{F}_n(Y) = \emptyset$. Since $HS_n(f)$ is joining, $HS_n(f)$ is surjective. Hence, by Theorem 8.2.8, f is weakly confluent. Thus, there exists a component R of $f^{-1}(Q)$ such that $f(R) = Q$. Let P_j be a component of $f^{-1}(y_j)$ for all $j \in \{2, \ldots, n-1\}$. Let $\mathcal{D}$ and $\mathcal{E}$ be the components of $\mathcal{C}_n(f)^{-1}(\mathcal{K})$ such that $R \cup P_2 \cup \cdots \cup P_{n-1} \cup D \in \mathcal{D}$ and $\{R \cup P_2 \cup \cdots \cup P_{n-1} \cup E \mid E \in \mathcal{F}_1(D)\} \subset \mathcal{E}$. In a similar way as it is done in Theorem 8.3.7, we have that $\mathcal{C}_n(f)(\mathcal{D}) \cap \mathcal{C}_n(f)(\mathcal{E}) = \emptyset$. Since $\mathcal{K} \cap \mathcal{F}_n(Y) = \emptyset$, $q_X^n(\mathcal{D})$ and $q_X^n(\mathcal{E})$ are components of $HS_n(f)^{-1}(\mathcal{K})$. Hence,

$$HS_n(f)(q_X^n(\mathcal{D})) \cap HS_n(f)(q_X^n(\mathcal{E})) = \emptyset.$$

A contradiction to the fact that $HS_n(f)$ is joining.

Q.E.D.

From Theorem 8.3.9 and Diagram I, we have:

8.3.10 Corollary *Let $f\colon X \to Y$ be a map between continua.*

(1) If $HS_n(f)$ is confluent, then f is confluent.
(2) If $HS_n(f)$ is semi-confluent, then f is semi-confluent.
(3) If $HS_n(f)$ is joining, then f is joining.

8.4 Monotone Maps

We study monotone maps. We prove the equivalence of the monotoneity of a map f, $\mathcal{C}_n(f)$ and $HS_n(f)$. We present some relationships between a monotone map and the fact that its induced maps are CE-maps, near-homeomorphisms, and universal. For this purpose, we use [2, 11, 14, 16, 24, 25, 27, 28, 31, 33, 34, 37, 38].

We begin with the following:

8.4.1 Lemma *Let $f\colon X \twoheadrightarrow Y$ be a surjective map between continua. If the induced map $HS(f)\colon HS(X) \twoheadrightarrow HS(Y)$ is monotone, then f is confluent.*

Proof Let Q be a subcontinuum of Y, and let K be a component of $f^{-1}(Q)$. We show that $f(K) = Q$. If Q is degenerate, clearly $f(K) = Q$. Suppose Q is nondegenerate, and $f(K) \neq Q$.

Let $\alpha\colon [0, 1] \to \mathcal{C}(Y)$ be an order arc such that $\alpha(0) = f(K)$ and $\alpha(1) = Q$, Theorem 1.8.20. Then $\alpha\left(\left[\frac{1}{2}, 1\right]\right) \cap \mathcal{F}_1(Y) = \emptyset$. Hence, $q_Y \circ \alpha\colon \left[\frac{1}{2}, 1\right] \to HS(Y)$ is an arc such that $q_Y \circ \alpha(1) = q_Y(Q)$, and $F_Y \notin q_Y \circ \alpha\left(\left[\frac{1}{2}, 1\right]\right)$ (Remark 7.1.5). Since $HS(f)$ is monotone, $HS(f)^{-1}\left(q_Y \circ \alpha\left(\left[\frac{1}{2}, 1\right]\right)\right)$ is a subcontinuum of $HS(X)$, and $F_X \notin HS(f)^{-1}\left(q_Y \circ \alpha\left(\left[\frac{1}{2}, 1\right]\right)\right)$, (∗) of Theorem 8.2.6. Thus, $q_X^{-1}\left(HS(f)^{-1}\left(q_Y \circ \alpha\left(\left[\frac{1}{2}, 1\right]\right)\right)\right)$ is a subcontinuum of $\mathcal{C}(X)$ and

$$\mathcal{F}_1(X) \cap q_X^{-1}\left(HS(f)^{-1}\left(q_Y \circ \alpha\left(\left[\frac{1}{2}, 1\right]\right)\right)\right) = \emptyset.$$

Let $x \in \bigcup q_X^{-1}\left(HS(f)^{-1}\left(q_Y \circ \alpha\left(\left[\frac{1}{2}, 1\right]\right)\right)\right)$. Then there exists an element A of $q_X^{-1}\left(HS(f)^{-1}\left(q_Y \circ \alpha\left(\left[\frac{1}{2}, 1\right]\right)\right)\right)$ such that $x \in A$. Thus, $HS(f)(q_X(A)) \in q_Y \circ \alpha\left(\left[\frac{1}{2}, 1\right]\right)$, and there exists $t \in \left[\frac{1}{2}, 1\right]$ such that $HS(f)(q_X(A)) = q_Y \circ \alpha(t)$. This implies that $\mathcal{C}(f)(A) = \alpha(t)$, and $f(A) \subset Q$. Hence, $x \in f^{-1}(Q)$. Therefore,

$$\bigcup q_X^{-1}\left(HS(f)^{-1}\left(q_Y \circ \alpha\left(\left[\frac{1}{2}, 1\right]\right)\right)\right) \subset f^{-1}(Q).$$

Note that, by construction, $K \subset \bigcup q_X^{-1}\left(HS(f)^{-1}\left(q_Y \circ \alpha\left(\left[\frac{1}{2}, 1\right]\right)\right)\right)$. Let us recall that $q_X^{-1}\left(HS(f)^{-1}\left(q_Y \circ \alpha\left(\left[\frac{1}{2}, 1\right]\right)\right)\right)$ is a subcontinuum of $\mathcal{C}(X)$ (see above). Thus, $\bigcup q_X^{-1}\left(HS(f)^{-1}\left(q_Y \circ \alpha\left(\left[\frac{1}{2}, 1\right]\right)\right)\right)$ is a subcontinuum of X, Corollary 6.1.2. Since K is a component of $f^{-1}(Q)$,

$$K = \bigcup q_X^{-1}\left(HS(f)^{-1}\left(q_Y \circ \alpha\left(\left[\frac{1}{2}, 1\right]\right)\right)\right).$$

Observe that $q_X^{-1}\left(HS(f)^{-1}\left(q_Y(Q)\right)\right) = q_X^{-1}\left(HS(f)^{-1}\left(q_Y \circ \alpha(1)\right)\right) \subset K$ and that:

$$Q = \mathcal{C}(f)\left(q_X^{-1}\left(HS(f)^{-1}\left(q_Y(Q)\right)\right)\right) =$$

$$f\left(q_X^{-1}\left(HS(f)^{-1}\left(q_Y(Q)\right)\right)\right) \subset f(K),$$

a contradiction. Therefore, $f(K) = Q$, and f is confluent.

Q.E.D.

8.4.2 Theorem *Let $f \colon X \to Y$ be a map between continua and let $n \in \mathbb{N}$. Then the following are equivalent:*

(1) f is monotone;
(2) $\mathcal{C}_n(f)$ is monotone;
(3) $HS_n(f)$ is monotone.

Proof Suppose f is monotone. Let $B \in \mathcal{C}_n(Y)$. Since f is monotone, $f^{-1}(B)$ has the same number of components as B. Let $A \in \mathcal{C}_n(f)^{-1}(B)$. Since $\mathcal{C}_n(f)(A) = f(A) = B$ and $A \subset f^{-1}(B)$, by Theorem 6.1.4, A intersects each component of $f^{-1}(B)$. Hence, by Theorem 1.8.20, there exists an order arc $\alpha \colon [0, 1] \to \mathcal{C}_n(X)$ such that $\alpha(0) = A$ and $\alpha(1) = f^{-1}(B)$. Since $B = f(A) \subset f(\alpha(t)) \subset f(f^{-1}(B)) = B$ for each $t \in [0, 1]$, $\alpha([0, 1]) \subset \mathcal{C}_n(f)^{-1}(B)$. Hence, $\mathcal{C}_n(f)^{-1}(B)$ is arcwise connected. Therefore, $\mathcal{C}_n(f)$ is monotone.

Suppose $\mathcal{C}_n(f)$ is monotone. Since the quotient map q_Y^n is monotone, by $(*)$ of Theorem 8.2.6 and Theorem 8.1.13, $HS_n(f)$ is monotone.

Suppose $HS_n(f)$ is monotone. We consider first the case $n = 1$. Let $y \in Y$. We show that $f^{-1}(y)$ is connected. Let $\{K_m\}_{m=1}^{\infty}$ be a decreasing sequence of nondegenerate subcontinua of Y such that $\bigcap_{m=1}^{\infty} K_m = \{y\}$. Hence, $\bigcap_{m=1}^{\infty} f^{-1}(K_m) = f^{-1}(\{y\})$.

Since for each positive integer m, K_m is a nondegenerate subcontinuum of Y and $HS(f)$ is monotone, we have that $\mathcal{C}(f)^{-1}(K_m)$ is connected (Remark 7.1.5 and $(*)$ of Theorem 8.2.6). Thus, we have that $\bigcup \mathcal{C}(f)^{-1}(K_m)$ is a subcontinuum of X (Corollary 6.1.2).

Since $HS(f)$ is monotone, by Lemma 8.4.1, f is confluent. Hence, by Theorem 8.3.1, $\bigcup \mathcal{C}(f)^{-1}(K_m) = f^{-1}(K_m)$ for every positive integer m. Therefore, $f^{-1}(K_m)$ is connected, and $f^{-1}(y)$ is connected also, Theorem 1.7.2. Thus, f is monotone.

Next, suppose $n \geq 2$. Let $B \in \mathcal{C}(Y) \setminus \mathcal{F}_1(Y)$. Then, by $(*)$ of Theorem 8.2.6 and our hypothesis, $\mathcal{C}_n(f)^{-1}(B)$ is connected. Since f is weakly confluent (by Theorem 8.2.8), there exists $A \in \mathcal{C}(X)$ such that $f(A) = B$. Hence, $A \in \mathcal{C}_n(f)^{-1}(B)$. This implies that $\mathcal{C}_n(f)^{-1}(B) \cap \mathcal{C}(X) \neq \emptyset$. Thus, $\bigcup \mathcal{C}_n(f)^{-1}(B) \in \mathcal{C}(X)$, Corollary 6.1.2. Note that $\bigcup \mathcal{C}_n(f)^{-1}(B) \subset f^{-1}(B)$. Let $x \in f^{-1}(B)$. Then $A \cup \{x\} \in \mathcal{C}_n(f)^{-1}(B)$ and $x \in \bigcup \mathcal{C}_n(f)^{-1}(B)$. Therefore, $\bigcup \mathcal{C}_n(f)^{-1}(B) = f^{-1}(B)$. In particular, $f^{-1}(B)$ is connected. Repeating the argument given for the case $n = 1$, we obtain that for each $y \in Y$, $f^{-1}(y)$ is connected. Therefore, f is monotone.

Q.E.D.

8.4.3 Lemma *Let Z be a compactum, let $n \in \mathbb{N}$ and let $P, Q \in C_n(Z)$. Then P and Q belong to the same component of $C_n(Z)$ if and only if for each component C of Z, $P \cap C \neq \emptyset$ is equivalent to $Q \cap C \neq \emptyset$.*

Proof Suppose that there exists a component C of Z such that $P \cap C \neq \emptyset$ and $Q \cap C = \emptyset$. Let U be an open and closed subset of Z such that $C \subset U$ and $U \cap Q = \emptyset$. Then the sets $\mathcal{U} = \{A \in C_n(Z) \mid A \cap U \neq \emptyset\}$ and $\mathcal{U}' = \{A \in C_n(Z) \mid A \subset Z \setminus U\}$ are disjoint open subsets of $C_n(Z)$, $C_n(Z) = \mathcal{U} \cup \mathcal{U}'$, $P \in \mathcal{U}$ and $Q \in \mathcal{U}'$. Therefore, P and Q do not belong to the same component of $C_n(Z)$.

Assume that for each component C of Z, $P \cap C \neq \emptyset$ if and only if $Q \cap C \neq \emptyset$. Let $R = \bigcup\{C \mid C$ is a component of Z and $P \cap C \neq \emptyset\}$. Note that $P \cup Q \subset R$ and each component of R intersects both P and Q. Hence, by Theorem 1.8.20, there exist order arcs from P to R and from Q to R. Therefore, P and Q belong to the same component of $C_n(Z)$.

Q.E.D.

8.4.4 Theorem *Let $f : X \twoheadrightarrow Y$ be a map between continua, and let $n \geq 3$. Then $C_n(f)$ is confluent if and only if f is monotone.*

Proof If f is monotone, then $C_n(f)$ is monotone (Theorem 8.4.2).

Suppose f is not monotone. Then there exist two points $p, q \in X$ such that $f(p) = f(q)$ and p and q lie in distinct components, say P and Q, of $f^{-1}(f(p))$, with $p \in P$ and $q \in Q$. Let $r \in X$ be such that $f(r) \neq f(p)$, and let R be a subcontinuum of X such that $r \in R$, $f(R)$ is nondegenerate, $f(p) \in Y \setminus f(R)$ and R is a component of $f^{-1}(f(R))$. Define a subcontinuum $\mathcal{K}$ of $C_n(Y)$ by

$$\mathcal{K} = \{\{f(p)\} \cup K \mid K \in C_{n-1}(f(R))\}.$$

Let $\mathcal{D}$ be the component of $C_n(f)^{-1}(\mathcal{K})$ containing $\{p, q, r\}$. Since $\mathcal{D}$ is a continuum, its union $Z = \sigma(\mathcal{D}) \in C_n(X)$, by Corollary 6.1.2. Note that $P \cup Q \cup R \subset Z$ and $\mathcal{D} \subset C_n(Z)$. We assert that P is a component of Z. To see this, let P' be the component of Z that contains P. Suppose that $P' \neq P$. Since P is a component of $f^{-1}(f(p))$, there exists a point $z \in P'$ such that $f(z) \neq f(p)$. Since $f(P') \subset \{f(p)\} \cup f(R)$, we have that $f(z) \in f(R)$. Since $f(p) \in f(P')$, we have that $f(P')$ is not connected, a contradiction. Hence, P is a component of Z. Similarly, Q is also a component of Z. Since $\mathcal{D}$ is a subcontinuum of $C_n(Z)$, $\mathcal{D}$ is contained in one of the components of $C_n(Z)$. Thus, by Lemma 8.4.3, for each element of $D \in \mathcal{D}$, $D \cap P \neq \emptyset$ and $D \cap Q \neq \emptyset$. Also, at most $n - 2$ components of D intersect R. Then $C_n(f)(\mathcal{D}) = \{\{f(p)\} \cup K \mid K \in C_{n-2}(R)\}$ is a proper subcontinuum of $\mathcal{K}$. Therefore, $C_n(f)$ is not confluent

Q.E.D.

The following example shows that Theorem 8.4.4 is the best possible. It also shows that for $n \geq 3$, $C_n(f)$ may not be confluent even when f is confluent and its range is locally connected.

8.4.5 Example Let $f \colon [-1, 1] \twoheadrightarrow [0, 1]$ be given by $f(t) = |t|$, where $|\cdot|$ denotes the absolute value. Then f is an open map; hence, confluent [37, (7.5), p. 148], which is not monotone. Note that, since $C_n(f)^{-1}(\{1\}) = \{\{1\}, \{-1\}, \{1, -1\}\}$, $C_n(f)$ is not monotone for $n \geq 2$. By Theorem 8.3.5, $C_2(f)$ is confluent and, by Theorem 8.4.4, $C_n(f)$ is not confluent for $n \geq 3$.

8.4.6 Theorem Let $f \colon X \twoheadrightarrow Y$ be a map between continua, and let $n \geq 3$. Then $C_n(f)$ is confluent if and only if $C_n(f)$ is monotone.

Proof If $C_n(f)$ is monotone, then clearly $C_n(f)$ is confluent. If $C_n(f)$ is confluent, then, by Theorem 8.4.4, f is monotone. Therefore, by Theorem 8.4.2, $C_n(f)$ is monotone.

Q.E.D.

8.4.7 Theorem Let $f \colon X \twoheadrightarrow Y$ be a map between continua and let $n \geq 3$. Then $HS_n(f)$ is confluent if and only if f is monotone.

Proof If f is monotone, by Theorem 8.4.2, $HS_n(f)$ is monotone; thus, confluent.

Suppose $HS_n(f)$ is confluent and f is not monotone. Hence, there exists a nondegenerate subcontinuum Q of Y such that $f^{-1}(Q)$ is not connected. Let D and E be two different components of $f^{-1}(Q)$. Let $p \in Y \setminus Q$. By Theorem 1.7.27, there exists a subcontinuum P of Y such that $\{p\} \subsetneq P \subset Y \setminus Q$. Let

$$\mathcal{K} = \{A \cup Q \mid A \in C_{n-1}(P)\}.$$

Then $\mathcal{K}$ is a subcontinuum of $C_n(Y)$. Let R be a component of $f^{-1}(P)$. By Corollary 8.3.10, f is confluent. Thus, $f(R) = P$.

Let $\mathcal{D}$ be the component of $C_n(f)^{-1}(\mathcal{K})$ such that $D \cup E \cup R \in \mathcal{D}$. By Corollary 6.1.2, $\bigcup \mathcal{D} \in C_n(X)$. Since $D \cup E \cup R \subset \bigcup \mathcal{D}$, there exists a component J of $\bigcup \mathcal{D}$ such that $D \subset J$. Note that $f(J) \subset P \cup Q$. Since $f(J)$ is connected and $f(J) \cap Q \neq \emptyset$, $f(J) \subset Q$. Hence, since D is a component of $f^{-1}(Q)$, $J = D$. Thus, D is a component of $\bigcup \mathcal{D}$. Similarly, one shows that E is a component of $\bigcup \mathcal{D}$.

Now, if $A \in \mathcal{D}$, then, by Lemma 6.1.4, each component of $\bigcup \mathcal{D}$ intersects A. Hence, $f(A)$ has at most $n - 2$ components in P. Thus, $C_n(f)(\mathcal{D}) \subset \{A \cup Q \mid A \in C_{n-2}(P)\}$ and $C_n(f)(\mathcal{D}) \neq \mathcal{K}$. Since $\mathcal{K} \cap \mathcal{F}_n(Y) = \emptyset$, $q_X^n(\mathcal{D})$ is a component of $HS_n(f)^{-1}(q_Y^n(\mathcal{K}))$ and $HS_n(f)(q_X^n(\mathcal{D})) \neq q_Y^n(\mathcal{K})$, a contradiction to the fact that $HS_n(f)$ is confluent.

Q.E.D.

8.4.8 Definition A surjective map $f \colon X \twoheadrightarrow Y$ between continua is a CE-map provided that $f^{-1}(y)$ has trivial shape for each $y \in Y$.

8.4.9 Remark Note that, by [24, 2.1 (A)], a compactum has trivial shape if and only if it is the intersection of a decreasing sequence of absolute retracts. Since absolute retracts are connected [25, Theorem 7, p. 341], a space with trivial shape is connected (Theorem 1.7.2). Hence, CE-maps are monotone.

8.4.10 Theorem *Let $f: X \twoheadrightarrow Y$ be a map between continua and let $n \in \mathbb{N}$. Then f is monotone if and only if $C_n(f)$ is a CE-map.*

Proof Assume f is monotone. Then $C_n(f)$ is monotone, Theorem 8.4.2. Let $B \in C_n(Y)$, and let $\Gamma = C_n(f)^{-1}(B)$. Thus, Γ is a subcontinuum of $C_n(X)$. By Theorem 6.2.3, $C(\Gamma)$ has trivial shape. Recall that the union map σ is continuous (Lemma 1.8.11). In particular, $\sigma: C(\Gamma) \twoheadrightarrow \Gamma$ is continuous. Hence, given $\mathcal{A} \in C(\Gamma)$, we have that $\sigma(\mathcal{A}) \in C_n(X)$, Corollary 6.1.2. In fact, $\sigma(\mathcal{A}) \in \Gamma$. Observe that for each $G \in \Gamma$, $\sigma(\{G\}) = G$. Thus, σ is an r-map from $C(\Gamma)$ onto Γ. Since trivial shape in an invariant under r-maps [2, (5.11), p. 103], Γ has trivial shape. Therefore, $C_n(f)$ is a CE-map.

Suppose $C_n(f)$ is a CE-map. Then, by Remark 8.4.9, $C_n(f)$ is monotone. Therefore, f is monotone, by Theorem 8.4.2.

Q.E.D.

The following example presents a monotone map r whose induced map to the hyperspace suspensions is an CE-map. It is not known if this is the case in general.

8.4.11 Example Let $\mathcal{S}^1$ denote the unit circle in the Euclidean plane $\mathbb{R}^2$. Let $r: \mathcal{S}^1 \times [0, 1] \twoheadrightarrow \mathcal{S}^1$ be given by $r((z, t)) = z$. Note that for each $z \in \mathcal{S}^1$, $r^{-1}(z) = \{z\} \times [0, 1]$. Hence, r is a monotone map. It is easy to see that, in this case, $C(r)^{-1}(\{z\}) = C(\{z\} \times [0, 1])$. Hence, $C(r)^{-1}(\mathcal{F}_1(\mathcal{S}^1)) = \bigcup \{C(r)^{-1}(\{z\}) \mid z \in \mathcal{S}^1\} = \bigcup \{C(\{z\} \times [0, 1]) \mid z \in \mathcal{S}^1\}$. Thus, $C(r)^{-1}(\mathcal{F}_1(\mathcal{S}^1))$ is homeomorphic to a solid torus. Observe that $\mathcal{F}_1(\mathcal{S}^1 \times [0, 1])$ is in the manifold boundary of $C(r)^{-1}(\mathcal{F}_1(\mathcal{S}^1))$. This implies that $q^1_{\mathcal{S}^1 \times [0,1]}(C(r)^{-1}(\mathcal{F}_1(\mathcal{S}^1))) = HS(r)^{-1}(F^1_{\mathcal{S}^1})$ is homeomorphic to the solid of revolution obtained by rotating the set $\{(x, y) \in \mathbb{R}^2 \mid (x - 1)^2 + y^2 \leq 1\}$ about the y axis. Note that this space is contractible. In particular, it has trivial shape [24, 2.1 (A)]. Therefore, by Theorem 8.4.10, $HS(r)$ is a CE-map.

8.4.12 Theorem *If $f: X \twoheadrightarrow Y$ is a monotone map between locally connected continua without free arcs, and $n \in \mathbb{N}$, then $C_n(f)$ is a near-homeomorphism.*

Proof First note that $C_n(X)$ and $C_n(Y)$ are homeomorphic to the Hilbert cube, by Theorem 6.1.8 . Since f is monotone, $C_n(f)$ is a CE-map, by Theorem 8.4.10. Hence, by [31, 7.8.4 and 7.5.7], $C_n(f)$ is a near-homeomorphism.

Q.E.D.

8.4.13 Remark Observe that [16, Example 4] shows that an induced map $C_n(f)$ may be a near-homeomorphism while f is not.

A proof of the following result may be found in [34, (2.2)].

8.4.14 Lemma *Let $f: X \twoheadrightarrow Y$ be a monotone map between continua. Then, for each $m \in \mathbb{N}$, there exist locally connected continua X_m and Y_m and a map $F: X_1 \to Y_1$ such that*

(1) $X_{m+1} \subset X_m$ and $Y_{m+1} \subset Y_m$ for each $m \in \mathbb{N}$;
(2) $\bigcap_{m=1}^{\infty} X_m = X$ and $\bigcap_{m=1}^{\infty} Y_m = Y$;
(3) the restriction $F|_{X_m}: X_m \to Y_m$ is a monotone surjection for each $m \in \mathbb{N}$;
(4) $F|_X = f$.

8.4.15 Theorem *Let $f: X \twoheadrightarrow Y$ be a monotone map from a continuum X onto a locally connected continuum Y, and let $n \in \mathbb{N}$. Then $\mathcal{C}_n(f)$ is universal.*

Proof Let $G: \mathcal{C}_n(X) \to \mathcal{C}_n(Y)$ be any map. Let X_m, Y_m and F be as in Lemma 8.4.14. Since Y is a locally connected continuum, $\mathcal{C}_n(Y)$ is an absolute retract [38, Théorème II$_m$]. Thus, there exists a map $\mathfrak{G}: \mathcal{C}_n(X_1) \to \mathcal{C}_n(Y)$ extending G. Fix $m \in \mathbb{N}$. According to Lemma 8.4.14, the continua X_m and Y_m are locally connected, and $F|_{X_m} = F_m$ is a monotone map from X_m onto Y_m. Hence, $\mathcal{C}_n(X_m)$ and $\mathcal{C}_n(Y_m)$ are absolute retracts [38, Théorème II$_m$] and, by Theorem 8.4.10, the induced map $\mathcal{C}_n(F_m): \mathcal{C}_n(X_m) \twoheadrightarrow \mathcal{C}_n(Y_m)$ is a surjective CE-map. By [33, Corollary 3.10] $\mathcal{C}_n(F_m)$ is universal. Since, for each $m \in \mathbb{N}$, the restriction $\mathfrak{G}|_{\mathcal{C}_n(X_m)}$ maps $\mathcal{C}_n(X_m)$ into $\mathcal{C}_n(Y_m)$, there exists $A_m \in \mathcal{C}_n(X_m)$ such that $\mathcal{C}_n(F_m)(A_m) = \mathfrak{G}(A_m)$, $m \in \mathbb{N}$. Note that for each $m \in \mathbb{N}$, $X_m \subset X_1$, $A_m \in \mathcal{C}_n(X_1)$, and $\mathcal{C}_n(F)(A_m) = \mathfrak{G}(A_m)$. Since $\mathcal{C}_n(X_1)$ is compact, the sequence $\{A_m\}_{m=1}^{\infty}$ has a convergent subsequence $\{A_{m_k}\}_{k=1}^{\infty}$. Let $A = \lim_{k \to \infty} A_{m_k}$. Since $\mathcal{C}_n(F)(A_{m_k}) = \mathfrak{G}(A_{m_k})$, by the continuity of $\mathcal{C}_n(F)$ and $\mathfrak{G}$, we obtain that $\mathcal{C}_n(F)(A) = \mathfrak{G}(A)$. Since $A_{m_k} \in \mathcal{C}_n(X_{m_k})$ for each $k \in \mathbb{N}$, and since $\bigcap_{k=1}^{\infty} X_{m_k} = X$, we have $A \in \mathcal{C}_n(X)$. Hence, $\mathfrak{G}(A) = G(A)$ and, since $F|_X = f$, it follows that $\mathcal{C}_n(F)(A) = \mathcal{C}_n(f)(A)$, $\mathcal{C}_n(f)(A) = G(A)$. Therefore, $\mathcal{C}_n(f)$ is universal.

$$\textbf{Q.E.D.}$$

The next results are applications of Theorem 8.4.15 and inverse limits.

8.4.16 Theorem *Let $\{X_m, f_m^{m+1}\}$ be an inverse sequence of locally connected continua whose inverse limit is X_∞. Suppose that for each $m \in \mathbb{N}$, there exists a subcontinuum Y_{m+1} of X_{m+1} with the property that $f_m^{m+1}|_{Y_{m+1}}$ maps Y_{m+1} monotonely onto X_m. Then, for each $n \in \mathbb{N}$, $\mathcal{C}_n(X_\infty)$ has the fixed point property.*

Proof By the hypothesis, for each $j \leq k$, it is not difficult to find a subcontinuum Z_k of X_k such that $f_j^k|_{Z_k}$ maps Z_k monotonely onto X_j. By Theorem 8.4.15, each $\mathcal{C}_n(f_j^k|_{Z_k}): \mathcal{C}_n(Z_k) \twoheadrightarrow \mathcal{C}_n(X_j)$ is universal. Thus, since each $\mathcal{C}_n(f_j^k|_{Z_k}) = \mathcal{C}_n(f_j^k)|_{\mathcal{C}_n(Z_k)}$, we obtain that, each $\mathcal{C}_n(f_j^k)|_{\mathcal{C}_n(Z_k)}: \mathcal{C}_n(Z_k) \to \mathcal{C}_n(X_j)$ is universal. Hence, each $\mathcal{C}_n(f_j^k): \mathcal{C}_n(X_k) \twoheadrightarrow \mathcal{C}_n(X_j)$ is universal (Proposition 2.6.4). Thus, since each $\mathcal{C}_n(X_m)$ is an absolute retract [38, Théorème II$_m$], $\varprojlim\{\mathcal{C}_n(X_m), \mathcal{C}_n(f_m^{m+1})\}$ has the fixed point property, Theorem 2.6.14. Thus, since

$\varprojlim\{\mathcal{C}_n(X_m), \mathcal{C}_n(f_m^{m+1})\}$ is homeomorphic to $\mathcal{C}_n(X_\infty)$, Theorem 2.3.4, $\mathcal{C}_n(X_\infty)$ has the fixed point property.

Q.E.D.

8.4.17 Definition A *dendrite* is a locally connected dendroid.

8.4.18 Lemma *Let X be a locally connected continuum and let D be a dendrite. If $f\colon X \twoheadrightarrow D$ is a quasi-monotone map, then there exists a subcontinuum Z of X such that $f|_Z$ maps Z monotonely onto D.*

Proof Let $f\colon X \twoheadrightarrow D$ be a quasi-monotone map. By Theorem 8.1.4, there exist a continuum Z, a monotone map $g\colon X \twoheadrightarrow W$ and a light open map $h\colon W \twoheadrightarrow D$ such that $f = h \circ g$. By [37, (2.4), p. 188], there exists a dendrite E of W such that $h|_E\colon E \twoheadrightarrow D$ is a homeomorphism. Let $Z = g^{-1}(E)$. Then $f|_Z$ maps Z monotonely onto D.

Q.E.D.

8.4.19 Theorem *Let $\{D_m, f_m^{m+1}\}$ be an inverse sequence of dendrites with quasi-monotone bonding maps whose inverse limit is D_∞. Then, for each $n \in \mathbb{N}$, $\mathcal{C}_n(D_\infty)$ has the fixed point property.*

Proof Observe that, by Lemma 8.4.18, for each $m \in \mathbb{N}$, there exits a subcontinuum Z_{m+1} of D_{m+1} such that $f_m^{m+1}|_{Z_{m+1}}$ maps Z_{m+1} monotonely onto D_m. Thus, the result now follows from Theorem 8.4.16.

Q.E.D.

8.4.20 Definition A surjective map $f\colon X \twoheadrightarrow Y$ between continua is said to be *monotonely refinable* if for each $\varepsilon > 0$, there exists a monotone and ε-map $g\colon X \twoheadrightarrow Y$ such that $d(f(x), g(x)) < \varepsilon$ for every $x \in X$.

8.4.21 Theorem *Let X and Y be continua and let $n \in \mathbb{N}$. If $f\colon X \twoheadrightarrow Y$ is monotonely refinable, then both $\mathcal{C}_n(f)$ and $HS_n(f)$ are monotonely refinable.*

Proof Note that, by Theorem 8.2.8, $\mathcal{C}_n(f)$ and $HS_n(f)$ are surjective. Let $\varepsilon > 0$. Since f is monotonely refinable, there exists a monotone and ε-map $g\colon X \twoheadrightarrow Y$ such that $d(f(x)), g(x)) < \varepsilon$ for each $x \in X$. Then, by Theorem 8.4.2 and Lemma 8.2.11, $\mathcal{C}_n(g)$ is a monotone and ε-map.

By Lemma 8.2.13, $\mathcal{H}(\mathcal{C}_n(f)(A), \mathcal{C}_n(g)(A)) < \varepsilon$ for all $A \in \mathcal{C}_n(X)$. Therefore, $\mathcal{C}_n(f)$ is monotonely refinable.

By Theorems 8.4.2 and 8.2.12, $HS_n(g)$ is a monotone and ε-map. Now, by Lemma 8.2.13, $\rho_Y^n(HS_n(f)(\chi), HS_n(g)(\chi)) < \varepsilon$ for every $\chi \in HS_n(X)$. Therefore, $HS_n(f)$ is monotonely refinable.

Q.E.D.

8.5 Open Maps

We show that if the induced map to the n-fold hyperspace of a map f is open, then f itself is open. The reverse implication is not true. We study relationships between an open map and the fact that its induced maps are monotone or r-maps. We present some fixed point theorems. We study OM-maps. We prove that for map f, if $n \geq 3$ and $HS_n(f)$ is open, then f is a homeomorphism. Also, if $n \geq 2$ and $C_n(f)$ is open, then, again, f is a homeomoprhism. To this end, we follow [1, 4, 5, 11, 13, 21, 32, 35, 37].

8.5.1 Theorem *Let X be a map between continua, and let $n \in \mathbb{N}$. If $C_n(f)$ is an open map, then f is open.*

Proof Let U be an open subset of X and let $x \in U$. Then $\langle U \rangle_n$ is an open subset of $C_n(X)$. Since $C_n(f)$ is an open map, $C_n(f)(\langle U \rangle_n)$ is an open subset of $C_n(Y)$. Observe that $C_n(f)(\langle U \rangle_n) \subset \langle f(U) \rangle_n$. Since $\{f(x)\} \in C_n(f)(\langle U \rangle_n)$, there exist open subsets $V_1, \ldots, V_m$ of Y such that $\{f(x)\} \in \langle V_1, \ldots, V_m \rangle_n \subset \langle f(U) \rangle_n$. Note that $\bigcap_{j=1}^{m} V_j \subset f(U)$. Hence, $f(x)$ is an interior point of $f(U)$. Therefore, $f(U)$ is an open subset of Y, and f is an open map.

Q.E.D.

The following example shows that the converse of Theorem 8.5.1 is not true.

8.5.2 Example Let $f \colon [-1, 1] \twoheadrightarrow [0, 1]$ be given by $f(t) = |t|$, where $|\cdot|$ is the absolute value. Then f is an open map. Let $U_1 = \left[-1, -\frac{1}{3}\right)$, let $U_2 = \left(-\frac{1}{2}, \frac{1}{2}\right)$ and let $U_3 = \left(\frac{1}{3}, 1\right]$. Observe that $[0, 1] \in C(f)(\langle U_1, U_2, U_3 \rangle_1)$. We prove that $[0, 1]$ is not an interior point of $C(f)(\langle U_1, U_2, U_3 \rangle_1)$. To this end, let $\delta \in \left(0, \frac{1}{3}\right)$. Then $\left[\frac{\delta}{2}, 1\right] \in C([0, 1])$ and $\mathcal{H}\left(\left[\frac{\delta}{2}, 1\right], [0, 1]\right) < \delta$, but there does not exist an element $K \in \langle U_1, U_2, U_3 \rangle_1$ such that $f(K) = \left[\frac{\delta}{2}, 1\right]$. Hence, $[0, 1]$ is not an interior point of $C(f)(\langle U_1, U_2, U_3 \rangle_1)$. A similar argument proves that $C_n(f)$ is not open for any $n \geq 2$.

8.5.3 Theorem *Let a continuum X be locally connected, and let a map $f \colon X \twoheadrightarrow Y$ be such that for some $n \in \mathbb{N}$ the induced map $C_n(f)$ is open. Then f is monotone.*

Proof Suppose the result is not true; i.e., f is not monotone. Thus, there exist two points p and q of X satisfying $f(p) = f(q)$ which belong to distinct components of $f^{-1}(f(p))$. By continuity of f, there exists a positive ε such that for every $L \in C_n(Y)$ with $f(p) \in L$ and $\mathcal{H}(L, \{f(p)\}) < \varepsilon$, the components of $f^{-1}(L)$ containing p and q, respectively, are distinct. By local connectedness of Y [37, (1.51), p. 26], there exists a subcontinuum W of Y such that $f(p) \in Int(W)$ and $\mathcal{H}(W, \{f(p)\}) < \varepsilon$. Let U_p and U_q be the components of $f^{-1}(W)$ containing p and q, respectively. Since in locally connected continua components of open sets are open (Lemma 1.7.11), we see that $p \in Int(U_p)$ and $q \in Int(U_q)$. Let $\delta > 0$ be such that $\mathcal{V}_\delta^d(p) \subset U_p$ and $\mathcal{V}_\delta^d(q) \subset U_q$. Let $\mathcal{B} = \alpha([0, 1])$ be an order arc in

$\mathcal{C}(Y)$ from $\{f(p)\}$ to Y through W (Theorem 1.8.20). Define $\mathcal{A}$ as the subset of $\mathcal{B}$ composed of all the elements L of $\mathcal{B}$ such that the component of $f^{-1}(L)$ containing p is distinct from the component of $f^{-1}(L)$ containing q. Observe that $W \in \mathcal{A}$ and that if L and L' belong to $\mathcal{B}$, L belongs to $\mathcal{A}$, and $L' \subset L$, then L' also belongs to $\mathcal{A}$. Hence, $\mathcal{A}$ is a connected subset of $\mathcal{B}$ containing $\{f(p)\}$ and W. Since $\mathcal{B} \setminus \mathcal{A}$ is closed, $\mathcal{A}$ is open in $\mathcal{B}$. Let $t_0 = \sup\{t \in [0,1] \mid \alpha(t) \in \mathcal{A}\} = \inf\{t \in [0,1] \mid \alpha(t) \in \mathcal{B} \setminus \mathcal{A}\}$, and let $R = \alpha(t_0)$. Then $R \in Cl_{\mathcal{C}_n(Y)}(\mathcal{A}) \setminus \mathcal{A}$. Let P denote the component of $f^{-1}(R)$ containing both p and q. Since $\mathcal{C}_n(f)$ is open, f is open (Theorem 8.5.1). Hence, it follows that $f(P) = R$ (each open map is confluent, Diagram I).

We show that $\mathcal{C}_n(f)\left(\mathcal{V}_\delta^{\mathcal{H}}(P) \cap \mathcal{C}_n(X)\right)$ is not open in $\mathcal{C}_n(Y)$. Assume this is not true. Let $t_1 \in [0,1]$ be such that $\alpha(t_1) = W$. Then $t_1 < t_0$. Choose $\eta > 0$ such that

$$\mathcal{V}_\eta^{\mathcal{H}}(R) \cap \mathcal{C}_n(Y) \subset \mathcal{C}_n(f)\left(\mathcal{V}_\delta^{\mathcal{H}}(P) \cap \mathcal{C}_n(X)\right) \setminus \alpha([0, t_1]).$$

Let $R_1 \in \mathcal{V}_\eta^{\mathcal{H}}(R) \cap \mathcal{C}_n(Y) \cap \mathcal{A}$. Then $W \subset R_1 \subset R \neq R_1$. Fix different points $y_1, \ldots, y_{n-1} \in R \setminus R_1$, and let $R_2 = R_1 \cup \{y_1, \ldots, y_{n-1}\} \in \mathcal{C}_n(Y)$. By the definition of the Hausdorff metric, we have $\mathcal{H}(R_2, R) < \eta$. Thus, there exists $K \in \mathcal{V}_\delta^{\mathcal{H}}(P) \cap \mathcal{C}_n(X)$ such that $f(K) = R_2$. Then K has n components $K_1, \ldots, K_n$ satisfying $f(K_1) = \{y_1\}, \ldots, f(K_{n-1}) = \{y_{n-1}\}$, and $f(K_n) = R_1$. Since $\mathcal{H}(K, P) < \delta$, there exists a point $x \in K$ such that $d(x, q) < \delta$. Then $x \in U_q$ and $f(x) \in W \subset R_1$. Hence, $x \in K_n$. Therefore, $U_q \cap K_n \neq \emptyset$. Similarly, $U_p \cap K_n \neq \emptyset$. Thus, $U_p \cup K_n \cup U_q$ is a connected subset of $f^{-1}(R_1)$ containing p and q. This contradicts the fact that $R_1 \in \mathcal{A}$.

Therefore, f is monotone.

Q.E.D.

As a consequence of Theorems 8.4.2 and 8.5.3, we have the following:

8.5.4 Corollary *Let X be a locally connected continuum, and let $f \colon X \twoheadrightarrow Y$ be a map such that for some $n \in \mathbb{N}$ the induced map $\mathcal{C}_n(f)$ is open. Then f is both monotone and open.*

8.5.5 Theorem *Let $f \colon X \twoheadrightarrow Y$ be an open map between compacta. Then the function $\Im(f) \colon 2^Y \to 2^X$ given by $\Im(f)(B) = f^{-1}(B)$ is continuous.*

Proof To see that $\Im(f)$ is continuous, let $\{B_m\}_{m=1}^\infty$ be a sequence of points of 2^Y converging to an element $B \in 2^Y$. Since f is open, 2^f is open [21, Theorem 4.3]. Thus, the sequence $\{(2^f)^{-1}(B_m)\}_{m=1}^\infty$ converges to $(2^f)^{-1}(B)$ (Theorem 8.1.18). Hence, the sequence $\{\sigma((2^f)^{-1}(B_m))\}_{m=1}^\infty$ converges to $\sigma((2^f)^{-1}(B))$, Lemma 1.8.11. It is easy to see that for each $D \in 2^Y$, the equality $\sigma((2^f)^{-1}(D)) = \Im(f)(D)$ holds. Thus, the sequence $\{\Im(f)(B_m)\}_{m=1}^\infty$ converges to $\Im(f)(B)$. Therefore, $\Im(f)$ is continuous.

Q.E.D.

8.5.6 Theorem *Let $f \colon X \twoheadrightarrow Y$ be an open and monotone surjective map between continua, and let $n \in \mathbb{N}$. Then the induced map $\mathcal{C}_n(f)$ is an r-map.*

Proof Let $h_n = \Im(f)|_{C_n(Y)}$. Then, by Theorem 8.5.5, h_n is continuous. Since f is monotone, $h_n(C_n(Y)) \subset C_n(X)$. Hence, h_n is well defined. Observe that $C_n(f) \circ h_n = 1_{C_n(Y)}$, where $1_{C_n(Y)}$ is the identity map on $C_n(Y)$. Therefore, $C_n(f)$ is an r-map.

Q.E.D.

Since the image of a contractible space under an r-map is contractible [1, Theorem 13.2, p. 26], we have the following result:

8.5.7 Corollary *Let X and Y be continua, and let $n \in \mathbb{N}$. If $f: X \twoheadrightarrow Y$ is a monotone and open map and $C_n(X)$ is contractible, then $C_n(Y)$ is contractible.*

Since the fixed point property is an r-invariant [1, Theorem 7.1, p. 18], we obtain:

8.5.8 Corollary *Let X and Y be continua, and let $n \in \mathbb{N}$. If $f: X \twoheadrightarrow Y$ is a monotone and open map and $C_n(X)$ has the fixed point property, then $C_n(Y)$ has the fixed point property.*

The two assumptions on the map f (i.e., monotoneity and openness) are essential in Theorem 8.5.6, even for $n = 1$. This can be seen by the next examples.

8.5.9 Example There exists a monotone and not open map $f: [0, 1] \twoheadrightarrow [0, 1]$ such that the induced map $C(f)$ is not an r-map. Define $f: [0, 1] \to [0, 1]$ by

$$f(t) = \begin{cases} \frac{3}{2}t, & \text{if } t \in [0, \frac{1}{3}); \\ \frac{1}{2}, & \text{if } t \in [\frac{1}{3}, \frac{2}{3}]; \\ \frac{3}{2}t - \frac{1}{2}, & \text{if } t \in (\frac{2}{3}, 1]. \end{cases}$$

Suppose that $C(f)$ is an r-map. Then there exists a map

$$\mathfrak{G}: C([0, 1]) \to C([0, 1])$$

such that $C(f) \circ \mathfrak{G} = 1_{C([0,1])}$. Let $B \in C([0, 1])$. Since $C(f) \circ \mathfrak{G}(B) = B$, $\mathfrak{G}(B) \subset f^{-1}(B)$. In particular, $\mathfrak{G}(\{\frac{1}{2}\}) \in C([\frac{1}{3}, \frac{2}{3}])$. Also, the partial maps $C(f)|_{\mathcal{F}_1([0,\frac{1}{3}])}: \mathcal{F}_1([0, \frac{1}{3}]) \twoheadrightarrow \mathcal{F}_1([0, \frac{1}{2}])$ and $C(f)|_{\mathcal{F}_1([\frac{2}{3},1])}: \mathcal{F}_1([\frac{2}{3}, 1]) \twoheadrightarrow \mathcal{F}_1([\frac{1}{2}, 1])$ are homeomorphisms. Hence, we have that the partial maps $\mathfrak{G}|_{\mathcal{F}_1([0,\frac{1}{2}])}: \mathcal{F}_1([0, \frac{1}{2}]) \twoheadrightarrow \mathcal{F}_1([0, \frac{1}{3}])$ and $\mathfrak{G}|_{\mathcal{F}_1([\frac{1}{2},1])}: \mathcal{F}_1([\frac{1}{2}, 1]) \twoheadrightarrow \mathcal{F}_1([\frac{2}{3}, 1])$ are also homeomorphisms. In the domain space, $\mathcal{F}_1([0, 1])$, of $\mathfrak{G}$, consider the two sequences $\{\{\frac{1}{2} - \frac{1}{2n}\}\}_{n=1}^{\infty}$ and $\{\{\frac{1}{2} + \frac{1}{2n}\}\}_{n=1}^{\infty}$. Note that the sequences converge to $\{\frac{1}{2}\}$, from the left and from the right, respectively. Thus, by the continuity of $\mathfrak{G}$, the value of $\mathfrak{G}(\{\frac{1}{2}\})$ must be $\{\frac{1}{3}\}$ and $\{\frac{2}{3}\}$ simultaneously, a contradiction.

8.5.10 Example There exists an open and not monotone map $f: S^1 \twoheadrightarrow S^1$ such that $C(f)$ is not an r-map. Define : $S^1 \twoheadrightarrow S^1$ by $f(z) = z^2$. Note that, by [32, (0.55)], $C(S^1)$ may be seen as the unit disk $\mathcal{B} = \{z \in \mathbb{C} \mid ||z|| \leq 1\}$, where its manifold boundary, $\partial(\mathcal{B}) = \{z \in \mathbb{C} \mid ||z|| = 1\}$, corresponds to $\mathcal{F}_1(S^1)$ and its centre, $\{0\}$, corresponds to $\mathcal{S}^1$. Let $\mathfrak{D} = \{A \in \mathcal{C}(\mathcal{S}^1) \mid \text{length}(A) \geq \pi\}$. Then $\mathfrak{D} = \{z \in \mathbb{C} \mid \|z\| \leq \frac{1}{2}\}$, $\mathfrak{D} = \mathcal{C}(f)^{-1}(\{0\})$ and $\mathcal{C}(f)|_{\mathcal{C}(\mathcal{S}^1)\setminus\mathfrak{D}} \colon \mathcal{C}(\mathcal{S}^1)\setminus\mathfrak{D} \twoheadrightarrow \mathcal{C}(\mathcal{S}^1)\setminus\{\{0\}\}$ is a 2-to-1 map. Suppose $\mathcal{C}(f)$ is an r-map. Then there exists a map $\mathfrak{G} \colon \mathcal{C}(\mathcal{S}^1) \to \mathcal{C}(\mathcal{S}^1)$ such that $\mathcal{C}(f) \circ \mathfrak{G} = 1_{\mathcal{C}(\mathcal{S}^1)}$. Let $B \in \mathcal{C}(\mathcal{S}^1)$. Since $\mathcal{C}(f) \circ \mathfrak{G}(B) = B$, $\mathfrak{G}(B) \subset f^{-1}(B)$. In particular, we have that $\mathfrak{G}(\{0\}) \in \mathfrak{D}$. In the domain space of $\mathfrak{G}$, consider the sequence $\{\{\frac{1}{n}\}\}_{n=1}^{\infty}$, which converges to $\{0\}$ from the right. Hence, by the continuity of $\mathfrak{G}$, $\mathfrak{G}(\{0\}) \in \{\frac{1}{2}, -\frac{1}{2}\}$. Meanwhile, the sequence $\{\{-\frac{1}{n}\}\}_{n=1}^{\infty}$, also in the domain of $\mathfrak{G}$, converges to $\{0\}$ from the left. Since $\mathcal{C}(f)^{-1}(\{-\frac{1}{n}\}) \in \{ti \mid t \in [\frac{1}{2}, 1]\} \cup \{-ti \mid t \in [\frac{1}{2}, 1]\}$, we obtain that $\mathfrak{G}(\{0\}) \in \{\frac{1}{2}i, -\frac{1}{2}i\}$. Hence, we have a contradiction.

The next result is consequence of Theorem 8.4.19.

8.5.11 Theorem *Let $\{D_m, f_M^{m+1}\}$ be an inverse sequence of dendrites with open bonding maps whose inverse limit is D_∞. Then, for each $n \in \mathbb{N}$, $\mathcal{C}_n(D_\infty)$ has the fixed point property.*

Proof Since, by [37, (8.11), p. 152], any open map defined on a locally connected continuum is quasi-monotone, the theorem follows from Theorem 8.4.19.

                                                                            Q.E.D.

8.5.12 Definition A continuum X is a *Knaster type continuum* if X is the inverse limit of an inverse sequence of arcs with surjective open bonding maps.

The next theorem shows that the n-fold hyperspace of a Knaster type continuum has the fixed point property and it follows from Theorem 8.5.11.

8.5.13 Theorem *Let $\{X_m, f_m^{m+1}\}$ be an inverse sequence of arcs with open bonding maps, whose inverse limit is X_∞. Then, for each $n \in \mathbb{N}$, $\mathcal{C}_n(X_\infty)$ has the fixed point property.*

It is known from [21, Theorem 5.2] that for each continuum X and a map $f \colon X \to Y$ the induced map $\mathcal{C}(f) \colon \mathcal{C}(X) \to \mathcal{C}(Y)$ is an OM-map if and only if f is an OM-map. For $n \geq 2$ we only have one implication (Theorem 8.5.14), the proof of which is patterned after the corresponding part of the proof of [21, Theorem 5.2]. It is important to remark that the class of OM-maps is the first known class of maps for which $n \leq 2$ matters in the implication: *If f is an OM-map, then $\mathcal{C}_n(f)$ is an OM-map.* As we show in Theorem 8.5.15 and Example 8.5.16, this implication holds for $n = 2$ and it does not hold for $n \geq 3$. We start with the two theorems.

8.5.14 Theorem *Let $f \colon X \twoheadrightarrow Y$ be a map between continua, and let $n \in \mathbb{N}$. If $\mathcal{C}_n(f)$ is an OM-map, then f is an OM-map.*

Proof Let $n \in \mathbb{N}$ and let $\{y_m\}_{m=1}^{\infty}$ be a sequence of points in Y that converges to a point $y \in Y$. Clearly the sequence of singletons $\{\{y_m\}\}_{m=1}^{\infty}$, considered as a sequence of elements of $\mathcal{C}_n(Y)$, converges to $\{y\} \in \mathcal{C}_n(Y)$. Let K be a component of $f^{-1}(y)$. Then $\mathcal{C}_n(K)$, considered as a subset of $\mathcal{C}_n(X)$, is a component of $\mathcal{C}_n(f)^{-1}(\{y\})$, Theorem 8.2.4. Since $\mathcal{C}_n(f)$ is an OM-map, by Theorem 8.1.17, we have that, for each $m \in \mathbb{N}$, there exists $K_m \in C_n(f)^{-1}(\{y_m\})$ such that some subsequence of the sequence $\{K_m\}_{m=1}^{\infty}$ converges to an element of $C_n(K)$. Since $K_m \subset f^{-1}(y_m)$, we obtain that $\limsup f^{-1}(y_m) \cap K \neq \emptyset$. Therefore, applying Theorem 8.1.17 once more, we conclude that f is an OM-map.

Q.E.D.

8.5.15 Theorem *Let $f \colon X \twoheadrightarrow Y$ be an OM-map between continua. Then the induced map $C_2(f)$ is an OM-map.*

Proof We use the characterization mentioned in Theorem 8.1.17. By [21, Theorem 5.2], the map $C(f) \colon C(X) \to C(Y)$ is an OM-map.

Claim 1 Let $P \in C(Y)$, let E be a component of $f^{-1}(P)$, and let $\{P_m\}_{m=1}^{\infty}$ be a sequence in $C(Y)$ converging to P. Then

$$\left(\limsup C(f)^{-1}(P_m) \right) \cap C(f)^{-1}(P) \cap C(E) \neq \emptyset.$$

To show Claim 1, note that, since f is a composition of two confluent maps, f itself is confluent (Diagram I). Thus, $f(E) = P$. Let $\mathcal{K}$ be the component of $C(f)^{-1}(P)$ containing E. Since $C(f)$ is an OM-map, by Theorem 8.1.17, there exists $A \in \left(\limsup C(f)^{-1}(P_m) \right) \cap \mathcal{K}$. Let $K = \bigcup \mathcal{K}$. By Corollary 6.1.2, K is a subcontinuum of X. Clearly $f(K) = P$ and $E \subset K$. This implies that $K = E$. Hence, $A \in \left(\limsup C(f)^{-1}(P_m) \right) \cap C(f)^{-1}(P) \cap C(E)$. Therefore, $\left(\limsup C(f)^{-1}(P_m) \right) \cap C(f)^{-1}(P) \cap C(E) \neq \emptyset$.

Claim 2 Let $P, Q \in C(Y)$, let E and F be components of $f^{-1}(P)$ and $f^{-1}(Q)$, respectively, and let $\{P_m\}_{m=1}^{\infty}$ and $\{Q_m\}_{m=1}^{\infty}$ be sequences in $C(Y)$ converging to P and to Q, respectively. Then there exist $A \in C(E)$ and $B \in C(F)$ such that $f(A) = P$, $f(B) = Q$, and $A \cup B \in \limsup C_2(f)^{-1}(P_m \cup Q_m)$.

In order to prove Claim 2, first apply Claim 1 to obtain an element $A \in \left(\limsup C(f)^{-1}(P_m) \right) \cap C(f)^{-1}(P) \cap C(E)$. Then take a sequence $\{m_k\}_{k=1}^{\infty}$ of positive integers and elements $A_k \in C(f)^{-1}(P_{m_k})$ such that $\{A_k\}_{k=1}^{\infty}$ converges to A. Apply Claim 1 again to obtain an element $B \in \left(\limsup C(f)^{-1}(Q_{m_k}) \right) \cap C(f)^{-1}(Q) \cap C(F)$. Hence, there exist a sequence $\{k_r\}_{r=1}^{\infty}$ positive integers and elements $B_r \in C(f)^{-1}(Q_{m_{k_r}})$ such that $\{B_r\}_{r=1}^{\infty}$ converges to B. Note that $\{A_{k_r} \cup B_r\}_{r=1}^{\infty}$ converges to $A \cup B$ (Lemma 1.8.11) and $f(A_{k_r} \cup B_r) = f(A_{k_r}) \cup f(B_r) = P_{m_{k_r}} \cup Q_{m_{k_r}}$. Therefore A and B have the required properties.

To prove the theorem, by Theorem 8.1.17, we need to show that if $P \in C_2(Y)$, $\mathcal{K}$ is a component of $C_2(f)^{-1}(P)$ and $\{P_m\}_{m=1}^{\infty}$ is a sequence in $C_2(Y)$ converging to P, then $\limsup C_2(f)^{-1}(P_m) \cap \mathcal{K} \neq \emptyset$. Let $D \in \mathcal{K}$. Then $f(D) = P$. We consider two cases.

Case (1) *P is not connected.*

Let $P = R_1 \cup R_2$, where R_1 and R_2 are the components of P. Then D is not connected. Thus, it can be written as $D = D_1 \cup D_2$, where D_1 and D_2 are the components of D, with $f(D_1) = R_1$ and $f(D_2) = R_2$. We may assume that each P_m can be written as $P_m = R_1^{(m)} \cup R_2^{(m)}$, where $R_1^{(m)}$ and $R_2^{(m)}$ are the components of P_m. Observe that $\left\{R_1^{(m)}\right\}_{m=1}^{\infty}$ converges to R_1 and $\left\{R_2^{(m)}\right\}_{m=1}^{\infty}$ converges to R_2. For each $j \in \{1, 2\}$, let E_j be the component of $f^{-1}(R_j)$ containing D_j. Let $D_0 = E_1 \cup E_2$. Then $f(D_0) = P$. If $\mathcal{A}$ is an order arc from D to D_0, then $\mathcal{A} \subset C_2(X)$ (Theorem 1.8.20). Since $f(D) = P = f(D_0)$, it follows that $f(L) = P$ for each $L \in \mathcal{A}$. Thus, $\mathcal{A} \subset \mathcal{K}$. In particular, $D_0 \in \mathcal{K}$. Applying Claim 2 to R_1 and R_2, we see that for each $j \in \{1, 2\}$, there exists $A_j \in C(E_j)$ such that $f(A_j) = R_j$ and $A_1 \cup A_2 \in \lim \sup C_2(f)^{-1}\left(R_1^{(m)} \cup R_2^{(m)}\right)$. Let $D' = A_1 \cup A_2$. Then $f(D') = P$ and $D' \subset D_0$. Let $\mathcal{B}$ be an order arc from D' to D_0 (Theorem 1.8.20). Note that $\mathcal{B} \subset \mathcal{K}$. Therefore, $D' \in \lim \sup C_2(f)^{-1}(P_m) \cap \mathcal{K}$.

Case (2) *P is connected.*
Let $D = D_1 \cup D_2$, where D_1 and D_2 are components of D (if D is connected, let $D_1 = D_2 = D$). For $j \in \{1, 2\}$, let E_j be the component of $f^{-1}(P)$ containing D_j. Since f is confluent, $f(E_j) = P$. Let $D_0 = E_1 \cup E_2$. Let $\mathcal{A}$ be an order arc from D to D_0 (Theorem 1.8.20). Then $f(L) = P$ for each $L \in \mathcal{A}$. Thus, $\mathcal{A} \subset \mathcal{K}$. In particular, $D_0 \in \mathcal{K}$.

For each $m \in \mathbb{N}$, let $P_m = R_m \cup S_m$, where R_m and S_m are components of P_m (if P_m is connected, let $R_m = S_m = P_m$). Since $C(X)$ is compact (Theorem 1.8.5), there exist subsequences $\{R_{m_k}\}_{k=1}^{\infty}$ and $\{S_{m_k}\}_{k=1}^{\infty}$ of the sequences $\{R_m\}_{m=1}^{\infty}$ and $\{S_m\}_{m=1}^{\infty}$, respectively, such that $\{R_{m_k}\}_{k=1}^{\infty}$ converges to R and $\{S_{m_k}\}_{k=1}^{\infty}$ converges to S for some $R, S \in C(Y)$. Note that $R \cup S = P$. Since $f(E_1) = P$, we can choose a component E of $f^{-1}(R)$ such that $E \subset E_1$. Similarly, there exists a component F of $f^{-1}(S)$ such that $F \subset E_2$.
By Claim 2, there exist $A \in C(E)$ and $B \in C(F)$ such that $f(A) = R$, $f(B) = S$ and $D' = A \cup B \in \lim \sup C_2(f)^{-1}(R_{m_k} \cup S_{m_k}) \subset \lim \sup C_2(f)^{-1}(P_m)$. Then $f(D') = P$ and $D' \subset D_0$. Let $\mathcal{B}$ be an order arc from D' to D_0 (Theorem 1.8.20). Note that $\mathcal{B} \subset \mathcal{K}$. Therefore, $D' \in \mathcal{K} \cap \lim \sup C_2(f)^{-1}(P_m)$.

Q.E.D.

8.5.16 Example There exists an open map $f : X \twoheadrightarrow Y$ between 2-cells X and Y such that for each integer $n \geq 3$, the induced map $C_n(f)$ is not an OM-map. Let $X = [-1, 1] \times [0, 1]$, $Y = [0, 1] \times [0, 1]$, and let $f : X \twoheadrightarrow Y$ be defined by $f((x, y)) = (|x|, y)$. We show that $C_3(f) : C_3(X) \to C_3(Y)$ is not an OM-map. The argument is similar for $n > 3$. Define

$$A_{-1} = \{-1, 0\} \times [0, 1] \subset X, \ A_1 = \{0, 1\} \times [0, 1] \subset X \text{ and}$$

$$B = \{0, 1\} \times [0, 1] \subset Y.$$

Note that $f(A_{-1}) = f(A_1) = B$ and that A_{-1} and A_1 are isolated elements of the preimage $C_3(f)^{-1}(B)$. Indeed, if $A \in C_3(f)^{-1}(B)$, then A has at most 3 components and $A \subset A_{-1} \cup A_1 = \{-1, 0, 1\} \times [0, 1]$. Therefore, if $A \in C_3(X)$

satisfies $f(A) = B$ and A is close enough to A_{-1} (or to A_1), then $A = A_{-1}$ (or $A = A_1$, respectively). Consequently, if

$$\mathcal{K} = \{A \in \mathcal{C}_3(X) \mid f(A) = B \text{ and } A_{-1} \neq A \neq A_1\},$$

then $\mathcal{K}$ is a component of $\mathcal{C}_3(f)^{-1}(B)$, and we have that

$$\mathcal{C}_3(f)^{-1}(B) = \{A_{-1}\} \cup \{A_1\} \cup \mathcal{K}.$$

For each $m \in \mathbb{N}$, define

$$B_m = (\{0\} \times ([0, \tfrac{1}{2} - \tfrac{1}{4m}] \cup [\tfrac{1}{2} + \tfrac{1}{4m}, 1])) \cup (\{1\} \times [0, 1]) \subset B \subset Y.$$

Thus, each B_m has 3 components, and the sequence $\{B_m\}_{m=1}^{\infty}$ converges to B. To prove that $\mathcal{C}_3(f)$ is not an OM-map, it is enough, by Theorem 8.1.17, to verify that

$$\mathcal{K} \cap \limsup \mathcal{C}_3(f)^{-1}(B_m) = \emptyset.$$

Let $A \in \limsup \mathcal{C}_3(f)^{-1}(B_m)$. Then there exists a sequence $\{A_{m_k}\}_{k=1}^{\infty}$ converging to A such that $A_{m_k} \in \mathcal{C}_3(X)$ and $f(A_{m_k}) = B_{m_k}$ for all $k \in \mathbb{N}$. Therefore, $A_{m_k} \subset f^{-1}(B_{m_k})$ for each $k \in \mathbb{N}$. Also, if $S = \{0\} \times [0, 1]$, then $S \subset A$ and, since A_{m_k} has at most 3 components the union of some two of which is close to S for sufficiently large k, there must be a third component which equals to either $\{-1\} \times [0, 1]$ or $\{1\} \times [0, 1]$. In any case, either $A = A_{-1}$ or $A = A_1$. Thus, $A \notin \mathcal{K}$. Therefore, $\mathcal{K} \cap \limsup \mathcal{C}_3(f)^{-1}(B_m) = \emptyset$. For an arbitrary $n > 3$, we define B_m as the union of the segment $\{1\} \times [0, 1]$ and of $n - 1$ mutually disjoint subsegments of $\{0\} \times [0, 1]$ such that B is the limit of the sequence $\{B_m\}_{m=1}^{\infty}$. Then the rest of the argument remains the same, with $\mathcal{C}_n(f)$ in place of $\mathcal{C}_3(f)$.

8.5.17 Remark Observe that in Example 8.5.16, we have Y is a subset of X and the map f is a retraction. In fact, f is a simple map. Thus, Example 8.5.16 shows that even if f is an open simple retraction between 2-cells, it is not necessarily true that $\mathcal{C}_n(f)$ is an OM-map for $n \geq 3$.

8.5.18 Remark Concerning the class of MO-maps, let us recall that there exist open surjective maps $f : [0, 1] \twoheadrightarrow [0, 1]$ such that $\mathcal{C}(f)$ are not MO-maps, see [13, Proposition 9]. The implication: *If $\mathcal{C}_n(f)$ is an MO-map, then f is an MO-map* is still open even for $n = 1$ (Question 9.6.1).

8.5.19 Theorem *Let $f : X \twoheadrightarrow Y$ be a map between continua. If $HS_n(f)$ is open, then f is light.*

Proof Suppose f is not light. Then there exists a nondegenerate subcontinuum A of X such that $f(A) \in \mathcal{F}_1(Y)$.

Let y_1 and y_2 be two distinct points of Y and let $\{D_k\}_{k=1}^{\infty}$ and $\{E_k\}_{k=1}^{\infty}$ be two sequences in $\mathcal{C}_n(Y) \setminus (\mathcal{C}_{n-1}(Y) \cup \mathcal{F}_n(Y))$ such that

$$\lim_{k\to\infty} D_k = \{y_1\} \text{ and } \lim_{k\to\infty} E_k = \{y_2\}.$$

For each $k \in \mathbb{N}$, let

$$L_k = \begin{cases} D_{\frac{k+1}{2}}, & \text{if } k \text{ is odd}; \\ L_{\frac{k}{2}}, & \text{if } k \text{ is even}. \end{cases}$$

Then $\{L_k\}_{k=1}^{\infty}$ is a divergent sequence of elements of $\mathcal{C}_n(Y)$ such that $\{L_k\}_{k=1}^{\infty} \cap \mathcal{F}_n(Y) = \emptyset$. Note that $\{q_Y^n(L_k)\}_{k=1}^{\infty}$ is a sequence of elements of $HS_n(Y)$ converging to F_Y^n. Since $HS_n(f)$ is open and $HS_n(f)(q_X^n(A)) = F_Y^n$, by Theorem 8.1.18, we have that $q_X^n(A) \in \liminf HS_n(f)^{-1}(q_Y^n(L_k))$. Hence, for each $k \in \mathbb{N}$, there exists $\chi_k \in HS_n(f)^{-1}(q_Y^n(L_k))$ such that $\{\chi_k\}_{k=1}^{\infty} \subset HS_n(X) \setminus \{F_X^n\}$ and converges to $q_X^n(A)$. For each $k \in \mathbb{N}$, $\chi_{2k-1} \in HS(f)^{-1}(q_Y^n(L_m))$ for some odd m and $\chi_{2k} \in HS(f)^{-1}(q_Y^n(L_\ell))$ for some even ℓ. Since $\{\chi_k\}_{k=1}^{\infty} \cap \{F_X^n\} = \emptyset$ and $q_X^n|_{\mathcal{C}_n(X)\setminus\mathcal{F}_n(X)}$ is a homeomorphism (Remark 7.1.5), we have that $\{(q_X^n)^{-1}(\chi_k)\}_{k=1}^{\infty}$ is a sequence of elements of $\mathcal{C}_n(X)$ converging to A. Since $\mathcal{C}_n(f)$ is continuous, we obtain:

$$\lim_{k\to\infty} \mathcal{C}_n(f)((q_X^n)^{-1}(\chi_k)) = \mathcal{C}_n(f)(A).$$

Observe that, by the construction of $\{\chi_k\}_{k=1}^{\infty}$, both of the sets $\{\mathcal{C}_n(f)((q_X^n)^{-1}(\chi_k))\}_{k=1}^{\infty} \cap \{D_k\}_{k=1}^{\infty}$ and $\{\mathcal{C}_n(f)((q_X^n)^{-1}(\chi_k))\}_{k=1}^{\infty} \cap \{E_k\}_{k=1}^{\infty}$ are infinite, contradicting the fact that the sequence $\{\mathcal{C}_n(f)((q_X^n)^{-1}(\chi_k))\}_{k=1}^{\infty}$ converges. Therefore, f is light.

Q.E.D.

8.5.20 Theorem *Let $f: X \twoheadrightarrow Y$ be a light map between continua. If $\mathcal{C}_n(f)$ is open, then $HS_n(f)$ is open.*

Proof Let $\mathcal{U}$ be an open subset of $HS_n(X)$. We consider two cases.

Case (1) $\mathcal{U} \cap \{F_X^n\} = \emptyset$.
Since f is light and the following restrictions $q_X^n|_{\mathcal{C}_n(X)\setminus\mathcal{F}_n(X)}$ and $q_Y^n|_{\mathcal{C}_n(Y)\setminus\mathcal{F}_n(Y)}$ are homeomorphisms (Remark 7.1.5), we have that $HS_n(f)(\mathcal{U})$ and $\mathcal{C}_n(f)((q_X^n)^{-1}(\mathcal{U}))$ are homeomorphic. Since $\mathcal{C}_n(f)$ is open, $HS_n(f)(\mathcal{U})$ is an open subset of $HS_n(Y)$.

Case (2) $F_X^n \in \mathcal{U}$.
Since $\mathcal{U}$ is open in $HS_n(X)$ and q_X^n is continuous, $(q_X^n)^{-1}(\mathcal{U})$ is open in $\mathcal{C}_n(X)$ and $\mathcal{F}_n(X) \subset (q_X^n)^{-1}(\mathcal{U})$. Since $\mathcal{C}_n(f)$ is open $\mathcal{C}_n(f)((q_X^n)^{-1}(\mathcal{U}))$ is open in $\mathcal{C}_n(Y)$. Note we also have that $\mathcal{F}_n(Y) \subset \mathcal{C}_n(f)((q_X^n)^{-1}(\mathcal{U}))$. Hence, $HS_n(f)(\mathcal{U}) = q_Y^n(\mathcal{C}_n(f)((q_X^n)^{-1}(\mathcal{U})))$ is an open subset of $HS_n(Y)$.

Q.E.D.

8.5.21 Theorem *Let $f: X \twoheadrightarrow Y$ be a map between continua and let $n \geq 3$. If $HS_n(f)$ is open, then f is a homeomorphism.*

Proof Since $HS_n(f)$ is open, $HS_n(f)$ is confluent (see Diagram I). Hence, since $n \geq 3$, f is monotone by Theorem 8.4.7. Also, since $HS_n(f)$ is open, by Theorem 8.5.19, f is light. Therefore, f is a homeomorphism (Theorem 8.1.29).

Q.E.D.

8.5.22 Theorem *Let $f: X \twoheadrightarrow Y$ be a map between continua and let $n \in \mathbb{N}$ be such that $C_n(f): C_n(X) \twoheadrightarrow C_n(Y)$ is open. If there exists a point $y \in Y$ such that $|f^{-1}(y)| > 1$, then there exist two subcontinua D and E of X such that $D \cap E = \emptyset$ and $f(D) = f(E) = L$, where L is a proper and irreducible subcontinuum between y and some point $y_0 \in Y \setminus \{y\}$.*

Proof Let $f: X \twoheadrightarrow Y$ be a map between continua and let $n \in \mathbb{N}$ be such that $C_n(f)$ is open. Let $b_1, \ldots, b_{n-1}$ and y be different points of Y such that $|f^{-1}(y)| > 1$. Let $a_1, \ldots, a_{n-1}$, x_1 and x_2 be points in X such that $f(a_j) = b_j$, for each $j \in \{1, \ldots, n-1\}$, and $x_1, x_2 \in f^{-1}(y)$, where $x_1 \neq x_2$. Let $B_m = V_{\frac{1}{m}}(y)$. Let C_m be the component of $Cl(B_m)$ such that $y \in C_m$. Note that $C_m \cap Bd(B_m) \neq \emptyset$, Theorem 1.7.27. Let D_m be an irreducible subcontinuum of C_m between y and some point of $Bd(B_m)$ [35, Proposition 11.30]. Observe that $\lim_{m \to \infty} D_m = \{y\}$. Hence, we have that:

$$\lim_{m \to \infty} (D_m \cup \{b_1, \ldots, b_{n-1}\}) = \{y, b_1, \ldots, b_{n-1}\}.$$

Since $\{x_1, a_1, \ldots, a_{n-1}\}$ and $\{x_2, a_1, \ldots, a_{n-1}\}$ both belong to $C_n(f)^{-1}(\{y, b_1, \ldots, b_{n-1}\})$, by Theorem 8.1.19, there exist two sequences $\{E_m\}_{m=1}^{\infty}$ and $\{F_m\}_{m=1}^{\infty}$ of points of $C_n(X)$ such that $\lim_{m \to \infty} E_m = \{x_1, a_1, \ldots, a_{n-1}\}$, $\lim_{m \to \infty} F_m = \{x_2, a_1, \ldots, a_{n-1}\}$ and:

$$E_m, F_m \in C_n(f)^{-1}(D_m \cup \{b_1, \ldots, b_{n-1}\}), \text{ for each } m \in \mathbb{N}.$$

Let $U_1, U_2, V_1, \ldots, V_{n-1}$ be open and pairwise disjoint subsets of X such that $x_1 \in U_1$, $x_2 \in U_2$ and $a_j \in V_j$ for every $j \in \{1, \ldots, n-1\}$. Thus, there exists $\ell \in \mathbb{N}$ such that $E_k \subset U_1 \cup V_1 \cup \cdots \cup V_{n-1}$ and $F_k \subset U_2 \cup V_1 \cup \cdots \cup V_{n-1}$, for each $k \geq \ell$ [37, Theorem (7.2), p. 12]. Hence, both E_k and F_k have exactly n components, for all $k \geq \ell$. Let k_0 be big enough such that if E and F are the components of E_{k_0} and F_{k_0}, respectively, such that $E \subset U_1$ and $F \subset U_2$, then $f(E) = f(F) = D_{k_0}$. Observe that $E \cap F = \emptyset$, Also, D_{k_0} is irreducible between y and some point $y_0 \in Y \setminus \{y\}$ and, by construction, $D_{k_0} \neq Y$.

Q.E.D.

8.5.23 Theorem *Let $f: X \twoheadrightarrow Y$ be a map between continua and let $n \geq 2$. If $C_n(f): C_n(X) \twoheadrightarrow C_n(Y)$ is open, then f is a homeomorphism.*

Proof Let $f: X \twoheadrightarrow Y$ be a map between continua and let $n \geq 2$ be such that $C_n(f): C_n(X) \to C_n(Y)$ is open. Suppose f is not a homeomorphism. Hence, there exists a point $y \in Y$ such that $|f^{-1}(y)| > 1$. Thus, there exist two subcontinua D and E of X such that $D \cap E = \emptyset$ and $f(D) = f(E) = L$, where L is an irreducible continuum between y and some different point $y_0 \in Y \setminus \{y\}$, Theorem 8.5.22. Note that L is a nondegenerate subcontinuum of Y and $D \setminus (f^{-1}(y) \cup f^{-1}(y_0)) \neq \emptyset$. Let $d \in D \setminus (f^{-1}(y) \cup f^{-1}(y_0))$. We consider two cases:

Case (1) $n = 2$.

Observe that $\{d\}$ and $f^{-1}(y) \cup E \cup f^{-1}(y_0)$ are disjoint closed subsets of X. Thus, there exist two open and disjoint sets U and V of X such that $d \in U$ and $f^{-1}(y) \cup E \cup f^{-1}(y_0) \subset V$. We show that $C_2(f)(\langle U, V \rangle_2)$ is not open. Note that $\{d\} \cup E \in \langle U, V \rangle_2$. Therefore, $C_2(f)(\{d\} \cup E) = L \in C_2(f)(\langle U, V \rangle_2)$. Since $\{y\}$ and $\{y_0\}$ are subsets of L, there exist two order arcs (Theorem 1.8.20) $\alpha_1, \alpha_2 : [0, 1] \to C(X)$ such that $\alpha_1(0) = \{y\}$, $\alpha_2(0) = \{y_0\}$ and $\alpha_1(1) = \alpha_2(1) = L$. Since $\alpha_1(1) = \alpha_2(1) = L$, it is not difficult to prove that there exists a point $s \in [0, 1]$ such that $\alpha_1(s) \cap \alpha_2(s) \neq \emptyset$ and $\alpha_1(t) \cap \alpha_2(t) = \emptyset$, for each $t < s$. Observe that $\alpha_1(0) \cup \alpha_2(0) = \{y, y_0\}$. Hence, $s > 0$. Since L is irreducible between y and y_0, and $y, y_0 \in \alpha_1(s) \cup \alpha_2(s)$, we have that $\alpha_1(s) \cup \alpha_2(s) = L$. Let $\{t_m\}_{m=1}^{\infty}$ be an increasing sequence in $[0, 1]$ such that $\lim_{m \to \infty} t_m = s$. Clearly, $\alpha_1(t_m) \cup \alpha_2(t_m) \in C_2(Y) \setminus C(Y)$, for each $m \in \mathbb{N}$. Let $L_m = \alpha_1(t_m) \cup \alpha_2(t_m)$. Observe that $\lim_{m \to \infty} L_m = L$.

We show that $L_m \notin C_2(f)(\langle U, V \rangle_2)$, for any $m \in \mathbb{N}$. Suppose that there exists $G \in \langle U, V \rangle_2$ such that $C_2(f)(G) = L_k$, for some $k \in \mathbb{N}$. Since $U \cap V = \emptyset$, $G \cap U \neq \emptyset$ and $G \cap V \neq \emptyset$, we obtain that G has two components. Let H and F be the components of G such that $H \subset U$ and $F \subset V$. Note that y and y_0 both belong to L_k. Thus, $G \cap f^{-1}(y) \neq \emptyset$ and $G \cap f^{-1}(y_0) \neq \emptyset$. Since $U \cap (f^{-1}(y) \cup f^{-1}(y_0)) = \emptyset$ and $H \subset U$, we have that $F \cap f^{-1}(y) \neq \emptyset$ and $F \cap f^{-1}(y_0) \neq \emptyset$. Hence, $f(F)$ is connected and $y, y_0 \in f(F) \subset L_k$, but this contradicts the fact that y and y_0 belong to different components of L_k. Thus, $L_m \notin C_2(f)(\langle U, V \rangle_2)$, for any $m \in \mathbb{N}$. Since $\lim_{m \to \infty} L_m = L$ and $L \in C_2(f)(\langle U, V \rangle_2)$, we have that L is not an interior point of $C_2(f)(\langle U, V \rangle_2)$.

Case (2) $n \geq 3$.

Observe that $L \neq Y$, Theorem 8.5.22. Hence, $X \setminus f^{-1}(L) \neq \emptyset$. Let $x_1, \ldots, x_{n-2}$ be different points in $X \setminus f^{-1}(L)$. Let $U, V, W_1, \ldots, W_{n-2}$ be open and pairwise disjoint subsets of X, such that $d \in U$, $f^{-1}(y) \cup E \cup f^{-1}(y_0) \subset V$ and $x_j \in W_j$, for each $j \in \{1, \ldots, n - 2\}$. Note that each point in $\langle U, V, W_1, \ldots, W_{n-2} \rangle_n$ has n components. Therefore, using a similar argument to the one given in Case (1), we may conclude that $C_n(f)(\langle U, V, W_1, \ldots, W_{n-2} \rangle_n)$ is not an open set.

Thus, $C_n(f)$ is not open, for any $n \geq 2$, by Cases (1) and (2). Hence, $|f^{-1}(y)| = 1$ for each $y \in Y$. Since f is define between continua, f is closed. Therefore, f is a homeomorphism.

Q.E.D.

8.6 Light Maps

Now we study light maps. The lightness of the induced maps imply the lightness of the original map. We prove that equivalence of the lightness between the induced maps on the n-fold hyperspaces and the induced maps on the n-fold hyperspace suspensions. We characterize the lightness of the induced maps in terms of the original map. We show that for $n \geq 2$ the lightness of the induced maps imply that the original map is a homeomorphism. We start by noting the following lemma:

8.6.1 Lemma *Let $f \colon X \twoheadrightarrow Y$ be a surjective map between continua, and let $n \in \mathbb{N}$. Then the following are equivalent:*

(1) f is light;
(2) $C_n(f)^{-1}(\mathcal{F}_n(Y)) = \mathcal{F}_n(X)$;
(3) $HS_n(f)^{-1}(F_Y^n) = \{F_X^n\}$.

Proof Suppose f is light. Let $y \in Y$ and let $A \in C_n(f)^{-1}(\{y\})$. Then $C_n(f)(A) = \{y\}$; i.e., $f(A) = y$. Since f is light, A is totally disconnected. Since A has at most n components, $A \in \mathcal{F}_n(X)$. It is clear that $C_n(f)(\mathcal{F}_n(X)) \subset \mathcal{F}_n(Y)$. Therefore, $C_n(f)^{-1}(\mathcal{F}_n(Y)) = \mathcal{F}_n(X)$.

Assume that $C_n(f)^{-1}(\mathcal{F}_n(Y)) = \mathcal{F}_n(X)$. Let $\chi \in HS_n(X) \setminus \{F_X^n\}$. Then $(q_X^n)^{-1}(\chi) \in C_n(X) \setminus \mathcal{F}_n(X)$. Hence, by hypothesis, $C_n(f)((q_X^n)^{-1}(\chi)) \in C_n(Y) \setminus \mathcal{F}_n(Y)$. Thus, $q_Y^n(C_n(f)((q_X^n)^{-1}(\chi))) \in HS_n(Y) \setminus \{F_Y^n\}$. This implies that $HS_n(f)(\chi) \neq F_Y^n$. Therefore, $HS_n(f)^{-1}(F_Y^n) = \{F_X^n\}$.

Suppose f is not light. Then there exists a point $y \in Y$ such that $f^{-1}(y)$ has a nondegenerate component, say A. Note that $A \in C_n(X) \setminus \mathcal{F}_n(X)$ and $C_n(f)(A) = \{y\} \in \mathcal{F}_n(Y)$. Hence, we have that $q_X^n(A) \in HS_n(X) \setminus \{F_X^n\}$ and $HS_n(f)(q_X^n(A)) = F_Y^n$. Therefore, $HS_n(f)^{-1}(F_Y^n) \neq \{F_X^n\}$.

Q.E.D.

8.6.2 Theorem *Let $f \colon X \twoheadrightarrow Y$ be a map between continua, and let $n \in \mathbb{N}$. If either $C_n(f)$ or $HS_n(f)$ is light, then f is light.*

Proof Suppose f is not light. Then there exist $x \in X$ and a nondegenerate subcontinuum A of X such that $f(A) = \{f(x)\}$. By Theorem 1.8.20, we can find an order arc such that its image, γ, is contained in $C(A)$ and $\gamma \cap \mathcal{F}_n(X) = \emptyset$. Hence, since $f(A) = \{f(x)\}$, $C_n(f)(\gamma) = \{\{f(x)\}\}$. Therefore, $C_n(f)$ is not light.

Since $\gamma \cap \mathcal{F}_n(X) = \emptyset$, $F_X^n \not\in q_X^n(\gamma)$ (Remark 7.1.5). Hence, $HS_n(f)(q_X^n(\gamma)) = \{F_Y^n\}$, by $(*)$ of Theorem 8.2.6. Therefore, $HS_n(f)$ is not light.

Q.E.D.

8.6.3 Remark Note that the converse implications of Theorem 8.6.2 are not true, use [12, 3.8].

As a consequence of Lemma 8.6.1 and Theorem 8.6.2 we have the following:

8.6.4 Theorem *Let $f \colon X \twoheadrightarrow Y$ be a surjective map between continua, and let $n \in \mathbb{N}$. Then $C_n(f)$ is light if and only if $HS_n(f)$ is light.*

Proof Let $A \in C_n(X) \setminus \mathcal{F}_n(X)$. Note that if $C_n(f)(A) \in C_n(Y) \setminus \mathcal{F}_n(Y)$, then $C_n(f)^{-1}(C_n(f)(A))$ is totally disconnected if and only if $HS_n(f)^{-1}(q_Y^n(C_n(f(A))))$ is totally disconnected, by Remark 7.1.5 and $(*)$ of Theorem 8.2.6.

Next, suppose $C_n(f)$ is light. Then, by Theorem 8.6.2, f is light. Hence, $HS_n(f)^{-1}(F_Y^n) = \{F_X^n\}$ (Lemma 8.6.1). Therefore, $HS_n(f)$ is light.

Finally, suppose $HS_n(f)$ is light. Then f is light (Theorem 8.6.2). Hence, $C_n(f)^{-1}(\mathcal{F}_n(Y)) = \mathcal{F}_n(X)$ (Lemma 8.6.1). Therefore, $C_n(f)$ is light.

Q.E.D.

The following theorem characterizes the lightness of the induced maps on n-fold hyperspaces.

8.6.5 Theorem *Let $f\colon X \to Y$ be a map between continua, and let $n \in \mathbb{N}$. Then $C_n(f)$ is light if and only if for each pair of elements A and B of $C_n(X)$ satisfying property (OA), $f(A) \subsetneqq f(B)$.*

Proof Suppose $C_n(f)$ is not light. Then there exist $D \in C_n(X)$ and a nondegenerate subcontinuum Γ of $C_n(X)$ such that $C_n(f)(\Gamma) = \{C_n(f)(D)\}$. Observe that $\bigcup \Gamma \in C_n(X)$ (Corollary 6.1.2), and $f\left(\bigcup \Gamma\right) = f(D)$. Since Γ is nondegenerate, there exists $A \in \Gamma$ such that $A \neq \bigcup \Gamma$. Hence, A and $\bigcup \Gamma$ satisfy property (OA) (Lemma 6.1.4), and $f(A) = f\left(\bigcup \Gamma\right) = f(D)$.

Next, suppose there exist $A, B \in C_n(X)$ satisfying property (OA) and $f(A) = f(B)$. By Theorem 1.8.20, there exists an order arc $\alpha\colon [0, 1] \to C_n(X)$ such that $\alpha(0) = A$ and $\alpha(1) = B$. Note that for every $t \in [0, 1]$, $f(\alpha(t)) = f(A)$. Hence, $\alpha([0, 1]) \subset C_n(f)^{-1}(C_n(f)(A))$. Therefore, $C_n(f)$ is not light.

Q.E.D.

As a consequence of Theorems 8.6.4 and 8.6.5, we have:

8.6.6 Theorem *Let $f\colon X \to Y$ be a map between continua, and let $n \in \mathbb{N}$. Then $HS_n(f)$ is light if and only if for each pair of elements A and B of $C_n(X)$ satisfying property (OA), $f(A) \subsetneqq f(B)$.*

8.6.7 Corollary *Let $f\colon X \twoheadrightarrow Y$ be a map between continua, and let $n \in \mathbb{N}$. If either $C_n(f)$ or $HS_n(f)$ is a k-to-1 map, for some positive integer k, then $k = 1$ and both $C_n(f)$ and $HS_n(f)$ are homeomorphisms.*

Proof Note that both $C_n(f)$ and $HS_n(f)$ are light maps (Theorem 8.6.4). Suppose $k \geq 2$. Then there exists $A \in C_n(X) \setminus \{X\}$ ($\chi \in HS_n(X) \setminus \{T_X^n\}$) such that $C_n(f)(A) = Y$ ($HS_n(f)(\chi) = T_Y^n$). Hence, $A \subsetneqq X$ ($(q_X^n)^{-1}(\chi) \subsetneqq X$) and $f(A) = f(X) = Y$ ($f\left((q_X^n)^{-1}(\chi)\right) = f(X) = Y$). Therefore, $C_n(f)$ is not light (Theorem 8.6.5) ($HS_n(f)$ is not light by Theorem 8.6.6), a contradiction. Thus, $k = 1$ and both $C_n(f)$ and $HS_n(f)$ are homeomorphisms.

Q.E.D.

The following example shows a light map f such that $C(f)$ is light but $C_2(f)$ is not light (compare with Corollary 8.6.10).

8.6.8 Example Let $A = Cl\left(\left\{\left(x, \sin\left(\frac{1}{x}\right)\right) \in \mathbb{R}^2 \mid 0 < x \le 1\right\}\right)$, let $B =$ $\{(x, y) \in \mathbb{R}^2 \mid (2, 0) + (-x, y) \in A\}$ and let $X = A \cup B$. Note that $A \cap B = \{(1, \sin(1))\}$. For each $y \in [-1, 1]$, identify the point $(0, y)$ with the point with the point $(2, y)$. Let Y be the quotient space and let $q: X \twoheadrightarrow Y$ be the quotient map. Using Theorem 8.6.5, one may show that $\mathcal{C}(q)$ is light. Let $P = \{(0, y) \mid -1 \le y \le 1\} \cup \{(2, 0)\}$ and let $Q = \{(0, y) \mid -1 \le y \le 1\} \cup \{(2, y) \mid -1 \le y \le 1\}$. Then $P, Q \in \mathcal{C}_2(X)$ and satisfy property (OA). But $q(P) = q(Q)$. Hence, by Theorem 8.6.5, $\mathcal{C}_2(q)$ is not light. By Theorem 8.6.4, $HS(f)$ is light, but $HS_2(f)$ is not light.

8.6.9 Theorem *Let $f: X \to Y$ be a map between continua and let $n \ge 2$. Then $\mathcal{C}_n(f)$ is light if and only if for each two disjoint subcontinua P and Q of X, we have that $f(P) \setminus f(Q) \ne \emptyset$ and $f(Q) \setminus f(P) \ne \emptyset$.*

Proof Suppose that for each two disjoint subcontinua P and Q of X, we have that $f(P) \setminus f(Q) \ne \emptyset$ and $f(Q) \setminus f(P) \ne \emptyset$. We prove that $\mathcal{C}_n(f)$ is light. Let $A, B \in \mathcal{C}_n(X)$ be such that they satisfy property (OA). Let $A_1, \dots, A_m$ be the components of A. Since $A \subsetneq B$, without loss of generality, we assume that $A_1 \subsetneq B_1$ for some component B_1 of B. We show that $f(B_1) \setminus f(A) \ne \emptyset$. To this end, we make an inductive construction. Since $A_1 \subsetneq B_1$, by Theorem 1.7.27, there exists a nondegenerate subcontinuum L_1 of X such that $L_1 \subset B_1 \setminus A_1$. Since $A_1 \cap L_1 = \emptyset$, by hypothesis, $f(L_1) \setminus f(A_1) \ne \emptyset$. Hence, $f(B_1) \setminus f(A_1) \ne \emptyset$. Suppose that, for some k, $f(B_1) \setminus f(A_1 \cup \cdots \cup A_k) \ne \emptyset$. We now prove that $f(B_1) \setminus f(A_1 \cup \cdots \cup A_{k+1}) \ne \emptyset$. In order to do this, we show

$$B_1 \setminus (f^{-1}(f(A_1 \cup \cdots \cup A_k)) \cup A_{k+1}) \ne \emptyset. \quad (*)$$

Suppose $B_1 \subset f^{-1}(f(A_1 \cup \cdots \cup A_k)) \cup A_{k+1}$. Since $A_1 \subsetneq B_1$ and B_1 is connected, we have that $B_1 \setminus (A_1 \cup \cdots \cup A_k) \ne \emptyset$ (each A_j is a component of A). Thus, by Theorem 1.7.27, there exists a nondegenerate subcontinuum K of X such that $K \subset B_1 \setminus (A_1 \cup \cdots \cup A_k)$. Since $B_1 \subset f^{-1}(f(A_1 \cup \cdots \cup A_k)) \cup A_{k+1}$, we have

$$K \subset f^{-1}(f(A_1 \cup \cdots \cup A_k \cup \cdots \cup A_k)) \setminus (A_1 \cup \cdots \cup A_{k+1}). \quad (**)$$

From here we obtain that

$$K = (f^{-1}(f(A_1)) \cap K) \cup \cdots \cup (f^{-1}(f(A_k)) \cap K). \quad (***)$$

Note that if $j \in \{1, \dots, k\}$, then $f^{-1}(f(A_j)) \cap K$ is a closed and totally disconnected subset of K. To see that it is totally disconnected, suppose it has a nondegenerate component R. Then, by $(**)$, $R \cap A_j = \emptyset$. But $f(R) \subset f(A_j)$, a contradiction to our general assumption. Therefore, each $f^{-1}(f(A_j)) \cap K$ is totally disconnected. Hence, we have written the continuum K as a finite union of totally disconnected closed sets, this is not possible. Therefore, $(*)$ is true.

By Theorem 1.7.27, there exists a nondegenerate subcontinuum L_k of X such that $L_k \subset B_1 \setminus (f^{-1}(f(A_1 \cup \cdots \cup A_k))) \cup A_{k+1})$. Hence, $f(L_k) \cap f(A_1 \cup \cdots \cup A_k) = \emptyset$. Since $L_k \cap A_{k+1} = \emptyset$, by our general assumption, $f(L_k) \setminus f(A_{k+1}) \neq \emptyset$. Thus, $f(L_k) \setminus f(A_1 \cup \cdots \cup A_{k+1}) \neq \emptyset$. Since $f(L_k) \subset f(B_1)$, we have that

$$f(B_1) \setminus f(A_1 \cup \cdots \cup A_{k+1}) \neq \emptyset.$$

Therefore, by mathematical induction, $f(B_1) \setminus f(A_1 \cup \cdots \cup A_m) \neq \emptyset$. Since $f(B_1) \subset f(B)$, we obtain that $f(A) \subsetneqq f(B)$. Therefore, by Theorem 8.6.5, $C_n(f)$ is light.

Next, suppose that there exist two disjoint nondegenerate subcontinua A and B of X such that $f(A) \subset f(B)$; i.e., $f(A) \setminus f(B) = \emptyset$. Let $a \in A$. Define $P = \{a\} \cup B$ and $Q = A \cup B$. Then $P, Q \in C_n(X)$, with $n \geq 2$. P and Q satisfy property (OA) and $f(P) = f(Q)$. Hence, by Theorem 8.6.5, $C_n(f)$ is not light.

Q.E.D.

8.6.10 Corollary *Let $f : X \to Y$ be a map between continua and let n and m be integers greater than or equal to 2. Then $C_n(f)$ is light if and only if $C_m(f)$ is light.*

As a consequence of Corollary 8.6.10 and Theorem 8.6.4, we obtain:

8.6.11 Corollary *Let $f : X \to Y$ be a map between continua and let n and m be integers greater than or equal to 2. Then $HS_n(f)$ is light if and only if $HS_m(f)$ is light.*

8.6.12 Theorem *Let $f : X \twoheadrightarrow Y$ be a weakly confluent map between continua and let $n \geq 2$. If $C_n(f)$ is light, then f is a homeomorphism.*

Proof Suppose $C_n(f)$ is light. Then, by Theorem 8.6.2, f is light. We prove that f is monotone. Being light and monotone, f is a homeomorphism (Theorem 8.1.29).

Suppose f is not monotone. Then there exists a point $y \in Y$ such that $f^{-1}(y)$ is not connected. Let C_1 and C_2 be two different components of $f^{-1}(y)$. Since f is light, $C_1 = \{x_1\}$ and $C_2 = \{x_2\}$ for some $x_1, x_2 \in X$. Thus, by Theorem 8.1.31, there exists an open subset W of Y such that $y \in W$ and x_1 and x_2 belong to different components of $f^{-1}(W)$. By Theorem 1.7.27, there exist two nondegenerate subcontinua P_1 and P_2 of X such that $x_1 \in P_1 \subset f^{-1}(W)$ and $x_2 \in P_2 \subset f^{-1}(W)$. Let $K = f(P_1) \cup f(P_2)$. Since $y \in f(P_1) \cap f(P_2)$, K is a subcontinuum of Y. Note that $f(K) \subset W$. Since f is weakly confluent, there exists a subcontinuum L of X such that $f(L) = K$. Since P_1 and P_2 are contained in different components of $f^{-1}(W)$, we have that either $P_1 \cap L = \emptyset$ or $P_2 \cap L = \emptyset$. Observe that $f(P_1) \subset f(L)$ and $f(P_2) \subset f(L)$. Hence, if $P_1 \cap L = \emptyset$, then $f(P_1) \setminus f(L) = \emptyset$. This implies that $C_n(f)$ is not light by Theorem 8.6.9. Similarly, if $P_2 \cap L = \emptyset$, then $C_n(f)$ is not light. Therefore, f is monotone.

Q.E.D.

As a consequence of Theorems 8.6.4 and 8.6.12, we obtain:

8.6.13 Corollary *Let $f : X \to Y$ be a weakly confluent map between continua and let $n \geq 2$. If $HS_n(f)$ is light, then f is a homeomorphism.*

To prepare this section we use [4, 8, 12, 27].

8.7 Freely Decomposable and Strongly Freely Decomposable Maps

We study induced freely decomposable and strongly freely decomposable maps and their relation with almost monotone and monotone maps. The material of this section, is based on [10, 15].

8.7.1 Theorem *There exists an almost monotone map* $f \colon X \twoheadrightarrow Y$ *between continua such that* $C_n(f)$ *is not strongly freely decomposable, for any* $n \in \mathbb{N}$. *Hence,* $C_n(f)$ *is not almost monotone, for any* $n \in \mathbb{N}$.

Proof Let $X = J \cup R$, where $J = \{(0, y) \in \mathbb{R}^2 \mid -1 \leq y \leq 3\}$ and $R = \left\{(x, \sin(\frac{1}{x})) \in \mathbb{R}^2 \mid 0 < x \leq \frac{2}{\pi}\right\}$. Define $Y = X/\sim$, where $(x_0, y_0) \sim (x_1, y_1)$ if and only if $x_0 = x_1 = 0$ and $e^{\pi y_0 i} = e^{\pi y_1 i}$. Let $f \colon X \twoheadrightarrow Y$ be the quotient map. Note that f is an almost monotone map. Observe that $Y = S \cup R'$, where $S = f(J)$ and $R' = f(R)$. Without loss of generality we assume that $S = \{(x, y) \in \mathbb{R}^2 \mid \|(x, y)\| = 1\}$.

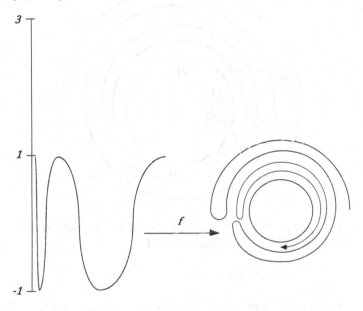

Let:

$$L_1 = \left\{(0, y) \in \mathbb{R}^2 \,\middle|\, \frac{1}{2} \leq y \leq \frac{3}{2}\right\},$$

$$L_2 = \left\{(0, y) \in \mathbb{R}^2 \,\middle|\, \frac{5}{2} \leq y \leq 3\right\},$$

$$L_3 = \left\{(0, y) \in \mathbb{R}^2 \,\middle|\, -1 \leq y \leq -\frac{1}{2}\right\},$$

and

$$L_4 = \left\{ \left(x, \sin\left(\frac{1}{x}\right) \right) \in \mathbb{R}^2 \ \middle| \ \frac{1}{\pi} \le x \le \frac{2}{\pi} \right\}.$$

Define $A = f(L_1) = \left\{ e^{\pi i t} \in S \mid \frac{1}{2} \le t \le \frac{3}{2} \right\}$ and $A' = f(L_4)$. Note that A' is an arc in R'. Then $\mathcal{A} = \langle A, A' \rangle_n$ is a subcontinuum of $\mathcal{C}_n(Y)$, Proposition 6.1.5.

We show that $Int_{\mathcal{C}_n(Y)}(\mathcal{A}) \ne \emptyset$. Let $p = f((0, -1)) \in S$ and let $W = \{ x \in \mathbb{R}^2 \mid \|x - p\| < 1 \}$. It is not difficult to show that there exist three open subsets U_1, U_2 and U_3 of $\mathbb{R}^2$ such that:

(1) $U_1 \cup U_2 \cup U_3 \subset W$;
(2) U_j is connected for each $j \in \{1, 2, 3\}$;
(3) $p \in U_2 \setminus (U_1 \cup U_3)$ and $U_1 \cap U_3 = \emptyset$;
(4) $(U_1 \cup U_2 \cup U_3) \cap A$ is connected.

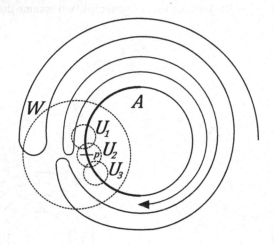

Let $V_1, V_2, \ldots, V_{n-1}$ be open, pairwise disjoint and connected subsets of $Y \setminus Cl_Y(W)$ such that $\cup_{i=1}^{n-1} V_i \subset A'$. Observe that if

$$E \in \langle U_1, U_2, U_3, V_1, \ldots, V_{n-1} \rangle_n,$$

then there exists a component E_0 of E such that $E_0 \subset A$ and $E \setminus E_0 \subset \cup_{i=1}^{n-1} V_i \subset A'$. Therefore, $\langle U_1, U_2, U_3, V_1, \ldots, V_{n-1} \rangle_n \subset \mathcal{A}$ and $Int_{\mathcal{C}_n(Y)}(\mathcal{A}) \ne \emptyset$.

Let $\mathcal{B} = Cl_{\mathcal{C}_n(Y)}(\mathcal{C}_n(Y) \setminus \mathcal{A})$. We prove that $\mathcal{B}$ is a proper subcontinuum of $\mathcal{C}_n(Y)$. Clearly, $\mathcal{B}$ is compact and since $Int_{\mathcal{C}_n(Y)}(\mathcal{A}) \ne \emptyset$, we have that $\mathcal{B} \ne \mathcal{C}_n(Y)$. Hence, we only need to show that $\mathcal{B}$ is connected.

Let $B \in \mathcal{C}_n(Y) \setminus \mathcal{A}$. Then there exists an order arc α from B to Y, Theorem 1.8.20. It is not difficult to see that $\alpha \subset \mathcal{C}_n(Y) \setminus \mathcal{A}$. Therefore, $\mathcal{C}_n(Y) \setminus \mathcal{A}$ is arcwise connected and $\mathcal{B}$ is a proper subcontinuum of $\mathcal{C}_n(Y)$. Clearly, $\mathcal{C}_n(Y) = \mathcal{A} \cup \mathcal{B}$.

Observe that $\langle L_1, L_4 \rangle_n$ is a continuum, Proposition 6.1.5. Also note that $f^{-1}(A) = L_1 \cup L_2 \cup L_3$ and $f^{-1}(A') = L_4$. Hence, $\langle L_1, L_4 \rangle_n \subset C_n(f)^{-1}(A)$. In fact, $\langle L_1, L_4 \rangle_n$ is an open and closed subset of $C_n(f)^{-1}(A)$. We already know that $\langle L_1, L_4 \rangle_n$ is closed. Let $D \in \langle L_1, L_4 \rangle_n$ and let $\varepsilon > 0$ be such that $\varepsilon < \frac{1}{2}$. Thus, clearly, if $E \in C_n(f)^{-1}(A)$ and $\mathcal{H}(D, E) < \varepsilon$, then $E \in \langle L_1, L_4 \rangle_n$ $(E \cap (L_2 \cup L_3) = \emptyset)$. Hence, $\langle L_1, L_4 \rangle_n$ is an open subset of $C_n(f)^{-1}(A)$. Therefore, $C_n(f)^{-1}(A)$ is not connected, and $C_n(f)$ is not strongly freely decomposable. By Diagram I, $C_n(f)$ is not almost monotone.

Q.E.D.

The map defined in the following theorem is taken from [15, Example 4.12, p. 138].

8.7.2 Theorem *There exists a map $f : X \twoheadrightarrow Y$ between continua such that both f and $C(f)$ are almost monotone and $C_n(f)$ is not strongly freely decomposable, for any $n \geq 2$.*

Proof Define H and S in polar coordinates (r, θ) by:

$$H = \{(r, \theta) \mid r = 1 + \theta^{-1} \text{ and } \theta \in [1, \infty)\} \text{ and}$$

$$S = \{(1, \theta) \mid \theta \in [0, 2\pi]\}.$$

Let $X = H \cup S$. Consider the equivalence relation on X given by: $(r, \theta) \sim (r', \theta')$ if and only if $(r, \theta) = (r', \theta')$ or, $r = r' = 1$ and $\theta = \theta' + \pi$. Denote by Y the quotient space and let $f : X \twoheadrightarrow Y$ be the quotient map.

We show that $C(f)$ is almost monotone. Let $\mathcal{D}$ be a subcontinuum of $C(Y)$ such that $Int_{C(Y)}(\mathcal{D}) \neq \emptyset$. It is not difficult to prove that there exists a point $D \in \mathcal{D}$ such that $D \subset f(H)$. Hence, either $\mathcal{D} \subset f(H)$ or, both $\mathcal{D} \cap f(H) \neq \emptyset$ and $\mathcal{D} \cap f(S) \neq \emptyset$. Thus, $C(f)^{-1}(\mathcal{D})$ is connected, by cases (1) and (3) of [15, Example 4.12, p. 139]. Therefore, $C(f)$ is almost monotone.

Assume that $n \geq 2$. Let $p = f((2, 1))$ in Y. Since Y is locally connected at p, there exists a subcontinuum K of Y such that $p \in Int_Y(K) \subset K \subset Y \setminus f(S)$. Let $\mathcal{K} = \langle K \rangle_n \cup \mathcal{F}_1(Y)$. Clearly, $\mathcal{K}$ is a subcontinuum of $C_n(Y)$ and $Int_{C_n(Y)}(\mathcal{K}) \neq \emptyset$. Moreover, $C_n(f)^{-1}(\mathcal{K})$ has two components: $\mathcal{F}_1(X) \cup \langle f^{-1}(K) \rangle_n$ and $\{\{(1, \theta), (1, \theta + \pi)\} \mid \theta \in [0, \pi]\}$. Thus, $C_n(f)^{-1}(\mathcal{K})$ is not connected. Finally, it is not difficult to see that $C_n(Y) \setminus \mathcal{K}$ is connected. Hence, $\mathcal{L} = Cl_{C_n(Y)}(C_n(Y) \setminus \mathcal{K})$ is a proper subcontinuum of $C_n(Y)$. Clearly, $C_n(Y) = \mathcal{K} \cup \mathcal{L}$. Therefore, $C_n(f)$ is not strongly freely decomposable.

Q.E.D.

8.7.3 Theorem *Let $f : X \twoheadrightarrow Y$ be a map between continua and let $n \in \mathbb{N}$. Then $C_n(f)$ is almost monotone if and only if $C_n(f)$ is strongly freely decomposable.*

Proof Since every almost monotone map is strongly freely decomposable (Diagram I), we have that the almost monotoneity of $C_n(f)$ implies their strongly freely decomposability. Since $C_n(Y)$ is unicoherent for each $n \in \mathbb{N}$, Theorem 6.2.5, the converse implication follows from Theorem 8.1.42.

Q.E.D.

The next result follows from Theorems 8.2.18 and 8.7.3.

8.7.4 Corollary *Let $f : X \to Y$ be a map between continua and let $n \in \mathbb{N}$. If $C_n(f)$ is strongly freely decomposable, then f is an almost monotone map.*

Now, we consider maps whose range are locally connected continua.

8.7.5 Theorem *Let $f : X \twoheadrightarrow Y$ be a map between continua, where Y is locally connected and let $n \in \mathbb{N}$. The following are equivalent:*

(1) $C_n(f)$ is freely decomposable;
(2) $C_n(f)$ is monotone;
(3) f is monotone.

Proof Suppose $C_n(f)$ is freely decomposable, we prove that $C_n(f)$ is monotone. Since Y is locally connected, $C_n(Y)$ is locally connected, Theorem 6.1.6. Since $C_n(Y)$ is colocally connected, Theorem 6.3.1, it follows easily that $C_n(Y) \setminus \{A\}$ is connected, for each $A \in C_n(Y)$. Thus, $C_n(f)^{-1}(A)$ is connected, Theorem 8.1.39. Therefore, $C_n(f)$ is monotone. Recall that the map $C_n(f)$ is monotone if and only if f is monotone, Theorem 8.4.2. Hence, since every monotone map is freely decomposable (Diagram I), we have that (1), (2) and (3) are equivalent.

Q.E.D.

Since every monotone map is strongly freely decomposable and every strongly freely decomposable map is freely decomposable (Diagram I), we have that the next corollary is a consequence of Theorems 8.7.5 and 8.7.3.

8.7.6 Corollary *Let $f : X \twoheadrightarrow Y$ be a map between continua, where Y is locally connected, and let $n \in \mathbb{N}$. The following are equivalent:*

(1) $C_n(f)$ is freely decomposable;
(2) $C_n(f)$ is strongly freely decomposable;
(3) $C_n(f)$ is almost monotone;
(4) $C_n(X)$ is monotone;
(5) f is monotone.

In the next example, we give a strongly freely decomposable map f such that $C_n(f)$ is not strongly freely decomposable, for any $n \in \mathbb{N}$, not even, when f is defined between locally connected continua.

8.7.7 Example There exists a strongly freely decomposable map f between locally connected continua such that $C_n(f)$ is not strongly freely decomposable, for any $n \in \mathbb{N}$. We denote $\mathcal{S}^1 = \{(x, y) \in \mathbb{R}^2 \mid x^2 + y^2 = 1\}$. Let $f : \mathcal{S}^1 \twoheadrightarrow [-1, 1]$ be the map defined by $f((x, y)) = x$. Clearly, f is strongly freely decomposable and it is not monotone. Hence, $C_n(f)$ is not freely decomposable, Corollary 8.7.6.

References

1. K. Borsuk, *Theory of Retracts*, Monografie Mat., Vol. 44, PWN (Polish Scientific Publishers), Warszawa, 1967.
2. K. Borsuk, *Theory of Shape*, Monografie Mat., Vol. 59 PWN (Polish Scientific Publishers), Warszawa, 1975.
3. J. Camargo, On the Semi-open Induced Mappings, Topology Proc., 32 (2008), 145–152.
4. J. Camargo, *Funciones Inducidas Entre Hiperespacios de Continuos*, Tesis Doctoral, Facultad de Ciencias, U. N. A. M., 2009. (Spanish)
5. J. Camargo, Openness of the Induced Map $C_n(f)$, Bol. Mat., 16(2) (2009), 115–123.
6. J. Camargo, Openness of Induced Maps and Homeomorphisms, Houston J. Math., 36 (2010), 199–213.
7. J. Camargo, Some Relationships Between Induced Mappings, Topology Appl., 157 (2010), 2038–2047.
8. J. Camargo, Lightness of Induced Mappings and Homeomorphisms, Canad. Math. Bull., 54 (2011), 607–618.
9. J. Camargo and S. Macías, On Freely Decomposable Maps, Topology Appl., 159 (2012), 891–899.
10. J. Camargo and S. Macías, On Strongly Freely Decomposable and Induced Maps, Glasnik Math., 48(68) (2013), 429–442.
11. J. J. Charatonik, A. Illanes and S. Macías, Induced Mappings on the Hyperspace $C_n(X)$ of a Continuum X, Houston J. Math., 28 (2002), 781–805.
12. J. J. Charatonik and W. J. Charatonik, Lightness of Induced Mappings, Tsukuba J. Math., 22 (1998), 179–192.
13. J. J. Charatonik and W. J. Charatonik, Induced MO-mappings, Tsukuba J. Math., 23 (1999), 245–252.
14. W. J. Charatonik, R^i-continua and Hyperspaces, Topology Appl., 23 (1986), 207–276.
15. W. J. Charatonik, Arc Approximation Property and Confluence of Induced Mappings, Rocky Mountain J. Math., 28 (1998), 107–154.
16. W. J. Charatonik, Induced Near-homeomorphisms, Comment. Math. Univ. Carolinae, 41 (2000), 133–137.
17. J. Dugundji, *Topology*, Allyn and Bacon, Inc., Boston, 1966.
18. A. Emeryk and Z. Horbanowicz, On Atomic Mappings, Colloq. Math., 27 (1973), 49–55.
19. G. R. Gordh, Jr. and C. B. Hughes, On Freely Decomposable Mappings of Continua, Glasnik Math., 14 (34) (1979), 137–146.
20. H. Hosokawa, Mappings of Hyperspaces Induced by Refinable Mappings, Bull. Tokyo Gakugei Univ., (4) 42 (1990), 1–8.
21. H. Hosokawa, Induced Mappings on Hyperspaces, Tsukuba J. Math., 21 (1997), 239–250.
22. H. Hosokawa, Induced Mappings on Hyperspaces II, Tsukuba J. Math., 21 (1997)773–783.
23. J. G. Hocking and G. S. Young, *Topology*, Dover Publications, Inc., New York, 1988.
24. J. Krasinkiewicz, Curves which are Continuous Images of Tree-like Continua are Movable, Fund. Math., 89 (1975), 233–260.
25. K. Kuratowski, *Topology*, Vol. II, Academic Press, New York, N. Y., 1968.
26. A. Lelek and D. R. Read, Composition of Confluent Mappings and Some Other Classes of Functions, Colloq. Math., 29 (1974), 101–112.
27. M. de J. López and S. Macías, Induced Maps on n-fold Hyperspaces, Houston J. Math., 33 (2007), 1047–4057.
28. S. Macías, Induced Maps on n-fold Hyperspace Suspensions, Topology Proc., 28 (2004), 143–152.
29. T. Maćkowiak, Continuous Mappings on Continua, Dissertationes Math. (Rozprawy Mat.), 158 (1979), 1–91.
30. T. Maćkowiak, Singular Arc-like Continua, Dissertationes Math. (Rozprawy Mat.), 257 (1986), 1–35.

31. J. van Mill, *Infinite-Dimensional Topology*, North Holland, Amsterdam, 1989.
32. S. B. Nadler, Jr., *Hyperspaces of Sets,* Monographs and Textbooks in Pure and Applied Math., Vol. 49, Marcel Dekker, New York, Basel, 1978. Reprinted in: Aportaciones Matemáticas de la Sociedad Matemática Mexicana, Serie Textos # 33, 2006.
33. S. B. Nadler, Jr., Universal Mappings and Weakly Confluent Mappings, Fund. Math., 221–235.
34. S. B. Nadler, Jr., Induced Universal Maps and Some Hyperspaces with the Fixed Point Property, Proc. Amer. Math. Soc., 100 (1987), 749–754.
35. S. B. Nadler, Jr., *Continuum Theory: An Introduction,* Monographs and Textbooks in Pure and Applied Math., Vol. 158, Marcel Dekker, New York, Basel, Hong Kong, 1992.
36. A. D. Wallace, Quasi-monotone Transformations, Duke Math. J., 7 (1940), 136–145.
37. G. T. Whyburn, *Analytic Topology,* Amer. Math. Soc. Colloq. Publ., Vol. 28, Amer. Math. Soc., Providence, R. I., 1942.
38. M. Wojdisławski, Rétractes Absolus et Hyperespaces des Continus, Fund. Math., 32 (1939), 184–192.
39. H. Xianjiu, Z. Fanping and Z. Gengrong, Semi-openness and Almost-openness of Induced Mappings, Appl. Math. J. Chinese Univ. Ser. B., 20(1) (2005), 21–26.

Chapter 9
Questions

In this small chapter we include some open questions related to the topics of the rest of the book.

9.1 Inverse Limits

9.1.1 Question Is it true that X is the inverse limit of graphs with monotone simplicial retractions as bonding maps if and only if X is locally connected and each cyclic element of X is a graph? (B. B. Epps [11])

Comment: If M is a semi-locally connected continuum a *cyclic element* of M is either a cut point of M, an end point of M or a nondegenerate subset of M which is maximal with respect to being a connected subset without cut points. A map $f \colon X \to Y$ between graphs is *simplicial* if it is a map between the set of vertexes of the graph and it is a linear extension on the edges.

9.1.2 Question Let $\{X_n, f_n^{n+1}\}$ be an inverse sequence of polyhedra with bonding maps $f_n^{n+1} \colon X_{n+1} \to X_n$ such that the inverse limit is a hereditarily indecomposable continuum. Let $g_n \colon S^2 \to X_n$ be a map such that g_n is homotopic to $f_n^{n+1} \circ g_{n+1}$ for each $n \in \mathbb{N}$. Is g_1 homotopic to a constant map? (J. Krasinkiewicz [11])

9.1.3 Question If X and Y are continua and $X \times Y$ is disk-like, then must X be arc-like? (C. L. Hagopian [14])

9.1.4 Question If a product of two continua is disk-like, must it be embeddable in $\mathbb{R}^3$? (C. L. Hagopian [15])

9.1.5 Question Does every disk-like continuum have the fixed point property? (S. Mardešić [33] and C. L. Hagopian [14], independently)

© Springer International Publishing AG, part of Springer Nature 2018
S. Macías, *Topics on Continua*, https://doi.org/10.1007/978-3-319-90902-8_9

Comment: An affirmative answer to this question would imply that every nonseparating plane continuum has the fixed point property. (C. L. Hagopian)

Comment: A partial positive answer is given in Theorem 7.7.2. (S. Macías)

9.1.6 Question Does each tree-like homogeneous continuum have the fixed point property? (C. L. Hagopian)

Comment: For the plane, Oversteegen and Tymchatyn have shown that this question has a positive answer. (W. Lewis)

9.1.7 Question Does every triod-like continuum have the fixed point property? (D. Bellamy)

9.1.8 Question Is there a map $f : [0, 1] \to [0, 1]$ such that f has a periodic orbit of odd period larger than one, f has no periodic orbit of period three and $\varprojlim\{[0, 1], f\}$ is homeomorphic to the pseudo-arc? (L. Block, J. Keesling and V. V. Uspenskij [5])

Comment: Let $f : X \to X$ be a map. A point $x \in X$ is said to be *periodic* provided that there exists $m \in \mathbb{N}$ such that $f^m(x) = x$. Now, x is said to have *period n* if $n = \min\{m \in \mathbb{N} \mid f^m(x) = x\}$. We say that f has a *periodic orbit of period n* provided that there exists a point $x \in X$ such that x has period n.

9.1.9 Question Suppose $n \geq 4$ is a positive integer and H and K are chainable continua having the property that each is an indecomposable continuum with only n end points and every nondegenerate proper subcontinuum is an arc. Are H and K homeomorphic? (W. T. Ingram)

9.1.10 Question Given $\lambda \in \left[\frac{1}{2}, 1\right]$, define $T_\lambda : [0, 1] \to [0, 1]$ by

$$
T_\lambda(x) =
\begin{cases}
2\lambda x, & \text{if } x \in \left[0, \dfrac{1}{2}\right]; \\
2\lambda(1 - x), & \text{if } x \in \left[\dfrac{1}{2}, 1\right].
\end{cases}
$$

If $\frac{1}{2} < \lambda_1 < \lambda_2 \leq 1$, then is $\varprojlim\{[0, 1], T_{\lambda_1}\}$ topologically different from $\varprojlim\{[0, 1], T_{\lambda_2}\}$? (W. T. Ingram [21])

Comment: This question has been solved by Barge and Diamond for $n = 5$. Kailhoffer and Raines have further partial solutions. (W. T. Ingram)

Comment: Sonja Štimac has announced at the meeting *Geometric Topology II*, celebrated at the Inter-University Centre, Dubrovnik, Croatia (September 29 through October 5, 2002), that in her dissertation (in Croatian) she has a positive answer to this question. No paper has appeared yet. (S. Macías)

Comment: A positive answer to this question is given by M. Barge, H. Bruin and S. Šitmac in [2]. (S. Macías)

9.2 The Set Function $\mathcal{T}$

9.2.1 Question If the set function $\mathcal{T}$ is continuous for the continuum X, is it true that X is $\mathcal{T}$-additive? (D. Bellamy [11])

Comment: A bushel of *Extra Fancy Stayman Winesap apples* for the solution. (D. Bellamy)

Comment: See Remark 3.4.18. (S. Macías)

Comment: A positive answer is given by J. Camargo and C. Uzcátegui in [9]. (S. Macías)

9.2.2 Question If $\mathcal{T}$ is continuous for the continuum X, is it true that the collection $\{\mathcal{T}(\{p\}) \mid p \in X\}$ is a continuous decomposition of X such that the quotient space is locally connected? (D. Bellamy [11])

Comment: A partial answer to this question is presented in Theorem 3.4.14 and in Corollary 5.1.23. A positive answer is given by S. Macías in [27] (S. Macías)

9.2.3 Question If X and Y are indecomposable continua, is $\mathcal{T}$ idempotent on $X \times Y$? Even for only closed sets in $X \times Y$? (D. Bellamy [11])

Comment: A negative answer to the first question is presented in Corollary 3.2.7, see also [28]. (S. Macías)

9.2.4 Question If $\mathcal{T}$ is continuous for X and X is a decomposable continuum, is it true that for each $p \in X$, $Int(\mathcal{T}(\{p\})) = \emptyset$? (D. Bellamy [11])

Comment: A partial answer to this question is presented in Theorem 3.4.14. A positive answer is given by J. Camargo and C. Uzcátegui in [9]. (S. Macías)

9.2.5 Question Let X be a continuum. If $X/\mathcal{T}$ denotes the finest decomposition space of X which shrinks each $\mathcal{T}(\{p\})$ to a point, is $X/\mathcal{T}$ locally connected? (D. Bellamy [22])

Comment: The answer is affirmative if it is assumed that X is $\mathcal{T}$-additive.

9.2.6 Question If X is an indecomposable continuum and W is a subcontinuum of $X \times X$ with nonempty interior, is $\mathcal{T}(W) = X \times X$? (F. B. Jones [22])

Comment: C. L. Hagopian [13] has shown that this question has an affirmative answer if X is chainable. (S. Macías)

9.2.7 Question If X is an atriodic continuum (or X contains no uncountable collection of pairwise disjoint triods) and X does not have a weak cut point, then is there a continuum $W \subset X$ such that $Int(W) \neq \emptyset$ and $\mathcal{T}(W) \neq X$? (H. Cook [22])

Comment: Let X be a continuum and let x be a point of X. We say that x is a *weak cut point* of X provided that there exist two points y and z in X such that if W is a subcontinuum of X and $\{y, z\} \subset W$, then $x \in W$.

9.2.8 Question If $\mathcal{T}$ is continuous for the continuum X and $f : X \twoheadrightarrow Z$ is a continuous and monotone surjection, is $\mathcal{T}$ continuous for Z also? (D. Bellamy [22])

Comment: A negative answer is given by A. Illanes and R. Leonel in [20]. (S. Macías)

9.2.9 Question If X is a strictly point $\mathcal{T}$-asymmetric dendroid, then is X smooth? (D. Bellamy [22])

Comment: A continuum X is *strictly point $\mathcal{T}$-asymmetric* if for any two distinct points p and q of X with $p \in \mathcal{T}(\{q\})$, we have that $q \notin \mathcal{T}(\{p\})$.
Comment: L. Fernández [12] has given a negative answer, see Example 3.1.96. (S. Macías)

9.2.10 Question Let X be a continuum. Suppose the restriction of $\mathcal{T}$ to the hyperspace of subcontinua of X is continuous. Does this imply that $\mathcal{T}$ is continuous for X? (D. Bellamy [3])

Comment: A partial answer to this question is presented in Theorem 3.1.32. (S. Macías)
Comment: Related results to Question 9.2.10 are in section 4 of [29].

9.2.11 Question Do open maps preserve $\mathcal{T}$-additivity? $\mathcal{T}$-symmetry? (D. Bellamy [22])

9.2.12 Question Is Theorem 3.7.12 true when K is closed and $\dim(K) = 0$? (S. Macías)

9.2.13 Question Is Theorem 3.7.13 true when K is closed and $\dim(K) = 0$? (S. Macías)

9.2.14 Question If X is a continuum, then is $\mathfrak{T}(X)$ an F_σ subset of 2^X? (D. Bellamy, L. Fernández and S. Macías [4])

9.2.15 Question Let X and Y be continua and let $f : X \twoheadrightarrow Y$ be a monotone map. When is it true that $|\mathfrak{T}(Y)| = |\mathfrak{T}(X)|$? (D. Bellamy, L. Fernández and S. Macías [4])

Comment: A positive answer for atomic maps is given by S. Macías in [30]. See Theorem 3.6.23. (S. Macías)

9.2.16 Question Is the converse of Theorem 3.6.31 true for a continuum X for which $\mathcal{T}$ is idempotent? (D. Bellamy, L. Fernández and S. Macías [4])

Comment: A characterization for the components of $\mathcal{T}$-closed sets is given by H. Abobaker and W. J. Charatonik in [1], and they present a negative answer. (S. Macías)

9.3 Homogeneous Continua

9.3.1 Question Is every homogeneous, hereditarily indecomposable, nondegenerate continuum a pseudo-arc? (F. B. Jones [22])

Comment: The *pseudo-arc* is the only nondegenerate, hereditarily indecomposable chainable continuum.
Comment: J. T. Rogers, Jr. has shown that it must be tree-like.

9.3.2 Question Is each nondegenerate, homogeneous, nonseparating plane continuum a pseudo-arc? (F. B. Jones [22])

Comment: F. B. Jones and C. L. Hagopian have shown that it must be hereditarily indecomposable. J. T. Rogers, Jr., has shown that it must be tree-like.

9.3.3 Question If a homogeneous continuum X contains an arc must it contain a solenoid or a simple closed curve? (C. L. Hagopian [22])

9.3.4 Question Is the simple closed curve the only nondegenerate, homogeneous, hereditarily decomposable continuum? (P. Minc [22])

Comment: Several partial answers to this question are given in [32]. (S. Macías)

9.3.5 Question Is every decomposable, homogeneous continuum of dimension greater than one aposyndetic? (J. T. Rogers, Jr. [22])

9.3.6 Question Is every nondegenerate, homogeneous, indecomposable continuum one-dimensional? (F. B. Jones [23])

9.3.7 Question Is each atriodic, homogeneous continuum circle-like? (C. L. Hagopian)

9.3.8 Question Is each aposyndetic, nonlocally connected, one-dimensional, homogeneous continuum an inverse limit of Menger curves? Menger curves and covering maps? (J. T. Rogers, Jr.)

9.3.9 Question Let X be an arcwise connected homogeneous continuum which is not S^1. Must X contain a simple closed curve of arbitrary small diameter? (L. Lum)

9.3.10 Question Let X be a homogeneous arcwise connected continuum which is not S^1. Let U be an open set in X and let M be an arc component of U. Is M cyclicly connected? (D. Bellamy)

Comment: An arcwise connected space Z is *cyclicly connected* provided that each pair of points of Z lie on a simple closed curve.

9.3.11 Question Let X be a homogeneous arcwise connected continuum which is not a simple closed curve. Is every arcwise connected open subset of X also cyclicly connected? (D. Bellamy)

9.3.12 Question Can Jones's Aposyndetic Decomposition Theorem be strengthened to give decomposition elements which are hereditarily indecomposable? (J. T. Rogers, Jr.)

9.3.13 Question If X is an arcwise connected homogeneous continuum other than a simple closed curve, must each pair of points of X be the vertices of a θ-curve in X? (D. Bellamy)

Comment: D. Bellamy and L. Lum have shown that each pair of points of X must lie on a simple closed curve.

9.3.14 Question Does each finite subset of a nondegenerate arcwise connected homogeneous continuum lie on a simple closed curve? (D. Bellamy)

9.3.15 Question Does there exist a homogeneous one-dimensional continuum with no nondegenerate chainable subcontinuum? (W. Lewis)

Comment: If there exists a nondegenerate, homogeneous, hereditarily indecomposable continuum other than the pseudo-arc, the answer is yes.

9.3.16 Question Is every continuum a continuous image of a homogeneous continuum? In particular, is the spiral around the triod such an image? (L. Fearnley)

9.3.17 Question Is there a nondegenerate homogeneous plane continuum X not homeomorphic to a simple closed curve, the pseudo-arc or the circle of pseudo-arcs?

Comment: L. C. Hoen and L. G. Oversteegen [17] recently showed that this question has a negative answer.

9.4   *n*-Fold Hyperspaces

9.4.1 Question Let X be a continuum. If $C_n(X)$ contains $(n+1)$-cells, then does X contain a decomposable subcontinuum? (S. Macías)[1]

Comment: J. Camargo, D. Herrera and S. Macías [7] have given a positive answer to this question. (S. Macías)

9.4.2 Question Is $C_3([0, 1])$ homeomorphic to $[0, 1]^6$? (R. Schori)

Comment: J. Camargo, D. Herrera and S. Macías [7] have given a negative answer to this question, see Theorem 6.9.22. (S. Macías)

9.4.3 Question If $n \geq 2$ is an integer and X is a continuum, then is $C_n(X)$ countable closed aposyndetic? 0-dimensional closed aposyndetic? (S. Macías [25])

Comment: J. M. Martínez-Montejano [34] has given a positive answer to this question, see Theorem 6.3.12. (S. Macías)

---

[1]We changed the original question because it had a positive answer by Theorem 6.1.12.

9.4.4 Question Let *n* be an integer greater than two. Let *X* be a continuum and let $x \in X$. If $\{x\}$ is arcwise accessible from $C_n(X) \setminus C(X)$, then is $\{x\}$ arcwise accessible from $C_2(X) \setminus C(X)$? (S. Macías)

9.4.5 Question Does there exist an indecomposable continuum *X* such that $C_n(X)$ is homeomorphic to the cone over a finite-dimensional continuum for some integer $n \geq 2$? (S. Macías and S. B. Nadler, Jr.)

9.4.6 Question Does there exist a hereditarily decomposable continuum *X* that is neither an arc nor a simple *m*-od such that $C_n(X)$ is homeomorphic to the cone over a finite-dimensional continuum for some $n \geq 3$? (S. Macías [26] and S. B. Nadler, Jr. [31])

9.4.7 Question Find a geometric model for $C_2(X)$, where *X* is a simple triod. (A. Illanes [19])

9.5 *n*-Fold Hyperspace Suspensions

9.5.1 Question For what continua *X* does the natural embedding in the proof of Theorem 7.1.29 embed $HS_m(X)$ as a retract of $HS_n(X)$? In particular, what about the case when *X* is S^1? (S. Macías)

9.5.2 Question For what continua *X*, can $HS_m(X)$ be embedded in $HS_n(X)$ as a retract $(m < n)$? (S. Macías)

9.5.3 Question What continua *X* have the property that $HS_n(X)$ is contractible for each $n \in \mathbb{N}$? (S. Macías)

Comment: Section 7.2 contains a couple of partial answers to this question.

9.5.4 Question If *X* is a locally connected continuum without free arcs and $n \in \mathbb{N}$, then is $HS_n(X)$ dimensionally homogeneous? (S. Macías)

9.5.5 Question Let *X* be a graph and let $n \in \mathbb{N}$. If *Y* is a continuum and $HS_n(Y)$ is homeomorphic to $HS_n(X)$, then is *Y* homeomorphic to *X*? (S. Macías and S. B. Nadler, Jr.)

Comment: By Theorem 7.4.13, we know that *Y* must be a graph. (S. Macías)
Comment: D. Herrera-Carrasco, A. Illanes, F. Macías-Romero and F. Vázquez-Juárez have given a positive answer to this question [16]. (S. Macías)

9.5.6 Question Is $HS_n([0, 1])$ or $HS_n(S^1)$ an absolute *n*-fold hyperspace suspension for each $n \geq 3$? (S. Macías and S. B. Nadler, Jr.)

Comment: D. Herrera-Carrasco, A. Illanes, F. Macías-Romero and F. Vázquez-Juárez have given a negative answer to this question [16]. (S. Macías)

9.5.7 Question Does there exist a compact *Q*-manifold, other than *Q*, that is an absolute *n*-fold hyperspace suspension? (S. Macías and S. B. Nadler, Jr.)

9.6   Induced Maps on n-Fold Hyperspaces

9.6.1 Question Let $f: X \twoheadrightarrow Y$ be a map between continua and let $n \in \mathbb{N}$. If $C_n(f)$ is and MO-map, then is f and MO-map? (H. Hosokawa [18] for $n = 1$, J. J. Charatonik, A. Illanes and S. Macías, for $n \geq 2$)

9.6.2 Question Let $f: X \to Y$ be a map between continua. If 2^f is almost monotone, then does it follow that $C_n(f)$ is almost monotone, for some $n \in \mathbb{N}$? (J. Camargo [6])

Comment: Observe that, by [8, Corollary 4.10], the answer to Question 9.6.2 is positive for maps whose range is locally connected.

9.6.3 Question Let $f: X \to Y$ be a map between continua and let $n \in \mathbb{N}$. If $HS_n(f)$ is OM, then is f OM? is $C_n(f)$ OM?. In particular, if $n \in \{1, 2\}$ and $HS_n(f)$ is OM, then is f OM? (J. Camargo [6])

9.6.4 Question Let $f: X \to Y$ be a map between continua and let $n \in \mathbb{N}$. Then:

(1) If f is semi-confluent (joining), then is $HS_n(f)$ semi-confluent (joining)?
(2) If $HS_n(f)$ is semi-confluent (joining), then is $C_n(f)$ semi-confluent (joining)?
(3) If $HS_n(f)$ is semi-confluent (joining), then is 2^f semi-confluent (joining)?
(4) If $C_n(f)$ is semi-confluent (joining), then is 2^f semi-confluent (joining)?
(5) If 2^f is semi-confluent (joining), then is $C_n(f)$ and/or $HS_n(f)$ semi-confluent (joining)? (J. Camargo [6])

9.6.5 Question Let $f: X \to Y$ be a map between continua and let $n \in \mathbb{N}$. If 2^f is confluent, then is $C_n(f)$ and/or $HS_n(f)$ confluent? (J. Camargo [6])

9.6.6 Question Do there exist a nonlocally connected continuum Y and a confluent map $f: X \to Y$ such that $C_2(f)$ is confluent? (M. de J. López and S. Macías [24])

9.6.7 Question Let $f: X \to Y$ be a map between continua. If f is almost monotone (weakly monotone), then is 2^f is almost monotone (weakly monotone)? (J. Camargo [6])

9.6.8 Question Let $f: X \to Y$ be a map between continua and let $n \in \mathbb{N}$. If $HS_n(f)$ is almost monotone (weakly monotone), then is 2^f and/or $C_n(f)$ almost monotone (weakly monotone)? (J. Camargo [6])

9.6.9 Question Let $f: X \to Y$ be a map between continua and let $n \in \mathbb{N}$. If 2^f is almost monotone (weakly monotone), then is $HS_n(f)$ almost monotone (weakly monotone)? (J. Camargo [6])

9.6.10 Question Let $f: X \to Y$ be a map between continua. What are the relationships between the following statements:

(1) 2^f is almost monotone (weakly monotone);
(2) $C_n(f)$ is almost monotone (weakly monotone) for each $n \in \mathbb{N}$? (J. Camargo [6])

9.6.11 Question Let $f: X \to Y$ be a map between continua. What properties of the spaces X or Y, or of the map f are necessary in order that whenever the induced map $C_n(f)$ or 2^f is light, we have that f is a homeomorphism? (J. Camargo [6])

9.6.12 Question Does there exist a map f, between indecomposable continua, such that $C_n(f)$ is light for $n \geq 2$ and f is not a homeomorphism? (J. Camargo [6])

9.6.13 Question Let $f: X \to Y$ be a map between hereditarily indecomposable continua and let $n \in \mathbb{N}$. If $C_n(f)$ is light, then is f a homeomorphism? (J. Camargo [6])

9.6.14 Question Let Y be a continuum and let $n \geq 2$. If $f: [0, 1] \times Y \to [0, 1]$ is the projection map, then is $C_n(f)$ open? (J. Camargo [6])

Comment: It is known that the answer to Question 9.6.14 is positive for $n = 1$ [10, Corollary 19]. (S. Macías)

9.6.15 Question Does there exist a map f between locally connected continua such that $HS_2(f)$ is open and f is not a homeomorphism? (J. Camargo [6])

Comment: It is known that for map f between locally connected continua and $n \neq 2$, $HS_n(f)$ is open if and only if f is a homeomorphism (for the "only if" part: for $n \geq 3$ see Theorem 8.5.21, and for $n = 1$ see [6, Proposición 4.24]) (S. Macías).

References

1. Hussam Abobaker and Włodzimierz J.Charatonik, $\mathcal{T}$-closed Sets, Topology Appl., 222 (2017), 274–277.
2. M. Barge, H. Bruin and S. Štimac, The Ingram Conjecture, Geometry & Topology, 16 (2012), 2481–2516.
3. D. P. Bellamy, Questions In and Out of Context, talk given in the VI Joint Meeting AMS-SMM, Houston, TX, May 13–15, 2004 (LaTeX edition by Jonathan Hatch).
4. D. P. Bellamy, L. Fernández and S. Macías, On $\mathcal{T}$-closed Sets, Topology Appl., 195 (2015), 209–225.
5. L. Block, J. Keesling and V. V. Uspenskij, Inverse Limits Which are the Pseudo-arc, Houston J. Math., 26 (2000), 629–638.
6. J. Camargo, *Funciones Inducidas Entre Hiperespacios de Continuos*, Tesis de Doctorado, Facultad de Ciencias, U. N. A. M., 2009. (Spanish)
7. J. Camargo, D. Herrera and S. Macías, Cells and n-fold Hyperspaces, Colloq. Math., 145 (2016), 157–166.
8. J. Camargo and S. Macías, On Strongly Freely Decomposable and Induced Maps, Glasnik Math., 48(68) (2013), 429–442.
9. J. Camargo and C. Uzcátegui, Continuity of the Jones' set function $\mathcal{T}$, Proc. Amer. Math. Soc., 145 (2017), 893–899.
10. J. J. Charatonik, W. J. Charatonik and A. Illanes, Openness of Induced Maps, Topology Appl., 98 (1999), 67–80.
11. H. Cook, W. T. Ingram and A. Lelek, *A List of Problems Known as Houston Problem Book*, in *Continua with the Houston Problem Book*, Lectures Notes in Pure and Applied Mathematics, Vol. 170, Marcel Dekker, Inc., New York, Basel, Hong Kong, 1995. (eds. H. Cook, W. T. Ingram, K. T. Kuperberg, A. Lelek, and P. Minc)

12. L. Fernández, On strictly point $\mathcal{T}$-asymmetric continua, Topology Proc., 35 (2010), 91–96.
13. C. L. Hagopian, Mutual Aposyndesis, Proc. Amer. Math. Soc., 23 (1969), 615–622.
14. C. L. Hagopian, Disk-like Products of λ Connected Continua, I, Proc. Amer. Math. Soc., 51 (1975), 448–452.
15. C. L. Hagopian, Disk-like Products of λ Connected Continua, II, Proc. Amer. Math. Soc., 52 (1975), 479–484.
16. D. Herrera-Carrasco, A. Illanes, F. Macías-Romero and F. Vázquez-Juárez, Finite Graphs have Unique $HS_n(X)$, Topology Proc., 44 (2014), 75–95.
17. L. C. Hoen and L. G. Oversteegen, A Complete Classification of Homogeneous Plane Continua, Acta Math., 216 (2016), 177–216.
18. H. Hosokawa, Induced Mappings on Hyperspaces, Tsukuba J. Math., 21 (1997), 239–250.
19. A. Illanes, A Model for the Hyperspace $C_2(S^1)$, Q. & A. in General Topology, 22 (2004), 117–130.
20. A. Illanes and Rocío Leonel, Continuity of Jones' function is not preserved under monotone mappings, Topology Appl., 231 (2017), 136–158.
21. W. T. Ingram, Inverse Limits on [0, 1] Using Tent Maps and Certain Other Piecewise Linear Bonding Maps, in *Continua with the Houston Problem Book*, Lectures Notes in Pure and Applied Mathematics, Vol. 170, Marcel Dekker, Inc., New York, Basel, Hong Kong, 1995, 253–258 (eds. H. Cook, W. T. Ingram, K. T. Kuperberg, A. Lelek, and P. Minc)
22. W. Lewis, Continuum Theory Problems, Topology Proc., 8 (1983), 361–394.
23. W. Lewis, The Classification of Homogeneous Continua, Soochow J. of Math., 18 (1992), 85–121.
24. M. de J. López and S. Macías, Induced Maps on n-fold Hyperspaces, Houston J. Math., 33 (2007), 1047–1057.
25. S. Macías, On the Hyperspaces $C_n(X)$ of a Continuum X, Topology Appl., 109 (2001), 237–256.
26. S. Macías, Fans Whose Hyperspaces Are Cones, Topology Proc., 27 (2003), 217–222.
27. S. Macías, A Decomposition Theorem for a Class of Continua for Which the Set Function $\mathcal{T}$ is Continuous, Colloq. Math., 109 (2007), 163–170.
28. S. Macías, On the Idempotency of the Set Function $\mathcal{T}$, Houston J. Math., 37 (2011), 1297–1305.
29. S. Macías, A Note On The Set Functions $\mathcal{T}$ and $\mathcal{K}$, Topology Proc., 46 (2015), 55–65.
30. S. Macías, Atomic Maps and $\mathcal{T}$-closed Sets, Topology Proc., 50 (2017), 97–100.
31. S. Macías and S. B. Nadler, Jr., n-fold Hyperspaces, Cones and Products, Topology Proc., 26 (2001–2002), 255–270.
32. S. Macías and S. B. Nadler, Jr., On Hereditarily Decomposable Homogeneous Continua, Topology Proc., 34 (2009), 131–145.
33. S. Mardešić, Mappings of Inverse Systems, Glasnik Mat. Fiz. Astronom. Ser. II, 18 (1963), 195–205.
34. J. M. Martínez-Montejano, Zero-dimensional Closed Set Aposyndesis and Hyperspaces, Houston J. Math., 32 (2006), 1101–1105.

Index

Printed in the United States
By Bookmasters